ts and
Monographs
in Physics

Springer

New York
Berlin
Heidelberg
Barcelona
Hong Kong
London
Milan
Paris
Singapore
Tokyo

W. Beiglböck
J.L. Birman
R.P. Geroch
E.H. Lieb
T. Regge
W. Thirring

Series Editors

Texts and Monographs in Physics

S. Albeverio, F. Gesztesy, R. Høegh-Krohn, and H. Holden: **Solvable Models in Quantum Mechanics** (1988).

R. Bass: **Nuclear Reactions with Heavy Ions** (1980).

A. Bohm: **Quantum Mechanics: Foundations and Applications,** Third Edition (1993).

O. Bratelli and D.W. Robinson: **Operator Algebras and Quantum Statistical Mechanics.** Volume I: C*- and W*-Algebras. Symmetry Groups. Decomposition of States (1979). Volume II: Equilibrium States. Models in Quantum Statistical Mechanics (1981).

K. Chadan and P.C. Sabatier: **Inverse Problems in Quantum Scattering Theory** (1977).

M. Chaichian and N.F. Nelipa: **Introduction to Gauge Field Theories** (1984).

G. Gallavotti: **The Elements of Mechanics** (1983).

W. Glöckle: **The Quantum Mechanical Few-Body Problem** (1983).

W. Greiner, B. Müller, and J. Rafelski: **Quantum Electrodynamics of Strong Fields** (1985).

J.M. Jauch and F. Rohrlrich: **The Theory of Photons and Electrons: The Relativistic Quantum Field Theory of Charge Particles with Spin One-half,** Second Expanded Edition (1980).

J. Kessler: **Polarized Electrons** (1976). Out of print. (Second Edition available as Springer Series in Atoms and Plasmas, Vol. 1.)

G. Ludwig: **Foundations of Quantum Mechanics I** (1983).

G. Ludwig: **Foundations of Quantum Mechanics II** (1985).

O. Nachtmann: **Elementary Particle Physics: Concepts and Phenomena** (1990).

R.G. Newton: **Scattering Theory of Waves and Particles,** Second Edition (1982).

A. Perelomov: **Generalized Coherent States and Their Applications** (1986).

H. Pilkuhn: **Relativistic Particle Physics** (1979).

R.D. Richtmyer: **Principles of Advanced Mathematical Physics.** Volume I (1978). Volume II (1981).

W. Rindler: **Essential Relativity: Special, General, and Cosmological,** Revised Second Edition (1980).

P. Ring and P. Schuck: **The Nuclear Many-Body Problem** (1980).

R.M. Santilli: **Foundations of Theoretical Mechanics.** Volume I: The Inverse Problem in Newtonian Mechanics (1978). Volume II: Birkhoffian Generalization of Hamiltonian Mechanics (1983).

M.D. Scadron: **Advanced Quantum Theory an Its Applications Through Feynman Diagrams** (1979).

N. Straumann: **General Relativity and Relativistic Astrophysics** (1984).

C. Truesdell and S. Bharatha: **The Concepts and Logic of Classical Thermodynamics as a Theory of Heat Engines: Rigorously Constructed upon the Foundation Laid by S. Carnot and F. Reech** (1977).

F.J. Ynduráin: **Quantum Chromodynamics: An Introduction to the Theory of Quarks and Gluons** (1983).

Arno Bohm

Quantum Mechanics: Foundations and Applications

Third Edition, Revised and Enlarged
Prepared with Mark Loewe

With 96 Illustrations

 Springer

Arno Bohm
Department of Physics
Center for Particle Theory
The University of Texas at Austin
Austin, TX 78712
USA

Mark Loewe
Department of Physics
Southwest Texas State University
San Marcos, TX 78666
USA

Series Editors

Wolf Beiglböck
Institut für Angewandte
  Mathematik
Universität Heidelberg
D-6900 Heidelberg 1
Federal Republic of
  Germany

Joseph L. Birman
Department of Physics
The City College of the
  City University of
  New York
New York, NY 10031
USA

Robert R. Geroch
Enrico Fermi Institute
University of Chicago
Chicago, IL 60637
USA

Elliott H. Lieb
Department of Physics
Joseph Henry Laboratories
Princeton University
Princeton, NJ 08540
USA

Tullio Regge
Istituto de Fisica Teorica
Universita di Torino
I-10125 Torino
Italy

Walter Thirring
Institut für Theoretische Physik
  der Universität Wien
A-1090 Wien
Austria

Library of Congress Cataloging-in-Publication Data
Bohm, Arno, 1936–
    Quantum mechanics : foundations and applications / Arno Bohm. —
3rd ed., rev. and enl. / prepared with M. Loewe.
        p. cm. — (Texts and monographs in physics)
    Includes bibliographical references and index.
    ISBN 0-387-95330-2
    1. Quantum theory. I. Loewe, M. II. Title. III. Series.
QC174.12.B63   1993
530.1′2—dc20   92-44947

Printed on acid-free paper.

First softcover printing 2001.
The first edition of this book appeared as: Arno Bohm, *Quantum Mechanics*. Springer-Verlag, New York, Heidelberg, Berlin, 1979.

Typeset by Composition House Ltd., Salisbury, England and Asco Trade Typesetting Ltd., Hong Kong.
Printed and bound by R.R. Donnelley & Sons, Harrisonburg, Virginia.
Printed in the United States of America.

9 8 7 6 5 4 3 2

ISBN 0-387-95330-2   SPIN 10843492

Springer-Verlag   New York Berlin Heidelberg
A member of BertelsmannSpringer Science+Business Media GmbH

To my students, colleagues, and friends
without whose help
this edition would not have been possible.

Preface to the Third Edition

This edition differs from the second chiefly in the addition of about 100 pages devoted to the quantum (or geometric, or Berry) phase, a subject that did not exist when this book was written. The changes in the remainder of the book consist of corrections of a small number of misprints.

While it may seem that adding two chapters on the quantum phase is overemphasizing a currently fashionable subject, they actually complete the development of quantum theory as given in this book. We start with simple models, synthesizing them into complicated "molecules." With the new chapters, we end with complicated "molecules," dividing them into simpler parts. This process of dividing a complex system into parts quite naturally gives rise to a gauge theory, of which the geometric phase is a manifestation — with consequences not only in theory, but observable in experiments. For this reason, the geometric phase is not a mere fashion, but a discovery that will retain its importance forever and must be discussed in textbooks on quantum mechanics.

I would like to acknowledge help and advice from Mark Loewe with the writing and also of the new part of the book. In addition, I would like to express my gratitude to J. Anandan, M. Berry, and C.A. Mead, who have read parts or all of the new material and have provided valuable advice.

Preface to the Second Edition

The first edition of this book was written as a text and has been used many times in a one-year graduate quantum mechanics course. One of the reviewers has made me aware that the book can also serve as, "... in principle, a handbook of nonrelativistic quantum mechanics." In the second edition we have therefore added material to enhance its usefulness as a handbook. But it can still be used as a text if certain chapters and sections are ignored. We have also revised the original presentation, in many places at the suggestion of students or colleagues. As a consequence, the contents of the book now exceed the material that can be covered in a one-year quantum mechanics course on the graduate level. But one can easily select the material for a one-year course omitting—according to one's preference—one or several of the following sets of sections: {I.7, XXI}, {X, XI} or just {XI}, {II.7, XIII}, {XIV.5, XV}, {XIX, XX}. Also the material of Sections I.5–I.8 is not needed to start with the physics in Chapter II. Chapters XI, XIII, XIX, and XX are probably the easiest to dispense with and I was contemplating the deletion of some of them, but each chapter found enthusiastic supporters among the readers who advised against it.

Chapter I—augmented with some applications from later chapters—can also be used as a separate introductory text on the mathematics of quantum mechanics.

The book is self-contained and does not require any prior knowledge of quantum mechanics, but it is a difficult book, because it is so concise. It offers a huge amount of material, more than one can find in texts with twice the number of pages. Consequently, some familiarity with the subject would be very helpful. Prerequisites are a knowledge of calculus, vector algebra, and analysis. Most physical examples are taken from the fields of atomic and molecular physics, as it is these fields that are best known to students at the stage when they learn quantum mechanics.

Texts on a subject established a half-century before are often written using the material and presentation established by the first generation of books on that subject. New applications, deeper insights, and unifying formulations that subsequently develop are easily overlooked. That has not been done in this text. I have presented a unified theoretical formulation which was made possible by later developments, and have included examples from more recent papers.

The changes incorporated in the second edition provide an easier access to the material, but leave the general idea unchanged. It is therefore fitting to quote from the preface of the first edition:

> ... in contrast to what one finds in the standard books, quantum mechanics is more than the overemphasized wave-particle dualism presented in the familiar mathematics of differential equations. "This latter dualism is only part of a more general pluralism" (Wigner) because, besides momentum and position, there is a plurality of other observables not commuting with position and momentum. As there is no principle that brings into prominence the position and momentum operators, a general formalism of quantum mechanics, in which every observable receives the emphasis it deserves for the particular problem being considered, is not only preferable but often much more practical.... It is this general form of quantum theory that is presented here.
>
> I have attempted to present the whole range from the fundamental assumptions to the experimental numbers. To do this in the limited space available required compromises. My choice ... was mainly influenced by what I thought was needed for modern physics and by what I found, or did not find, in the standard textbooks. Detailed discussions of the Schrödinger differential equation for the hydrogen atom and other potentials can be found in many good books.[1] On the other hand, the descriptions of the vibrational and rotational spectra of molecules are hardly treated in any textbooks of quantum mechanics, though they serve as simple examples for the important procedure of quantum-mechanical model building.... So I have treated the former rather briefly and devoted considerable space to the latter.
>
> Groups have not been explicitly made use of in this book. However, the reader familiar with this subject will see that group theory is behind most of the statements that have been cast here in terms of algebras of observables.

---

[1] The subject also is usually adequately treated in undergraduate courses.

This is a physics book, and though mathematics has been used extensively, I have not endeavored to make the presentation mathematically rigorous.... Except in the mathematical inserts, which are given in openface brackets ⟦M: ⟧, the reader will not even be made aware of these mathematical details.

The mathematical inserts are of two kinds. The first kind provides the mathematics needed, and the second kind indicates the underlying mathematical justification....

Quantum mechanics starts with Chapter II, where the most essential basic assumptions (axioms) of quantum mechanics are made plausible from the example of the harmonic oscillator as realized by the diatomic molecule. Further basic assumptions are introduced in later chapters when the scope of the theory is extended. These basic assumptions ("postulates") are not to be understood as mathematical axioms from which everything can be derived without using further judgment and creativity. An axiomatic approach of this kind does not appear to be possible in physics. The basic assumptions are to be considered as a concise way of formulating the quintessence of many experimental facts.

The book consists of two clearly distinct parts, Chapters II–XI and Chapters XIV–XXI, with two intermediate chapters, Chapters XII and XIII. The first part is more elementary in presentation, though more fundamental in subject matter .... The second part, which starts with Chapter XIV, treats scattering and decaying systems. The presentation there is more advanced.

Chapter XIV gives a derivation of the cross section under very general conditions.... Two different points of view—one in which the Hamiltonian time development is assumed to exist, and the other making use of the S-matrix—are treated in a parallel fashion. The required analyticity of the S-matrix is deduced from causality. One of the main features of the presentation is to treat discrete and continuous spectra from the same point of view. For this the rigged Hilbert space is needed, which provides not only a mathematical simplification but also a description which is closer to physics.

Major changes to the book have been made in Chapters I, XIII, and XXI, which were almost totally rewritten. Chapter XXI discusses the new notion of Gamow vectors for the description of decaying states. They were created when the first edition was written in order to achieve the desired unity of description of all of quantum mechanics. Chapter I had to be expanded to provide the mathematical background for Chapter XXI. To start a physics book with a mathematical introduction may create an incorrect impression. I therefore want to emphasize that the book contains many more experimental numbers than mathematical theorems. Extensive revisions have also been made in Chapters II, IV, XIV, XVI, XVII, and XVIII; and many improvements were made in Chapters III, V, VIII, and IX. The appendix to Section

V.3 has been rewritten to provide a simple but typical example for the construction of noncompact group representations. Not all chapters could be revised because of time limitations. Chapters VII, X, XI, and XIX have been scrutinized only a little and Chapters VI, XII, XV, and XX remain essentially as they were in the first edition.

Acknowledgments

For the second edition, as for the first edition, I am indebted to many for their help, encouragement, and advice. Chapter XIII was rewritten with K. Kraus, who together with A. Peres also suggested improvements to Chapter II. On the material of Chapter XXI, I received advice from L. Khalfin and M. Gadella. The new version of Chapter I grew out of a joint project with G. B. Mainland. The revisions of the first part of the book were made together with M. Loewe. For the revisions of the second part of the book I was assisted by J. Morse. P. Busch proofread Section XIII.1. I received many letters pointing out misprints and inadequacies, suggesting improvements, and encouraging me through the tedious task of preparing a new edition. I would like to thank R. Scalettar, A. Y. Klimik, L. Fonda, and T. Mertelmeier for pointing out errors in the first edition. The numerous misprints could not have been purged without the help of students in my classes. Support from D.O.E. and the Alexander von Humboldt Foundation is gratefully acknowledged. I am particularly grateful to M. Loewe who proofread the entire book and made many improvements. If the second edition is better than the first, it is mainly due to him.

Contents

Mathematical Preliminaries

The mathematical language of quantum mechanics is introduced in this chapter. It does not contain any physics.

I.1 The Mathematical Language of Quantum Mechanics

To formulate Newtonian mechanics, the mathematical language of differential and integral calculus was developed. Though one can get some kind of understanding of velocity, acceleration, etc., without differential calculus (in particular for special cases), the real meanings of these physical notions in their full generality become clear only after one is familiar with the idea of the derivative. On the other hand, though, the abstract mathematical definitions of calculus become familiar to us only if we visualize them in terms of their physical realizations. Nowadays, no one would attempt to understand classical mechanics without knowing calculus.

Quantum mechanics, too, has its mathematical language, whose development went parallel to the development of quantum mechanics and whose creation in its full generality was inspired by the needs of quantum physics. This is the language of linear spaces, linear operators, associative algebras, etc., which has meanwhile grown into one of the main branches of mathematics—linear algebra and functional analysis. Although one might obtain some sort of understanding of quantum physics without knowing its mathematical language, the precise and deep meaning of the physical notions in their full generality will not reveal themselves to anyone who does not understand its mathematical language.

Therefore we shall start the quantum-mechanics course with some of the vocabulary and grammar of this language. We shall not try to be mathematically rigorous, since one can still communicate in a language that one does not speak completely correctly. We shall also not give all the mathematics that is needed at the beginning, and you need not be worried if you do not understand everything right away; one learns a language best by using it. We shall give in this chapter not much more mathematics than is needed to make the initial statements about physics. We shall then have to learn new mathematical notions whenever they arise, while we proceed with the development of the physical ideas.

Before we start to study the mathematical structures that are employed in quantum mechanics, we should make the following observation: A mathematical structure is not something real—it only exists in our mind and is created by our mind (though often inspired by outside influences). It is obtained by taking a set of objects and equipping this set with a structure by defining relations between these objects. Modern mathematics distinguishes three basic kinds of structures: algebraic, topological, and ordering. The mathematical structures we use are complicated combinations of these three. For example, the real numbers have an algebraic structure given by the usual laws of addition and multiplication; they have a topological structure given by the meaning of the usual limiting process for an infinite series of numbers, and they have an ordering structure given by the relations expressed by $<$.

We shall use predominantly algebraic structures, although in order to speak the mathematical language of quantum mechanics correctly, topological structures are essential. We shall start with the definition of a linear space, and linear operators and give the definition of an associative algebra. That will provide us with enough vocabulary and grammar to enable us to start communicating physics. We will then give an intuitive description of some fundamental mathematical properties needed throughout the book. In Section I.4 we present the eigenvector decomposition. The underlying mathematics is the nuclear spectral theorem, one of the greatest achievements of mathematics. We shall not prove it, we shall not even precisely state it, but explain it by analogy as a generalization of the coordinate expansion in the three-dimensional case. Section I.5 describes the spaces of sequences and functions as the coordinate representation of linear spaces (in which the nuclear spectral theorem holds). The generalized eigenvectors, of which the reader should have acquired an intuitive understanding in Sections I.4 and I.5, will then be defined as continuous functionals in Section I.7, which is the only place in which a brief discussion of some topological notions is given.

I.2 Linear Spaces, Scalar Product

A linear space Φ is a set of elements $\phi, \psi, \chi, \ldots$ which is given an algebraic structure that is a generalization of certain aspects of the three-dimensional real space $\mathbb{R}^3$. The elements, also called vectors, are defined to obey rules which are well-known properties of vectors in $\mathbb{R}^3$. The linear spaces which we define are in general not three-dimensional, but can have any dimension N, often infinite, and use, in general, the complex numbers $\mathbb{C}$ rather than the real numbers $\mathbb{R}$. The rules that define the linear space are:

(a) For any two elements $\phi, \psi \in \Phi$ there is defined an element $\phi + \psi \in \Phi$, the sum or addition of ϕ and ψ, which obeys the following rules:

$$\phi + \psi = \psi + \phi, \tag{2.1a}$$

$$(\phi + \psi) + \chi = \phi + (\psi + \chi). \tag{2.1b}$$

There exists in Φ an element 0—called the zero vector—with the property $\phi + 0 = \phi$ for every $\phi \in \Phi$. $\tag{2.1c}$

(b) For every complex number, $a \in \mathbb{C}$, and every element $\phi \in \Phi$, there exists an element $a\phi \in \Phi$, called the product of the vector ϕ with the number a, which obeys the following rules:

$$1\phi = \phi, \tag{2.1d}$$

$$0\phi = 0. \tag{2.1e}$$

(0 on the left is the number zero, 0 on the right is the zero vector of (2.1c).)

$$a(b\phi) = (ab)\phi, \quad a, b \in \mathbb{C}, \quad \phi \in \Phi, \tag{2.1f}$$

$$a(\phi + \psi) = a\phi + a\psi, \quad\quad\quad \psi \in \Phi, \tag{2.1g}$$

$$(a + b)\phi = a\phi + b\phi. \tag{2.1h}$$

The element $(-1)\phi$ is usually denoted $-\phi$; because of (2.1d), (2.1h), and (2.1e) one obtains the following result:

$$\phi + (-\phi) = (1 + (-1))\phi = 0\phi = 0.$$

The above collection of relations between elements of Φ and complex numbers defines the linear space and every set of objects that fulfills these relations is called a linear space. Often such objects have more properties than those stated above.

Clearly the vectors $\mathbf{a}, \mathbf{b}, \ldots$ in the three-dimensional space $\mathbb{R}^3$ fulfill the above relations. However, the set of complex infinitely differentiable continuous functions which vanish rapidly at infinity also fulfills these relations. One often says that the abstract linear space defined by the above rules is realized by other mathematical objects like functions, if these other objects appear to us more "real" than the "abstract" vectors. Thus if one feels more familiar with functions one may prefer the "realization" of Φ by a space of functions over the space Φ itself.

In physics the abstract mathematical objects are realized by physical objects. Thus a physicist's "realization" of a linear space is not by other more familiar or more interesting mathematical objects, but by physical objects. In particular, in quantum physics, the elements of the space Φ will be the mathematical images of pure physical states; thus, the linear space is "realized" by the physical states of a quantum system.

A linear space is a set with very little mathematical structure. We will equip it with more structure by defining a scalar product. This notion is again a generalization from the three-dimensional real space $\mathbb{R}^3$.

A linear space is called a *scalar product space* (or *Euclidean space* or *pre-Hilbert space*) if in it a function (ϕ, ψ) of the two vectors $\phi, \psi \in \Phi$ is defined which is a complex number and has the following properties

$$(\phi, \phi) \geq 0 \quad \text{and} \quad (\phi, \phi) = 0 \quad \text{iff} \quad \phi = 0. \tag{2.2a}$$

$$(\psi, \phi) = \overline{(\phi, \psi)} \tag{2.2b}$$

(the bar denotes complex conjugate).

$$(\phi, a\psi) = a(\phi, \psi) \tag{2.2c}$$

($a \in \mathbb{C}$, the set of complex numbers).

$$(\phi_1 + \phi_2, \psi) = (\phi_1, \psi) + (\phi_2, \psi). \tag{2.2d}$$

This function is called the *scalar product* of the elements ϕ and ψ. The usual scalar product in $\mathbb{R}^3$, $(\mathbf{a}, \mathbf{b}) = \mathbf{a} \cdot \mathbf{b}$ clearly fulfills the conditions (2.2a)–(2.2d) with all numbers being real instead of complex.

As in $\mathbb{R}^3$ one calls two vectors ϕ and ψ *orthogonal* if

$$(\phi, \psi) = 0. \tag{2.3}$$

With the scalar product defined by (2.2a)–(2.2d) one defines the *norm* $\|\phi\|$ of a vector ϕ by

$$\|\phi\| = +\sqrt{(\phi, \phi)}. \tag{2.4}$$

For any vector ψ different from the zero vector one can always define a vector $\hat{\psi} = \psi/\|\psi\|$, which has the property $\|\hat{\psi}\| = 1$ and is called a *normalized* vector.

Sometimes one needs in a linear space a more general notion than the scalar product, the bilinear Hermitian form.

A complex-valued function $h(\phi, \psi)$ of two vector arguments is a *Hermitian form* if it satisfies

$$h(\phi, \psi) = \overline{h(\psi, \phi)}, \tag{2.5b}$$

$$h(\phi, a\psi) = ah(\phi, \psi) \quad (a \in \mathbb{C}), \tag{2.5c}$$

$$h(\phi_1 + \phi_2, \psi) = h(\phi_1, \psi) + h(\phi_2, \psi). \tag{2.5d}$$

If in addition h satisfies

$$h(\phi, \phi) \geq 0 \tag{2.5a}$$

for every vector ϕ, then h is said to be a *positive Hermitian form*. A positive Hermitian form is called *positive definite* if

$$\text{from} \quad h(\phi, \phi) = 0 \quad \text{follows} \quad \phi = 0 \quad \text{for every vector } \phi. \tag{2.6}$$

Thus a Hermitian form fulfills the conditions (2.2b), (2.2c), and (2.2d), but not the condition (2.2a) for a scalar product. However, a positive definite Hermitian form is, by (2.6), a scalar product.

Positive Hermitian forms, which are not necessarily scalar products, satisfy the *Cauchy-Schwarz-Bunyakovski inequality*:

$$|h(\phi, \psi)|^2 \le h(\phi, \phi)h(\psi, \psi). \tag{2.7}$$

If h is positive definite, equality holds iff $\phi = a\psi$ for some $a \in \mathbb{C}$.

A set M in the linear space Φ is called a *subspace* of Φ if M is a linear space under the same definitions of the operations of addition and multiplication by a number as given for Φ, i.e., if it follows from $\phi, \psi \in M$ that $a\phi \in M$ and $\phi + \psi \in M$.

An expression of the form $a_1\phi_1 + a_n\phi_n$ is called a *linear combination* of the vectors $\phi_1, \phi_2, \ldots, \phi_n$; the vectors $\phi_1, \phi_2, \ldots, \phi_n$ are said to be *linearly dependent* if there exist numbers $a_1, a_2, \ldots, a_n$, not all zero, for which $a_1\phi_1 + a_2\phi_2 + \cdots + a_n\phi_n = 0$. If the equation $a_1\phi_1 + a_2\phi_2 + \cdots + a_n\phi_n = 0$ holds only for $a_1 = a_2 = \cdots = a_n = 0$, then the vectors $\phi_1, \phi_2, \ldots, \phi_n$ are called *linearly independent*. A space Φ is said to be *finite-dimensional* and, more precisely, *n-dimensional* if there are n and not more than n linearly independent vectors in Φ. If the number of linearly independent vectors in Φ is arbitrarily great, then Φ is said to be *infinite-dimensional*. Every system of n linearly independent vectors in an n-dimensional space Φ is called a *basis* for Φ.

An *isomorphism* between two algebraic structures $\mathscr{A}$ and $\mathscr{B}$ is a one-to-one correspondence between the sets $\mathscr{A}$ and $\mathscr{B}$ (i.e., to every $a \in \mathscr{A}$ there corresponds exactly one $b \in \mathscr{B}$ and vice versa: $a \leftrightarrow b$), which preserves the algebraic operations.

Two linear scalar product spaces Φ and Φ' are, thus, isomorphic if from

$$\Phi \ni f \leftrightarrow f' \in \Phi',$$
$$\Phi \ni g \leftrightarrow g' \in \Phi' \tag{2.8}$$

it follows that

$$\alpha f + \beta g \leftrightarrow \alpha f' + \beta g', \qquad \alpha, \beta \in \mathbb{C} \tag{2.9}$$

and

$$(f, g)_\Phi = (f', g')_{\Phi'}. \tag{2.10}$$

Scalar product spaces (and in particular Hilbert spaces) for which (2.9) and (2.10) are fulfilled are also called isometric. It often happens that two scalar product spaces are isomorphic as linear spaces, i.e., are in a one-to-one correspondence which fulfills (2.9), but are not isomorphic as scalar product spaces, i.e., the correspondence does not fulfill (2.10).

I.3 Linear Operators

Vectors in $\mathbb{R}^3$ can be transformed into each other. One example is the rotation R of a vector $\mathbf{a}$ into a vector $R\mathbf{a} = \mathbf{b}$. In analogy to this one defines transformations or *linear operators* on a linear space Φ. A function A, $A: \Phi \to \Phi$, that maps each vector ϕ in a linear space Φ into a vector $\psi \in \Phi$,

$$A\phi = \psi,$$

is called a linear operator if for all $\phi, \psi \in \Phi$ and $a \in \mathbb{C}$ it fulfills the conditions

$$A(\phi + \psi) = A\phi + A\psi, \tag{3.1}$$

$$A(a\phi) = aA\phi. \tag{3.2}$$

It is called *antilinear* if it fulfills instead of (3.2) the relation

$$A(a\phi) = a^*A\phi, \tag{3.3}$$

where a^* is the complex conjugate of a.

The operations of addition of two operators $A + B$, multiplication of an operator by a complex number aA, and multiplication of two operators AB, are defined in the following way:

$$(A + B)\phi = A\phi + B\phi, \quad (aA)\phi = a(A\phi), \quad (AB)\phi = A(B\phi), \tag{3.4}$$

for all $\phi \in \Phi$. It is easily verified that $A + B$, aA and AB are linear operators if A and B are linear operators.

Of special interest are the zero operator, denoted 0, and the unit operator or identity operator, denoted I, which are defined by

$$0\psi = 0, \quad I\psi = \psi \tag{3.5}$$

for every $\psi \in \Phi$. Note that 0 on the left is the zero operator, while 0 on the right is the zero vector of (2.1c).

For every linear operator A defined on *all* of Φ, one can define an operator $A^\dagger$ by $(A^\dagger\phi, \psi) = (\phi, A\psi)$ for every $\phi, \psi \in \Phi$. The operator $A^\dagger$ is called the *adjoint operator* of A. An operator for which $A^\dagger = A$ is called *self-adjoint* or Hermitian.[1]

The definition of linear operators was inspired by the properties of transformations on the three-dimensional space. Linear operators on a linear space Φ may represent such transformations on the physical three-dimensional space, but they can also have other physical interpretations. In particular, in quantum physics they represent physical observables.

An important notion is the notion of eigenvalue and eigenvector. A symmetric tensor $\mathbb{I}$ in three dimensions can be diagonalized by transforming to its principal axis:

$$\mathbb{I} \cdot \mathbf{a} = I_{(a)}\mathbf{a}, \tag{3.6a}$$

where $I_{(a)}$ is the eigenvalue and $\mathbf{a}$ the corresponding eigenvector. Similarly one defines eigenvalues and eigenvectors in a linear space Φ. A nonzero vector $\psi \in \Phi$ is called an eigenvector of the linear operator A if

$$A\psi = \lambda\psi \quad \text{with} \quad \lambda \in \mathbb{C}. \tag{3.6b}$$

λ is called the eigenvalue of A corresponding to the eigenvector ψ. For a given operator A there may be many (perhaps infinitely many) different

---

[1] We will usually use the term Hermitian if we do not want to distinguish between the mathematically precisely defined notions *self-adjoint*, *essentially self-adjoint*, and *symmetric*.

eigenvectors with different eigenvalues. There are also linear operators on linear spaces which do not have one single eigenvector in that space.

If A is a Hermitian operator, $A^\dagger = A$, then eigenvectors and eigenvalues have the following properties:

(1) All eigenvalues are real.
(2) If ϕ_1 and ϕ_2 are eigenvectors of A with eigenvalues λ_1 and λ_2, respectively, and if $\lambda_1 \neq \lambda_2$, then ϕ_1 and ϕ_2 are orthogonal to each other, $(\phi_1, \phi_2) = 0$.

The notion of eigenvalue is important in quantum physics: As mentioned above, the operators represent the observables of a physical system. The eigenvalues then represent the numbers which are obtained in a measurement of one of these observables.

An operator B is called the inverse of an operator A if

$$BA = AB = I. \tag{3.7a}$$

It is denoted by

$$B = A^{-1}. \tag{3.7b}$$

A linear operator U is called a unitary operator if

$$U^\dagger U = UU^\dagger = I. \tag{3.8a}$$

Because of (3.7) one defines a unitary operator also by the condition:

$$U^\dagger = U^{-1}. \tag{3.8b}$$

Let A and B be two operators. A and B are said to *commute* if

$$[A, B] \equiv AB - BA = 0. \tag{3.9}$$

$[A, B]$ is called the *commutator* of A and B.

A set $\mathscr{A}$ is an (*associative*) *algebra with unit element* iff

(a) $\mathscr{A}$ is a linear space.
(b) For every pair $A, B \in \mathscr{A}$, there is defined a product $AB \in \mathscr{A}$ such that

$$(AB)C = A(BC), \tag{3.10a}$$
$$A(B + C) = AB + AC, \tag{3.10b}$$
$$(A + B)C = AC + BC, \tag{3.10c}$$
$$(aA)B = A(aB) = aAB. \tag{3.10d}$$

(c) There exists an element $I \in \mathscr{A}$ such that

$$IA = AI = A \tag{3.11}$$

for all $A \in \mathscr{A}$.

A subset $\mathscr{A}_1$ of an algebra is called a *subalgebra of* $\mathscr{A}$ if $\mathscr{A}_1$ is an algebra with the same definitions of the operations of addition, multiplication by a number, and multiplication as given for $\mathscr{A}$; i.e., if from $A, B \in \mathscr{A}_1$, it follows that $A + B \in \mathscr{A}_1$, $aA \in \mathscr{A}_1$, and $AB \in \mathscr{A}_1$.

An algebra $\mathscr{A}$ is called a *-*algebra* if we have on the algebra a *-*operation* (involution) $A \mapsto A^\dagger$ that has the following defining properties:

(d)
$$(aA + bB)^\dagger = \bar{a}A^\dagger + \bar{b}B^\dagger,$$
$$(AB)^\dagger = B^\dagger A^\dagger,$$
$$(A^\dagger)^\dagger = A, \tag{3.12}$$
$$I^\dagger = I,$$

where $A, B \in \mathscr{A}$ and $a, b \in \mathbb{C}$. From the definitions of the sum and product of operators with a number—given in (3.4)—and from the definition of the formal adjoint operator, one can see that the set of linear operators fulfills all the axioms (a), (b), (c), and (d) of a *-algebra. Thus the set of linear operators in a linear space forms a *-algebra. A *-subalgebra of this algebra is called an *operator* *-*algebra*. It can be shown that in a certain sense every *-algebra can be realized as an operator *-algebra in a scalar-product space (generalization of the Gelfand–Naimark–Segal reconstruction theorem). In quantum mechanics physical systems are assumed to be described by operator *-algebras.

A set $X_1, X_2, \ldots, X_n$ of elements of $\mathscr{A}$ is called a *set of generators*, and $\mathscr{A}$ is said to be *generated* by the X_i ($i = 1, 2, \ldots, n$) iff each element of $\mathscr{A}$ can be written

$$A = cI + \sum_{i=1}^{n} c^i X_i + \sum_{i,j}^{n} c^{ij} X_i X_j + \cdots, \tag{3.13}$$

where $c, c^i, c^{ij}, \ldots \in \mathbb{C}$.

Defining algebraic relations are relations among the generators

$$P(X_i) = 0, \tag{3.14}$$

where $P(X_i)$ is a polynomial with complex coefficients of the n variables X_i. An element $B \in \mathscr{A}$,

$$B = bI + \sum b^i X_i + \sum b^{ij} X_i X_j + \cdots, \tag{3.15}$$

where $b, b^i, \ldots \in \mathbb{C}$, is *equal* to the element A iff (3.15) can be brought into the same form (3.13) with the same coefficients $c, c^i, c^{ij}, \ldots$ by the use of the defining relations (3.14).

I.4 Basis Systems and Eigenvector Decomposition

As in the three-dimensional space $\mathbb{R}^3$, one can also introduce a system of basis vectors in a general linear space Φ. In $\mathbb{R}^3$ one conveniently chooses a system of three normalized vectors

$$\mathbf{e}_1, \mathbf{e}_2, \mathbf{e}_3, \qquad \|\mathbf{e}_i\| = 1,$$

which are orthogonal to each other

$$\mathbf{e}_i \cdot \mathbf{e}_j = \delta_{ij}, \qquad i, j = 1, 2, 3.$$

Such an *orthonormal basis system* can also always be chosen in a linear scalar product space Φ. We denote these basis vectors in various ways by

$$e_i = |e_i) = |i\rangle, \qquad i = 1, 2, 3, \ldots, N. \qquad (4.1)$$

Their scalar products are written in one of the following ways:

$$(e_i, e_j) \equiv (e_i|e_j) \equiv \langle i|j\rangle = \delta_{ij}, \qquad i, j = 1, 2, \ldots. \qquad (4.2)$$

If Φ is N-dimensional then there are N linearly independent vectors in this orthonormal basis system. N can be infinite.

The basis system fulfilling (4.2) can be chosen arbitrarily, but it is convenient to choose it such that the particular physical problem under investigation takes its simplest mathematical form. For example, if one describes a rigid body with moment of inertia tensor $\mathbb{I}$ in the usual three-dimensional space, then it is useful to choose the basis system, $e_i, i = 1, 2, 3$ such that

$$\mathbf{e}_i \cdot \mathbb{I} \cdot \mathbf{e}_j = \mathrm{I}_{(j)} \delta_{ij},$$

i.e., choose—according to (3.6a)—eigenvectors of the tensor $\mathbb{I}$.

Similarly, it is extremely useful to choose as the basis system for the space of physical states, Φ, eigenvectors of an operator A which represents an important observable of the physical system under investigation (most frequently one chooses for A the energy operator H, the position operator Q or the momentum operator P). Thus one would want to choose for the basis system a set of vectors e_i which fulfill

$$A e_i = a_i e_i, \qquad a_i \in \mathbb{C}, \qquad (4.3)$$

where the a_i are the eigenvalues. These normalized eigenvectors are often denoted by

$$e_i = |a_i) \quad \text{or} \quad e_i = |a_i\rangle.$$

The symbol $|\,\rangle$ with a letter ϕ, a or i in it denoting a vector ϕ, an eigenvalue a or an index i of an eigenvalue is called a *ket*; the symbol $\langle\,|$ with a letter in it is called a *bra*.[2]

In the three-dimensional space $\mathbb{R}^3$ every vector $\mathbf{v}$ can be expanded with respect to a basis system of eigenvectors of any symmetric tensor, i.e.,

$$\mathbf{v} = \sum e_i v_i, \quad \text{where the} \quad v_i = \mathbf{e}_i \cdot \mathbf{v} \qquad (4.4a)$$

are the coordinates or components of the vector $\mathbf{v}$ with respect to the basis of eigenvectors $\{e_i\}$ of the symmetric tensor $\mathbb{I}$ (3.6a). The same can be proven

[2] P. A. M. Dirac (1958).

for any finite-dimensional linear scalar product space Φ. This means that: For every Hermitian operator A in a finite-dimensional space Φ there exists a system of eigenvectors

$$Ae_i = a_i e_i, \quad i = 1, 2, \ldots, N = \text{finite}, \tag{4.3}$$

such that every vector $\phi \in \Phi$ can be written as

$$\phi = \sum_{i=1}^{N} e_i(e_i, \phi) \equiv \sum_{i=1}^{N} e_i(a_i | \phi), \tag{4.4b}$$

where the complex numbers

$$c_i = (e_i, \phi) \equiv (a_i | \phi) \tag{4.5}$$

are the components of the vector ϕ with respect to the basis $\{e_i\}$.

The set of a_i's (which are real if A is Hermitian) is called the spectrum of A and the above statement is called the spectral theorem for the operator A in a finite-dimensional space. Equation (4.4b) is called the *spectral decomposition* of the vector ϕ or the *eigenvector expansion* of ϕ.

For infinite-dimensional spaces the above statement is in general not correct. Though there is always a countably infinite set of orthogonal basis vectors, every self-adjoint operator need not have a countably infinite set of eigenvectors which form a basis. Furthermore, quantum physics also requires operators whose set of eigenvalues is a continuous set, or even a union of a discrete set and a continuous set.

It could have been that all operators which appear in quantum physics have the property that the set of their eigenvalues is discrete. That would have been the case if the measurement for every observable in quantum physics could only lead to a discrete set of numbers. Then only an infinite-dimensional generalization of (4.3) and (4.4) would be needed. However, there are observables in physics whose measurement can lead to any number out of a continuous set of numbers (e.g., the observables momentum and position can in many cases take any value x with $-\infty < x < +\infty$). Therefore we need not only the infinite-dimensional generalization (which can be obtained in the Hilbert space) but also the continuous, infinite-dimensional generalization of (4.3) and (4.4). There exist in fact spaces Φ—or more precisely, there exist topologies for infinite-dimensional linear spaces—for which the generalization of the finite-dimensional spectral decomposition can be proven for all self-adjoint operators needed in physics.[3] This generalization is the nuclear spectral theorem. As the eigenvector decomposition is so important for physics we will only use spaces for which this theorem is valid.

We cannot present the mathematics here and will explain the discrete and continuous eigenvector decomposition in an infinite-dimensional space in analogy to the finite-dimensional case (4.4b) or (4.4a). We consider the discrete and continuous cases separately; the general case of an arbitrary self-adjoint operator of physics will be a combination of these two cases.

---

[3] These spaces Φ and their conjugates $\Phi^\times$ (cf. Section I.7 below) (and the closely related theory of distributions) and the triplet $\Phi \subset \mathscr{H} \subset \Phi^\times$ they form with the Hilbert space $\mathscr{H}$ did in fact not exist when they were needed for quantum mechanics. The creation of these mathematical structures was inspired by the development of quantum theory.

We will call the self-adjoint operator with a discrete set of eigenvalues H and the operator with a continuous set of eigenvalues Q. Then the spectral theorem asserts:

There exists a system of eigenvectors $|E_n)$ in the discrete case and $|x\rangle$ in the continuous case;

$$H|E_n) = E_n|E_n), \qquad E_n = E_0, E_1, E_2, \ldots, \tag{4.3d}$$

$$Q|x\rangle = x|x\rangle; \qquad -\infty \leq m \leq x \leq M \leq +\infty, \tag{4.3c}$$

such that *every* $\phi \in \Phi$ can be expanded in terms of these eigenvectors:

$$\phi = \sum_{n=0}^{\infty} |E_n)(E_n|\phi), \tag{4.4d}$$

$$\phi = \int_m^M dx|x\rangle\langle x|\phi\rangle, \tag{4.4c}$$

and $\phi = 0$ if and only if all its components are zero, i.e., $(E_n|\phi) = 0$ for all E_n and $\langle x|\phi\rangle = 0$ for all x.[4]

A system of eigenvectors $|E_n)$ or $|x\rangle$ with these properties is called *complete* or a *basis system*. Thus the spectral theorem asserts the existence of a complete system of eigenvectors of a self-adjoint operator. $(E_n|\phi)$ or $\langle x|\phi\rangle$ are called the coordinates or components of ϕ with respect to the basis system $\{|E_n)\}$ or $\{|x\rangle\}$, respectively. They are, as in the three-dimensional case, the scalar products of the eigenvectors with ϕ:

$$\langle x|\phi\rangle = (|x\rangle, \phi), \tag{4.5c}$$

$$(E_n|\phi) = (|E_n), \phi). \tag{4.5d}$$

Thus $(E_n|\phi)$ is the infinite-dimensional generalization of the v_i in (4.4a), and $\langle x|\phi\rangle$ is the continuous infinite-dimensional generalization of v_i.

Whereas the $|E_n)$ are proper eigenvectors, the $|x\rangle$ are called generalized eigenvectors, or, *eigenkets*. Though we can manipulate them as if they were proper eigenvectors, mathematically there is an important difference between the discrete basis vectors $|E_n)$ and the continuous basis vectors $|x\rangle$: the $|E_n)$ are in Φ while the $|x\rangle$ are in $\Phi^\times$, the space of continuous antilinear functionals over Φ. We shall define and explain these mathematical notions in Section I.7. Here we shall try to convey the meaning of these generalized eigenvectors by analogy to the finite-dimensional case.

The set of eigenvalues E_n in (4.3d) is called the spectrum of the operator H. If H has a discrete set of eigenvalues, the spectrum is called discrete. All the corresponding eigenvectors $|E_n)$ enter into the discrete basis vector expansion (4.4d) and there are no further eigenvectors with discrete eigenvalue that enter into the basis vector expansion (4.4d). The set of continuous eigenvalues x, whose eigenvectors enter into the generalized eigenvector expansion (4.4c), is called the continuous spectrum of the operator Q.

---

[4] The simple nondegenerate form (4.4d), (4.4c) is valid if the operator A (H or Q) is cyclic, i.e., if there exists an $f \in \Phi$ such that $A^n f \equiv f_{(n)}$ generate the entire space Φ, which means that any $\phi \in \Phi$ can be written as $\phi = \sum_n f_{(n)} c_{(n)}$ where $c_{(n)}$ are complex numbers. Degenerate spectra, which occur when more than one quantum number is needed, will be discussed later in the text.

In general—and this depends upon the properties of the space Φ—there are more generalized eigenvectors of Q—i.e., kets which fulfill (4.3c) and whose precise definition will be given in Section I.7—than enter in the eigenvector expansion (4.4c). Their generalized eigenvalues we will not include in the definition of the continuous spectrum. Whereas the discrete eigenvalues of a self-adjoint operator are always real, the generalized eigenvalues need not be real; they can be real or complex, and even if they are real they need not necessarily belong to the spectrum, i.e., appear in the integral (4.4c). But for a self-adjoint operator there is always a real subset of the set of generalized eigenvalues whose eigenvectors are complete.

The most general form of the spectral theorem for an operator A representing a physical observable is a combination of (4.4d) and (4.4c):

$$\phi = \sum |a_i)(a_i|\phi\rangle + \int da\, |a\rangle\langle a|\phi\rangle, \qquad (4.4g)$$

where the sum is over the discrete spectrum and the integral is over the (absolutely) continuous spectrum of A. It can happen that some or all values a_i appearing in the sum also appear in the integral. Then they are called discrete eigenvalues in the continuous spectrum. If this happens for a_k then $|a_k)$, is still orthogonal to all $|a\rangle$ including $|a_k\rangle$, i.e.,

$$(a_k|\phi^c) = 0 \quad \text{for} \quad \phi^c = \int da\, |a\rangle\langle a|\phi^c\rangle.$$

The spectral theorem (4.4b), (4.4d), (4.4c), and (4.4g)—which we have to accept here without proof[5]—is the central statement, from which the other results of this section follow.

To see that the coordinates $(E_n|\phi)$ are indeed what their notation indicates, namely the scalar product of the vector with the basis vector $|E_n)$, we calculate the scalar product of (4.4d) with the eigenvector $|E_m)$. From (4.4d), it follows that

$$(|E_m), \phi) = \sum_{n=0}^{\infty} (|E_m), |E_n))(E_n|\phi). \qquad (4.6d)$$

Since $|E_m)$ and $|E_n)$ are eigenvectors of the same Hermitian operator H, if $E_n \neq E_m$ they must be orthogonal to each other,

$$(|E_m), |E_n)) = 0. \qquad (4.7')$$

For $E_n = E_m$ we normalize them:

$$(|E_n), |E_n)) = \||E_n)\|^2 = 1. \qquad (4.7'')$$

---

[5] The finite-dimensional case (4.4b) is easily reduced to the problem of the number of roots of a polynomial of degree N, as shown in Section I.5 below in particular by Equation (5.17). Equations (4.4d) and (4.4c) need much more mathematical preparation, cf. I. M. Gelfand *et al.* (1964), Vol. 4, or K. Maurin (1968).

This we combine and write

$$(|E_m), |E_n)) = (E_m|E_n) = \delta_{E_n E_m} = \delta_{nm}; \qquad n, m = 1, 2, \ldots, \qquad (4.7d)$$

where the Kronecker δ is defined by

$$\delta_{nm} = \delta_{E_n E_m} = \begin{cases} 1 & \text{for } n = m; \quad E_n = E_m, \\ 0 & \text{for } n \neq m; \quad E_n \neq E_m. \end{cases} \qquad (4.8)$$

So these eigenvectors of the self-adjoint operator H have the property (4.2) as required of orthonormal basis vectors. Inserting (4.7d) into (4.6d) one obtains:

$$(|E_m), \phi) = \sum_{n=0}^{\infty} \delta_{mn}(E_n|\phi) = (E_m|\phi). \qquad (4.5)$$

This is the expected identity (4.5d).

The spectral theorem (4.4d) can be written in different forms: one can omit the arbitrary vector $\phi \in \Psi$ on both sides of (4.4d) and obtain the *spectral resolution of the identity operator I*:†

$$I = \sum_{n=0}^{\infty} |E_n)(E_n|. \qquad (4.9d)$$

One can multiply both sides of (4.4d) with the operator H,

$$H\phi = \sum_{n=0}^{\infty} H|E_n)(E_n|\phi) = \sum_{n=0}^{\infty} E_n|E_n)(E_n|\phi),$$

and then omit the arbitrary vector ϕ on both sides to obtain:

$$H = \sum_{n=0}^{\infty} E_n|E_n)(E_n|. \qquad (4.10d)$$

This identity between the operator H and the weighted sum of operators $|E_n)(E_n|$ is called the *spectral resolution of the self-adjoint operator H with discrete spectrum*.

One can take the scalar product of (4.4d) with another $\psi \in \Phi$, then one obtains

$$(\psi, \phi) = \sum_{n=0}^{\infty} (\psi|E_n)(E_n|\phi) = \sum_{n=0}^{\infty} (E_n|\psi)^*(E_n|\phi). \qquad (4.11d)$$

In particular if one chooses $\psi = \phi$ one obtains

$$\|\phi\|^2 = (\phi, \phi) = \sum_{n=0}^{\infty} (\phi|E_n)(E_n|\phi) = \sum_{n=0}^{\infty} |(E_n|\phi)|^2. \qquad (4.12d)$$

Equation (4.11d) is the analogue of the formula

$$\mathbf{v} \cdot \mathbf{x} = \sum_{i,j=1}^{3} v_i \mathbf{e}_i \cdot \mathbf{e}_j x_j = \sum_{i=1}^{3} v_i x_i$$

for the ordinary scalar product in $\mathbb{R}^3$.

† The quantities $\Lambda_n = |E_n)(E_n|$ are projection operators as defined in the mathematical insert on p. 58.

As in the three-dimensional space $\mathbb{R}^3$, a vector ϕ is completely specified by its components $(E_n|\phi)$ with respect to a given basis $|E_n)$. But unlike the three-dimensional space $\mathbb{R}^3$, where any sequence of three real numbers (x_1, x_2, x_3) defines a vector x, or unlike the N-dimensional complex space $\mathbb{C}^N$ where any sequence of N complex numbers $(\xi_1, \xi_2, \ldots, \xi_N)$ defines a vector x, an arbitrary infinite sequence of complex numbers

$$(\phi_1, \phi_2, \phi_3, \ldots, \phi_n, \ldots)$$

does not in general define a vector in Φ. As can be seen from (4.12d), in order that the scalar product be defined, the infinite sequence that defines a vector must at least fulfill the condition

$$\sum_{i=1}^{\infty} |\phi_i|^2 < \infty, \tag{4.12d'}$$

i.e., it must be *square summable*. If one further wants to demand that every operator A of the set of operators which represents physical observables be defined in the whole space Φ, one has to require that all $A\phi$ are well defined in Φ, which means that $(A\phi, A\phi)$ must be finite. Choosing $A = H^p$ where $p = 0, 1, 2, \ldots$ is any power, one obtains for this requirement the following condition:

$$(H^p\phi, H^p\phi) = \sum_{n=0}^{\infty} (\phi|H^p|E_n)(E_n|H^p|\phi)$$

$$= \sum_{n=0}^{\infty} E_n^{2p}|(E_n|\phi)|^2 < \infty \quad \text{for any} \quad p = 0, 1, 2, \ldots. \tag{4.13d}$$

Thus not only will $\{(E_n|\phi)|n = 0, 1, 2, \ldots\}$ have to be square summable, but also $\{E^p(E_n|\phi)\}$ has to be a square summable sequence for any $p = 0, 1, 2, \ldots$.

Fortunately these (topological) questions of what happens at infinity are not very relevant for physics as only a finite number of the $|(E_n|\phi)|$ can be determined experimentally.

We now turn to the continuous spectrum and repeat the above considerations for the continuous case. We calculate the scalar product of ϕ with the generalized eigenvector $|x\rangle$[6] using Equation (4.4c):

$$(|x\rangle, \phi) \equiv \int dy\, (|x\rangle, |y\rangle)\langle y|\phi\rangle.$$

This we rewrite with the following definition of the new symbol

$$\langle x|y\rangle \equiv (|x\rangle, |y\rangle)$$

---

[6] More precisely, using the notions that will be introduced in Section I.7, we should say that we calculate the value $(|x\rangle, \phi)$ of the functional $|x\rangle$ at the vector $\phi \in \Phi$, $(|x\rangle, \phi) = \langle x|\phi\rangle$ is a generalization of the usual scalar product.

as

$$(|x\rangle, \phi) = \int dy \, \langle x|y\rangle\langle y|\phi\rangle. \tag{4.6c}$$

$\langle x|\phi\rangle$ is the coordinate of the vector ϕ along the direction of the basis vector $|x\rangle$. $(|x\rangle, \phi)$ is the scalar product of ϕ with the basis vector $|x\rangle$. These two quantities should be the same, as stated by (4.5). Therefore (4.6c) has the form

$$\langle x|\phi\rangle = \int_{-\infty}^{\infty} dy \, \langle x|y\rangle\langle y|\phi\rangle. \tag{4.6c'}$$

The scalar products $\langle y|\phi\rangle$ are functions of the continuous variable y in the same way as the scalar products $(E_n|\phi)$ are functions of the discrete variable E_n. Equation (4.6c′) therefore says that the mathematical quantity $\langle x|y\rangle$ has the property that it maps the function $\phi(y) = \langle y|\phi\rangle$ by integration into its value at the point x: $\phi(x) = \langle x|\phi\rangle$. There exist no well-behaved and not even a locally integrable function which has this property.

Such a quantity is called a distribution or a generalized function. The distribution $\langle x|y\rangle$ *defined* by (4.6c′) for a class of well-behaved functions $\phi(x) = \langle x|\phi\rangle$ is called the Dirac δ-function and is denoted in analogy to (4.7d) as

$$\langle x|y\rangle = \delta(x - y). \tag{4.7c}$$

$\delta(x - y)$ is a generalization of the Kronecker $\delta_{E_n E_m}$ which is usually defined by (4.8) but which could as well have been defined by

$$(E_m|\phi) = \sum_{n=0}^{\infty} \delta_{E_m E_n}(E_n|\phi) \tag{4.6d'}$$

for a class of infinite sequences $\{(E_n|\phi)\}$. When we will use function sequences of the δ-type in Section II.8, we will see in which sense $\delta(x - y)$ can be considered as a generalization of the right-hand side of (4.8).

The eigenvectors $|E_n)$ are normalized to 1 by (4.7″); the generalized eigenvectors $|x\rangle$ fulfilling (4.7c) are called δ-function normalized. They are not dimensionless, but have the dimension $1/\sqrt{\dim dx}$. For example, if dx has the dimension cm, then $\langle x'|x\rangle$ has the dimension cm^{-1}, and $|x\rangle$ has the dimension cm$^{-1/2}$.

Instead of the generalized eigenvectors with δ-function normalization (4.7c) one could also choose generalized eigenvectors of Q with a different normalization. Instead of (4.4c) one writes

$$\phi = \int d\mu(x)|x\}_{\rho\rho}\{x|\phi\rangle, \tag{4.4c_\mu}$$

where the $|x\}_\mu$ are again eigenvectors of Q:

$$Q|x\}_\rho = x|x\}_\rho, \tag{4.3c_\mu}$$

and $\rho(x) = d\mu(x)/dx$ is a real nonnegative and integrable function.

In order that the new components of ϕ, the $_\rho\{x|\phi\rangle$, be the scalar product of ϕ with the new eigenvectors $|x\}_\rho$, i.e., in order that

$$_\rho\{x|\phi\rangle = \int \rho(y) \, dy \, _\rho\{x|y\}_{\rho\rho}\{y|\phi\rangle, \tag{4.6c'_ρ}$$

one has to demand

$$d\mu(y) \, _\rho\{x|y\}_\rho = dy \, \langle x|y\rangle = dy \, \delta(x - y).$$

So the normalization of the new generalized eigenvectors is

$$_\rho\{x|y\}_\rho = \left(\frac{d\mu(y)}{dy}\right)^{-1} \delta(x - y) = \rho^{-1}(y) \, \delta(x - y). \tag{4.7c_ρ}$$

Thus when the integration contains the weight function $\rho(x)$, the generalized eigenvector normalization contains the factor $\rho^{-1}(x)$.

The transformation between the two basis systems is

$$|x\}_\rho = |x\rangle \frac{1}{\sqrt{\rho(x)}}. \tag{4.14}$$

The most appropriate choice for $\rho(x)$ depends upon the property of the operator Q and its relation to the other operators of the problem.[7]

As in the case of the discrete spectrum the spectral theorem (4.4c) can be written in different forms. Omitting the arbitrary vector ϕ one obtains the resolution of the identity:

$$I = \int dx \, |x\rangle\langle x| = \int \rho(x) \, dx \, |x\}_{\rho\rho}\{x|. \tag{4.9c}$$

Multiplying both sides of (4.4c) with the operator Q and then omitting the arbitrary vector ϕ one obtains after (4.3c) has been used

$$Q = \int dx \, x|x\rangle\langle x| \tag{4.10c}$$

which we call the spectral resolution of the self-adjoint operator Q with (absolutely[7]) continuous spectrum.

The scalar product of two vectors $\phi, \psi \in \Phi$, obtained from (4.4c), is

$$(\psi, \phi) = \int dx \, \langle\psi|x\rangle\langle x|\phi\rangle. \tag{4.11c}$$

---

[7] The nuclear spectral theorem for an arbitrary self-adjoint operator does in fact not assert (4.4g) but (4.4c$_\rho$) with a general measure $d\mu(x)$, and it does not say anything about the spectral measure $d\mu(x)$ in addition to the assertion of its existence. However, all operators used in physics are of the special kind for which either $d\mu(x) = \rho(x) \, dx$ with $\rho(x)$ as described above (such operators are said to have an absolutely continuous spectrum and for them one can always make the transformation (4.14) from $|x\}_\rho$ to $|x\rangle$), or they have the property that $d\mu(x) = \sum_{x_i} \delta(x - x_i) \, dx$ (these are the operators with discrete spectrum), or they have both an absolutely continuous and a discrete spectrum. So (4.4g) is the most general form needed for physics.

From $(\psi, \phi) = (\phi, \psi)^*$ and from $(\phi, \psi) = \int dx \langle\phi|x\rangle\langle x|\psi\rangle$ for any $\phi, \psi \in \Phi$ one concludes that for the generalized scalar product one has the same relation as for the scalar product:

$$\langle\psi|x\rangle = \langle x|\psi\rangle^*.$$

Using the notation

$$\langle x|\phi\rangle = \phi(x), \qquad \langle x|\psi\rangle^* = \psi^*(x),$$

one can write (4.11c) in the form

$$(\psi, \phi) = \int dx \, \psi^*(x)\phi(x). \qquad (4.11c')$$

In particular, if one chooses $\psi = \phi$ one obtains

$$\|\phi\|^2 = (\phi, \phi) = \int dx \, \phi^*(x)\phi(x) = \int dx \, |\phi(x)|^2. \qquad (4.12c)$$

From this we see that not any arbitrary function $\phi(x)$ can give the components of a vector $\phi \in \Phi$ with respect to the continuous basis system $|x\rangle$, but only those functions for which the integral on the right-hand side of Equation (4.12c) exists, i.e., the square integrable functions.[8]

If one demands of the space Φ that on all its elements the operator Q and any arbitrary power thereof, Q^p $(p = 0, 1, 2, \ldots)$, be well defined, then one must have that

$$\|Q^p\phi\|^2 = (Q^p\phi, Q^p\phi) = \int dx \, x^{2p}|\phi(x)|^2 < \infty. \qquad (4.13c)$$

Thus $|\phi(x)|$ must decrease faster than any power of x. If other operators are also to be defined everywhere in Φ further conditions will have to be imposed on the components $\langle x|\phi\rangle$ of $\phi \in \Phi$. Thus the realization of Φ must be much better than L^2. An example of a realization of Φ is the *Schwartz-space S. S* is defined as the space of infinitely differentiable complex-valued functions which together with their derivatives vanish at infinitely more rapidly than any power of $1/x$. We call those functions well behaved.

It can be that the space Φ is such that the components of all vectors $\psi \in \Phi$ with respect to the continuous basis $|x\rangle$, $\psi(x) = \langle x|\psi\rangle$, are boundary values of analytic functions $\psi(z)$ on the complex plane or a domain of the complex

---

[8] The scalar product space in which the scalar product is realized by the integral (4.11c') is called the space of square integrable functions L^2. If the scalar product space is complete then it is called a Hilbert space. For this reason a scalar product space is also called a pre-Hilbert space. The space of square (Lebesgue) integrable functions is one realization of the Hilbert space. Another realization of the Hilbert space is the space of square summable infinite sequences, cf. (4.12d').

plane, e.g., the lower half plane. Then the contour of integration in (4.11c′) can be deformed and one obtains

$$(\psi, \phi) = \int_{-\infty}^{+\infty} dx \; \psi^*(x)\phi(x) = \int_{\mathscr{C}} dz \; \psi^*(z)\phi(z)$$

$$= \int_{\mathscr{C}} dz \; \langle \psi | z \rangle \langle z | \phi \rangle. \tag{4.15}$$

$\mathscr{C}$ can be any contour which is obtained from the contour along the real axis by deforming it into the domain of analyticity without passing over a singularity. This defines the generalized eigenvectors $|z\rangle$ of the self-adjoint operator Q:

$$Q|z\rangle = z|z\rangle \tag{4.16}$$

with complex eigenvalue z. These generalized eigenvectors $|z\rangle$ can also be used in a generalized basis vector expansion. This is obtained from (4.15) by omitting the arbitrary vector $\psi \in \Phi$

$$\phi = \int_{\mathscr{C}} dz \; |z\rangle \langle z | \phi \rangle. \tag{4.4cc}$$

Thus instead of using the generalized basis vector expansion (4.4c) with respect to the generalized eigenvectors $|x\rangle$ with $x \in \mathbb{R}$, one can in this case as well use the generalized basis vector expansion (4.4cc) with respect to the generalized eigenvectors $|z\rangle$ with complex eigenvalues $z \in \mathbb{C}$. This possibility will be used in the description of decaying states.

I.5 Realizations of Operators and of Linear Spaces

We have already mentioned that the components of all vectors $\phi \in \Phi$ with respect to a basis system constitute a realization of the space Φ. A three-dimensional vector $\mathbf{x}$ in R^3 is thus realized by the sequence (x_1, x_2, x_3) of its components $x_i = \mathbf{e}_i \cdot \mathbf{x}$. The components, of course, depend upon the basis system chosen. If one takes another basis system $\mathbf{e}_i'$ the components of the same vector $\mathbf{x}$ change: $x_i' = \mathbf{e}_i' \cdot \mathbf{x}$.

In a finite-dimensional linear scalar product space the components are finite sequences of complex numbers (4.5). In an infinite-dimensional space they are either infinite sequences $\phi_i = (E_i | \phi) \equiv (i | \phi)$ which fulfill certain conditions such as (4.12d′) and (4.13d), or continuous infinite sequences $\phi(x) = \langle x | \phi \rangle$, i.e., functions of a continuous variable x, which also fulfill additional conditions like (4.12c) and (4.13c).

As in the three-dimensional case, if one changes the basis system of the space, the components of a given vector ϕ change too. Thus, if we take in addition to H of (4.3d) another operator A which has, like H, a discrete spectrum

$$A|a_i\rangle = a_i|a_i\rangle, \qquad a_i = a_0, a_1, a_2, \ldots, \tag{5.1}$$

then the same vector ϕ of (4.4d) can also be expanded as

$$\phi = \sum_{i=0}^{\infty} |a_i\rangle\langle a_i|\phi\rangle \equiv \sum_i |i\rangle\langle i|\phi\rangle, \tag{5.2}$$

where we defined for writing convenience: $|i\rangle \equiv |a_i\rangle$.

If A and H do not commute, then the components $\langle a_i|\phi\rangle$ and $(E_n|\phi)$ are entirely different. The components can be written as a column matrix

$$\phi \leftrightarrow (E_n|\phi) = \begin{pmatrix} (E_0|\phi) \\ (E_1|\phi) \\ \vdots \end{pmatrix}, \tag{5.3}$$

$$\phi \leftrightarrow \langle a_i|\psi\rangle = \begin{pmatrix} \langle a_0|\phi\rangle \\ \langle a_1|\phi\rangle \\ \langle a_2|\phi\rangle \\ \vdots \end{pmatrix}. \tag{5.4}$$

Column matrices can be added, by adding each of their components, they can be multiplied by a number by multiplying each component and a scalar product between column matrices can be defined as by the expression furthest to the right in Equation (4.11d). With these definitions it is easy to see that the set of column matrices: $\langle i|\phi\rangle$, $\langle i|\psi\rangle$, $\langle i|\chi\rangle$, ..., $i = 1, 2, 3, ...$ form a linear scalar product space $\mathbb{C}_{(i)}^{\infty}$ (cf. also Problem 1). In the same way the set of column matrices of components with respect to the basis system $\{|E_n\rangle | E_n = E_1, E_2, ...\}$, the $(E_n|\phi)$, $(E_n|\psi)$, (E_n, χ), ..., form a linear scalar product space $\mathbb{C}_{(E_n)}^{\infty}$. It is easy to see that $\mathbb{C}_{(E_n)}^{\infty}$ and $\mathbb{C}_{(i)}^{\infty}$ are isomorphic to each other and isomorphic to the space Φ:

$$(\langle i|\phi\rangle) \leftrightarrow ((E_n|\phi)) \leftrightarrow \phi,$$
$$(\langle i|\psi\rangle) \leftrightarrow ((E_n|\psi)) \leftrightarrow \psi.$$

Therewith we can define the space Φ by specifying the set of all infinite column matrices. For instance we shall define $\mathbb{C}_{(E_n)}^{\infty}$ as the space of all $(E_n|\phi)$ which fulfill (4.13d) with $E_n = (n + \frac{1}{2})$, $n = 0, 1, 2, ...$. Φ can then be defined as the space isomorphic to it.[9] Thus the same vector can be represented by entirely different column matrices.

When vectors are realized by column matrices, operators are realized by quadratic matrices. We obtain these matrices of an operator B in the following way:

Calculate

$$B\phi = \sum_i B|i\rangle\langle i|\phi\rangle, \tag{5.5}$$

---

[9] $\mathbb{C}_{(E_n)}^{\infty}$ has a natural topological structure which Φ then inherits. Φ is then the largest space in which all the operators H^p, $p = 0, 1, 2, ...$ are continuous operators, and its topological structure could have been defined by this requirement.

and take the scalar product of this equation with $|a_j\rangle$

$$\langle j|B\phi\rangle = \sum_i \langle j|B|i\rangle\langle i|\phi\rangle. \qquad (5.6)$$

This can be written in matrix notation as

$$\begin{pmatrix} \langle 1|B\phi\rangle \\ \langle 2|B\phi\rangle \\ \langle 3|B\phi\rangle \\ \vdots \end{pmatrix} = \begin{pmatrix} \langle 1|B|1\rangle & \langle 1|B|2\rangle & \langle 1|B|3\rangle \cdots \\ \langle 2|B|1\rangle & \langle 2|B|2\rangle & \langle 2|B|3\rangle \\ \langle 3|B|1\rangle & \langle 3|B|2\rangle & \cdots & \cdots \\ \vdots & \vdots & \vdots \end{pmatrix} \begin{pmatrix} \langle 1|\phi\rangle \\ \langle 2|\phi\rangle \\ \langle 3|\phi\rangle \\ \vdots \end{pmatrix}. \qquad (5.7)$$

The right-hand side of (5.6) gives the definition of multiplication of a quadratic matrix with the column matrix on the right-hand side of (5.7). Thus we see that the linear operator B is realized by the infinite matrix

$$B \leftrightarrow \langle j|B|i\rangle = \begin{pmatrix} \langle 1|B|1\rangle & \langle 1|B|2\rangle & \cdots \\ \langle 2|B|1\rangle & \langle 2|B|2\rangle & \cdots \\ \langle 3|B|1\rangle & & \cdots \\ \vdots & \end{pmatrix}, \qquad (5.8)$$

which is called the matrix of the operator B with respect to the basis system $\{|i\rangle | i = 1, 2, 3, \ldots\}$.

Now, let D be a second operator and apply $D + B$ to the basis vector $|i\rangle$ using (3.4)

$$(B + D)|i\rangle = B|i\rangle + D|i\rangle.$$

Taking the scalar product of both sides of this equation with the basis vector $|j\rangle$ one obtains (using (2.2d)):

$$\langle j|(B + D)|i\rangle = \langle j|B|i\rangle + \langle j|D|i\rangle. \qquad (5.9)$$

Thus the matrix of the sum of two linear operators is equal to the sum of the matrices, as defined by the right-hand side of (5.9). Let us now consider the matrix element of DB and use the resolution of the identity (4.9d) for the basis system (5.1):

$$\langle j|D\,B|i\rangle = \sum_{n=0}^{\infty} \langle j|D|n\rangle\langle n|B|i\rangle. \qquad (5.10)$$

Written in matrix notation this has the form:

$$\begin{pmatrix} \langle 1|DB|1\rangle & \langle 1|DB|2\rangle & \langle 1|DB|3\rangle & \cdots \\ \langle 2|DB|1\rangle & \langle 2|DB|2\rangle & \langle 2|DB|3\rangle & \cdots \\ & \cdots\cdots\cdots & \end{pmatrix}$$

$$= \begin{pmatrix} \langle 1|D|1\rangle & \langle 1|D|2\rangle & \langle 1|D|3\rangle & \cdots \\ \langle 2|D|1\rangle & \langle 2|D|2\rangle & \langle 2|D|3\rangle & \cdots \\ & \cdots\cdots\cdots & \end{pmatrix} \begin{pmatrix} \langle 1|B|1\rangle & \langle 1|B|2\rangle & \cdots \\ \langle 2|B|1\rangle & \langle 2|B|2\rangle & \cdots \\ \langle 3|B|1\rangle & \langle 3|B|2\rangle & \cdots \\ \vdots & \end{pmatrix}$$

$$(5.10')$$

where the right-hand side of (5.10) gives the definition of the multiplication of two (infinite-dimensional) quadratic matrices.

Thus in the realization of the space Φ by the space of column vectors, the operators are realized by matrices in such a way that the sum of two operators corresponds to the sum of their matrices and the product of two operators corresponds to the product of their matrices.

If the basis system is chosen to be a system of eigenvectors of the operator A as in (5.1), then the matrix of A with respect to this basis system is diagonal as from (5.1) follows

$$\langle j|A|i\rangle = a_i\langle j|i\rangle = a_i\delta_{ij}, \tag{5.11}$$

or, written as a matrix,

$$\begin{pmatrix} \langle 1|A|1\rangle & \langle 1|A|2\rangle & \langle 1|A|3\rangle & \cdots \\ \langle 2|A|1\rangle & \langle 2|A|2\rangle & \cdots \\ \langle 3|A|1\rangle & \cdots \\ \vdots & \vdots \end{pmatrix} = \begin{pmatrix} a_1 & 0 & 0 & 0 & 0 & \cdots \\ 0 & a_2 & 0 & 0 & 0 & \cdots \\ 0 & 0 & a_3 & 0 & 0 & \cdots \\ \vdots & \vdots & \vdots & \vdots & \vdots \end{pmatrix}. \tag{5.11'}$$

The matrix elements with respect to the basis system of eigenvectors of A of all other operators that do not commute with A have nonzero off-diagonal matrix elements.

As the proper eigenvectors of H are also elements of Φ one can use (4.4d) and expand them with respect to the basis system $|i\rangle$ of (5.1) and vice versa

$$|E_n) = \sum_{i=1}^{\infty} |i\rangle\langle i|E_n); \tag{5.12}$$

$$|i\rangle = \sum_{n=1}^{\infty} |E_n)(E_n|i\rangle. \tag{5.12'}$$

Taking the scalar product of this equation with ϕ and then taking the complex conjugate one obtains

$$(E_n|\phi) = \sum_{i=1}^{\infty} (E_n|i\rangle\langle i|\phi). \tag{5.13}$$

Thus $(E_n|i\rangle$ are the matrix elements of a matrix that transforms the infinite-column matrix (5.4) into the infinite-column matrix (5.3). It is called the transition matrix (or transformation matrix) between the two basis systems $\{|i\rangle|i=1, 2, \ldots$ and $\{|E_n)|n=1, 2, \ldots\}$. Its elements $(E_n|i\rangle$ are also called transition coefficients. Taking the scalar product of (5.12') with $|j\rangle$ gives

$$\langle j|i\rangle = \delta_{ij} = \sum_{n=1,2,\ldots}^{\infty} \langle j|E_n)(E_n|i\rangle = \sum_n (E_n|j\rangle^*(E_n|i\rangle. \tag{5.14'}$$

Similarly one obtains

$$(E_m|E_n) = \delta_{mn} = \sum_{j=1}^{\infty} (E_m|j\rangle\langle j|E_n) = \sum_j (E_m|j\rangle(E_n|j\rangle^*. \tag{5.14}$$

A matrix whose matrix elements fulfill the condition (5.14) is called a unitary matrix. If the matrix elements are also real, which happens sometimes in Φ (and always in a real linear space), then the matrix is called orthogonal.

If the matrix elements of the operator A in the basis $|E_n\rangle$ are given, $(E_n|A|E_m)$, then one can obtain the matrix elements $\langle i|A|j\rangle$ by using the transformation (5.12'):

$$a_j\delta_{ij} = \langle i|A|j\rangle = \sum_{n=1}^{\infty}\sum_{m=1}^{\infty} \langle i|E_n\rangle(E_n|A|E_m)(E_m|j\rangle. \qquad (5.15)$$

One says, the transition matrix transforms the matrix $(E_n|A|E_m)$ into its diagonal form. Multiplying both sides of (5.15) by $\langle E_r|i\rangle$, summing over i and using (5.14') one obtains:

$$\sum_{m=1}^{\infty} (E_r|A|E_m)(E_m|j\rangle = a_j(E_r|j\rangle. \qquad (5.16)$$

If the eigenvalues a_j are unknown, then this is the eigenvalue problem in matrix notation. For the case that the space is N-dimensional (instead of ∞-dimensional) this is a system of N linear homogeneous equations[10] for the N unknown $(E_m|j\rangle$:

$$\sum_{m=1}^{N} [(E_n|A|E_m) - a\delta_{nm}](E_m|j\rangle = 0. \qquad (5.17)$$

It has nonzero solutions for $(E_m|j\rangle$ if and only if

$$\det[(E_n|A|E_m) - a\delta_{nm}] = 0. \qquad (5.18)$$

The eigenvalues of the matrix $(E_n|A|E_m)$ are the N solutions of this equation. These solutions are not necessarily distinct. Equation (5.18) is a polynomial of degree N and, by a famous theorem of algebra,[10] every polynomial of degree N has N (in general complex but if A is Hermitian then real) solutions.

We now turn to the case that one of the basis systems,

$$\{|x\rangle| - \infty < x < + \infty\}, \qquad (5.19)$$

is continuous. Then the vector ϕ is realized by a function of a continuous variable x, rather than by a function of a discrete variable as in (5.3) and (5.4):

$$\phi \leftrightarrow \langle x|\phi\rangle = \phi(x). \qquad (5.20)$$

As in the discrete case, the space Φ can be defined by its isomorphism (5.20) to a space of functions. We want to consider the particular case where Φ is the space of vectors whose components $\langle x|\phi\rangle$ are elements of the function space S.

[10] See, e.g., A Lichnerowicz (1967).

The eigenvector $|E_n\rangle$ is a particular vector of the space Φ; so one can use the continuous basis system expansion (4.4c) for it:

$$|E_n\rangle = \int dx\, |x\rangle\langle x|E_n\rangle. \tag{5.21}$$

The $\langle x|E_n\rangle$ are the analogue of the transition matrix elements $\langle i|E_n\rangle$ between two discrete basis systems. They are the transition coefficients between the discrete basis system $|E_n\rangle$ and the continuous basis system $|x\rangle$.

For fixed value of x the $\langle x|E_n\rangle$ are functions of the discrete variable E_n and for fixed value of E_n they are functions of the continuous variable x. They also occur if one takes the scalar product of the basis vector expansion (4.4d) with the continuous basis vector $|x\rangle$:

$$\langle x|\phi\rangle = \sum_{n=0}^{\infty} \langle x|E_n\rangle(E_n|\phi\rangle. \tag{5.22}$$

The $\langle x|E_n\rangle$ constitute a particular set of functions in the space S, which, as a consequence of (4.3d), have the property:

$$\langle x|H|E_n\rangle = E_n\langle x|E_n\rangle. \tag{5.23}$$

Because of this property they are called eigenfunctions of the operator H. Equation (5.22), which is an immediate consequence of the eigenvector expansion (4.4d), is called the eigenfunction expansion of the function $\langle x|\phi\rangle \in S$.

We want to illustrate this on a well-known example. We choose for the space of functions the subspace $K(a) \subset S$ of all functions $\phi(x) = \langle x|\phi\rangle$ which are identically zero outside the domain $|x| < a$.

Let the operator Q have the continuous spectrum $\{x| -a < x < +a\}$. So the spectral representation of an arbitrary vector $\phi \in \Phi$ is

$$\phi = \int_{-a}^{+a} dx\, |x\rangle\langle x|\phi\rangle. \tag{5.24}$$

Let the self-adjoint operator $H = P^2$ be defined by

$$\langle x|H|\phi\rangle = -\frac{d^2}{dx^2}\langle x|\phi\rangle \tag{5.25}$$

for every component $\langle x|\phi\rangle$ of any $\phi \in \Phi$. In order that Q, P and any power of these operators be defined in Φ, the space of components $\langle x|\phi\rangle$ must be the space of continuous infinitely differentiable functions for which

$$\int_{-a}^{a} dx\, \left| x^n \frac{d^m}{dx^m} \langle x|\phi\rangle \right|^2 < \infty.$$

Further

$$\langle x \geq a|\phi\rangle = \langle x \leq -a|\phi\rangle = 0. \tag{5.26}$$

(This boundary condition is important to establish the self-adjointness of H, cf. Problem 14.) Let the eigenvector of H be denoted by $|n\rangle$:

$$H|n\rangle = E_n|n\rangle. \tag{5.27}$$

We want to find all eigenvalues E_n and the transition coefficients $\langle x|n\rangle$. From (5.25) and (5.27) follows

$$-\frac{d^2}{dx^2}\langle x|n\rangle = E_n\langle x|n\rangle. \tag{5.28}$$

The solutions of this differential equation which fulfill the boundary condition (5.26) are

$$\langle x|n\rangle = \frac{1}{\sqrt{a}}\cos\left(\frac{n\pi}{2a}x\right) \quad \text{for } n = 1, 3, 5, 7, \ldots,$$

$$\langle x|n\rangle = \frac{1}{\sqrt{a}}\sin\left(\frac{n\pi}{2a}x\right) \quad \text{for } n = 2, 4, 6, \ldots. \tag{5.29}$$

The eigenvalues are

$$E_n = \frac{n^2\pi^2}{4a^2}. \tag{5.30}$$

The normalization factor $1/\sqrt{a}$ in (5.29) has been chosen so that $\langle n|n\rangle = 1$, which according to (5.24) is given by

$$\langle n|n\rangle = \int_{-a}^{+a} dx \, \langle n|x\rangle\langle x|n\rangle$$

$$= \frac{1}{a}\int_{-a}^{+a} dx \cos\left(\frac{n\pi}{2a}x\right)\cos\left(\frac{n\pi}{2a}x\right) = 1, \tag{5.31e}$$

or

$$\langle n|n\rangle = \frac{1}{a}\int_{-a}^{+a} dx \sin\left(\frac{n\pi}{2a}x\right)\sin\left(\frac{n\pi}{2a}x\right) = 1. \tag{5.31o}$$

In the last equalities of (5.31e) and (5.31o) properties of integrals over the trigonometric functions have been used. For $n' \neq n$ one should have as a consequence of the orthogonality of eigenvectors of self-adjoint operators

$$\langle n'|n\rangle = \int_{-a}^{+a} dx \, \langle n'|x\rangle\langle x|n\rangle = 0. \tag{5.32}$$

But this can also be obtained by inserting (5.29) into (5.32) and calculating the integrals over the trigonometric functions. Because of (5.32), one says that the eigenfunctions $\langle n'|x\rangle$ and $\langle n|x\rangle$ are orthogonal.

The eigenfunction expansion (5.22) for this particular case is

$$\langle x|\phi\rangle = \phi(x) = \sum_{n=1,3,5}^{\infty} a_n \cos\left(\frac{n\pi}{2a}x\right) + \sum_{n=2,4,6}^{\infty} b_n \sin\left(\frac{n\pi}{2a}x\right) \quad \text{for } |x| < a,$$

$$\phi(x) = 0 \quad \text{for } |x| \geq 0. \tag{5.33}$$

The coordinates of the vector ϕ are given by:

$$\sqrt{a}\, b_n = \langle n|\phi\rangle = \int_{-a}^{+a} dx \, \langle n|x\rangle\langle x|\phi\rangle$$

$$= \frac{1}{\sqrt{a}} \int_{-a}^{+a} dx \, \sin\left(\frac{n\pi}{2a} x\right)\phi(x); \qquad n = 2, 4, 6\ldots$$

and

$$\sqrt{a}\, a_n = \langle n|\phi\rangle = \int_{-a}^{+a} dx \, \langle n|x\rangle\langle x|\phi\rangle$$

$$= \frac{1}{\sqrt{a}} \int_{-a}^{+a} dx \, \cos\left(\frac{n\pi}{2a} x\right)\phi(x); \qquad n = 1, 3, 5,\ldots\ .$$

Equation (5.33) is called a Fourier series representation of the arbitrary function $\phi(x) \in K(a)$. The coordinates of the vector ϕ with respect to the basis system $|n\rangle$, a_n and b_n, are called the Fourier coefficients of the function $\phi(x) = \langle x|\phi\rangle$. Because (5.22) is just a generalization of a classical Fourier series (5.33) one also calls (5.22) and even (4.4d) often a Fourier series representation or Fourier expansion. In general the eigenfunctions $\langle x|E_n\rangle$ will not be trigonometric functions, however (5.32) must always be fulfilled, i.e., the $\langle x|E_n\rangle$ always form an orthogonal system of basis functions. Well-known examples of orthogonal basis functions are the Hermite Polynomials, the Legendre Polynomials, and the Laguerre Polynomials. In particular, for the space S, when the spectrum of Q is the entire real line $\mathbb{R}$, the Hermite polynomials are an appropriate choice of basis functions.

So far we have considered transition coefficients between two discrete basis systems $\langle i|E_n\rangle$, and the transition coefficients $\langle x|E_n\rangle$ between a discrete and a continuous basis system. Now we want to go one step further and consider transition coefficients between two continuous basis systems. We choose for Φ the space realized by S and consider the operator P defined by (cf. Problem 7)

$$\langle x|P|\phi\rangle = \frac{1}{i}\frac{d}{dx}\langle x|\phi\rangle \quad \text{for every} \quad \phi \in \Phi, \qquad \text{i.e.,} \quad \langle x|\phi\rangle \in S. \quad (5.34)$$

We want to consider (5.34) also for the case that ϕ is an eigenvector of the operator P (extending the definition of P if this eigenvector is not in Φ):

$$P|p\rangle = p|p\rangle. \qquad (5.35)$$

Equation (5.34) then reads

$$\langle x|P|p\rangle = p\langle x|p\rangle = \frac{1}{i}\frac{d}{dx}\langle x|p\rangle. \qquad (5.36)$$

It is well known that this differential equation has solutions for any complex value of p given by

$$\langle x|p\rangle = \frac{1}{\sqrt{2\pi}} e^{ixp}, \qquad p \in \mathbb{C}. \qquad (5.37)$$

None of these solutions, however, is square integrable so that $|p\rangle \notin \Phi$ (cf. Problem 9).

We can expand any $\phi \in \Phi$ with respect to the basis system $|p\rangle$ of eigenvectors of P:

$$\phi = \int_{\text{Spectr. } P} dp \, |p\rangle\langle p|\phi\rangle, \tag{5.38}$$

where the integral has to extend over the spectrum of the operator P. We want to determine the spectrum of P. It is contained in the set of generalized eigenvalues which is identical with all of $\mathbb{C}$. We take the scalar product[11] of (5.38) with $|x\rangle$

$$\langle x|\phi\rangle = \int_{\text{Spectr. } P} dp \, \langle x|p\rangle\langle p|\phi\rangle = \frac{1}{\sqrt{2\pi}} \int dp \, e^{ixp}\langle p|\phi\rangle. \tag{5.39}$$

If the p integration in (5.39) extends over the real line, $-\infty < p < +\infty$, then (5.39) says that $\phi(x) = \langle x|\phi\rangle$ is the Fourier transform of the function $\tilde{\phi}(p) = \langle p|\varphi\rangle$. We therefore want to give a brief review of the properties of the Fourier transform.[12]

Let $\phi(x)$ be an element of the space S, then the (inverse) Fourier transform,† which we denote by $F_p^{-1}[\phi(x)] \equiv \tilde{\phi}(p)$, is defined by

$$F_p^{-1}[\phi(x)] \equiv \tilde{\phi}(p) = \frac{1}{\sqrt{2\pi}} \int_{-\infty}^{+\infty} dx \, e^{-ixp}\phi(x). \tag{5.40a}$$

The Fourier transform has the following properties:

(a) F^{-1} maps S onto itself; that means $\tilde{\phi}(p)$ also belongs, as a function of p, to the space S.

(b) The Fourier transform which we denote by $F_x[\tilde{\phi}(p)]$ and which is defined by

$$\phi(x) \equiv F_x[\tilde{\phi}(p)] = \frac{1}{\sqrt{2\pi}} \int_{-\infty}^{+\infty} dp \, e^{ixp}\tilde{\phi}(p) \tag{5.40b}$$

also belongs to the space S if $\tilde{\phi}(p) \in S$.

(c)

$$\left(\frac{1}{i}\frac{d}{dx}\right)^n \phi(x) = \frac{1}{\sqrt{2\pi}} \int_{-\infty}^{+\infty} dp \, p^n e^{ixp}\tilde{\phi}(p) = F_x[p^n\tilde{\phi}(p)] \tag{5.40c}$$

for any power $n = 0, 1, 2, 3, \ldots$, which we also write as

$$\left(\frac{1}{i}\frac{d}{dx}\right)^n F_x[\tilde{\phi}(p)] = F_x[p^n\tilde{\phi}(p)],$$

---

[11] Section I.4, footnote 6.

[12] For the proofs and details see I. M. Gelfand et al. (1964), Vol. 2, Chapter III, or E. J. Beltrami and M. R. Wohlers, *Distributions and the Boundary Values of Analytic Functions*, Academic Press, New York, 1966.

† One sometimes calls inverse Fourier transform what we call Fourier transform here and vice versa; we may do the same.

(d)

$$\left(-\frac{1}{i}\frac{d}{dp}\right)^n \tilde{\phi}(p) = \frac{1}{\sqrt{2\pi}} \int_{-\infty}^{+\infty} dx \, x^n e^{-ixp} \phi(x) = F_p^{-1}[x^n \phi(x)] \quad (5.40d)$$

for any power $n = 0, 1, 2, 3, \ldots$, which we also write as

$$\left(-\frac{1}{i}\frac{d}{dp}\right)^n F_p^{-1}[\phi(x)] = F_p^{-1}[x^n \phi(x)],$$

(e)

$$\int dx \, \psi^*(x)\phi(x) = \int dp \, \tilde{\psi}^*(p)\tilde{\phi}(p), \quad (5.40e)$$

where $\tilde{\psi}(p)$, $\tilde{\phi}(p) \in S$ are the Fourier transforms of $\psi(x)$, $\phi(x) \in S$.

Thus the Fourier transform F and its inverse F^{-1} establish two reciprocal isomorphisms between the Schwartz space functions S_x of the variable x and the Schwartz space functions S_p of the variable p.

We can now return to the problem of the spectrum of P. As every $\langle x|\phi\rangle = \phi(x) \in S$ can be written according to (5.40b) and (5.37) in the form (5.39) with the integration extending over the real axis $-\infty < p < +\infty$ with $\langle p|\phi\rangle$ being again an element of S, we conclude that

$$\text{Spectr. } P = \{p \,|\, -\infty < p < \infty\},$$

i.e., the real axis $\mathbb{R}$. Thus we have found the spectral resolution (5.38) of ϕ with respect to the basis system of eigenvectors of P. The spectrum of P is continuous and the $|p\rangle$ are generalized eigenvectors (not elements of Φ). The transition coefficients between the generalized basis systems $\{|x\rangle\}$ and $\{|p\rangle\}$ are given by (5.37) with $p \in \mathbb{R}$ and $x \in \mathbb{R}$. For fixed value of x they are functions of the continuous variable p, and for fixed value of p they are functions of the continuous variable x. They satisfy the differential equation (5.36), but they are not well-behaved functions, i.e., elements of the function space S. Therefore they are called generalized eigenfunctions or distributions. It is easily seen by exchanging x and p in the preceding arguments and using (5.40a) in place of (5.40b) that

$$\langle p|x\rangle = \frac{1}{\sqrt{2\pi}} e^{-ipx}. \quad (5.41)$$

Consequently we have shown that for this special case of transition coefficients between two continuous basis systems the following relation holds:

$$\langle p|x\rangle = \langle x|p\rangle^*. \quad (5.42)$$

The exponential function is just one example of transition coefficients between two continuous basis systems. Other examples that we will meet in the second part of the book are the spherical Bessel functions.

Finally, in this section we want to consider the realization of operators in terms of a continuous basis system. This is again done in analogy to the discrete case. To obtain the analogue of (5.5) we apply the operator B to the continuous basis system expansion (4.4c)

$$B\phi = \int dx\, B|x\rangle\langle x|\phi\rangle. \tag{5.43}$$

Taking the scalar product[13] with the basis vector $|y\rangle \in \{|x\rangle | \infty < x < \infty\}$ we obtain

$$\langle y|B\phi\rangle = \int dx\, \langle y|B|x\rangle\langle x|\phi\rangle \tag{5.44}$$

which is the continuous analogue of (5.6). This is an integral transform transforming the function $\langle x|\phi\rangle$ into the function $\langle y|\psi\rangle = \langle y|B\phi\rangle$. The continuous analogue of the matrix element (5.8), $\langle y|B|x\rangle$, is called the kernel of the integral transform. Choosing for $B = A$ and for $\phi = |i\rangle$, the eigenvector (5.1) of A we obtain from (5.44)

$$\int dx\, \langle y|A|x\rangle\langle x|i\rangle = a_i\langle y|i\rangle. \tag{5.45}$$

This is the analogue of (5.16). If the eigenvalues a_i and eigenfunctions $\langle y|i\rangle$ are unknown this is a homogeneous integral equation for the determination of these values.

I.6 Hermite Polynomials as an Example of Orthonormal Basis Functions

We start with the operator Q realized by the multiplication operator and the operator P realized by the differential operator (Equation (5.34)) on the space of functions S:

$$\langle x|Q|\phi\rangle = x\langle x|\phi\rangle, \qquad -\infty < x < +\infty, \tag{6.1}$$

$$\langle x|P|\phi\rangle = \frac{1}{i}\frac{\partial}{\partial x}\langle x|\phi\rangle, \qquad \langle x|\phi\rangle \in S. \tag{6.2}$$

Another operator H is defined by

$$H = \tfrac{1}{2}(P^2 + Q^2), \tag{6.3}$$

and its eigenvectors are denoted by $|n\rangle$:

$$H|n\rangle = E_n|n\rangle. \tag{6.4}$$

We denote the space in which Q, P, H and any element of the algebra generated by them act by Φ. Thus $\phi \in \Phi \Leftrightarrow \langle x|\phi\rangle \in S$.[14]

---

[13] Section I.4, footnote 6.

[14] One can also construct Φ in the following manner (cf. Chapter II): Start from the algebra of operators P, Q, and H which fulfills in addition to (6.3) the relation $PQ - QP = (1/i)I$, and construct the space Φ as the largest space in which this algebra is given (represented) as an algebra of (continuous) operators and in which there exists at least one eigenvector of H.

The discreteness of the spectrum of H, which is implied by (6.4), is in fact a consequence of the assumption that the eigenvectors $|n\rangle$ of H lie in the space Φ or that $\langle x|n\rangle \in S$, as will be shown below.

The transition coefficients between the Q-eigenvectors and the H-eigenvectors, $\langle x|n\rangle$, are orthonormal basis functions of the space S in the sense of (5.22). We want to determine $\langle x|n\rangle$ and the eigenvalues E_n explicitly.

We denote

$$\lambda = 2E_n. \tag{6.5}$$

Using (6.3) in (6.4) we obtain with (6.2) and (6.1):

$$2\langle x|H|n\rangle = \left(-\frac{d^2}{dx^2} + x^2\right)\langle x|n\rangle = \lambda\langle x|n\rangle. \tag{6.6}$$

Thus we have to find the solutions

$$\psi(x) = \langle x|n\rangle \tag{6.7}$$

of the differential equation

$$\psi''(x) + (\lambda - x^2)\psi(x) = 0. \tag{6.8}$$

In addition we will demand that the $\psi(x) \in S$. Then they must in particular fulfill the boundary conditions:

$$\psi(x) \to 0 \quad \text{faster than any polynomial of } 1/x \text{ as } x \to \pm\infty \tag{6.9}$$

Equation (6.8) is highly singular at $|x| \to \infty$ and for large values of $|x|$ is given approximately by $\psi'' - x^2\psi = 0$, which has the solutions $\psi \sim e^{\pm x^2/2}$. One therefore makes the ansatz

$$\psi(x) = c_1 e^{-x^2/2}y(x) + c_2 e^{x^2/2}y_2(x) \tag{6.10}$$

with y_2 and y expanded as power series in x:

$$y = \sum_{k=0}^{\infty} a_k x^k. \tag{6.11}$$

The second term in (6.10) increases rapidly and cannot lead to an element of S, therefore c_2 must be zero. Inserting

$$\psi(x) = c_1 e^{-x^2/2}y(x) \tag{6.12}$$

into (6.8) one obtains the differential equation

$$y'' - 2xy' + (\lambda - 1)y = 0. \tag{6.13}$$

Substituting (6.11) into (6.13) and collecting powers of x one obtains

$$(k + 2)(k + 1)a_{k+2} - 2ka_k + (\lambda - 1)a_k = 0$$

which leads to the recurrence formula

$$a_{k+2} = \frac{2k - (\lambda - 1)}{(k + 2)(k + 1)}a_k. \tag{6.14}$$

This determines the coefficients of the higher powers in (6.11) from those of the lower powers. By comparing the ratio a_{k+2}/a_k for large values of k with the ratios b_{k+2}/b_k in a power series expansion of e^{x^2}, we can show that the function (6.11) with (6.14) will be divergent like e^{x^2}, as $x \to \pm\infty$. Thus, the only way that we can construct solutions $y(x)$ of the form (6.12) which converge in the desired manner at $\pm\infty$ is to choose values for λ such that the series (6.11) terminates. This we can accomplish by letting $\lambda - 1 = 2n$ for $n = 0, 1, 2, \ldots$, and we thus see that $\psi(x)$ can be an element of S only if

$$\lambda = 2n + 1 \quad \text{or} \quad E_n = n + \tfrac{1}{2}, \qquad n = 0, 1, 2, \ldots. \tag{6.15}$$

The solutions of the differential equation (6.13) corresponding to the eigenvalues (6.15) are also labelled by n and denoted by $y(x) = H_n(x)$. Thus, we obtain a family of equations

$$H_n'' - 2xH_n' + 2nH_n = 0, \qquad n = 0, 1, 2, \ldots, \tag{6.16}$$

which are called the Hermite differential equations. Their solutions $H_n(x)$ are called the Hermite polynomials. It is easy to check that the polynomials defined by:

$$H_n(x) = (-1)^n e^{x^2} \frac{d^n}{dx^n} (e^{-x^2}) \tag{6.17}$$

fulfill (6.16). Equation (6.17) is called the Rodrigues formula for the Hermite polynomials.

Therewith we have found for the transition coefficient (6.7):

$$\langle x|n \rangle = \psi_n(x) = \frac{1}{N_n} e^{-x^2/2} H_n(x). \tag{6.18}$$

The factor N_n will be chosen so that

$$(n|n) = \int dx \, (n|x)\langle x|n \rangle = \left| \frac{1}{N_n} \right|^2 \int dx \, e^{-x^2} H_n(x) H_n(x) = 1. \tag{6.19}$$

The orthogonality of the eigenvectors of the self-adjoint operator H can now be written in the form

$$N_{n'}^* N_n (n'|n) = \int dx \, e^{-x^2} H_{n'}(x) H_n(x) = 0 \quad \text{for} \quad n \neq n', \tag{6.20}$$

The second equality of (6.20), which we obtained as a consequence of the orthogonality of the $|n)$, is called the orthogonality relation for the Hermite polynomials. The normalization factor N_n has to be calculated using the special properties of the Hermite polynomials defined by (6.17). This is done using standard methods in the Appendix to this section. From (A.12) one obtains

$$|N_n|^2 = \sqrt{\pi} n! \, 2^n. \tag{6.21}$$

N_n and therewith the transition coefficients $\langle x|n \rangle$ in (6.16) are only determined up to a phase factor, i.e., a factor $e^{i\alpha(n)}$ of modulus one, which can still vary with n. It is standard to choose these factors equal to $(+1)$. Thus we obtain

$$\langle x|n \rangle = \frac{1}{\sqrt{2^n n!\sqrt{\pi}}} e^{-x^2/2} H_n(x). \tag{6.22}$$

Appendix to Section I.6

The normalization factor N_n in (6.19) is calculated in the following way: We consider the function $f(t) = e^{-t^2 + 2tx}$. Considered as a function of the complex variable t, $f(t)$ is an entire function and can be expanded into a Taylor series of t:

$$e^{-t^2 + 2tx} = \sum_{n=0}^{\infty} a_n(x)t^n \quad \text{for all} \quad |t| < \infty, \tag{A.1}$$

where the expansion coefficients depend upon the parameter x. These coefficients are given by

$$a_n = \frac{1}{n!} \frac{d^n f(t)}{dt^n}\bigg|_{t=0}. \tag{A.2}$$

As $f(t)$ is an analytic function for every value of the parameter x we can use the Cauchy theorem which states that

$$\frac{d^n f(t)}{dt^n} = \frac{n!}{2\pi i} \oint \frac{f(z)}{(z-t)^{n+1}} dz, \tag{A.3}$$

where the integration is along a contour enclosing the point $z = t$. Applying this to $f(z) = e^{-z^2 + 2zx}$ and using (A.2) we obtain at $t = 0$

$$a_n = \frac{1}{2\pi i} \oint \frac{e^{-z^2 + 2zx}}{z^{n+1}} dz. \tag{A.4}$$

On the other hand, we can use (A.3) for the function $f(z) = e^{-z^2}$ and obtain

$$\frac{d^n}{dx^n} e^{-x^2} = \frac{n!}{2\pi i} \oint \frac{e^{-z^2}}{(z-x)^{n+1}} dz, \tag{A.5}$$

where the contour of integration encloses the point $z = x$, which can be in particular any point on the real axis.

On the left-hand side of (A.5) we use (6.17). On the right-hand side we perform a change of integration variable $z = x - t$, $dz = -dt$. Then (A.5) goes over into

$$e^{-x^2}(-1)^n H_n(x) = \frac{n!}{2\pi i} \oint \frac{1}{(-1)^n} \frac{e^{-(x-t)^2}}{t^{n+1}} dt, \tag{A.6}$$

where the contour of integration encloses the point $t = 0$. Dividing both sides by $e^{-x^2}(-1)^n$ one obtains

$$H_n(x) = \frac{n!}{2\pi i} \oint \frac{e^{-t^2 + 2tx}}{t^{n+1}} \, dt. \tag{A.7}$$

Comparison of (A.4) and (A.7) shows that the coefficients in (A.1) are given by

$$a_n(x) = \frac{1}{n!} H_n(x)$$

and (A.1) can be written

$$e^{-t^2 + 2tx} = \sum_{n=0}^{\infty} \frac{1}{n!} H_n(x)t^n \quad \text{for} \quad |t| < \infty. \tag{A.8}$$

The function $f(t) = e^{-t^2 + 2tx}$ is called the generating function of the Hermite polynomials. Many properties of orthogonal polynomials are easily derived using their generating function.

We are now ready to calculate the normalization factor N_n. Using (A.8) we obtain

$$\int_{-\infty}^{+\infty} dx \, e^{-t^2 + 2tx} e^{-s^2 + 2sx} e^{-x^2} = \sum_{n,m=0}^{\infty} \frac{t^n s^m}{n! \, m!} \int_{-\infty}^{+\infty} dx \, e^{-x^2} H_n(x) H_m(x). \tag{A.9}$$

The left-hand side of (A.9) can be written as

$$e^{2st} \int_{-\infty}^{+\infty} dx \, e^{-(x-(s+t))^2} = e^{2st} \int_{-\infty}^{+\infty} d\xi \, e^{-\xi^2} = e^{2st} \sqrt{\pi}. \tag{A.10}$$

Expanding e^{2st} into a power series one obtains

$$\text{left-hand side (A.9)} = \sqrt{\pi} \sum_{v=0}^{\infty} \frac{2^v (st)^v}{v!} \tag{A.11}$$

Equating equal powers of s and t on the right-hand side of (A.9) and (A.11) one obtains

$$\int_{-\infty}^{+\infty} dx \, e^{-x^2} H_n(x) H_n(x) = \sqrt{\pi} n! \, 2^n. \tag{A.12}$$

For $n \neq m$ one again obtains (6.20). Because of the property (6.20) and (A.12) the $H_n(x)$ are called orthonormal on the interval $-\infty < x < +\infty$ with respect to the weight function $(\sqrt{\pi} n! \, 2^n)^{-1} e^{-x^2}$.

Other important properties obtained from the generating function are the recurrence relations. To obtain the recurrence relation for the Hermite polynomials we differentiate (A.8) with respect to t:

$$\sum_{n=0}^{\infty} \frac{2x}{n!} H_n(x)t^n - \sum_{n=0}^{\infty} \frac{2}{n!} H_n(x)t^{n+1} = \sum_{n=1}^{\infty} \frac{1}{(n-1)!} H_n(x)t^{n-1}.$$

Equating powers of t^n one obtains:

$$2xH_n(x) = H_{n+1}(x) + 2nH_{n-1}(x). \tag{A.13}$$

This recurrence relation can be used to calculate the Hermite polynomials step by step starting from $H_0(x) = 1$, $H_1(x) = 2x$.

In Chapter II we will also need some properties of the Fourier transforms of the $H_n(x)$ which are easily derived from the generating function. Multiplying (A.8) by $e^{iyx - x^2/2}$, where $-\infty < y < +\infty$, and integrating we obtain:

$$\int_{-\infty}^{+\infty} e^{2xt - t^2 + iyx - x^2/2}\, dx = \int e^{iyx - x^2/2}\, dx \sum_n \frac{1}{n!} H_n(x) t^n$$

$$= \sum_{n=0}^{\infty} \frac{t^n}{n!} \int_{-\infty}^{+\infty} e^{iyx - x^2/2} H_n(x)\, dx. \tag{A.14}$$

The integral on the left-hand side is calculated using the substitution $\eta = x - 2t$ and gives (cf. Problem 11):

$$e^{t^2} e^{i2yt} \int e^{-\eta^2/2} e^{iy\eta}\, d\eta = e^{-(it)^2} e^{2y(ti)} \sqrt{2\pi}\, e^{-y^2/2}. \tag{A.15}$$

For the right-hand side of this equality we use (A.8) with t replaced by (it). Then we obtain from (A.14):

$$\sqrt{2\pi}\, e^{-y^2/2} \sum_{n=0}^{\infty} \frac{(it)^n}{n!} H_n(y) = \sum_{n=0}^{\infty} \frac{t^n}{n!} \int_{-\infty}^{+\infty} e^{iyx} e^{-x^2/2} H_n(x)\, dx. \tag{A.16}$$

Comparing coefficients of identical powers of t we obtain

$$i^n e^{-y^2/2} H_n(y) = \frac{1}{\sqrt{2\pi}} \int_{-\infty}^{+\infty} dx\, e^{iyx} e^{-x^2/2} H_n(x). \tag{A.17}$$

I.7 Continuous Functionals[15]

In the previous sections of this chapter we presented the mathematics of quantum mechanics as an algebraic structure. Generalized eigenvectors were introduced by analogy to the basis vectors of a three-dimensional space and it was asserted that they could be treated very much like ordinary vectors. However, they are not vectors of the space Φ but are in fact continuous functionals on the space. In this section we define them and explain their distinction from ordinary vectors.

Let Φ be a complex linear space. A functional on Φ is a mapping F from the space Φ into the complex numbers $\mathbb{C}$, $F: \Phi \to \mathbb{C}$. (If Φ is a real space then

---

[15] This section can be omitted in first reading. It gives a mathematical explanation of notions used in Sections I.4 and I.5 and is not essential for an understanding of the following chapters except for Chapter XXI.

the mapping is into the real numbers $\mathbb{R}$.) It is thus the analog of a complex-valued function of a real variable $x \in \mathbb{R}$, $F: \mathbb{R} \to \mathbb{C}$, only now the variable is not a real number but a vector $\phi \in \Phi$. If F satisfies

$$F(\alpha\phi + \beta\psi) = \alpha^*F(\phi) + \beta^*F(\psi) \quad \text{for all} \quad \phi, \psi \in \Phi \quad \text{and} \quad \alpha, \beta \in \mathbb{C},$$

$$(7.1)$$

then F is called an antilinear functional. If F satisfies

$$F(\alpha\phi + \beta\psi) = \alpha F(\phi) + \beta F(\psi), \qquad (7.1a)$$

then F is called a linear functional. (If Φ is a real space there is no distinction between linear and antilinear mappings.) We will consider here antilinear functionals rather than linear ones (note that in the mathematical literature one usually considers linear functionals).

An example of an antilinear functional on a linear space is given by

$$F(\phi) = (\phi, \psi), \qquad (7.2)$$

where ψ is a *fixed* element $\psi \in \Phi$ and (ϕ, ψ) is the scalar product of ψ with ϕ where ϕ varies over Φ (cf. Problem 12). Because of this example and because in the general case we want to consider a functional to be a generalization of the scalar product, one uses for the antilinear functional $F(\phi)$ the symbol

$$F(\phi) = \langle \phi | F \rangle. \qquad (7.3)$$

Any two antilinear functionals, F_1 and F_2, on a linear space Φ may be added and multiplied by numbers according to

$$(\alpha F_1 + \beta F_2)(\phi) = \alpha F_1(\phi) + \beta F_2(\phi), \qquad \alpha, \beta \in \mathbb{C}, \qquad (7.4')$$

or, using the notation (7.3),

$$\langle \phi | \alpha F_1 + \beta F_2 \rangle = \alpha \langle \phi | F_1 \rangle + \beta \langle \phi | F_2 \rangle. \qquad (7.4)$$

The functional $\alpha F_1 + \beta F_2$ defined by (7.4) is again an antilinear functional over Φ (Problem 13). Thus the set of antilinear functionals on a linear space Φ forms a linear space.[16] This space is called the conjugate space or dual space (precisely, the algebraic dual or algebraic conjugate space) to the space Φ and is denoted by $\Phi^\times$.

Let Φ be finite-dimensional, dim $\Phi = n$, and let e_i, $i = 1, 2, \ldots, n$ be a basis for Φ; let F be an arbitrary antilinear functional on Φ. Let us denote by f_i the complex numbers

$$f_i = F(e_i), \qquad i = 1, 2, \ldots, n. \qquad (7.5)$$

The functional F at an arbitrary $\phi \in \Phi$, $\phi = \sum_{i=1}^{n} \phi^i e_i$, can then be written as

$$F(\phi) = \sum_{i=1}^{n} F(\phi^i e_i) = \sum_{i} \phi^{*i} F(e_i) = \sum_{i=1}^{n} \phi^{*i} f_i. \qquad (7.6)$$

---

[16] This is the analog to the statement that the set of linear operators on a linear space forms an algebra (i.e., a linear space in which a multiplication is defined). (7.4) is the analog to the first two equations of (3.4).

One has thus a one-to-one correspondence between the set of antilinear functionals on a finite-dimensional space and the sequence of numbers $(f_1, f_2, \ldots, f_n)$. With these numbers one can now define a vector f in Φ by

$$f = \sum_{i=1}^{n} f_i e_i. \tag{7.7}$$

Then the right-hand side of (7.6) is the scalar product of the vectors ϕ and f and we have shown that in a finite-dimensional linear scalar product space there is a one-to-one correspondence between the functionals F and the $f \in \Phi$:

$$F \leftrightarrow f \tag{7.8a}$$

such that the value of the functional $F(\phi)$ at the element $\phi \in \Phi$ is given by the scalar product:

$$F(\phi) = (\phi, f). \tag{7.8b}$$

This shows that in the finite-dimensional case the symbol (7.3) can always be identified with the scalar product[17]

$$\langle \phi | F \rangle = (\phi, f). \tag{7.8c}$$

When the identification (7.8) can be made, then Φ is said to be self-dual[17] and we write

$$\Psi^{\times} = \Phi. \tag{7.8d}$$

For infinite-dimensional scalar product spaces such an identification is in general not possible. As we have already seen that the scalar product always defines an antilinear functional, we can identify those antilinear functionals F_f, whose values at any $\phi \in \Phi$ can be written as the scalar product $F_f(\phi) = (\phi, f)$ with a fixed $f \in \Phi$, with the vectors f: $F_f \leftrightarrow f$. Then we have

$$\Phi^{\times} \supset \Phi. \tag{7.9}$$

For the understanding of functionals on infinite-dimensional spaces topological notions are of great importance. So far we have only discussed how algebraic structures are imposed upon a set. The linear scalar product space defined in Section I.2, which we want to call Ψ if it has no other structure, is a purely algebraic notion. One can now impose upon it a topological structure. Topological structures are defined by giving a specific meaning to the notion of convergence of infinite sequences.[18]

The Hilbert space convergence of a sequence of vectors $\phi_1, \phi_2, \phi_3, \ldots,$ $\phi_\nu, \ldots$ to the vector $\phi \in \mathscr{H}$, denoted by

$$\phi_\nu \overset{\mathscr{H}}{\to} \phi \quad \text{for} \quad \nu \to \infty, \tag{7.10'}$$

---

[17] This identification is not necessary and often it is useful not to identify the functionals with elements of Φ.

[18] There are topological spaces for which the definition of convergence of sequences is not sufficient to define the topology. But for the spaces Φ which we shall consider here (with first axiom of countability) this definition is sufficient.

is defined as

$$\|\phi_v - \phi\| \to 0 \quad \text{for} \quad v \to \infty. \tag{7.10}$$

The Φ-space convergence is defined with the help of the algebra of operators and is therefore dependent upon it. For the algebra generated by P, Q, H of (6.3) and (6.4) it is defined by[19]

$$\phi_v \overset{\Phi}{\to} \phi \quad \text{for } v \to \infty \qquad \text{iff} \tag{7.11'}$$

$$\|H^p(\phi_v - \phi)\| \to 0 \quad \text{for } v \to \infty \qquad \text{for every } p = 0, 1, 2, 3, \ldots . \tag{7.11}$$

As (7.11) includes (for $p = 0$) the property (7.10) it is clear that from

$$\phi_v \overset{\Phi}{\to} \phi \quad \text{follows} \quad \phi_v \overset{\mathscr{H}}{\to} \phi \tag{7.12}$$

but not vice versa. The convergence (a topology) defined by (7.11) is called *stronger* (finer) than the convergence (topology) defined by (7.10), and the convergence defined by (7.10) is called weaker (coarser) than the convergence defined by (7.11). $\mathscr{H}$ is the space which contains in addition to the elements of Ψ all limit elements of $\overset{\mathscr{H}}{\to}$-convergent sequences. Φ is the space which contains in addition to the elements of Ψ all limit elements of $\overset{\Phi}{\to}$-convergent sequences. As because of (7.12) every Φ-convergent sequence is also $\mathscr{H}$-convergent but not vice versa, we have

$$\Phi \subset \mathscr{H}. \tag{7.13}$$

The antilinear functionals that we will consider in the case of infinite-dimensional spaces will always be *continuous* functionals. An antilinear functional is continuous iff from

$$\phi_v \to \phi \quad \text{follows} \quad F(\phi_v) \overset{C}{\to} F(\phi) \quad \text{for} \quad v \to \infty, \tag{7.14}$$

where $\overset{C}{\to}$ denotes convergence for complex numbers.[20] $\Phi^\times$, $\mathscr{H}^\times$, etc. will always denote spaces of continuous functionals.

We can now consider the set of continuous linear functionals on $\mathscr{H}$ and on Φ.

$\Phi^\times$ is the set of all F^Φ with the property that

$$F^\Phi(\phi_v) \overset{C}{\to} F^\Phi(\phi) \quad \text{for all} \quad \phi_v \overset{\Phi}{\to} \phi.$$

$\mathscr{H}^\times$ is the set of all $F^{\mathscr{H}}$ with the property that

$$F^{\mathscr{H}}(\phi_v) \overset{C}{\to} F^{\mathscr{H}}(\phi) \quad \text{for all} \quad \phi_v \overset{\mathscr{H}}{\to} \phi.$$

The condition fulfilled by $F \in \mathscr{H}^\times$ is more stringent than the condition fulfilled by $F \in \Phi^\times$ because according to (7.12) there are more sequences for which $\phi_v \overset{\mathscr{H}}{\to} \phi$ than sequences for which $\phi_v \overset{\Phi}{\to} \phi$. Therefore

$$\mathscr{H}^\times \subset \Phi^\times. \tag{7.15}$$

---

[19] This is the weakest topology or convergence that makes all elements of the algebra continuous operators.

[20] A sequence $c_1, c_2, \ldots$ of $c_v \in \mathbb{C}$ converges to a number c iff $|c_v - c| \to 0$ for $v \to \infty$.

As mentioned above, infinite-dimensional linear spaces do not in general have the property (7.8d). However, the infinite-dimensional Hilbert space has the following remarkable property expressed by the Fréchet–Riesz theorem: For every $\mathcal{H}$-continuous functional $F^{\mathcal{H}}$ there exists an $f \in \mathcal{H}$ which is uniquely determined, such that

$$F^{\mathcal{H}}(\phi) \equiv \langle \phi | F^{\mathcal{H}} \rangle = (\phi, f). \tag{7.16}$$

Thus for $\mathcal{H}$ one can make the identification (7.8) as for finite-dimensional spaces and obtains from (7.13) and (7.15) the triplet of spaces:

$$\Phi \subset \mathcal{H} = \mathcal{H}^{\times} \subset \Phi^{\times} \tag{7.17}$$

called a *Gelfand triplet* or *rigged Hilbert space*. The symbol $\langle \phi | F \rangle$ is therefore an extension of the scalar product to those $F \in \Phi^{\times}$ which are not in $\mathcal{H}$.

One can now consider antilinear continuous functionals $\tilde{\phi}$ on $\Phi^{\times}$ and denote the space of all $\tilde{\phi}$ by $\Phi^{\times\times}$. For a large class of linear topological spaces Φ (called reflexive) there is a natural one-to-one correspondence between a $\phi \in \Phi$ and $\tilde{\phi} \in \Phi^{\times\times}$ given by

$$\tilde{\phi}(F) = \overline{F(\phi)} = \overline{\langle \phi | F \rangle}. \tag{7.18}$$

As $\overline{\langle \phi | F \rangle}$ is an extension of $(f, \phi) = \overline{(\phi, f)}$ we can consider the functional $\tilde{\phi}$ at the vector $F \in \Phi^{\times}$ as an extension of the scalar product (f, ϕ). It is therefore convenient to write $\tilde{\phi}(F) \equiv \langle F | \phi \rangle$, where $F \in \Phi^{\times}$ is now the variable, i.e., $\langle F | \phi \rangle$ is the value of the functional $\tilde{\phi}$ on the space $\Phi^{\times}$ at the vector F. With this definition of $\langle F | \phi \rangle$, (7.18) reads

$$\langle F | \phi \rangle = \overline{\langle \phi | F \rangle} \tag{7.18'}$$

giving an extension of the property (2.2b) of the scalar product.

In order to define the generalized eigenvectors which appear in the spectral resolution (4.4c) we will have to consider the linear operators on Φ. We will only consider operators which are continuous on Φ. In analogy to the definition of a continuous functional we say that an operator $A: \Phi \to \Phi$ is continuous iff for all convergent sequences

$$\phi_{\nu} \overset{\Phi}{\to} \phi \quad \text{follows that} \quad A\phi_{\nu} \overset{\Phi}{\to} A\phi.$$

The notion of continuity thus depends—like all topological notions—on the definition of convergence. An operator which is a continuous operator with respect to the Φ-convergence need not be a continuous operator with respect to the $\mathcal{H}$-convergence. (Many physical observables cannot be represented by continuous operators on the Hilbert space $\mathcal{H}$ but they can still be continuous operators on Φ.)

For every continuous linear operator A on Φ one can define an adjoint operator $A^{\times}$ on $\Phi^{\times}$ by

$$\langle \phi | A^{\times} | F \rangle = \langle A\phi | F \rangle. \tag{7.19}$$

One can prove that $A^{\times}$ is a linear continuous operator on $\Phi^{\times}$. This is the extension of the Hilbert space-adjoint operator defined by (cf. Section I.3):

$$(\phi, A^{\dagger}f) = (A\phi, f) \quad \text{for all } \phi \in \Phi \qquad (7.20)$$

(which in general is not defined for all $f \in \mathcal{H}$).

If one has two operators A and $\bar{A}$ with A defined on a subspace of the space on which $\bar{A}$ is defined and with the property $A\phi = \bar{A}\phi$ for all $\phi \in$ subspace, then one writes $A \subset \bar{A}$ and calls $\bar{A}$ an *extension of the operator A*.* For an operator A on Φ which is self-adjoint and has a unique self-adjoint[21] extension $\bar{A}$ to $\mathcal{H}$ one has therefore the following triplet of operators:

$$A \subset \bar{A} = A^{\dagger} \subset A^{\times} \qquad (7.21)$$

on the triplet of spaces

$$\Phi \subset \mathcal{H} \subset \Phi^{\times}.$$

A functional F over Φ is called a generalized eigenvector of the operator A on Φ with eigenvalue ω iff

$$\langle A\phi | F \rangle = \omega \langle \phi | F \rangle \quad \text{for all} \quad \phi \in \Phi. \qquad (7.22)$$

The generalized eigenvector F with eigenvalue ω is also denoted

$$|F\rangle = |\omega\rangle. \qquad (7.23)$$

Another precise form of writing (7.22) is

$$A^{\times}|\omega\rangle = \omega|\omega\rangle, \qquad (7.24)$$

which is often just written as

$$A|\omega\rangle = \omega|\omega\rangle. \qquad (7.25)$$

If $F \leftrightarrow f \in \mathcal{H}$ then (7.22) becomes

$$(A\phi, f) = (\phi, A^{\dagger}f) = \omega(\phi, f) \quad \text{for all} \quad \phi \in \Phi, \qquad (7.26)$$

which for A self-adjoint (as Φ is dense in $\mathcal{H}$) is identical with the definition of an eigenvector (cf. (3.6b)):

$$Af = \omega f. \qquad (7.27)$$

The definition (7.22) of a generalized eigenvector and its generalized eigenvalue is thus a generalization of the definition (7.27) of a proper eigenvector and its proper eigenvalue. This generalization need not have all the properties of ordinary eigenvectors and eigenvalues. In particular, ω in (7.22) need not be real even if A is (essentially) self-adjoint.

The eigenvectors $|x\rangle$ that appear in the continuous eigenvector expansion are generalized eigenvectors in the sense of (7.22). The coordinate $\langle x|\phi\rangle = \phi(x)$ of the vector $\phi \in \Phi$ along the basis vector $|x\rangle$, or the value of the function

---

[21] An operator which is defined on Φ, is Hermitian on Φ as defined in Section I.3 and has a unique self-adjoint extension to $\mathcal{H}$ is called essentially self-adjoint.
* For operators the symbol $\subset$ does not mean inclusion as for spaces.

ϕ at the number x, is thus the complex conjugate of the value of the functional $|x\rangle \in \Phi^\times$ at the vector $\phi \in \Phi$: $\langle x|\phi\rangle = \langle \phi|x\rangle^*$. Using (7.18), one can also say that $\langle x|\phi\rangle$ is the value of the functional $\phi \in \Phi^{\times\times} = \Phi$ at the vector $\langle x| \in \Phi^\times$.

I.8 How the Mathematical Quantities Will Be Used

We conclude this chapter by a table which gives the correspondences between the mathematical objects introduced in this chapter and the physical quantities which they will represent. To establish, explain, and utilize this correspondence will be the subject of the following chapters of this book.

Mathematical Image	Physical Quantity
vector $\phi \in \Phi$ (modulo a phase factor)	pure physical state
linear operator A on Φ	physical observable
eigenvalues (spectrum) of A	values obtained in a measurement of the observable A
eigenvector $|\lambda\rangle$ of A with eigenvalue λ	state in which measurement of A gives the value λ, eigenstate, bound state
generalized eigenvector $|\lambda\rangle$ of A with real eigenvalue λ	scattering state
generalized eigenvector $|\omega\rangle$ of A with complex eigenvalue $\omega = E - i\Gamma/2$	resonance state with resonance energy E and width Γ
scalar product $(\lambda|\phi)$ with $|\lambda\rangle \in \Phi$, $\phi \in \Phi$	probability amplitude
modulus squared of it $|(\lambda|\phi)|^2$	probability to obtain the value λ in a measurement of the observable A on a physical system in the state ϕ
functional $\langle\lambda|\phi\rangle$	wave function
modulus squared of it $|\langle\lambda|\phi\rangle|^2$	probability density for measuring the value λ for the observable A in the state ϕ
diagonal matrix element or expectation value $(\phi, A\phi)$	average value for the measurement of A in in the state ϕ
matrix element $(\psi, A\phi)$	transition amplitude
its modulus square $|(\psi, A\phi)|^2$	transition probability for transition caused by A from state ϕ to state ψ

Problems

1. Let $\mathbb{C}^n$ denote the set of all sequences $x = (\xi_1, \xi_2, \ldots, \xi_n)$ with $\xi_i \in \mathbb{C}$ (complex numbers). Define addition and multiplication with elements of $\mathbb{C}$ by the formulas

$$(\xi_1, \xi_2, \ldots, \xi_n) + (\eta_1, \eta_2, \ldots, \eta_n) = (\xi_1 + \eta_1, \xi_2 + \eta_2, \ldots, \xi_n + \eta_n).$$
$$\alpha(\xi_1, \xi_2, \ldots, \xi_n) = (\alpha\xi_1, \alpha\xi_2, \ldots, \alpha\xi_n).$$

(a) Show that $\mathbb{C}^n$ is a linear space.

(b) Let $x = (\xi_1, \ldots, \xi_n)$, $y = (\eta_1, \ldots, \eta_n)$ be two elements of $\mathbb{C}^n$. Show that (x, y) defined by

$$(x, y) = \sum_{i=1}^{n} \bar{\xi}_i \eta_i$$

fulfills all conditions of a scalar product.

2. Let Φ be $\mathbb{C}^n$ of Problem 1 and let a map A be defined by

$$A(\xi_1, \xi_2, \ldots, \xi_n) = (\xi_1', \xi_2', \ldots, \xi_n'),$$

where

$$\xi'_i = \sum_{k=1}^{n} a_{ik} \xi_k,$$

where a_{ik} $(i, k = 1, 2, \ldots, n)$ are fixed numbers.

$$a \equiv \begin{pmatrix} a_{11} & a_{12} & \cdots & a_{1n} \\ a_{21} & a_{22} & \cdots & a_{2n} \\ \vdots & & & \\ a_{n1} & a_{n2} & \cdots & a_{nn} \end{pmatrix}$$

is called the matrix of the map A. Define another map B in the same way. Show that these maps A and B are linear operators on the space $\mathbb{C}^n$ (i.e., fulfill the relations for linear operators). Show that the set of operators form an algebra if multiplication and addition of their matrices and multiplication of their matrices by a number are defined by the usual rules for matrix multiplication.

3. Show that af $(a \in \mathbb{C}, f \in \Phi)$ is an eigenvector of a linear operator A if f is an eigenvector of A with eigenvalue λ. What is the eigenvalue of the vector af?

4. Show that a Hermitian operator A has the following properties:
   (a) All eigenvalues are real.
   (b) Two eigenvectors ϕ_1 and ϕ_2 of A are orthogonal to each other if the eigenvalues corresponding to them are different from each other.

5. Show that two vectors ϕ and ψ are equal iff all their components with respect to a basis system are equal.

6. Show that the Cauchy–Schwarz–Bunyakovski inequality, Equation (2.7), follows from the definition (2.5) of a positive Hermitian form.

7. Let Q be the operator with continuous spectrum $\{x \mid -\infty < x < +\infty\}$ and let $|x\rangle$ denote its generalized eigenvector. Define another operator P by:

$$\langle x|P|\psi\rangle = \frac{1}{i}\frac{d}{dx}\langle x|\psi\rangle \quad \text{for all} \quad \psi \in \Phi \qquad (i^2 = -1).$$

(a) Show that P is a linear operator.
(b) Show that in order to have the operators Q and P well defined, i.e., $\|P\phi\|$ and $\|Q\phi\|$ be finite for all vectors $\phi \in \Phi$, the components of every vector $\langle x|\phi\rangle$ must be infinitely differentiable continuous functions which together with their derivatives vanish at infinity more rapidly than any polynomial of x (i.e., $\langle x|\phi\rangle$ must be functions of the Schwartz space S).

(c) Show that the operators P and Q fulfill the commutation relation

$$PQ - QP = \frac{1}{i} 1.$$

8. Let Φ be a linear scalar product space with elements $\phi, \psi, \chi, \dots$. Let $|i\rangle, |E_n\rangle$, $i = 1, 2, \dots, E_n = E_1, E_2, \dots$, be two discrete basis systems in Φ and $|x\rangle, -\infty < x < +\infty$ a continuous basis system of Φ. Show that the spaces of components $\mathbb{C}_{(i)}^\infty$ with elements $\langle i|\phi\rangle, \langle i|\psi\rangle, \langle i|\chi\rangle, \dots$ and $\mathbb{C}_{(E_n)}^\infty$ with elements $(E_n|\phi), (E_n|\psi), (E_n|\chi), \dots$ and $S_{(x)}$ with elements $\langle x|\phi\rangle, \langle x|\psi\rangle, \langle x|\chi\rangle, \dots$ are all isomorphic to the space Φ and isomorphic to each other.

9. Show that the generalized function $\langle x|p\rangle = (1/\sqrt{2\pi})e^{ixp}$ is a generalized eigenfunction of the operator defined in Problem 7:

$$\langle x|P|p\rangle = \frac{1}{i}\frac{d}{dx}\langle x|p\rangle = p\langle x|p\rangle$$

if p is real, but that it is not a well-behaved function, i.e., that $|p\rangle \notin \Phi$.

10. Show that the polynomials defined by

$$H_n(\xi) = (-1)^n e^{\xi^2} \frac{d^n}{d\xi^n}(e^{-\xi^2}), \qquad n = 0, 1, 2, \dots$$

satisfy the differential equation

$$H_n'' - 2\xi H_n' + 2nH_n = 0.$$

11. Calculate the Fourier transform

$$\phi(x) = \frac{1}{\sqrt{2\pi}} \int_{-\infty}^{+\infty} dp\, e^{ixp}\tilde{\phi}(p)$$

of the function $\tilde{\phi}(p) = e^{-ap^2} \in S\ (a > 0)$.

12. Show that the scalar product (ϕ, ψ) with $\psi \in \Phi$ fixed and $\phi \in \Phi$ variable defines an antilinear functional $F_\psi(\phi) \equiv (\phi, \psi)$ on the space Φ.

13. Let F_1 and F_2 be linear functions over Φ, α, β be complex numbers, show that $(\alpha F_1 + \beta F_2)$ defined by

$$(\alpha F_1 + \beta F_2)(\phi) = \alpha F_1(\phi) + \beta F_2(\phi)$$

for all $\phi \in \Phi$ is again a linear functional over Φ.

14. Let P^2 be the operator defined by Equation (5.25) and P the operator defined by

$$\langle x|P|\phi\rangle = \frac{1}{i}\frac{d}{dx}\langle x|\phi\rangle$$

with the same requirements for the functions $\langle x|\phi\rangle$. Show that as a consequence of (5.26) P and P^2 are self-adjoint operators. Are P and P^2 self-adjoint operators if (5.26) is not fulfilled? *Hint*: Use integration by parts.

15. In Section I.7 it was shown that

$$\mathcal{H} \subset \Phi^\times \quad \text{by proving} \quad \mathcal{H}^\times \subset \Phi^\times$$

and using the Fréchet–Riesz theorem $\mathscr{H} = \mathscr{H}^\times$. Give an alternate proof of $\mathscr{H} \subset \Phi^\times$ without using $\mathscr{H} = \mathscr{H}^\times$ by proving the following:

(a) Show that from

$$h_n \overset{\mathscr{H}}{\to} h \quad \text{follows} \quad (h_n, f) \to (h, f)$$

for all $f \in \mathscr{H}$. *Hint*: Use the Cauchy–Schwarz–Bunyakowski inequality.

(b) Show that every $f \in \mathscr{H}$ defines an element of $F \in \Phi^\times$ by $F(\varphi) = (\varphi, f)$ for every $\varphi \in \Phi$. E.g., show that F, which is antilinear by Problem 12, is continuous.

Hint: Show that from $g_n \overset{\Phi}{\to} g$ follows $F(g_n) \to F(g)$ using the result of part (a).

CHAPTER II

Foundations of Quantum Mechanics—The Harmonic Oscillator

This chapter, the longest in the book, introduces three of the basic assumptions of quantum mechanics and then illustrates them, using mainly the example of the harmonic oscillator. Though some historical remarks are included, neither the historical development nor any other heuristic way towards quantum mechanics is followed. The basic assumptions are formulated, explained, and applied. In Sections II.2, II.4, the basic assumptions are introduced; in Sections II.3, II.5, II.7 the harmonic oscillator is used to illustrate them. Section II.6 contains the derivations of some general consequences and might be omitted in first reading. The discussion for the continuous spectra, important for the description of the scattering and decay phenomena in the second part of the book, is given in Section II.8. Several remarks throughout this chapter emphasize the particular problems connected with generalized eigenvalues and eigenvectors and our unified treatment of continuous and discrete spectra. In Section II.9 we are ready to explain the physical meaning of the quantum-mechanical constant of nature, $\hbar$.

II.1 Introduction

Physicists believe that there is something in nature, or in each restricted domain of it, that may be "understood"; that there is a structure in nature. To "understand" means to bring this structure into congruence with some structure in our mind, with a structure of thought objects, with a structure that has been created by our minds. For physics this structure of thought objects is a mathematical structure. So to understand part of physical nature means to map its structure on a mathematical structure. To obtain a physical theory, then, means to obtain a mathematical image of a physical system, by which we mean any suitably restricted domain of physical nature (e.g., just an atom, or even just a set of substates of it, or all atoms and molecules).

For the domains of quantum physics the mathematical structures are algebras of linear operators in linear spaces. The discovery of this, the fundamental properties of the algebra, and the other basic assumptions of quantum mechanics was a very difficult process, the history of which we do not want to describe here. It is much easier to start with the basic assumptions and to show that conclusions derived from them really describe some parts of nature. This is the path we want to follow here.

II.2 The First Postulate of Quantum Mechanics

The first basic assumption or postulate ("axiom") of Quantum Mechanics is:

I.   A physical observable, defined by the prescription for its measurement, is represented by a linear operator[1] in a linear space. The mathematical image of a physical system is an operator *-algebra in a linear scalar-product space $\mathcal{H}$. The algebra is generated by some basic physical quantities, and the multiplication is defined by algebraic relations.

The content of this postulate will only become clear after one has seen it in operation in a number of concrete physical examples. This postulate is the result of a long and tedious process of abstraction and one of the most glorious achievements of science.

The elements of the algebra, the linear operators representing a physical observable, are not to be considered as anything in particular—like differential operators acting on well-behaved functions. Rather, the algebraic symbol stands for something which is defined only in its relationship to other such symbols through mathematical operations in which these symbols take part. A single algebraic symbol is thus empty and its properties are assigned to it by its relation to all other symbols. And the more (independent) relations one imposes, the more specific becomes the property of the mathematical entity. The meaning of a single algebraic symbol is determined by the way in which the algebraic language as a whole is used.

To illustrate axiom I we use one of the simplest examples, the one-dimensional harmonic oscillator. The one-dimensional harmonic oscillator can

---

[1] In Chapter XIX we shall also have to admit semilinear operators.

be defined as the physical system whose mathematical image is the algebra of operators that is generated by the following Hermitian operators:

H, representing the observable energy.
P, representing the observable momentum.
Q, representing the observable position.

The defining algebraic relations that are fulfilled by these operators are

$$[P, Q] \equiv PQ - QP = \frac{\hbar}{i} I, \tag{2.1a}$$

$$H = \frac{P^2}{2\mu} + \frac{\mu\omega^2}{2} Q^2, \tag{2.1b}$$

where $\hbar$, μ, and ω are numbers, I is the unit operator, and $i = \sqrt{-1}$. The physical meaning of the numbers μ and ω will be found by correspondence to the classical system to be the mass and angular frequency of the oscillator, which are system constants. The physical meaning of the universal constant[2] $\hbar$ will be discussed in Section II.9.

The justification for the above statements is that there exist in nature physical systems on which one can perform measurements that lead to numbers that can be calculated from the above assumptions. Therefore it is reasonable to say that these physical systems, which are as always "idealized physical systems," have as their mathematical image the above mathematical structure. The name "harmonic oscillator" for this physical system arises from the correspondence to a classical system that has the same name.

The classical-mechanical system called a one-dimensional harmonic oscillator has the following structure (Figure 2.1). Two mass points, with masses m_1 and m_2, are acted upon by a force F along the line joining the two points. The force F is proportional to the difference between the distance r and its equilibrium length r_0:

$$F = -k(r - r_0) = -kx, \tag{2.2}$$

where $x = r - r_0$. The potential energy for this system is then

$$V = +\tfrac{1}{2}kx^2, \tag{2.3}$$

and the kinetic energy is

$$E_{\text{kin}} = \frac{1}{2\mu} p^2, \tag{2.4}$$

[2] $2\pi\hbar = h$ is called Planck's constant.

Figure 2.1 Model of a one-dimensional oscillator.

Figure 2.2 Classical one-dimensional oscillator.

where we have introduced the reduced mass

$$\mu \equiv \frac{m_1 m_2}{m_1 + m_2} \tag{2.5}$$

and the momentum

$$p = \mu \frac{dx}{dt}. \tag{2.6}$$

We can therefore define a classical harmonic oscillator as a system whose potential energy is proportional to the square of the distance from its equilibrium position, or as a system whose total energy is

$$E = \frac{1}{2\mu} p^2 + \tfrac{1}{2}kx^2. \tag{2.7}$$

As is well known, such a physical system is an idealization of the mechanical system in nature that consists of two masses connected by a spring (Figure 2.2). Such a mechanical system performs oscillations with an angular frequency

$$\omega = \sqrt{\frac{k}{\mu}}. \tag{2.8}$$

If we write Equation (2.7) in terms of ω by solving Equation (2.8) for k, we have

$$E = \frac{1}{2\mu} p^2 + \frac{\mu\omega^2}{2} x^2. \tag{2.9}$$

This expression looks exactly like one of the relations defining the quantum-mechanical harmonic oscillator. The important difference is that in the classical case the energy E, the momentum p, and the position x are real numbers, whereas in the quantum-mechanical case these quantities are represented by the operators H, P, and Q, respectively.

For real numbers a and b one always has the relation $ab - ba = 0$; thus

$$px - xp = 0. \tag{2.10}$$

Generally, for operators A and B, $AB - BA \neq 0$, i.e., operators do not, in general, commute. For the particular case of the momentum and position operators one has the relation (2.1a).

Fundamental relations like (2.1a) and (2.1b) cannot be derived; they can only be conjectured. This is true not only for quantum mechanics but also for classical physics. Thus the relation (2.9) is essentially the result of conjectures that go back to Newton.

The process of conjecture in physical theory can be described simply as follows: First one considers the experimental data in a certain domain of physics, e.g., the atomic domain. Then one tries to construct a mathematical structure from which one can calculate numbers agreeing with the experimental data. The algebra of operators defined by the relations (2.1) is such a mathematical structure for the quantum-mechanical harmonic oscillator. The conjecture (2.1b) was simple for us, since it was obtained by correspondence with the classical case. The physical meaning of the system constants μ and ω follows from the classical analogue. The conjecture (2.1a), however, was not so simple. The replacement of the classical relation (2.10) by the quantum-mechanical relation (2.1a), and thus the realization that the quantum-mechanical observables are not represented by real numbers but by operators, was one of the greatest achievements of science. The relation (2.1a) is called the Heisenberg (or canonical) commutation relation. The physical meaning of this relation will become clear when we study its consequences in the following sections.

> Since we do not want to develop quantum mechanics along its historical path, we shall give only a very brief and partial sketch of the motivation for (2.1a). It was known from experiment that light is emitted from atoms at discrete frequencies. At the time the developments leading to (2.1a) took place, the atom was pictured as consisting of a positively charged nucleus with a very small radius and negatively charged electrons revolving around it at a distance of about 10^{-8} cm. Describing this picture in terms of classical physics lead to a dilemma. A revolving and therefore accelerating electron would emit radiation, leading to a continuous loss of its energy and a continuous change in the frequency of the emitted radiation until the electron would finally, and very rapidly, fall into the nucleus. This contradicted the experimental situation that atoms are quite stable, and the frequency of the emitted radiation can only take certain definite values (spectral lines), which are characteristic of the atom. Hence classical theory could not describe the observations in the atomic domain, if the above picture of the atom was kept. One did not wish to abandon the picture of a miniature planetary system for the atoms, since replacing the Hamiltonian by another classical Hamiltonian that would take only discrete values would be virtually impossible. To obtain only discrete values for the Hamiltonian, some extra conditions were imposed on the classical theory that were not only supplementary to, but also in contradiction with the classical theory. This was the so-called "old quantum theory" (Bohr).

Quantum mechanics originated in the idea of Werner Heisenberg and Max Born (1925) and Erwin Schrödinger (1926) to interpret the

Hamiltonian as a mathematical quantity that would take only discrete values. From an analysis of the harmonic oscillator, Heisenberg and Born and Jordan concluded that in order for the energy to be a system of discrete values, the momentum and position should be given by the matrices[3]

$$P = -i\sqrt{\frac{\mu\omega\hbar}{2}}\begin{bmatrix} 0 & \sqrt{1} & 0 & 0 & \cdots \\ -\sqrt{1} & 0 & \sqrt{2} & 0 & \cdots \\ 0 & -\sqrt{2} & 0 & \sqrt{3} & \cdots \\ 0 & 0 & -\sqrt{3} & 0 & \cdots \\ \vdots & \vdots & \vdots & \vdots & \ddots \end{bmatrix}, \quad (2.11)$$

$$Q = \sqrt{\frac{\hbar}{2\mu\omega}}\begin{bmatrix} 0 & \sqrt{1} & 0 & 0 & \cdots \\ \sqrt{1} & 0 & \sqrt{2} & 0 & \cdots \\ 0 & \sqrt{2} & 0 & \sqrt{3} & \cdots \\ 0 & 0 & \sqrt{3} & 0 & \cdots \\ \vdots & \vdots & \vdots & \vdots & \ddots \end{bmatrix}, \quad (2.12)$$

where $2\pi\hbar = h$ is Planck's constant. Schrödinger built on a partial theory due to de Broglie, according to which free particles should diffract like light and should therefore be described by a wave function that satisfies a wave equation. From this he concluded that the momentum should be the differential operator

$$P = \frac{\hbar}{i}\frac{\partial}{\partial x} \quad (2.13)$$

applied to the wave function. Both the Heisenberg and the Schrödinger *Ansatz* lead to the relation (2.1a); and soon after the two formulations were made known, it was shown that they were two different realizations of the same mathematical entities, namely linear operators in a linear space. We shall give a brief description of subsequent historical developments in Section II.5.[4]

The quantum-mechanical harmonic oscillator is an idealization of microphysical systems in nature; e.g., diatomic molecules are, under certain conditions, quantum-mechanical harmonic oscillators. The simplest possible geometrical (classical) picture of the diatomic molecule is two atoms, of masses m_1 and m_2, bound together by an elastic force.

There is an essential difference between the descriptions, in everyday language, of the physical systems whose idealizations are the classical

---

[3] This particular representation of the P and Q as matrices of operators with respect to a certain basis of vectors in an inner-product space will be derived in the next section (Section II.3).

[4] A few historical remarks will appear throughout the book. For a description of the historical developments see, e.g., Jammer (1966).

harmonic oscillator and the quantum-mechanical harmonic oscillator. Everyday language gives quite an adequate description of classical systems, e.g., two carts with masses connected by a spring. Everyday language is, however, insufficient to describe quantum-mechanical systems. The quantum-mechanical harmonic oscillator is best described by the mathematical structure given by (2.1a, b). This is the justification for the above definition and the conjectures that led to it.

The procedure of conjecturing the mathematical relations outlined above for the harmonic oscillator works for many microphysical systems, so we summarize it again:

1. Obtain a classical picture of the microphysical system, and express the observables as functions of position and momentum coordinates.
2. Replace the real numbers corresponding to momentum p and position x in the expressions for other classical observables by the Hermitian operators P and Q, respectively. The operators P and Q satisfy the Heisenberg commutation relation, and from them we obtain the algebra of observables for a quantum-mechanical system.

Not all classical observables can be expressed in terms of momentum and position: the intrinsic angular momentum or spin of a top, for example, cannot be expressed as a function of the momentum and position of the top. For these observables the above procedure of correspondence between the classical and the quantum-mechanical observables has to be generalized. We can give a vague formulation of this conjecturing procedure: in the transition from the mathematical description of the classical system to the mathematical description of the corresponding quantum system, we replace the numbers representing classical observables by operators that represent the corresponding quantum observables. The mathematical properties of these operators are specified by the mathematical (in particular algebraic) relations between them. These relations between the observables are obtained from the relations between the classical observables and some additional relations that specify the property of noncommutativity. These relations can often be obtained from analogy to known relations between similar quantities.

There exist (in elementary-particle physics for example) many quantum-mechanical observables without known classical counterparts. The defining relations between these observables can then be obtained only phenomenologically from a comparison between the consequences of these relations and the experimental data.

We shall discuss many examples of this conjecturing procedure. Rarely is the choice of the relations unambiguous, and the ultimate justification of the defining relations, like that of any fundamental theoretical assumption, is their success in providing agreement between theoretical consequences and experimental data.

The representation of physical observables by operators poses the problem of relating these mathematical quantities to experimental data, which are numbers. The solution of this problem is given by a number of further basic assumptions or axioms, which provide the actual *interpretation* of quantum mechanics. We shall introduce these axioms at suitable places when they are needed.

II.3 Algebra of the Harmonic Oscillator

We now give a mathematical description of the quantum-mechanical harmonic oscillator. To find the mathematical properties of the algebra of the harmonic oscillator means to find how the operators of this algebra act on the vectors of the space $\mathcal{H}$. In order to do this we introduce a new operator a and the operator $a^\dagger$ (which is its adjoint):

$$a = \frac{1}{\sqrt{2}}\left(\sqrt{\frac{\mu\omega}{\hbar}}\, Q + \frac{i}{\sqrt{\mu\omega\hbar}}\, P\right),$$

$$a^\dagger = \frac{1}{\sqrt{2}}\left(\sqrt{\frac{\mu\omega}{\hbar}}\, Q - \frac{i}{\sqrt{\mu\omega\hbar}}\, P\right). \tag{3.1}$$

Then we introduce the new operator

$$N = a^\dagger a, \tag{3.2}$$

which is Hermitian. Inserting (3.1) into (3.2) we get

$$N = \frac{\mu\omega}{2\hbar}\, Q^2 + \frac{1}{2\mu\omega\hbar}\, P^2 - \frac{i}{2\hbar}\ (PQ - QP). \tag{3.3}$$

Using the relations (2.1) we obtain

$$N = \frac{1}{\omega\hbar}\, H - \frac{1}{2}\, I. \tag{3.4}$$

Similarly we calculate

$$aa^\dagger = \frac{1}{\omega\hbar}\, H + \frac{1}{2}\, I. \tag{3.5}$$

Therefore as a consequence of the canonical commutation relation (2.1a) we obtain for a and $a^\dagger$ the commutation relation

$$[a, a^\dagger] \equiv aa^\dagger - a^\dagger a = I. \tag{3.6}$$

Furthermore, we can express the energy operator H in terms of a and $a^\dagger$ to obtain

$$H = \hbar\omega a^\dagger a + \frac{\hbar\omega}{2}\, I = \hbar\omega(N + \tfrac{1}{2}I). \tag{3.7}$$

Now we want to find the eigenvalues and eigenvectors of H and N. As the eigenvectors of N are eigenvectors of H, it is sufficient to find all the eigenvectors of N.

Let us assume that there is at least one eigenvector of N in $\mathscr{H}$, and call its eigenvalue λ. (This is not a trivial assumption, since there are many operators that do not have eigenvectors in $\mathscr{H}$. In fact it is an additional assumption without which Equation (2.1a) is ambiguous.) Let φ_λ be this eigenvector; thus

$$N\varphi_\lambda = \lambda\varphi_\lambda. \tag{3.8}$$

Then calculate $Na\varphi_\lambda$ using Equation (3.6),

$$Na\varphi_\lambda = (a^\dagger a)a\varphi_\lambda = (aa^\dagger - I)a\varphi_\lambda = a(a^\dagger a - I)\varphi_\lambda$$
$$= a(N - I)\varphi_\lambda = a(\lambda - 1)\varphi_\lambda = (\lambda - 1)a\varphi_\lambda.$$

Thus

$$N(a\varphi_\lambda) = (\lambda - 1)(a\varphi_\lambda), \tag{3.9}$$

so that either $a\varphi_\lambda = 0$ or $a\varphi_\lambda$ is an eigenvector of N with eigenvalue $\lambda - 1$. Suppose the latter is the case. Then we can define

$$\varphi_{\lambda-1} \equiv a\varphi_\lambda. \tag{3.10}$$

Repeating the above calculation on $\varphi_{\lambda-1}$, we find that

$$\varphi_{\lambda-2} \equiv aa\varphi_\lambda \tag{3.11}$$

is either zero or an eigenvector with eigenvalue $\lambda - 2$. We continue in this way and obtain a sequence of vectors

$$\varphi_{\lambda-m} \equiv a^m\varphi_\lambda \qquad (m = 0, 1, 2, \ldots), \tag{3.12}$$

which are eigenvectors of N with eigenvalue $(\lambda - m)$, as long as $\varphi_{\lambda-m} \neq 0$. In a similar way one calculates $N(a^\dagger\varphi_\lambda)$ to get

$$N(a^\dagger\varphi_\lambda) = (\lambda + 1)(a^\dagger\varphi_\lambda). \tag{3.13}$$

Hence $\tilde{\varphi}_{\lambda+1} \equiv a^\dagger\varphi_\lambda$ is either the zero vector or an eigenvector of N with eigenvalue $\lambda + 1$. It is easily shown that always $\tilde{\varphi}_{\lambda+1} \neq 0$. Suppose $a^\dagger\varphi_\lambda = 0$. That means

$$\|a^\dagger\varphi_\lambda\| = 0. \tag{3.14}$$

But

$$\|a^\dagger\varphi_\lambda\|^2 = (a^\dagger\varphi_\lambda, a^\dagger\varphi_\lambda), \tag{3.15}$$

and by definition,

$$(a^\dagger\varphi_\lambda, a^\dagger\varphi_\lambda) = (\varphi_\lambda, aa^\dagger\varphi_\lambda). \tag{3.16}$$

By the commutation relation for a and $a^\dagger$ we have

$$(\varphi_\lambda, aa^\dagger\varphi_\lambda) = (\varphi_\lambda, a^\dagger a\varphi_\lambda) + (\varphi_\lambda, \varphi_\lambda)$$
$$= \|a\varphi_\lambda\|^2 + \|\varphi_\lambda\|^2 > 0, \tag{3.17}$$

since $\varphi_\lambda \neq 0$. Thus $\tilde{\varphi}_{\lambda+1}$ is always an eigenvector of N with eigenvalue $\lambda + 1$. Again by repeating the above process one obtains a sequence of vectors $\tilde{\varphi}_{\lambda+n}$ $(n = 0, 1, 2, \ldots)$, which are eigenvectors of N with eigenvalue $\lambda + n$. These are always nonzero, as follows from the above proof.

We now determine when $\varphi_{\lambda-m}$ can be zero. We calculate

$$(\varphi_{\lambda-m}, N\varphi_{\lambda-m}) = (\lambda - m)(\varphi_{\lambda-m}, \varphi_{\lambda-m}) = (\lambda - m)\|\varphi_{\lambda-m}\|^2$$

and

$$(\varphi_{\lambda-m}, N\varphi_{\lambda-m}) = (\varphi_{\lambda-m}, a^\dagger a\varphi_{\lambda-m}) = \|a\varphi_{\lambda-m}\|^2.$$

Since the norm is always nonnegative, it follows that

$$\lambda - m = \frac{\|a\varphi_{\lambda-m}\|^2}{\|\varphi_{\lambda-m}\|^2} \geq 0. \tag{3.18}$$

Therefore the sequence of eigenvectors $\varphi_{\lambda-m}$ must terminate after a finite number of steps, and there must exist one vector φ_0 such that

$$a\varphi_0 = 0.$$

The vector φ_0 is an eigenvector of N with eigenvalue zero, since $N\varphi_0 = a^\dagger a\varphi_0 = a^\dagger 0 = 0$. Now we define the normalized vectors

$$\phi_0 = \frac{\varphi_0}{\|\varphi_0\|},$$

$$\phi_1 = C_1 a^\dagger \phi_0, \tag{3.19}$$

$$\vdots$$

$$\phi_n = C_n (a^\dagger)^n \phi_0$$

$$\vdots$$

where C_n are complex numbers. From the considerations above we know that the vectors ϕ_n are eigenvectors of N, with eigenvalue n, and the C_n are chosen such that the ϕ_n are normalized. Thus we have

$$N\phi_n = n\phi_n, \qquad \|\phi_n\| = 1. \tag{3.20}$$

The C_n are calculated as follows: From Equations (3.20) and (3.19),

$$1 = \|\phi_n\|^2 = (a^{\dagger n}\phi_0, a^{\dagger n}\phi_0)|C_n|^2,$$

but

$$(a^{\dagger n}\phi_0, a^{\dagger n}\phi_0) = \frac{1}{|C_{n-1}|^2}(a^\dagger\phi_{n-1}, a^\dagger\phi_{n-1}),$$

so

$$1 = \frac{|C_n|^2}{|C_{n-1}|^2}(\phi_{n-1}, aa^\dagger\phi_{n-1}).$$

Using Equation (3.6),

$$1 = \frac{|C_n|^2}{|C_{n-1}|^2} (\phi_{n-1}, (a^\dagger a + 1)\phi_{n-1})$$

$$= \frac{|C_n|^2}{|C_{n-1}|^2} (\phi_{n-1}, (N + 1)\phi_{n-1})$$

$$= n \frac{|C_n|^2}{|C_{n-1}|^2} (\phi_{n-1}, \phi_{n-1})$$

$$= n \frac{|C_n|^2}{|C_{n-1}|^2};$$

hence C_n must be chosen so that

$$n|C_n|^2 = |C_{n-1}|^2. \tag{3.21}$$

Since ϕ_0 is normalized, $C_0 = 1$, and one solution of Equation (3.21) is then

$$C_n = \sqrt{\frac{1}{n!}}. \tag{3.22}$$

There are other solutions of (3.21) which differ from (3.22) by a factor of modulus 1; we choose (3.22).

We may summarize the procedure as follows: Start with a normalized vector $\phi_0 \in \mathcal{H}$ that has the following property:

$$a\phi_0 = 0. \tag{3.23}$$

Then apply to ϕ_0 the operator $a^\dagger$ to obtain a system of eigenvectors of N,

$$\phi_n = \frac{1}{\sqrt{n!}} a^{\dagger n} \phi_0, \tag{3.24}$$

with eigenvalues $n = 0, 1, 2, \ldots$. These eigenvectors are normalized by the construction above, and they are orthogonal to each other since they are eigenvectors of a Hermitian operator with different eigenvalues (cf. Problem I.4). Thus we have the orthonormality relations

$$(\phi_n, \phi_{n'}) = \delta_{nn'}, \tag{3.25}$$

and the ϕ_n form an orthonormal system in $\mathcal{H}$. The operators a and $a^\dagger$ are defined on this orthonormal system by

$$a\phi_n = \sqrt{n}\phi_{n-1},$$
$$a^\dagger\phi_n = \sqrt{n+1}\phi_{n+1}. \tag{3.26}$$

Because of this property, a is called the annihilation operator and $a^\dagger$ the creation operator. Both are often referred to as ladder operators. All the elements of the algebra of the harmonic oscillator that are functions of a and $a^\dagger$ are defined on the elements of this orthonormal system. Consider the set of all vectors

$$\psi = \sum_n \alpha_n \phi_n, \tag{3.27}$$

where the α_n are complex numbers and the sum runs through an arbitrarily large but finite number of n. This set of vectors forms a linear space called the space spanned by the vectors ϕ_n. We will denote this space by $\mathscr{H}$ or Φ when we ignore details.

[More precisely, the space of all the $\psi = \sum_{n=0}^{\infty} \alpha_n \phi_n$ for which $\sum_{n=0}^{\infty} |\alpha_n|^2 < \infty$ will be denoted by $\mathscr{H}$ (for Hilbert space), and the space of all ψ for which $\sum_{n=0}^{\infty} |\alpha_n|^2 (n+1)^p < \infty$ for all $p = 0, 1, 2, 3, \ldots$ will be denoted by Φ. Obviously

$$\Phi \subset \mathscr{H}. \tag{3.28}$$

The advantage of Φ over $\mathscr{H}$ is that all operators representing observables can be defined on the whole space Φ, but not on all of $\mathscr{H}$. Though both spaces Φ and $\mathscr{H}$ are mathematical generalizations, Φ is more convenient for physics.]

As all observables are functions of a and $a^\dagger$, and a and $a^\dagger$ are known by (3.26) and (3.27) for all ψ, all observables for the quantum-mechanical harmonic oscillator are in principle known on Φ. So the task of determining the mathematical properties of the algebra of the harmonic oscillator is in principle completed.

Let us calculate the diagonal matrix element of the energy operator H between the vector ϕ_n for a fixed value of n. Using Equations (3.7) and (3.20), we have

$$H\phi_n = \hbar\omega(N + \tfrac{1}{2})\phi_n = \hbar\omega(n + \tfrac{1}{2})\phi_n \tag{3.29}$$

and

$$\begin{aligned}(\phi_n, H\phi_n) &= \hbar\omega(\phi_n, N\phi_n) + \hbar\omega(\phi_n, \tfrac{1}{2}\phi_n) \\ &= \hbar\omega(n + \tfrac{1}{2})|\phi_n|^2 = \hbar\omega(n + \tfrac{1}{2}).\end{aligned}$$

We call this number E_n:

$$E_n = \hbar\omega(n + \tfrac{1}{2}). \tag{3.30}$$

E_n are the eigenvalues of the operator H. The set of all eigenvalues is called the *spectrum* of an operator. The spectrum $\{E_n | n = 1, 2, \ldots\}$ of the operator H is the energy spectrum of the harmonic oscillator.

II.4 The Relation Between Experimental Data and Quantum-Mechanical Observables

We now wish to formulate the basic axiom of quantum mechanics that gives the connection between the operator representing an observable and the experimentally measured values for this observable. Before doing this, we observe some fundamental differences in the experimental properties of the classical harmonic oscillator and the quantum-mechanical oscillator.

Consider the classical picture in Figure 2.1. Assume that the system has no energy—i.e., does not perform any vibrations, or vibrates with zero amplitude.

The system can be excited into vibration if, for example, one of the masses is struck by another massive projectile. The oscillator can perform vibrations of any amplitude, depending on the amount of energy transferred to it, which in turn depends on the momentum of the projectile. Thus for the classical oscillator one can prepare a projectile that excites the oscillator to vibrate with any desired energy E. The energy is given by

$$E = \tfrac{1}{2}ka^2, \tag{4.1}$$

where a is the amplitude of the vibrations

$$x(t) = a \sin(\omega t + \theta_0).$$

In the realization of the classical harmonic oscillator as two massive carts (see Figure 2.2), the system may be excited, for example, by hitting one of the carts with a third cart of suitable momentum.

The situation is quite different for the quantum-mechanical harmonic oscillator. To see this we will consider an analogous experiment with microphysical systems. As is typical for experiments with microphysical systems, one does not have a single physical system but an *ensemble* of systems, i.e., a collection of systems that are in a certain sense identical. The ensemble of microphysical harmonic oscillators that is used in this experiment is a gas of CO molecules, and an electron beam serves as an ensemble of structureless projectiles.

> Remark: It should be remembered that the statement that a physical system is structureless has no universal meaning. Any system can be considered structureless under certain conditions but not under others. For instance, in the domain of physics for which the kinetic theory of gases is valid, the molecules can be considered as structureless systems, whereas when higher energies are considered, they are seen to have structure. The electron is a system that can be considered to be without "energy structure" for all currently available energies.

Experiments of this type are called *energy-loss experiments*, and are based on the original experiment performed by Franck and Hertz in 1914.

Figure 4.1 shows a schematic diagram of such experiments. A beam of electrons leaves a monochromator[5] with a very narrow energy band of

[5] A monochromator is an apparatus that prepares a beam with very well-defined momentum and, therefore, with very well-defined kinetic energy. An analyzer is the same kind of apparatus used for a different purpose.

Figure 4.1 Schematic diagram of an energy-loss experiment.

(a)

(b)

Figure 4.2 (a) Schematic diagram of double electrostatic analyzer. Electrons are emitted from the thoria-coated iridium filament. They pass between the cylindrical grids at an energy about 2.05 eV, and are accelerated into the collision chamber, where they are crossed with a molecular beam. Those electrons scattered into the acceptance angle of the second electrostatic analyzer pass between the cylindrical grids, again at an energy from 0 to ≈ 2 eV. The electrons pass the exit slit into the second chamber and impinge onto an electron multiplier. (b) Energy spectrum of scattered electrons in CO at an incident electron energy of 2.05 eV.

energy E_e, and enters a collision chamber filled with an ensemble of CO molecules (or any other physical system whose structure is to be investigated) kept at a low temperature. Some of the electrons scatter into an analyzer which focuses only those electrons with energy $E_{e'}$ onto the detector. (The energy resolution is typically 0.005 to 0.05 eV; in the particular experiment described in Figure 4.2(a), it is 0.06 eV.) The energy $E_{e'}$ that is selected by the analyzer can be varied, so that one can measure the intensity I (the electron current at the detector) of the electrons as a function of the energy loss $E = E_e - E_{e'}$.

The result of an actual experiment, performed with the apparatus schematically depicted in Figure 4.2(a) on the diatomic molecule CO, is shown in Figure 4.2(b).[6] We see a maximum for an energy loss of $E_0 = E_e - E_{e'} = 0$; i.e., a major fraction of the electron current e does not lose any energy (elastically scattered electrons). Then there is a relative maximum of intensity for the electrons that have lost the energy E_1 (≈ 0.26 eV), which means that a large portion of the electrons of energy E_e lose energy inelastically to the CO molecules. A third bump in electron intensity occurs at the energy loss $E_2 = 2E_1$, and so on. As a whole, Figure 4.2 shows us that the scattered electron current e' is a mixture of electrons that have lost one of the eight discrete amounts of energy $E_0, E_1, E_2, \ldots, E_7$ at $v = 0, v = 1, \ldots, v = 7$, respectively, to the CO molecules.

We conclude from this experimental situation[7] that the CO molecules cannot be excited to any arbitrary energy value. Only a discrete number of energy values are possible; i.e., the physical system has a discrete number of energy levels. This situation is represented by an energy-level diagram (Figure 4.3). In this particular case of the vibrating molecule, the energy-level diagram consists of a series of equidistant levels, as can be seen from the experimental results in Figure 4.2(b).

From Figure 4.2(b), one sees that this is only approximately true and that the distance between energy levels actually decreases with increasing energy. This discrepancy indicates the limitation of the

[6] From G. J. Schulz, *Phys. Rev.* **135**, A988 (1964), with permission.

[7] Multiple scattering of electrons and CO molecules is negligible because the intensity of the electron current e and the density of CO molecules are sufficiently low.

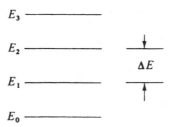

Figure 4.3 Energy-level diagram of the harmonic oscillator.

harmonic oscillator as the model for the vibrating molecule. We will discuss the corrections to this model later.

Comparing this experimental energy spectrum with the result in Equation (3.29), we see that they agree if we take $\Delta E = \hbar\omega$ for the energy difference between two neighboring energy levels. Thus the diagonal matrix elements $(\phi_n, H\phi_n)$ give the possible energy values of the quantum-mechanical oscillator (e.g., vibrating molecule), apart from an overall constant. We interpret the situation as follows: The projectile (electron) excites the harmonic oscillator (vibrating molecule) into any one of a discrete number of *states* described by ϕ_n. If no excitation takes place, the harmonic oscillator remains in the ground state, described by ϕ_0. If all the electrons that passed through the gas had lost the same amount of energy, i.e., if there were only one bump in the experimental curve at $v = n_0$, the ensemble of molecules struck by these electrons would be described by one of the ϕ_n, say ϕ_{n_0}. In this case the ensemble of CO molecules would be said to be in a *pure state* ϕ_{n_0}. This is often expressed by saying that "all molecules have the energy E_{n_0}." The results in Figure 4.2(b) show that the ensemble of molecules considered there (i.e., those molecules that have collided with the electrons that are scattered into the acceptance angle of the analyzer, which in the experiment of Figure 4.2(a) is about 72° from the electron beam axis) is not in a pure state, since the electrons have not lost one particular value E_{n_0}, but rather a variety of energies $E_0, E_1, E_2, \ldots, E_7$. Such an ensemble of molecules is said to be in a *mixture of states*, or the state of the ensemble is said to be a *mixture*. This mixture can be described by the set of vectors $\phi_0, \phi_1, \phi_2, \ldots,$ $\phi_n, \ldots$ and a set of numbers (relative probabilities) $w_0, w_1, \ldots, w_n, \ldots$ where w_n is chosen to be proportional to the height h_n of the bump corresponding to the energy E_n, and they are normalized so that

$$\sum_n w_n = 1. \tag{4.2}$$

Thus one may form a mental picture of the mixture as a collection of N molecules, where $N_n = w_n N$ is the number of molecules with energy E_n in the ensemble. If it were possible to pick one molecule out of the ensemble, w_n would be the probability that this molecule has energy E_n. However, a complete dynamical description of a single specimen picked from a microphysical system is not possible. w_n is measured as the relative intensity of the electron current that has lost the energy E_n and therefore represents a statement about a large number of single specimens.

Before we introduce a concise description of the state of a mixture, we shall give an alternate description of a pure state. To do this we shall need some more mathematics, which we now introduce.

⟦An operator W is called *positive* if $(\phi, W\phi) \geq 0$ for all $\phi \in \mathcal{H}$. A Hermitian operator P is called a *projection operator* or *projector* iff $P^2 = P$.

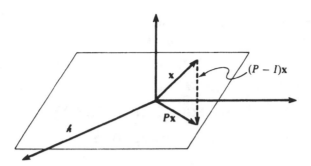

Figure 4.4 Projection operator onto a plane.

It is easily seen that the set $h = \{\phi \in \mathcal{H} : P\phi = \phi\}$ is a subspace $h \subseteq \mathcal{H}$; for if $\phi, \psi \in h$ and $\alpha, \beta \in \mathbb{C}$, then $P(\alpha\phi + \beta\psi) = \alpha P\phi + \beta P\psi = \alpha\phi + \beta\psi$, i.e., h is closed under multiplication by a number and under vector addition. Alternatively, $h = \{P\phi : \phi \in \mathcal{H}\}$, and hence we write $h = P\mathcal{H}$. As an example, let $\mathcal{H} = \mathbb{R}^3$, the three-dimensional Euclidean space, and let h be a plane passing through the origin. Then for any $\mathbf{x} \in \mathbb{R}^3$ let $P\mathbf{x}$ be the ordinary projection of $\mathbf{x}$ onto the plane (see Figure 4.4).

Two subspaces $h_1, h_2 \subseteq \mathcal{H}$ are said to be *mutually orthogonal* if for each $\phi \in h_1$ and for each $\psi \in h_2$, we have $(\phi, \psi) = 0$. The set $h = \{\phi + \psi : \phi \in h_1 \text{ and } \psi \in h_2\}$ is called the *direct sum* of h_1 and h_2, and is denoted $h = h_1 \oplus h_2$. Similarly, given a collection $h_1, h_2, h_3, \ldots,$ of mutually orthogonal subspaces, one can form their direct sum, $h = \sum_i \oplus h_i = \{\sum_i \varphi_i : \varphi_i \in h_i\}$.

Given a projector P that projects onto the subspace h, the operator $I - P$ is also a projector. [*Proof*: Since $P^2 = P$, it follows that $(I - P)^2 = I - IP - PI + P^2 = I - P - (P - P^2) = I - P.$] $I - P$ thus determines another subspace, which we will denote by $h^{\perp}$. All vectors in $h^{\perp}$ are orthogonal to those in h; for if $\phi \in h$ and $\psi \in h^{\perp}$, then $\phi = P\phi$ and $\psi = (I - P)\psi$; using the definition ($P^{\dagger} = P$ and $P^2 = P$) of a projector, we have $(\phi, \psi) = (P\phi, (I - P)\psi) = (\phi, P(I - P)\psi) = (\phi, (P - P^2)\psi) = (\phi, 0\psi) = 0$. Since for any $\psi \in \mathcal{H}$ we may write $\psi = P\psi + (I - P)\psi$, any projector P gives a *decomposition* of $\mathcal{H}$ into orthogonal subspaces, $\mathcal{H} = h \oplus h^{\perp}$.

For any $\phi \in \mathcal{H}$, the set $\{\alpha\phi : \alpha \in \mathbb{C}\}$ is a one-dimensional subspace of $\mathcal{H}$, called the *space spanned by* ϕ.

Finally, given a basis $\{\phi_{\gamma}\}$ of $\mathcal{H}$, we define the *trace of the operator* A by

$$\operatorname{Tr} A \equiv \sum_{\gamma} (\phi_{\gamma}, A\phi_{\gamma}). \tag{4.3}$$

It can be shown (Problem 19) that $\operatorname{Tr} A$ is independent of the orthonormal basis chosen. $\operatorname{Tr} A$ need not be finite. We will in general assume that it is finite for those operators whose trace we take. The trace has the following properties:

$$\text{Tr}(AB) = \text{Tr}(BA). \tag{4.4a}$$

In general, $\text{Tr}(A_1 A_2 \ldots A_n) = \text{Tr}(A_2 A_3 \ldots A_n A_1)$.

$$\text{Tr}((A + B)C) = \text{Tr}(AC) + \text{Tr}(BC).] \tag{4.4b}$$

We now use these new mathematical concepts to describe the harmonic oscillator. Let

h_0 be the space spanned by ϕ_0,

h_1 be the space spanned by ϕ_1,

h_2 be the space spanned by ϕ_2,

$$\vdots$$

h_n be the space spanned by ϕ_n.

$$\vdots$$

Then the space $\mathcal{H}$ is the orthogonal sum

$$\sum_{n=0}^{\infty} \oplus h_n = \mathcal{H}. \tag{4.5}$$

Let Λ_n $(n = 0, 1, 2, \ldots)$ denote the projection operators that project on the spaces h_n. Then the operator form of (4.5) is written

$$\sum_{n=0}^{\infty} \Lambda_n = I. \tag{4.6}$$

It says that every $f \in \mathcal{H}$ can be written as the sum of its components $\Lambda_n f$ in the subspaces h_n:

$$f = \sum_{n=0}^{\infty} \Lambda_n f. \tag{4.7}$$

This is equivalent to writing

$$f = \sum_{n=0}^{\infty} \phi_n (\phi_n, f) = \sum_{n=0}^{\infty} |\phi_n\rangle\langle\phi_n|f\rangle, \tag{4.8}$$

which is, according to (I.2.15b), the statement that ϕ_n is a basis. Thus Λ_n may also be written $\Lambda_n = |\phi_n\rangle\langle\phi_n|$, and (4.6) may be written

$$I = \sum_{n=0}^{\infty} |\phi_n\rangle\langle\phi_n|. \tag{4.9}$$

We will often denote the projection operator Λ on the one-dimensional subspace spanned by a unit vector ψ by

$$\Lambda = |\psi\rangle\langle\psi|. \tag{4.10}$$

If we apply the operator H to the arbitrary vector f and use (4.9), we have, because of (3.29) and (3.30),

$$Hf = \sum_{n=0}^{\infty} H|\phi_n\rangle\langle\phi_n|f\rangle = \sum_{n=0}^{\infty} E_n|\phi_n\rangle\langle\phi_n|f\rangle.$$

Ás f was arbitrary, we may as well omit it and write

$$H = \sum_{n=0}^{\infty} E_n|\phi_n\rangle\langle\phi_n| = \sum_{n=0}^{\infty} E_n\Lambda_n. \tag{4.11}$$

This is called the *spectral representation* of the operator H. With the notion of the trace we can express the energy values $E_n = (\phi_n, H\phi_n)$ in the form

$$E_n = \text{Tr}(H\Lambda_n). \tag{4.12}$$

PROOF. Since

$$\Lambda_n\phi_\gamma = \begin{cases} \phi_n & \text{if } n = \gamma, \\ 0 & \text{if } n \neq \gamma, \end{cases}$$

one obtains

$$\text{Tr}(H\Lambda_n) = \sum_\gamma (\phi_\gamma, H\Lambda_n\phi_\gamma)$$

$$= \sum_\gamma (\phi_\gamma, H\phi_n)\delta_{\gamma n} = E_n. \tag{4.13}$$

Thus the state of the harmonic oscillator in which the energy is E_n can be described by the projection operator Λ_n in place of ϕ_n. Clearly Λ_n and ϕ_n do not represent the same mathematical objects. Λ_n represents the subspace $\hbar_n = \Lambda_n\mathcal{H} = \{\Lambda_n f \mid f \in \mathcal{H}\}$, which is the one-dimensional subspace spanned by ϕ_n (a one-dimensional subspace is often called a *ray*). The space $\hbar_n$ contains many normalized vectors, namely, all those that can be obtained from ϕ_n by multiplication with a phase factor $e^{i\alpha}$, where $0 \leq \alpha \leq 2\pi$. (The set of all $e^{i\alpha}\phi$ with $\|\phi\| = 1$ is called a *unit ray*.)

It is generally assumed that a pure state of a physical system is described by a projection operator Λ on a one-dimensional subspace (or by a unit ray) and not by a vector. However, it is very common to call not only Λ but also any vector $\phi \in \Lambda\mathcal{H}$ a "state" or "state vector."

The result described by (4.12) is a special one because the one-dimensional projection operator Λ_n is in the spectral representation of H. Situations do occur where an observable A is measured in a state described by a one-dimensional projection operator Λ not in the spectral representation of A (e.g., it could be that $[A, \Lambda] \neq 0$). Then the "value" of the observable A in the state Λ is denoted by $\langle A\rangle_\Lambda$, or just by $\langle A\rangle$, and it is determined from measurements using the experimental set-up that defines the observable A. $\langle A\rangle$ is obtained by the following prescription: One performs a series of N measurements of the observable A; i.e., one either has N identical systems on which one performs the measurement, or one performs the same measurement N times on a physical system, making sure that the system is in the same state for each measurement. Let the numbers measured for the observable A in this experiment be denoted $a_0, a_1, \ldots, a_n, \ldots$, and let N_n be the number of measurements that gave the result a_n. Then $\langle A \rangle$ is the average of these measurements,

$$\langle A \rangle = \sum_n \frac{N_n}{N} a_n, \tag{4.14a}$$

for sufficiently large N [i.e., N must be large enough so that the right-hand side of (4.14a) would change only slightly if N were to be increased]. The ratio $w_n = N_n/N$ is the probability of obtaining the value a_n in one measurement. Thus $\langle A \rangle$ is an experimentally determined quantity.

The operators A and Λ are mathematical quantities. A is the mathematical image for the observable, and Λ is the mathematical image for the state. If the mathematical image of the system (i.e., the theory of the system) is known, then in particular one can calculate $\mathrm{Tr}(A\Lambda)$; i.e., $\mathrm{Tr}(A\Lambda)$ is a mathematical quantity that follows from the theory.

The second basic assumption or postulate of quantum mechanics relates the experimentally determined quantity $\langle A \rangle$ to the theoretically calculated quantity $\mathrm{Tr}(A\Lambda)$. It is formulated now only for a pure state:

II′.   A pure state of a quantum-mechanical physical system is characterized by a projection operator Λ on a one-dimensional subspace $\Lambda\mathcal{H}$ (or by the one-dimensional subspace itself). The average value $\langle A \rangle$, determined in an experiment performed on the physical system in the state Λ, is related to $\mathrm{Tr}(A\Lambda)$ by

$$\langle A \rangle = \mathrm{Tr}(A\Lambda). \tag{4.14b}$$

$\mathrm{Tr}(A\Lambda)$ [and because of (4.14b) also $\langle A \rangle$] is called the *expectation value* of the observable A in the state Λ.

For the harmonic oscillator in a pure state Λ_n, say Λ_1 (the projection operator on the subspace spanned by ϕ_1), measurements of H will always yield the same value. Consequently

$$\langle H \rangle = \sum_n \frac{N_n}{N} E_n = \frac{N}{N} E_1. \tag{4.15}$$

Hence, in the pure state Λ_1, the probability w_1 for the value E_1 is $w_1 = 1$ and all other $w_i = 0$.

A pure state Λ is called an *eigenstate of the observable A*, or *pure eigenstate of A*, if the one-dimensional space $\Lambda\mathcal{H}$ is spanned by an eigenvector ϕ of A. (This definition of eigenstates of A will be generalized in Section II.6 to include states which are not necessarily pure, i.e., not described by one-dimensional projection operators.) Thus, the pure state Λ_1 of the harmonic oscillator is also an eigenstate of the energy observable H, i.e., a pure eigenstate of H. In general, as we shall see later, a physical system can be in a pure state described by a one-dimensional space, and the measurement of some observable A, of which this state is not an eigenstate, will still yield a set of different numbers $a_1, a_2, \ldots, a_n, \ldots$. This is the point at which quantum mechanics deviates drastically from classical theories. We shall discuss this matter in detail later.

We now discuss the description of a mixture rather than a pure state. Consider the CO molecules struck by the electrons in the energy loss experiments of Figures 4.1 or 4.2. If we measure the energy of the molecules N times, we expect

$$N_0 = w_0 N \text{ measurements will give the value } E_0,$$
$$N_1 = w_1 N \text{ measurements will give the value } E_1,$$
$$\vdots$$
$$N_n = w_n N \text{ measurements will give the value } E_n,$$

because w_i is the ratio of molecules in the ensemble, with energy E_i. Thus the average value $\langle H \rangle$ of the energy operator H is

$$\langle H \rangle = \sum_n w_n E_n = \sum_n w_n(\phi_n, H\phi_n), \tag{4.16}$$

which is to be equal to the expectation value of H in the mixed state. We want to write this expectation value as a trace of some operator; to do this we introduce the positive Hermitian operator

$$W \equiv \sum_n w_n \Lambda_n, \tag{4.17}$$

where each Λ_n is the projector on the eigenspace $\hbar_n$ of H spanned by ϕ_n. Now we calculate

$$\text{Tr}(HW) = \sum_\gamma (\phi_\gamma, HW\phi_\gamma)$$
$$= \sum_\gamma \sum_n w_n(\phi_\gamma, H\Lambda_n \phi_\gamma).$$

Since

$$\Lambda_n \phi_\gamma = \delta_{n\gamma} \phi_n,$$

we have

$$\text{Tr}(HW) = \sum_\gamma \sum_n w_n(\phi_\gamma, H\phi_n)\delta_{n\gamma}$$
$$= \sum_n w_n(\phi_n, H\phi_n). \tag{4.18}$$

Thus if the state of the mixture is described by the operator W defined in Equation (4.17), then the expectation value of the energy is given by

$$\langle H \rangle = \text{Tr}(HW). \tag{4.19}$$

It is important to note that in Equation (4.19), W is not a projection operator as would be the case for a pure state.

The statement contained in Equation (4.19) is conjectured to be the general expression for quantum mechanics, and we formulate it as the second postulate of quantum mechanics:

II. The state of a quantum-mechanical system is characterized by a positive Hermitian operator W. The expectation value of an observable A, determined in an experiment as the average value $\langle A \rangle$ according to (4.14a), is given by

$$\langle A \rangle = \text{Tr}(AW) \quad \text{if Tr } W = 1$$

or by

$$\langle A \rangle = \frac{\text{Tr}(AW)}{\text{Tr } W}.$$

The operator W is called the *statistical operator*, and its matrix is called the *density matrix*. The state of the system is called a pure state iff W is the projection operator on a one-dimensional subspace.

It is customary to normalize the probabilities w_n so that

$$\sum_{n=0}^{\infty} w_n = 1, \tag{4.20}$$

i.e., to normalize the statistical operator so that

$$\text{Tr } W = 1. \tag{4.21}$$

Note, however, that Equation (4.20) as well as Equation (4.21) is only a convention. Instead of w_n we could have used the height of the bumps h_n to characterize the state. In that case we would get, in place of Equation (4.17),

$$\tilde{W} = \sum_n h_n \Lambda_n \tag{4.22}$$

and

$$\text{Tr } \tilde{W} = \sum_n h_n.$$

It is easy to see that

$$W = \frac{\tilde{W}}{\text{Tr } \tilde{W}}.$$

W and $\tilde{W}$ describe the same state of the harmonic oscillator. However, if one uses $\tilde{W}$ to characterize the state, then the expectation value of the observable H is given by

$$\langle H \rangle = \frac{\text{Tr}(H\tilde{W})}{\text{Tr } \tilde{W}} \quad \text{or in general} \quad \langle A \rangle = \frac{\text{Tr}(A\tilde{W})}{\text{Tr } \tilde{W}}. \tag{4.23}$$

Thus, whenever possible, it is more convenient to use the normalized statistical operator for the description of the state.

The second basic assumption of quantum mechanics cannot be derived; it can only be conjectured. It has been conjectured, in the long process of discovering quantum mechanics, from a variety of experimental results. We have in our above considerations made this axiom plausible for the particular case of the energy measurement of the harmonic oscillator.

Consider the state of the harmonic oscillator given by

$$W = \sum_n w_n \Lambda_n.$$

The projection operator Λ_p on the eigenspace of H with eigenvalue E_p is also an observable. Thus we can calculate the expectation value of Λ_p in the state W:

$$\langle \Lambda_p \rangle = \text{Tr}(\Lambda_p W) = \sum_n \text{Tr}(\Lambda_p \Lambda_n) w_n. \tag{4.24}$$

From the properties of projection operators we know that

$$\Lambda_p \Lambda_n = 0 \quad \text{for} \quad p \neq n,$$
$$\Lambda_n \Lambda_n = \Lambda_n; \tag{4.25}$$

consequently

$$\langle \Lambda_p \rangle = \sum_n \delta_{np} \, \text{Tr}(\Lambda_n) w_n \tag{4.26}$$

$$= \text{Tr}(\Lambda_p) w_p, \tag{4.27}$$

so

$$\langle \Lambda_p \rangle = w_p, \tag{4.28}$$

because $\text{Tr } \Lambda_p = 1$, since Λ_p is the projector on a one-dimensional subspace. Therefore Λ_p is the observable whose expectation value is the probability of obtaining the value E_p in a measurement of the energy, i.e., the probability of finding the system in the state Λ_p. In particular, for the pure state $W = \Lambda_n$ we have

$$\langle \Lambda_p \rangle = \text{Tr}(\Lambda_p \Lambda_n) = \begin{cases} 0 & \text{for } p \neq n, \\ 1 & \text{for } p = n. \end{cases} \tag{4.29}$$

Generally, projection operators represent a particular kind of observable called a *property* or *proposition*, whose expectation value tells us something about the state.

The postulate II states how the expectation value is calculated if the statistical operator W is known. We now discuss how the statistical operator has to be chosen in any particular case. To do this we use our belief, which is supported by all experimental evidence, that *a repetition of a measurement always leads to the same result,* or, in other words, that all experiments performed under the same conditions give the same experimental numbers. It is assumed that generally a certain physical system may be in a number of states $1, 2, \ldots$, which in quantum mechanics are described by statistical operators $W_1, W_2, \ldots$, and that these states can be determined by the measurements of observables $A, B, \ldots$. Thus every state of the physical system is the result of a *preparation* of the system. A preparation is a series of manipulations with physical equipment that leads to the measurement of an observable (or several observables). These measurements may or may not affect the state of the system. In classical physics the effect of a measurement upon the state of the physical system is always neglected; the essential difference in quantum physics is that here this effect is taken into account. Thus in principle, and also in practice, not all measurements can be performed simultaneously. Therefore one must always be able to reproduce a certain physical state, so that it can be determined by a series of successive measurements. After the measurement of an observable, the system will be in a state that has been prepared by this measurement. This state must be described by a statistical operator for which the calculated expectation value of the observable agrees with the average value just measured. That is, the W must be chosen so that the calculated value agrees with the measured value that one will obtain in an "immediate" repetition of the measurement, i.e., in a measurement that is performed before the state has a chance to change (develop in time). In fact, this principle was used when we wrote down the state of the CO molecules. In the example of the ensemble of CO molecules, the preparation consists of setting up the apparatus and bombarding the gas with monochromatic electrons. This prepares the CO gas in a state that is a mixture of energy eigenstates Λ_n, with the mixing ratios w_n. The statistical operator W has to express what has been measured on the physical system, or what preparations have been performed. Therefore we postulate the statistical operator to be

$$W = \sum_n w_n \Lambda_n. \tag{4.30}$$

With this W, one calculates for the outcome of the measurement of the energy H the numbers

$$E_0, E_1, E_2, \ldots,$$
$$w_0, w_1, w_2, \ldots.$$

Thus the original values obtained in the preparation of the state are predicted for an immediate repetition of the measurement (i.e., a measurement before the state W has changed), which in the present case may be any finite amount of time later, since the situation in this experiment is stationary, but before another measurement is made or before the setup is altered (e.g., by switching off the electron beam).

Equation (4.30) is the third postulate of quantum mechanics for the case that the spaces $\Lambda_n \mathcal{H}$ are one-dimensional. In general the space of eigenvectors of an observable A with eigenvalue a_i is not one-dimensional. Then the eigenvalue a_i is called degenerate and the spectrum $\{a_1, a_2, \ldots\}$ is called a degenerate spectrum. The space $\Lambda_{a_i} \mathcal{H} = \{f_{a_i} | f_{a_i} = \Lambda_{a_i} f, f \in \mathcal{H}\}$ is the space of all eigenvectors of A with eigenvalue a_i. Λ_{a_i} is the projection operator on the subspace $\Lambda_{a_i} \mathcal{H}$ of the space $\mathcal{H}$. The dimension of this subspace, $\dim(\Lambda_{a_i} \mathcal{H})$, can have any finite value or can be infinite, and can be different for different eigenvalues a_i. The space $\Lambda_{a_i} \mathcal{H}$ is called the eigenspace of the operator A with eigenvalue a_i.

After these preparations we are ready to formulate the third postulate of quantum mechanics for the general case of an observable with degenerate spectrum.

IIIa. Given that an observable A has been measured with the following results

$$a_1, a_2, a_3, \ldots,$$

$$w_{a_1}, w_{a_2}, w_{a_3}, \ldots,$$

where w_{a_i} gives the relative frequency with which the value a_i was measured (i.e., after N measurements the value a_i occurred $w_{a_i} N$ times), then the state of the system immediately following the measurement is represented by the statistical operator W given by

$$W = \sum_{a_i} w_{a_i} \Lambda_{a_i} \frac{1}{\dim(\Lambda_{a_i} \mathcal{H})}, \qquad (4.31)$$

where Λ_{a_i} are the projection operators on the eigenspaces of A with eigenvalues a_i, and where $\dim(\Lambda_{a_i} \mathcal{H}) = \mathrm{Tr}\, \Lambda_{a_i}$ denotes the dimension of the space $\Lambda_{a_i} \mathcal{H}$.

Before we discuss the general case, let us consider the particular situation that a_n has been measured with the probability $w_n = 1$ and all other $w_i (i \neq n) = 0$. Then the statistical operator is chosen as

$$W = \Lambda_{a_n} \frac{1}{\dim(\Lambda_{a_n} \mathcal{H})}. \qquad (4.32)$$

[*Note.* The Λ_{a_i} are not necessarily projection operators on one-dimensional spaces. If Λ_{a_n} is the projector on a one-dimensional space, the state (4.32) is said to be a pure state.] As we shall prove below, one obtains for the expectation value

$$\langle A \rangle = \mathrm{Tr}(AW) = \frac{1}{\dim(\Lambda_{a_n} \mathcal{H})} \mathrm{Tr}(A \Lambda_{a_n}) = a_n, \qquad (4.33)$$

and for the probability

$$w_{a_n} = \langle \Lambda_{a_n} \rangle = \mathrm{Tr}(\Lambda_{a_n} W) = \frac{1}{\dim(\Lambda_{a_n} \mathcal{H})} \mathrm{Tr}\, \Lambda_{a_n} = 1. \qquad (4.34)$$

In other words, this statistical operator (4.32) predicts, for an immediate repetition of the measurement of A, the value a_n with certainty. Before we prove (4.33) and (4.34), it is convenient to present some more mathematics.

[[If the space $\Lambda_a \mathcal{H}$ of eigenvectors of A with eigenvalue a has the dimension $\dim(\Lambda_a \mathcal{H})$, then there exist $\dim(\Lambda_a \mathcal{H})$ linearly independent eigenvectors. One may then choose an arbitrary basis of

$\dim(\Lambda_a \mathcal{H})$ mutually orthonormal eigenvectors of A with eigenvalue a, i.e., a set of vectors

$$\phi_a^1, \phi_a^2, \ldots, \phi_a^\kappa, \ldots, \phi_a^{\dim(\Lambda_a \mathcal{H})}$$

that fulfill

$$A\phi_a^\kappa = a\phi_a^\kappa \tag{4.35}$$

and

$$(\phi_a^\kappa, \phi_a^{\kappa'}) = \delta_{\kappa\kappa'}. \tag{4.36}$$

The basis for the whole space is given by

$$\phi_{a_1}^1, \phi_{a_1}^2, \ldots, \phi_{a_1}^{\dim(\Lambda(a_1)\mathcal{H})}$$
$$\vdots$$
$$\phi_{a_j}^1, \phi_{a_j}^2, \ldots, \phi_{a_j}^{\dim(\Lambda(a_j)\mathcal{H})} \tag{4.37}$$
$$\vdots$$

In general, $\dim(\Lambda_{a_i} \mathcal{H}) \neq \dim(\Lambda_{a_j} \mathcal{H})$. Any of the $\dim(\Lambda_{a_j} \mathcal{H})$ may or may not be infinite.

An arbitrary vector $f \in \mathcal{H}$ is then written, according to (I.4.4d)

$$f = \sum_{a_i} \sum_{\kappa=1}^{\dim(\Lambda_{a_i}\mathcal{H})} |\phi_{a_i}^\kappa\rangle\langle\phi_{a_i}^\kappa|f\rangle \tag{4.38}$$

This statement and the resulting form (4.42), given here without proof, is the *spectral theorem* for an operator with discrete *degenerate* spectrum.[8]

If we apply A to the arbitrary vector f written in this form we obtain

$$Af = \sum_{a_i\kappa} a_i|\phi_{a_i}^\kappa\rangle\langle\phi_{a_i}^\kappa|f\rangle, \tag{4.39}$$

or, if we omit the arbitrary f,

$$A = \sum_{a_i\kappa} a_i|\phi_{a_i}^\kappa\rangle\langle\phi_{a_i}^\kappa| = \sum_{a_i} a_i \sum_\kappa |\phi_{a_i}^\kappa\rangle\langle\phi_{a_i}^\kappa|. \tag{4.40}$$

The operator

$$\sum_\kappa |\phi_{a_i}^\kappa\rangle\langle\phi_{a_i}^\kappa| = \Lambda_{a_i} \tag{4.41}$$

is the projection operator on the subspace $\Lambda_{a_i} \mathcal{H}$ of eigenvectors of A with eigenvalue a_i, since

$$\Lambda_{a_i}\Lambda_{a_j} = \sum_\kappa \sum_{\kappa'} |\phi_{a_i}^\kappa\rangle\langle\phi_{a_i}^\kappa|\phi_{a_j}^{\kappa'}\rangle\langle\phi_{a_j}^{\kappa'}| = \sum_\kappa |\phi_{a_i}^\kappa\rangle\langle\phi_{a_j}^\kappa|\delta_{a_i a_j} = \Lambda_{a_i}\delta_{a_i a_j}$$

$$A(\Lambda_{a_i} f) = \sum_\kappa a_i|\phi_{a_i}^\kappa\rangle\langle\phi_{a_i}^\kappa|f\rangle = a_i(\Lambda_{a_i} f).$$

---

[8] For operators with continuous spectra one has to start with (I.4.4c) instead of (I.4.4d); the spectral theorem looks similar to (4.42) with the sum replaced by an integral, cf. K. Maurin (1967), p. 131.

Therewith (4.40) can be written

$$A = \sum_{a_i} a_i \Lambda_{a_i}. \tag{4.42}$$

This is called the *spectral representation of the operator A* (with discrete spectrum). The set of Λ_{a_i} is called the *spectral family*.

If one omits the arbitrary f in (4.38), one obtains the resolution of the identity operator I with respect to the spectral family of the operator A:

$$I = \sum_{a_i \kappa} |\phi_{a_i}^\kappa\rangle\langle\phi_{a_i}^\kappa| = \sum_{a_i} \Lambda_{a_i}.$$

We now prove (4.33). We use the basis (4.37) and calculate

$$\mathrm{Tr}(A\Lambda_{a_n}) = \sum_{a_j \kappa} (\phi_{a_j}^\kappa, A\Lambda_{a_n}\phi_{a_j}^\kappa) = \sum_\kappa (\phi_{a_n}^\kappa, A\phi_{a_n}^\kappa)$$

$$= \sum_\kappa a_n(\phi_{a_n}^\kappa, \phi_{a_n}^\kappa) = a_n \dim(\Lambda_{a_n}\mathcal{H}). \tag{4.43}$$

Thus

$$\mathrm{Tr}(AW) = \mathrm{Tr}(A\Lambda_{a_n}) \frac{1}{\dim(\Lambda_{a_n}\mathcal{H})} = a_n.$$

Equation (4.34) is established in the same way.

We now proceed with the introduction of some further mathematical notions: two Hermitian operators A, B are called *commuting operators* iff

$$AB = BA \quad \text{or} \quad [A, B] = 0. \tag{4.44}$$

Let Λ_{a_i} be the spectral family of A, and P_{b_i} the spectral family of B. Then the statement that A and B commute is equivalent[9] to

$$\Lambda_{a_i} P_{b_j} = P_{b_j} \Lambda_{a_i} \quad \text{for every } a_i, b_j. \tag{4.45a}$$

Thus, using (4.42), an operator A commutes with B if

$$AP_{b_j} = P_{b_j}A \quad \text{for every } b_j. \tag{4.45b}$$

The spectral representation (4.42), of the Hermitian operator A or any other operator for which (4.42) holds, can be used to define *functions of operators*.[10] Let $f(a_i)$ be a function of the discrete variable a_i; then, we define the operator $f(A)$ by

$$f(A) = \sum_{a_i} f(a_i)\Lambda_{a_i}. \tag{4.46}$$

---

[9] In fact (4.45a) is the precise statement of commutativity; in order to make (4.44) precise one has to add the requirement that $A^2 + B^2$ be essentially self-adjoint (its Hilbert space closure is self-adjoint).

[10] For the mathematical problems connected with this definition see K. Maurin (1967), Section V.5.

Note that $f(A)$ might not be defined on the same subspace Φ of the Hilbert space $\mathscr{H}$ on which A is defined; however, if $f(a_i)$ is a polynomial in a_i, then $f(A)$ is defined everywhere in Φ.
From (4.45a) it follows that if $[A, B] = 0$, then $[f(A), B] = 0.]$

A pure eigenstate of A (e.g., the energy eigenstate $W = \Lambda_n$ of the one-dimensional harmonic oscillator) is a special case of (4.32).

We return now to the general expression (4.31). Comparing it with (4.42), we recognize that (4.31) is the spectral representation of a Hermitian operator W with respect to the spectral family $\{\Lambda_{a_i}\}$ of the observable A. The $w_{a_i}/\dim(\Lambda_{a_i}\mathscr{H})$ are the eigenvalues of W. For the expectation value of A, we obtain from (4.31)

$$\mathrm{Tr}(AW) = \sum_{a_i} w_{a_i} \frac{1}{\dim(\Lambda_{a_i}\mathscr{H})} \mathrm{Tr}(A\Lambda_{a_i}). \tag{4.47}$$

Inserting (4.43) into this, we obtain

$$\langle A \rangle = \mathrm{Tr}(AW) = \sum_{a_i} w_{a_i} a_i, \tag{4.48}$$

as it ought to be for the expectation value, i.e., the average value in a series of measurements of the observable A in the state of the system described by W.

The projection operator Λ_a on the eigenspace of A with eigenvalue a represents the observable "probability of a," as it should. To see this we calculate

$$\mathrm{Tr}(\Lambda_a W) = \sum_{a_i} w_{a_i} \mathrm{Tr}(\Lambda_a \Lambda_{a_i}) \frac{1}{\dim(\Lambda_{a_i}\mathscr{H})}$$

$$= \sum_{a_i} w_{a_i} \delta_{a_i a} \mathrm{Tr}(\Lambda_a) \frac{1}{\dim(\Lambda_{a_i}\mathscr{H})} = w_a.$$

Here we have used

$$\Lambda_a \Lambda_{a_i} = 0 \quad \text{for} \quad a_i \neq a,$$
$$\Lambda_a \Lambda_a = \Lambda_a.$$

Thus the statistical operator W of (4.31) leads to the prediction we wanted, namely that an immediate repetition of the measurement of an observable gives the same results as the original measurement that constituted the preparation of the state.

If no measurement has been performed that constitutes the preparation of a state, and no information concerning the state has been obtained, then every pure state is assumed to be equally probable. The statistical operator for such a state must, therefore, be taken to be proportional to the unit operator

$$\tilde{W} = I. \tag{4.49}$$

This hypothesis—that properties not affected in a measurement are assumed to occur with the same probability—has already been incorporated into basic assumption IIIa, i.e., into equations (4.31) and (4.32). There, no observable that distinguishes between the different values $\kappa = 1, 2, \ldots$, $\dim(\Lambda_{a_i}\mathcal{H})$ had been measured. Thus, it is assumed in (4.31) and (4.32) that for each a_i any of the pure states $|\phi_{a_i}^1\rangle\langle\phi_{a_i}^1|, |\phi_{a_i}^2\rangle\langle\phi_{a_i}^2|, \ldots, |\phi_{a_i}^\kappa\rangle\langle\phi_{a_i}^\kappa|, \ldots$ appears with the same probability, namely the probability (cf. (4.34)):

$$w_{a_i}^\kappa = \text{Tr}(|\phi_{a_i}^\kappa\rangle\langle\phi_{a_i}^\kappa|W) = \sum_{a_j} \text{Tr}(|\phi_{a_i}^\kappa\rangle\langle\phi_{a_i}^\kappa|\Lambda_{a_j}) \frac{w_{a_i}}{\dim(\Lambda_{a_j}\mathcal{H})}$$

$$= \sum_{a_j\kappa'} \text{Tr}(|\phi_{a_i}^\kappa\rangle\langle\phi_{a_i}^\kappa|\phi_{a_j}^{\kappa'}\rangle\langle\phi_{a_j}^{\kappa'}|) \frac{w_{a_j}}{\dim(\Lambda_{a_j}\mathcal{H})} = \frac{w_{a_i}}{\dim(\Lambda_{a_i}\mathcal{H})}$$

Further, any other pure state of the space $\Lambda_{a_i}\mathcal{H}$ which is described by a vector $\psi \in \Lambda_{a_i}\mathcal{H}$ also appears with this probability, because this probability is

$$w_{a_i}^\psi = \text{Tr}(|\psi\rangle\langle\psi|W) = \sum_{\kappa'} \text{Tr}(|\psi\rangle\langle\psi|\phi_{a_i}^{\kappa'}\rangle\langle\phi_{a_i}^{\kappa'}|) \frac{w_{a_i}}{\dim(\Lambda_{a_i}\mathcal{H})}$$

$$= \left(\sum_{\kappa'} |\langle\psi|\phi_{a_i}^{\kappa'}\rangle|^2\right) \frac{w_{a_i}}{\dim(\Lambda_{a_i}\mathcal{H})} = \frac{w_{a_i}}{\dim(\Lambda_{a_i}\mathcal{H})}.$$

The last equality follows because $\|\psi\|^2 = 1$ and

$$\|\psi\|^2 = \sum_{a_j\kappa} |\langle\psi|\phi_{a_j}^\kappa\rangle|^2 = \sum_\kappa |\langle\psi|\phi_{a_i}^\kappa\rangle|^2,$$

as $\langle\psi|\phi_{a_j}^\kappa\rangle = 0$ for $a_j \neq a_i$.

The operator Λ_{a_i} is the unit operator in the space $\Lambda_{a_i}\mathcal{H}$ of eigenvectors of A with eigenvalue a_i, which one sees immediately if one compares (4.41) with (4.9). (4.32) then says that W is given by the unit operator in the subspace of eigenvectors of A with eigenvalue a_i, except for the normalization, which is chosen such that $\text{Tr } W = 1$. Thus the statement (4.32) is the same as the statement (4.49) restricted to the finite-dimensional subspace $\Lambda_{a_i}\mathcal{H}$. The general form (4.49) is of little practical use, because states about which one has no information at all are of no interest. But (4.49) is frequently used when restricted to a subspace as in (4.32).

If the system is known to be in thermodynamic equilibrium, then the statistical operator W is chosen to be the Gibbs distribution

$$\tilde{W} = e^{-H/kT}, \tag{4.50}$$

where H is the energy operator, and k and T are the Boltzmann constant and absolute temperature, respectively. A justification of (4.50) can be found in books on statistical physics.

One of the consequences of the basic assumption IIIa is that the possible values of an observable that can be measured in any experiment are the eigenvalues of that observable's operator. This means that *in the process of conjecturing the theory, the operator that represents an observable must be chosen in such a way that all the values obtained in the experiment that measures the observable in question are eigenvalues of the operator.*

The measurement of an observable also constitutes a preparation of the system. Generally, this means that the measurement of an observable causes alterations in the state of the system. These statements are now formulated in the second part of the postulate III.

IIIb. Let a system be in the state W, and let an observable B be measured on the system. Let the eigenvalues of B be $b_1, b_2, b_3, \ldots$, and let $\Lambda_{b_1}, \Lambda_{b_2}, \ldots$ be the projectors on the corresponding eigenspaces. Then the state of the system after the measurement is given by

$$W' = \sum_{b_i} \Lambda_{b_i} W \Lambda_{b_i}. \tag{4.51}$$

If this measurement is used to select a certain subensemble with a definite value b of B, then the state is given by

$$W' = \frac{1}{\text{Tr}(\Lambda_b W \Lambda_b)} \Lambda_b W \Lambda_b. \tag{4.52}$$

There is a special case in which the measurement of an observable does not alter the state W; this is the case if and only if the operator B that represents the observable commutes with W, because only then Equation (4.51) becomes, according to (4.45b),

$$W' = \sum_i \Lambda_{b_i} W \Lambda_{b_i} = \sum_i \Lambda_{b_i} \Lambda_{b_i} W$$

$$= \sum_i \Lambda_{b_i} W;$$

and, since

$$\sum_i \Lambda_{b_i} = I,$$

we get

$$W' = \sum_i \Lambda_{b_i} W = W.$$

Two observables are called *compatible* if the measurement of one does not alter the state that has been prepared by the measurement of the other. The preceding consideration then shows that compatible observables are represented by commuting operators. Observables represented by noncommuting operators are *incompatible*.

If one has two pure states, $W_1 = \Lambda_\psi = |\psi\rangle\langle\psi|$ and $W_2 = \Lambda_\phi = |\phi\rangle\langle\phi|$, one can mix them to obtain a new state

$$W = \lambda W_1 + (1 - \lambda)W_2; \qquad 0 > \lambda > 1. \tag{4.53}$$

W describes the state which is a mixture of the pure states W_1 and W_2 with mixing ratio $\lambda:(1 - \lambda)$. It is not a pure state.

If ϕ and ψ are orthogonal to each other, i.e., $\Lambda_\psi \Lambda_\phi = 0$, and one mixes them with equal ratios, $\lambda = \frac{1}{2}$, then one obtains a state

$$W = \tfrac{1}{2}(\Lambda_\psi + \Lambda_\phi) = \tfrac{1}{2}\Lambda \qquad (4.54)$$

which is described by a projection operator Λ. Λ projects on the two-dimensional subspace spanned by ϕ and ψ. The probability to find the property Λ_ψ, i.e., the property of the pure state W_1, in the state W is according to (4.28)

$$w_\psi = \mathrm{Tr}(\Lambda_\psi W) = \tfrac{1}{2}\,\mathrm{Tr}(\Lambda_\psi(\Lambda_\psi + \Lambda_\phi)) = \tfrac{1}{2}, \qquad (4.55)$$

as it should be.

With the two pure states Λ_ψ and Λ_ϕ one can define new state vectors

$$\alpha_1 \psi + \alpha_2 \phi, \qquad (4.56)$$

where α_1, α_2 are arbitrary complex numbers. All these vectors also represent pure states. For instance one can choose $\alpha_1 = 1/\sqrt{2} = \alpha_2$. Then

$$\chi = \frac{1}{\sqrt{2}}(\psi + \phi) \qquad (4.57)$$

is again a normalized vector. The projection operator

$$\Lambda_\chi = |\chi\rangle\langle\chi| \qquad (4.58)$$

on the one-dimensional space spanned by χ again represents a pure state. The state Λ_χ is called a *superposition* of the states Λ_ψ and Λ_ϕ or the state vector χ given by (4.56) is called the superposition of the vectors ψ and ϕ. To specify the superposition of two states not only is the relative magnitude of α_1 and α_2 important, but so also is their relative phase. With the choice $\alpha_1 = 1/\sqrt{2} = -\alpha_2$ in (4.56) one obtains the normalized vector

$$\eta = \frac{1}{\sqrt{2}}(\psi - \phi) \qquad (4.59)$$

which is orthogonal to χ in (4.57). The projection operator

$$\Lambda_\eta = |\eta\rangle\langle\eta| \qquad (4.60)$$

represents a different pure state than Λ_χ.

In contrast to the combination (4.53) of the two states Λ_ψ and Λ_ϕ which one also calls an incoherent combination, the combination (4.56) is called a *coherent* combination of the two state vectors ψ and ϕ. The most general incoherent combination of the energy eigenstates $\Lambda_n = |\phi_n\rangle\langle\phi_n| (n = 1, 2, \ldots)$ of the harmonic oscillator is (4.17) and the most general coherent combination of the energy eigenvectors ϕ_n is (3.27). Any change of the magnitudes or phases of the α_n except for an overall magnitude or phase change of ψ in (3.27) gives a vector which describes a different pure state Λ_ψ. Any state vector may be considered as the result of superposition of two or more state vectors and this can be done in an infinite number of ways since instead of the eigenvectors ϕ_n of H one can choose any other basis of the space Φ for the expansion of the vector ψ. Conversely, any two or more state vectors can be superposed to give a new state vector. This *superposition principle* is a consequence of the property (I.2.1a) of the linear space Φ. If one postulates that any vector in Φ represents a pure, physically preparable state, then any pure state can be viewed as a linear combination of an arbitrary number of other states. Such a postulate cannot be checked because for every vector in Φ one would have to build an apparatus that prepares it. Certainly, the energy eigenstates of the harmonic oscillator $\Lambda_n = |\phi_n\rangle\langle\phi_n|$ are of greater importance than a pure state Λ_ψ where ψ is any arbitrary linear combination (3.27). Whether with any two energy eigenvectors ϕ_n and ϕ_m the vector $\psi_n = \phi_n + \alpha\phi_m$, $\alpha \in \mathbb{C}$ also represents a physically preparable state is very doubtful (and certainly impossible if one considers the oscillator as a stationary system, see Chapter XII below). However, that the linear combination $\psi_1 + \psi_2$ of a certain pair of state vectors ψ_1 and ψ_2 also represents a pure physical state is an experimentally proven fact (interference, superposition of polarized light, superposition of K-mesons).

The superpositions χ and η span the same two-dimensional space as the original vectors ψ and ϕ and the projection operator Λ on this two-dimensional space, (4.54), can also be written as

$$\Lambda = \Lambda_\chi + \Lambda_\eta. \tag{4.61}$$

Thus the mixed state W is given by

$$W = \tfrac{1}{2}(\Lambda_\psi + \Lambda_\phi) = \tfrac{1}{2}(\Lambda_\chi + \Lambda_\eta)$$

and can, therefore, as well be viewed as a mixture of the pure state Λ_χ and the pure state Λ_η with the mixing ratios $w_\chi = w_\eta = \tfrac{1}{2}$. Whether this is of practical importance depends upon the particular situation. If the superpositions χ and η represent physically preparable pure states then the state W can be obtained by mixing the pure states Λ_ψ and Λ_ϕ as well as by mixing the pure states Λ_χ and Λ_η.

II.5 The Basic Assumptions Applied to the Harmonic Oscillator, and Some Historical Remarks

We now return to the discussion of the harmonic oscillator. Let us assume that its state has been found to be

$$W = \Lambda_n,$$

where Λ_n is the projection operator on the space spanned by ϕ_n. This is the case if a measurement of the energy always gives the result E_n. According to the basic assumption II, the expectation value of the momentum P in the pure state Λ_n is given by

$$\langle P \rangle = \mathrm{Tr}(PW) = \mathrm{Tr}(P\Lambda_n)$$

$$= \sum_m (\phi_m, P\Lambda_n \phi_m)$$

$$= (\phi_n, P\phi_n), \tag{5.1}$$

so

$$\langle P \rangle = \langle n|P|n \rangle. \tag{5.2}$$

We leave it as an exercise for the reader to show that as a consequence of (3.26) and (3.1),

$$\langle P \rangle = 0. \tag{5.3}$$

Similarly the expectation value of the position Q in the pure state $W = \Lambda_n$ is

$$\langle Q \rangle = \langle n|Q|n \rangle = 0. \tag{5.4}$$

We now calculate the expectation value of P^2 (i.e., the kinetic energy $P^2/2\mu$) and the expectation value of Q^2 (or the potential energy $\omega^2 \mu Q^2/2$):

$$\langle P^2 \rangle = (\phi_n, P^2 \phi_n).$$

Using Equations (3.1), we have

$$P^2 = -\frac{\mu\omega\hbar}{2}(a^\dagger - a)^2$$

so

$$\langle P^2 \rangle = -\frac{\mu\omega\hbar}{2}(\phi_n, (a^\dagger - a)^2 \phi_n).$$

Inserting for a and $a^\dagger$ from Equation (3.26), we get

$$\langle P^2 \rangle = \mu\omega\hbar(n + \tfrac{1}{2}) \tag{5.5}$$

In a similar fashion we calculate

$$\langle Q^2 \rangle = (\phi_n, Q^2 \phi_n).$$

Again from (3.1),

$$Q^2 = \frac{\hbar}{2\mu\omega}(a^\dagger + a)^2;$$

so

$$(\phi_n, Q^2 \phi_n) = \frac{\hbar}{2\mu\omega}(\phi_n, (a^\dagger + a)^2 \phi_n),$$

and finally

$$\langle Q^2 \rangle = \frac{\hbar}{\mu\omega}(n + \tfrac{1}{2}). \tag{5.6}$$

Although the expectation values of the observables P, Q, P^2, and Q^2 are uniquely determined, it does not follow that one should expect to obtain the same value a_i for any measurement of an observable A (although such is the case in the measurement[11] of the observable H in the pure state Λ_n). What does follow—according to the definition of $\langle A \rangle$—is that the average of the values obtained in a series of many repeated measurements of A on the state Λ_n is $\langle A \rangle$, which of course is uniquely determined. In general it is possible that the values obtained in each measurement (a_1 for the first measurement, a_2 for the second, a_3 for the third, etc.) are entirely different from each other and widely dispersed.

In order to describe the dispersion of the measured values quantitatively, let us define, for any observable A and state W, a new observable given by

$$(A - \langle A \rangle I)^2.$$

To see that this observable is suitable for the description of the dispersion of the measured values $a_1, a_2, a_3, \ldots$, we consider the expectation value of $(A - \langle A \rangle I)^2$ in the state W, which is called the *dispersion of A in the state W*:

$$
\begin{aligned}
\operatorname{disp}_{(W)} A &\equiv \langle (A - \langle A \rangle_W I)^2 \rangle_W \\
&= \operatorname{Tr}(A^2 W) - 2\operatorname{Tr}(A\langle A \rangle_W W) + \operatorname{Tr}(\langle A \rangle_W^2 W) \\
&= \operatorname{Tr}(A^2 W) - (\operatorname{Tr}(AW))^2 = \langle A^2 \rangle_W - \langle A \rangle_W^2 \qquad (5.7)
\end{aligned}
$$

(unless otherwise stated we always assume that $\operatorname{Tr} W = 1$); thus

$$\operatorname{disp}_{(W)} A = \langle A^2 \rangle_W - \langle A \rangle_W^2. \qquad (5.8)$$

For the special case of the observable H for the harmonic oscillator in the state Λ_n we obtain

$$
\begin{aligned}
\operatorname{disp}_{(\Lambda_n)} H &= \operatorname{Tr}(H^2 \Lambda_n) - (\operatorname{Tr}(H\Lambda_n))^2 \\
&= (\phi_n, H^2 \phi_n) - E_n^2.
\end{aligned}
$$

However, since ϕ_n is the eigenvector of H with eigenvalue E_n, we have

$$
\begin{aligned}
H^2 \phi_n &= HE_n \phi_n = E_n H \phi_n \\
&= E_n^2 \phi_n;
\end{aligned}
$$

thus

$$\operatorname{disp}_{(\Lambda_n)} H = E_n^2 - E_n^2 = 0.$$

This shows that the dispersion of H in the state Λ_n is zero. More generally, the dispersion of an arbitrary observable A in a pure eigenstate Λ of A is zero. However, the converse of this statement is not true; i.e., if the dispersion of an observable A is zero, then the state W need not be a pure eigenstate of A (in particular, it need not be a pure state). For this reason, the notion of *eigenstate* will be generalized.

---

[11] We remark that by measurement we always mean an idealized measurement in which the measuring apparatus does not lead to uncertainties; i.e., the experimental error is zero.

Let A be an observable with degenerate spectrum. A state W is called an eigenstate of the observable A if W can be represented as

$$W = \sum_\kappa w^\kappa |\phi_a^\kappa\rangle\langle\phi_a^\kappa|, \tag{5.9}$$

where $w^\kappa \geq 0$, $\sum_\kappa w^\kappa = 1$ (W is normalized), and ϕ_a^κ, $\kappa = 1, 2, \ldots$, $\dim \Lambda_a \mathcal{H}$ is any basis of the eigenspace $\Lambda_a \mathcal{H}$ of A with eigenvalue a.

In Section II.6 it will be shown that the eigenstates of an observable are precisely those states in which the dispersion of A is zero. A special case of an eigenstate is (5.9) with $w^\kappa = w = (\dim \Lambda_a \mathcal{H})^{-1}$ for which W is given by

$$W = \Lambda_a \frac{1}{\text{Tr}(\Lambda_a)}. \tag{5.10}$$

The energy eigenstate Λ_n for the harmonic oscillator is not an eigenstate of either of the observables P and Q, so its dispersion is not zero. The dispersion is easily calculated:

$$\text{disp}_{(\Lambda_n)} P = \langle P^2\rangle - \langle P\rangle^2.$$

From Equations (5.3) and (5.5),

$$\langle P^2\rangle - \langle P\rangle^2 = \hbar\mu\omega(n + \tfrac{1}{2}). \tag{5.11}$$

Similarly

$$\text{disp}_{(\Lambda_n)} Q = \langle Q^2\rangle - \langle Q\rangle^2 = \frac{\hbar}{\mu\omega}(n + \tfrac{1}{2}). \tag{5.12}$$

The square root of the dispersion is called the *uncertainty of A in the state W*, and is denoted by ΔA:

$$\Delta A = \sqrt{\text{disp } A}. \tag{5.13}$$

From Equations (5.11) and (5.12) it follows that

$$\Delta P \, \Delta Q = \hbar(n + \tfrac{1}{2}). \tag{5.14}$$

Thus for any pure energy eigenstate of the harmonic oscillator,

$$\Delta P \, \Delta Q \geq \hbar\tfrac{1}{2}. \tag{5.15}$$

This relation is true in general, as we shall see later, and is a consequence of the Heisenberg commutation relations $[P, Q] = (\hbar/i)I$. It is called the *Heisenberg uncertainty relation*.

As shown above for the particular case of the pure eigenstates of the observable H, the dispersion of an observable in an eigenstate of that observable is zero. Physically, this means that a measurement of the observable on an eigenstate gives one value (which must be an eigenvalue of the corresponding operator) with certainty, i.e., with unit probability.

Now consider the mixture W of the energy-loss experiment described by (4.17). By definition

$$\text{disp}_{(W)} H = \langle H^2\rangle - \langle H\rangle^2. \tag{5.16}$$

Using (4.18) and the corresponding expression for $\langle H^2 \rangle$, we have

$$\langle H^2 \rangle - \langle H \rangle^2 = \sum_\gamma E_\gamma^2 w_\gamma - \left(\sum_\gamma w_\gamma E_\gamma \right)^2$$

$$= \left(\sum_\gamma w_\gamma \right) \left(\sum_\gamma E_\gamma^2 w_\gamma \right) - \left(\sum_\gamma w_\gamma E_\gamma \right)^2 \geq 0. \qquad (5.17)$$

ΔE is called standard deviation or root-mean-square (r.m.s) deviation. For the last inequality we have used the Cauchy inequality for sequences:

$$\sum_{n=0}^{\infty} x_n^2 \sum_{n=0}^{\infty} y_n^2 \geq \left(\sum_{n=0}^{\infty} x_n y_n \right)^2 \quad \text{with} \quad x_n = \sqrt{w_n}, \ y_n = E_n \sqrt{w_n}, \qquad (5.18)$$

which is a consequence of (I.2.7). Thus $\text{disp}_{(W)} H$ in a mixture is always greater than or equal to zero. It is equal to zero if

$$w_\gamma = \delta_{n\gamma}, \quad \text{i.e.,} \quad W = \Lambda_n,$$

and it deviates from zero according to the number of and values of w_γ that differ from zero. Thus disp H, and therewith ΔH, is indeed a quantitative expression for the uncertainty in the measurement of the observable H. From the first line in Equation (5.17) we see that the uncertainty ΔH of the observable H corresponds in the measurement of this observable to the standard deviation or r.m.s. deviation of the series of measurements described above Equation (4.16).

It is no surprise that disp H or ΔH was greater than zero for the mixed state W of CO molecules in the collision chamber, since that was a mixture of CO molecules of different energies. The reason that the outcome of the measurement of H in W is not fully determined is that not enough preparations had been performed to separate different systems, or more precisely, to separate systems in different states, from each other.

The harmonic oscillators in the collision chamber can be separated, at least in a *gedanken* experiment, into eight sets, each one being a set of N_n molecules having the same energy E_n, i.e., such that a measurement of H on the nth such set gives with certainty the value E_n. The state of each set is then Λ_n. It is impossible to further separate (even in a *gedanken* experiment) any of these sets into two or more subsets of different systems or systems in different states. This is the reason for calling Λ_n a pure state. If we now perform the measurement[12] of an observable on the system in a pure state, one would expect from classical considerations that this should lead with certainty to a well-defined number, as a set in a pure state consists of only one species. If we choose for this observable H or N, this is indeed the case. However, if we choose P or Q, we see from (5.11), (5.12) that it is not. Even in pure states the outcome of the measurement of an observable is in general not uniquely determined. *Thus statistics cannot in principle be eliminated from quantum mechanics*; systems in the "same pure state" do not give

---

[12] An idealized measurement; see footnote 11.

identical values in a measurement. All that can be said is with what probability w_n a certain value a_n can be expected in the measurement of an observable.

Let us illustrate this with a *gedanken* experiment. We consider the measurement of the observable Q in the state Λ_n, and ask for the probability with which we can expect the value x in a measurement of Q. By definition, the expectation value of an observable A is

$$\langle A \rangle = \sum_i w_i a_i, \tag{5.19}$$

where a_i are the eigenvalues of A, and w_i are the probabilities for finding the values a_i in a measurement of A. We calculate $\langle Q \rangle$ as follows:

$$\langle Q \rangle = \text{Tr}(Q\Lambda_n) = \sum_i \langle i | Q\Lambda_n | i \rangle, \tag{5.20}$$

where $\psi_i = |i\rangle$ is any complete basis system and we have used the following notation:

$$(\phi, \psi) = \langle \phi | \psi \rangle \quad \text{or} \quad (\psi_i, \psi_j) = \langle i | j \rangle,$$
$$(\phi, A\psi) = \langle \phi | A | \psi \rangle. \tag{5.21}$$

For $|i\rangle$ we may take any complete basis system; let us therefore choose the system of eigenvectors of the operator Q, which we denote by $|x\rangle$, where x is the eigenvalue of Q:

$$Q|x\rangle = x|x\rangle. \tag{5.22}$$

Then

$$\langle Q \rangle = \sum_x \langle x | Q\Lambda_n | x \rangle$$
$$= \sum_x x \langle x | \Lambda_n | x \rangle$$
$$= \sum_x x \langle x | n \rangle \langle n | x \rangle.$$

The last step follows from writing $\Lambda_n = |n\rangle\langle n|$.

We introduce new notation, defining

$$\phi_n(x) \equiv \langle x | n \rangle. \tag{5.23}$$

Then

$$\langle Q \rangle = \sum_x x \phi_n(x) \overline{\phi_n(x)}$$
$$= \sum_x x |\phi_n(x)|^2. \tag{5.24}$$

If we now compare this with the general expression (5.19) for the expectation value of an operator, we conclude that

$$|\langle x | n \rangle|^2 = |\phi_n(x)|^2 = w_n(x) \tag{5.25}$$

is the probability for obtaining the value x in the measurement of the observable Q for an oscillator in the state Λ_n. $\phi_n(x)$ is called the *wave function of the state* with energy E_n. Thus we have found that the probability for finding the value x for the position operator Q is the absolute value squared of the wave function. It was in this form that the probability interpretation was originally introduced in quantum mechanics by M. Born [influenced by an earlier suggestion of Albert Einstein (1916)].

As we shall discuss later, the $\sum_x$ is in fact an integral, and $w_n(x) = |\langle x|n\rangle|^2$ is a quantity of dimension cm^{-1}, i.e., a *probability density*. The probability is the integral of $w_n(x)$ over a certain interval.

We have derived the fact that $|\phi(x)|^2$ describes the probability of finding the physical system at x from the basic assumption II. Historically the events took place in reverse order, and II is a generalization of the discovery by Max Born in June 1926 that the wave function $\phi(x)$, which—as we shall see in Section II.7—is a solution of the Schrödinger equation, is connected with probability.

Born was at that time investigating the collision process between free particles and an atom, using the formalism of Schrödinger's wave mechanics, which for this kind of process he found superior to the matrix mechanics of Heisenberg, Born, and Jordan. However, he could not accept Schrödinger's interpretation that $\phi(x)$ represents matter waves. Recalling these events, Born said in his Nobel Prize Lecture in 1954: "On this point [Schrödinger's matter-wave interpretation] I could not follow him. This was connected with the fact that my Institute and that of James Franck were housed in the same building in Göttingen University. Every experiment by Franck and his assistants on electron collision appeared to me as a new proof of the corpuscular nature of the electron." He calculated the formula for the scattered wave in the Schrödinger formalism (Born approximation[13]) and concluded that if this formula is to allow for a corpuscular interpretation, then the most natural interpretation of $\phi(x)$ is that $|\phi(x)|^2$ describes the probability density for detecting the corpuscle. He summarized his conclusion in 1926: "The motion of particles conforms to the law of probability but the probability itself is propagated in accordance with causality."

Born's interpretation was soon generalized to include other observables in addition to the position. And the time was ripe for the general formulation of quantum mechanics. The starting point for this was the discovery that the canonical transformations can be carried over to matrix mechanics and wave mechanics (F. London, May 1926; P. Jordan, July 1926). In this way the canonical transformations were recognized as transformations of an abstract space. Transformations of objects with continuous indices were introduced by Dirac and Jordan in December 1926. The mathematical elaboration of the London–Jordan–Dirac transformation theory was undertaken in Göttingen by David Hilbert together with L. Nordheim and John von Neumann (1926–1927). Notions like state and observable evolved.

---

[13] See Section XIV.5 below.

Subsequently, von Neumann developed an axiomatic theory of Hilbert spaces and established the association between physical states and Hilbert-space vectors as well as the correspondence between observables and linear operators. He also introduced the statistical operator.

Parallel to the mathematical formulation went the development of the physical interpretation, which was discussed principally in Copenhagen (1926–1927). Whereas Bohr wanted to make the wave-particle duality the starting point of the physical interpretation, Heisenberg relied upon the transformation theory. His basic approach was to regard something as defined if it is measurable. In his famous *gedanken* experiment in which he discussed the position measurement of the electron with a microscope, he explained the uncertainty of position and momentum (5.15), and therewith the probabilistic nature of quantum-mechanical predictions, as the effect of the measurement apparatus. Object and observer have no independent reality; the measurement affects the state of the object (basic assumption III).

While Heisenberg derived his uncertainty principle from the transformation theory, Bohr developed his conception of complementarity from the wave-particle duality (see Section II.9): There exist complementary properties—like position and momentum—and the exact measurement of one precludes the possibility of obtaining information of the other. Properties are not actualities; they are only possibilities for the physical system. These developments formed the basis of the so-called Copenhagen interpretation of quantum mechanics.[14] Its physical content is formulated in the basic assumptions of this chapter.

Von Neumann's Hilbert-space formulation could not accommodate continuous eigenvalues. Since that time, stimulated by the Dirac formalism, a new branch of mathematics—the theory of distributions—was developed and for the first time systematically presented by L. Schwartz (1950–1951). On this basis I. M. Gelfand and collaborators introduced around 1960 the rigged Hilbert space. Its use for quantum mechanics was suggested around 1965.

II.6 Some General Consequences of the Basic Assumptions of Quantum Mechanics

After this detailed discussion of the case of the harmonic oscillator, let us investigate some general consequences of the basic assumptions of quantum mechanics. Let us consider an arbitrary state W of an arbitrary microphysical system and see what we can predict for the measurement of the

---

[14] The interpretation of quantum mechanics suggested by Born and established by Heisenberg and the Copenhagen school represented such a drastic departure from the well-established ideas of their time that decades had to pass before it could gain general acceptance, and even today new attempts are being undertaken to return to a deterministic interpretation. Born's statistical interpretation of quantum mechanics was one of the most far-reaching contributions of physics to contemporary thinking.

values of some arbitrary Hermitian observables $A, B, \ldots$ of this system. These predictions are contained in the following two statements, which we will prove below.

1. The dispersion of any observable A in an arbitrary state W fulfills

$$\mathrm{disp}_{(W)}\, A \geq 0 \qquad\qquad (6.1)$$

with

$$\mathrm{disp}_{(W)}\, A = 0$$

if and only if W is an eigenstate of A.
2. The uncertainties of two observables A, B in an arbitrary state W fulfill the uncertainty relation

$$\Delta A\, \Delta B \geq \tfrac{1}{2}|\langle[A, B]\rangle_W|. \qquad\qquad (6.2)$$

A special case of this uncertainty relation is the Heisenberg uncertainty relation; from

$$[P, Q] = \frac{\hbar}{i}I \quad \text{and} \quad \left\langle\frac{\hbar}{i}I\right\rangle = \frac{\hbar}{i}\mathrm{Tr}(WI) = \frac{\hbar}{i}$$

it follows that

$$\Delta P\, \Delta Q \geq \frac{\hbar}{2} \qquad\qquad (6.3)$$

for any state W. The equality sign in the Heisenberg uncertainty relation holds if $W = \Lambda_0$, the ground state of the one-dimensional harmonic oscillator, and for certain other pure states as well (cf. Problem 4e).

The uncertainty relation is a very general property of two operators, and just depends upon the assumption that the observables are represented by linear operators in a linear space. It says that unless A and B are compatible observables, there are always states for which at least one of the observables cannot be measured with exact accuracy (since for every operator $C = i[A, B]$ there is always a W such that

$$|\langle C\rangle_W| = |\mathrm{Tr}(WC)| \neq 0).$$

In the case that C (or $-C$) is positive definite (as, e.g., for $A = P$ and $B = Q$), $|\mathrm{Tr}(WC)| > 0$ for every state W, and therefore the two observables A and B can never be measured simultaneously with unrestricted accuracy. If the system has been prepared in an eigenstate of one of the observables A, $\Delta A = 0$ (i.e., A has been measured to give accurately one value), then according to (6.2) the values for B are completely uncertain. This is a consequence of the previously discussed fact that in quantum physics a measurement changes the state, and the measurement of one observable interferes with the measurement of another observable unless these two observables are compatible, i.e., are represented by operators that commute with each other. Thus only compatible observables can be simultaneously measured, and only for compatible observables do there exist common eigenstates.

PROOF OF STATEMENT 1.   A variant of the spectral theorem (4.42) states that every statistical operator can be written in the form

$$W = \sum_i w_i \Lambda_i \tag{6.4}$$

with real numbers w_i, and one-dimensional projection operators Λ_i which project onto spaces $\Lambda_i \mathcal{H}$ spanned by single vectors ϕ_i such that the vectors $\phi_1, \phi_2, \ldots$, together form an orthonormal basis:

$$(\phi_i, \phi_j) = \delta_{ij}. \tag{6.5}$$

The numbers w_i satisfy

$$w_i \geq 0 \tag{6.6}$$

[which follows from the positivity of W since by (6.4) and (6.5) we have $w_i = (\phi_i, W\phi_i)$], and

$$\sum_i w_i = \mathrm{Tr}\ W = 1. \tag{6.7}$$

Since $\mathrm{Tr}(BW)$ for an arbitrary operator B can be evaluated with any basis, we choose the particular basis $\{\phi_i\}$ from (6.5) and obtain

$$\mathrm{Tr}(BW) = \sum_j \left(\phi_j, B \sum_i w_i \Lambda_i \phi_j \right) = \sum_i w_i (\phi_i, B\phi_i), \tag{6.8}$$

where we have used $\Lambda_i \phi_j = \delta_{ij} \phi_i$. Taking $B = I$ in (6.8) leads to (6.7) because $(\phi_i, I\phi_i) = (\phi_i, \phi_i) = 1$ for all i.

From the definition (5.7), we have

$$\mathrm{disp}_{(W)} A = \langle (A - \alpha I)^2 \rangle_W = \mathrm{Tr}((A - \alpha I)^2 W) \tag{6.9}$$

where $\alpha = \langle A \rangle_W = \mathrm{Tr}(AW)$. Thus, taking $B = (A - \alpha I)^2$ in (6.8) yields

$$\mathrm{disp}_{(W)} A = \sum_i w_i (\phi_i, (A - \alpha I)^2 \phi_i)$$

$$= \sum_i w_i ((A - \alpha I)\phi_i, (A - \alpha I)\phi_i)$$

$$= \sum_i w_i \|(A - \alpha I)\phi_i\|^2. \tag{6.10}$$

Therefore $\mathrm{disp}_{(W)} A \geq 0$, since $w_i \geq 0$ by (6.9) and obviously $\|(A - \alpha I)\phi_i\|^2 \geq 0$ for all i. Furthermore, $\mathrm{disp}_{(W)} A = 0$ implies

$$w_i \|(A - \alpha I)\phi_i\|^2 = 0 \quad \text{for all } i. \tag{6.11}$$

For a given i, this in turn implies that either $w_i = 0$ or $\|(A - \alpha I)\phi_i\|^2 = 0$. If $w_i = 0$, then $w_i \Lambda_i$ is also zero, and this term can be omitted from the sum in (6.4). Thus the sum in (6.4) can be taken to run only over those values of i having $w_i > 0$. For the values of i which remain, $\|(A - \alpha I)\phi_i\|^2 = 0$ is required by (6.11); i.e.,

$$(A - \alpha I)\phi_i = A\phi_i - \alpha\phi_i = 0. \tag{6.12}$$

Therefore, if $\mathrm{disp}_{(W)} A = 0$, then all ϕ_i for the remaining values of i must be eigenvectors of A belonging to the same eigenvalue α. But then the state W in (6.4) describes an eigenstate of A as defined in (5.9).

The converse, i.e., that $\mathrm{disp}_{(W)} A = 0$ for W given by (5.9), is proven by a straightforward calculation (see Problem 10).   □

PROOF OF STATEMENT 2.   For two operators A and B we define

$$h(A, B) \equiv \mathrm{Tr}(WA^\dagger B) = \mathrm{Tr}(WI(A^\dagger B)) = h(I, A^\dagger B) = \langle A^\dagger B \rangle. \qquad (6.13)$$

One easily shows from (4.3) and (4.4) that

(a) $h(A, A) \geq 0,$
(b) $h(A, B) = \overline{h(B, A)},$
(c) $h(aA, B) = \bar{a}h(A, B),$
(d) $h(A + C, B) = h(A, B) + h(C, B),$

i.e., $h(A, B)$ defined by (6.13) fulfills all the conditions of (I.2.5) that define a positive Hermitian form. Since $h(A, B)$ is a positive Hermitian form, the Schwarz inequality (I.2.7) holds:

$$|h(A, B)|^2 \leq h(A, A)h(B, B). \qquad (6.14)$$

From this, statement 2 is easily proven (we assume A, B to be Hermitian):

$$|\langle [A, B] \rangle|^2 = |h(I, [A, B])|^2 = |h(A, B) - h(B, A)|^2 \leq 4h(A, A)h(B, B)$$

so if we replace

$$A \to A - \alpha I, \qquad \alpha = \mathrm{Tr}(WA),$$
$$B \to B - \beta I, \qquad \beta = \mathrm{Tr}(WB),$$

we obtain

$$h(A - \alpha I, A - \alpha I)h(B - \beta I, B - \beta I) \geq \tfrac{1}{4}|h(I, [A - \alpha I, B - \beta I])|^2$$

or (for A and B Hermitian)

$$\mathrm{disp}\, A \cdot \mathrm{disp}\, B \geq \tfrac{1}{4}|h(1, [A, B])|^2 = \tfrac{1}{4}|\langle [A, B] \rangle|^2. \qquad (6.15)$$

The square root of this gives (6.2), and therefore statement 2 has been proven.   □

II.7 Eigenvectors of Position and Momentum Operators; the Wave Functions of the Harmonic Oscillator

In this section we shall investigate the properties of the wave functions. We will begin by examining the properties of the operators P and Q; in particular, we wish to know what are the possible eigenvalues x of Q. That is, for which values x can

$$Q|x\rangle = x|x\rangle \qquad (7.1)$$

be fulfilled? Then we wish to find the transition coefficient $(x|n)$ which transforms between a vector $|n\rangle = \phi_n$ (an eigenstate of H and N) and the vectors $|x\rangle$ (eigenstates of Q):

$$|n\rangle = \sum_x |x\rangle(x|n). \qquad (7.2)$$

From (3.1) and (3.26) it follows that

$$Q|n) = \sqrt{\frac{\hbar}{2\mu\omega}} (a + a^\dagger)|n)$$

$$= \sqrt{\frac{\hbar}{2\mu\omega}} (\sqrt{n}|n - 1) + \sqrt{n + 1}|n + 1)). \tag{7.3}$$

Taking the scalar product of this equation with $|x)$, we obtain

$$(n|Q|x) = \sqrt{\frac{\hbar}{2\mu\omega}} (\sqrt{n}(n - 1|x) + \sqrt{n + 1}(n + 1|x)). \tag{7.4}$$

On the other hand, taking the scalar product of (7.1) with $|n)$, we obtain

$$(n|Q|x) = x(n|x). \tag{7.5}$$

Comparing (7.4) with (7.5) yields

$$x(n|x) = \sqrt{\frac{\hbar}{2\mu\omega}} (\sqrt{n}(n - 1|x) + \sqrt{n + 1}(n + 1|x)), \tag{7.6}$$

or, denoting $n + 1 = m$,

$$\sqrt{m}(m|x) = \sqrt{\frac{2\mu\omega}{\hbar}} x(m - 1|x) - \sqrt{m - 1}(m - 2|x). \tag{7.7}$$

Since (7.3) is valid for $n = 0, 1, 2, \ldots,$ (7.7) is valid for $m = 1, 2, 3, \ldots$. For $n = 0$ (i.e., $m = 1$), we obtain instead of (7.3)

$$Q|0) = \sqrt{\frac{\hbar}{2\mu\omega}} \sqrt{0 + 1}|0 + 1) = \sqrt{\frac{\hbar}{2\mu\omega}}|1); \tag{7.3a}$$

and instead of (7.7),

$$\sqrt{1}(1|x) = \sqrt{\frac{2\mu\omega}{\hbar}} x(0|x). \tag{7.7a}$$

Thus we see that (7.7) is a recurrence relation for $(m|x)$; if $(0|x)$ is known, we can determine $(1|x)$ by (7.7a) and then determine $(2|x)$ by (7.7). With $(1|x)$ and $(2|x)$ we can determine $(3|x)$ by (7.7), and so on.

To find out what the transition coefficients $(m|x)$ are, we introduce

$$y \equiv \sqrt{\frac{\mu\omega}{\hbar}} x \tag{7.8}$$

and

$$f_n(y) \equiv \sqrt{2^n n!} \frac{(n|x)}{(0|x)}, \tag{7.9}$$

which is defined for all x with $(0|x) \neq 0$ (if $(0|x) = 0$, then by (7.7) and (7.7a) $(n|x) = 0$ for all n). Then from (7.7) it follows that

$$\sqrt{\frac{m}{2^m m!}}\, f_m(y) = \sqrt{\frac{2}{2^{m-1}(m-1)!}}\, y f_{m-1}(y) - \sqrt{\frac{m-1}{2^{m-2}(m-2)!}}\, f_{m-2}(y),$$

$$(7.10)$$

or

$$f_m(y) = 2y f_{m-1}(y) - 2(m-1) f_{m-2}(y). \tag{7.11}$$

From (7.7a) we have

$$f_1(y) = 2y f_0(y), \tag{7.12}$$

and from (7.9)

$$f_0(y) = 1. \tag{7.13}$$

Equations (7.11), (7.12), and (7.13) are the recurrence relations for the Hermite functions and have solutions for any complex number y.[15] Thus for any complex value x there is a solution $(n|x)$ of the recurrence relation (7.7). For physical reasons we need to consider only real values of x, because x is the value that is obtained in the position measurement and must be real. For real values of y, the solutions $f_m(y)$ of (7.11), (7.12), (7.13) are the Hermite polynomials:

$$f_n(y) = H_n(y) = (-1)^n e^{y^2} \frac{d^n(e^{-y^2})}{dy^n}. \tag{7.14}$$

Thus from (7.9) we may obtain the transition coefficient $(n|x)$ for every real value of x for which $(0|x)$ is defined. We restrict ourselves to those solutions of (7.7) for which $(0|x)$ is finite, because $|(0|x)|^2$, the probability for obtaining the value x in a measurement of Q in the ground state Λ_0, is assumed to be finite.

Combining (7.8), (7.9), and (7.14), we have

$$(n|x) = \frac{1}{\sqrt{2^n n!}}\,(0|x) H_n\!\left(\sqrt{\frac{\mu\omega}{\hbar}}\, x\right) \tag{7.15}$$

for $-\infty < x < +\infty$.

[The set $(\psi|x) = \sum_n \bar{a}_n(\phi_n|x)$, considered for a fixed value x as a function of ψ, is an antilinear functional over the space Φ of the ψ in Equation (3.27) [i.e., it fulfills Equation (I.7.1)]. Thus $|x)$ is a continuous functional, cf. Section I.7]. As was mentioned in Chapter I, the set of continuous linear functionals on a space Φ, which we denote by $\Phi^\times$, is in general larger than the space Φ, except for the

---

[15] Cf. Appendix of Section I.6 or N. N. Lebedev, *Special Functions and Their Applications*, Prentice Hall, Englewood Cliffs, N.J., 1965, Section 10.4.

Hilbert space $\mathcal{H}$, where $\mathcal{H}^\times = \mathcal{H}$ (Frechet–Riesz theorem.) Taking into account (3.28), we obtain $\Phi \subset \mathcal{H} \subset \Phi^\times$, the *Gelfand triplet* or *rigged Hilbert space* for the harmonic oscillator. The $|x)$ that obey (7.1) are elements of $\Phi^\times$ and are the *generalized eigenvectors* of the operator Q; the set

$$|x) \qquad -\infty < x < +\infty$$

is a *generalized basis system*. The spectrum of Q is continuous and is the whole real axis $\mathbb{R}$. The spectrum of the operator N, (3.2), or H, (3.7), is *discrete*. As x can be any real value, the formal sum in (7.2) is actually an integral. (7.2) is a special case of the *nuclear spectral theorem*, according to which every $\varphi \in \Phi$ can be written as

$$\varphi = \int dx\,|x\rangle\langle x|\varphi\rangle = \int dx\,|x\rangle\varphi(x), \qquad (7.16)$$

as was explained in detail in Section I.4. Here we denote the generalized eigenvectors of Q,

$$Q|x\rangle = x|x\rangle, \qquad (7.17)$$

by $|x\rangle$ instead of $|x)$ to emphasize the difference between them and the eigenvectors belonging to the discrete spectrum, $|n)$. If we take the "scalar product" of φ in (7.16) with $|x'\rangle$, or more precisely the functional $|x'\rangle$ at the element φ, we have

$$\langle x'|\varphi\rangle = \int dx\,\langle x'|x\rangle\langle x|\varphi\rangle. \qquad (7.18)$$

$\langle x'|x\rangle$ is therefore that mathematical object which by integration (7.18) maps the "well-behaved" function $\varphi(x) = \langle x|\varphi\rangle$ into its value at the point x': $\varphi(x') = \langle x'|\varphi\rangle$. It is the *Dirac δ-function* or *δ-distribution*

$$\langle x'|x\rangle = \delta(x' - x), \qquad (7.19)$$

so (7.18) is written

$$\varphi(x') = \langle x'|\varphi\rangle = \int dx\,\delta(x' - x)\langle x|\varphi\rangle$$

$$= \int dx\,\delta(x' - x)\varphi(x). \qquad (7.20)$$

In the transition from the symbolic (7.2) to (7.16),

$$\sum_x |x\rangle\langle x| \quad \text{was replaced by} \quad \int dx\,|x\rangle\langle x|.$$

Consequently, the vectors $|x)$ and the generalized vectors $|x\rangle$ must have different dimensions in order that

$$|x)(x| \quad \text{and} \quad dx\,|x\rangle\langle x|$$

may have the same dimension.

The $|x\rangle\langle x|\,dx$ are the analogues of the projection operators; however, they do not project on any vector of the Hilbert space $\mathcal{H}$ (or Φ), because the $|x\rangle$ are not elements of these spaces. Projection operators are $\int_{\Delta x} dx\,|x\rangle\langle x|$ where the integration extends over a finite interval Δx. As in the case for operators with discrete spectra, one can define functions of the operator Q with continuous spectrum by

$$f(Q) = \int dx\, f(x)|x\rangle\langle x|,$$

where $f(x)$ is a function defined on the spectrum of Q with $\int dx\, f(x)|\psi(x)|^2 < \infty$ for all $\psi \in \Phi$.]

We now return to the evaluation of the harmonic-oscillator function and calculate $(0|x\rangle$. We write $\phi_n = |n)$ in the form (7.16),

$$|n) = \int dx\,|x\rangle\langle x|n), \tag{7.16'}$$

and take the scalar product of this with ϕ_m,

$$\delta_{mn} = (\phi_m, \phi_n) = (m|n) = \int dx\,(m|x\rangle\langle x|n). \tag{7.21}$$

We shall make use of

$$\langle x|n) = \overline{(n|x\rangle}. \tag{7.22}$$

[If $|n\rangle = \phi_n$ and $|x\rangle = \chi$ both were vectors in the Hilbert space then (7.22) would be the condition (I.2.2b) of the definition of the scalar product. However $(n|x\rangle$ is the functional $|x\rangle \in \Phi^\times$ at the element $|n) \in \Phi$ and $\langle x|n)$ is a functional $|n) \in \Phi^{\times\times} = \Phi$ at the element $|x\rangle \in \Phi^\times$, so (7.22) is not obvious. But it can be shown that (7.22) is generally true for transition coefficients between a discrete basis and a continuous basis. Hereafter we shall drop the typographical distinction $|\ )$ as opposed to $|\ \rangle$, and shall use generally $|\ \rangle$, unless we want to stress the distinction between generalized and ordinary vectors.]

We insert (7.15) into (7.21) and obtain

$$\sqrt{\frac{\hbar}{2^n 2^m n!\, m!\, \mu\omega}} \int dy \left| \left\langle 0 \left| \sqrt{\frac{\hbar}{\mu\omega}}\, y \right\rangle \right|^2 H_m(y) H_n(y) = \delta_{mn}. \tag{7.23}$$

If we compare this with the orthogonality relations for the Hermite poly-nomials,[16]

$$\int dy \, e^{-y^2} H_m(y) H_n(y) = \begin{cases} 0 & \text{for } m \neq n, \\ n! \, 2^n \sqrt{\pi} & \text{for } m = n, \end{cases} \qquad (7.24)$$

we find

$$\left| \left\langle 0 \left| \sqrt{\frac{\hbar}{\mu\omega}} \, y \right\rangle \right|^2 = \sqrt{\frac{\mu\omega}{\pi\hbar}} \, e^{-y^2}. \qquad (7.25)$$

Thus, up to an arbitrary phase factor (which we choose to be unity),

$$\langle 0 | x \rangle = \left(\frac{\mu\omega}{\pi\hbar} \right)^{1/4} e^{-(\mu\omega/2\hbar)x^2}. \qquad (7.26)$$

With this and (7.15) we obtain the transition coefficients $\langle n | x \rangle$ between the x- and n-bases, i.e., the harmonic-oscillator wave functions $\phi_n(x)$:

$$\langle n | x \rangle = \left(\frac{\mu\omega}{\pi\hbar} \right)^{1/4} \frac{1}{\sqrt{2^n n!}} H_n \left(\sqrt{\frac{\mu\omega}{\hbar}} x \right) e^{-(\mu\omega/2\hbar)x^2} = \langle x | n \rangle = \phi_n(x). \quad (7.27)$$

A few of the wave functions are plotted in Figure 7.1.

We now repeat the procedure for the operator P that we have already gone through for the operator Q. The eigenvectors of P will be denoted by $|p\rangle$:

$$P | p \rangle = p | p \rangle \qquad (7.28)$$

The action of P on $|n\rangle$ (cf. Problem 13) is

$$P | n \rangle = -i \sqrt{\frac{\hbar\mu\omega}{2}} (a - a^\dagger) | n \rangle$$

$$= -i \sqrt{\frac{\hbar\mu\omega}{2}} (\sqrt{n} | n - 1 \rangle - \sqrt{n + 1} | n + 1 \rangle). \qquad (7.29)$$

[16] Lebedev, *ibid.*, Section 4.13.

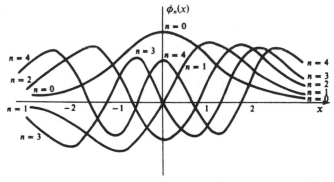

Figure 7.1 Wave functions of the harmonic oscillator [from Ludwig (1954), with permission].

If we take the "scalar product" with $|p\rangle$ and use (7.28), we obtain

$$p\langle p|n\rangle = -i\sqrt{\frac{\hbar\mu\omega}{2}}\,(\sqrt{n}\langle p|n-1\rangle - \sqrt{n+1}\langle p|n+1\rangle), \qquad (7.30)$$

or

$$p\langle n|p\rangle = i\sqrt{\frac{\hbar\mu\omega}{2}}\,(\sqrt{n}\langle n-1|p\rangle - \sqrt{n+1}\langle n+1|p\rangle). \qquad (7.31)$$

If we introduce the new quantities $(n|p)$ defined by

$$\langle n|p\rangle = i^n(n|p), \qquad (7.32)$$

then

$$\begin{aligned} i\langle n-1|p\rangle &= ii^{n-1}(n-1|p), \\ -i\langle n+1|p\rangle &= -ii^{n+1}(n+1|p), \end{aligned} \qquad (7.33)$$

so (7.31) may be written

$$p(n|p) = \sqrt{\frac{\hbar\mu\omega}{2}}\,(\sqrt{n}(n-1|p) + \sqrt{n+1}(n+1|p)). \qquad (7.34)$$

We see that this is exactly the same recurrence relation as in (7.6), with $x\sqrt{\mu\omega/\hbar}$ replaced by $p/\sqrt{\mu\omega\hbar}$. Thus by the same argument as for $\langle n|x\rangle$, we find [using (7.32)] that

$$\langle n|p\rangle = i^n\left(\frac{1}{\pi\mu\omega\hbar}\right)^{1/4}\frac{1}{\sqrt{2^n n!}}H_n\left(\frac{1}{\sqrt{\hbar\mu\omega}}p\right)e^{-p^2/2\mu\omega\hbar}. \qquad (7.35)$$

We have seen that the eigenvectors $|n\rangle$ of the energy operator H for the harmonic oscillator have the very particular property that the transition coefficients (7.27) between these vectors and the x-basis are the same as the transition coefficients (7.35) between these vectors and the p-basis except for a phase factor.

By the same argument as above for the operator Q, we conclude that the spectrum of P is continuous,

$$\text{spectrum } P = \{p | -\infty < p < \infty\}, \qquad (7.36)$$

and the $|p\rangle$ are generalized eigenvectors.

The transition coefficients $\langle p|n\rangle$ in

$$|n\rangle = \int dp\,|p\rangle\langle p|n\rangle \qquad (7.37)$$

are called the *wave functions in the momentum representation* and are denoted

$$\phi_n(p) \equiv \langle p|n\rangle. \qquad (7.38)$$

The probability density of finding the value p in a measurement of the momentum P in the energy eigenstate Λ_n of the oscillator is $w_n(p) = |\langle p|n \rangle|^2$. Also, for any arbitrary vector φ the transition coefficient

$$\varphi(p) = \langle p|\varphi \rangle \tag{7.39}$$

in

$$\varphi = \int dp \, |p\rangle\langle p|\varphi\rangle \tag{7.40}$$

is called the *momentum-space wave function* or the *wave function in the p-representation* for the state φ.

So far, we have obtained the matrix element of Q in the x-representation,

$$\langle n|Q|x \rangle = x\langle n|x \rangle; \tag{7.41}$$

or more generally, for every well-behaved vector φ,

$$\langle \varphi|Q|x \rangle = x\langle \varphi|x \rangle, \qquad \langle x|Q|\varphi \rangle = x\langle x|\varphi \rangle. \tag{7.42}$$

Likewise, we have obtained the matrix element of P in the p-representation,

$$\langle n|P|p \rangle = p\langle n|p \rangle; \tag{7.43}$$

or more generally,

$$\langle \varphi|P|p \rangle = p\langle \varphi|p \rangle, \qquad \langle p|P|\varphi \rangle = p\langle p|\varphi \rangle. \tag{7.44}$$

We now want to calculate the matrix elements of P in the basis of generalized eigenvectors of Q and the matrix elements of Q in the basis of generalized eigenvectors of P. We do this in two steps: First we find out what the symbol $\langle x|p \rangle$, the transition matrix element between the x-basis and the p-basis, represents, and then we determine $\langle x|P|n \rangle$.

We begin by taking the scalar product of (7.16′) with $|p\rangle$ [or, more precisely, we will consider ϕ_n as a functional at the generalized eigenvector $|p\rangle \in \Phi^{\times}$, $p \in$ spectrum P, and use (7.16′)]:

$$\langle p|n \rangle = \int dx \, \langle p|x \rangle\langle x|n \rangle; \tag{7.45}$$

and then we take the scalar product of (7.37) with $\langle x|$ [or more precisely, consider ϕ_n as a functional at the generalized eigenvector

$$|x\rangle \in \Phi^{\times}, \qquad x \in \text{spectrum } Q,$$

and use (7.37)]:

$$\langle x|n \rangle = \int dp \, \langle x|p \rangle\langle p|n \rangle. \tag{7.46}$$

In (7.45) and (7.46), $\langle x|n \rangle$ and $\langle p|n \rangle$ are given by (7.27) and (7.35), respectively. The Hermite polynomials have the property [Equation (I.A.17)][17]:

$$i^n e^{-\eta^2/2} H_n(\eta) = \int_{-\infty}^{+\infty} d\xi \, \frac{e^{i\xi\eta}}{\sqrt{2\pi}} \, e^{-\xi^2/2} H_n(\xi). \tag{7.47}$$

Inserting (7.35) and (7.27) into this relation, it follows that

$$\langle n|p \rangle = \int_{-\infty}^{+\infty} dx \, \frac{e^{ixp/\hbar}}{\sqrt{2\pi\hbar}} \, \langle n|x \rangle, \tag{7.48}$$

or, taking the complex conjugate,

$$\langle p|n \rangle = \int dx \, \frac{e^{-ixp/\hbar}}{\sqrt{2\pi\hbar}} \, \langle x|n \rangle. \tag{7.49}$$

Comparing (7.49) with (7.45), we find that the $\langle p|x \rangle$ are given by

$$\langle p|x \rangle = \frac{1}{\sqrt{2\pi\hbar}} \, e^{-ixp/\hbar}. \tag{7.50}$$

In the same way one obtains from (7.46) and (7.47)

$$\langle x|p \rangle = \frac{1}{\sqrt{2\pi\hbar}} \, e^{ixp/\hbar}. \tag{7.51}$$

(7.50) and (7.51) together give

$$\langle x|p \rangle = \overline{\langle p|x \rangle}. \tag{7.52}$$

It is now simple to calculate the matrix element of P in the basis of generalized eigenvectors of Q, using (7.51):

$$\langle x|P|\varphi \rangle = \int dp \, p \langle x|p \rangle \langle p|\varphi \rangle = \int dp \, p \, \frac{e^{ixp/\hbar}}{\sqrt{2\pi\hbar}} \, \langle p|\varphi \rangle$$

$$= \int dp \, \frac{\hbar}{i} \frac{d}{dx} \, \langle x|p \rangle \langle p|\varphi \rangle = \frac{\hbar}{i} \frac{d}{dx} \int dp \, \langle x|p \rangle \langle p|\varphi \rangle.$$

Thus

$$\langle x|P|\varphi \rangle = \frac{\hbar}{i} \frac{d}{dx} \, \langle x|\varphi \rangle. \tag{7.53}$$

In the same way, using (7.50), one obtains

$$\langle p|Q|\varphi \rangle = -\frac{\hbar}{i} \frac{d}{dp} \, \langle p|\varphi \rangle. \tag{7.54}$$

[17] Lebedev, *ibid.*, Section 4.12.

Each vector in the linear space can be fully characterized by its components with respect to a basis system. In Sections I.4 and I.5 it was discussed in detail that in the same way as one represents the vector $\mathbf{x}$ in the three-dimensional space by its component x_i with respect to the three basis vectors $\mathbf{e}_i$, according to $\mathbf{x} = \sum_{i=1}^{3} \mathbf{e}_i x_i$, one can represent the vector φ by its components $\langle n|\varphi \rangle$ with respect to the discrete basis $\varphi_n = |n\rangle$ of eigenvectors of H according to $\varphi = \sum_{n=0}^{\infty} |n\rangle\langle n|\varphi\rangle$ [Equation (4.8)]. This is also true if the basis is a continuous basis, according to

$$\varphi = \int dx \, |x\rangle\langle x|\varphi\rangle. \tag{7.16}$$

Thus to the vector φ corresponds the function $\langle x|\varphi\rangle = \varphi(x)$ and to the vector $P\varphi$ corresponds the function $\langle x|P|\varphi\rangle$. Equation (7.53) then states that in the realization of the space of vectors φ by the space of wave functions $\langle x|\varphi\rangle = \varphi(x)$, the momentum operator is realized by the differential operator times $\hbar/i$, and (7.42) states that the position operator is realized by the operator of multiplication with the number x. This *realization* is called the *Schrödinger representation*.

[It should be noted that the Schrödinger representation could not be derived from the commutation relation (2.1a) alone, but required an additional assumption: that the operator H in (2.1b) has at least one proper eigenvector. This assumption is one of several equivalent forms of the additional requirement. Other forms are that $P^2 + Q^2$ be essentially self adjoint or that the representation of the Heisenberg algebra be integrable to the representation of the Weyl group (Weyl's form of the canonical commutation relation).]

The Schrödinger representation of the energy eigenvalue equation

$$H\phi_n = E_n \phi_n \tag{7.55}$$

is the time-independent Schrödinger equation which we shall derive now for the harmonic oscillator. The energy operator for the oscillator was given by (2.1b):

$$H = \frac{1}{2\mu} P^2 + \frac{\mu\omega^2}{2} Q^2. \tag{7.56}$$

Let us take the matrix element of H between $\langle x|$ and $|n\rangle$:

$$\langle x|H|n\rangle = \frac{1}{2\mu} \langle x|P^2|n\rangle + \frac{\mu\omega^2}{2} \langle x|Q^2|n\rangle. \tag{7.57}$$

We compute

$$\langle x|P^2|n\rangle = \sum_{n'} \langle x|P|n'\rangle\langle n'|P|n\rangle$$

$$= \sum_{n'} \frac{\hbar}{i}\frac{d}{dx}\langle x|n'\rangle\langle n'|P|n\rangle$$

$$= \frac{\hbar}{i}\frac{d}{dx}\sum_{n'}\langle x|n'\rangle\langle n'|P|n\rangle$$

$$= \frac{\hbar}{i}\frac{d}{dx}\langle x|P|n\rangle$$

$$= \left(\frac{\hbar}{i}\right)^2 \frac{d^2}{dx^2}\langle x|n\rangle. \tag{7.58}$$

Moreover, from (7.1),

$$\langle x|Q^2|n\rangle = x^2\langle x|n\rangle. \tag{7.59}$$

Inserting these in (7.57), we have for the matrix element of the energy operator

$$\langle x|H|n\rangle = -\frac{\hbar^2}{2\mu}\frac{d^2}{dx^2}\langle x|n\rangle + \frac{\mu\omega^2}{2}x^2\langle x|n\rangle$$

$$= \left(-\frac{\hbar^2}{2\mu}\frac{d^2}{dx^2} + \frac{\mu\omega^2}{2}x^2\right)\langle x|n\rangle. \tag{7.60}$$

Therefore, in the Schrödinger representation the eigenvalue equation (7.55) is

$$\left(-\frac{\hbar^2}{2\mu}\frac{d^2}{dx^2} + \frac{\mu\omega^2}{2}x^2\right)\langle x|n\rangle = E_n\langle x|n\rangle. \tag{7.55'}$$

Summarizing, we have seen that in the x-representation, the operator P is given by $-i\hbar\, d/dx$,

$$P \leftrightarrow -i\hbar\frac{d}{dx}, \tag{7.61}$$

and the energy operator by

$$H \leftrightarrow \left(-\frac{\hbar}{2\mu}\frac{d^2}{dx^2} + \frac{\mu\omega^2}{2}x^2\right).$$

II.8 Postulates II and III for Observables with Continuous Spectra

The description of the basic assumptions II and III in Section II.4 was given in terms of operators with discrete spectra. In this section we shall give the description for the case that the observables have continuous spectra. With the help of the formalism introduced here we shall be able to discuss in more detail the problems connected with the measurement of position and momentum, the subject of the next section.

The statistical operator of a pure state,

$$W = \Lambda_{b_0} = |b_0\rangle\langle b_0|,$$

where Λ_{b_0} is the projection operator onto the one-dimensional subspace spanned by the (proper) eigenvector $|b_0\rangle$ of an observable B, is written in an arbitrary basis system $\{\phi_n\}$ as

$$W = \sum_{mn} c_m \overline{c}_n |\phi_m\rangle\langle\phi_n|, \tag{8.1a}$$

where

$$c_n = \langle\phi_n|b_0\rangle. \tag{8.1b}$$

If $\{\phi_n\}$ is the basis system $\{\phi_b - |b\rangle\}$ in which B and therefore W is diagonal, then

$$c_b = \langle\phi_b|b_0\rangle = \langle b|b_0\rangle = \delta_{bb_0}. \tag{8.1b'}$$

For a mixture the statistical operator is

$$W = \sum_{mn} w_{mn} |\phi_m\rangle\langle\phi_n|, \tag{8.2}$$

where w_{mn} are the matrix elements of W in the basis $\{\psi_n\}$; the collection $\{w_{mn}\}$ is the *density matrix*.

In the case of a continuous spectrum Equation (8.1a) has to be replaced by

$$W = \int dx' \, dx'' \, f(x')\overline{f}(x'')|x'\rangle\langle x''|, \tag{8.3a}$$

where $\{|x\rangle\}$ is the basis system of generalized eigenvectors of an observable, say Q, with a continuous spectrum. The (well-behaved test) function $f(x)$ of the continuous variable x for the pure state $W = \Lambda_{b_0}$ is given by

$$f(x) = \langle x|b_0\rangle = \langle x|b_0). \tag{8.3b}$$

Here we have used the second notation to emphasize that $|x\rangle$ is a generalized eigenvector and $|b_0)$ is a proper eigenvector. For a mixture the statistical operator is in analogy to (8.2):

$$W = \int dx' \, dx'' \, F(x', x'')|x'\rangle\langle x''|, \tag{8.4}$$

where $F(x', x'')$ is a function of the two continuous variables x' and x'', the generalized eigenvalues of Q.

Let us now assume that the observable B that was used to prepare the state W is Q, and that we want to prepare and describe a pure state. Naively one would want to choose

$$f(x) = \langle x|x_0\rangle = \delta(x - x_0),$$

in analogy to the discrete case (8.1b'), so that

$$W = \Lambda_{x_0} = |x_0\rangle\langle x_0|.$$

However, it does not make sense to ask for a measurement of a particular point of an observable with a continuous spectrum. Any measurement apparatus, a macroscopic system, is of finite size, not infinitesimally small. Counters have finite extension and need a finite amount of energy; slits—as narrow as they may be—have finite extension. Therefore one cannot ask for a state of a physical system in which the measurement of an observable Q with a continuous spectrum [say spectrum $Q = (-\infty, +\infty)$] gives exactly one particular value, because such a state can be neither measured nor prepared. It only makes sense to ask for a state of a physical system in which the measurement of the observable Q gives values x lying in some particular interval $x_0 - \epsilon < x < x_0 + \epsilon$ around a mathematically defined point x_0. (Physically neither an exact point nor an exact interval, i.e., an interval with precisely determined endpoints, can be defined.) Thus for physical reasons, the analogue to (8.1a) for an observable Q with continuous spectrum is

$$W = \int dx' \, dx'' \, f_\epsilon(x', x_0) \tilde{f}_\epsilon(x'', x_0) |x'\rangle\langle x''|, \tag{8.5}$$

where $f_\epsilon(x, x_0)$ is a function that has its main support in the neighborhood $(x_0 - \epsilon, x_0 + \epsilon)$ of the point x_0 and describes in a more or less direct way the resolution of the measurement apparatus by which the state was prepared.

For the state W given by (8.1a), the *probability* that a measurement of B will yield the value b is

$$\text{Tr}(\Lambda_b W) = (b|W|b) = (b|b_0)(b_0|b) = \delta_{bb_0}\delta_{b_0b} = \delta_{bb_0}. \tag{8.6}$$

For the state W given by (8.5), the likelihood that a measurement of Q will yield the value x is described not by a probability but by the probability distribution (probability density)

$$\langle x|W|x\rangle = \int dx' \, dx'' \, f_\epsilon(x'; x_0) \tilde{f}_\epsilon(x''; x_0) \langle x|x'\rangle\langle x''|x\rangle$$

$$= \int dx' \, dx'' \, f_\epsilon(x'; x_0) \tilde{f}_\epsilon(x''; x_0) \delta(x - x') \delta(x'' - x)$$

$$= f_\epsilon(x; x_0) \tilde{f}_\epsilon(x; x_0) = |f_\epsilon(x; x_0)|^2 = F_\epsilon(x; x_0). \tag{8.7}$$

Here we have used (7.19) [or (I.4.7c$_\rho$) if dx is symbolic for $d\mu(x)$].

Symbolically we may write, in analogy to (8.6), $\langle x|W|x\rangle = \text{Tr}(\Lambda_x W)$ with $\Lambda_x = |x\rangle\langle x|$; however, one has to keep in mind that Λ_x is *not* a projection operator onto a subspace of the space of physical states. The physical question, i.e., the question that has an answer that can be compared with an experimental determination, is: What is the probability of obtaining a value x in the interval between x_1 and x_2 when a measurement of the observable Q is made on the state W? Here the interval (x_1, x_2) is determined by the resolution of the apparatus with which the measurement is made, and again cannot be infinitesimally small and exactly determined.

The operator which represents finding the position in the interval $x_1 < x < x_2$ is the projection operator

$$\Lambda(x_1, x_2) = \int_{x_1}^{x_2} dx \, |x\rangle\langle x|; \tag{8.8}$$

the probability of finding the position anywhere in $x_1 < x < x_2$ when the system is in the state W of (8.5) is then

$$\text{Tr}(\Lambda(x_1, x_2)W) = \int dx' \, \langle x'|\Lambda(x_1, x_2)W|x'\rangle = \int_{x_1}^{x_2} dx \, \langle x|W|x\rangle$$

$$= \int_{x_1}^{x_2} dx \, |f_\epsilon(x; x_0)|^2 = \int_{x_1}^{x_2} dx \, F_\epsilon(x; x_0). \tag{8.9}$$

The function $F_\epsilon(x; x_0) = |f_\epsilon(x; x_0)|^2$ is determined by the experimental situation. It describes the resolution of the apparatus by which the state W is prepared. The better the resolution, the smaller is ϵ, and in the idealized but unphysical limit $\epsilon \to 0$,

$$|f_\epsilon(x; x_0)|^2 = F_\epsilon(x; x_0) \to \delta(x - x_0). \tag{8.10}$$

The limit (8.10) would correspond to an apparatus with infinitely sharp resolution, which cannot exist in reality. Thus the state W in (8.5) can be only as "pure" or as sharp as the experimental resolution allows, and that can never be "ideally sharp" as described by a δ-function.

In order that the state W given in (8.5) be normalized, $f_\epsilon(x, x_0)$ must fulfill

$$\text{Tr } W = \int_{-\infty}^{+\infty} dx \, \langle x|W|x\rangle = \int_{-\infty}^{+\infty} dx \, |f_\epsilon(x, x_0)|^2 = \int_{-\infty}^{+\infty} dx \, F_\epsilon(x, x_0) = 1.$$

$$\tag{8.11}$$

Examples for $F_\epsilon(x) = F_\epsilon(x, 0)$ are (function sequences of δ-type):

$$F_\epsilon(x) = \frac{1}{\pi} \frac{\epsilon}{x^2 + \epsilon^2} \qquad \text{(Figure 8.1)}, \tag{8.12}$$

$$F_\epsilon(x) = \frac{1}{\sqrt{2\pi}} \frac{1}{\epsilon} \exp\left(-\frac{1}{2} \frac{x^2}{\epsilon^2}\right) \qquad \text{(Figure 8.2)}, \tag{8.13}$$

$$F_\epsilon(x) = \begin{cases} 0 & \text{if } x < -\epsilon/2, \\ 1/\epsilon & \text{if } -\epsilon/2 < x < +\epsilon/2, \\ 0 & \text{if } +\epsilon/2 < x. \end{cases} \qquad \text{(Figure 8.3)}, \tag{8.14}$$

$$F_\epsilon(x) = \begin{cases} 0 & \text{if } x \le -\epsilon, \\ \dfrac{x + \epsilon}{\epsilon^2} & \text{if } -\epsilon \le x \le 0, \\ \dfrac{-x + \epsilon}{\epsilon^2} & \text{if } 0 \le x \le \epsilon, \\ 0 & \text{if } +\epsilon \le x, \end{cases} \qquad \text{(Figure 8.4)}. \tag{8.15}$$

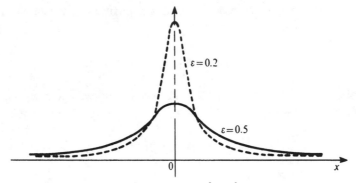

Figure 8.1 Graph of $F_\epsilon(x) = (1/\pi)\epsilon/(x^2 + \epsilon^2)$ for two values of ϵ.

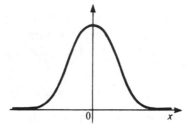

Figure 8.2 Graph of $F_\epsilon(x) = (1/\sqrt{2\pi})(1/\epsilon) \exp(-\tfrac{1}{2}x^2/\epsilon^2)$.

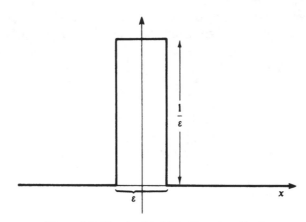

Figure 8.3 Graph of $F_\epsilon(x)$ in Equation (8.14).

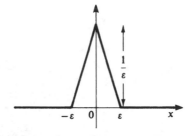

Figure 8.4 Graph of $F_\epsilon(x)$ in Equation (8.15).

All these functions have their main support in the interval $-\epsilon < x < +\epsilon$ and have the property

$$F_\epsilon(x) \to \delta(x) \quad \text{as } \epsilon \to 0.$$

Thus we have arrived at the *basic assumption IIIa for continuous spectrum*: The statistical operator W given by (8.5) with a distribution function $F_\epsilon(x)$ as narrow as experimentally possible is the continuous-spectrum analog of a pure eigenstate.

We shall call such a state an "almost sharp" state and an "almost eigenstate" of the observable with continuous spectrum.

An arbitrary state, in whose preparation no attempt at sharpness may have been made, is described by (8.4). $F(x', x'')$ is connected with the probability distribution function of the observable Q by

$$\operatorname{Tr}(\Lambda_x W) - \langle x|W|x\rangle = \int dx'\, dx''\, F(x', x'')\langle x|x'\rangle\langle x''|x\rangle = F(x, x) \quad (8.16)$$

This is the appropriate place to mention the interrelationship between the physical measurability and preparability of a state and the mathematical description of quantum physics in a rigged Hilbert space $\Phi \subset \mathcal{H} \subset \Phi^\times$, though we cannot give a detailed description here.[18]

The resolution function $F_\epsilon(x; x_0)$ or $f_\epsilon(x; x_0)$ that corresponds to a realistic experimental apparatus must be a well-behaved function; e.g., it can be something like that given in (8.12)–(8.15) with a small but finite ϵ for the description of a state as "sharp" as the preparation allows. One can make this precise by requiring that $f_\epsilon(x; x_0)$ be an element of the Schwartz space, which is a realization of Φ. More appropriate still may be to require that $F_\epsilon(x; x_0)$ have compact support. Those $f_\epsilon(x; x_0)$'s that are either distributions or elements of the Hilbert space (i.e., Lebesgue square-integrable functions that are determined only up to their values on a set of measure zero) cannot describe the resolution of a realistic apparatus (e.g., the detector efficiency).

If $f_\epsilon(x, x_0)$ is an element of the Schwartz space, then the state one obtains by (8.5) or (8.4) is in the space Φ. More precisely, W is a positive definite Hilbert-space-bounded trace-class operator with $W\mathcal{H} \subset \Phi$. If W is a pure state, $W = |f\rangle\langle f|$, then $f(x) = \langle x|f\rangle$ given by (8.3) is an element of the Schwartz space, and $|f\rangle$ is an element of Φ. Thus the requirement that the space of physical states consist of only those states that can be prepared in a realistic experiment means that in a precise formulation of the preceding basic assumptions the space of physical states should not be the Hilbert space $\mathcal{H}$, but the nuclear topological space Φ of the rigged Hilbert

---

[18] An elementary discussion of this subject is given in A. Bohm, *The Rigged Hilbert Space and Quantum Mechanics*, Springer Lecture Notes in Physics, vol. 78, 1978.

space $\Phi \subset \mathscr{H} \subset \Phi^\times$. Though the elements of $\Phi^\times$ which are not in Φ do not represent physical states, $\Phi^\times$ is of immense value, as it contains the generalized eigenvectors of self-adjoint operators with continuous spectra, which are not elements of $\mathscr{H}$; cf. Sections I.4 and I.7.

The Hilbert space $\mathscr{H}$ is infinite-dimensional, and it is infinite-dimensional in a very particular sense; namely, it is complete with respect to a particular topology, that is, with respect to a particular meaning of convergence of infinite sequences. Since an infinite number of states can never be prepared, physical measurements cannot tell us anything about infinite sequences, but can at most give us information about arbitrarily large but finite sequences. Therefore, physics cannot give us sufficient information to tell how to take the limit to infinity, i.e., how to choose the topology. The choice of the topology will be a mathematical generalization, but then one should choose the topology such that it is most convenient.

With the choice of Φ as the space of physical states, one does not only obtain a description which is closer to reality for the above-mentioned reasons connected with measurement and preparation of physical states, but one also obtains a tremendous simplification of the mathematical description. Even the simplest algebra of operators that appears in quantum mechanics, the one fulfilling the relation (2.1a), cannot be represented by $\mathscr{H}$-continuous operators. Consequently there are vectors that lie outside the domain of the definition of an observable that is represented by an unbounded operator in $\mathscr{H}$, e.g., the vector that would correspond to a state of infinite energy. In contrast to this, it seems possible that for all physical systems the observables can be represented by Φ-continuous operators. Thus in the rigged-Hilbert-space description the mathematical image of a physical system is an algebra of continuous operators, domain questions do not arise and all the mathematical manipulations that physicists perform are rigorously justified.

After having considered the continuous-spectrum analogue (8.5) of (4.31) in the axiom IIIa we shall now consider the basic assumption IIIb for a continuous spectrum. Let the physical system be in a state W; this state may have been prepared by measurements of an observable with a continuous spectrum or with a discrete spectrum, or may even be completely unspecified. If a measurement of an observable Q with continuous spectrum $-\infty < x < +\infty$ is made on the state W and is used to select out a subensemble with values x of Q lying in the range $x_1 < x < x_2$, then the (unnormalized) state W' that describes this subensemble is

$$W' = \Lambda(x_1, x_2) W \Lambda(x_1, x_2)$$

$$= \int_{x_1}^{x_2} dx' \int_{x_1}^{x_2} dx'' \, |x'\rangle\langle x'|W|x''\rangle\langle x''|. \tag{8.17}$$

This is the continuous spectrum analog of (4.52). Note that for $x_1 \to -\infty$, $x_2 \to -\infty$ we have $W' = W$, which says that no measurement has been performed. Again, $x_1 < x < x_2$ must be a finite-size interval; it cannot be infinitesimally small, due to the finite resolution and finite extension of the measuring apparatus.

To show that (8.17) represents a state where measurement of Q must yield a value x in the interval $x_1 < x < x_2$, we compute the probability density for getting the value x, which according to (8.17) is given by

$$\langle \Lambda_x \rangle' = \frac{1}{\mathrm{Tr}\, W'} \mathrm{Tr}(\Lambda_x W') = \frac{1}{\mathrm{Tr}\, W'} \langle x | W' | x \rangle$$

$$= \frac{1}{\mathrm{Tr}\, W'} \int_{x_1}^{x_2} dx' \int_{x_1}^{x_2} dx'' \, \langle x | x' \rangle \langle x' | W | x'' \rangle \langle x'' | x \rangle$$

$$= \begin{cases} 0 & \text{if } x < x_1 \text{ or } x > x_2, \\ (\mathrm{Tr}\, W')^{-1} \langle x | W | x \rangle & \text{if } x_1 < x < x_2. \end{cases} \tag{8.18}$$

An actual position measurement is over some range of values in a small interval $x_0 - \epsilon < x < x_0 + \epsilon$. The probability for measuring Q to be in such an interval is

$$\langle \Lambda(x_0 - \epsilon, x_0 + \epsilon) \rangle'$$

$$= \int_{x_0 - \epsilon}^{x_0 + \epsilon} dx \, \langle \Lambda_x \rangle'$$

$$= \begin{cases} 0 & \text{if } (x_0 - \epsilon, x_0 + \epsilon) \cap (x_1, x_2) = \emptyset, \\ \dfrac{1}{\mathrm{Tr}\, W'} \displaystyle\int_{x_0 - \epsilon}^{x_0 + \epsilon} dx \, \langle x | W | x \rangle & \text{if } (x_0 - \epsilon, x_0 + \epsilon) \subseteq (x_1, x_2). \end{cases}$$

$$\tag{8.19}$$

Thus W' represents a state where the position is in the interval (x_1, x_2).

II.9 Position and Momentum Measurements— Particles and Waves

The material of Section II.8 has provided us with a formalism in which we can discuss problems connected with the measurement of position and momentum in more detail. Let us return to the classical-mechanical model of the harmonic oscillator, which consisted of two mass points vibrating around an equilibrium point. For simplicity we assume that $m_2 \gg m_1$, so we can consider m_2 to be at rest. The position x and momentum p are then essentially the position and momentum of mass point 1.

How can a measurement of the position be made on the quantum-mechanical harmonic oscillator? We have considered the diatomic molecule as a realization of this system. The question is, then, how do we measure the position, i.e., the separation between the two atoms of the CO molecules in the collision chamber? There seems to be no way one can think of to set up a measurement apparatus that would do so. The same argument also applies to a momentum measurement on this ensemble. Thus the above discussion in Section II.7 of the probabilities $w(x)$ and $w(p)$ is not of much practical value if one restricts oneself to the quantum-mechanical harmonic-oscillator system. With the formalism developed in Section II.8 we can give a theoretical explanation of this situation.

According to (8.17), after a measurement of the position operator Q, the physical system is in an almost eigenstate of the position operator; i.e., all molecules in the ensemble have a position localized in the interval $(x - \epsilon, x + \epsilon)$. For our classical model of the CO molecule as two oscillating mass points (with $m_2 \gg m_1$ for simplicity), the above statement means that m_1 is located at a distance $\bar{x}$ from m_2 that lies in the interval $(x - \epsilon, x + \epsilon)$. However, a classical system for which m_1 is located in a restricted domain of space is not an oscillator. It is a localized mass point, which we call a particle. A measurement of the position operator Q on the harmonic oscillator would destroy the oscillator and result in a physical system that has as its classical image a *particle*. Thus the fact that it is impossible in practice to set up an experiment that would measure the observable Q on the ensemble of CO molecules in the collision chamber has a theoretical description in the basic assumption IIIb as described by (8.17).

The above arguments also exemplify the fact that for a particular physical system, such as the vibrating CO molecule, there are different kinds of observables. Of one kind (e.g., the energy operator of the vibrating CO molecule) one can easily prepare eigenstates of the particular physical system. Of the other kind (e.g., the relative-position operator of the vibrating CO molecule) the preparation of eigenstates will destroy the system.

The measurement of this second kind of observable can only be described if one enlarges the physical system. Thus in our example we need not restrict ourselves to CO molecules, but can consider the larger physical system that consists of CO molecules, O atoms, and C atoms. Then the CO molecules are just particular states of this larger physical system. Other states are the states in which the C and O atoms are not bound. In this larger system, the measurement of the position (the distance between C and O atoms) is always possible at least in principle.

For the almost eigenstate of the position operator Q, we have

$$W(x) = \iint dx'\, dx''\, f_\epsilon(x' - x)\overline{f_\epsilon}(x'' - x)|x'\rangle\langle x''|, \qquad (9.1x)$$

where $Q|x\rangle = x|x\rangle$ and $|f_\epsilon(x' - x)|^2$ has the property of the functions (8.12)–(8.15). One can define the almost eigenstate of the momentum operator P,

$$W(p) = \int_{-\infty}^{\infty} dp' \int_{-\infty}^{\infty} dp''\, f_\epsilon(p' - p)\tilde{f}_\epsilon(p'' - p)|p'\rangle\langle p''|, \qquad (9.1p)$$

where $P|p\rangle = p|p\rangle$ and $|f_\epsilon(p' - p)|^2$ is again a function of the kind given in (8.12)–(8.15). Q and P here are related by (2.1a). In such a state the measurement of the momentum P gives a value $\bar{p}$ in the interval $p - \epsilon < \bar{p} < p + \epsilon$. By the same arguments as above, such a state cannot be realized by the oscillator; therefore we do not want to require (2.1b), and shall just require that Q and P be related by (2.1a) and that as a consequence[19] of this $\langle p|x\rangle$ is given by (7.50). A possible relation between the energy operator H and the P and Q that facilitates the preparation of almost momentum eigenstates is $H = P^2/2m$, where m is a system constant, the mass. A quantum physical system obeying this relation is called a free nonrelativistic elementary particle.

A quantum-mechanical system with well-defined momentum has as its classical image a *wave*.

Let us discuss this last point a bit more. The characteristic or defining property of a *particle* is that it is localized in space, i.e., it can be assigned a definite position. In the same way, the characteristic or defining property of a *wave*, in particular a plane wave, is that it has infinite periodic spatial extension. The mathematical description of a plane wave is

$$\phi(x, t) = Ae^{ikx}e^{-i\omega t}, \tag{9.2}$$

so the periodic spatial extension is described by

$$u(x) = Ae^{ikx}, \tag{9.3}$$

where A is the amplitude and k is the wave number. The fundamental period or wavelength is $\lambda = 2\pi/k$. If the wave is not a plane wave, i.e., does not have a well-defined wave number or wavelength but has wave numbers k' inside a certain interval $(k - \Delta k, k + \Delta k)$, then one has a wave packet described by

$$\phi(x) = \int_{-\infty}^{+\infty} dk' \, A(k')e^{ik'x}, \tag{9.4}$$

where $A(k')$ is nonzero only in the interval $2\,\Delta k$ about the value k.

To see that a quantum-mechanical system with well-defined momentum has as its classical image a wave, let us calculate the spatial distribution of a state with well-defined momentum. Let us first do this simple-mindedly, working with "states" having exact momentum $|p\rangle$ and "states" having exact position $|x\rangle$ (which are "described" by $W(p) = dp\,|p\rangle\langle p|$ and $W(x) = dx\,|x\rangle\langle x|$, respectively), though we should keep in mind that such "states" are not physical states.

For a pure state Λ_ψ with "well-behaved" state vector ψ, the spatial distribution is given by the wave function $\psi(x) = \langle x|\psi\rangle$; according to (5.25) the probability density for measuring the value x in a position measurement of the state ψ is $w_\psi(x) = |\langle x|\psi\rangle|^2$. Now let ψ be the "not well-behaved" and

---

[19] As remarked in ⟦ ⟧ at the end of Section II.7, (2.1a) is not sufficient to derive (7.50). One has to make an additional assumption, e.g., that $P^2 + Q^2$ is essentially self-adjoint.

therefore unphysical "state" vector $|p\rangle$. The spatial distribution is then given by $\langle x|p\rangle$, which is, according to (7.51),

$$\langle x|p\rangle = \frac{1}{\sqrt{2\pi\hbar}} e^{ixp/\hbar}. \tag{9.5}$$

If we compare this with the spatial distribution of the plane wave given by (9.3), we see that an "exact momentum eigenstate" of a quantum-mechanical system has the same spatial distribution as the classical system called a plane wave, with the wave number and wavelength given by the de Broglie relation:

$$k = \frac{1}{\hbar}p \quad \text{and} \quad \lambda = \frac{2\pi\hbar}{p} = \frac{h}{p}. \tag{9.6a}$$

This is the justification for calling an ensemble of quantum-mechanical systems with definite momentum p a wave with wavelength $\lambda = h/p$.

Let us now return to the physical case where the state is given by (9.1p) and is an almost momentum eigenstate with as narrow a momentum distribution $|f_\epsilon(p' - p)|^2$ as possible. The probability density for obtaining the value x in a measurement of the observable Q in the state $W(p)$ is given according to (8.7) by

$$\langle x|W(p)|x\rangle = \int_{-\infty}^{+\infty} dp' \int_{-\infty}^{+\infty} dp'' \, f_\epsilon(p' - p)\bar{f}_\epsilon(p'' - p)\langle x|p'\rangle\langle p''|x\rangle$$

$$= \int_{-\infty}^{+\infty} dp' \, f_\epsilon(p' - p)\frac{e^{ixp'/\hbar}}{\sqrt{2\pi\hbar}} \int_{-\infty}^{+\infty} dp'' \, \bar{f}_\epsilon(p'' - p)\frac{e^{-ixp''/\hbar}}{\sqrt{2\pi\hbar}} \tag{9.7}$$

In order to compare this with the wave function for a (classical) wave packet (9.4), we define

$$\phi(x) \equiv \int_{-\infty}^{+\infty} dp' \, f_\epsilon(p' - p)\frac{e^{ixp'/\hbar}}{\sqrt{2\pi\hbar}}$$

$$= \int_{-\infty}^{+\infty} dk' \, \hbar \frac{f_\epsilon(k' - k)}{\sqrt{2\pi\hbar}} e^{ixk'}, \tag{9.8}$$

where $p' = \hbar k'$. With

$$A(k') = \sqrt{\frac{\hbar}{2\pi}} \, f_\epsilon(k' - k),$$

$\phi(x)$ indeed describes a wave packet (9.4), and the intensity of this wave gives, according to

$$\langle x|W(p)|x\rangle = |\phi(x)|^2, \tag{9.9}$$

the *probability density* for the value x.

Thus a quantum-mechanical system in a state with well-defined momentum (in a narrow momentum interval) has as its classical image a wave packet with wavelength λ in a narrow interval:

$$\frac{2\pi\hbar}{p - \epsilon} > \lambda > \frac{2\pi\hbar}{p + \epsilon}. \tag{9.10}$$

The intensity of this wave corresponds to the probability density for the measurement of the position. $\phi(x)$ describes the probability and is called the *probability amplitude*.

We now understand why $\psi(x) = \langle x | \psi \rangle$ has been called a wave function: Let W_ψ be a pure state, i.e., W_ψ be a projector onto the space spanned by ψ. Then the probability density for obtaining the value x in a measurement of the position is

$$w_\psi(x) = \langle x | W_\psi | x \rangle = \langle x | \psi \rangle \langle \psi | x \rangle$$
$$= \psi(x)\overline{\psi}(x) = |\psi(x)|^2. \tag{9.11}$$

On the other hand, we can calculate $\langle x | W_\psi | x \rangle$ by inserting a complete system of generalized states

$$I = \int dp' \, |p'\rangle \langle p'|,$$

to obtain

$$\langle x | W_\psi | x \rangle = \int dp' \int dp'' \, \langle x | p' \rangle \langle p' | \psi \rangle \langle \psi | p'' \rangle \langle p'' | x \rangle$$

$$= \int dp' \frac{e^{ixp'/\hbar}}{\sqrt{2\pi\hbar}} \psi(p') \int dp'' \frac{e^{-ixp''/\hbar}}{\sqrt{2\pi\hbar}} \psi(p''). \tag{9.12}$$

If we compare (9.11) and (9.12), we see that the wave function, whose physical interpretation was that its absolute value squared is the probability distribution, is given by

$$\psi(x) = \int dp' \frac{e^{ixp'/\hbar}}{\sqrt{2\pi\hbar}} \psi(p') \tag{9.13}$$

with $\psi(p') = \langle p' | \psi \rangle$. Thus the wave function $\psi(x)$ may be considered as a superposition of plane waves, $\psi(x)$ comes closer to being a plane wave as the transition coefficient $\langle p' | \psi \rangle$ between the pure state ψ and the generalized momentum eigenvector $|p'\rangle$ comes closer to being a δ-function $\delta(p' - p)$.

The wave character of the quantum system can be experimentally displayed by diffraction experiments. Let us consider a system that has been prepared in an almost eigenstate of momentum, i.e., a monochromatic beam of quantum particles. Such a monochromatic beam is easy to prepare if the particles are charged, but harder to prepare if they are neutral (like atoms or molecules). Let us therefore consider first a monochromatic beam of electrons, such as that prepared for the electron-loss experiment that we considered before. The electron beam is produced by a filament, then accelerated by an electric potential, and then passed through crossed electric and magnetic fields of very well-defined field strengths. Only electrons with very well-defined momenta are not deflected. If $f_\epsilon(p' - p)$ describes the momentum spread of this beam, then the spatial distribution is given by (9.8):

$$\phi(x) = \int dk' \, \hbar \frac{f_\epsilon(k' - k)}{\sqrt{2\pi\hbar}} \, e^{ixk'}, \tag{9.14}$$

which becomes closer to being a plane wave,

$$\phi(x) \approx A e^{ixk},$$

as the momentum spread about the value $\hbar k$ becomes narrower. This approximate plane wave may be diffracted by a grid, and one obtains a typical diffraction pattern.

Related to Equation (9.6a) by special relativity is another equation:

$$\omega = \frac{1}{\hbar} E, \tag{9.6b}$$

where ω is the angular frequency of light, $\omega = 2\pi\nu$. The connection between (9.6a) and the relation (9.6b) is best obtained from the dispersion law for our waves, $\omega = f(k)$, and the relation between energy and momentum, $E = f(p)$. Thus, for the system "electromagnetic wave–photon" the dispersion law is $\omega = ck$ ($\nu = c/\lambda$), and the relation between energy and momentum is $E = cp$ [$E^2 - (cp)^2 = (mc^2)^2$ with rest mass $m = 0$]. From these classical relations (the first for a wave, the second for a massless relativistic particle) it follows immediately that (9.6b) is a consequence of (9.6a) and vice versa. For the system "electron de Broglie wave–nonrelativistic electron" the dispersion law is $\omega = k^2/2(m/\hbar)$ and the relation between energy and momentum is $E = p^2/2m$, which again shows that (9.6a) and (9.6b) are consequences of each other.

The two physical systems "electromagnetic wave–photon" and "electron de Broglie wave–electron" are two examples of physical systems called elementary particles.[20] The historical development of the ideas about these two systems was, however, quite distinct: To begin with, all known experimental facts about light could be explained by a particle theory (Newton, 1663) as well as a wave theory (C. Huygens, 1678). The same domain of physical knowledge could be described by two different pictures. Then, diffraction of light was discovered (Young, 1801), which could only be explained by the wave picture, not the particle picture. The cathode rays, on the other hand, though considered by some to be, like light, due to a process in the ether, were soon proven to behave like negatively charged particles in an electric and magnetic field (J. J. Thomson,

---

[20] It would have been better to coin a new name for these physical systems (e.g., "quanta") to avoid any reference to Newtonian particles.

1897; J. Perrin, 1897). Thus light was a wave and electrons were particles. Had the photoelectric effect (P. Lenard, 1902) been discovered before the diffraction phenomena of light, and had the electron diffraction (Davisson and Germer, G. P. Thomson, 1927) been the first effect observed with the cathode rays, the situation might have been just the reverse.

As light was considered a wave, one had the dispersion law $\omega = ck$, or $\omega = c/\lambda$, and Einstein (1905) showed that in order to explain the photoelectric effect one has to associate with light, besides the energy given by Planck's (1900) formula $E = \hbar\omega$, the momentum $p = E/c$.

Electrons, on the other hand, were considered particles, and one had the energy-momentum relation $E = p^2/2m$. But de Broglie (1924) postulated that they are waves, with a wave number given by (9.6a). Then he derived the dispersion law for these waves from the Planck formula (9.6b) to be $\omega = k^2/2(m/\hbar)$.

The two relations

$$E = \hbar\omega, \qquad \mathbf{p} = \hbar\mathbf{k} \tag{9.6}$$

are justly considered the gate to quantum theory. We derived at least the second equation (9.6a) from the defining relation (2.1a). But historically they were the starting point of the long conjecturing process that led to (2.1a) and ultimately to quantum mechanics in general. The relations (9.6) connect two different classical pictures (often called dual).

The one picture is a particle, i.e., a localizable mass point. Classically, such a particle can be given definite values of energy and momentum (e.g., the momentum and the energy measured by an apparatus that moves with a velocity $-\mathbf{v}$ relative to such a localized mass point are $\mathbf{p} = m\mathbf{v}$ and $E = p^2/2m$).

The other picture is a wave, i.e., a more or less periodic disturbance of infinite spatial extension. A wave has a frequency ω and a wave vector $\mathbf{k}$.

In quantum mechanics the physical system (quantum, elementary particle), of which we should always imagine an ensemble, can occur in different states. For instance, it can be in an almost momentum eigenstate described by (9.1p). We may then ask about the position of such a state, e.g., we may want to know the probability of finding the position anywhere in an interval $x - \Delta/2 < x' < x + \Delta/2$. This is according to the basic assumption II, expressed with the projection operator of (8.8) as:

$$\mathrm{Tr}\left(\Lambda\left(x - \frac{\Delta}{2}, x + \frac{\Delta}{2}\right)W(p)\right) = \int_{x-\Delta/2}^{x+\Delta/2} dx' \langle x'|W(p)|x'\rangle.$$

For the sake of definiteness we may assume that the almost momentum eigenstate is given by (9.1p) with $F_\epsilon(p' - p)$ given by (9.14) or Figure 8.3, so that $f_\epsilon(p' - p) = 1/\sqrt{\epsilon}$ in the interval between $p - \epsilon/2$ and $p + \epsilon/2$. The probability density of the position operator is then obtained using (9.7) as

$$\langle x | W(p) | x \rangle = \int_{p-\epsilon/2}^{p+\epsilon/2} dp' \, \frac{1}{\sqrt{\epsilon}} \frac{e^{ixp'/\hbar}}{\sqrt{2\pi\hbar}} \int_{p-\epsilon/2}^{p+\epsilon/2} dp'' \, \frac{1}{\sqrt{\epsilon}} \frac{e^{-ixp''/\hbar}}{\sqrt{2\pi\hbar}}$$

$$= \frac{1}{2\pi\hbar} \epsilon \quad \text{for} \quad \epsilon x/\hbar \ll 1. \tag{9.15}$$

This means that the probability density of the position operator is independent of the position. Thus the probability of finding for the physical system a value in the interval of length Δ around x is

$$\int_{x-\Delta/2}^{x+\Delta/2} dx' \, \langle x' | W(p) | x' \rangle = \frac{1}{2\pi\hbar} \epsilon \Delta, \tag{9.16}$$

i.e., the system is located with the same small probability anywhere in space.

In the limiting case of exact momentum when the momentum spread $\epsilon \to 0$, the probability of finding the physical system in an interval Δ anywhere would be zero. However, this limiting case is unphysical, because there is no apparatus that could prepare such a state. For example, an almost momentum eigenstate of electrons, a monochromatic electron beam, can be prepared by the apparatus used in the energy-loss experiment of Section II.4. The electrons produced by a filament are accelerated by an electrical potential. Even if this potential is very well defined, the electrons will have still the same momentum spread with which they were emitted by the filament. This electron beam then passes through a monochromator (an electric or magnetic field, or in some cases crossed electric and magnetic fields), and only the electrons with a definite velocity or momentum will arrive at the position of the slit. But again, even if the field is very well defined, this slit has a finite opening, so that electrons with slightly different momenta can pass through. In general, because of the necessarily finite size of the macroscopic measurement apparatus, the preparation of a monochromatic beam with zero spread is impossible.

As we have seen above, for an almost momentum eigenstate every position has the same probability, i.e., the position is undetermined. The physical system is in a wave state with a rather well-defined wave number and wavelength; thus it appears as a wave.

Thus, according to the formulation of quantum mechanics based upon the basic assumptions I, II, and III, a quantum physical system in a momentum eigenstate behaves like a wave—it "is" a wave—until a measurement is performed that alters the state (e.g., a position measurement). *As long as such a measurement is not performed, the theoretical predictions are exactly the same as in a classical wave theory.*

We can go through the same arguments starting with an almost position eigenstate $W(x)$ described by (9.1x). The probability density of the momentum operator is then obtained as

$$\langle p | W(x) | p \rangle = \frac{1}{2\pi\hbar} \xi, \tag{9.17}$$

and the probability of finding the momentum in an interval between $p - \epsilon/2$ and $p + \epsilon/2$ is

$$\int_{p-\epsilon/2}^{p+\epsilon/2} dp' \, \langle p' | W(x) | p' \rangle = \frac{1}{2\pi\hbar} \, \xi\epsilon. \tag{9.18}$$

As this probability does not depend upon the central value p, any value of p has the same probability. For an almost position eigenstate the momentum is undetermined. Such an almost position eigenstate we have called a particle, but we see now that, in distinction to a classical (Newtonian) particle, the quantum physical system in a particle state cannot have a definite value of momentum.

The property described by

$$W(p) = \Lambda\left(p - \frac{\epsilon}{2}, p + \frac{\epsilon}{2}\right), \tag{9.19}$$

i.e., that the physical system is in a state with well-defined momentum, and the property

$$W(x) = \Lambda\left(x - \frac{\xi}{2}, x + \frac{\xi}{2}\right), \tag{9.20}$$

i.e., that the physical system is in a state with well-defined position, are called *complementary* to each other. The realization of one property precludes any prediction for any complementary property. If a physical system is in the state (9.19) of well-defined momentum then nothing can be said about its position; any position x is equally probable or improbable as expressed by (9.15). If a quantum physical system is prepared to occur with the classical attributes of a wave then it cannot bear the classical attributes of a particle and vice versa.

If the properties $\tilde{\Lambda}$ and Λ are complementary then the properties $1 - \tilde{\Lambda}$ and Λ, $\tilde{\Lambda}$ and $1 - \Lambda$, and $1 - \tilde{\Lambda}$ and $1 - \Lambda$ are also complementary. $1 - \Lambda$ is the negation of the property Λ: If Λ is the property that the quantum physical system is in the space domain between $x - \xi/2$ and $x + \xi/2$ then $1 - \Lambda$ is the property that the system is outside this domain. However, in distinction to the classical case, the quantum physical system need not have the property $1 - \Lambda$ if it does not have the property Λ; e.g., it could be in the state $\tilde{\Lambda}$, i.e., it could occur as a wave and have no position at all or, instead of $\tilde{\Lambda}$, it could as well have any other property incompatible with Λ and $1 - \Lambda$. There are thus an infinite number of possibilities for a quantum system to have *neither* the property Λ *nor* its negation $1 - \Lambda$, whereas a classical system would always have either the property Λ or its negation $1 - \Lambda$.

The constant $\hbar$ was introduced in the defining relation (2.1a) without much discussion. With the consequences of (2.1a) derived so far we can already understand the meaning of $\hbar$.

$\hbar$ is a conversion factor for the values of one and the same physical quantity in different units. Momentum and energy can be measured in the conventional ways. One prepares an electron beam with sharp momentum by, e.g., using the apparatus of Figure 4.2(a): Electrons, first treated as particles, are emitted by the filament and accelerated by the potential difference of an electric field to the velocity v which can be calculated from $v^2 = 2(e/m)V$. With V measured in volts and with the usual value

$$\frac{e}{m} = 1.7588 \cdot 10^{11} \frac{\text{Coulomb}}{\text{kg}} = 1.7588 \cdot 10^5 \frac{\text{cm}^2}{\text{sec}^2\,\text{V}}$$

one obtains v in cm/sec or $p = mv$ in erg sec/cm.

This electron beam is then treated as a wave. It is diffracted by a grid (grids with the right grid constants of approximately $10^{-8}\,\text{cm} = 1\,\text{Å}$ are crystals) and its wave number k is measured. Using our present knowledge that particles and waves are just two different appearances (of many possible appearances) of the quantum physical system "electron" for which the particle property $v = dx/dt =$ velocity of the particle is related to the wave property $\lambda = 2\pi/k =$ wavelength of the wave by the relation (9.6a):

$$v = \frac{\hbar}{m} k = \frac{\hbar}{m} \frac{2\pi}{\lambda}, \tag{9.21}$$

one can determine $\hbar/m$ from the value of v measured with the electric field and from the value of λ measured with the grid. One need not restrict oneself to this particular experiment only, but one can do many more experiments with electrons measuring independently v from the particle property of the electron and λ from its wave property. The value of $\hbar/m$ that one obtains using (9.19) will be the same. Thus $\hbar/m$ is a constant as long as one uses only electrons.

Now one can repeat the same procedure with other physical systems, e.g., protons, and one can again determine the value of $\hbar/m$ from Equation (9.21). One finds that this leads to a different value of $\hbar/m$ than in the experiments with electrons. Thus the conversion factor $\hbar/m$ between k and v depends upon the nature of the microsystem.

But m can be measured independently—and it is conventional to use for it the independent unit g—by first measuring e/m in deflection experiments with cathode rays and proton beams and then by measuring e in a Millikan experiment. In this way one finds that $\hbar$, the conversion factor between k in cm^{-1} and $p = mv$ in g cm/sec,

$$mv = \hbar k, \tag{9.6a'}$$

has the same value for all physical systems, namely[21]

$$\hbar = 1.0546 \cdot 10^{-28}\,\text{g}\,\frac{\text{cm}}{\text{sec}}\,\text{cm}$$

$$= 1.0546 \cdot 10^{-28}\,\text{erg sec} = 6.58218 \cdot 10^{-16}\,\text{eV sec}. \tag{9.22}$$

The conversion factor $\hbar$ between wave number k and momentum p has nothing to do with the structure of the microphysical system. It is not a system

---

[21] $1\,\text{eV} = 1.6 \times 10^{-12}\,\text{erg} = 1.6 \times 10^{-12}\,\text{dyn cm}$; $1\,\text{dyn} = 1\text{g cm/sec.}^2$

constant like μ and ω in the microphysical system of Section II.2, or like $(\hbar/m)$; it is a universal constant.

Why is the conversion factor between k and v a system-dependent constant and the conversion factor between k and p an universal constant of nature? The reason for this lies in the structure of the symmetry group of space–time, the Galilei group (similar arguments also apply to the symmetry group of relativistic space–time, the Poincaré group). The observables $G_i/\hbar = mQ_i/\hbar$ and $P_i/\hbar$—obeying the commutation relations $[G_i/\hbar, P_j/\hbar] = i\delta_{ij}(m/\hbar)$ $(i, j = 1, 2, 3)$, from which the three-dimensional version of (2.1a) follows— are not just observables of a particular microsystem or of various micro-systems but they represent the generators of the Galilei transformations: $(1 + ix^i P_i/\hbar)$ represents the displacement by an infinitesimal vector x^i (in cm), and $(1 + iv^i G_i/\hbar)$ represents an infinitesimal proper Galilei transforma- tion (to a frame moving with velocity v^i in cm/sec). Thus $P_i/\hbar$ and $mQ_i/\hbar$ (but not $Q_i/\hbar$) have a universal meaning which comes from the structure of the symmetry group in nature and not from the structure of a particular physical system. $m/\hbar$ is an invariant of the group of Galilei transformations, whose value characterizes the representation of the (extended) Galilei group and, therewith, the physical system that this representation describes.

The conversion factor $\hbar$ appears in the expression for the Galilei trans- formation because x^i and P_i have in the past been chosen to have independent units: x^i is measured in cm and P_i is measured in erg sec/cm, a unit which is derived from the fundamental unit g for the mass

$$\left(1 \text{ erg } \frac{\text{sec}}{\text{cm}} = 1 \text{ dyn cm} \frac{\text{sec}}{\text{cm}} = 1 \text{ g} \frac{\text{cm}}{\text{sec}^2} \text{ cm} \frac{\text{sec}}{\text{cm}}\right).$$

But this choice is no longer necessary now that one knows quantum mechan- ics and the fundamental physical meaning of the representations of the Galilei group. One can decide to choose the units such that $\hbar = 1$ because all one needs is that $x^i P_i/\hbar$ and $v^i mQ_i/\hbar$ be dimensionless quantities. Taking for x^i the conventional units cm, the choice $\hbar = 1$ means that P_i has the units cm^{-1} and that the units of the mass are then no longer the independent units g but are the derived units sec/cm^2. The value of $\hbar$ in the conventional c.g.s. system, which we denote by $\{\hbar\}$ and which according to (9.22) is given by $\{\hbar\} = 1.0546 \cdot 10^{-28}$, is therefore nothing other than the conversion factor between g and sec/cm^2. Thus choosing the conversion factor $\hbar = 1$, only cm and sec remain as fundamental units.

For relativistic space–time one has an additional conversion factor between a distance x measured in cm and the time t (in sec) that light takes to traverse this distance, $t = (1/c)x$, where c is the velocity of light. Choosing $c = 1$ for this conversion factor eliminates sec as an independent unit and 1 sec $= \{1/c\}$ cm, where $\{c\} = 2.9979 \cdot 10^{10}$ denotes the value of c in the c.g.s. system. Then cm remains as the only fundamental unit.

Instead of cm^{-1} one usually takes in relativistic particle physics the units of 1 eV or 1 MeV $= 10^6$ eV for energy, mass and momentum. The distance is then measured in MeV^{-1}. The value $\{\hbar c\}$ of $\hbar c$ in the c.g.s. system of units is then the conversion factor between the units of energy, mass, and momentum in MeV and in cm^{-1}:

$$1 \text{ cm}^{-1} = \{\hbar c\} \text{ MeV} = 1.97328 \cdot 10^{-11} \text{ MeV}.$$

Thus we have seen that $\hbar$, *like any constant of nature, is just a conversion factor for the values of one and the same physical quantity in different units.* Before a constant of nature is discovered, the values of the same quantity in different units (e.g., the momentum $\mathbf{p}$ and the wave vector $\mathbf{k}$) are thought to be different physical quantities. After the constant has been discovered and the different physical quantities have been related by a new theory, one can measure these quantities in the same units and eliminate in this way one of the previously independent units.

Problems

1. Show that the operators a and $a^\dagger$, defined by Equations (3.1), are the adjoints of each other.

2. Show that the collection of all linear operators on a (complex) linear space R constitutes an associative algebra with unit element under the operations of scalar multiplication of an operator, addition of operators, and multiplication (composition) of operators.

3. For the quantum-mechanical harmonic oscillator, calculate the diagonal matrix elements of the energy operator H between the states

$$\tilde{\phi}_n = e^{i\alpha_n}\phi_n, \qquad \alpha_n = \text{real},$$

where ϕ_n is the eigenvector belonging to the eigenvalue n of the operator $N = a^\dagger a$, according to Equation (3.20).

4. Let ϕ_n be the vectors of Equation (3.19). Define for each complex number z the *coherent state vector*[22]

$$|z\rangle \equiv e^{-|z|^2/2} \sum_{n=0}^{\infty} \frac{z^n}{\sqrt{n!}}\, \phi_n.$$

(a) Show that the coherent state $|z\rangle$ is an eigenstate of the operator $a^\dagger$ with eigenvalue z.

(b) Show that the scalar product of ϕ_n with $|z\rangle$ has the property

$$|\langle \phi_n|z\rangle|^2 = \frac{|z|^{2n}}{n!}\, e^{-|z|^2}.$$

(c) Show that the scalar product $\langle z|z'\rangle$ of two coherent states $|z\rangle$ and $|z'\rangle$ is given by

$$\langle z|z'\rangle = \exp(-\tfrac{1}{2}|z|^2 + \bar{z}\, z' - \tfrac{1}{2}|z'|^2).$$

(d) Calculate the expectation value of the operators H, Q, and P in the coherent state $|z\rangle$, i.e., calculate $\langle z|H|z\rangle$, $\langle z|Q|z\rangle$, and $\langle z|P|z\rangle$.

(e) Show that for the pure coherent state $W = |z\rangle\langle z|$ the equality holds in (6.3):

$$\Delta P\, \Delta Q = \frac{\hbar}{2}.$$

---

[22] For the applications of coherent states consult Sargent Scully-Lamb (1974), Ch. XV.

5. Figure PS.1(a) shows a schematic diagram of a slightly different experimental setup than the one of Figure 4.2(a). Results of an energy-loss experiment performed on N_2 with this apparatus are shown in Figure PS.1(b). The bump at $v = 0$ in Figure PS.1(b) is approximately 30 times larger than the bumps at $v = 1, 2, 3, \ldots$; this is in contrast to the situation depicted in Figure 4.2(b), where the $v = 0$ bump is approximately three times larger than the succeeding bumps. Explain the difference. Determine from Figure

Figure PS.1 (a) Schematic diagram of the double electrostatic analyzer. Electrons are emitted by the filament, deflected by the cylindrical grids (4A and 4B) with radii of 1.0 and 1.5 cm, respectively, injected into the collision chamber 5, and analyzed by sweeping the voltage between electrodes 6 and 3. S_1, S_2, S_3, S_4, S_5 are shields to collect unwanted electrons; 4C and 7C are top and bottom grids. Typical operating voltages between the electrodes indicated: $(4A-4B) = 1.2\,V$; $(7A-B) = 1.2\,V$; $(Fil-3) = 1.4\,V$; $(F-S_1) = 20\,V$; $(F-S_2) = (F-S_3) = 20\,V$. The electron collector is grounded. All slits are 0.5×4 mm. (b) Energy spectrum of forward-scattered electrons at an incident electron energy of 2.6 eV. [From Schulz, G. J: *Physical Review* **125**, 229 (1962), with permission.]

PS.1(b) the statistical operator W that describes the state of the ensemble of N_2 molecules of the experiment of Figure PS.1(a).

6. Suppose we have a quantum-mechanical oscillator.

  (a) Let the state of the system be given by (4.17). Show that (4.21) follows from (4.20).
  (b) Show that $W = \tilde{W}/(\mathrm{Tr}\ \tilde{W})$ is the normalized statistical operator, where $\tilde{W}$ is given by (4.22).
  (c) Let Λ be a projection operator. Show that $\mathrm{Tr}\ \Lambda = \dim(\Lambda\mathscr{H})$.

7. Calculate the expectation values of the operators Q, Q^2, P, and P^2 when the quantum-mechanical oscillator is in the state $W = \Lambda_n$, where Λ_n is the projection operator onto the energy eigenspace corresponding to the eigenvalue $E_n = \hbar\omega(n + \frac{1}{2})$.

8. Let Λ be the projector onto the one-dimensional subspace spanned by ψ ($\|\psi\| = 1$). Show that the dispersion $\mathrm{disp}_{(\Lambda)} H$ of H when the system is in the state $W = \Lambda$ satisfies

$$\mathrm{disp}_{(\Lambda)} H \geq 0.$$

9. Calculate $\mathrm{disp}_{(W)} H$ for the state of the CO molecules in the collision chamber of the energy-loss experiment described in Section II.4. Show that $\mathrm{disp}_{(W)} H \geq 0$.

10. Let W be given by Equation (5.9). Show that $\mathrm{disp}_{(W)} A = 0$.

11. Let W be given by (4.31). Calculate $\langle A \rangle$ and show that this is the result we would expect if $\langle A \rangle$ is to represent the average value of A. Is W' as given by (4.51) normalized, i.e., does $\mathrm{Tr}\ W' = 1$?

12. Let H and B be two observables with spectra E_n and b_i, respectively. Let Λ_n and Λ_{b_i} be the projection operators on the eigenspaces of H with eigenvalue E_n and of B with eigenvalue b_i, respectively. Let the quantum-mechanical system be in a state in which an energy measurement has always resulted in the value E_n.

  (a) What is the statistical operator W for this state?
  (b) In this state W a measurement of the observable B is performed. What is the statistical operator W' after the B measurement, without selection of subsystems?
  (c) What is the probability of finding the value b_j in a measurement of B on the physical system in the state W? in the state W'? Compare these two probabilities.
  (d) The observable H is now measured again. What is the probability of obtaining E_n in this measurement of H? Compare this probability with the probability of obtaining E_n in the original state W.
  (e) Let A be a third observable with eigenvalues a_i and projectors Λ_{a_i} onto the eigenspaces. Calculate the probabilities of obtaining the value a_k in a measurement of A when the system is in the state W and when the system is in the state W'. When are these probabilities equal?

13. Calculate the matrix elements $\langle n|P|m \rangle$ and $\langle n|Q|m \rangle$ of the operators P and Q between the energy eigenvectors $\phi_n = |n\rangle$ of the harmonic oscillator.

14. Give explicit expressions for the first six Hermite polynomials

$$H_n(y) \qquad (n = 0, 1, 2, 3, 4, 5)$$

and discuss the probability distribution $|\langle x|n \rangle|^2$ for the harmonic oscillator in the six lowest energy eigenstates.

15. In the energy eigenstate $W = \Lambda_1$ corresponding to the energy eigenvalue $E_1 = \frac{3}{2}\hbar\omega$, the probability $w_1(x)\,\Delta x$ of measuring the position to be within an interval

$(x - \Delta x/2, x + \Delta x/2)$ about the value x is nonzero for $x > a = \sqrt{2E_1/(\mu\omega^2)}$. From this it follows that the potential energy V may be larger than the total energy E_1 because

$$V = \tfrac{1}{2}\mu\omega^2 x^2 > \tfrac{1}{2}\mu\omega^2 a^2 = E_1.$$

Point out what is wrong with this argument, and give a correct interpretation of the situation.

16. Show that the operators $P^{(S)} = (\hbar/i)\, d/dx$ and $Q^{(S)} = x$, considered as operators on the wave function, satisfy the Heisenberg commutation relation.

17. It was derived in Section II.7 that, on the wave function, the momentum operator P and the position operator Q are represented by the operators $P^{(S)} = (\hbar/i)\, d/dx$ and $Q^{(S)} = x$. Which assumption besides the Heisenberg commutation relation has been used in this derivation?

18. Let a quantum-mechanical particle with mass m be confined by impenetrable walls to the region $-a < x < +a$ (a one-dimensional rectangular well with infinitely high walls). This means that the expectation value of the potential-energy operator V between the generalized position eigenstates $|x\rangle$ is

$$\langle V\rangle_x = \frac{\langle x|V|x\rangle}{\langle x|x\rangle} = -c,$$

where c is a constant; and that the probability that a measurement of the position Q will give the value x is nonzero only for $-a < x < +a$. Denote by $|n\rangle$ the eigenvectors of the energy operator

$$H = \frac{P^2}{2m} + V,$$

and let Λ_n denote the projector onto the space spanned by $|n\rangle$. Calculate the eigenvalues of H, and compute the expectation values of Q, Q^2, P, and P^2 when the system is in an energy eigenstate Λ_n.

19. Show that for any linear operators A, B and for $\lambda \in \mathbb{C}$, the trace as defined by Equation (4.3) has the following properties:

(a) It is independent of the choice of basis.
(b) $\mathrm{Tr}(\lambda A) = \lambda\, \mathrm{Tr}\, A$.
(c) $\mathrm{Tr}(A + B) = \mathrm{Tr}\, A + \mathrm{Tr}\, B$.
(d) $\mathrm{Tr}(AB) = \mathrm{Tr}(BA)$.

20. Let $W(p)$ be an almost momentum eigenstate described by the statistical operator

$$W(p) = \int_{-\infty}^{+\infty} dp' \int_{-\infty}^{+\infty} dp''\; f_\epsilon(p' - p)\bar{f}_\epsilon(p'' - p)|p'\rangle\langle p''|$$

with

$$f_\epsilon(p' - p) = \frac{1}{\sqrt{\pi}}\frac{\sqrt{\epsilon}}{p' - p - i\epsilon}$$

(Lorentzian momentum distribution). Calculate the probability density of the position operator and the probability of finding the position in the interval $x_0 - \Delta x/2 < x < x_0 + \Delta x/2$. Discuss the result.

21. Davisson and Germer scattered low-energy electrons from metal targets. For 45-eV electrons incident normally on a crystal face, compute the angle between the incident beam and the scattering maximum if the metal is assumed to be of simple cubic structure with a lattice constant of 3.52 Å.

22. Has it been derived that electrons are waves with wavelength $\lambda = h/p$? If so, then explain how this has been done. If not, explain what *has* been derived.

CHAPTER III

Energy Spectra of Some Molecules

Section III.1 discusses how the energy levels of a quantum-mechanical system emitting dipole radiation are observed. A derivation of the transition probability is not given. In Section III.2 the defining relations of angular momentum are established. Section III.3 derives the representations of the algebra of angular momentum. In Section III.4 the energy spectrum of a rotator is derived and compared with the experimental spectrum of diatomic molecules. Section III.5 contains the basic assumption about the physical combination of two nonidentical quantum-mechanical systems and the application of this assumption to the description of vibrating and rotating diatomic molecules.

III.1 Transitions Between Energy Levels of Vibrating Molecules—The Limitations of the Oscillator Model

A quantum-mechanical system in a certain stationary state [e.g., the energy eigenstate Λ_n of the diatomic molecule (oscillator)] will remain in that state so long as it is not acted upon by outside forces. In practice, any quantum-mechanical system is acted upon by weak external forces, such as external electromagnetic fields or internal electromagnetic fields that arise from the motion of charges within the system. Under the influence of such forces, the state is liable to change. If the system has a discrete set of states (e.g., the energy eigenstates of the oscillator), then a weak external disturbance does not change these states (or, more precisely, it changes the energy levels by a negligibly small amount), but the system may jump from one state to another.

The theory of such transitions, which can be developed as a consequence of the basic assumptions of quantum mechanics, will be presented later in Chapters XIV and XXI. For the moment we shall just give some semi-classical arguments and state the result, which we shall use to obtain transition frequencies and selection rules.

We accept here as an empirical fact that under the influence of ever-present external disturbances, the quantum system may perform transitions from one energy eigenstate with energy E_n to another with energy E_m, and emit or absorb the energy difference

$$E_n - E_m$$

as electromagnetic radiation in the form of a light quantum or photon of frequency

$$v_{nm} = \frac{E_n - E_m}{h} = \frac{E_n - E_m}{2\pi\hbar}. \tag{1.1}$$

If the electromagnetic field has the frequency v_{nm}, then the quantum-mechanical system can absorb a photon of this frequency and jump from a state of energy E_n to a state of higher energy E_m. On the other hand, if a quantum system is in an excited state E_n (a state of higher energy than the ground state), it can emit a photon of frequency v_{nm} and drop to a state of lower energy E_m.

Transitions between two states cannot occur under the influence of electromagnetic radiation if the matrix element of the total electric displacement operator $\mathbf{D}$ of the system vanishes between these two states.[1] Also, the probability for such a transition, and thus the intensity of the emitted (or absorbed) electromagnetic radiation, is proportional to the square of the modulus of this matrix element.

To illustrate, let us return to our classical picture of the diatomic molecule. If the molecule consists of unlike atoms (e.g., CO) then it has an electric dipole moment, since the centers of the positive and negative charges do not coincide. The dipole moment is the vector directed from the center of negative charges to the center of positive charges and is given by

$$\mathbf{D} = q\mathbf{d},$$

where q is the charge and d is the distance between the centers of the charges. The permanent dipole moment $\mathbf{D}_0$ of the molecule lies along the internuclear axis.

If the interatomic (or internuclear) distance changes, the dipole moment will change, and to a good approximation it may be assumed that the dipole moment is a linear function of the deviation from the equilibrium position of the interatomic distance:

$$\mathbf{D} = \mathbf{D}_0 + q\mathbf{x}. \tag{1.2}$$

---

[1] This statement applies only to dipole radiation. There may be quadrupole or higher multipole radiation even for molecules for which the dipole moment is zero; however, the magnitude of the higher multipole radiation is exceedingly small.

Therefore the dipole moment changes with the frequency of the mechanical vibration. Oscillating charges radiate an electromagnetic field, and on the basis of classical electrodynamics the emitted light should have a frequency equal to the frequency of the oscillator, i.e.,

$$v = \frac{\omega}{2\pi}, \tag{1.3}$$

where $\omega = \sqrt{k/m}$ is the angular frequency of the classical oscillator [see (II.2.8)].

If the molecule consists of two like atoms (e.g., O_2, N_2), then the dipole moment is zero, because the centers of positive and negative charge coincide and oscillations of the molecule about its equilibrium position do not lead to oscillations of the center of charge. No emission or absorption of electromagnetic radiation occurs.

Let us now turn to the quantum-mechanical molecule. Quantum-theoretically, the emission of radiation takes place as a result of a transition of the oscillator from a higher to a lower energy state, and absorption takes place by the reverse process. The frequency of the emitted light is given by

$$v_{nm} = \frac{E_n - E_m}{h}. \tag{1.1}$$

The intensity of the emission, classically proportional to the time-averaged value (over one period) of the square of the dipole moment $\mathbf{D}$, is in quantum theory proportional to the absolute value squared of the transition matrix elements

$$\langle m|\mathbf{D}|n \rangle = \mathbf{D}_{mn}, \tag{1.4}$$

where $\mathbf{D}$ is the dipole operator, obtained from (1.2) by the usual procedure of replacing the number x with the operator $\mathbf{Q}$:

$$\mathbf{D} = \mathbf{D}_0 + q\mathbf{Q}. \tag{1.5}$$

The transition probability per unit time (intensity) A_{nm} for spontaneous dipole emission[2] in the transition from an energy state with energy value E_n to a state with energy E_m is given by[3]

$$A_{nm} = \frac{4}{3\hbar c^3} \omega_{nm}^3 |\mathbf{D}_{mn}|^2, \tag{1.6}$$

where $\omega_{nm} = (E_n - E_m)/\hbar$, c is the velocity of light, and

$$|\mathbf{D}_{mn}|^2 = \frac{1}{\dim(\Lambda_n \mathcal{H})} \mathrm{Tr}(\Lambda_n \mathbf{D} \cdot \Lambda_m \mathbf{D}) = \frac{1}{\dim(\Lambda_n \mathcal{H})} \sum_{i=1}^{3} \sum_{v,\mu} |\langle \mu, m|D_i|n, v \rangle|^2. \tag{1.7}$$

---

[2] The probabilities for induced emission and absorption are also proportional to $|\mathbf{D}_{mn}|^2$ and to the intensity of the incident radiation.

[3] This formula can be obtained using semiclassical arguments, or it can be derived by applying the results of Chapter XXI.

D is given by (1.5), Λ_n is the projection operator on the energy eigenspace with eigenvalue E_n, and $\dim(\Lambda_n \mathscr{H})$ is the dimension of this energy eigenspace. μ and ν are the same kind of index as κ in (II.4.35); they label the different vectors within the energy eigenspaces $\Lambda_m \mathscr{H}$ and $\Lambda_n \mathscr{H}$.

Since we are now not interested so much in knowing the intensity as we are in knowing when this intensity is zero, the precise form of the transition probability is not of primary interest to us at the moment. Equation (1.6) can be derived using the general formalism developed in Chapter XXI. For the special case of the one-dimensional oscillator we replace the dipole and position vectors by the one-dimensional quantities D and Q and (1.7) goes over into $|\langle m|D|n\rangle|^2$.

For many quantum-mechanical systems, a great majority of the matrix elements of D vanish, so there is a severe limitation on the possibilities for transitions. The rules that express this limitation are called *selection rules*. In order to determine which particular transitions can actually occur for the harmonic oscillator, we have to calculate the matrix elements

$$\langle m|D|n\rangle = q\langle m|Q|n\rangle. \tag{1.8}$$

The matrix elements of the position operator between energy eigenstates have already been calculated [see Equations (II.7.3)] and are given by

$$\langle m|Q|n\rangle = \sqrt{\frac{\hbar}{2\mu\omega}}(\sqrt{n}\,\langle m|n-1\rangle + \sqrt{n+1}\,\langle m|n+1\rangle). \tag{1.9}$$

Thus we see that the transition probability and hence the emission and absorption intensity of light are zero except when the quantum numbers n and m differ by unity. Thus the selection rule for the harmonic oscillator is

$$n - m = \pm 1. \tag{1.10}$$

Transitions in the harmonic oscillator are possible only between neighboring energy levels. The frequency of light that is emitted (for $E_n > E_m$) or absorbed (for $E_n < E_m$) is given according to (1.1) and (1.10) by

$$\nu_{nm} = \frac{E_n - E_m}{h} = \frac{\hbar\omega}{h}[(n + \tfrac{1}{2}) - (m + \tfrac{1}{2})] = \frac{\omega}{2\pi}. \tag{1.11}$$

Thus quantum-theoretically the frequency of the radiated light is equal to the frequency $\omega/2\pi$ of the oscillator, and is independent of the energy level n. Similar arguments apply for absorption. Thus we have seen that for the particular case of the quantum-mechanical harmonic oscillator, the frequency of emitted and absorbed light is the same as it would be for the classical oscillator.

If we recall the energy-level diagram for the harmonic oscillator (Figure II.4.3) we can indicate the allowed transitions by vertical lines (see Figure 1.1). All these transitions give rise to the same frequency. This is a consequence of the equal spacing between energy levels.

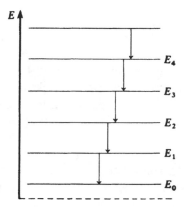

Figure 1.1 Dipole transitions between the energy levels of the harmonic oscillator.

For a diatomic molecule consisting of two like atoms (e.g., O_2), the dipole moment operator (1.5) is the zero operator, and therefore no transitions between different energy levels occur.

Let us now turn to the comparison of our theoretical results with the experimental situation.

In order to find out what frequency we should expect, we turn first to the energy-loss spectrum of the CO molecules (Figure II.4.2). From the distance between the bumps in the energy-loss spectrum we find that the difference between the various energy levels of the vibrating CO molecule is

$$\Delta E = 0.265 \, \text{eV}. \tag{1.12a}$$

If we calculate the frequency from this according to (1.11) we find

$$\nu = \frac{\Delta E}{2\pi\hbar} = \frac{0.265 \, \text{eV}}{2\pi \times 6.58 \times 10^{-16} \, \text{eV sec}} = 6.4 \times 10^{13} \, \text{sec}^{-1} \tag{1.12b}$$

and

$$\lambda = \frac{c}{\nu} = 0.466 \times 10^{-3} \, \text{cm} = 4.66 \, \mu \tag{1.12c}$$

$(1\mu = 10^{-4} \, \text{cm}; 1 \, \text{Å} = 10^{-8} \, \text{cm} = 10^{-4} \, \mu)$.

In molecular spectroscopy it is customary to give the frequency not in sec^{-1} but in cm^{-1}, i.e., give instead of the frequency ν the wave number $\nu/c = 1/\lambda$, which indicates the number of waves per cm. We shall not introduce a new symbol for it but also call it ν, and the unit next to the number will then tell us what is meant. The frequency in cm^{-1}, or wave number, of the radiation emitted by the transition between the vibrational levels of CO is then

$$\nu = 2140 \, \text{cm}^{-1}. \tag{1.12d}$$

Thus we expect, from the energy-loss spectrum of CO, that the vibrating CO molecules emit or absorb electromagnetic radiation only with the frequency given by (1.12), i.e., we expect one spectral line in the near infrared region.[4]

---

[4] In order to give a feeling for the orders of magnitude of various quantities involved in molecular spectroscopy, we show a table of the regions of the electromagnetic spectrum in Figure 1.2.

If we compare this with the absorption or emission spectrum, we find that it is indeed correct. If the absorption spectrum is obtained with a thin layer of absorbing gas, one finds only a single, broad, intense absorption line (or band) in the near infrared region, with a wavelength around $\lambda = 4.66\,\mu$. For other diatomic molecules consisting of unlike atoms, one finds the same situation; e.g., for HCl this band lies at $\lambda = 2.46\,\mu$. One also finds that such bands do not appear for molecules consisting of like atoms, such as O_2, N_2, H_2.

If the absorption is observed with thicker layers of gas, the intensity of absorption of the fundamental band naturally increases, and in addition a second band of similar form appears quite weakly, at approximately half the

THE ELECTROMAGNETIC SPECTRUM

Energy	Frequency	Transitions	Radiation	Wave Number	Wave Length
E (ev)	ν (sec^{-1})			ν (cm^{-1})	λ (cm)
5×10^{-6}	1.2×10^{9}	Nuclear Magnetic Resonance Spin Orientations in Magnetic Field Electron Spin Resonance	Radio Waves / Micro-waves (radar)	4×10^{-2}	25
3.1×10^{-3}	7.5×10^{11}	Molecular Rotations		25	4×10^{-2}
5×10^{-2}	1.2×10^{13}			400	2.5×10^{-3}
		Molecular Vibrations	Infrared Region		
0.5	1.2×10^{14}			4000	2.5×10^{-4}
1.55	3.8×10^{14}		Visible	12.5×10^{3}	8×10^{-5}
3.1	7.5×10^{14}			25×10^{3}	4×10^{-5}
		Valence Electronic Transitions			
6.2	1.5×10^{15}		Ultra-violet	50×10^{3}	2×10^{-5}
1240	3×10^{17}			10^{7}	10^{-7}
		Inner Shell Electronic Transitions	X-rays		
1.24×10^{4}	3×10^{18}			10^{8}	10^{-8}
		Nuclear Transitions	Gamma Rays		

Figure 1.2 Schematic diagram of the electromagnetic spectrum. Note that the scale is nonlinear. Boundaries between regions are generally quite arbitrary.

Figure 1.3 Coarse structure of the infrared spectrum of HCl (schematic). The intensity actually falls off five times faster than indicated by the height of the vertical lines. Herzberg (1966), vol. 1.

wavelength or double the frequency (wave number). If the thickness of the layer is still further increased (up to several meters at atmospheric pressure), a third and possibly even a fourth and a fifth band appear whose wavelengths are approximately a third, a fourth, and a fifth, respectively, of that of the first band; that is to say, their frequencies are three, four, and five times as great. Figure 1.3 gives schematically the complete infrared spectrum of HCl. In this figure the lengths of the vertical lines that represent the bands give an indication of their intensity. However, the actual decrease in intensity is five times as fast as is indicated in the drawing.

The explanation of these additional bands with lower intensity is that the diatomic molecule is not quite a harmonic oscillator. In a harmonic oscillator the restoring force increases indefinitely with increasing distance from the equilibrium point. However, it is clear that in an actual molecule, when the atoms are at a great distance from one another, the attractive force is zero. Thus the quantum-mechanical harmonic oscillator is only a simplified model of the vibrating molecule, and if one wants to describe the finer details of vibrating molecules, then the anharmonic forces also have to be taken into account. The energy levels of the anharmonic oscillator are not equidistant like those of the harmonic oscillator, but rather their separation decreases slowly with increasing n.

The energy levels and absorption spectrum for an anharmonic but almost harmonic oscillator are shown in Figure 1.4. (For the sake of clarity a faster decrease of ΔE is drawn than is actually found in most observed cases.) The selection rule (1.10),

$$n - m = \pm 1,$$

holds only approximately for the anharmonic oscillator and applies only to the most intense transitions. But now transitions with $n - m = \pm 2, \pm 3, \ldots$ can also appear—though with rapidly decreasing intensity. All these results can be calculated using perturbation theory, which we shall introduce in Chapter VIII. We describe these facts here to demonstrate that the simple soluble quantum-mechanical models like the harmonic oscillator describe only the principal structures of a microphysical system in nature, and cannot be expected to describe all the details. This is not a deficiency of the harmonic

Figure 1.4 Energy levels and infrared transitions of the anharmonic oscillator. The absorption spectrum is given schematically beneath.

oscillator model but a general property of physical theories. Models are only idealizations and cannot be expected to reproduce the experimental results up to the last digit. The explanation of a new decimal place in an experimental number often requires a new model, and sometimes a completely new theory.

We shall see this presently when we consider the transition frequencies in the near infrared region in more detail, as obtained with a spectrometer of sufficiently high resolution. The broad spectral lines for the CO molecule around $v = 2140$ cm^{-1} is then resolved into a number of individual narrow lines, as shown in Figure 1.5. That is, around $v = 2140$ cm^{-1} one does not have a single line but a band, called the vibration–rotation band. As one sees from this figure, this band consists of a series of almost equidistant lines, with one line missing in the center of the band. Going out from the gap, there are two branches, which are called the P branch (towards longer wavelengths) and the R branch (towards smaller wavelengths). Figure 1.6 shows the same effect for the $n = 1$ line of Figure 1.3 for HCl.

One would expect such fine structure in the absorption or emission spectrum of electromagnetic radiation for the CO molecule if the energy levels of the vibrating molecule of Figure 1.1 were split into a series of sublevels as shown in Figure 1.7, which shows only any two neighboring energy levels of the energy spectrum of the vibrating molecule as given in Figure 1.1.

The description of such a splitting lies outside the capability of an oscillator model. It can only mean that a state characterized by the quantum number n is not a pure state but is in fact a mixture of states with different energies. In the oscillator, however, the state characterized by n was a pure state described by a projection operator Λ_n on a one-dimensional subspace spanned by ϕ_n, namely $\Lambda_n \mathcal{H}$. The state of the diatomic molecule characterized by the quantum number n must have at least as many dimensions as energy levels (when the number of energy levels equals the dimension then to any energy value there belongs a one-dimensional subspace or a projector on a one-dimensional subspace). Consequently the oscillator model alone describes only part of the properties of a diatomic molecule, and in order to describe the finer details of the spectrum we have to combine the oscillator model with

Figure 1.5 The vibration–rotation band of carbon monoxide.

Figure 1.6 The fundamental absorption band for HCl under high resolution. (The lines are doubled due to the presence of the two isotopes Cl^{35} and Cl^{37} in the ratio 3:1; we do not discuss this effect here.) [From Alpert Keiser and Szymanski (1970), with permission.]

Figure 1.7 Origin and appearance of rotational structure. P and R branches are shown to the left and right, respectively, on the spectrometer tracing of the CO fundamental absorption band at 2144 cm^{-1}. The Q branch (dashed line) is missing. Energy levels are shown to scale, except that the distance between upper and lower vibrational states (2144 cm^{-1}) should be about five times as great as in the figure. [From Bauman (1962), with permission.]

another model, which describes these finer details and which reflects additional features of the diatomic molecule that we have so far not taken into account. This new model is the *rotator model*.

If we return to our classical picture of the CO molecule, considering it as two atoms of mass m_1 and m_2 at a distance x, then we see that this classical object not only can perform vibrations along the x-axis, but also can perform rotations in three-dimensional space around its center of mass. As long as it is in the vibrational ground state, that is, as long as the energy involved is less than 0.26 eV for CO, it is a rigid rotator, i.e., it can be considered as two pointlike masses m_1, m_2 fastened to the ends of a weightless rigid rod of length x. We shall therefore first study the rigid–rotator model by itself. This will provide us with a description of the CO states that are characterized by the quantum number $n = 0$, and will also approximately describe each set of states with a given vibrational quantum number n. Then we shall see how these two models are combined to form the vibrating rotator or the rotating vibrator.

III.2 The Rigid Rotator

To conjecture the mathematical image (the algebra of operators) for the rotator, we proceed in the same way as for the oscillator: We consider the classical rotator, and we replace the three coordinates of momentum p_i and the three coordinates of position x_i in all expressions for the observables with operators P_i and Q_i that fulfill the canonical commutation relations

$$[P_i, Q_j] = \frac{\hbar}{i}\delta_{ij}I, \qquad [Q_i, Q_j] = 0, \qquad [P_i, P_j] = 0 \qquad (2.1)$$

($\delta_{ij} = 1$ for $i = j$ and $\delta_{ij} = 0$ for $i \neq j; i, j = 1, 2, 3$)—the obvious generalization of (II.2.1a) to the three-dimensional case.

In classical mechanics the energy E of rotation of a rigid body is given by

$$E = \tfrac{1}{2}I\omega^2. \qquad (2.2)$$

Here ω is the angular velocity of the rotation,[5] and I is the moment of inertia of the system about the axis of rotation. The angular velocity is related to the number of rotations per second, v_{rot} (the rotational frequency), by

$$\omega = 2\pi v_{rot}. \qquad (2.3)$$

The angular momentum of the system is given by $l = I \cdot \omega$. Introducing this into (2.2), the energy may also be expressed by

$$E = \frac{l^2}{2I}. \qquad (2.4)$$

---

[5] We use the same symbol ω for the rotational angular velocity that we used for the vibrational angular velocity in Chapter II.

Figure 2.1 Dumbbell model of a diatomic molecule.

The moment of inertia about the axis of rotation for the rotator model is given by

$$I = m_1 r_1^2 + m_2 r_2^2,$$

where

$$r_1 = \frac{m_2}{m_1 + m_2} x \quad \text{and} \quad r_2 = \frac{m_1}{m_1 + m_2} x \tag{2.5}$$

are the distances of the masses m_1 and m_2 from the center of mass C, and x is the distance between the two mass points (see Figure 2.1). Substitution gives

$$I = \frac{m_1 m_2}{m_1 + m_2} x^2, \tag{2.6}$$

that is, the moment of inertia about the axis of rotation is the same as that of a mass point of mass

$$\mu = \frac{m_1 m_2}{m_1 + m_2} \tag{2.7}$$

at a distance x from the axis; μ is called the *reduced mass* of the molecule.

Thus, instead of considering the rotation of the rigid rotator, we can equally well consider the rotation of a single mass point of mass μ with coordinates x_i, where the vector $\mathbf{x} = (x_1, x_2, x_3)$ is the position vector for the mass point μ with the center of mass C as the origin. If we denote the momentum of the mass point μ in this coordinate system by $\mathbf{p} = (p_1, p_2, p_3)$, then the angular momentum about the point C is given by

$$\mathbf{l} = \mathbf{x} \times \mathbf{p}, \tag{2.8}$$

so its components are given by

$$l_i = \sum_{j,k} \epsilon_{ijk} x_j p_k \equiv \epsilon_{ijk} x_j p_k. \tag{2.9}$$

In this equation $\epsilon_{ijk} = +1$ for $(ijk) = (123)$ and every even permutation thereof, $\epsilon_{ijk} = -1$ for (ijk) an odd permutation of (123), and $\epsilon_{ijk} = 0$ otherwise. It is understood from now on that two identical indices in a term indicate a summation from 1 to 3 on those indices unless stated otherwise.

The operators that represent the corresponding quantum-mechanical observables are then obtained by our general principle of replacing x_i, p_j by Q_i, P_j fulfilling (2.1). Thus the angular-momentum operator $\mathbf{L}$ is given by

$$\mathbf{L} = \mathbf{Q} \times \mathbf{P},$$

or

$$L_i = \epsilon_{ijk} Q_j P_k, \tag{2.10}$$

and the energy operator is given in correspondence to (2.4) by

$$H = \frac{1}{2I} \mathbf{L}^2 = \frac{1}{2I} \sum_{i=1}^{3} L_i L_i = \frac{1}{2I} L_i L_i. \tag{2.11}$$

As Q_j and P_k are Hermitian operators, it follows that L_i and H are Hermitian:

$$L_i^\dagger = \epsilon_{ijk} P_k^\dagger Q_j^\dagger = \epsilon_{ijk} P_k Q_j = \epsilon_{ijk} Q_j P_k = L_i. \tag{2.11a}$$

From the Heisenberg commutation relations (2.1) one obtains the commutation relations of the operators L_i that represent the components of angular momentum. The calculation goes as follows:

$$
\begin{aligned}
[L_i, L_l] &= \epsilon_{ijk} \epsilon_{lmn} [Q_j P_k, Q_m P_n] \\
&= \epsilon_{ijk} \epsilon_{lmn} (Q_j [P_k, Q_m P_n] + [Q_j, Q_m P_n] P_k) \\
&= \epsilon_{ijk} \epsilon_{lmn} (Q_j \{ [P_k, Q_m] P_n + Q_m [P_k, P_n] \} \\
&\quad + \{ [Q_j, Q_m] P_n + Q_m [Q_j, P_n] \} P_k) \\
&= \epsilon_{ijk} \epsilon_{lmn} \left(Q_j P_n \frac{\hbar}{i} \delta_{km} - Q_m P_k \frac{\hbar}{i} \delta_{jn} \right) \\
&= \frac{\hbar}{i} (\epsilon_{ijk} \epsilon_{lkn} Q_j P_n - \epsilon_{ijk} \epsilon_{lmj} Q_m P_k).
\end{aligned}
$$

By changing the summation indices we can write this as

$$\frac{\hbar}{i} (\epsilon_{imk} \epsilon_{lkn} Q_m P_n - \epsilon_{ikn} \epsilon_{lmk} Q_m P_n) = \frac{\hbar}{i} (\epsilon_{imk} \epsilon_{knl} + \epsilon_{ink} \epsilon_{klm}) Q_m P_n. \tag{2.12}$$

We now make use of the following property of the ϵ_{imk}, which is easily verified:

$$\epsilon_{imk} \epsilon_{knl} = \delta_{in} \delta_{ml} - \delta_{mn} \delta_{il}. \tag{2.13}$$

Writing the second term in (2.12) in the form (2.13):

$$\epsilon_{ink} \epsilon_{klm} = \delta_{il} \delta_{nm} - \delta_{im} \delta_{nl},$$

and adding it to (2.13) gives

$$\epsilon_{imk} \epsilon_{knl} + \epsilon_{ink} \epsilon_{klm} = \delta_{in} \delta_{ml} - \delta_{im} \delta_{nl} = -\epsilon_{ilk} \epsilon_{kmn}.$$

The last equality again follows from (2.13). Substitution of this expression into (2.12) gives

$$[L_i, L_l] = i\hbar \epsilon_{ilk} \epsilon_{kmn} Q_m P_n.$$

If we insert the definition (2.10) of L_i on the right-hand side, we obtain the commutation relation of the angular-momentum operators L_i:

$$[L_i, L_l] = i\hbar\epsilon_{ilk}L_k. \qquad (2.14)$$

We have obtained the commutation relations (2.14) of the angular-momentum operators (2.10) from the Heisenberg commutation relations (2.1). The expression of the energy operator (2.11) does not contain the P_i and Q_i explicitly; this is true for all physical observables of the rotator. In fact the operators P_i and Q_i are unphysical observables for the quantum-mechanical rotator system in the following sense: It is not possible to prepare the quantum-mechanical rotator so that it is in a generalized eigenstate of momentum or position. Therefore, for the rotator, the L_i that obey the commutation relations (2.14) are the fundamental physical observables, and the fact that they may be obtained from the momentum and position operators can be ignored. In fact, the operators L_i defined by (2.10) or the classical quantities l_i defined by (2.9) are a particular case of an observable that is connected with new degrees of freedom of physical systems in the three-dimensional space.

A physical object in the three-dimensional physical space has six degrees of freedom: three translational degrees of freedom described by the three coordinates x_i, and three rotational degrees of freedom, described by a rotation $R(\alpha, \beta, \gamma)$ that depends upon three angles α, β, γ (e.g., the three Euler angles, or the three angles of rotation around three fixed coordinate axes). The momentum p_i is the canonical variable conjugate to the coordinate x_i, and the canonical variable conjugate to the angular coordinate α_i is the angular momentum l_i.

If we forget about the constituents of the diatomic molecule and consider the dumbbell as an entire physical system whose center of mass is fixed at a particular point in space, then the remaining degrees of freedom are the rotations by the three angles, and the dynamical variables are these angles and their conjugates, the angular momenta l_i. In our particular case above the l_i are obtained from the momenta and coordinates of the constituents, which are thought of as mass points, i.e., not possessing rotational degrees of freedom.

In general, a classical extended particle, in the three-dimensional physical space, has as variables the momenta p_i and the angular momenta s_i (spin) associated with it, which are the conjugates of the linear coordinates x_i and the angular coordinates α_i respectively. For the case that the extended physical object is a dumbbell the spin s_i is the orbital angular momentum $l_i = \epsilon_{ijk}x_jp_k$ of its two constituents.

For the quantum physical extended particle, the momentum is represented by the operator P_i, and the spin by the operator S_i. It is now easy to conjecture that the defining commutation relations of the spin operators S_i are

$$[S_i, S_j] = i\hbar\epsilon_{ijk}S_k. \qquad (2.15)$$

For the special case of the spin of the dumbbell, these relations were derived in (2.14) from the relations (2.1), which had already proven to be successful.

Equation (2.15) can also be derived from the properties of the group of rotations, if one assumes that the rotation $R(\alpha, \beta, \gamma)$ of the physical object is represented (continuously) by a (unitary) operator $U(\alpha, \beta, \gamma)$ in the space of physical states of this quantum-physical object. This latter assumption can be derived from the fact that the rotations are symmetry transformations and that the rotation group is a symmetry group of the physical system (Wigner Theorem). We will not discuss this connection here, but we take the algebra of observables defined by (2.15) as the starting point.

We will now investigate the properties of the algebra of operators generated by the J_i, which fulfill the commutation relations

$$[J_i, J_k] = i\hbar\epsilon_{ikl}J_l \qquad (i, k, l = 1, 2, 3), \qquad (2.16)$$

where J_i may be either an L_i as given by (2.10) or an S_i. In particular we shall find the properties of all the J_i, which are linear Hermitian operators in a linear space. It will turn out that the set of all J_i is richer than the set of the L_i given by (2.10). The algebra generated by the J_i is called the *enveloping algebra* of the group SU(2) and is denoted by $\mathscr{E}(\mathrm{SU}(2))$.[6]

III.3 The Algebra of Angular Momentum

We shall now find all possible solutions of the commutation relation (2.16) that fulfill $J_i^\dagger = J_i$. This means that we shall construct all linear spaces in which the J_i that fulfill (2.16) act as linear Hermitian operators. We assume that there exists at least one eigenvector of J_3 in these spaces.

Instead of working with the J_i $(i = 1, 2, 3)$, we introduce the following linear combinations:

$$H_3 = \hbar^{-1}J_3, \quad H_+ = \hbar^{-1}(J_1 + iJ_2), \quad H_- = \hbar^{-1}(J_1 - iJ_2), \quad (3.1)$$

The Hermiticity condition $J_i^\dagger = J_i$ is expressed by

$$H_3^\dagger = H_3, \qquad H_+^\dagger = H_-, \qquad H_-^\dagger = H_+. \qquad (3.2)$$

From (2.16) it then follows that

$$[H_3, H_\pm] = \pm H_\pm,$$

$$[H_+, H_-] = 2H_3. \qquad (3.3)$$

The operator $\mathbf{J}^2$ can be written

$$\mathbf{J}^2 = \hbar^2\mathbf{H}^2 \qquad (3.4)$$

with

$$\mathbf{H}^2 = H_+H_- + H_3^2 - H_3 = H_-H_+ + H_3^2 + H_3. \qquad (3.5)$$

---

[6] J_i are called the generators of the group SU(2), meaning that every element of the group is given by $e^{i\omega J_j}$. The commutation relations of the group SO(3) and the group SU(2) are the same. We therefore use the symbol $\mathscr{E}(\mathrm{SO}(3))$ and $\mathscr{E}(\mathrm{SU}(2))$ interchangeably. In general, the enveloping algebra of any Lie group (of operators) is defined as the associative algebra whose generators—in the sense of the definition of Equation (I.3.13)—are the generators of the Lie group.

$\mathbf{H}^2$ and therfore $\mathbf{J}^2$ have the property that

$$[\mathbf{H}^2, H_3] = 0, \qquad [\mathbf{H}^2, H_\pm] = 0, \tag{3.6a}$$

or in general

$$[\mathbf{H}^2, A] = 0, \tag{3.6b}$$

where A is any element of the form

$$A = aI + a^i J_i + a^{ij} J_i J_j + a^{ijk} J_i J_j J_k + \cdots, \tag{3.7}$$

where a, a^i, a^{ij}, a^{ijk}, ..., are complex numbers. Because of the property (3.6), $\mathbf{H}^2$ and $\mathbf{J}^2$ are called *invariant operators* (*Casimir operators*) of the algebra $\mathcal{E}(SU(2))$.

We now find all *ladder representations* of $\mathcal{E}(SU(2))$, i.e., all solutions of the commutation relations (2.16) by linear operators in linear spaces that are obtained by applying to one eigenvector of H_3 the whole algebra $\mathcal{E}(SU(2))$. (This one eigenvector is assumed to exist.)

We choose a vector $f = f^c$ which is an eigenvector of $\mathbf{H}^2$ with eigenvalue c:

$$\mathbf{H}^2 f = cf. \tag{3.8}$$

As $\mathbf{H}^2$ commutes with every A of the form (3.7), Af is again an eigenvector of $\mathbf{H}^2$ with eigenvalue c.

PROOF. $\mathbf{H}^2 Af = A\mathbf{H}^2 f = cAf.$ □

We now choose f such that it is also an eigenvector of H_3 and we call the eigenvalue m. The normalized eigenvector we denote by $f_m = f_m^c$:

$$H_3 f_m = m f_m, \qquad (f_m, f_m) = 1. \tag{3.8'}$$

(f_m are called *weight vectors*; m is called a *weight*.)

To choose a vector such that it is simultaneously an eigenvector of two operators is possible iff these two operators commute, because then

$$\mathbf{H}^2 H_3 f_m = H_3 \mathbf{H}^2 f_m$$

which by (3.8) and (3.8') becomes an identity. Note that if two operators do not commute then, in general, they have no common eigenvector (see Problem 7).

Now we define

$$f_{m\pm} = H_\pm f_m$$

and calculate using (3.3)

$$H_3 f_{m\pm} = H_3 H_\pm f_m = (H_\pm H_3 \pm H_\pm) f_m$$
$$= (H_\pm m \pm H_\pm) f_m = (m \pm 1) H_\pm f_m$$
$$= (m \pm 1) f_{m\pm}.$$

$$\tag{3.9}$$

Thus $f_{m\pm}$ is, if it is different from 0, an eigenvector of H_3 with eigenvalue $m \pm 1$.

We note some properties of f_m:

1. The number c is nonnegative.

PROOF.

$$c = (f_m, \mathbf{H}^2 f_m) = \hbar^{-2}(f_m, (J_1^2 + J_2^2 + J_3^2)f_m)$$

$$= \hbar^{-2}(\|J_1 f_m\|^2 + \|J_2 f_m\|^2 + \|J_3 f_m\|^2) \geq 0. \qquad \square$$

2. Any eigenvalue m of H_3 fulfills

$$m^2 \leq c. \tag{3.10}$$

PROOF.

$$0 \leq \hbar^{-2}\{(J_1 f_m, J_1 f_m) + (J_2 f_m, J_2 f_m)\}$$

$$= \hbar^{-2}\{(f_m, \mathbf{J}^2 f_m) - (f_m, J_3^2 f_m)\} = \hbar^{-2}\hbar^2(c - m^2)\|f_m\|^2,$$

from which follows $(c - m^2) \geq 0$. $\qquad \square$

Note that in the proofs the Hermiticity property of J_i has been used.

If we start with an arbitrary eigenvector $f_{m_0}^c$ of H_3 and $\mathbf{H}^2$ and apply H_+ successively, then we obtain according to (3.9) new eigenvectors f_m^c of H_3 with ever-increasing eigenvalues. Because of (3.10), after a finite number of steps we must reach the eigenvector f_i^c with the largest eigenvalue l of H_3, i.e., with

$$f_{l+}^c = H_+ f_l^c = 0. \tag{3.11}$$

From (3.5) it then follows that

$$\mathbf{H}^2 f_l^c = (H_3^2 + H_3)f_l^c = l(l + 1)f_l^c. \tag{3.12}$$

Consequently the eigenvalue c of $\mathbf{H}^2$ and the largest eigenvalue l of H_3 (highest weight) are connected by

$$c = l(l + 1).$$

Instead of characterizing the eigenvectors of $\mathbf{H}^2$ and H_3 by c and m, we can characterize them by l and m.

If we apply H_- successively to f_m^l, then after a finite number of steps, because of (3.10), we must reach the vector f_μ^l with the lowest eigenvalue μ of H_3, i.e., with

$$f_{\mu-}^l = H_- f_\mu^l = 0. \tag{3.13}$$

From (3.5) it then follows that

$$\mathbf{H}^2 f_\mu^l = (H_3^2 - H_3)f_\mu^l = \mu(\mu - 1)f_\mu^l. \tag{3.14}$$

Comparing (3.14) with (3.12) we find

$$l(l + 1) = \mu(\mu - 1),$$

and the only solution of this equation for μ that fulfills (3.10) is

$$\mu = -l. \tag{3.15}$$

Thus if we start with the vector f_l^l, apply H_- successively to it and normalize, we obtain the sequence of vectors

$$f_{l-1}^l = (\alpha_l)^{-1} H_- f_l^l,$$
$$f_{l-2}^l = (\alpha_{l-1})^{-1} H_- f_{l-1}^l, \tag{3.16}$$
$$\vdots$$
$$f_{m-1}^l = (\alpha_m)^{-1} H_- f_m^l,$$

where the α_m are

$$\alpha_m = \sqrt{(H_- f_m^l, H_- f_m^l)}.$$

We shall finally reach the vector with the smallest weight μ:

$$f_\mu^l = (\alpha_{\mu+1})^{-1} H_- f_{\mu+1}^l.$$

As $\mu = -l$, we have $2l + 1$ vectors in the sequence (3.16).

$$f_m^l \quad \text{with} \quad m = l, l-1, l-2, \ldots, -l+1, -l, \tag{3.17}$$

which fulfill

$$(f_m^l, f_{m'}^l) = \delta_{mm'}. \tag{3.18}$$

As $2l + 1$ is a number of vectors, it must be an integer; consequently l can be only one of the following numbers:

$$l = 0, \tfrac{1}{2}, 1, \tfrac{3}{2}, \ldots . \tag{3.19}$$

Thus for a given number l, which can be one of the numbers in (3.19), we have $2l + 1$ vectors f_m^l that are orthogonal to each other and span a space, which we shall call $\mathscr{R}^l$:

$$\mathscr{R}^l = \left\{ f : f = \sum_{m=-l}^{+l} a^m f_m^l \right\}. \tag{3.20}$$

We shall now determine the value of the normalization constant α_m:

$$\alpha_m \overline{\alpha_m} = (H_- f_m, H_- f_m) = (f_m, H_+ H_- f_m)$$
$$= (f_m, (\mathbf{H}^2 - H_3^2 + H_3) f_m) = l(l+1) - m^2 + m.$$

Thus, except for a phase factor (which remains undetermined),

$$\alpha_m = \sqrt{l(l+1) - m^2 + m} = \sqrt{(l+m)(l-m+1)}, \tag{3.21}$$

and we have

$$H_- f_m^l = \sqrt{(l+m)(l-m+1)}\, f_{m-1}^l = \alpha_m f_{m-1}^l. \tag{3.22}$$

It remains to determine $H_+ f_m^l$. We know already that $H_+ f_m^l \propto f_{m+1}^l$. We set $H_+ f_m^l = \beta_m f_{m+1}^l$ and calculate

$$\overline{\beta_m}(f_{m+1}^l, f_{m+1}^l) = (H_+ f_m^l, f_{m+1}^l)^- = (f_m^l, H_- f_{m+1}^l) = \alpha_{m+1}(f_m^l, f_m^l).$$

$$l = 2$$

Figure 3.1 Example of a weight diagram of an irreducible representation of SU(2).

Therefore

$$\overline{\beta_m} = \alpha_{m+1} = \sqrt{(l+m+1)(l-m)} = \beta_m,$$

and we have

$$H_+ f^l_m = \sqrt{(l-m)(l+m+1)} \, f^l_{m+1} = \alpha_{m+1} f^l_{m+1}. \qquad (3.23)$$

Summarizing, we have found the following: For every integer or half-integer value l there is a space $\mathcal{R}^l$ spanned by $2l + 1$ orthogonal vectors f^l_m $(m = -l, \ldots, l)$. In this space the operators H_+, H_-, H_3 are given by (3.8'), (3.22), (3.23); and thus the action of any element $A \in \mathcal{E}(SU(2))$ given by (3.7) on any vector $f \in \mathcal{R}^l$ given by (3.20) is determined. To indicate that for every l one obtains a different operator, one could also write $H^{(l)}_+, H^{(l)}_-$, $H^{(l)}_3$ for the operators in (3.8'), (3.22), (3.23). The space $\mathcal{R}^l$ is called an *irreducible representation space* of $\mathcal{E}(SU(2))$. In this space the elements H_+, H_-, H_3 defined by (3.3) are represented by the operators given in (3.8'), (3.22), (3.23). These operators are called the $(2l + 1)$-*dimensional irreducible representation* of the H_+, H_-, H_3. As can be seen, they depend upon l; for each $l = 0$, $\frac{1}{2}, 1, \ldots$, there is one "different" set of operators. All of the vectors in $\mathcal{R}^l$ are eigenvectors of $\mathbf{H}^2$ with the same eigenvalue $l(l + 1)$. There is no element $A \in \mathcal{E}(SU(2))$ that can transform from a vector $f^l \in \mathcal{R}^l$ to a vector $f^{l'} \in \mathcal{R}^{l'}$ with $l \neq l'$. This fact is expressed by the statement that $\mathcal{R}^l$ is "invariant under all $A \in \mathcal{E}(SU(2))$." In particular, $\mathcal{R}^l$ remains invariant under the action of the J_i $(i = 1, 2, 3)$ and the $H_\pm, H_3$.

If we plot the possible values of m in an irreducible representation along a line, then we obtain the weight diagram of the representation characterised by l, of SU(2). This is the simplest example of a weight diagram for Lie groups. For $l = 2$ this is shown in Figure 3.1. To each dot there corresponds a basis vector f^l_m in the representation space $\mathcal{R}^l$, or equivalently there corresponds the one-dimensional subspace spanned by f^l_m. Each such subspace (or basis vector) represents a (pure) physical state. Thus to each point in the weight diagram there corresponds a pure physical state.

The smallest space is $\mathcal{R}^0$, which is one-dimensional, as shown in Figure 3.2.

In general, we have a weight diagram for each representation, and the action of the operators H_+, H_- can be represented in this diagram, as shown

$$l = 0$$

Figure 3.2 Weight diagram of the one-dimensional representation of SU(2).

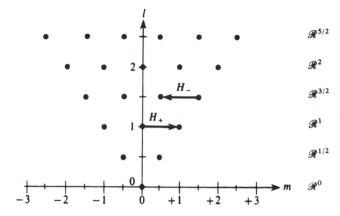

Figure 3.3 Weight diagrams of irreducible representations of SU(2).

in Figure 3.3. For each weight diagram, there is a space $\mathcal{R}^l$, and for each space $\mathcal{R}^l$ there is a possible state (or set of states) of the quantum-mechanical system characterized by the value l. Because of the correspondence between the operator $\mathbf{L}$ (2.10) and the classical angular momentum $\mathbf{l}$ (2.8), the number l is called the *angular momentum quantum number*:

$$\mathbf{L}^2 f^l = \hbar^2 l(l + 1) f^l.$$

Thus the quantum-mechanical angular momentum can take only a discrete number of values. The physical state that corresponds to a given space $\mathcal{R}^l$, and which is described by the statistical operator $W = (\dim \mathcal{R}^l)^{-1} \Lambda^l = (2l + 1)^{-1} \Lambda^l$ where Λ^l is the projector onto $\mathcal{R}^l$, has a definite angular momentum l. Except for $l = 0$ this is not a pure state but a mixture. The preparation of a pure state requires not only a measurement of $\mathbf{L}^2$ but also a measurement of L_3 (or any other component of the vector $\mathbf{L}$). If the value of L_3 is always measured to be m then the state is the pure state represented by the projection operator $\Lambda^l_m = |f^l_m\rangle\langle f^l_m|$ on the one-dimensional subspace $\mathcal{R}^l_m$.

The space $\mathcal{R}^l$ is the direct sum of one-dimensional spaces $\mathcal{R}^l_m$,

$$\mathcal{R}^l = \sum_{m=-l}^{+l} \oplus \mathcal{R}^l_m, \tag{3.24}$$

and each $\mathcal{R}^l_m$ is spanned by the vector f^l_m.

It has already been mentioned that not all sets of linear operators J_i ($i = 1, 2, 3$) that fulfill (2.16) can be given by (2.10). It can be proven (Problem 1) that for L_i given by (2.10) the number l can only take the values $l = 0, 1, 2, \ldots$; in other words, the operators L_i given by (2.10) can be represented by operators in the spaces $\mathcal{R}^l$ with $l = 0, 1, 2, \ldots$ only. Thus for the operators given by (2.10) there are a countable number of representatives $L_i^{(l)}$ in spaces $\mathcal{R}^l$ that remain invariant under the action of the $L_i^{(l)}$. These spaces $\mathcal{R}^l$ are not, however, invariant under the action of the operators Q_j and P_j. Using (2.1) and the definition (2.10), one calculates immediately

$$[L_i, Q_j] = i\hbar\epsilon_{ijk}Q_k,$$

and

$$[L_i, P_j] = i\hbar\epsilon_{ijk}P_k. \tag{3.25}$$

Operators like **P** and **Q**, which fulfill these commutation relations with **L**, are called *vector operators*. From (3.25) it follows that

$$[\mathbf{L}^2, Q_j] = i\hbar\epsilon_{ijk}(L_iQ_k + Q_kL_i) \neq 0,$$

so that Q_j will change the eigenvalue $l(l + 1)$ of $\mathbf{L}^2$. This means that Q_j can transform from one $\mathcal{R}^l$ to another $\mathcal{R}^{l'}$ with $l' \neq l$.

The operators L_i in $\mathcal{R}^l$ with $l = \frac{1}{2}, \frac{3}{2}, \frac{5}{2}, \ldots$ or any direct sum of them,

$$\sum_{\substack{l = \text{half-odd} \\ \text{integer}}} \oplus \mathcal{R}^l$$

cannot be expressed as functions of the operators Q_i and P_i. For $l = \frac{1}{2}$ the operators $L_i^{(l=1/2)}$ are called spin operators. The 2×2 matrices σ_i with the matrix elements $2(f_{m'}^{l=1/2}, L_i f_m^{l=1/2})$ are called Pauli matrices (Problem 5).[7]

III.4 Rotation Spectra

As mentioned before, the algebra $\mathscr{E}(SU(2))$ does not contain operators that transform out of a given $\mathcal{R}^l$. The algebra of observables of the quantum-mechanical rotator is, however, larger than $\mathscr{E}(SU(2))$; additional elements can be formed, e.g., algebraic functions of J_i and P_i or of J_i and Q_i. The observables Q_i, for instance, have the property that they transform from a given $\mathcal{R}^l$ into the neighboring $\mathcal{R}^{l+1}$ and $\mathcal{R}^{l-1}$:

$$Q_i : \mathcal{R}^l \rightarrow \mathcal{R}^{l-1} \oplus \mathcal{R}^{l+1}, \tag{4.1}$$

but not into $\mathcal{R}^{l\pm n}$ for $n > 1$. We shall prove this in the section on parity.

The space of physical states $\mathcal{R}$ of the quantum-mechanical rotator is the direct sum of the spaces $\mathcal{R}^l$:

$$\mathcal{R} = \sum_{l=0}^{\infty} \oplus \mathcal{R}^l. \tag{4.2}$$

---

[7] The representations of the J_i given by (3.8'), (3.22), (3.23), and (3.1) for integer values of l are related to group representations of the rotation group SO(3) by taking $e^{i\omega^1 J_1}$, $e^{i\omega^2 J_2}$, $e^{i\omega^3 J_3}$ and products thereof, where $e^{i\omega^3 J_3}$ represents the rotation around the third axis by an angle ω^3. For half-odd integer values of l the J_i are not related to group representations of SO(3); they are, however, still ray representations of SO(3). As physical states are represented by rays and not by vectors—according to hypothesis II—not only proper group representations but also ray representations are relevant for quantum physics. Therefore representations with any integer or half-odd integer value of l are realized in quantum physics. All these representations are group representations of the "covering group" SU(2) which, therefore, has also been called (by E. P. Wigner) the quantum mechanical rotation group. The occurrence of half-odd integer angular momenta is, therefore, a natural consequence of the basic assumptions of quantum mechanics. In terms of the algebra of observables this means that all representations of the algebra of angular momentum (2.16) are realized in nature.

Before we justify this statement, we want to give a brief description of the properties of $\mathscr{R}$.

$\mathscr{R}$ is not an irreducible representation space of $\mathscr{E}(SO(3))$, the algebra of angular momenta. It is called a *reducible representation space*, and it reduces as given in (4.2) into a direct sum of irreducible representation spaces $\mathscr{R}^l$ ($l = 0, 1, 2, \ldots$). The operators H_+, H_-, H_3, now considered as operators in the big space $\mathscr{R}$, transform every element of a given space $\mathscr{R}^l$ into an element that is again in the same $\mathscr{R}^l$. Thus the subspaces $\mathscr{R}^l$ of $\mathscr{R}$ are left "invariant" by the H_+, H_-, H_3, and consequently by any $A \in \mathscr{E}(SU(2))$. The operator $\mathbf{H}^2$—which is a number in the space $\mathscr{R}^l$, namely $l(l+1)$—has in $\mathscr{R}$ a notrivial spectrum, namely

$$\text{spectrum } \mathbf{H}^2 = \{l(l+1), l = 0, 1, 2, 3, \ldots\} \tag{4.3}$$

The weight diagram for the representation in $\mathscr{R}$ is shown in Figure 4.1.

We shall now justify the statement (4.2). We assume that the J_i are the angular momenta $L_i = \epsilon_{ijk} Q_j P_k$; then, according to the result stated in Section III.3 and proved in Problem 1, only integer values of l are allowed, i.e., $\mathscr{R}$ contains only $\mathscr{R}^l$ with $l = 0, 1, 2, \ldots$. That $\mathscr{R}$ contains all the $\mathscr{R}^l$ ($l = 0, 1, 2, \ldots$) follows from the fact that there are observables for the rotator (e.g., the operators Q_i) that transform from a given $\mathscr{R}^l$ to the neighboring $\mathscr{R}^{l-1}$ and $\mathscr{R}^{l+1}$ according to (4.1). That each $\mathscr{R}^l$ appears only once follows from the fact that for the rotator no additional quantum number is necessary; if one $\mathscr{R}^{l_0}$ were to appear twice or more, then there would be two or more vectors $f_m^{l_0}(1)$, $f_m^{l_0}(2)$, $\ldots$ with the same quantum numbers l_0, m, and a new quantum number would be necessary to distinguish between these two or more vectors. But the rotator is just that model for which there is no other diagonal observable besides angular momentum ($\mathbf{L}^2$ and L_3). (In other words, a real physical system can be a rotator only to the extent that no other quantum numbers are necessary for the description of its properties; e.g., polyatomic (symmetric top) molecules cannot in general be described by the rotator model, and even for the diatomic molecule, the rotator model is only an approximate description that neglects all but the rotational properties of

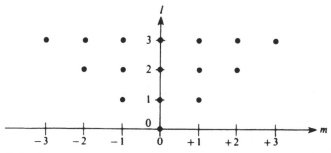

Figure 4.1 Collection of SO(3) weight diagrams belonging to an irreducible representation of SO(3,1) or E(3).

the dumbbell.) Thus, as always, the justification of (4.2) is that in nature there are physical systems whose physical states are (up to a certain limitation) described by $\mathscr{R}$.

[Mathematically this can be stated more briefly: "The spectrum-generating algebra of the rotator is $\mathscr{E}(E_3)$." $\mathscr{E}(E_3)$ is generated by L_i, Q_i that fulfill the commutation relations

$$[L_i, L_j] = i\hbar\epsilon_{ijk}L_k, \qquad [L_i, Q_j] = i\hbar\epsilon_{ijk}Q_k, \qquad [Q_i, Q_j] = 0,$$

and $\mathscr{R}$ is a particular irreducible representation space of $\mathscr{E}(E_3)$.]

Each dot on the weight diagram of $\mathscr{R}$ represents the pure state that is described by the one-dimensional subspace $\mathscr{R}_m^l$ (l, m fixed) spanned by f_m^l. The (normalized) statistical operator for the pure state $W = \Lambda_m^l$, where Λ_m^l is the projector on $\mathscr{R}_m^l$, represents a quantum-mechanical system for which the angular momentum has the definite value l and the 3-component of angular momentum, H_3, has a definite value m. As there is no distinguished direction in space and the coordinate system has been chosen arbitrarily, H_3 represents the angular momentum around an arbitrarily choosable direction; it is also called the helicity.

The values of the energy operator in $\mathscr{R}$, i.e., the energy spectrum of the rotator is obtained from (2.11) as

$$\text{spectrum } H = E_l = \frac{1}{2I}\hbar^2 l(l + 1). \tag{4.4}$$

Thus we see that the energy levels depend upon l, as represented in the diagram of Figure 4.2. If we compare this with Figure 1.7, we see that the

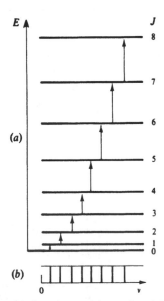

Figure 4.2 Energy levels and infrared transitions of a rigid rotator: (a) The energy-level diagram, (b) the resulting spectrum (schematic). [From Herzberg (1966), with permission.]

rotator has indeed the energy spectrum required to interpret the infrared spectrum of diatomic molecules.

In contrast to the oscillator, the energy eigenspaces (i.e., the spaces of vectors with the same energy eigenvalue) for the rotator are not one-dimensional, except for $l = 0$. Therefore the state of a rotator with a definite energy value E_{l_0} ($l_0 \neq 0$) need not be a pure state. If only energy measurements have been made with the result E_{l_0}, the statistical operator is given by

$$W = \Lambda^{l_0} \quad \text{(unnormalized)}, \tag{4.5}$$

or

$$W = (\text{Tr } \Lambda^{l_0})^{-1}\Lambda^{l_0} = (2l_0 + 1)^{-1}\Lambda^{l_0} \quad \text{(normalized)}, \tag{4.5'}$$

where Λ^{l_0} is the projector on the $(2l_0 + 1)$-dimensional space $\mathscr{R}^{l_0}$. By an energy measurement alone it is not possible to prepare a pure state of the rotator. Only under certain additional conditions—if a direction in space is distinguished (e.g., by an external magnetic field)—can one prepare a state with definite helicity, i.e., a pure state $\Lambda^{l_0}_{m_0}$. If only the energy of the rotator has been measured but not the helicity, then the states with different helicities are assumed to appear with equal weight, which is why one chooses

$$W = \Lambda^{l_0} = \Lambda^{l_0}_{-l_0} + \Lambda^{l_0}_{-l_0+1} + \cdots + \Lambda^{l_0}_m + \cdots + \Lambda^{l_0}_{l_0} \tag{4.6}$$

for the (unnormalized) statistical operator. Equation (4.6) is a special case of (II.4.32) and (II.4.49).

In order to calculate the frequencies that can be emitted and absorbed by a rotator, we have to know the selection rules. In our classical picture of the rotator we can consider it as a rotating dipole moment $\mathbf{D}$, with

$$\mathbf{D} = \text{const } \mathbf{Q}, \tag{4.7}$$

where $\mathbf{Q}$ is the vector between the centers of positive and negative charge. Classically the radiation is then a consequence of the rotation of this electrical dipole moment. Quantum-mechanically, the intensity of the absorbed or emitted radiation is proportional to the absolute value squared of the matrix element of the operator $\mathbf{D}$, i.e., proportional to

$$|\langle f^{l'}_{m'}|\mathbf{Q}|f^l_m\rangle|^2. \tag{4.8}$$

Thus dipole radiation will only be obtained in transitions from states $f^{l'}_{m'}$ to f^l_m between which the matrix element (4.8) is nonzero (quadrupole and higher-order radiation is negligibly small). It will be shown in the section on parity that [as was stated in (4.1)]

$$\langle f^{l'}_{m'}|Q_i|f^l_m\rangle = 0 \quad \text{unless} \quad l = l' \pm 1. \tag{4.9}$$

That is, the selection rule for dipole radiation of the rotator is

$$\Delta l = l - l' = \pm 1. \tag{4.10}$$

If we compare this result with the experimental situation for CO, depicted in Figure 1.7, we observe complete agreement. Figure 1.7 shows the transitions between states not only with different values of angular momentum l but also with different values of the vibrational quantum number n.

We also expect radiation from transitions between different rotator states of the diatomic molecule that belong to the same oscillator state $n = 0$. These transitions (absorption) are indicated by the ↑ in Figure 4.2. The frequency of this radiation, given in wave number units cm^{-1} obtained by dividing (1.1) by c, is

$$\nu_{l'l} = \frac{E_{l'} - E_l}{2\pi\hbar c}.$$

With (4.4) and (4.10) we calculate

$$\nu_{l+1,l} = \frac{\hbar^2}{2I} \frac{(l+1)(l+2) - (l+1)l}{2\pi\hbar c}$$

$$= \frac{h}{4\pi cI} 2(l+1) = B2(l+1), \qquad (4.11)$$

where

$$B = \frac{h}{8\pi^2 cI}. \qquad (4.11')$$

Thus the spectrum of a simple rigid rotator consists of a series of equidistant lines, as schematically drawn at the bottom of Figure 4.2.

We expect the frequency for pure rotational transitions to be much smaller than the vibrational frequency because the spacing between the rotational energy levels is much smaller than between the vibrational energy levels as seen in Figure 1.7 (note the scale factor of $\frac{1}{5}$ there).

The pure rotation spectrum lies in the far infrared. The absorption spectrum of HCl in the far infrared has been measured, and the experimental results are given in the second column of Table 4.1. From (4.11) we expect that the frequencies will have an equidistant spacing. Therefore in the third column of the table the differences between the successive frequencies are given. According to (4.11) this difference must be

$$\Delta\nu = \nu_{l+1,l} - \nu_{l,l-1} = 2B. \qquad (4.12)$$

The first eleven frequencies have roughly equal spacing and fitting them with (4.11) we obtain

$$B_{HCl} = \frac{h}{8\pi^2 cI_{HCl}} \approx 10.35 \, cm^{-1}. \qquad (4.13)$$

The fourth column of the table gives the values calculated from (4.11) with the value (4.13). We observe fairly good agreement between the calculated and observed values if we compare only the first eleven frequencies in column two with column four. The agreement becomes worse however at the higher values of l. The differences between successive frequencies at the higher values of l are smaller than at the lower values of l. The frequencies are not equally spaced but the trend suggests that (4.11) can be modified in order to obtain better agreement at the higher values of l as well. The last column of the table gives a fit to the expansion

$$\nu_{l+1,l} = 2b(l+1) - 4d(l+1)^3 \qquad (4.14)$$

Table 4.1 Absorption spectrum of HCl in the far infrared. [Data for $l = 1, 2, 3$ from McCubbin, *J. Chem. Phys.* **20**, 668 (1952); for $l = 4, \ldots, 11$ from R. L. Hansler and R. A. Oetjen, *J. Chem. Phys.* **21**, 1340 (1953); for $l = 17, \ldots, 33$ from J. Strong, *Phys. Rev.* **45**, 877 (1934).] The units of ν are cm^{-1}.

l	$\nu_{l-1, l\text{obs}}$	$\Delta\nu_{\text{obs}}$	$\nu_{\text{calc}} = 20.70\,l$	$\nu_{\text{calc}} = 20.88\,l$ $-0.001837\,l^3$
1	20.8		20.70	20.87
		20.8		
2	41.6		41.41	41.74
		20.9		
3	62.5		62.11	62.58
		20.53		
4	83.03		82.82	83.39
		21.10		
5	104.13		103.52	104.15
		20.60		
6	124.73		124.22	124.86
		20.64		
7	145.37		144.93	145.50
		20.52		
8	165.89		165.63	166.07
		20.34		
9	186.23		186.33	186.54
		20.37		
10	206.60		207.04	206.92
		22.26		
11	228.86		227.74	227.19
17	345.6		351.96	345.86
		19.0		
18	364.6		372.67	365.05
		19.6		
19	384.2		393.37	384.04
		19.2		
20	403.4		414.07	402.82
		18.7		
21	422.1		434.78	421.37
		18.0		
22	440.1		455.48	439.70
		18.2		
23	458.3		476.18	457.79
		18.1		
24	474.4		496.89	475.62
		19.9		
25	494.3		517.59	493.18
		16.9		
26	511.2		538.30	510.48
		16.5		
27	527.7		559.00	527.48
		15.2		
28	542.9		579.70	544.19
		18.2		
29	561.1		600.41	560.59
		15.3		
30	576.4		621.11	576.66
		13.2		
31	589.6		641.81	592.41
		19.8		
32	609.4		662.52	607.82
		14.0		
33	623.4		683.22	622.87

(b, d are constants). Comparing this last column with the observed values in the second column, we see that the agreement of (4.14) with the experimental data is far better than that of (4.11). The energy spectrum that corresponds to (4.14) is given by[8]

$$E_l = [bl(l + 1) - dl^2(l + 1)^2]2\pi\hbar c \qquad (4.15)$$

(b, d are constants). The energy levels (4.15) have been drawn in Figure 4.3 with an exaggerated value of d.

[8] The energy spectrum

$$E_l = 2\pi\hbar c[b' - d'l]l(l + 1),$$

which agrees well with nuclear rotation spectra, gives poorer agreement with molecular rotation spectra than (4.15).

Figure 4.3 Energy levels of the nonrigid rotator. For comparison, the energy levels of the corresponding rigid rotator are indicated by broken lines (for $J < 6$, they cannot be drawn separately). [From Herzberg (1966), with permission.]

The explanation for the better fit of (4.15) to the experimental values is that the diatomic molecule HCl is not exactly a rigid rotator. The bonds between atoms are not rigid, and the interatomic distance varies with the speed of rotation, giving rise to a centrifugal distortion. Equation (4.15) can be obtained if we return to the classical picture in which the molecule is considered as two hard spheres (atoms) joined, not as in Figure 2.1 by a rigid rod, but by a spring. If the molecule rotates about an axis perpendicular to this spring, then at equilibrium the centrifugal force $l^2/(\mu x^3)$ equals the centripetal force $k(x - x_e)$, where k is the spring constant and x_e the interatomic distance of the stationary molecule. Thus

$$k(x - x_e) = \frac{l^2}{\mu x^3}. \tag{4.16}$$

The energy of this system is [cf. (II.2.3) and (2.4)]

$$E \doteq \frac{l^2}{2\mu x^2} + \tfrac{1}{2}k(x - x_e)^2. \tag{4.17}$$

Making use of the expansion

$$x^2 = x_e^2\left(1 + 2\frac{x - x_e}{x_e} + \cdots\right) \tag{4.18}$$

and (4.16), one obtains for E

$$E = \frac{1}{2\mu x_e^2}l^2 - \frac{1}{2\mu^2 k x_e^6}(l^2)^2 + O((l^2)^3). \tag{4.19}$$

The first term is the energy of the rigid rotator, and the second term is the contribution due to the centrifugal forces. Going to the quantum system operator by replacing the number l^2 with the operator $\mathbf{L}^2$, one obtains the energy operator

$$H = \frac{1}{2\mu x_e^2}\mathbf{L}^2 - \frac{1}{2\mu^2 k x_e^6}(\mathbf{L}^2)^2, \qquad (4.20)$$

from which the spectrum (4.14) follows. The better fit of (4.14) to the experimental values confirms the above classical consideration. But we also observe that the empirical value of d is orders of magnitude smaller than that of b, obtained from the above fit:

$$b_{HCl} = 10.438 \text{ cm}^{-1}, \qquad d_{HCl} = 0.00046 \text{ cm}^{-1}. \qquad (4.21)$$

This shows that the rigid rotator is a remarkably good model of the rotating diatomic molecule.[9]

As we shall see below, the spacings between the levels of the rotating CO molecule are considerably smaller than for the HCl molecule. Therefore the pure rotation spectrum of CO lies at a considerably longer wavelength, where experimental investigation is very difficult.

We now want to obtain some quantitative features of the classical picture for the diatomic molecule. From the value (4.13) we calculate the moment of inertia of HCl:

$$I_{HCl} = 2.71 \times 10^{-40} \text{ g cm}^2.$$

With

$$m_{Cl} = \frac{35.45 \text{ g}}{N_A} = 6.0 \times 10^{-23} \text{ g}$$

and

$$m_H = \frac{1.008}{N_A} = 0.167 \times 10^{-23} \text{ g},$$

one calculates (2.7):

$$\mu_{HCl} = \frac{m_{Cl} m_H}{m_{Cl} + m_H} = 1.63 \times 10^{-24} \text{ g}$$

From (2.6) we may calculate the internuclear distance of the HCl molecules using the values I_{HCl} and μ_{HCl}:

$$x_{HCl} = 1.29 \times 10^{-8} \text{ cm}$$

---

[9] The data of Table 4.1 are accurate enough to test the correction term $d(l(l+1))^2$. We have tried to find other empirical formulas like $E_l = 2\pi\hbar c(b' - d'l)l(l+1)$ and found that it gives a much poorer agreement with the data of Table 4.1 than (4.15).

Thus we have calculated from the infrared absorption spectrum that the size of the molecule is of the order of 10^{-8} cm. This order of magnitude agrees very well with the values of atomic and molecular radii obtained from other classical considerations. We want to stress, however, that x is the value for the classical picture of the quantum-mechanical system and is not the expectation value of a quantum-mechanical observable.

III.5 Combination of Quantum Physical Systems— The Vibrating Rotator

We shall now combine the quantum-mechanical rotator model with the quantum-mechanical oscillator model to form the quantum-mechanical vibrating rotator (or rotating oscillator) model. This will provide a description of the experimental situation shown in Figures 1.5, 1.6, and 1.7. We first discuss the general case of a combination of two quantum physical systems. For this we require a new mathematical notion: the *direct product* or *tensor product* of linear spaces.

⟦Let R_1 and R_2 be two linear spaces, let $u_i \in R_1$ and $v_j \in R_2$, and let $a_{ij} \in \mathbb{C}$ (complex numbers). The set of all (arbitrarily large but finite) sums

$$f = \sum_{i,j} a_{ij} u_i v_j, \qquad (5.1)$$

where a_{ij} takes any value in $\mathbb{C}$, forms a linear space, which is called the *direct-product space* and is denoted $R_1 \otimes R_2$. $u_i v_j$ is the formal product, which is also written $u_i v_j = u_i \otimes v_j$. If $(u, u')_1$ denotes the scalar product in R_1, and if $(v, v')_2$ denotes the scalar product in R_2, then the scalar product in $R_1 \otimes R_2$ is defined by

$$\left(\sum_{ij} a_{ij} u_i v_j, \sum_{lm} b_{lm} u'_l v'_m \right) = \sum_{ijlm} \overline{a_{ij}} b_{lm} (u_i, u'_l)_1 (v_j, v'_m)_2. \qquad (5.2)$$

(*Remark*: If R_1 and R_2 are Hilbert spaces, then the "completion" of $R_1 \otimes R_2$ with respect to this scalar product is the direct Hilbert-space product.)

If ϕ_ν is a basis in R_1 and ψ_μ is a basis in R_2, then

$$f_{\nu\mu} = \phi_\nu \otimes \psi_\mu = \phi_\nu \psi_\mu \qquad (5.3)$$

is a basis in $R_1 \otimes R_2$.

If A_1 is a linear operator in R_1 and A_2 is a linear operator in R_2, then the operators

$$C = A_1 \otimes I, \qquad B = I \otimes A_2 \qquad (5.4)$$

in $R_1 \otimes R_2$ are defined in the following way:

$$Cf = \sum_{ij} a_{ij}(A_1 u_i) \otimes v_j, \qquad (5.4')$$

$$Bf = \sum_{ij} a_{ij} u_i \otimes (A_2 v_j).$$

The linear operator $A = A_1 \otimes A_2$ in $R_1 \otimes R_2$ is defined by

$$A(u_j \otimes v_k) = (A_1 \otimes A_2)(u_j \otimes v_k) \underset{\text{def}}{=} (A_1 u_j) \otimes (A_2 v_k). \quad (5.5)$$

It is easily seen that if A_1, B_1 are linear operators in R_1 and A_2, B_2 are linear operators in R_2, then

$$A_1 B_1 \otimes A_2 B_2 = (A_1 \otimes A_2)(B_1 \otimes B_2). \qquad (5.6)$$

Every operator A in the direct-product space is a linear combination of direct products of operators, i.e.,

$$A = \sum_i A_1^{(i)} \otimes A_2^{(i)}, \qquad (5.7)$$

with $A_1^{(i)}$ linear operators in R_1 and $A_2^{(i)}$ linear operators in R_2.]

With the notion of the direct product of spaces we can formulate the basic assumption about the physical combination of two quantum-mechanical systems:

IVa. Let one physical system be described by an algebra of operators, $\mathscr{A}_1$, in the space R_1, and the other physical system by an algebra $\mathscr{A}_2$ in R_2. The direct product space $R_1 \otimes R_2$ is then the space of physical states of the physical combination of these two systems, and its observables are operators in the direct-product space [given in the form (5.7)]. The particular observables of the first system alone are given by $A_1 \otimes I$, and the observables of the second system alone by $I \otimes A_2$ (I = identity operator).

We reemphasize that IVa is a basic assumption of quantum mechanics and can only be justified by the fact that such physical systems exist.

We shall now apply this basic assumption IVa to the diatomic molecule that vibrates and rotates.

We called the space of physical states of the oscillator $\mathscr{H}$. In $\mathscr{H}$ we introduced a basis of eigenvectors of the operator N or H_{osc}:

$$\text{basis of } \mathscr{H}: \quad \phi_n = |n\rangle \qquad (n = 0, 1, 2, \dots). \qquad (5.8)$$

The action of all observables (all elements of the algebra of observables of the quantum-mechanical oscillator) of the harmonic oscillator on the basis vectors $|n\rangle$ is known from Chapter II.

We called the space of physical states of the rotator $\mathscr{R}$. In $\mathscr{R}$ we introduced a basis of eigenvectors of the operators L_3 and $\mathbf{L}^2$ or H_{rot}:

$$\text{basis of } \mathscr{R}: \quad f_m^l = |lm\rangle$$

$$(l = 0, 1, 2, \dots; m = \text{integer with } -l \le m \le +l). \qquad (5.9)$$

The space of physical states of the vibrating rotator is, according to IVa, the direct-product space

$$\mathfrak{S} = \mathscr{H} \otimes \mathscr{R}, \tag{5.10}$$

and the observables are the operators $\sum_i A_{osc}^{(i)} \otimes A_{rot}^{(i)}$, where $A_{osc}^{(i)}$ is any observable of the oscillator and $A_{rot}^{(i)}$ is any observable of the rotator. The basis system in $\mathfrak{S}$ is obtained as the direct product of the basis systems ϕ_n in $\mathscr{H}$ and f_m^l in $\mathscr{R}$, and is denoted by $|nlm\rangle$:

$$|n\,l\,m\rangle = |n\rangle \otimes |l\,m\rangle = \phi_n \otimes f_m^l. \tag{5.11}$$

We have already mentioned that the rotating diatomic molecule is not a rigid rotator and the vibrating diatomic molecule is not a harmonic oscillator. Furthermore, the rotations and vibrations are not independent motions of the molecule: In our classical picture the diatomic molecule is a system of two mass points which are connected by a massless spring. Consequently there are interactions between the vibration and rotation caused, e.g., by the fact that during the vibration the internuclear distance $x = (x_i x_i)^{1/2}$ changes, and consequently the moment of inertia $I = \mu x^2$ changes.

For the moment we want to neglect all these finer details and consider the idealized system that is simultaneously a rigid rotator and a harmonic oscillator—keeping in mind, however, that this is an idealized system, which can at best be only approximately correct.

The energy operator of this idealized physical combination of the harmonic oscillator and rigid rotator is given by

$$H = H_{osc} \otimes I + I \otimes H_{rot}, \tag{5.12}$$

where

$$H_{osc} = \hbar\omega(N + \tfrac{1}{2}I) = \frac{1}{2\mu}P^2 + \frac{\mu\omega^2}{2}Q^2$$

and

$$H_{rot} = \frac{1}{2I}\mathbf{L}^2.$$

If the interactions between the two systems are neglected, all observables are given by

$$A = A_{osc} \otimes I + I \otimes A_{rot}. \tag{5.13}$$

From (5.12) we obtain the energy spectrum of the idealized vibrating rotator:

$$\text{spectrum } H = E_{nl} = \hbar\omega(n + \tfrac{1}{2}) + \frac{1}{2I}\hbar^2 l(l + 1). \tag{5.14}$$

The experiments show that the system constants ω for the oscillator and I for the rotator fulfill $\hbar\omega > \hbar^2/(2I)$ (the pure vibrational transitions are in the near infrared and the pure rotation transitions are in the far infrared).

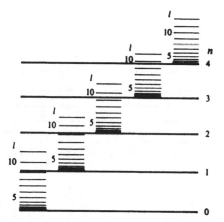

Figure 5.1 Energy levels of the vibrating rotator. For each of the first five vibrational levels, a number of rotational levels are drawn (short horizontal lines.) [From Herzberg (1966), with permission.]

The energy-level diagram that we obtain under these conditions from (5.14) is shown in Figure 5.1. To obtain the transition frequencies we use the selection rules

$$\Delta n = \pm 1 \quad \text{and} \quad \Delta l = \pm 1 \tag{5.15}$$

given by (1.10) and (4.10). If we consider a particular vibrational transition from n to $n + 1$ (absorption) or $n + 1$ to n (emission), we obtain from (5.14) the frequencies (in cm^{-1})

$$\nu_R = \frac{E_{n+1,l+1} - E_{n,l}}{2\pi\hbar c}$$

$$= \nu_0 + 2B + 2Bl \qquad (\Delta l = +1), \tag{5.16R}$$

$$\nu_P = \frac{E_{n+1,l-1} - E_{n,l}}{2\pi\hbar c}$$

$$= \nu_0 - 2Bl \qquad (\Delta l = -1) \tag{5.16P}$$

for absorption, where B is given by (4.11'):

$$B = \frac{h}{8\pi^2 cI} = \frac{\hbar}{4\pi cI} \tag{4.11'}$$

and

$$\nu_0 = \frac{\omega}{2\pi c}.$$

Thus we have two series of equidistant lines, which are called the R and P branches, with a gap at ν_0 (as $\Delta l = 0$ is excluded by the selection rule). The corresponding transitions are indicated in the energy-level diagram, Figure 5.2. The frequency spectrum calculated from (5.16) is depicted in strip (b).

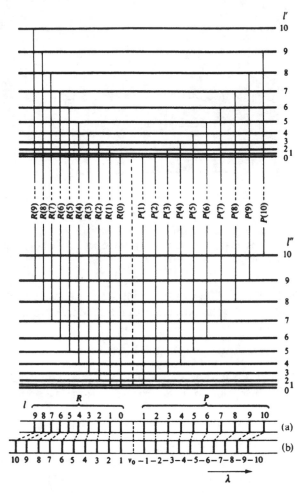

Figure 5.2 Energy-level diagram explaining the fine structure of a rotation-vibration band. In general, the separation of the two vibrational levels is considerably larger compared to the spacing of the rotational levels than shown in the figure (indicated by the broken parts of the vertical lines representing the transitions). The schematic spectrograms (*a*) and (*b*) give the resulting spectrum with and without allowance for the interaction between rotation and vibration. In these spectrograms, unlike most of the others, short wavelengths are at the left. [From Herzberg (1966), with permission.]

The observed spectrum from Figure 1.7 is depicted in strip (a). Thus the prediction of the vibrating-rotator model without interaction between vibration and rotation agrees rather well, but is not quite accurate. The observed lines in the *R* branch draw closer together, and those in the *P* branch draw farther apart, than the predicted equidistant lines. This is due to the interaction of rotation and vibration.

If one assumes that the moment of inertia is different in different vibrational states, then one obtains

$$E_{nl} = \hbar\omega(n + \tfrac{1}{2}) + \frac{1}{2I_n}\hbar^2 l(l + 1) \qquad (5.17)$$

instead of (5.14), and for the wave numbers of the resulting lines

$$v = \frac{E_{n'l'} - E_{n''l''}}{2\pi\hbar c} = v_0(n' - n'') + B_{n'}l'(l' + 1) - B_{n''}l''(l'' + 1), \quad (5.18)$$

where

$$B_n = \frac{h}{8\pi^2 c I_n}.$$

From this one obtains the absorption frequencies (in wave number units cm^{-1}) for transitions $n''l'' \to n'l'$ between neighboring vibrational levels:

$$v_R = v_0(n' - n'') + 2B_{n'} + (3B_{n'} - B_{n''})l + (B_{n'} - B_{n''})l^2$$

$$(l' = l + 1, l'' = l, \Delta l = +1), \quad (5.19R)$$

$$v_P = v_0(n' - n'') - (B_{n'} + B_{n''})l + (B_{n'} - B_{n''})l^2$$

$$(l' = l - 1, l'' = l, \Delta l = -1). \quad (5.19P)$$

Equations (5.19) given excellent agreement with the empirical fine structure of the infrared bands.

For the HCl molecule the values of B_n have been obtained for the various bands $n' \leftrightarrow n''$:

$$0 \leftrightarrow 1 \qquad 0 \leftrightarrow 2 \qquad 0 \leftrightarrow 3 \qquad 0 \leftrightarrow 4 \qquad 0 \leftrightarrow 5$$

(transitions with $\Delta n > 1$ occur as a consequence of the small anharmonicity; cf. Figure 1.3). The results are summarized in Table 5.1. The difference ΔB_n between successive values is very nearly a constant, so that B_n can be fitted by the formula

$$B_n = B_e - \alpha_e(n + \tfrac{1}{2}), \quad (5.20)$$

where α_e is a constant small compared to $B_e = 10.5909\ cm^{-1}$, the equilibrium value of B_n. The value of B_n given in the table for the rotation vibration spectrum agree within the accuracy of the measurements with the value $b_{HCl} = 10.438\ cm^{-1}$ obtained from the pure rotation spectrum of HCl, Equation (4.21).

Table 5.1 Rotational constants of HCl in the different vibrational levels of the electronic ground state. [From Herzberg (1966), p. 800, with permission.]

n	B_n (cm^{-1})	ΔB_n (cm^{-1})
0	10.4400	
		0.3034
1	10.1366	
		0.3037
2	9.8329	
		0.2986
3	9.5343	
		0.302
4	9.232	
		0.299
5	8.933	

Experimental values for the vibration-rotation spectra of the CO molecule are not as numerous or as accurate. From the spectrum depicted in Figure 1.5 one obtains

$$B^{CO} = 1.96 \text{ cm}^{-1}. \tag{5.21}$$

This value, and hence the fine structure in the energy spectrum, is considerably smaller than that for HCl.

The diatomic molecule with the largest rotational constant B_e, and thus the largest energy difference between rotational levels, is the H_2 molecule, for which $B_e^{H_2} = 60.80 \text{ cm}^{-1}$.

A qualitative theoretical explanation of (5.17) with (5.20) follows from the classical picture of the diatomic molecule as two rigid spheres connected by a spring. When this system is in a state of higher vibrational energy, it has a larger amplitude and consequently a larger moment of inertia. Consequently I_n^{-1} decreases with increasing n, as expressed by (5.20).

For the quantum-mechanical observables, the empirical formula (5.20) means that in the presence of an interaction between vibrational and rotational degrees of freedom, the form (5.13) is not sufficient. For the energy operator one has in addition to (5.12) an interaction term, for which one may try as a first guess

$$H_{\text{int}} = g H_{\text{osc}} \otimes H_{\text{rot}}, \tag{5.22}$$

where g is a coupling constant of dimension $(\text{eV})^{-1}$. Thus the energy operator for the vibrating, rotating, interacting diatomic molecule in this approximation will be given by

$$H = H_{\text{osc}} \otimes I + I \otimes H_{\text{rot}} + g H_{\text{osc}} \otimes H_{\text{rot}}, \tag{5.23}$$

where

$$H_{\text{osc}} = \hbar\omega(N + \tfrac{1}{2}I) \tag{5.24}$$

and

$$H_{\text{rot}} = \frac{1}{2I_e} \mathbf{L}^2. \tag{5.25}$$

I_e is the moment of inertia that corresponds to the equilibrium separation x_e: $I_e = \mu x_e^2$.

The energy values of the diatomic molecule with vibration-rotation interaction are the expectation values of H in the physical states. It is, of course, not obvious that the $|n\,l\,m\rangle = \phi_n \otimes f_m^l$ of (5.11), where ϕ_n are eigenvectors of N and f_m^l are eigenvectors of $\mathbf{L}^2$ and L_3, represent the pure states of this physical system. However, as they happen to be also eigenstates of the energy operator H of (5.23), they are the obvious choice for states in an energy measurement. Thus the energy values are the eigenvalues of H in the basis $|n\,l\,m\rangle$ given by (5.11) which are calculated to be:

$$E_{nl} = \hbar\omega(n + \tfrac{1}{2}) + \frac{1}{2I_e}\hbar^2 l(l+1) + g\hbar\omega \frac{\hbar^2}{2I_e}(n+\tfrac{1}{2})l(l+1). \tag{5.26}$$

The wave number of the radiation quantum corresponding to the energy value, i.e.,

$$v_{nl} = \frac{E_{nl}}{hc} = \frac{E_{nl}}{2\pi\hbar c} \qquad (5.27)$$

is called the *term value*[10] [cf. Equation (1.12)]. The term values of the vibrating rotator are therefore, by (5.26),

$$v_{nl} = v_0(n + \tfrac{1}{2}) + (B_e - \alpha_e(n + \tfrac{1}{2}))l(l + 1), \qquad (5.28)$$

where we have used the standard notation of molecular spectroscopy:

$$B_e = \frac{\hbar}{4\pi c I_e} = \frac{\hbar}{4\pi c \mu x_e^2}, \qquad (5.29)$$

$$\alpha_e = -\frac{g v_0 \hbar^2}{2 I_e}, \qquad (5.30)$$

$$B_n = B_e - \alpha_e(n + \tfrac{1}{2}). \qquad (5.31)$$

According to the above qualitative considerations I_n^{-1} and therefore B_n decrease with n, thus α_e should be larger than zero, which is experimentally always fulfilled. Equation (5.28) with (5.31) gives for the wave numbers of the transitions in the R branch

$$v_R = v_{n'l+1} - v_{n''l}$$

$$= v_0(n' - n'') + 2B_{n'} + (3B_{n'} - B_{n''})l + (B_{n'} - B_{n''})l^2, \quad (5.32R)$$

and for the wave numbers in the P branch

$$v_P = v_{n'l-1} - v_{n''l}$$

$$= v_0(n' - n'') - (B_{n'} + B_{n''})l + (B_{n'} - B_{n''})l^2, \qquad (5.32P)$$

i.e., the empirically well-established formulas (5.19). This also shows that our first guess (5.22) was a good guess.

Two previously mentioned effects have not been taken into account in (5.23) with (5.24) and (5.25). These are the anharmonicity of the oscillator and the influence of centrifugal forces. Thus (5.28) is not the end of the story of the vibrating and rotating diatomic molecule. If these effects are also taken into account, then up to a certain degree of accuracy one obtains for the term values of a vibrating rotator

$$v_{nl} = \omega_e(n + \tfrac{1}{2}) - \omega_e \xi_e(n + \tfrac{1}{2})^2 + B_n l(l + 1) - D_n l^2(l + 1)^2, \quad (5.33)$$

where ξ_e is a small parameter expressing the anharmonicity,[11] and

$$B_n = B_e - \alpha_e(n + \tfrac{1}{2}), \qquad (5.34)$$

$$D_n = D_e + \beta_e(n + \tfrac{1}{2}). \qquad (5.35)$$

---

[10] Note that we use the same symbol v for the frequency (in $\sec^{-1}$) and the wave number (in cm^{-1}); cf. statement following Equation (1.12c).

[11] $\omega_e \xi_e$ is given by the anharmonicity of the oscillator (terms proportional to Q^3, Q^4, etc.) and can be calculated by perturbation theory. See Problem VIII.1.

ω_e is the standard notation for

$$\omega_e = \frac{\omega}{2\pi c} = \frac{1}{2\pi c}\sqrt{\frac{k}{\mu}}. \qquad (5.36)$$

B_e is given by (5.29):

$$B_e = \frac{\hbar}{4\pi c I_e} = \frac{\hbar}{4\pi c \mu x_e^2}. \qquad (5.29)$$

According to the semiclassical consideration leading to (4.20), D_e may be expressed in terms of the reduced mass μ, equilibrium separation x_e and spring constant k as

$$D_e = \frac{\hbar^3}{4\pi c k \mu^2 x_e^6}. \qquad (5.37)$$

From (5.36), (5.29), and (5.37) it follows that the three system parameters D_e, B_e, ω_e are not independent but are related by

$$D_e = \frac{4B_e^3}{\omega_e^2}. \qquad (5.38)$$

The parameters ξ_e, α_e, β_e, expressing the degree of anharmonicity, are known empirically to be small:

$$\xi_e \ll 1, \qquad \frac{\alpha_e}{B_e} \ll 1, \qquad \frac{\beta_e}{D_e} \ll 1,$$

which must be the case, as they represent the effect of corrections to models which are very well realized by microphysical systems in nature. The system parameters ω_e, ξ_e, B_e, D_e, α_e, and β_e have been experimentally determined for many diatomic molecules and are collected in tables [cf. Herzberg (1966)]. Equation (5.33) gives a very good description of the vibration-rotation spectra of diatomic molecules, and only in exceptional cases are higher correction terms needed.

Diatomic molecules are vibrating rotators only as long as the internal energy is sufficiently low—roughly, in the region of energy of infrared radiation. For higher energies (1–20 eV) which correspond to the visible and ultraviolet regions, the molecules are no longer just vibrating rotators, since new degrees of freedom become accessible to electronic transitions. In each electronic state the molecule is, however, still a vibrator, in the same way as in each vibrational state the molecule is a rotator. This leads to energy spectra as depicted schematically in Figure 5.3 for two electronic states. We shall not describe the electronic structure of molecules here, they are of the same nature as the electronic structure of atoms which will be discussed in later chapters of the book.

Figure 5.3 Vibrational and rotational levels of two electronic states A and B of a molecule (schematic). Only the first few rotational and vibrational levels are drawn in each case. [From Herzberg (1966), with permission.]

Problems

1. Let P_i and Q_j $(i, j = 1, 2, 3)$ satisfy the canonical communication relation $[P_i, Q_j] = (\hbar/i)\delta_{ij}I$. Define the orbital angular momentum $L_i \equiv \epsilon_{ijk}Q_jP_k$ $(\mathbf{L} = \mathbf{Q} \times \mathbf{P})$. The component operators are Hermitian [Equation (2.11a)] and satisfy the commutation relation (2.14). Define H_3, H_+, and H_- as follows:

$$H_3 \equiv \hbar^{-1}L_3 \quad \text{and} \quad H_\pm \equiv \hbar^{-1}(L_1 \pm iL_2).$$

(a) Show that

$$[H_3, H_\pm] = \pm H_\pm$$

and that

$$[H_+, H_-] = 2H_3.$$

Show that

$$\mathbf{L}^2 \equiv L_iL_i = \hbar^2(H_+H_- + H_3^2 - H_3)$$
$$= \hbar^2(H_-H_+ + H_3^2 + H_3).$$

(b) Show that $\mathbf{H}^2 \equiv \hbar^{-2}\mathbf{L}^2$ is an invariant operator of $\mathscr{E}(SU(2))$; i.e., show that

$$[\mathbf{H}^2, H_3] = 0, \qquad [\mathbf{H}^2, H_+] = 0$$

and

$$[\mathbf{H}^2, H_-] = 0,$$

and consequently that

$$[\mathbf{H}^2, A] = 0,$$

where A is any element of the algebra generated by the L_i,

$$A = aI + a^i L_i + a^{ij} L_i L_j + \cdots$$

$(a, a^i, a^{ij}, \ldots \in \mathbb{C})$.

(c) Prove that the spectrum of $\mathbf{L}^2$ is $\hbar^2 l(l + 1)$ $(l = 0, 1, 2, 3, \ldots)$. *Hint*: Express the operator L_3 in terms of the annihilation operators a_i and creation operators $a_i^\dagger$:

$$a_i = \frac{1}{\sqrt{2\hbar}}(Q_i + iP_i),$$

$$a_i^\dagger = \frac{1}{\sqrt{2\hbar}}(Q_i - iP_i)$$

(units chosen so that $\mu\omega = 1$) of the three-dimensional harmonic oscillator, and show that as a consequence of the spectrum of the harmonic oscillator L_3 can have only integral eigenvalues.

2. Calculate the internuclear distance for the CO molecule using the absorption spectrum given in Figure 1.7.

3. The energy-loss spectrum of vibrating H_2 molecules (Figure PS.1) shows two bumps at 0 and at 0.52 eV with respective intensities 3.5 and $7.8 \times \frac{1}{30} = 0.26$ (arbitrary units), respectively. What is the statistical operator W for the ensemble of H_2 molecules

Figure PS.1 Energy-loss spectrum of H_2 [From G. J. Schultz, *Phys. Rev.* **135**, A988 (1964) with permission.]

in this energy-loss experiment? Can dipole transitions occur in this H_2 gas? What would be the frequency for these dipole transitions?

4. In an infrared absorption experiment with HCl molecules in the ground state it is observed that the following frequencies $v(cm^{-1})$ are absorbed:

20.68	82.72	144.76	206.80
41.36	103.40	165.44	227.48
62.04	124.08	186.12	

Assume that in this absorption experiment only transitions between neighboring energy levels take place (dipole transitions). What are the energy levels of this (rotating) HCl molecule if the zero of the energy scale is fixed by $E_0 = 0$?

5. Let $\{f_{-1/2}^{j=1/2}, f_{1/2}^{j=1/2}\}$ be the orthonormal basis for the space $\mathscr{R}^{1/2}$, defined in Section III.3.

(a) Show that the matrices σ_i with matrix elements $\sigma_i = (2/\hbar)(f_{m'}^{j=1/2}, J_i f_m^{j=1/2})$ are given by

$$\sigma_1 = \begin{bmatrix} 0 & 1 \\ 1 & 0 \end{bmatrix}, \qquad \sigma_2 = \begin{bmatrix} 0 & -i \\ i & 0 \end{bmatrix}, \qquad \sigma_3 = \begin{bmatrix} 1 & 0 \\ 0 & -1 \end{bmatrix}$$

with respect to the basis $\{f_{-1/2}^{j=1/2}, f_{1/2}^{j=1/2}\}$. The σ_i are the Pauli spin matrices.

(b) Show that the Pauli spin matrices together with the unit matrix are a basis for the vector space of operators on $\mathscr{R}^{1/2}$.

6. Let $\mathscr{R}_1$ and $\mathscr{R}_2$ be two finite-dimensional linear scalar-product spaces and $\mathscr{R} = \mathscr{R}_1 \oplus \mathscr{R}_2$ be their direct sum. Let A_1 be a linear operator in $\mathscr{R}_1$ and A_2 be a linear operator in $\mathscr{R}_2$. The direct sum $A_1 \oplus A_2$ is defined as the map

$$(A_1 \oplus A_2)\psi = (A_1 \oplus A_2)(\psi_1 \oplus \psi_2) = (A_1\psi_1 \oplus A_2\psi_2)$$

for $\psi_i \in \mathscr{R}_i$ and $\psi = \psi_1 \oplus \psi_2 \in \mathscr{R}$. Let $\mathscr{H} = \mathscr{R}_1 \otimes \mathscr{R}_2$ be the direct product of $\mathscr{R}_1$ and $\mathscr{R}_2$. The direct product $A_1 \otimes A_2$ is defined as the map

$$(A_1 \otimes A_2)h = (A_1 \otimes A_2)(h_1 \otimes h_2) = A_1 h_1 \otimes A_2 h_2$$

for $h_i \in \mathscr{R}_i$ and $h = h_1 \otimes h_2 \in \mathscr{H}$.

(a) Show that in an appropriate basis the matrix a of the operator $A_1 \oplus A_2$ has the form

$$\begin{pmatrix} a_1 & 0 \\ 0 & a_2 \end{pmatrix},$$

where a_1 is the matrix of A_1 and a_2 is the matrix of A_2.

(b) Show that the matrix of the operator $A_1 \otimes A_2$ when written in an appropriate basis has the property

$$a_{jk, il} = a_{ji}^{(1)} a_{kl}^{(2)},$$

i.e., may be written as a block matrix

$$\begin{pmatrix} a_{11}^{(1)}a^{(2)} & \cdots & a_{1n}^{(1)}a^{(2)} \\ a_{21}^{(1)}a^{(2)} & \cdots & a_{2n}^{(1)}a^{(2)} \\ \vdots & & \vdots \\ a_{n1}^{(1)}a^{(2)} & \cdots & a_{nn}^{(1)}a^{(2)} \end{pmatrix}$$

or

$$\begin{pmatrix} a_{11}^{(2)}a^{(1)} & \cdots & a_{1m}^{(2)}a^{(1)} \\ \vdots & & \vdots \\ a_{m1}^{(2)}a^{(1)} & \cdots & a_{mm}^{(2)}a^{(1)} \end{pmatrix}$$

where $a^{(i)}$ is the matrix of A_i ($i = 1, 2$) and $a_{rs}^{(i)}$ are the matrix elements.

(c) Show that

(i) $\mathrm{Tr}(A_1 \oplus A_2) = \mathrm{Tr}\, A_1 + \mathrm{Tr}\, A_2$,

(ii) $\mathrm{Tr}(A_1 \otimes A_2) = \mathrm{Tr}\, A_1 \,\mathrm{Tr}\, A_2$.

7. Let A, B, and C be three linear operators which fulfill $[A, B] = C$. Under what condition can a vector f be simultaneously an eigenvector of A and an eigenvector of B?

CHAPTER IV

Complete Systems of Commuting Observables

In this chapter it is explained that the question of what constitutes a complete system of commuting observables is not a mathematical question but can only be answered by experiment.

For the one-dimensional oscillator we needed one number to label the basis vectors of the space of physical states. We used the energy (or principal) quantum number n, which is the eigenvalue of the observable N of Equations (II.3.2)–(II.3.7) and which is related to the energy by (II.3.30):

$$N|n\rangle = n|n\rangle \quad \text{or} \quad H_{\text{osc}}|n\rangle = E_n|n\rangle, \qquad n = 0, 1, 2, \ldots. \qquad (1.1)$$

Instead of the quantum number n one could have used the "continuous quantum number" x, i.e., the generalized eigenvalue of the position operator (II.7.1). Instead of the basis vectors $|n\rangle$ one could have used the generalized basis vectors $|x\rangle$ which are connected with the $|n\rangle$ by the transformation (II.7.16'). One could as well have used the generalized basis vectors $|p\rangle$ of (II.7.28) and (II.7.37), or even other basis systems. All these basis systems for the one-dimensional oscillator are labeled by one quantum number. One says that the spectra of N, H, Q, and P are simple or nondegenerate.[1]

For the rotator we needed two quantum numbers for which we had chosen, in Section III.3, the integer or half-integer numbers j and j_3, which are connected to the eigenvalues of the square of the angular momentum operator $\mathbf{J}^2$ and the arbitrarily chosen component J_3:

$$\mathbf{J}^2|j,j_3\rangle = j(j+1)\hbar^2|j,j_3\rangle; \qquad J_3|j,j_3\rangle = j_3\hbar|j,j_3\rangle. \qquad (1.2)$$

---

[1] These observables are cyclic operators for the space of physical states of the one-dimensional oscillator. Cf. footnote 4 of Section I.4.

The spectrum of $\mathbf{J}^2$ is not simple, it is degenerate, and in order to label the basis vectors one needs in addition to j the quantum number j_3 (the "magnetic" quantum number).[2]

Instead of the basis vectors (1.2) one could use another system of basis vectors for the space of physical states $\mathscr{R}$ of the rotator, e.g., the $|j j_2\}$ defined by

$$\mathbf{J}^2|j j_2\} = j(j + 1)\hbar^2|j j_2\}; \qquad J_2|j j_2\} = j_2 \hbar|j j_2\}. \qquad (1.3)$$

Equations (1.2) or (1.3), along with normalization, completely define the basis vectors up to a phase factor (i.e., the subspace spanned by the basis vectors). This means: If $\mathbf{J}^2\psi = a\psi$ and $J_3\psi = b\psi$ for $\psi \in \mathscr{R}$, then $a = j'(j' + 1)\hbar^2$ where j' is an integer or half-integer, $b = j_3'$ where j_3' is an integer or half-integer with $-j' \le j_3' \le j'$, and $\psi = \alpha|j', j_3'\rangle$ with $\alpha \in \mathbb{C}$. The two basis systems (1.2) and (1.3) are related by a basis transformation, (I.5.12), with transformation coefficients $\langle j j_3 | j' j_2'\}$ which are completely determined by the algebra of observables, except for phase factors (which are usually fixed by a convention).

Whereas for the oscillator one needs one observable and one quantum number related to it (e.g., H^{osc} and n), for the rotator one needs always two commuting observables and two quantum numbers related to them (e.g., $\mathbf{J}^2, J_3$ and j, j_3). To a certain extent one has a choice of which observables one wants to take, but the number of labels needed to specify the basis vectors of the space of physical states is dictated by the physical system.[3]

A system of commuting Hermitian operators that specifies the (generalized) basis system completely is called (following Dirac) a *complete system of commuting observables* (c.s.c.o.). The (generalized) eigenvalues of a c.s.c.o. are called quantum numbers.

For the one-dimensional harmonic oscillator, the c.s.c.o. consists of one operator, either the operator H, or the operator Q, or some other operator. For the rotator the c.s.c.o. consists of two operators; $\mathbf{J}^2$ and J_3 are a convenient choice, but others are also possible, even operators with continuous spectrum.[4] For different physical systems and, therewith, different algebras of observables one usually has different c.s.c.o.'s. The larger the algebra (i.e., the more complicated the physical system), the larger is the number of operators in a c.s.c.o.

From what is said above and the fact that in a direct-product space a basis system is obtained as the direct product of the basis vectors in the two factor

---

[2] None of the observables J_i or $\mathbf{J}^2$ are cyclic operators in the space of physical states of the rotator.

[3] It can happen that in the transition from a discrete quantum number to a continuous quantum number a new label—which usually takes a finite number of values—will appear so that the set of discrete quantum numbers for a given physical system may contain fewer quantum numbers than the set containing one or more continuous quantum numbers. Thus the number of observables for the complete systems of commuting observables of a given physical system is not necessarily a fixed number.

[4] Such a basis system is given in (3.76) of the Appendix to Section V.3.

spaces, it is clear that one c.s.c.o. of the combination of two systems is given by the combination of the two c.s.c.o.'s for each subsystem.

For example, a c.s.c.o. for the vibrating rotator is given by

$$N, \mathbf{J}^2, J_3. \tag{1.4}$$

But there are other possible c.s.c.o.'s for the vibrating rotator. Some of them may gain physical importance when the interaction between rotator and vibrator becomes strong and the assumption that the Hamiltonian of the combined system commutes with (1.4) becomes a poor approximation.

[The problem of determining for which operator *-algebras there exists a c.s.c.o. is unsolved; the requirement of the existence of a c.s.c.o. is certainly a restrictive condition. For certain classes of operator *-algebras $\mathscr{A}$ there do exist c.s.c.o.s (e.g., if $\mathscr{A}$ is the enveloping algebra of a nilpotent or semisimple group).]

The physicist's problem is not to find a c.s.c.o. for a given algebra, but is usually the reverse: From the experimental data he finds out how many quantum numbers are required, and what are the possible values of these quantum numbers. This gives him, according to the basic assumption IIIa, the complete commuting system $\{A_k\}$ and its spectrum. He then conjectures the total algebra $\mathscr{A}$ by adding to $\{A_k\}$ a minimum of other operators such that the matrix elements of elements of $\mathscr{A}$ calculated from the properties of this algebra agree with the experimental values of the corresponding observables.

Thus, the question of what is a c.s.c.o. for a particular physical system and the question of when a system of commuting operators is complete are physical questions. If an experiment gives more values than can be supplied by a given system of commuting operators, then this system is not complete; one has to introduce a new quantum number, i.e., enlarge the system of commuting operators. This usually requires a further enlargement of the algebra.

An example of this procedure is the transition from the rotator model of the diatomic molecule to the vibrating rotator model: As long as the energy is small ($\lesssim 10^{-2}$ eV) only rotational states can be excited. The diatomic molecule is described by the algebra of operators of the rotator and the quantum numbers are j and j_3; $\mathbf{J}^2, J_3$ is a c.s.c.o. If the energy is increased one observes new levels, each of which are the starting points of a whole rotational band, as depicted in Figure 5.1 of Chapter III. To describe these levels one introduces a new quantum number n and assigns the values 0, 1, 2, etc. of this new quantum number to them. With the new quantum number n one has introduced a new observable N so that the c.s.c.o. is now (1.4). Then one adds to this set of operators further operators $a^\dagger$ and a which together with N form the algebra of observables such that the other features of the physical system (e.g., the equal spacing between the levels) are correctly described.

Not only can an increase in energy, i.e., an extension of the domain of applicability of the theory, require an enlargement of the c.s.c.o. and of the algebra of observables, but so also can an increase in the precision of measurement. This can also be illustrated by the vibrating rotator model:

As long as one could not resolve the fine structure of the order of 10^{-3} eV, the quantum number n was sufficient to describe the infrared spectrum of the diatomic molecules and the oscillator model provided an approximate description which was accurate enough for coarse-resolution experiments (every theoretical description is an approximate description). However, when the results of high-resolution experiments had to be explained, new quantum numbers j and j_3 (to explain splitting in a magnetic field) had to be introduced. The c.s.c.o. $\mathbf{J}^2$, J_3 corresponding to these new quantum numbers is then augmented by the transition operators to form the whole new rotator algebra which is then combined with the oscillator algebra to give the algebra of observables of the vibrating rotator.

Because of the above described procedure for obtaining the algebra of observables $\mathscr{A}$ for a given physical system, we can assume that for the algebras in quantum mechanics one always has a c.s.c.o. Thus for the algebra $\mathscr{A}$ there exists a set of commuting operators

$$A_1, A_2, \ldots, A_N \in \mathscr{A} \qquad (1.5)$$

which have a set of (generalized) eigenvectors,[5]

$$A_k|\lambda_1, \ldots, \lambda_N\rangle = \lambda_k|\lambda_1, \ldots, \lambda_N\rangle, \qquad (1.6)$$

such that every physical state vector ϕ can be represented as

$$\phi = \int_\Lambda d\mu(\lambda)|\lambda_1 \cdots \lambda_N\rangle\langle\lambda_1 \cdots \lambda_N|\phi\rangle. \qquad (1.7)$$

Here the set $\Lambda = \{(\lambda_1\lambda_2 \cdots \lambda_N)\}$ is the spectrum of the c.s.c.o. (1.5) with $\Lambda_k = \{\lambda_k\}$ being the spectrum of the observable A_k. This means that Λ_k is a set of generalized eigenvalues if A_k has a continuous spectrum, it is the set of all discrete eigenvalues if A_k has a discrete spectrum, and it is the set of all discrete and continuous eigenvalues if A_k has a discrete and continuous spectrum. $\int_\Lambda d\mu(\lambda) = \int_{\Lambda_1}\int_{\Lambda_2}\cdots\int_{\Lambda_N} d\mu(\lambda_1, \ldots, \lambda_N)$ means summation over the discrete spectrum and integration over the continuous spectrum. Thus (1.7) is the generalization of the spectral theorem (I.4.4g) from the one variable a to the N variables $\lambda_1, \ldots, \lambda_N$.

It often happens that some of the observables $A_1 A_2, \ldots, A_M$ of the c.s.c.o. have only a (absolutely[6]) continuous spectrum $\Lambda_C = \{(\lambda_1, \ldots, \lambda_M)\}$ and the other observables $A_{M+1}, \ldots, A_N$ of the c.s.c.o. have only a discrete spectrum $\Lambda_D = \{(\lambda_{M+1}, \ldots, \lambda_N)\}$. Then (1.7) takes the form

$$\phi = \int_{\Lambda_C} \cdots \int d\mu(\lambda_1, \lambda_2, \ldots, \lambda_M) \sum_{\Lambda_D} \cdots \sum |\lambda_1, \ldots, \lambda_M, \lambda_{M+1}, \ldots, \lambda_N\rangle$$
$$\times \langle\lambda_1, \ldots, \lambda_N|\phi\rangle, \quad (1.8)$$

---

[5] Mathematically this assumption can be stated: the conditions of the nuclear spectral theorem are fulfilled.

[6] Cf. footnote 7, Section I.4.

where the summation runs over all discrete eigenvalues $\{\lambda_k\}$ $k = M + 1$, $M + 2, \ldots, N$ of all observables $A_{M+1}, \ldots, A_N$, and the integration extends over a set of generalized eigenvalues $\{\lambda_k\}$ $k = 1, 2, \ldots, M$ of all observables $A_1, \ldots, A_M$.

The statement (1.7) or (1.8) is the basis of the Dirac formulation of quantum mechanics.[7]

An immediate consequence of the statement (1.6), (1.7), or (1.8) is the following statement: Let $A_1, \ldots, A_N$ be a c.s.c.o. with (generalized) eigenvectors $|\lambda_1 \cdots \lambda_N\rangle = |\lambda\rangle$ such that

$$A_k|\lambda\rangle = \lambda_k|\lambda\rangle \qquad (k = 1, 2, \ldots, N).$$

Then from

$$A_i|a\} = a_i|a\} \qquad (i = 1, 2, \ldots, N) \tag{1.9}$$

it follows that the sets of eigenvalues are the same:

$$\{(a_1, a_2 \ldots a_N)\} = \{(\lambda_1, \lambda_2 \ldots \lambda_N)\} \tag{1.10a}$$

and that for the eigenvectors with the same eigenvalue $a = \lambda$ one has:

$$|a\} = \alpha|\lambda\rangle \tag{1.10b}$$

where $\alpha \in \mathbb{C}$.

For the case that the spectrum Λ is continuous, i.e., that one of the λ_i can take continuous values, the formulation of (1.10) has to be given a precise meaning. In the following we shall use (1.10) only for the case of a discrete spectrum.

---

[7] It is not a theorem, because the mathematical premises characterizing the algebra $\mathscr{A}$ cannot be stated. But it is based on the general nuclear spectral theorem, which is a generalization of the nuclear spectral theorem for a cyclic operator discussed in Section 1 4 to a strongly commuting family of operators $A_1, \ldots, A_N$. Note that, already for the enveloping algebra of a unitary irreducible representation of a noncompact group, the number N is not fixed.

CHAPTER V

Addition of Angular Momenta—
The Wigner–Eckart Theorem

In Section V.1 the elementary rotator is defined as the system described by an irreducible representation of the algebra of angular momentum. Section V.2 discusses then the direct product of two irreducible representations of angular momentum and its reduction with respect to the total angular momentum. The Clebsch–Gordan coefficients are introduced, their recursion relations are derived, and their most frequently used values are tabulated. In Section V.3 tensor operators are introduced and the Wigner–Eckart theorem for the rotation group is stated without derivation. In Section V.4 a new observable, parity, is introduced. Parity is then applied to discuss the spectrum of diatomic symmetric-top molecules. In an Appendix to Section V.3 the irreducible representations of the algebras of SO(3, 1), SO(4), and E(3) are derived.

V.1 Introduction—The Elementary Rotator

In Chapter III we have treated the quantum-mechanical rotator. Its space of physical states is given by (III.4.2):

$$\mathscr{R} = \sum_{l=0}^{\infty} \oplus \mathscr{R}^l, \tag{1.1}$$

where $\mathscr{R}^l$ is the irreducible-representation space of the algebra of angular momentum of the dumbbell, L_i, with

$$[L_i, L_j] = i\hbar\epsilon_{ijk}L_k. \tag{1.2}$$

$\mathscr{E}(SO(3)_{L_i})$ is a subalgebra of the algebra of observables for the rotator.[1]

Let us now consider the physical system whose space of physical states is $\mathscr{R}^l$, with an arbitrary fixed value $l = 0, \frac{1}{2}, 1, \frac{3}{2}, \ldots$, and whose algebra of observables is the angular momentum algebra. Every rotator can, under certain conditions, be considered as such a physical system—namely, if the energy E^{ext} that can be transferred to or from this rotator is small compared to the energy differences $E_l - E_{l-1}$ and $E_{l+1} - E_l$. In nature, this condition is rarely fulfilled, because a state of higher energy always tends to decay into a state of lower energy "spontaneously," and therefore the excited state has only a limited lifetime, unless E_l is the ground-state energy value. It is, nevertheless, customary to consider these unstable states often as independent physical systems, not only in nonrelativistic quantum mechanics, but also in the relativistic case of elementary-particle physics. We call a physical system whose space of states is $\mathscr{R}^l$ an *elementary rotator*. Elementary rotators are not just substructures of the diatomic molecule; they appear in other quantum-mechanical systems in all areas of physics.

V.2 Combination of Elementary Rotators

According to the basic assumption IV, the space of physical states of the combination of two elementary rotators with spaces $\mathscr{R}^{j_1}_{(1)}$ and $\mathscr{R}^{j_2}_{(2)}$ is the direct-product space $\mathscr{R}^{j_1}_{(1)} \otimes \mathscr{R}^{j_2}_{(2)} = \mathscr{R}$. Let $J_i^{(1)}$ and $J_i^{(2)}$ be the angular-momentum operators in $\mathscr{R}^{j_1}_{(1)}$ and $\mathscr{R}^{j_2}_{(2)}$, respectively, and let

$$|j_1 m_1\rangle \qquad (m_1 = -j_1, \ldots, +j_1)$$

and

$$|j_2 m_2\rangle \qquad (m_2 = -j_2, \ldots, +j_2)$$

be the basis systems in $\mathscr{R}^{j_1}_{(1)}$ and $\mathscr{R}^{j_2}_{(2)}$, respectively. Then a basis system in $\mathscr{R}^{j_1}_{(1)} \otimes \mathscr{R}^{j_2}_{(2)}$ is given by

$$|j_1 m_1\rangle \otimes |j_2 m_2\rangle = |j_1 m_1 j_2 m_2\rangle. \tag{2.1}$$

The $J_i^{(\alpha)}$ $(\alpha = 1, 2)$ fulfill the commutation relations

$$[J_i^{(\alpha)}, J_j^{(\alpha)}] = i\hbar\epsilon_{ijk} J_k^{(\alpha)}. \tag{2.2}$$

We define operators in $\mathscr{R} = \mathscr{R}^{j_1}_{(1)} \otimes \mathscr{R}^{j_2}_{(2)}$ by

$$J_i = J_i^{(1)} \otimes I + I \otimes J_i^{(2)}. \tag{2.3}$$

---

[1] Generators of $\mathscr{E}(SO(3))$ can represent various angular momenta. For example, they may be the orbital angular momenta of a mass point (III.2.10), or they may be the angular momenta of a dumbbell, or they may be the spin operator of an electron, etc. That is, the generators of $\mathscr{E}(SO(3))$ fulfilling (1.2) may have various different physical realizations. Though all $\mathscr{E}(SO(3))$ are mathematically equivalent, their generators may represent different physical observables. In order to specify which physical quantity the $\mathscr{E}(SO(3))$ represents, we write the observables as subscripts to the symbol for the algebra, e.g., as $\mathscr{E}(SO(3)_{L_i})$.

As a consequence of (2.2) and (2.3) it follows that the J_i fulfill the commutation relation

$$[J_i, J_j] = i\hbar\epsilon_{ijk}J_k, \qquad (2.4)$$

Thus the J_i of (2.3) fulfill the defining relations of the generators of the algebra of angular momentum, and are thus a representation of angular momentum in the space $\mathcal{R}^{j_1}_{(1)} \otimes \mathcal{R}^{j_2}_{(2)}$. The J_i are called the *total-angular-momentum operators* of the combined system.

The basis system (2.1) of $\mathcal{R}$ consists of eigenvectors of the following complete set of commuting operators (c.s.c.o.):

$$\mathbf{J}^{(1)2} \otimes I, \qquad J_3^{(1)} \otimes I, \qquad I \otimes \mathbf{J}^{(2)2}, \qquad I \otimes J_3^{(2)}, \qquad (2.5)$$

with the eigenvalues

$$
\begin{aligned}
\mathbf{J}^{(1)2} \otimes I | j_1 m_1 j_2 m_2 \rangle &= j_1(j_1 + 1)\hbar^2 | j_1 m_1 j_2 m_2 \rangle, \\
J_3^{(1)} \otimes I | j_1 m_1 j_2 m_2 \rangle &= m_1 \hbar | j_1 m_1 j_2 m_2 \rangle, \\
I \otimes \mathbf{J}^{(2)2} | j_1 m_1 j_2 m_2 \rangle &= j_2(j_2 + 1)\hbar^2 | j_1 m_1 j_2 m_2 \rangle, \\
I \otimes J_3^{(2)} | j_1 m_1 j_2 m_2 \rangle &= m_2 \hbar | j_1 m_1 j_2 m_2 \rangle.
\end{aligned}
\qquad (2.6)
$$

We now introduce the abbreviated notation for the operators in $\mathcal{R}$:

$$J_i^{(1)} = J_i^{(1)} \otimes I, \qquad J_i^{(2)} = I \otimes J_i^{(2)}. \qquad (2.7)$$

These $J_i^{(\alpha)}$ are operators in the space $\mathcal{R} = \mathcal{R}^{j_1}_{(1)} \otimes \mathcal{R}^{j_2}_{(2)}$ in contrast to the original $J_i^{(\alpha)}$, which were operators in $\mathcal{R}^{j_\alpha}_{(\alpha)}$ only. The $J_i^{(\alpha)}$ of (2.7) clearly fulfill the commutation relation (2.2), so we denote them by the same symbol. However, we should keep in mind that the old $J_i^{(\alpha)}$ are the restrictions of the new $J_i^{(\alpha)}$ to the space $\mathcal{R}^{j_\alpha}_{(\alpha)}$.

The basis (2.1) is in general not a physical basis for the combined system. A basis is physical if it consists of eigenstates in which the physical system can be prepared. If all physical states are eigenstates of the energy operator H of the physical system, then the physical basis must consist of eigenvectors of operators that commute with the energy operator. In general not all the $J_i^{(\alpha)}$ commute with the energy operator (in particular, not if there is an interaction term between the two angular momenta).

The operators that in general commute with the energy operator are the total angular momenta J_i (they commute with H whenever H is rotationally invariant). Therefore it is convenient to choose a basis in $\mathcal{R}$ that consists of eigenvectors of the system of commuting operators

$$\mathbf{J}^{(1)2} = \mathbf{J}^{(1)2} \otimes I, \qquad \mathbf{J}^{(2)2} = I \otimes \mathbf{J}^{(2)2}, \qquad \mathbf{J}^2 = \sum_i J_i^2, \qquad J_3. \quad (2.8)$$

We denote this basis by

$$| j_1 j_2 j m \rangle. \qquad (2.9)$$

It has the properties

$$\mathbf{J}^{(\alpha)2}|j_1 j_2 j\, m\rangle = j_\alpha(j_\alpha + 1)\hbar^2|j_1 j_2 j\, m\rangle,$$

$$J_3|j_1 j_2 j\, m\rangle = m\hbar|j_1 j_2 j\, m\rangle, \qquad (2.10)$$

$$\mathbf{J}^2|j_1 j_2 j\, m\rangle = j(j + 1)\hbar^2|j_1 j_2 j\, m\rangle.$$

Each vector, and consequently each basis vector of the basis (2.1), can be expanded according to (I.4.4d) with respect to the basis system (2.9):

$$|j_1 m_1 j_2 m_2\rangle = \sum_{j,\,m} |j_1 j_2 j\, m\rangle\langle j_1 j_2 j\, m|j_1 m_1 j_2 m_2\rangle; \qquad (2.11)$$

and each basis vector of the basis (2.9) can be expanded with respect to the basis system (2.1),

$$|j_1 j_2 j\, m\rangle = \sum_{m_1,\,m_2} |j_1 m_1 j_2 m_2\rangle\langle j_1 m_1 j_2 m_2|j_1 j_2 j\, m\rangle. \qquad (2.12)$$

The transition coefficients $\langle j_1 m_1 j_2 m_2|j_1 j_2 j\, m\rangle$ are called the *Clebsch–Gordan* or *Wigner coefficients*, and are denoted by

$$
\begin{aligned}
\langle j_1 m_1 j_2 m_2|j_1 j_2 j\, m\rangle &= \langle j_1 m_1 j_2 m_2|j\, m\rangle \\
&= C(j_1 j_2 j\, m_1 m_2 m) \\
&= \langle m_1 m_2|j\, m\rangle \\
&= C^{j_1 j_2 j}_{m_1 m_2 m}
\end{aligned} \qquad (2.13)
$$

for fixed values of j_1 and j_2. In (2.13) we have given various notations used in the literature.

Taking the scalar product of (2.11) with $|j_1 m_1' j_2 m_2'\rangle$ we obtain the orthogonality relation of the Clebsch–Gordon coefficients:

$$\delta_{m_1' m_1}\delta_{m_2' m_2} = \sum_{j,\,m} \langle j_1 m_1' j_2 m_2'|j\, m\rangle\langle j\, m|j_1 m_1 j_2 m_2\rangle, \qquad (2.14)$$

and taking the scalar product of (2.12) with $|j_1 j_2 j'\, m'\rangle$, we obtain

$$\delta_{jj'}\delta_{mm'} = \sum_{m_1,\,m_2} \langle j'\, m'|j_1 m_1 j_2 m_2\rangle\langle j_1 m_1 j_2 m_2|j\, m\rangle. \qquad (2.15)$$

We now want to find the spectrum of $\mathbf{J}^2$ and J_3 in $\mathscr{R} = \mathscr{R}_{(1)}^{j_1} \otimes \mathscr{R}_{(2)}^{j_2}$, i.e., the values that j and m in basis system (2.9) can take.

For the sake of clarity in the following discussion we use curly brackets for the direct product basis vectors,

$$|j_1 m_1 j_2 m_2\} = |j_1 m_1\rangle \otimes |j_2 m_2\rangle,$$

to distinguish them from the total angular momentum basis vectors $|j_1 j_2 j\, m\rangle$. Also, since j_1 and j_2 are fixed values, we just write

$$|m_1 m_2\} = |j_1 m_1 j_2 m_2\} \quad \text{and} \quad |j\, m\rangle = |j_1 j_2 j\, m\rangle.$$

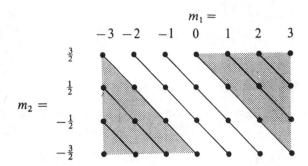

Figure 2.1 Dot pattern of direct product basis vectors $|m_1 m_2\} = |j_1 m_1\rangle \otimes |j_2 m_2\rangle$ in the space $\mathscr{R} = \mathscr{R}_{(1)}^{j_1=3} \otimes \mathscr{R}_{(2)}^{j_2=3/2}$. Shading indicates those basis vectors for which $|m| \geq |j_1 - j_2|$.

Since the vectors $|m_1 m_2\}$ are eigenvectors of $J_3 = J_3^{(1)} + J_3^{(2)}$ with eigenvalue $m_1 + m_2$, the spectrum of J_3 in $\mathscr{R}$ is given as all possible sums of $m_1 = j_1, j_1 - 1, \ldots, -j_1$ with $m_2 = j_2, j_2 - 1, \ldots, -j_2$:

$$\text{spectrum } J_3 = \{j_1 + j_2, j_1 + j_2 - 1, \ldots, -j_1 - j_2\}. \qquad (2.16)$$

Since the vectors $|j\,m\rangle$ are eigenvectors of J_3 with eigenvalue m, the quantum number m in the basis $|j\,m\rangle$ also ranges over the values (2.16). The subspace of $\mathscr{R}$ consisting of all the eigenvectors of J_3 with a particular value for m is denoted by $\mathscr{R}_m$. The dimension of $\mathscr{R}_m$ is obtained by counting the number of direct product basis vectors $|m_1 m_2\}$ for which $m_1 + m_2 = m$ and it is given by

$$\dim \mathscr{R}_m = \begin{cases} j_1 + j_2 + 1 - |m| & \text{for } |m| \geq |j_1 - j_2|, \\ j_1 + j_2 + 1 - |j_1 - j_2| & \text{for } |m| \leq |j_1 - j_2|. \end{cases} \qquad (2.17)$$

Figure 2.1 illustrates the case $j_1 = 3, j_2 = \frac{3}{2}$ and suggests how this counting is done for general j_1 and j_2. The dimension of a space $\mathscr{R}_m$ is obtained by counting the number of dots along the diagonal for which $m_1 + m_2 = m$.

The space $\mathscr{R}_{j_1 + j_2}$ is, according to (2.17), one-dimensional and it is spanned by the direct product basis vector $|j_1 j_2\}$. The vector $|j_1 j_2\}$ is also an eigenvector of $\mathbf{J}^2$ with eigenvalue $(j_1 + j_2)(j_1 + j_2 + 1)$:

PROOF.   We have in general

$$\begin{aligned} \mathbf{J}^2 &= (\mathbf{J}^{(1)} + \mathbf{J}^{(2)})^2 = \mathbf{J}^{(1)2} + 2\mathbf{J}^{(1)} \cdot \mathbf{J}^{(2)} + \mathbf{J}^{(2)2} \\ &= \mathbf{J}^{(1)2} + \mathbf{J}^{(2)2} + 2(J_1^{(1)} J_1^{(2)} + J_2^{(1)} J_2^{(2)} + J_3^{(1)} J_3^{(2)}) \\ &= \mathbf{J}^{(1)2} + \mathbf{J}^{(2)2} + 2J_3^{(1)} J_3^{(2)} + J_-^{(2)} J_+^{(1)} + J_-^{(1)} J_+^{(2)}. \end{aligned} \qquad (2.18)$$

Now,

$$(J_-^{(2)} J_+^{(1)} + J_-^{(1)} J_+^{(2)})|j_1 j_1\rangle \otimes |j_2 j_2\rangle = 0, \qquad (2.19)$$

since j_1 and j_2 are the highest possible values of m_1 and m_2, i.e., since $J^{(1)}_+|j_1j_1\rangle = 0$ and $J^{(2)}_+|j_2j_2\rangle = 0$. Thus

$$\mathbf{J}^2|j_1j_2\} = (\mathbf{J}^{(1)2} + \mathbf{J}^{(2)2} + 2J^{(1)}_3J^{(2)}_3)|j_1j_1\rangle \otimes |j_2j_2\rangle$$

$$= [j_1(j_1 + 1) + j_2(j_2 + 1) + 2j_1j_2]\hbar^2|j_1j_1\rangle \otimes |j_2j_2\rangle$$

$$= (j_1 + j_2)(j_1 + j_2 + 1)\hbar^2|j_1j_2\}. \qquad \square \qquad (2.20)$$

As $|j_1j_2\}$ is an eigenvector of the c.s.c.o. (2.8), it is related by a complex number to the total angular momentum basis vector $|j_1 + j_2\, j_1 + j_2\rangle$ which has the same eigenvalues and spans the same one-dimensional subspace of $\mathcal{R}$:

$$|j_1 + j_2\, j_1 + j_2\rangle = \alpha|j_1j_2\}, \qquad \alpha \in \mathbb{C}. \qquad (2.21)$$

Assuming both basis vectors are normalized, α is a phase factor and we may choose it to be unity so that

$$|j_1 + j_2\, j_1 + j_2\rangle = |j_1j_2\}, \qquad (2.22)$$

and

$$\{j_1j_2|j_1 + j_2\, j_1 + j_2\rangle = 1. \qquad (2.23)$$

Besides spanning $\mathcal{R}_{j_1 + j_2}$, $|j_1 + j_2\, j_1 + j_2\rangle$ belongs to the space $\mathcal{R}^{j_1 + j_2}$ of all eigenvectors of $\mathbf{J}^2$ with eigenvalue $j(j + 1) = (j_1 + j_2)(j_1 + j_2 + 1)$. Any such eigenspace $\mathcal{R}^j$ of the total angular momentum is invariant under the action of $J_i = J^{(1)}_i + J^{(2)}_i$ and has the properties of an elementary rotator space with respect to these total angular momentum operators. Thus, since $|j_1 + j_2\, j_1 + j_2\rangle$ has the highest value for m of any vector in $\mathcal{R}^{j_1 + j_2}$, a basis of total angular momentum vectors $|j\, m\rangle$ may be obtained in $\mathcal{R}^{j_1 + j_2}$ by applying

$$J_- = J^{(1)}_- + J^{(2)}_- \qquad (2.24)$$

successively to $|j_1 + j_2\, j_1 + j_2\rangle$ following the procedure of Section III.3. One obtains by this procedure the $2(j_1 + j_2) + 1$ basis vectors

$$|j_1 + j_2\, m\rangle, \qquad m = j_1 + j_2, j_1 + j_2 - 1, \dots, -(j_1 + j_2). \qquad (2.25)$$

Now consider the space $\mathcal{R}_{j_1 + j_2 - 1}$ of eigenvectors of J_3 with eigenvalue $m = j_1 + j_2 - 1$. According to (2.17), this space is two-dimensional (unless $j_1 = 0$ or $j_2 = 0$). One total angular momentum basis vector in $\mathcal{R}_{j_1 + j_2 - 1}$ is the vector $|j_1 + j_2\, j_1 + j_2 - 1\rangle$ which already appears in (2.25) as a basis vector in $\mathcal{R}^{j_1 + j_2}$. Due to the properties of the spaces $\mathcal{R}^j$, the other basis vector $|j\, m = j_1 + j_2 - 1\rangle$ in $\mathcal{R}_{j_1 + j_2 - 1}$ must belong to a space $\mathcal{R}^j$ for which $j = j_1 + j_2 - 1 + n$, $n =$ nonnegative integer. However, it certainly cannot belong to a space $\mathcal{R}^j$ for which $j > j_1 + j_2$ since this would imply that values of m larger than $j_1 + j_2$ occur in $\mathcal{R}$ and it also cannot belong to the space $\mathcal{R}^{j_1 + j_2}$ since it would then not be orthogonal to $|j_1 + j_2\, j_1 + j_2\rangle$. Thus, this new vector belongs to the space $\mathcal{R}^{j_1 + j_2 - 1}$. This determines the basis vector $|j_1 + j_2 - 1\, j_1 + j_2 - 1\rangle$ up to an arbitrary phase which is fixed by requiring

$$\{j_1j_2 - 1|j_1 + j_2 - 1\, j_1 + j_2 - 1\rangle \quad \text{be real and positive.} \qquad (2.26)$$

As in the case of $\mathcal{R}^{j_1 + j_2}$, since $|j_1 + j_2 - 1\, j_1 + j_2 - 1\rangle$ has the highest value for m of any vector in $\mathcal{R}^{j_1 + j_2 - 1}$, a basis in $\mathcal{R}^{j_1 + j_2 - 1}$ is generated by applying J_- successively to $|j_1 + j_2 - 1\, j_1 + j_2 - 1\rangle$ to give the $2(j_1 + j_2 - 1) + 1$ basis vectors

$$|j_1 + j_2 - 1\, m\rangle, \qquad m = j_1 + j_2 - 1, j_1 + j_2 - 2, \dots, -(j_1 + j_2 - 1).$$

$$(2.27)$$

According to (2.17), the next J_3 eigenspace, $\mathscr{R}_{j_1+j_2-2}$, is three-dimensional (unless $j_1 = 0, \frac{1}{2}$ or $j_2 = 0, \frac{1}{2}$). We have already found in $\mathscr{R}_{j_1+j_2-2}$ the orthogonal vectors $|j_1 + j_2 j_1 + j_2 - 2\rangle$ from (2.25) and $|j_1 + j_2 - 1 j_1 + j_2 - 2\rangle$ from (2.27) and need one more total angular momentum basis vector orthogonal to these to span $\mathscr{R}_{j_1+j_2-2}$. By an argument similar to that given in the previous paragraph, the new basis vector $|j m = j_1 + j_2 - 2\rangle$ must have $j = j_1 + j_2 - 2$ and so belongs to the total angular momentum space $\mathscr{R}^{j_1+j_2-2}$ spanned by the $2(j_1 + j_2 - 2) + 1$ basis vectors

$$|j_1 + j_2 - 2\ m\rangle, \qquad m = j_1 + j_2 - 2, j_1 + j_2 - 3, \ldots, -(j_1 + j_2 - 2).$$

$$(2.28)$$

The phases of these vectors are fixed by requiring

$$\{j_1 j_2 - 2|j_1 + j_2 - 2 j_1 + j_2 - 2\rangle \text{ be real and positive.} \qquad (2.29)$$

We continue in this way and consider in succession the J_3 eigenspaces $\mathscr{R}_m$, $m = j_1 + j_2, j_1 + j_2 - 1, \ldots, |j_1 - j_2|$. At each step a new total angular momentum space $\mathscr{R}^{j=m}$ must be introduced which contains the vector $|m\ m\rangle$ needed to complete, along with previously determined vectors

$$|j_1 + j_2\ m\rangle, |j_1 + j_2 - 1\ m\rangle, \ldots, |m + 1\ m\rangle,$$

the basis for the space $\mathscr{R}_m$. This is so because, by (2.17), the dimensionality of the successive spaces $\mathscr{R}_m$ increase by one until $\mathscr{R}_{|j_1-j_2|}$ is reached. Once $\mathscr{R}_{|j_1-j_2|}$ has been considered, and the last space $\mathscr{R}^{j=|j_1-j_2|}$ introduced, none of the remaining J_3 eigenspaces $\mathscr{R}_m$, $m = |j_1 - j_2| - 1, |j_1 - j_2| - 2, \ldots, -(j_1 + j_2)$, have dimension greater than that of $\mathscr{R}_{|j_1-j_2|}$ and they are in fact spanned by total angular momentum basis vectors already found in the spaces $\mathscr{R}^{j_1+j_2}, \mathscr{R}^{j_1+j_2-1}, \ldots, \mathscr{R}^{|j_1-j_2|}$. Thus the values j may take are $j = j_1 + j_2, j_1 + j_2 - 1, \ldots, |j_1 - j_2|$. The phase choices (2.23), (2.26), (2.29), etc. are summarized in the convention

$$\{j_1 j - j_1|j j\rangle \geq 0, \qquad j = m = j_1 + j_2, j_1 + j_2 - 1, \ldots, |j_1 - j_2|. \quad (2.30)$$

This is the standard convention for the Clebsch–Gordan coefficients.

Therewith we have seen that the spectrum of $\mathbf{J}^2$ in $\mathscr{R}$ is given by

$$\text{spectrum } \mathbf{J}^2 = \{j(j + 1): j = j_1 + j_2, j_1 + j_2 - 1, \ldots, |j_1 - j_2|\}, \quad (2.31)$$

and the space $\mathscr{R}$ may be expressed as

$$\mathscr{R} \equiv \mathscr{R}^{j_1}_{(1)} \otimes \mathscr{R}^{j_2}_{(2)} = \mathscr{R}^{j_1+j_2} \oplus \mathscr{R}^{j_1+j_2-1} \oplus \cdots \oplus \mathscr{R}^{|j_1-j_2|} \qquad (2.32)$$

This is called the reduction of $\mathscr{R}$ into a sum of irreducible total angular momentum spaces.

Summarizing, we have seen that the space $\mathscr{R}^{j_1}_{(1)} \otimes \mathscr{R}^{j_2}_{(2)}$ is in general not an irreducible-representation space or ladder-representation space of the algebra of total angular momentum $\mathscr{E}(SO(3)_{J_i})$. That is, not all vectors of $\mathscr{R}^{j_1}_{(1)} \otimes \mathscr{R}^{j_2}_{(2)}$ can be obtained by applying the $J_\pm$ a sufficient number of times to one of its vectors, but rather it is the direct sum of several such irreducible representation spaces $\mathscr{R}^j$ as given in (2.32). We have seen in particular that the Clebsch–Gordan coefficients, which are the transition coefficients between the two basis systems (2.1) and (2.9), are zero unless j is as in (2.30) and $m = m_1 + m_2$:

$$\{m_1 m_2|j m\rangle = 0$$

$$\text{for} \quad m \neq m_1 + m_2, \qquad j \neq j_1 + j_2, j_1 + j_2 - 1, \ldots, |j_2 - j_1|. \quad (2.33)$$

The Clebsch–Gordan coefficients are calculated recursively.[2] The recursion relations are obtained if one takes the matrix element of $J_\pm = J_\pm^{(1)} + J_\pm^{(2)}$ between the states $|j\,m\rangle$ and $|m_1\,m_2\}$

$$
\begin{aligned}
\{m_1 m_2|J_\pm|j\,m\rangle &= \sqrt{j(j+1) - m(m\pm 1)}\,\hbar\{m_1 m_2|j\,m\pm 1\rangle \\
&= \{m_1 m_2|J_\pm^{(1)} + J_\pm^{(2)}|j\,m\rangle \\
&= ((J_\mp^{(1)} + J_\mp^{(2)})|m_1 m_2\}, |j\,m\rangle) \\
&= \sqrt{j_1(j_1+1) - m_1(m_1 \mp 1)}\,\hbar\{m_1 \mp 1\,m_2|j\,m\rangle \\
&\quad + \sqrt{j_2(j_2+1) - m_2(m_2 \mp 1)}\,\hbar\{m_1 m_2 \mp 1|j\,m\rangle.
\end{aligned} \quad (2.34\pm)
$$

Here we have used (III.3.21–23),

$$
L_\pm f_m^l = \sqrt{l(l+1) - m(m\pm 1)}\,\hbar f_{m\pm 1}^l
$$

for

$$
L_\pm = J_\pm,\, J_\pm^{(1)},\, J_\pm^{(2)}
$$

and

$$
J_\pm^{(\alpha)\dagger} = J_\mp^{(\alpha)}.
$$

In the first step we shall describe the calculation of the Clebsch–Gordan coefficients $\{m_1 m_2|j\,j\rangle$; in the second step we then turn to the calculation of Clebsch–Gordan coefficients for all values of m, $\{m_1 m_2|j\,m\rangle$.

Setting $m = j$ in (2.34+), we obtain

$$
-\sqrt{j_1(j_1+1) - m_1(m_1 - 1)}\{m_1 - 1\,m_2|j\,j\rangle
$$
$$
= \sqrt{j_2(j_2+1) - m_2(m_2 - 1)}\{m_1 m_2 - 1|j\,j\rangle. \quad (2.35)
$$

Because of (2.33) we can restrict our attention to the values $m_1 + (m_2 - 1) = (m_1\quad 1) \mid m_2 = j(=m)$. Then (2.35) can be written

$$
\{m_1 - 1\,j - m_1 + 1|j\,j\rangle = - \left[\frac{(j_2 + j - m_1 + 1)(j_2 - j + m_1)}{(j_1 + m_1)(j_1 - m_1 + 1)}\right]^{1/2} \{m_1 j - m_1|j\,j\rangle \quad (2.36)
$$

Starting from $m_1 = j_1$ one can calculate all $\{m_1 m_2|j\,j\rangle$ successively from (2.36). One obtains for an arbitrary m_1 and $m_2 = j - m_1$

$$
\begin{aligned}
\{m_1 m_2|j\,j\rangle &= (-1)^{j_1 - m_1} \\
&\times \left[\frac{(j_2 + j - j_1 + 1)(j_2 + j - j_1 + 2)\cdots(j_2 + j - m_1)(j_2 - j + j_1)}{2j_1(2j_1 - 1)\cdots(2j_1 - j_1 + m_1 + 1)}\right. \\
&\quad \times \left.\frac{(j_2 - j + j_1 - 1)\cdots(j_2 - j + m_1 + 1)}{1\cdot 2\cdots(j_1 - m_1)}\right]^{1/2} \{j_1 j - j_1|j\,j\rangle \\
&= (-1)^{j_1 - m_1}\left[\frac{(j_1 + m_1)!\,(j_2 - j + j_1)!\,(j_2 + j - m_1)!}{(2j_1)!\,(j_1 - m_1)!\,(j_2 - j + m_1)!\,(j_2 + j - j_1)!}\right]^{1/2} \\
&\quad \times \{j_1 j - j_1|j\,j\rangle. \quad (2.37)
\end{aligned}
$$

[2] In the following a brief outline of the calculation of the Clebsch–Gordan coefficients is given, only the results given in the table will be needed later and this part may, therefore, be omitted at a first reading.

From the orthogonality relation (2.15), it follows that for $j = j' = m = m'$

$$1 = \sum_{m_1=-j_1}^{+j_1} |\{m_1 m_2 | j j\}|^2. \tag{2.15'}$$

Inserting (2.37) into (2.15') and making use of the equality[3]

$$\sum_{m_1} \frac{(j_1 + m_1)!\,(j_2 + j - m_1)!}{(j_1 - m_1)!\,(j_2 - j + m_1)!} = \frac{(j_1 + j_2 + j + 1)!\,(-j_1 + j_2 + j)!\,(j_1 - j_2 + j)!}{(2j + 1)!\,(j_1 + j_2 - j)!}, \tag{2.38}$$

one obtains

$$|\{j_1 j - j_1 | j j\}|^2 = \frac{(2j_1)!\,(2j + 1)!}{(j_1 + j_2 + j + 1)!\,(j_1 - j_2 + j)!}. \tag{2.39}$$

And with the phase convention (2.30) one obtains

$$\{j_1 j - j_1 | j j\} = +\sqrt{\frac{(2j_1)!\,(2j + 1)!}{(j_1 + j_2 + j + 1)!\,(j - j_2 + j)!}}. \tag{2.40}$$

This equation, together with (2.37), gives all coefficients that occur on the right-hand side of Equation (2.34) for $m = j$. Equation (2.34−) then gives the values of

$$\{m_1 m_2 | j\, m = j - 1\}$$

and successively the values of $\{m_1 m_2 | j\, m\}$ for all $m = j - 1, j - 2, \ldots, -j$.

In this way a general formula for the Clebsch–Gordan coefficients can be derived.[4]

The result of the above described derivation is the following formula for the Clebsch–Gordon coefficients:

$$\langle j_1 m_1 j_2 m_2 | j\, m\rangle = \{m_1 m_2 | j\, m\}$$

$$= \delta_{m_1 + m_2, m} \left[\frac{(2j + 1)(j_1 + j_2 - j)!\,(j_1 - j_2 + j)!\,(-j_1 + j_2 + j)!}{(j_1 + j_2 + j + 1)!}\right]^{1/2}$$

$$\times \,[(j_1 + m_1)!\,(j_1 - m_1)!\,(j_2 + m_2)!\,(j_2 - m_2)!\,(j + m)!\,(j - m)!]^{1/2}$$

$$\times \sum_z \{(-1)^z/[z!\,(j_1 + j_2 - j - z)!\,(j_1 - m_1 - z)!\,(j_2 + m_2 - z)!$$

$$\times \,(j - j_2 + m_1 + z)!\,(j - j_1 - m_2 + z)!]\} \tag{2.41}$$

The summation index z ranges over all integral values for which the argument of every factorial involving z is nonnegative. In (2.41) we have returned to the notation (2.1) for the direct product basis, but we shall still use $|j\, m\rangle$ for the total angular momentum basis.

We also list some of the properties (symmetry relations) of the Clebsch–Gordan coefficients:

$$\langle j_1 m_1 j_2 m_2 | j_3 m_3\rangle = (-1)^{j_2 + m_2}\left[\frac{2j_3 + 1}{2j_1 + 1}\right]^{1/2}\langle j_2 - m_2 j_3 m_3 | j_1 m_1\rangle. \tag{2.42}$$

We now give explicit expressions for the Clebsch–Gordan coefficients for the cases $j_2 = 0, \frac{1}{2}, 1$.

[3] A derivation of (2.38) can be found in Edmonds (1957), Appendix 1.

[4] A derivation using the same notation as used here is given in Edmonds (1957), pp. 44–45.

Table 2.1 $\langle j_1\ m - m_2\ \frac{1}{2}\ m_2 | j\ m \rangle$

	$m_2 = \frac{1}{2}$	$m_2 = -\frac{1}{2}$
$j = j_1 + \frac{1}{2}$	$\sqrt{\dfrac{j_1 + m + \frac{1}{2}}{2j_1 + 1}}$	$\sqrt{\dfrac{j_1 - m + \frac{1}{2}}{2j_1 + 1}}$
$j = j_1 - \frac{1}{2}$	$-\sqrt{\dfrac{j_1 - m + \frac{1}{2}}{2j_1 + 1}}$	$\sqrt{\dfrac{j_1 + m + \frac{1}{2}}{2j_1 + 1}}$

For $j_2 = 0$,

$$\langle j_1 m_1 0 0 | j m \rangle = \delta_{j_1 j} \delta_{m_1 m}. \tag{2.43}$$

The values for $j_2 = \frac{1}{2}$ and $j_2 = 1$ are given in Tables 2.1 and 2.2.

Instead of the Clebsch–Gordan coefficients it is frequently more convenient to use the Wigner 3-j symbols, since these display symmetry properties more clearly. The 3-j symbol is defined by

$$\begin{pmatrix} j_1 & j_2 & j_3 \\ m_1 & m_2 & m_3 \end{pmatrix} = (-1)^{j_1 - j_2 - m_3}(2j_3 + 1)^{-1/2} \langle j_1 m_1 j_2 m_2 | j_3 - m_3 \rangle. \tag{2.44}$$

Its symmetry properties are given by:

$$\begin{pmatrix} j_1 & j_2 & j_3 \\ m_1 & m_2 & m_3 \end{pmatrix} = \begin{pmatrix} j_2 & j_3 & j_1 \\ m_2 & m_3 & m_1 \end{pmatrix} = \begin{pmatrix} j_3 & j_1 & j_2 \\ m_3 & m_1 & m_2 \end{pmatrix}, \tag{2.45}$$

$$(-1)^{j_1 + j_2 + j_3} \begin{pmatrix} j_1 & j_2 & j_3 \\ m_1 & m_2 & m_3 \end{pmatrix} = \begin{pmatrix} j_2 & j_1 & j_3 \\ m_2 & m_1 & m_3 \end{pmatrix}$$

$$= \begin{pmatrix} j_1 & j_3 & j_2 \\ m_1 & m_3 & m_2 \end{pmatrix} = \begin{pmatrix} j_3 & j_2 & j_1 \\ m_3 & m_2 & m_1 \end{pmatrix}, \tag{2.46}$$

and

$$\begin{pmatrix} j_1 & j_2 & j_3 \\ m_1 & m_2 & m_3 \end{pmatrix} = (-1)^{j_1 + j_2 + j_3} \begin{pmatrix} j_1 & j_2 & j_3 \\ -m_1 & -m_2 & -m_3 \end{pmatrix}. \tag{2.47}$$

Table 2.3[5] lists several values for the 3-j symbols from which many others can be calculated using the symmetry properties (2.45)–(2.47)

[5] From Edmonds (1957), with permission.

Table 2.2 $\langle j_1\ m - m_2\ 1\ m_2 | j\ m \rangle$

$j =$	$m_2 = 1$	$m_2 = 0$	$m_2 = -1$
$j_1 + 1$	$\sqrt{\dfrac{(j_1 + m)(j_1 + m + 1)}{(2j_1 + 1)(2j_1 + 2)}}$	$\sqrt{\dfrac{(j_1 - m + 1)(j_1 + m + 1)}{(2j_1 + 1)(j_1 + 1)}}$	$\sqrt{\dfrac{(j_1 - m)(j_1 - m + 1)}{(2j_1 + 1)(2j_1 + 2)}}$
j_1	$-\sqrt{\dfrac{(j_1 + m)(j_1 - m + 1)}{2j_1(j_1 + 1)}}$	$\dfrac{m}{\sqrt{j_1(j_1 + 1)}}$	$\sqrt{\dfrac{(j_1 - m)(j_1 + m + 1)}{2j_1(j_1 + 1)}}$
$j_1 - 1$	$\sqrt{\dfrac{(j_1 - m)(j_1 - m + 1)}{2j_1(2j_1 + 1)}}$	$-\sqrt{\dfrac{(j_1 - m)(j_1 + m)}{j_1(2j_1 + 1)}}$	$\sqrt{\dfrac{(j_1 + m + 1)(j_1 + m)}{2j_1(2j_1 + 1)}}$

Table 2.3

$$\begin{pmatrix} j_1 & j_2 & j_3 \\ 0 & 0 & 0 \end{pmatrix} = (-1)^{J/2} \left[\frac{(j_1+j_2-j_3)!\,(j_1+j_3-j_2)!\,(j_2+j_3-j_1)!}{(j_1+j_2+j_3+1)!} \right]^{1/2} \frac{(\tfrac{1}{2}J)!}{(\tfrac{1}{2}J-j_1)!\,(\tfrac{1}{2}J-j_2)!\,(\tfrac{1}{2}J-j_3)!}$$

if J is even.

$$\begin{pmatrix} j_1 & j_2 & j_3 \\ 0 & 0 & 0 \end{pmatrix} = 0 \text{ if } J \text{ is odd, where } J = j_1 + j_2 + j_3$$

$$\begin{pmatrix} J+\tfrac{1}{2} & J & \tfrac{1}{2} \\ M & -M-\tfrac{1}{2} & \tfrac{1}{2} \end{pmatrix} \quad (-1)^{J-M-1/2}\left[\frac{J-M+\tfrac{1}{2}}{(2J+2)(2J+1)} \right]^{1/2} \qquad (J+\tfrac{1}{2},\,J,\,\tfrac{1}{2})$$

$$\begin{pmatrix} J+1 & J & 1 \\ M & -M-1 & 1 \end{pmatrix} \quad (-1)^{J-M-1}\left[\frac{(J-M)(J-M+1)}{(2J+3)(2J+2)(2J+1)} \right]^{1/2} \qquad (J+1,\,J,\,1)$$

$$\begin{pmatrix} J+1 & J & 1 \\ M & -M & 0 \end{pmatrix} \quad (-1)^{J-M-1}\left[\frac{(J+M+1)(J-M+1)\cdot 2}{(2J+3)(2J+2)(2J+1)} \right]^{1/2}$$

$$\begin{pmatrix} J & J & 1 \\ M & -M-1 & 1 \end{pmatrix} \quad (-1)^{J-M}\left[\frac{(J-M)(J+M+1)\cdot 2}{(2J+2)(2J+1)(2J)} \right]^{1/2} \qquad (J,\,J,\,1)$$

$$\begin{pmatrix} J & J & 1 \\ M & -M & 0 \end{pmatrix} \quad (-1)^{J-M}\frac{M}{[(2J+1)(J+1)J]^{1/2}}$$

$$\begin{pmatrix} J+\tfrac{3}{2} & J & \tfrac{3}{2} \\ M & -M-\tfrac{3}{2} & \tfrac{3}{2} \end{pmatrix} \quad (-1)^{J-M+1/2}\left[\frac{(J-M-\tfrac{1}{2})(J-M+\tfrac{1}{2})(J-M+\tfrac{3}{2})}{(2J+4)(2J+3)(2J+2)(2J+1)} \right]^{1/2} \qquad (J+\tfrac{3}{2},\,J,\,\tfrac{3}{2})$$

$$\begin{pmatrix} J+\tfrac{3}{2} & J & \tfrac{3}{2} \\ M & -M-\tfrac{1}{2} & \tfrac{1}{2} \end{pmatrix} \quad (-1)^{J-M+1/2}\left[\frac{3(J-M+\tfrac{1}{2})(J-M+\tfrac{3}{2})(J+M+\tfrac{3}{2})}{(2J+4)(2J+3)(2J+2)(2J+1)} \right]^{1/2}$$

$$\begin{pmatrix} J+\frac{1}{2} & J & \frac{3}{2} \\ M & -M-\frac{3}{2} & \frac{3}{2} \end{pmatrix} = (-1)^{J-M-1/2}\left[\frac{3(J-M-\frac{1}{2})(J-M+\frac{1}{2})(J+M+\frac{3}{2})}{(2J+3)(2J+2)(2J+1)2J}\right]^{1/2}$$

$$\begin{pmatrix} J+\frac{1}{2} & J & \frac{3}{2} \\ M & -M-\frac{1}{2} & \frac{1}{2} \end{pmatrix} = (-1)^{J-M-1/2}\left[\frac{J-M+\frac{1}{4}}{(2J+3)(2J+2)(2J+1)2J}\right]^{1/2}(J+3M+\frac{3}{2})$$

$$\left(J+\frac{1}{2}, J, \frac{3}{2}\right)$$

$$\begin{pmatrix} J+2 & J & 2 \\ M & -M-2 & 2 \end{pmatrix} = (-1)^{J-M}\left[\frac{(J-M-1)(J-M)(J-M+1)(J-M+2)}{(2J+5)(2J+4)(2J+3)(2J+2)(2J+1)}\right]^{1/2}$$

$$\begin{pmatrix} J+2 & J & 2 \\ M & -M-2 & 2 \end{pmatrix} = 2(-1)^{J-M}\left[\frac{(J+M+2)(J-M+2)(J-M+1)(J-M)}{(2J+5)(2J+4)(2J+3)(2J+2)(2J+1)}\right]^{1/2}$$

$$\begin{pmatrix} J+2 & J & 2 \\ M & -M-1 & 1 \end{pmatrix} = (-1)^{J-M}\left[\frac{6(J+M+2)(J+M+1)(J-M+2)(J-M+1)}{(2J+5)(2J+4)(2J+3)(2J+2)(2J+1)}\right]^{1/2}$$

$$(J+2, J, 2)$$

$$\begin{pmatrix} J+1 & J & 2 \\ M & -M-2 & 2 \end{pmatrix} = 2(-1)^{J-M+1}\left[\frac{(J-M-1)(J-M)(J-M+1)(J+M+2)}{(2J+4)(2J+3)(2J+2)(2J+1)2J}\right]^{1/2}$$

$$\begin{pmatrix} J+1 & J & 2 \\ M & -M-1 & 1 \end{pmatrix} = (-1)^{J-M+1}2(J+2M+2)\left[\frac{(J-M+1)(J-M)}{(2J+4)(2J+3)(2J+2)(2J+1)2J}\right]^{1/2}$$

$$\begin{pmatrix} J+1 & J & 2 \\ M & -M & 0 \end{pmatrix} = (-1)^{J-M+1}2M\left[\frac{(6J+M+1)(J-M+1)}{(2J+4)(2J+3)(2J+2)(2J+1)2J}\right]^{1/2}$$

$$(J+1, J, 2)$$

$$\begin{pmatrix} J & J & 2 \\ M & -M-2 & 2 \end{pmatrix} = (-1)^{J-M}\left[\frac{6(J-M-1)(J-M)(J+M+1)(J+M+2)}{(2J+3)(2J+2)(2J+1)2J(2J-1)}\right]^{1/2}$$

$$\begin{pmatrix} J & J & 2 \\ M & -M-1 & 1 \end{pmatrix} = (-1)^{J-M}(1+2M)\left[\frac{6(J+M+1)(J-M)}{(2J+3)(2J+2)(2J+1)2J(2J-1)}\right]^{1/2}$$

$$\begin{pmatrix} J & J & 2 \\ M & -M & 0 \end{pmatrix} = (-1)^{J-M}\frac{2[3M^2 - J(J+1)]}{[(2J+3)(2J+2)(2J+1)2J(2J-1)]^{1/2}}$$

$$(J, J, 2)$$

V.3 Tensor Operators and the Wigner–Eckart Theorem

We have considered the angular-momentum operators in some detail in the previous section. Now we consider a more general class of operators, called *tensor operators*,[6] which are defined by their relationship with the angular-momentum operators. The simplest example of such an operator is a *scalar operator*, which is defined to be any operator S satisfying

$$[J_i, S] = 0. \tag{3.1}$$

Another example is any set of three operators V_i ($i = 1, 2, 3$) that satisfy

$$[J_i, V_j] = i\hbar\epsilon_{ijk}V_k \qquad (i, j, k = 1, 2, 3). \tag{3.2}$$

Such a set of operators is called a *vector operator* or *regular tensor operator*. We note that the angular momentum itself is a vector operator.

Rather than using the "Cartesian" components V_i, it is more convenient to use the "spherical" components V_0, $V_{\pm 1}$ defined by

$$V_0 = V_3, \qquad V_{\pm 1} = \frac{\mp 1}{\sqrt{2}}(V_1 \pm iV_2). \tag{3.3}$$

Note that the $H_\pm$ of (III.3.1) (or the $J_\pm = \hbar H_\pm$) are *not* the spherical components of a vector operator, but differ from them by the factor $+1/\sqrt{2}$ or $-1/\sqrt{2}$; therefore we use the notation $J_{\pm 1}$ for the spherical components of the angular momentum operator, in contrast to $J_\pm$ for the raising and lowering operators. Equation (3.2) becomes

$$[J_\kappa, V_\kappa] = 0 \qquad (\kappa = -1, 0, +1), \tag{3.4a}$$

$$[J_\pm, V_{\mp 1}] = \sqrt{2}\hbar V_0, \qquad [J_\pm, V_0] = \sqrt{2}\hbar V_{\pm 1}, \qquad [J_0, V_\kappa] = \kappa\hbar V_\kappa. \tag{3.4b}$$

In general, we define a *tensor operator* of rank j to be (in spherical components) a set of $2j + 1$ operators[7] $T_\kappa^{(j)}$ ($\kappa = -j, -j + 1, \ldots, +j$) that satisfy

$$[J_0, T_\kappa^{(j)}] = \kappa\hbar T_\kappa^{(j)},$$

$$[J_\pm, T_\kappa^{(j)}] = \sqrt{j(j + 1) - \kappa(\kappa \pm 1)}\hbar T_{\kappa \pm 1}^{(j)}. \tag{3.5}$$

This can be written in a more compact form using the Clebsch–Gordan coefficients: ($\mu = -1, 0, +1$)

$$[J_\mu, T_\kappa^{(j)}] = \sqrt{j(j + 1)}\langle j\,\kappa\,1\,\mu | j\,\kappa + \mu\rangle T_{\kappa + \mu}^{(j)} \tag{3.5'}$$

---

[6] We shall actually only consider what are called "irreducible tensor operators."

[7] We emphasize the use of spherical components by the use of Greek letters such as κ in labeling the components. While such emphasis is not needed for the general tensor operator, in the case that the tensor operator is a vector operator the use of Greek letters will enable us to distinguish between the Cartesian components (Latin letters) and the spherical components (Greek letters).

We note that scalar and vector operators are tensor operators of rank 0 and 1, respectively.

The matrix elements of tensor operators have an important property which is expressed by the *Wigner-Eckart theorem*:

Let $T_\kappa^{(J)}$ be a tensor operator. The matrix element of $T_\kappa^{(J)}$ between the angular momentum eigenstates may be written as

$$\langle j'm'|T_\kappa^{(J)}|j\,m\rangle = \langle j\,m\,J\,\kappa|j'm'\rangle\langle j'\|T^{(J)}\|j\rangle, \qquad (3.6)$$

where $\langle j\,m\,J\,\kappa|j'\,m'\rangle$ is a Clebsch–Gordan coefficient and the symbol $\langle j'\|T_\kappa^{(J)}\|j\rangle$ [defined by (3.6)] denotes a quantity that depends on j',j,J and the nature of the tensor operator $T_\kappa^{(J)}$ but does not depend on m, m', or κ. $\langle j'\|T^{(J)}\|j\rangle$ is called the reduced matrix element[8] *of the tensor operator.*

This theorem will not be proven here, since the proof is pure mathematics and does not give any additional insight into the physics. If in addition to j and m there are other quantum numbers, say $\eta = (a_1, a_2, \ldots, a_N)$, then the reduced matrix element will in general also depend upon these quantum numbers, i.e.,

$$\langle \eta'j'm'|T_\kappa^{(J)}|\eta\,j\,m\rangle = \langle j\,m\,J\,\kappa|j'm'\rangle\langle\eta'j'\|T^{(J)}\|\eta\,j\rangle. \qquad (3.6')$$

The additional observables $\eta^{\mathrm{op}} = (A_1, A_2, \ldots, A_N)$ whose eigenvalues are the quantum numbers $\eta = (a_1, a_2, \ldots, a_N)$ must have the property that $[\eta^{\mathrm{op}}, J_i] = 0$ or $[\eta^{\mathrm{op}}, \mathrm{SO}(3)_{J_i}] = 0$. It should be emphasized that the Wigner–Eckart theorem is both a theorem and a definition. It is a theorem in that (3.6) says that the matrix element of $T_\kappa^{(J)}$ can be factored such that the dependence on m, m', and κ is entirely contained in the Clebsch–Gordan coefficient, and it is a definition in that (3.6) defines the reduced matrix elements.

The Wigner–Eckart theorem has become one of the most important tools for the understanding of physics. Equation (3.6) is the Wigner–Eckart theorem for the rotation group, which is connected with the algebra of angular momenta J_i as an algebra of observables. Many physical systems have an (enveloping) algebra of a group as a subalgebra of the algebra of observables, and they have observables that are tensor operators with respect to this group. For those observables the Wigner–Eckart theorem, a generalization of (3.6), holds.

The Wigner–Eckart theorem expresses the matrix elements of the tensor operators, which are directly connected with the numbers measured in an experiment, in terms of Clebsch–Gordan coefficients and reduced matrix elements. The Clebsch–Gordan coefficients are known mathematical quantities and are calculated from the properties of the group. The reduced matrix elements are physical parameters whose values have to be obtained

---

[8] In some textbooks, a constant factor or function of j appears explicitly in the Wigner–Eckart theorem. In our notation these factors have been absorbed into the reduced matrix element.

from the experimental data. [If an observable has further additional properties—e.g., Equation (VI.3.12) of the Lenz vector in Chapter VI—then the
reduced matrix elements can be reduced to a still smaller number of parameters.] Thus the importance of the Wigner–Eckart theorem is that it allows
one to fit the large number of experimentally observable matrix elements in
terms of a much smaller number of more fundamental quantities, the reduced
matrix elements. The property of being a tensor operator is often all one
knows of an observable, and the Wigner–Eckart theorem then is the only
tool at hand.

A particular consequence of the Wigner–Eckart theorem (3.6) is

$$\langle j'm'|T^{(J)}_\kappa|j\,m\rangle = 0$$

$$\text{if} \quad m' \neq \kappa + m \quad \text{or} \quad j' \neq J + j, J + j - 1, \ldots, |J - j| \qquad (3.7)$$

which follows immediately from the properties of the Clebsch–Gordan
coefficients (2.33).

As an illustration of the Wigner–Eckart theorem, we note that for a scalar
operator S, Equation (3.6) becomes

$$\langle j'm'|S|j\,m\rangle = \delta_{m'm}\delta_{j'j}\langle j\|S\|j\rangle. \qquad (3.8)$$

This says that S cannot change the angular-momentum quantum number,
which is just as we would expect. For the angular-momentum operators
themselves, Equation (3.6) gives

$$\langle j'm'|J_\kappa|j\,m\rangle = \langle j\,m\,1\,\kappa|j'm'\rangle\langle j'\|J\|j\rangle, \qquad (3.9)$$

where the Clebsch–Gordan coefficients are listed in Table 2.2. By comparing
the Clebsch–Gordan coefficient in the table with the formula (III.3.8'),
(III.3.22), and (III.3.23), we see that

$$\langle j'\|J\|j\rangle = \delta_{j'j}\hbar\sqrt{j(j + 1)}. \qquad (3.10)$$

If $|j\,m\rangle$ is a basis in the space in which V_κ is a general vector operator, then
we can expand $V_\kappa|j\,m\rangle$ with respect to this basis:

$$V_\kappa|j\,m\rangle = \sum_{j'm'} |j'm'\rangle\langle j'm'|V_\kappa|j\,m\rangle$$

and obtain, according to the Wigner–Eckart theorem (3.6),

$$V_\kappa|j\,m\rangle = \sum_{j'm'} |j'm'\rangle\langle j\,m\,1\,\kappa|j'm'\rangle\langle j'\|V\|j\rangle. \qquad (3.11)$$

If the c.s.c.o. contains, in addition to $\mathbf{J}^2$ and J_3, N other operators $A_1, \ldots, A_N$
with spectrum $\eta = (a_1, a_2, \ldots, a_N)$, then the basis is $|\eta\,j\,m\rangle$, and instead of
(3.11) one has, according to (3.6'),

$$V_\kappa|\eta\,j\,m\rangle = \sum_{\eta'j'm'} |\eta'j'm'\rangle\langle j\,m\,1\,\kappa|j'm'\rangle\langle\eta'j'\|V\|\eta\,j\rangle \qquad (3.11')$$

We shall for the remainder of this section suppress the additional quantum
numbers η, but it should be understood that whenever additional quantum

numbers η are needed, the reduced matrix elements will depend upon these quantum numbers and summation over η' is implied.

According to (3.7), the only nonvanishing terms in (3.11) are those for which $j' = j + 1, j, j - 1$ and $m' = m + \kappa$:

$$
\begin{aligned}
V_\kappa | j\, m \rangle = &| j - 1\, m + \kappa \rangle \langle j\, m\, 1\, \kappa | j - 1\, m + \kappa \rangle \langle j - 1 \| V \| j \rangle \\
&+ | j\, m + \kappa \rangle \langle j\, m\, 1\, \kappa | j\, m + \kappa \rangle \langle j \| V \| j \rangle \\
&+ | j + 1\, m + \kappa \rangle \langle j\, m\, 1\, \kappa | j + 1\, m + \kappa \rangle \langle j + 1 \| V \| j \rangle.
\end{aligned}
\tag{3.12}
$$

This is the most general form possible for the action of a vector operator. According to (3.12), then, any vector operator can be completely specified by three reduced matrix elements (which in general may depend upon η, η'). Actually, as we shall show below, only two quantities are needed to determine a vector operator.

By using the Clebsch–Gordan coefficients of Table 2.2, we can write (3.12) explicitly. For the 0-component,

$$
\begin{aligned}
V_0 | j\, m \rangle = &| j - 1\, m \rangle \left(-\sqrt{\frac{(j - m)(j + m)}{j(2j + 1)}} \right) \langle j - 1 \| V \| j \rangle \\
&+ | j\, m \rangle \left(\frac{m}{\sqrt{j(j + 1)}} \right) \langle j \| V \| j \rangle \\
&+ | j + 1\, m \rangle \left(\sqrt{\frac{(j - m + 1)(j + m + 1)}{(2j + 1)(j + 1)}} \right) \langle j + 1 \| V \| j \rangle,
\end{aligned}
$$

or

$$
\begin{aligned}
V_0 | j\, m \rangle = &\sqrt{j^2 - m^2}\, c_j | j - 1\, m \rangle - m a_j | j\, m \rangle \\
&- \sqrt{(j + 1)^2 - m^2}\, d_j | j + 1\, m \rangle,
\end{aligned}
\tag{3.13}
$$

where the c_j, a_j, and d_j are defined by Equation (3.13) itself:

$$
c_j = -\frac{\langle j - 1 \| V \| j \rangle}{\sqrt{j(2j + 1)}},
$$

$$
a_j = -\frac{\langle j \| V \| j \rangle}{\sqrt{j(j + 1)}},
\tag{3.14}
$$

$$
d_j = -\frac{\langle j + 1 \| V \| j \rangle}{\sqrt{(2j + 1)(j + 1)}}.
$$

Likewise,

$$
\begin{aligned}
V_{+1} | j\, m \rangle = &-| j - 1\, m + 1 \rangle \sqrt{(j - m - 1)(j - m)/2}\, c_j \\
&+ | j\, m + 1 \rangle \sqrt{(j + m + 1)(j - m)/2}\, a_j \\
&- | j + 1\, m + 1 \rangle \sqrt{(j + m + 1)(j + m + 2)/2}\, d_j
\end{aligned}
\tag{3.15}
$$

and

$$V_{-1}|j\,m\rangle = -|j-1\,m-1\rangle\sqrt{(j+m)(j+m-1)/2}\,c_j$$
$$-|j\,m-1\rangle\sqrt{(j-m+1)(j+m)/2}\,a_j$$
$$-|j+1\,m-1\rangle\sqrt{(j-m+1)(j-m+2)/2}\,d_j. \quad (3.16)$$

Equations (3.13), (3.15), and (3.16) say that a vector operator is completely determined by the three functions c_j, a_j, and d_j (which may in general depend upon η', η) of the discrete parameter j. In the mathematical note below it will be shown that in fact only two such functions are needed, as it is possible to choose the basis $|j\,m\rangle$ in such a way that

$$d_j = c_{j+1}, \quad (3.17)$$

i.e.,

$$\sqrt{2j+3}\langle j+1\|V\|j\rangle = \sqrt{2j+1}\langle j\|V\|j+1\rangle. \quad (3.17')$$

⟦We define a new basis

$$|h_m^j\rangle = \omega(j)|j\,m\rangle, \quad (3.18)$$

where $\omega(j)$ is a complex number. The angular-momentum operators have the same form (Section III.3) in the $|h_m^j\rangle$ basis as they do in the $|j\,m\rangle$ basis. In general, the $|h_m^j\rangle$ are not normalized unless

$$|\omega(j)| = 1. \quad (3.19)$$

In this basis, Equation (3.13) becomes

$$V_0|h_m^j\rangle = \sqrt{j^2-m^2}\,c_j\,\frac{\omega(j)}{\omega(j-1)}|h_m^{j-1}\rangle - ma_j|h_m^j\rangle$$
$$-\sqrt{(j+1)^2-m^2}\,d_j\,\frac{\omega(j)}{\omega(j+1)}|h_m^{j+1}\rangle. \quad (3.20)$$

We define $A_j = a_j$, and we wish to choose $\omega(j)$ such that

$$c_j\frac{\omega(j)}{\omega(j-1)} = C_j \quad \text{and} \quad d_j\frac{\omega(j)}{\omega(j+1)} = D_j = C_{j+1}. \quad (3.21)$$

This will be possible if

$$c_j\frac{\omega(j)}{\omega(j-1)} = \frac{\omega(j-1)}{\omega(j)}d_{j-1}, \quad (3.22)$$

i.e., if

$$\omega^2(j) = \omega^2(j-1)\frac{d_{j-1}}{c_j}. \quad (3.23)$$

Suppose j_0 is the smallest value of j in the space, i.e.,

$$\langle j_0 - 1\|V\|j_0\rangle = 0. \quad (3.24)$$

Then (3.23) will be satisfied if we choose

$$\omega(j) = \sqrt{\omega^2(j_0)\, \frac{d_{j_0}}{c_{j_0+1}} \frac{d_{j_0+1}}{c_{j_0+2}} \cdots \frac{d_{j-1}}{c_j}} \tag{3.25}$$

so (3.20) becomes

$$V_0|h_m^j\rangle = \sqrt{j^2 - m^2}\,C_j|h_m^{j-1}\rangle - mA_j|h_m^j\rangle$$
$$- \sqrt{(j+1)^2 - m^2}\,C_{j+1}|h_m^{j+1}\rangle.] \tag{3.26}$$

Thus every vector operator can be written in the form (3.13), (3.15), (3.16) with (3.17), where the c_j and a_j are functions of j. It should be noted, however, that in general we may have $d_j = c_{j+1}$ for only one vector operator V_κ at a time. If two different vector operators are involved in one problem and the basis has been chosen so that for the reduced matrix element of one of them (3.17) holds, then in this basis, the other or any additional vector operator is expressed in terms of three independent reduced matrix elements.

So far we have not required anything of V_κ except that it be a vector operator. As a consequence the reduced matrix elements c_j and a_j are arbitrary functions of j. If we specify V_κ further we will obtain more specific information about the c_j and a_j. In the Appendix we will study three specific cases of vector operators which together with the J_κ will generate the algebra of a group, and we will see that in each of these cases the functions c_j and a_j are completely specified by a pair of two numbers. But if nothing else is known of V_κ (from physics) except that it is a vector operator then the c_j and a_j cannot be further calculated. They can only be determined phenomenologically from the experimental value of one particular matrix element of one component of V_κ, e.g., c_j can be obtained from $\langle j-1, j-1|V_0|j, j-1\rangle$ and a_j from $\langle j\,j|V_0|j\,j\rangle$. All other matrix elements $\langle j\,m|V|j\,m\rangle$ for $m = -j, -j+1, \ldots, j$ and $\kappa = 0, +1, -1$ can then be calculated from these two experimental values using the Clebsch–Gordan coefficients.

Often the vector operators have a specified Hermiticity property. They are either Hermitian, defined by

$$V_0^\dagger = V_0, \qquad V_{\pm 1}^\dagger = -V_{\mp 1},$$

or they are skew-Hermitian, defined by

$$V_0^\dagger = -V_0, \qquad V_{\pm 1}^\dagger = +V_{\mp 1}.$$

It is easy to see (Appendix) that for Hermitian V_κ the functions c_j are purely imaginary, $\bar{c}_j = -c_j$, and the functions a_j are real, $\bar{a}_j = a_j$. For skew-Hermitian V_κ one has $\bar{c}_j = c_j$ and $\bar{a}_j = -a_j$.

Appendix to Section V.3—Representations of the Enveloping Algebras SO(3,1), SO(4), and E_3

Algebras that are generated by the angular-momentum operators J_i and a particular vector operator are needed for various problems of quantum physics. We have already seen an example of such an algebra in Section III.4, where $\mathscr{E}(E_3)$ was the algebra generated by the angular momentum J_i and the dipole operator Q_i. In Chapter VI we shall encounter the algebra $\mathscr{E}(SO(4))$ generated by the angular-momentum operator L_i and the Lenz vector A_i. $\mathscr{E}(SO(3,1))$ is the algebra of the Lorentz group and has various applications in quantum physics.

In this appendix we give a derivation of the representations of these three algebras. These algebras are the enveloping algebras of the groups $SO(3,1)$ [the pseudo-orthogonal group in $(3 + 1)$ dimensions], $SO(4)$ [the orthogonal group in four dimensions], and E_3 [the three-dimensional Euclidean group], but we shall not discuss these group-theoretic connections here.

The defining relations of these algebras are given by

$$[H_i, H_j] = i\epsilon_{ijk}H_k, \tag{3.27}$$

$$[H_i, F_j] = i\epsilon_{ijk}F_k, \tag{3.28}$$

$$[F_i, F_j] = \lambda^2 i\epsilon_{ijk}H_k, \tag{3.29$_{\lambda^2}$}$$

where $\lambda^2 = -1$ for $\mathscr{E}(SO(3,1))$, $\lambda^2 = +1$ for $\mathscr{E}(SO(4))$, and $\lambda^2 = 0$ for $\mathscr{E}(E_3)$. Equation (3.27) defines the algebra $\mathscr{E}(SO(3))$ of angular momentum. Equation (3.28) specifies that F_i is a vector operator with respect to $\mathscr{E}(SO(3))$; the F_i, $i = 1, 2, 3$, are the Cartesian components. Equation (3.29) then specifies that this vector operator together with H_k generates one of the above algebras. If in addition to (3.27) and (3.28), (3.29) is also fulfilled, then **F** is a very particular vector operator and its reduced matrix elements have very particular values. We shall now determine these values. We shall derive the representations of (3.27)–(3.29) for $\lambda^2 = -1$ and obtain the other two cases $\lambda^2 = 1$ and $\lambda^2 = 0$ by a change of the Hermiticity property and by a limiting process, respectively.

To find all linear representations of the algebra (3.27)–(3.29) means to find all linear operators in all linear spaces that fulfill these commutation relations. We shall restrict ourselves here to the subclass of those representations whose representation space $\mathscr{R}$ contains each of the representation spaces $\mathscr{R}^l$ ($l = 0$, or $\frac{1}{2}$, or etc of $\mathscr{E}(SO(3))$ at most once.[9] In each $\mathscr{R}^l$ we have, according to Section III.3, the basis

$$f_m^l, \qquad m = -l, l + 1, \ldots, +l, \tag{3.30}$$

and the space $\mathscr{R}$ is then spanned by these basis vectors f_m^l, where l runs through a set of values that is to be determined. In other words,

$$\mathscr{R} = \sum_l \oplus \mathscr{R}^l. \tag{3.31}$$

If one of the $\mathscr{R}^l$, say for $l = l_0$, were to appear more than once, then f_m^l would not be a basis system of the space $\mathscr{R}$, and a new label (quantum

---

[9] These include all the representations that are connected with unitary representations of the groups $SO(3,1)$ and $SO(4)$.

number) would be needed to distinguish orthogonal vectors with the same values of l_0 and $m: f_m^{l_0\eta}$.

The action of the linear operators H_k on all f_m^l is already known from Section III.3, and we have now to determine the action of the F_k upon the f_m^l. We introduce the components [in analogy to (III.3.1)]

$$F_0 = F_3, \qquad F_\pm = F_1 \pm iF_2 = \mp\sqrt{2}F_{\pm 1}. \qquad (3.32)$$

F_0 and $F_{\pm 1}$ are the standard spherical components of the vector operator **F**. The commutation relations of the F_0, $F_\pm$ with $H_\pm$, $H_0 = H_3$ and with each other are then

$$\left.\begin{array}{ll} [H_+, F_-] = 2F_0, & [H_-, F_+] = -2F_0 \\[4pt] [F_\kappa, H_\kappa] = 0, & \kappa = 0, +, - \\[4pt] [H_0, F_\pm] = \pm F_\pm, & [H_\pm, F_0] = \mp F_\pm \end{array}\right\} \qquad (3.33)$$

$$[F_0, F_\pm] = \mp H_\pm, \qquad [F_+, F_-] = -2H_0. \qquad (3.34)$$

Equation (3.33) corresponds to (3.28) and expresses the vector-operator property of **F**, and (3.34) corresponds to (3.29) with $\lambda^2 = -1$ and specifies that this vector operator is a generator of $\mathscr{E}(SO(3,1))$. If we now use the vector-operator property and the Wigner–Eckart theorem, then we obtain for F_0 and $F_\pm$ the expressions (3.13), (3.15), and (3.16). If we also choose the phase factor of the basis vectors appropriately, as discussed in the mathematical note just before this appendix, then (3.17) can be used, and we obtain for the action of F_0 and $F_\pm$ on the basis vectors the expressions

$$F_0 f_m^l = \sqrt{l^2 - m^2}\, c_l f_m^{l-1} - m a_l f_m^l - \sqrt{(l+1)^2 - m^2}\, c_{l+1} f_m^{l+1}, \qquad (3.35_0)$$

$$F_+ f_m^l = \sqrt{(l-m)(l-m-1)}\, c_l f_{m+1}^{l-1} - \sqrt{(l-m)(l+m+1)}\, a_l f_{m+1}^l$$
$$+ \sqrt{(l+m+1)(l+m+2)}\, c_{l+1} f_{m+1}^{l+1}, \qquad (3.35_+)$$

$$F_- f_m^l = -\sqrt{(l+m)(l+m-1)}\, c_l f_{m-1}^{l-1} - \sqrt{(l+m)(l-m+1)}\, a_l f_{m-1}^l$$
$$- \sqrt{(l-m+1)(l-m+2)}\, c_{l+1} f_{m-1}^{l+1}. \qquad (3.35_-)$$

These have been obtained using only (3.33); and every vector operator, in a space in which $\mathscr{R}^l$ appears at most once, has this form.

The vectors f_m^l have, according to the results of Section III.3, the property

$$H_3 f_m^l = m f_m^l, \qquad \mathbf{H}^2 f_m^l = l(l+1) f_m^l,$$
$$H_\pm f_m^l = \sqrt{(l \mp m)(l \pm m + 1)}\, f_{m\pm 1}^l. \qquad (3.36)$$

The f_m^l have in addition been chosen so that (3.17) is fulfilled.

So far we have only exploited (3.27) and (3.28); we shall now make use of (3.29) or (3.34) to determine the unknown coefficients a_l and c_l [related to the reduced matrix elements by (3.14)]. It is sufficient to use just one of the three relations (3.34) because the other two are consequences of it and the relations (3.33). We choose

$$[F_+, F_0] = H_+. \qquad (3.34')$$

Applying both sides of (3.34′) to the vector f_m^l, using (3.35), and comparing coefficients of f_{m+1}^l, f_{m+1}^{l+1}, and f_{m+1}^{l-1} leads to the following equations:

$$[a_l(l+1) - (l-1)a_{l-1}]c_l = 0, \tag{3.37a}$$

$$[a_{l+1}(l+2) - la_l]c_{l+1} = 0, \tag{3.37b}$$

$$(2l-1)c_l^2 - (2l+3)c_{l+1}^2 - a_l^2 = 1. \tag{3.37c}$$

One starts with an arbitrary basis vector f_m^l which one assumes is contained in the particular representation space $\mathcal{R}$ that one wants to construct. Applying the operators H_κ to f_m^l a sufficient number of times, one obtains, as described in Section III.3, the whole space $\mathcal{R}^l$. Applying the vector operators F_κ to f_m^l a sufficient number of times, one reaches all the other $\mathcal{R}^j \subset \mathcal{R}$ where j differs from l by an integer, because F_κ changes l by 0, +1, −1. As $j > 0$, there must always exist a smallest value of j in any $\mathcal{R}$. This smallest j we call k_0; it can be any of the allowed values for j and will characterize the space $\mathcal{R}$. If k_0 is the smallest value of j, then according to (3.35)

$$c_{k_0} = 0.$$

If k_0 is integer, then $\mathcal{R}$ contains only integer $l \geq k_0$, and if k_0 is half-integer, then $\mathcal{R}$ contains only half-integer $l \geq k_0$:

$$\mathcal{R} = \sum_{l=k_0, k_0+1, \ldots} \oplus \mathcal{R}^l. \tag{3.38}$$

For all l for which $c_l \neq 0$ one obtains from (3.37a)

$$a_l(l+1) - (l-1)a_{l-1} = 0, \tag{3.39a}$$

and for all l for which $c_{l+1} \neq 0$ one obtains from (3.37b)

$$a_{l+1}(l+2) - la_l = 0. \tag{3.39b}$$

Defining

$$\rho_l = l(l+1)a_l, \tag{3.39c}$$

these two conditions lead to

$$\rho_l - \rho_{l-1} = 0,$$
$$\rho_{l+1} - \rho_l = 0, \tag{3.39d}$$

which means that ρ_l is independent of l. We call this arbitrary complex constant $\rho_l = ik_0 c$, where c is an arbitrary complex number, and we shall show below that the factoring out of k_0 is always possible. a_l can then be written

$$a_l = \frac{ik_0 c}{l(l+1)} \qquad \text{for any} \quad l \geq k_0, l \geq \tfrac{1}{2}. \tag{3.40}$$

For the case that both $c_l = 0$ and $c_{l+1} = 0$, it follows from (3.37c) that $a_l^2 = -1$, and the matrix elements (3.35) of F_κ are completely determined. With (3.40) this means that $k_0 = l$, $c = l+1$ and that $\mathcal{R} = \mathcal{R}^l$.

It remains to show that the factoring out of k_0 is always possible, i.e., that $a_l = 0$ for $k_0 = 0$. But for $k_0 = 0$, (3.39a) holds for $l = 1$ and leads to $a_1 = 0$. For $l = 2$ this in turn leads to $a_2 \cdot 3 - a_1 = a_2 \cdot 3 = 0$, and proceeding this way we obtain $a_l = 0$ for $l = 1, 2, 3, \ldots$. For $l = 0$ the factor in front of a_0 in (3.35) is zero so that the appearance of a_0 is meaningless. Thus (3.40) holds generally.

To determine c_l we use (3.37c). Defining

$$\sigma_l = (2l - 1)(2l + 1)c_l^2, \qquad (3.41)$$

we can write (3.37c) as

$$\sigma_l - \sigma_{l+1} - (2l + 1)a_l^2 = 2l + 1,$$

or using (3.40)

$$\sigma_l - \sigma_{l+1} = (2l + 1) - k_0^2 c^2 \left(\frac{1}{l^2} - \frac{1}{(l+1)^2} \right). \qquad (3.42)$$

We calculate now for any value $k \geq k_0, k \geq 1$

$$
\sigma_{k_0} - \sigma_k = \sum_{l=k_0}^{k-1} (\sigma_l - \sigma_{l+1}) = \sum_{l=k_0}^{k-1} (2l + 1) - k_0^2 c^2 \sum_{l=k_0}^{k-1} \left(\frac{1}{l^2} - \frac{1}{(l+1)^2} \right)
$$

$$
= k^2 - k_0^2 - k_0^2 c^2 \left(\frac{1}{k_0^2} - \frac{1}{k^2} \right)
$$

$$
= \frac{(k^2 - k_0^2)(k^2 - c^2)}{k^2}.
$$

As k_0 is the smallest value of l, $c_{k_0} = 0$, and consequently $\sigma_{k_0} = 0$, so that one obtains

$$\sigma_k = - \frac{(k^2 - k_0^2)(k^2 - c^2)}{k^2}, \qquad (3.43)$$

and by (3.41),

$$c_l = \frac{i}{l} \sqrt{\frac{(l^2 - k_0^2)(l^2 - c^2)}{4l^2 - 1}} \qquad \text{for any} \quad l \geq k_0, l \geq 1. \qquad (3.44)$$

Therewith, we have found all possibilities of how the operators H_κ and F_κ, that fulfill (3.33) and (3.34), can act upon the vectors f_m^l. These possibilities are characterized by the pair of numbers (k_0, c), where k_0 is an integer or half-integer and c is a complex number. For each choice of these two numbers the action of the operators on the f_m^l is given by (3.35) and (3.36), with a_l and c_l given by (3.40) and (3.44). In other words, for each pair (k_0, c) we have a representation of the algebra $\mathscr{E}(SO(3,1))$ generated by the H_κ and F_κ in the representation space $\mathscr{R}$ of (3.38). This space and the operators acting in it are characterized by the pair of numbers $(k_0, c): \mathscr{R}(k_0, c), H_\kappa^{(k_0, c)}$, and $F_\kappa^{(k_0, c)}$.

So far, we have not introduced into $\mathcal{R}(k_0, c)$ a scalar product and cannot speak of Hermiticity of the operators. Therefore, the symbol $\sum_{l=k_0, k_0+1, \ldots} \oplus$ defines the space of linear combinations

$$f = f^{k_0} + f^{k_0+1} + \cdots + f^l + \cdots + f^N,$$

with $f^l \in \mathcal{R}^l$ and N in general larger than any given number. For all of the pairs (k_0, c), the $H_\kappa^{(k_0, c)}$ and $F_\kappa^{(k_0, c)}$ are all of the linear representations of $\mathscr{E}(SO(3,1))$ that contain a given value of angular momentum l at most once. These representation spaces are in general infinite-dimensional; only for certain values of c are they finite-dimensional.

The finite-dimensional cases are obtained from (3.35) and (3.44). Looking at (3.35), we see that successive application of F_κ will lead to ever higher values of l unless for some value $l = k_1$ one has $c_{k_1+1} = 0$; k_1 is then the highest value that l can take in $\mathcal{R}$. According to (3.44), this means that the value of c must be such that $c^2 = (k_1 + 1)^2$. Thus for

$$k_0 = 0, \tfrac{1}{2}, 1, \ldots, \quad \text{and} \quad c = \pm(k_1 + 1), \quad \text{where} \quad k_1 = k_0, k_0 + 1, \ldots,$$
$$\text{(3.45)}$$

the representation space is finite-dimensional and is given by

$$\mathcal{R}(k_0, \pm(k_1+1)) = \sum_{l=k_0, k_0+1, \ldots}^{k_1} \oplus \mathcal{R}^l. \tag{3.46}$$

Though $\mathcal{R}(k_0, (k_1 + 1))$ and $\mathcal{R}(k_0, -(k_0 + 1))$ are the same direct sum of irreducible representation spaces of $\mathscr{E}(SO(3))$, the F_κ act differently in these two spaces, as can be seen from (3.44) and (3.35). They are inequivalent representation spaces of $\mathscr{E}(SO(3,1))$, but they have the same reduction with respect to $\mathscr{E}(SO(3))$. From (3.40) and (3.44) it follows that for the finite-dimensional cases (3.45) the a_l and c_l fulfill the conditions

$$a_l = -\bar{a}_l, \qquad c_l = \bar{c}_l. \tag{3.47b}$$

We will now introduce a scalar product into the spaces $\mathcal{R}(k_0, c)$, of which we must require

$$(f_m^l, f_{m'}^{l'}) = \delta^{ll'} \delta_{mm'} \tag{3.48}$$

(so that $\mathbf{H}^2$ is a Hermitian operator in $\mathcal{R}(k_0, c)$). Then from (3.35_0) one obtains

$$(F_0 f_m^{l-1}, f_m^l) = -\sqrt{l^2 - m^2}\ \bar{c}_l (f_m^l, f_m^l),$$
$$(f_m^{l-1}, F_0 f_m^l) = \sqrt{l^2 - m^2}\ c_l (f_m^{l-1}, f_m^{l-1}), \tag{3.49_1}$$

and

$$(F_0 f_m^l, f_m^l) = -m\bar{a}_l (f_m^l, f_m^l),$$
$$(f_m^l, F_0 f_m^l) = -m a_l (f_m^l, f_m^l). \tag{3.49_2}$$

With (3.48), this means that

$$(F_0 f_m^{l-1}, f_m^l) c_l = -(f_m^{l-1}, F_0 f_m^l) \bar{c}_l, \tag{3.50_1}$$

and

$$(F_0 f_m^l, f_m^l)a_l = (f_m^l, F_0 f_m^l)\bar{a}_l. \tag{3.50_2}$$

From (3.50) and (3.47b) it follows that for the finite-dimensional cases (3.45) the operator F_0 must be skew-Hermitian: $F_0^\dagger = -F_0$. By similar arguments, or by (3.34), it can also be shown that the other components of $\mathbf{F}$ are skew-Hermitian. Thus $\mathbf{F}$ is a skew-Hermitian vector operator,

$$F_0^\dagger = -F_0, \qquad F_\pm^\dagger = -F_\mp, \tag{3.47a}$$

if the representation is finite-dimensional.

The other very important case is that $\mathbf{F}$ is a Hermitian vector operator:

$$F_0^\dagger = F_0, \qquad F_\pm^\dagger = F_\mp, \tag{3.51a}$$

i.e., that we have a Hermitian representation of $\mathscr{E}(SO(3, 1))$ [which is connected with a unitary representation of the group $SO(3, 1)$]. In this case, it follows from (3.51a) and (3.50) that

$$a_l = \bar{a}_l, \qquad c_l = -\bar{c}_l. \tag{3.51b}$$

This means according to (3.40) that ic must be real unless $k_0 = 0$, and according to (3.44) that $(l^2 - k_0^2)(l^2 - c^2)/(4l^2 - 1)$ must be positive. The latter is possible only when c^2 is real, i.e., when c is real or purely imaginary.

Thus for any integer of half-integer value of k_0 and any purely imaginary value of c, (3.51) can be fulfilled. If c is real, then k_0 must be equal to zero in order that $a_l = \bar{a}_l$. Then the expression under the square root in (3.44) becomes $l^2(l^2 - c^2)/(4l^2 - 1)$, which will be positive for all $l = 1, 2, \ldots$ only if $c^2 < 1$.

Thus we have seen that (3.51) can be fulfilled in two cases:

a. For $k_0 = 0, \frac{1}{2}, 1, \ldots$ and $ic = $ real, $-\infty < ic < +\infty$
b. For $k_0 = 0, 0 \leq c < 1$.

In neither of these two cases can c_{l+1} ever become zero. Consequently, according to (3.35), one obtains from any given f_m^l also an f_m^{l+1}. One can apply F_κ an arbitrary number of times arriving at ever higher values for l. The representation space $\mathscr{R} = \mathscr{R}(k_0, c)$ with (k_0, c) fulfilling case a or b is infinite-dimensional:

$$\mathscr{R}(k_0, c) = \sum_{l=k_0, k_0+1, \ldots}^{\infty} \oplus \mathscr{R}^l. \tag{3.52}$$

It is the (algebraic)[10] direct sum of the irreducible representation spaces $\mathscr{R}^l(l = k_0, k_0 + 1, \ldots)$ of the algebra of angular momentum, $\mathscr{E}(SO(3))$.

$(k_0 = 0, c = 1)$ is the trivial case in which all H_κ and F_κ are represented by the zero operator.

---

[10] It can be completed with respect to various topologies, e.g., with respect to the Hilbert-space topology defined by the scalar product, or with respect to the topology defined by the countable number of scalar products $(\phi, \psi)_n = (\phi, (\mathbf{H}^2 + \mathbf{F}^2 + 1)^n \psi)$ leading to the nuclear space Φ. Cf. Böhm (1978).

Summarizing our results, we have found that the irreducible-representation[11] spaces of $\mathscr{E}(SO(3,1))$ [the algebra defined by (3.27), (3.28), $(3.29_{\lambda^2=-1})$] for which H_i and F_i are Hermitian are characterized by two numbers (k_0, c) that can take any of the values in case a or b. The reduction of these representation spaces $\mathscr{R}(k_0, c)$ with respect to the algebra of $\mathscr{E}(SO(3))$ is given by (3.52), and the operators H_κ and F_κ are given by (3.36) and (3.35) with (3.40) and (3.44). The irreducible representation spaces of $\mathscr{E}(SO(3,1))$ with skew-Hermitian F_i are characterized by the two numbers $(k_0, c) = (k_0, \pm(k_1 + 1))$ which can take any of the values (3.45). The reduction of these representation spaces with respect to $\mathscr{E}(SO(3))$ is given by (3.46), and H_κ and F_κ are again given by (3.36), (3.35), (3.40), (3.44).

It is now easy to obtain the representations of $\mathscr{E}(SO(4))$ and $\mathscr{E}(E_3)$. To obtain the representations of $\mathscr{E}(SO(4))$ we define

$$A_j = -iF_j, \qquad A_\kappa = -iF_\kappa \qquad (\kappa = 0, +, -). \tag{3.53}$$

If F_j fulfills the commutation relation $(3.29_{\lambda^2=-1})$ then A_i fulfills the commutation relation $(3.29_{\lambda^2=+1})$; i.e., it fulfills together with H_i the commutation relations of $\mathscr{E}(SO(4))$:

$$[A_i, A_j] = i\epsilon_{ijk}H_k \tag{$3.29_{\lambda^2=+1}$}$$

If F_j are skew-Hermitian $(F_j^\dagger = -F_j)$, then A_j are Hermitian $(A_j^\dagger = A_j)$. Therefore we conclude that the Hermitian irreducible representations of $\mathscr{E}(SO(4))$ are characterized by two numbers $(k_0, \pm(k_1 + 1))$ that can take the values (3.45). The irreducible-representation spaces are all finite-dimensional and given by (3.46). The operators A_κ are again given by (3.35) with (3.40) and (3.44), where F_κ is given by (3.53).

The algebra $\mathscr{E}(SO(3,1))$ has two independent invariant operators, i.e., two independent operators that commute with all generators H_i and with F_i. They are

$$C_1 = \mathbf{F}^2 - \mathbf{H}^2, \qquad C_2 = \mathbf{F} \cdot \mathbf{H}. \tag{3.54}$$

As they commute with all H_κ and F_κ their eigenvalues cannot change by applying H_κ, F_κ to the corresponding eigenvector. Therefore their eigenvalues must be the same in the whole irreducible representation space. One can calculate these eigenvalues applying C_1 and C_2 to a suitably chosen f_m^l using (3.36), (3.35) with (3.40) and (3.44). The result is

$$C_1 f_m^l(k_0, c) = (-k_0^2 - c^2 + 1)f_m^l(k_0, c), \tag{3.55_1}$$

$$C_2 f_m^l(k_0, c) = -ik_0 c f_m^l(k_0, c). \tag{3.55_2}$$

The two invariant operators for $\mathscr{E}(SO(4))$ are

$$C_1^{SO(4)} = \mathbf{A}^2 + \mathbf{H}^2, \qquad C_2^{SO(4)} = \mathbf{A} \cdot \mathbf{H}, \tag{3.56}$$

---

[11] The irreducible representations are also called ladder representations because they are obtained by climbing from one value of l to the next.

with the eigenvalues

$$C_1^{SO(4)} f_m^l(k_0, \pm(k_1 + 1)) = (k_0^2 + c^2 - 1) f_m^l(k_0, \pm(k_1 + 1))$$

$$= (k_0^2 + k_1^2 + 2k_1) f_m^l(k_0, \pm(k_1 + 1)), \qquad (3.57_1)$$

$$C_2^{SO(4)} f_m^l(k_0, \pm(k_1 + 1)) = -k_0 c f_m^l(k_0, \pm(k_1 + 1))$$

$$= \pm k_0 (k_1 + 1) f_m^l(k_0, \pm(k_1 + 1)). \qquad (3.57_2)$$

Thus, the eigenvalues of the invariant operators specify the irreducible representation completely; it is, however, more convenient to characterize the irreducible representations by the values of (k_0, c) or by the values of $(k_0, k_1, \pm)$, where $\pm$ refers to the two inequivalent representations with the same value of k_0 and k_1.

We finally mention the special case where $k_0 = 0$ or $C_2 = 0$. This is the case one encounters most frequently in the application to simple quantum-mechanical systems. We shall meet this case for $\mathscr{E}(SO(4))$ again in Chapter VI in the derivation of the hydrogen spectrum. It is the case that does not require parity doubling, as we shall see below in Section V.4 [in particular Equation (4.25)]. We have encountered it already in Section III.4; it leads to a spin spectrum given by (III.4.2) and depicted by Figure III.4.1. For $k_0 = 0$ it follows from (3.52) for Hermitian representations of $\mathscr{E}(SO(3, 1))$ that

$$\mathscr{R}(0, c) = \sum_{l = 0, 1, \ldots} \oplus \mathscr{R}^l, \qquad (3.58)$$

and from (3.46) for Hermitian representations of $\mathscr{E}(SO(4))$ that

$$\mathscr{R}(0, k_1 + 1 = n) = \sum_{l = 0, 1, \ldots}^{n-1} \oplus \mathscr{R}^l \qquad (n = 1, 2, 3, \ldots). \qquad (3.59)$$

From (3.40) follows that for this case

$$a_l = 0 \quad \text{for all } l \qquad (\text{or } \langle l \| F \| l \rangle = 0) \qquad (3.60)$$

and the F_κ and A_κ change the value of l by ± 1. Therefore, the representations $(0, k_1 + 1)$ and $(0, -k_1 - 1)$ are equivalent.

We shall now derive the irreducible representations of $\mathscr{E}(E_3)$ from the irreducible representations (k_0, c) of $\mathscr{E}(SO(3, 1))$ by a limiting process, called Inönü–Wigner contraction. We define the operators

$$P_i = \lambda F_i, \qquad J_i = H_i. \qquad (3.61)$$

If H_i and F_i fulfill the commutation relations (3.27), (3.28), (3.29$_{\lambda^2 = -1}$) of $\mathscr{E}(SO(3, 1))$, then P_i and J_i fulfill the commutation relations

$$[J_i, J_k] = i\epsilon_{ikl} J_l, \qquad (3.62)$$

$$[J_i, P_k] = i\epsilon_{ikl} P_l, \qquad (3.63)$$

$$[P_i, P_k] = -\lambda^2 \epsilon_{ikl} J_l, \qquad (3.64)$$

and P_i depends upon λ. The invariant operators are now

$$\lambda^2 C_1 = P_i P_i - \lambda^2 J_i J_i, \tag{3.65}$$

$$\lambda C_2 = P_i J_i. \tag{3.66}$$

In the limit $\lambda^2 \to 0$ the commutation relation (3.64) goes into the commutation relation

$$[P_i, P_k] = 0. \tag{3.67}$$

Equations (3.62), (3.63), and (3.67) are the commutation relations that define the algebra $\mathscr{E}(E_3)$. Thus in the limit $\lambda^2 \to 0$ the commutation relations of $\mathscr{E}(SO(3, 1))$ go into those of $\mathscr{E}(E_3)$. However, if we simply take the limit $\lambda \to 0$, then we see from (3.61) that $P_i \to 0$. In order to obtain the algebra of operators of $\mathscr{E}(E_3)$ from the algebra of operators $\mathscr{E}(SO(3, 1))$ in $\mathscr{R}(k_0, c)$ by the limit $\lambda \to 0$, we must simultaneously increase the F_κ so that P_κ does not go to zero. This can be done, as seen from (3.35), (3.40) and (3.44), by increasing the value of $|ic| \to \infty$ when $\lambda \to 0$. The contraction in the representation is therefore performed in the following way:[11a]

$$\lambda \to 0, \qquad |ic| \to \infty \qquad \text{such that } (-ic\lambda) \to \epsilon \tag{3.68}$$

where ϵ is a finite real value. From (3.35), (3.40), and (3.44) we see that, e.g., P_3 is then given by

$$
\begin{aligned}
P_3 f_m^l &= \lim_{(-ic\lambda) \to \epsilon} \lambda F_0 f_m^l \\
&= \sqrt{l^2 - m^2}\, \tilde{c}_l f_m^{l-1} - m\tilde{a}_l f_m^l - \sqrt{(l+1)^2 - m^2}\, \tilde{c}_{l+1} f_m^{l+1}, \quad (3.69)
\end{aligned}
$$

where

$$\tilde{c}_l = \lim_{(-ic\lambda) \to \epsilon} \lambda c_l = \lim_{(-ic\lambda) \to \epsilon} \frac{i}{l} \sqrt{\frac{(l^2 - k_0^2)(\lambda^2 l^2 - \lambda^2 c^2)}{4l^2 - 1}} = \frac{i}{l} \epsilon \sqrt{\frac{l^2 - k_0^2}{4l^2 - 1}} \tag{3.70}$$

and

$$\tilde{a}_l = \lim_{(-ic\lambda) \to \epsilon} \lambda a_l = \lim_{(-ic\lambda) \to \epsilon} \frac{\lambda i c k_0}{l(l+1)} = \frac{k_0 \epsilon}{l(l+1)}. \tag{3.71}$$

The invariant operators of $\mathscr{E}(E_3)$ are $P_i P_i$ and $P_i J_i$. From (3.65) we see that in this limiting process

$$P_i P_i = \lim_{(-ic\lambda) \to \epsilon} \lambda^2 C_1 = \lim_{(-ic\lambda) \to \epsilon} (-k_0^2 \lambda^2 - \lambda^2 c^2 + \lambda^2 1) = \epsilon^2, \tag{3.72}$$

where we have used (3.55$_1$) for the irreducible representation (k_0, c). In the same way it follows from (3.66), using (3.55$_2$), that

$$P_i J_i = \lim_{(-ic\lambda) \to \epsilon} \lambda C_2 = \lim_{(-ic\lambda) \to \epsilon} (-\lambda i c k_0) = \epsilon k_0. \tag{3.73}$$

[11a] The choice of the negative sign in the contraction limit (3.68) is irrelevant and has been made in order to get the conventional signs in (3.73) and (3.78) from the convention for sign c used in (3.35) and (3.40).

The representation of $\mathscr{E}(E_3)$ obtained in this contraction process (3.68) is therefore characterized by the two parameters (k_0, ϵ) with $-\infty < \epsilon < +\infty$.

The $\mathscr{E}(SO(3))$ subalgebra and the reduction with respect to this sub-algebra are not affected by the contraction process (3.68). Therefore the irreducible representation space $\mathscr{R}(k_0, \epsilon)$ of $\mathscr{E}(E_3)$ obtained from the representation space $\mathscr{R}(k_0, c)$ is again given by [cf. (3.52)]

$$\mathscr{R}(k_0, \epsilon) = \sum_{l=k_0, k_0+1, \ldots}^{\infty} \oplus \mathscr{R}^l. \tag{3.74}$$

Summarizing our results, we have found that the Hermitian irreducible-representation spaces of $\mathscr{E}(E_3)$, defined by (3.62), (3.63), (3.67), are characterized by two numbers (k_0, ϵ) which take the values $k_0 = 0, \frac{1}{2}, 1, \ldots, -\infty < \epsilon < +\infty$. The reduction of this space with respect to $\mathscr{E}(SO(3))$ is given by (3.74), and the operators $J_\kappa = H_\kappa$ and P_κ are given by (3.36) and (3.35) with F_κ replaced by P_κ and a_l, c_l replaced by $\tilde{a}_l$, and $\tilde{c}_l$ given in (3.70) and (3.71).

The basis system f_m^l of the representation space $\mathscr{R}(k_0, \epsilon)$ is the one in which the following complete system of commuting operators is diagonal:

$$J_3, \quad J_i J_i, \quad P_i P_i, \quad P_i J_i. \tag{3.75}$$

As the P_i commute [Equation (3.67)], one usually chooses a basis system of the representation space in which the P_i are diagonal. As a complete system of commuting operators one chooses therefore

$$P_1, P_2, P_3, \quad P_i J_i \tag{3.76}$$

with the corresponding basis vectors $|p_i, k_0\rangle$, which have the property

$$\langle k_0' p_i' | p_i k_0 \rangle = \delta_{k_0 k_0'} \delta^3(\mathbf{p} - \mathbf{p}'), \tag{3.77}$$

$$P_i | p_i k_0 \rangle = p_i | p_i k_0 \rangle, \qquad P_i J_i | p_i k_0 \rangle = k_0 \epsilon | p_i k_0 \rangle, \tag{3.78}$$

$$\epsilon^2 = p_i p_i.$$

Without a derivation,[12] we mention that the transformation matrices from the basis $f_m^l(k_0, \epsilon)$ to the basis $|p_i k_0\rangle$,

$$|p_i k_0\rangle = \sum_{l, m} f_m^l(k_0, \epsilon) \langle \epsilon \, k_0 l \, m | p_i k_0 \rangle, \tag{3.79}$$

are given by

$$\langle \epsilon \, k_0 l \, m | p_i k_0 \rangle = \sqrt{\frac{2l+1}{4\pi}} \frac{1}{|\epsilon|} D_{mk_0}^l(\phi, \theta, -\phi). \tag{3.80}$$

Here (ϕ, θ) are the spherical coordinates of $\mathbf{p}$; $p_1 = |\epsilon| \sin \theta \cos \phi$, $p_2 = |\epsilon| \sin \theta \sin \phi$, $p_3 = |\epsilon| \cos \theta$; and $D_{mk_0}^l(\phi, \theta, -\phi)$ is the rotation matrix.[13]

In this appendix we have derived a huge class of irreducible representations of $\mathscr{E}(SO(4))$, $\mathscr{E}(SO(3, 1))$, and $\mathscr{E}(E_3)$, many of which are used for applications in this book or for applications to other problems in quantum

---

[12] A derivation is given in Appendix I of A. Bohm and R. B. Teese, Spectrum generating group of the symmetric top molecule, *J. Math. Phys.* **17**, 94 (1976).

[13] The rotation matrices can be found in, e.g., L. C. Biedenharn and J. D. Louck [1979], Chapter 3, or A. R. Edmonds [1957].

physics. We have used here the same method that was used in Section III.3 for the derivation of the representation of $\mathscr{E}(SO(3))$. Though group theory has not been introduced, we want to mention that these are all the unitary representations of the groups SO(4), SO(3, 1) (the homogeneous Lorentz group), and E_3 (the Euclidean group in three dimensions).

V.4 Parity

Parity is a very important observable in quantum physics, as all quantum-mechanical systems are usually in states that are eigenstates of parity. The *parity operation P*, which is also called *space inversion*, is the operation of taking the mirror image. If a physical object has the position x_i and momentum p_i, then its mirror image has the position $x_i^R = -x_i$ and momentum $p_i^R = -p_i$. For a quantum physical system this transformation P is, according to the basic assumption I, represented by a linear operator in the space of physical states. We call this operator the parity operator U_P.

Some properties of U_P follow directly from the physical interpretation of the parity P; others are convention. As we shall justify below, U_P is required to fulfill the following defining relations:

$$U_P Q_i U_P^{-1} = -Q_i \quad (i = 1, 2, 3), \tag{4.1}$$

$$U_P P_i U_P^{-1} = -P_i \quad (i = 1, 2, 3), \tag{4.2}$$

$$U_P J_i U_P^{-1} = +J_i \quad (i = 1, 2, 3), \tag{4.3}$$

where Q_i, P_i, and J_i are the position, momentum, and angular-momentum operators, respectively. A further requirement on U_P is

$$U_P U_P^{\dagger} = I \tag{4.4}$$

and

$$U_P^2 = I. \tag{4.5}$$

A consequence of (4.4) and (4.5) is

$$U_P^{\dagger} = U_P^{-1}, \qquad U_P^{-1} = U_P, \qquad U_P^{\dagger} = U_P. \tag{4.6}$$

Linear operators that fulfill (4.4) are called *unitary* operators;[14] thus U_P is unitary and according to (4.6) also Hermitian.

Equation (4.3) follows from (4.1) and (4.2) if $J_i = L_i = \epsilon_{ijk} Q_j P_k$; for other angular momenta it is a postulate.

From (4.1), (4.2), and (4.3) we obtain the relation of U_P with any observable that is a function of momentum, position, and angular momentum. When new independent observables are introduced, their relations with U_P will have to be postulated.

---

[14] See also Section XV.3 for the definition of a unitary operator.

The parity provides a classification of tensor operators. A tensor operator $T_\kappa^{(k)}$ for which

$$U_P T_\kappa^{(k)} U_P^{-1} = (-1)^k T_\kappa^{(k)} \tag{4.7}$$

is called a *(proper) tensor operator*; if

$$U_P T_\kappa^{(k)} U_P^{-1} = -(-1)^k T_\kappa^{(k)}, \tag{4.8}$$

then $T_\kappa^{(k)}$ is called a *pseudotensor operator*. Thus the operators Q_i and P_i are, according to (4.1) and (4.2), proper vector operators, whereas the operator J_i is a pseudovector (also called an axial-vector) operator, according to (4.3).

All energy operators that we have met so far fulfill the condition

$$U_P H U_P^{-1} = H; \tag{4.9}$$

i.e., they are proper scalar operators. Equation (4.9) is, however, not universally fulfilled; the part of the energy operator that causes the weak decay is known not to fulfill (4.9) (parity nonconservation).

In the following we shall give some justification of the above postulates:

Let ψ be a vector in the space of states $\mathcal{H}$, and denote by ψ^R the vector we obtain from ψ by applying U_P:

$$\psi^R = U_P \psi. \tag{4.10}$$

Let us consider the pure physical state of which ψ is the representative vector, i.e., the state Λ_ψ, which is the projector on the one-dimensional space spanned by ψ. Let us denote by Λ_ψ^R the projector on the space spanned by ψ^R, and let us make the "physically reasonable" assumption that this is again a pure state (i.e., ψ^R spans a one-dimensional space); ψ^R is one of the vectors that represent the state obtained from Λ_ψ by a space reflection, because U_P is an operator that represents this space reflection. Thus Λ_ψ^R is the space-reflected state of Λ_ψ, and these two states are connected by

$$\Lambda_\psi^R = |\psi^R\rangle\langle\psi^R| = U_P|\psi\rangle\langle\psi|U_P^\dagger = U_P \Lambda_\psi U_P^\dagger. \tag{4.11}$$

As a generalization of (4.11) we obtain [cf. (II.4.31)] that for any arbitrary state W,

$$W^R = U_P W U_P^\dagger \tag{4.12}$$

will describe the state that is obtained by a space reflection of the state W. When W is normalized (Tr $W = 1$), the space-reflected state should also be normalized (Tr $W^R = 1$), or more generally,

$$\text{Tr } W^R = \text{Tr } W, \tag{4.13}$$

i.e., space reflection should not change the probability. Thus

$$\text{Tr } W = \text{Tr}(U_P W U_P^\dagger) = \text{Tr}(W U_P^\dagger U_P),$$

which is fulfilled if one requires (4.4).

If one takes the mirror image of an object two times, then one gets the original object back. Therefore

$$W = (W^R)^R = U_P W^R U_P^\dagger = U_P U_P W U_P^\dagger U_P^\dagger. \tag{4.14}$$

This is fulfilled if one requires (4.5), which is sufficient to ensure (4.14) but not necessary for (4.14); i.e., (4.5) is also partially a convention.

Let $\bar{x}_i$ be the expectation value of the position operator of a quantum-mechanical system in the state W,

$$\bar{x}_i = \mathrm{Tr}(Q_i W);$$

and let $\bar{x}_i^R$ be the expectation value in the state W^R of the quantum-mechanical system obtained from the state W by space reflection, $W^R = U_p W U_p^\dagger$:

$$\bar{x}_i^R = \mathrm{Tr}(Q_i W^R).$$

Then, because of the physical meaning of space inversion,

$$\bar{x}_i^R = -\bar{x}_i; \tag{4.15}$$

i.e., space reflection changes the value of $\bar{x}_i$ into the value $-\bar{x}_i$. Thus we should have

$$\mathrm{Tr}(Q_i W) = -\mathrm{Tr}(Q_i W^R) = -\mathrm{Tr}(Q_i U_P W U_P^\dagger)$$
$$= -\mathrm{Tr}(U_P^\dagger Q_i U_P W). \tag{4.16}$$

This is fulfilled if

$$Q_i = -U_P^\dagger Q_i U_P,$$

or, using (4.4), if

$$U_P Q_i U_P^{-1} = -Q_i,$$

which is the requirement (4.1). The other defining relations (4.2) and (4.3) can be justified by similar arguments.

We have therefore justified the defining relations (4.1)–(4.5) of the operator that represents the observable parity by showing that these relations lead to consequences that we expect of the physical operation of space reflection. Thus U_P defined by (4.1)–(4.5) can represent the observable parity. We have not, however, shown that this U_P is the only operator that can represent parity. (A few remarks concerning this question are given in Section XIX.2, in particular in the Appendix of that section.)

We shall now study the properties of the parity operator U_P in angular-momentum eigenstates—for example, as they occur in the rotator space (III.4.2):

$$\mathscr{R} = \sum_{l=0}^{\infty} \oplus \mathscr{R}^l. \tag{4.17}$$

As the system of commuting observables

$$\mathbf{J}^2, J_3 \tag{4.18}$$

commutes with U_P, it follows that

$$\mathbf{J}^2, J_3, U_P \tag{4.19}$$

is also a system of commuting observables. We have to distinguish two cases: (a) that (4.18) is already a c.s.c.o.—as is the case for the rotator—and (b) that (4.19) is the c.s.c.o. This latter case is called *parity doubling*.

Case (a): If (4.18) is the c.s.c.o., then $|j j_3\rangle$ is a basis system. Because of (4.3) it follows that

$$U_P|j j_3\rangle = \pi|j j_3\rangle, \tag{4.20}$$

$[\pi \neq 0$ because of (4.4)], i.e., $|j j_3\rangle$ are eigenvectors of U_P. The eigenvalue $\pi = \pi(j, j_3)$, which we call the parity of the state $|j j_3\rangle$, could in general depend upon j, j_3. We will now show that

$$\pi(j, j_3) = (-1)^j \eta, \tag{4.21}$$

where η, called the *intrinsic parity*, is independent of j and j_3, and where $|\eta| = 1$.

From (III.3.22) and (III.3.23) it follows that[15]

$$U_P J_\pm|j j_3\rangle = U_P \sqrt{j(j + 1) - j_3(j_3 \pm 1)}|j j_3 \pm 1\rangle$$
$$= \pi(j, j_3 \pm 1)\sqrt{j(j + 1) - j_3(j_3 \pm 1)}|j j_3 \pm 1\rangle.$$

Because of (4.3) the left-hand side of this must be equal to

$$J_\pm U_P|j j_3\rangle = \pi(j, j_3)J_\pm|j j_3\rangle$$
$$= \pi(j, j_3)\sqrt{j(j + 1) - j_3(j_3 \pm 1)}|j j_3 \pm 1\rangle.$$

Comparing the last two equations, we conclude that

$$\pi(j, j_3) = \pi(j, j_3 \pm 1), \tag{4.22}$$

i.e., that π is independent of j_3. To find the dependence of π upon j, we have to use (4.1) and the vector-operator property of Q_i. Writing

$$Q_{\kappa = \pm 1} = \frac{Q_1 \pm iQ_2}{\mp\sqrt{2}}, \qquad Q_{\kappa = 0} = Q_3,$$

and using (3.12), we have

$$Q_\kappa|j j_3\rangle = |j - 1 j_3 + \kappa\rangle\langle j j_3 1 \kappa|j - 1 j_3 + \kappa\rangle\langle j - 1\|Q\|j\rangle$$
$$+ |j j_3 + \kappa\rangle\langle j j_3 1 \kappa|j j_3 + \kappa\rangle\langle j\|Q\|j\rangle$$
$$+ |j + 1 j_3 + \kappa\rangle\langle j j_3 1 \kappa|j + 1 j_3 + \kappa\rangle\langle j + 1\|Q\|j\rangle, \tag{4.23}$$

---

[15] In this and succeeding sections we shall use units such that $\hbar \equiv 1$. We shall on occasion restore the $\hbar$'s to an equation, in which case we are using the usual (Gaussian) c.g.s. units, cf. the explanation on p. 111 and on p. 209.

where $\langle j' \| Q \| j \rangle$ are the reduced matrix elements of Q. If we apply U_P to both sides of (4.23), then from (4.1) it follows that

$$Q_\kappa U_P |j j_3 \rangle = \pi(j) Q_\kappa |j j_3 \rangle = -U_P Q_\kappa |j j_3 \rangle$$

$$= - |j-1 j_3 + \kappa \rangle \pi(j-1) \langle j j_3 1 \kappa | j-1 j_3 + \kappa \rangle \langle j-1 \| Q \| j \rangle$$

$$- |j j_3 + \kappa \rangle \pi(j) \langle j j_3 1 \kappa | j j_3 + \kappa \rangle \langle j \| Q \| j \rangle$$

$$- |j+1 j_3 + \kappa \rangle \pi(j+1) \langle j j_3 1 \kappa | j+1 j_3 + \kappa \rangle \langle j+1 \| Q \| j \rangle.$$

$$(4.24)$$

If we insert (4.23) into the second term of this equation and compare it with the last term we conclude

$$\langle j \| Q \| j \rangle = 0, \tag{4.25}$$

$$\pi(j-1) = -\pi(j) \quad \text{if} \quad \langle j-1 \| Q \| j \rangle \neq 0, \tag{4.26}$$

$$\pi(j+1) = -\pi(j) \quad \text{if} \quad \langle j+1 \| Q \| j \rangle \neq 0. \tag{4.27}$$

According to (3.17'),[16]

$$\sqrt{2j+3} \, \langle j+1 \| Q \| j \rangle = \sqrt{2j+1} \langle j \| Q \| j+1 \rangle, \tag{4.28}$$

so if $\langle j \| Q \| j+1 \rangle = 0$ for every j, then this and (4.25) would mean that Q_κ is the zero operator. Thus for some value(s) of j, $\langle j \| Q \| j+1 \rangle \neq 0$; for all such values we have by (4.26), (4.27)

$$\pi(j) = -\pi(j+1). \tag{4.29}$$

From this it follows that

$$\pi(j) = (-1)^j \eta \qquad (\eta = \text{const}). \tag{4.21}$$

To determine what η can be we use (4.5):

$$U_P U_P |j j_3 \rangle = \pi(j) \pi(j) |j j_3 \rangle = I |j j_3 \rangle, \tag{4.30a}$$

and (4.4):

$$\langle j j_3 | I | j j_3 \rangle = \langle j j_3 | U_P^\dagger U_P | j j_3 \rangle = \bar{\pi}(j) \pi(j) \langle j j_3 | j j_3 \rangle. \tag{4.30b}$$

Thus

$$\pi(j)\pi(j) = 1, \qquad \bar{\pi}(j)\pi(j) = 1, \tag{4.31}$$

and consequently $\pi(j) = \bar{\pi}(j)$, so $\pi(j)$ is real and $\pi(j)^2 = 1$, i.e., $\pi(j)$ is $+1$ or is -1. Thus

$$\eta = +1 \quad \text{or} \quad \eta = -1. \tag{4.32}$$

---

[16] If the basis has not been chosen such that (3.17) is fulfilled for the vector operator Q but if the Q_i are Hermitian then one can prove that

$$-\sqrt{2j+3} \langle j+1 \| Q \| j \rangle = \sqrt{2j+1} \langle j \| Q \| j+1 \rangle.$$

For skew-Hermitian Q_i, i.e., $Q_i^\dagger = -Q_i$, one can prove a similar relation with (-1) on the left-hand side replaced by $(+1)$.

We can now give a complete justification for the selection rule (III.4.10):

$$\Delta j = \pm 1. \tag{4.33}$$

It is a consequence of the fact that Q_κ is a proper vector operator.

From the property

$$[J_k, Q_l] = i\epsilon_{klm}Q_m \qquad (k, l, m = 1, 2, 3) \tag{4.34}$$

it follows that $\Delta j = \pm 1, 0$ [see Equation (3.7)]. From (4.25), i.e., the property (4.1) of the Q_i together with the assumption that there is no parity doubling, it follows that $\Delta j \neq 0$, which establishes (III.4.10).

Case (b): If (4.19) is the c.s.c.o., then the parity π is an additional label for the basis vectors:

$$U_P|j j_3 \pi\rangle = \pi|j j_3 \pi\rangle. \tag{4.35}$$

The question now is what is the spectrum of U_P, i.e., what are the possible values of π. From (4.4) and (4.5) it follows in the same way as above that (4.31) must hold. Thus

$$\pi = +1 \quad \text{or} \quad \pi = -1. \tag{4.36}$$

Instead of (4.23) we now have

$$\begin{aligned}
Q_\kappa|j j_3 \pi\rangle &= |j - 1\, j_3 + \kappa, -\pi\rangle\langle j j_3 1 \kappa|j - 1\, j_3 + \kappa\rangle\langle j - 1, -\pi\|Q\|j\,\pi\rangle \\
&+ |j\, j_3 + \kappa, -\pi\rangle\langle j j_3 1 \kappa|j\, j_3 + \kappa\rangle\langle j, -\pi\|Q\|j\,\pi\rangle \\
&+ |j + 1\, j_3 + \kappa, -\pi\rangle\langle j j_3 1 \kappa|j + 1\, j_3 + \kappa\rangle\langle j + 1, -\pi\|Q\|j\,\pi\rangle,
\end{aligned} \tag{4.37}$$

where in addition to the vector-operator property (4.34) of Q_κ we have taken (4.1) into account, i.e., that Q_κ changes the eigenvalue of U_P, and also (4.36). Thus unlike case (a), (4.1) with (4.36) does not lead to any conditions on the reduced matrix elements $\langle j' \pi'\|Q\|j\,\pi\rangle$ and parities. Consequently for every choice of j, j_3 there are two vectors

$$|j j_3 \pi = +1\rangle \quad \text{and} \quad |j j_3 \pi = -1\rangle. \tag{4.38}$$

That is the reason case (b) is called the parity-doubling case.

It is customary to obtain an analogy with case (a) by labeling the states not with π but rather with the intrinsic parity η, which is defined by

$$U_P|j j_3 \eta\rangle = \eta(-1)^j|j j_3 \eta\rangle \qquad (\eta = \pm 1). \tag{4.39}$$

The parity $\pi(j)$ in a state with angular momentum j and intrinsic parity η is then given by

$$\pi(j) = (-1)^j \eta. \tag{4.21}$$

The selection rule is now seen from (4.37) to be

$$j \to j + 1, j, j - 1, \tag{4.40}$$

$$\pi \to -\pi. \tag{4.41}$$

Equation (4.41), which is a consequence of (4.1), was also fulfilled in case (a), only there it was already implicit in the selection rule (4.33) as a consequence of (4.21).

If parity is an independent observable (case (b)) then, according to (4.38), for every value of j we have two angular momentum spaces

$$\mathscr{R}^{j^{+}} \quad \text{and} \quad \mathscr{R}^{j^{-}}. \tag{4.42}$$

As Q_{κ} transforms from a given value of j^{π} to $(j + 1)^{-\pi}, j^{-\pi},$ and $(j - 1)^{-\pi}$, one obtains all angular-momentum spaces unless there is a j_0 such that

$$\langle j_0 - 1\,\eta \| Q \| j_0\,\eta \rangle = 0 \tag{4.43a}$$

or a j_1 such that

$$\langle j_1 + 1\,\eta \| Q \| j_1\,\eta \rangle = 0. \tag{4.43b}$$

That is, j_0 is the smallest value of j that can be reached by repeated application of Q_{κ}, and j_1 is the largest. The space of all states may then be written

$$\mathscr{R} = \sum_{j=j_0}^{j_1} \oplus \mathscr{R}^{j}_{\eta=+1} \oplus \sum_{j=j_0}^{j_1} \oplus \mathscr{R}^{j}_{\eta=-1} = \mathscr{R}_{\eta=+1} \oplus \mathscr{R}_{\eta=-1}. \tag{4.44}$$

Are there physical systems in nature whose space of physical states is given by (4.44)?

When we considered the rotating diatomic molecule, we used as its classical picture the dumbbell, i.e., we assumed that the moment of inertia about the line joining the two atoms was zero. However, because there are a number of electrons revolving about the two nuclei, a better classical model in many cases is a dumbbell with a flywheel on its axis. Thus in many cases the diatomic molecule is not a simple quantum-mechanical rotator but a quantum-mechanical symmetric top; the total angular momentum $\mathbf{j}$ is no longer perpendicular to the direction of the figure axis $\mathbf{a} = \mathbf{x}/|\mathbf{x}|$, but instead has a constant component in the direction $\mathbf{a}$, i.e.,

$$\mathbf{j} \cdot \mathbf{a} = \text{const}, \tag{4.45}$$

due to the revolution of the electrons.

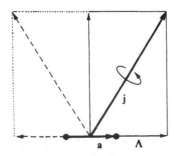

Figure 4.1 Vector diagram for the symmetric top. The curved arrow indicates the rotation of the whole diagram about $\mathbf{j}$. The dashed part of the figure gives the vector diagram when the sense of the direction of $\mathbf{a}$ is reversed.

To obtain the energy operator and the commutation relations of the quantum mechanical symmetric top we use again the correspondence between classical and quantum physics: We start from the classical non-relativistic symmetric top[17] with moment of inertia I_A about the symmetry axis given by the direction $\mathbf{a}$ and moment of inertia I_B about any axis perpendicular to $\mathbf{a}$. As $\mathbf{a}$ rotates with the angular frequency $\boldsymbol{\omega}$ one has $\dot{\mathbf{a}} = \boldsymbol{\omega} \times \mathbf{a}$. From this one obtains—by taking the cross product with $\mathbf{a}$ and using $\mathbf{a}^2 = 1$—that

$$\boldsymbol{\omega} = (\boldsymbol{\omega} \cdot \mathbf{a})\mathbf{a} + \mathbf{a} \times \dot{\mathbf{a}}. \tag{4.46}$$

Thus $\boldsymbol{\omega}$ has a component in the direction of $\mathbf{a}$ and one perpendicular to $\mathbf{a}$. The angular momentum $\mathbf{j} = \mathbb{I} \cdot \boldsymbol{\omega}$, where $\mathbb{I}$ is the tensor of inertia, can be written

$$\mathbf{j} = I_A(\boldsymbol{\omega} \cdot \mathbf{a})\mathbf{a} + I_B \mathbf{a} \times \dot{\mathbf{a}}.$$

Inserting again (4.46) this can be written

$$\mathbf{j} = I_B \boldsymbol{\omega} + (I_A - I_B)(\boldsymbol{\omega} \cdot \mathbf{a})\mathbf{a}.$$

Using

$$\mathbf{a} \cdot \mathbf{j} = \mathbf{a} \cdot \mathbb{I} \cdot \boldsymbol{\omega} = I_A \mathbf{a} \cdot \boldsymbol{\omega} \quad \text{and} \quad \mathbf{a} = \frac{\mathbf{x}}{|\mathbf{x}|}$$

one obtains from the equation above:

$$\boldsymbol{\omega} = \frac{1}{I_B}\mathbf{j} - \frac{I_A - I_B}{I_B I_A}\frac{1}{\mathbf{x}^2}(\mathbf{x} \cdot \mathbf{j})\mathbf{x}. \tag{4.47}$$

The energy of the top with angular frequency $\boldsymbol{\omega}$ and angular momentum $\mathbf{j}$ is

$$E = \tfrac{1}{2}\boldsymbol{\omega} \cdot \mathbf{j} = \tfrac{1}{2}\boldsymbol{\omega} \cdot \mathbb{I} \cdot \boldsymbol{\omega}.$$

Inserting (4.47) into this equation one obtains the energy of the classical symmetric top in terms of $\mathbf{j}$ and $\mathbf{x}$:

$$E = \frac{1}{2I_B}\left(\mathbf{j}^2 - \frac{I_A - I_B}{I_A}\frac{1}{\mathbf{x}^2}(\mathbf{x} \cdot \mathbf{j})^2\right). \tag{4.48}$$

For the quantum-mechanical system, the numbers j_i and x_i are replaced by the operators J_i and Q_i. The energy operator of the quantum-mechanical symmetrical top is therefore given by

$$H = \frac{1}{2I_B}\left(\mathbf{J}^2 - \frac{I_A - I_B}{I_A}\frac{1}{\mathbf{Q}^2}(\mathbf{Q} \cdot \mathbf{J})^2\right), \tag{4.49}[18]$$

[17] Corben (1968).

[18] $1/\mathbf{Q}^2$ is the operator defined as the inverse of $\mathbf{Q}^2 : 1/\mathbf{Q}^2 = (\mathbf{Q}^2)^{-1}$; see [[M]] in Section VI.3.

and the Q_i and J_i fulfill the commutation relations:

$$[J_i, J_j] = i\hbar\epsilon_{ijk}J_k, \qquad [J_i, Q_j] = i\hbar\epsilon_{ijk}Q_k, \qquad [Q_i, Q_j] = 0. \quad (4.50)$$

The solutions of these commutation relations, which are identical with (3.62), (3.63), and (3.67), have been derived in the Appendix to Section V.3. It has been shown that for every real number x and every integer or half-integer j_0 there exists an irreducible representation space which has the reduction with respect to angular momentum given by (4.44) with $j_1 = \infty$. The eigenvalues of the Casimir operators $\mathbf{Q}^2$ and $\mathbf{Q} \cdot \mathbf{J}$ are given by

$$\mathbf{Q}^2 = Q_i Q_i = x^2,$$
$$\mathbf{Q} \cdot \mathbf{J} = Q_i J_i = j_0 x. \qquad (4.51)$$

The irreducible representation spaces (4.44), in which the J_i, Q_j, and U_P fulfilling (4.50) and (4.1)–(4.3) act as operators, are therefore candidates for spaces of physical states of the symmetrical top. In these spaces x^2 and $j_0 x$ are constants; they would therefore describe symmetrical tops for which (4.45) is fulfilled, like the rotating diatomic molecule with electrons revolving about the two nuclei (as long as the state of the electron does not change, which would change the value of j_0). For the general symmetric top (4.45) is not fulfilled and its space of physical states is therefore a direct sum of the irreducible representation spaces (4.44) with different values of j_0. We will not consider the general symmetric top molecules here.[19]

The matrix elements of Q_i are given by (4.37) with the reduced matrix elements (determined from (3.14) (3.17) (3.70) and (3.71)) given by:

$$\langle j\,\eta'\|Q\|j\,\eta\rangle = -\frac{j_0 x}{[j(j+1)]^{1/2}}\,\delta_{\eta\eta'}, \qquad (4.52)$$

$$\langle j-1\,\eta'\|Q\|j\,\eta\rangle = -i\left[\frac{(j^2 - j_0^2)x^2}{j(2j-1)}\right]^{1/2}\delta_{\eta\eta'}, \qquad (4.53)$$

$$\langle j+1\,\eta'\|Q\|j\,\eta\rangle = -i\left[\frac{((j+1)^2 - j_0^2)x^2}{(j+1)(2j+3)}\right]^{1/2}\delta_{\eta\eta'}. \qquad (4.54)$$

The η-dependence follows from (4.1) and (4.3). These reduced matrix elements, together with (4.37) and the usual matrix elements of J_κ [as given by (III.3.8'), (III.3.22), and (III.3.23)], constitute a representation in $\mathcal{R}$ of the enveloping algebra $\mathscr{E}(E_3)$ of the three-dimensional Euclidean group extended by parity. When $j_0 = 0$, then $\langle j\|Q\|j\rangle = 0$ and the space is given by (4.17). Equation (4.54) shows that j can become arbitrarily large, i.e., $j_1 = \infty$.

[19] A. Bohm and R. B. Teese, J. Math. Phys. 17, 94 (1976).

The energy spectrum of the symmetrical top, i.e., the matrix elements of the energy operator H in the basis $|j\,j_3\,\eta\rangle$ of $\mathcal{R}$, is obtained from (4.49) using (4.51):

$$\text{spectrum } H = \frac{1}{2I_B}\left(j(j+1) - \frac{I_A - I_B}{I_A}\Lambda^2\right) \tag{4.55}$$

$$(j = \Lambda, \Lambda + 1, \Lambda + 2, \ldots), \tag{4.55a}$$

where $\Lambda = j_0$ is the conventional notation used in molecular spectroscopy. The angular momentum spectrum (4.55a) follows from (4.44) or (3.74). It is also intuitively (classically) clear, as the value of total angular momentum j must always be larger than its component Λ in the direction of $\mathbf{Q}$. The component along the direction $\mathbf{Q}$ (axis of the molecule) can take the two values $\pm\Lambda = j_0 x/|x|$ (the $-$ sign corresponding to the dashed line in Figure 4.1). As

$$Q_i J_i U_P - -U_P Q_i J_i, \tag{4.56}$$

parity states are not eigenstates of the operator $Q_i J_i$ that represents this component. Thus in parity eigenstates the molecule does not have definite values for the component of angular momentum along the molecular axis.[20]

From (4.55) it follows that the energy levels of the symmetric top, in which the component of angular momentum along the top axis is fixed, are the same as those of the simple rotator except that there is a shift of magnitude proportional to Λ^2, and except that levels with $j < \Lambda$ are absent. Each energy level has, in addition to the usual $(2j + 1)$-fold degeneracy, a twofold degeneracy because of the two possible values of $\eta = \pm 1$. The energy diagram of this kind of symmetric top for the case $\Lambda = 1$ is shown in Figure 4.2, with that of the rotator for comparison. The $+$ or $-$ sign gives the values of the

[20] This shows the inadequacy of the classical picture of a dumbbell with an electron revolving about its axis, on which Fig. 4.1 is based. In the quantum-mechanical state with definite parity the electron is neither revolving clockwise (positive value for $\mathbf{j} \cdot \mathbf{a}$) nor counterclockwise (negative value for $\mathbf{j} \cdot \mathbf{a}$) around the direction $\mathbf{a}$. It is also not in a mixture of states with clockwise and counterclockwise revolving electrons. But it is in a pure state which is incompatible with either of these eigenstates of $\mathbf{J} \cdot \mathbf{Q}$.

Figure 4.2 Energy diagram (a) of the rotator and (b) of the symmetric top. For the symmetric top $\Lambda = 1$ is assumed. The dashed level with $j = 0$ therefore does not occur. [From G. Herzberg, (1966), Vol. 1 with permission.]

parity π. The levels with a given value of j and $\pi = \pm$, which according to (4.55) should be degenerate, are drawn slightly separated. The selection rule $\Delta j = \pm 1, 0$ has already been derived. Thus for these kinds of molecules, in addition to the R and P branches given by (III.5.19), there also appear in the vibration-rotation bands transitions with $\Delta j = 0$, i.e., a series of lines given by

$$\nu_Q = \nu_0 + (B_{n'} - B_{n''})j + (B_{n'} - B_{n''})j^2 \tag{4.57}$$

(Here and also in the formula (III.5.19) for the R and P branches, ν_0 is not the oscillator's wave number but

$$\nu_0^{osc} - [(I_A - I_B)/I_A](B_{n'}\Lambda_{(n')}^2 - B_{n''}\Lambda_{(n'')}^2).$$

But as $\Lambda_{(n')}$ and $\Lambda_{(n'')}$ are constant, this last term can be absorbed in a suitably redefined ν_0.) This series of lines is called the Q-branch. The energy-level diagram and the transitions for the case that the upper rotational band has the value $\Lambda = 1$ and the lower band the value $\Lambda = 0$ are depicted in Figure 4.3.

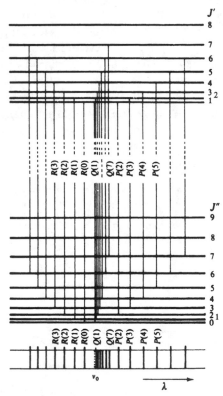

Figure 4.3 Energy level diagram for a band with P, Q, and R branches. For the sake of clarity, in the spectrogram below, the lines of the P and R branches which form a single series are represented by longer lines than those of the Q branch. The separation of the lines in the Q branch has been made somewhat too large in order that the lines might be drawn separately. The convergence in the P and R branches is frequently much more rapid than shown. [From Herzberg (1966), with permission.]

$\Delta j = 0$ transitions are not observed in the infrared spectrum of diatomic molecules. Therefore $\Lambda = 0$ for the electronic ground state of the diatomic molecules, i.e., the motion of the electrons in the electronic ground states is such that no angular momentum about the interatomic axis results, so the diatomic molecule in the ground state is a dumbbell. However, all three branches have been observed in transitions between different electronic levels of diatomic molecules. Therefore in excited electronic states Λ is in general different from zero but has a constant value as long as the electronic state does not change. Thus in an excited electronic state, the diatomic molecule is a dumbbell with a flywheel rotating around its axis. Figure 4.4(a) shows an example of the energy levels and transitions between an electronic state with the electronic angular-momentum component $\Lambda = 0$ and intrinsic parity $\eta = +1$ ($^1\Sigma^+$), and an electronic state with $\Lambda = 1$ ($^1\Pi$). Figure 4.4(b) shows the same for an electronic state with $\Lambda = 0$ and $\eta = -1$ ($^1\Sigma^-$) and the $^1\Pi$ state. The splitting between the two "degenerate" levels of opposite parity, which was indicated in Figure 4.2, and which should not be there according to (4.55), does in fact occur. It is a consequence of the interaction between the electronic motion and the rotation of the molecule (Λ-type doubling), which was not taken into account in the derivation of (4.55).

The transition frequencies given by (4.57) are in the visible or ultraviolet region, as ν_0 is in this region for transitions between different electronic states.

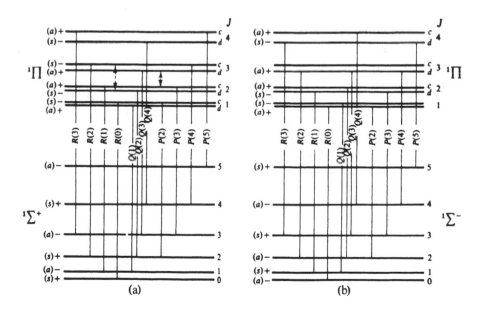

Figure 4.4 Energy-level diagram for the first lines of (a) a $^1\Pi - {}^1\Sigma^+$ transition and (b) a $^1\Pi - {}^1\Sigma^-$ transition. For the sake of clarity, the Λ-type doubling in the $^1\Pi$ state has been greatly exaggerated. The broken arrow to the left in (a) gives $R(2)-Q(2)$, and the one to the right gives $Q(3)-P(3)$. Their difference gives the sum of the Λ doublings for $J = 2$ and $J = 3$ in the upper state. [From Herzberg (1966), with permission.]

Problem

1. Determine the energy levels of the (isotropic) three-dimensional harmonic oscillator and the degree of degeneracy of each energy level. Determine the possible values of the orbital angular momentum in the nth energy level. To do this, proceed as follows:

(a) Use the basic assumption IV to obtain the space of physical states and the product basis from the combination of three one-dimensional oscillators. In this product basis the energy operator $H = P^2/(2m) + (m\omega^2/2)Q^2$ is diagonal. Use this fact to obtain the spectrum of H and the degeneracy of its eigenvalues.

(b) Introduce the angular momentum operator $L_i = \epsilon_{ijk}Q_jP_k$ and the c.s.c.o. H, $\mathbf{L}^2$, and L_3. This c.s.c.o. is not diagonal in the product basis obtained in (a). Introduce a basis of eigenvectors of the c.s.c.o. H, $\mathbf{L}^2$, and L_3 (the angular momentum basis), and determine the transition coefficients between the product basis and the angular-momentum basis. Determine the possible values of l for a given energy level E_n.

Hydrogen Atom— The Quantum-Mechanical Kepler Problem

The classical Kepler problem is described in Section VI.2, where the Lenz vector is introduced. Section VI.3 gives the algebra of angular momentum and the Lenz vector, and Section VI.4 describes the representation of this algebra. In Section VI.5 we present the algebraic derivation of the hydrogen spectrum and discuss, at the end, fine-structure effects.

VI.1 Introduction

An atom is described in everyday language as a nucleus with positive charge $+Ze$ and Z electrons, each with negative charge $-e$, that move around the nucleus in fixed, stable orbits. The simplest atom is the hydrogen atom, which —in this language—consists of the proton as the nucleus and one electron moving around it in the field of the Coulomb electrostatic force of the proton. Thus the hydrogen atom is considered to be the realization of the quantum-mechanical analog of the classical Kepler problem.

The quantum-mechanical Kepler problem was first solved by Pauli (in 1925) using the Heisenberg commutation relations. Later it became customary to treat the hydrogen atom as a solution of the Schrödinger equation for the $1/r$ potential, which was introduced in 1926. Our presentation is based on Pauli's and Bargmann's treatment.[1]

To conjecture the algebra of observables for this problem we shall follow our general procedure for quantum-mechanical systems that have classical

[1] W. Pauli, *Z. Phys.* **36**, 336 (1926); V. Bargmann, *Z. Phys.* **99**, 576 (1936).

analogs and obtain the relations between the quantum-mechanical observables by correspondence from the relations between the classical observables. We therefore recall some facts about the classical Kepler problem.

VI.2 Classical Kepler Problem

Let x_i be the coordinates of the electron (with charge $-e$), and let p_i be its momentum components. Let the proton (with charge $+e$) be located at the origin of the coordinate system. Then the Coulomb force between electron and proton is given by

$$\mathbf{F} = -\frac{e^2}{r^2}\mathbf{n},$$

where

$$\mathbf{n} = \mathbf{x}/r \quad \text{and} \quad r = |\mathbf{x}| = \sqrt{\sum_i x_i x_i},$$

and the potential energy (Coulomb potential) is given by

$$V(r) = -\frac{e^2}{r}.$$

[Here the charge e, the system constant, is measured in electrostatic units: $e = 4.803 \times 10^{-10}$ esu; 1 esu $= 1$ cm dyn$^{1/2}$ $= 1$ g$^{1/2}$ cm$^{3/2}$ sec^{-1}. If e is measured in the practical unit of charge, $e = 1.602 \times 10^{-19}$ coulomb, then $\mathbf{F}$ is given by $|\mathbf{F}| = (4\pi\epsilon_0)^{-1}(e^2/r^2)$ with $(4\pi\epsilon_0)^{-1} = 8.99 \times 10^9$ kg m^3 sec^{-2} C^{-2}]. Therefore the energy of the electron is given by

$$E = \frac{p^2}{2m_e} - \frac{e^2}{r}, \tag{2.1}$$

where m is the mass of the electron. [The electron mass $m_e = 9.11 \times 10^{-31}$ kg is much smaller than the proton mass $m_p = 1836m_e$; if one wants to take into account that m_p is not infinite, then m in (2.1) is given by the reduced mass

$$m_e \to m = \frac{m_e m_p}{m_e + m_p},$$

and $\mathbf{p}$ is not the momentum of the electron but rather

$$\mathbf{p} = \frac{-m_e \mathbf{p}_p + m_p \mathbf{p}_e}{m_e + m_p}$$

—the reduced momentum.] We will write below m_e even if we mean the reduced mass.

Besides the energy E, there are two other constants of the motion: the angular momentum

$$\mathbf{l} = \mathbf{x} \times \mathbf{p}, \tag{2.2}$$

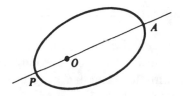

Figure 2.1

and the *Lenz vector* [or *Runge–Lenz vector*; cf. Barger and Olsson (1973)]

$$\tilde{\mathbf{a}} \equiv -\frac{\mathbf{p} \times \mathbf{l}}{m_e} + \frac{e^2}{r}\mathbf{x}. \tag{2.3}$$

The closed orbits of the classical Kepler problem are ellipses (Figure 2.1). $\mathbf{l}$ is an axial vector that is perpendicular to the plane of the orbit. As $\mathbf{l}$ is a constant of the motion, the orbit lies in some plane through O. For the Kepler problem, $V = -e^2/r$, not only is the plane of the orbit fixed, but the orbit itself is also fixed, i.e., the major axis PA is fixed and the ellipse does not precess. The constant of motion that is used to characterize the orientation of the major axis in the orbital plane is the Lenz vector $\tilde{\mathbf{a}}$, which is directed along the major axis from O to A.

The constant of motion $\tilde{\mathbf{a}}$ is a special feature of the $1/r$ potential. Whereas $\mathbf{l}$ is a constant of the motion for every spherically symmetric system [i.e., for $V(\mathbf{x}) = V(|\mathbf{x}|)$], $\tilde{\mathbf{a}}$ is a constant of the motion only for $V(\mathbf{x}) \propto 1/r$; a small deviation of the potential from this form causes the axis PA of the ellipse to precess slowly in the plane perpendicular to $\mathbf{l}$, so that the orbit is not closed.

For later comparison with the quantum-mechanical case, we list here some relations between the classical constants of the motion.[2] Let a particular orbit have the semimajor axis with length $\hat{a}$, the semiminor axis with length $\hat{b}$, and eccentricity $\epsilon = \sqrt{1 - \hat{b}^2/\hat{a}^2}$. Then

$$E = -\frac{e^2}{2\hat{a}}, \qquad \mathbf{l}^2 = m_e e^2 \hat{a}(1 - \epsilon^2),$$

$$\tilde{\mathbf{a}}^2 = \frac{2E}{m}\mathbf{l}^2 + e^4 = (e^2\epsilon)^2. \tag{2.4}$$

One defines

$$\mathbf{a} = \tilde{\mathbf{a}}\, m_e/\sqrt{-2m_e E}.$$

From (2.3) follows immediately

$$\mathbf{l} \cdot \tilde{\mathbf{a}} = \mathbf{l} \cdot \mathbf{a} = 0, \tag{2.5}$$

and a short calculation using (2.4) shows that

$$\mathbf{a}^2 + \mathbf{l}^2 = \frac{m_e^2 e^4}{-2m_e E}. \tag{2.6}$$

---

[2] See, e.g., Barger and Olsson (1973), Chapters 4–3.

VI.3 Quantum-Mechanical Kepler Problem

The quantum-mechanical observables of the quantum-mechanical Kepler problem for the hydrogen atom are obtained by replacing the numbers p_i, x_i in (2.1), (2.2), (2.3) with the Hermitian operators P_i and Q_i that fulfill the Heisenberg commutation relations (III.2.1).

For the angular momentum this has been done before; the angular momentum operators are given by

$$L_i = \epsilon_{ijk} Q_j P_k \qquad (i = 1, 2, 3). \tag{3.1}$$

The energy operator for the hydrogen atom is conjectured from (2.1) to be

$$H = \frac{\mathbf{P}^2}{2m_e} - e^2 Q^{-1}, \tag{3.2}$$

where Q^{-1} is the inverse of the operator $Q = \sqrt{Q_1^2 + Q_2^2 + Q_3^2}$.

⟦The mathematical meaning of Q^{-1} requires some explanation. As Q_i are Hermitian, $A = \sum_i Q_i^2$ is a positive-definite operator. An operator A is *positive definite* if $(f, Af) \geq 0$ and if from $(f, Af) = 0$ follows $f = 0$ for all $f \in \mathcal{H}$. In particular, A has the property that its spectrum $\{a\}$ (all its eigenvalues) is nonnegative. For every Hermitian positive-definite operator A, one uniquely defines a positive-definite Hermitian operator B such that $B^2 = A$. B is written as $B = \sqrt{A}$. It has the same system of eigenvectors as A, commutes with every operator that commutes with A, and has the eigenvalues $\{\sqrt{a}\}$. Often this definition of the operator $\sqrt{A}$ is inadequate, and one must allow for eigenvalues $\{-\sqrt{a}\}$.

The inverse operator B^{-1} of an operator B is the operator that (if it exists) has the property that $B^{-1}B = BB^{-1} = I$. If B is a Hermitian operator with spectrum $\{b\}$, then the spectrum of B^{-1} is $\{1/b\}$, and B^{-1} has the same system of eigenvectors as B.⟧

To conjecture the quantum-mechanical Lenz vector from (2.3), we have to agree first what the quantum-mechanical correspondence of the numbers $(\mathbf{p} \times \mathbf{l})_i = \epsilon_{ijk} p_j l_k$ should be. As the operators P_j and L_k do not commute, the operator $\epsilon_{ijk} P_j L_k$ is not Hermitian, but instead

$$(\epsilon_{ijk} P_j L_k)^\dagger = \epsilon_{ijk} L_k P_j = \epsilon_{ijk} P_j L_k - 2i\hbar P_i.$$

As we want the quantum-mechanical observables $\tilde{A}_i$ that correspond to the components of the Lenz vector to be Hermitian operators, we replace the classical quantity by the symmetrized product,

$$\epsilon_{ijk} p_j l_k \rightarrow \epsilon_{ijk} \tfrac{1}{2} \{P_j, L_k\},$$

where the *anticommutator* $\{A, B\}$ of two operators A and B is defined by

$$\{A, B\} \equiv AB + BA.$$

Thus the quantum-mechanical Lenz vector is

$$\tilde{A}_i \equiv \frac{1}{2m} \epsilon_{ikl}\{P_l, L_k\} + \frac{e^2 Q_i}{Q}. \tag{3.3}$$

For the sake of convenience we shall now introduce some new units. So far the momentum P has been measured in eV/c, where c is the velocity of light, and the position Q has been measured in cm. The canonical commutation relations read

$$[P_i, Q_j] = \frac{\hbar}{i} I \delta_{ij},$$

where $\hbar = 6.58 \times 10^{-18}$ eV sec $= 1.97 \times 10^{-5}$ eV cm/c with the velocity of light $c = 2.998 \times 10^{10}$ cm/sec. The Planck constant $2\pi\hbar$ was the conversion factor between the measurement of energy E in units of eV and the measurement of energy ν in units of sec^{-1} (frequency units): $E = (2\pi\hbar)\nu = h\nu$ (Planck–Einstein relation); or it is the conversion factor between the momentum in units of inverse length (cm^{-1})—i.e., as the reciprocal of the de Broglie wavelength, $1/\lambda$—and the momentum in standard units (erg sec/cm): $p = 2\pi\hbar/\lambda$ (the de Broglie relation). After we have convinced ourselves that frequency and energy or inverse wavelength and momentum are the same physical quantities measured in different units, it is no longer necessary to carry the conversion factor $\hbar$ between these different units.[3] Therefore we put $\hbar \equiv 1$ in all our calculations, which means that we replace the position Q measured in cm with the position $Q/\hbar$ measured in (eV/c)$^{-1}$. We call $Q/\hbar$ simply Q. Thus, the canonical commutation relations now read

$$[P_i, Q_j] = -iI\delta_{ij}. \tag{3.4}$$

For the particular case of the electron in a hydrogen atom we simplify the notation further by using $m_e H$ as the energy operator, i.e., instead of $\mathbf{P}^2/2m - e^2/Q$ we use the energy operator measured in units of eV $\cdot m_e$, which we simply call H:

$$H = \frac{\mathbf{P}^2}{2} - \frac{a}{Q}, \quad \text{where} \quad a = \frac{m_e e^2}{\hbar} = m_e c\alpha. \tag{3.5}$$

The quantity

$$\alpha \equiv \frac{e^2}{\hbar c}$$

is the *Sommerfeld fine-structure constant*. In these units the Lenz vector is given by

$$\tilde{A}_i = \tfrac{1}{2}\epsilon_{ikl}\{P_l, L_k\} + aQ_iQ^{-1}. \tag{3.6}$$

---

[3] That energy and momentum are also the same physical quantities measured in different units follows from relativity.

The commutation relation of the angular-momentum operators is given in these units by

$$[L_i, L_j] = i\epsilon_{ijk}L_k, \tag{3.7}$$

which is immediately seen from (III.2.14).

Now that we have conjectured the quantum-mechanical observables and thus the algebra of operators for the quantum-mechanical hydrogen-atom we want to derive the properties of this algebra and compare the results with the experimental data. In particular, we would like to know the spectrum of H, which will give us the energy levels of the hydrogen atom. It is clear that if we knew all algebras of operators in linear spaces generated by P_i, Q_j that fulfill (3.4), then the algebra of observables and space of states for the hydrogen atom would be one of them. For the hydrogen atom we need only one representation, i.e., one particular algebra of operators in one particular space of states. Instead of giving the precise mathematical specifications for this representation, we shall just go ahead and derive it. We remark that the space of physical states (subspace of the Hilbert space) for the hydrogen atom is different from the one for the three-dimensional oscillator, though the basic commutation relations (3.4) are the same.

As in the case of the oscillator, the momentum P_i and the position Q_i of the electron are not physical observables, in the sense that the physical system (hydrogen atom) cannot be prepared in approximate eigenstates of P_i or Q_i; such a preparation would break the hydrogen atom apart. The hydrogen atom, like every nonrelativistic quantum-mechanical system, appears in eigenstates (pure or mixtures) of the energy operator. The energy operator (3.2), (3.5) of the hydrogen atom differs from that of the oscillator, which is of the form $\mathbf{P}^2 + \mathbf{Q}^2$ [cf. (II.2.1b)]. The space of physical states is the space spanned by the physically preparable states. For the oscillator these are eigenstates of $\mathbf{P}^2 + \mathbf{Q}^2$, and for the hydrogen atom they are eigenstates of H in (3.5). Since the energy operators differ, the spaces of physical states differ.

In the previously considered cases of the oscillator and rotator, new observables were introduced, as functions of the P_i and Q_j, which were more directly related to the physically preparable states (e.g., the L_i for the rotator). After the relations between these new observables had been derived as consequences of (3.4), we could forget about the origin of these relations and consider them as new fundamental (defining) relations for the particular physical system under consideration. We shall follow the same procedure here for the hydrogen atom.

In order to decide what to choose for these fundamental observables in the case of the hydrogen atom, we recall that a state vector can be an eigenvector of two different operators only if these two operators commute. Physical states for stationary nonrelativistic quantum-mechanical systems appear to be in general eigenstates (or mixtures thereof) of the energy operator. Therefore we choose operators that commute with H as these new observables; these are the quantum-mechanical constants of the motion, as we shall discuss later when we consider time development (Chapter XII).

As l_i and $\tilde{a}_i$ are the classical constants of the motion, we expect that the angular-momentum operator L_i and the Lenz operator $\tilde{A}_i$ will be constants of the motion, i.e., commute with H, and are therefore to be chosen as our new fundamental observables for the hydrogen atom. A straightforward calculation (Problem 1) shows that as a consequence of the Heisenberg commutation relations (3.4) it follows that indeed

$$[H, L_i] = 0 \quad \text{and} \quad [H, \tilde{A}_i] = 0. \tag{3.8}$$

As H commutes with L_i and $\tilde{A}_i$, it will commute with the whole algebra generated by L_i and $\tilde{A}_i$. Consequently there is no operator in this algebra that transforms from a given eigenvector of H with eigenvalue E to a vector that is not an eigenvector of H with the same eigenvalue E. Let us denote the space of eigenvectors of H with eigenvalue E by $\mathcal{R}(E)$. Then all L_i, $\tilde{A}_i$ and all (well-defined) functions $A = A(\mathbf{L}, \tilde{\mathbf{A}})$ of them transform a vector $f \in \mathcal{R}(E)$ into a vector $Af = g$ that is again in $\mathcal{R}(E)$ with the same eigenvalue E. In other words, the algebra of L_i, $\tilde{A}_i$ leaves the eigenspaces $\mathcal{R}(E)$ *invariant*. Let us therefore first consider this algebra and investigate the structure of the eigenspaces $\mathcal{R}(E)$ of H.

To identify this algebra better (and to see that it is a simple, well-known mathematical structure) we define a new set of operators:

$$A_i = (\sqrt{-2H})^{-1}\tilde{A}_i = \tilde{A}_i(\sqrt{-2H})^{-1}. \tag{3.9}$$

We have to ask ourselves whether the definition (3.9) makes sense. On the space $\mathcal{R}(E)$ there is no problem, because on $\mathcal{R}(E)$ $\sqrt{-2H} = \sqrt{-2E}$, i.e., $\sqrt{-2H}$ is a number, so on $\mathcal{R}(E)$ A_i and $\tilde{A}_i$ differ just by a constant factor. Thus the restriction of A_i to $\mathcal{R}(E)$, which we again call A_i, is well defined. A_i is also well defined wherever $\sqrt{-2H}$ is well defined. This is the case for the space of vectors f for which $(f, (-2H)f) > 0$, according to the definition of the square root and inverse of an operator. This space is called the negative-energy space, or space of bound states. We restrict ourselves to this space. [For the space where $+2H$ is positive definite, instead of A_i one defines $A_i' = \sqrt{2H}\,\tilde{A}_i$ and finds that these A_i' fulfill commutation relations that differ from those of A_i by a factor of (-1), i.e., A_i and iA_i' fulfill the same commutation relations; see below.]

We denote the negative-energy space by $\mathcal{R}$. On $\mathcal{R}$, A_i given by (3.9) is well defined. From (3.8) it follows immediately that

$$[H, A_i] = 0. \tag{3.10}$$

By a straightforward but lengthy calculation (Problem 1) one can show that as a consequence of the Heisenberg commutation relation (3.4) the A_i fulfill the following commutation relations with L_i and with each other:

$$[L_i, A_j] = i\epsilon_{ijk}A_k, \tag{3.11}$$

$$[A_i, A_j] = i\epsilon_{ijk}L_k, \qquad A_i^\dagger = A_i. \tag{3.12}$$

With the L_i and A_i one defines the operators

$$C_1 \equiv \mathbf{A}^2 + \mathbf{L}^2 = A_1^2 + A_2^2 + A_3^2 + L_1^2 + L_2^2 + L_3^2, \tag{3.13}$$

$$C_2 \equiv \mathbf{A} \cdot \mathbf{L} = \mathbf{L} \cdot \mathbf{A} = A_1 L_1 + A_2 L_2 + A_3 L_3. \tag{3.14}$$

These operators have the property that they commute with the A_i and L_j (and consequently with the algebra generated by A_i, L_j) if A_i and L_j fulfill the commutation relations (3.7), (3.11), (3.12):

$$[C_1, L_i] = 0, \qquad [C_1, A_i] = 0, \tag{3.15}$$

$$[C_2, L_i] = 0, \qquad [C_2, A_i] = 0. \tag{3.16}$$

The algebra generated by the L_i and A_i that fulfill the commutation relations (3.7), (3.11), (3.12) is called $\mathscr{E}(SO(4))$.

[It is the enveloping algebra of the four-dimensional rotation group SO(4). The algebra generated by L_i and $A_i' = \sqrt{2H}\tilde{A}_i$, where $2H$ is positive definite, is the enveloping algebra $\mathscr{E}(SO(3,1))$ of the group SO(3, 1). cf. Appendix to Section V.3]

The operators C_1 and C_2 are the *invariant* or *Casimir operators* of $\mathscr{E}(SO(4))$. In an irreducible representation space of $\mathscr{E}(SO(4))$, i.e., a space that is obtained by applying every element of $\mathscr{E}(SO(4))$ to one angular momentum eigenvector (ladder representation), the operators C_1 and C_2 are multiples of the identity operator. That is, they each have only one eigenvalue, c_1 and c_2 respectively, and these values characterize the representation space [in the same way as the value $j(j + 1)$ characterizes the representation space of $\mathscr{E}(SO(3))$]. For the particular case where L_i and A_i are defined by (3.1), (3.6), (3.9), one can calculate (Problem 1) that as a consequence of (3.4),

$$C_1 = a^2(-2H)^{-1} - I, \tag{3.17}$$

$$C_2 = 0; \tag{3.18}$$

i.e., the operator C_1 is related to the energy operator (3.5), and C_2 is the zero operator. The relations (3.17) and (3.18) are the quantum-mechanical analogue of the classical relations (2.6) and (2.5) between a_i and l_i. Thus the space $\mathscr{R}(E)$ is a particular representation space of $\mathscr{E}(SO(4))$, namely the one in which $c_2 = 0$ and $c_1 = a^2(-2E)^{-1} - 1$.

If we know all possible representation spaces of $\mathscr{E}(SO(4))$ that fulfill (3.18), then we know the properties of all the operators L_i, A_i given by (3.1), (3.3), (3.9) in the spaces $\mathscr{R}(E)$, for all possible values E with $E < 0$. It will turn out that c_1 cannot take any arbitrary real value, but rather that the spectrum of C_1 is discrete. [For the invariant operator $\mathbf{J}^2$ of $\mathscr{E}(SO(3))$ we found the same result, namely that only the discrete set of eigenvalues $j(j + 1)$ with $j = 0, \frac{1}{2}, 1, \frac{3}{2}, \ldots$ is possible.] Consequently from (3.17) it will follow that the spectrum of the energy operator H is discrete and the spectrum of H is obtained by (3.17) from the spectrum of C_1.

Our task is therefore to find all the representation spaces of $\mathscr{E}(SO(4))$, in particular those that fulfill (3.18). This can be done in essentially the same way as it was done for $\mathscr{E}(SO(3))$ in Section III.3, except that the calculations for $\mathscr{E}(SO(4))$ are more involved. In the next section we shall give a description of the properties of the representation spaces of $\mathscr{E}(SO(4))$ which should suffice for the understanding of the subsequent material. A derivation of the representations of $\mathscr{E}(SO(4))$ has been given in the Mathematical Appendix to Section V.3.

VI.4 Properties of the Algebra of Angular Momentum and the Lenz Vector[4]

We choose as basis vectors in $\mathscr{R}(E)$ the angular-momentum vectors $|l\,m\rangle$. This is possible because

$$H, \mathbf{L}^2, L_3$$

is a system of commuting operators. Whether this is a c.s.c.o. for the hydrogen atom can only be decided by comparison with experiment. Therefore we make the assumption that this is a c.s.c.o.—i.e., that l, m and the energy are the only quantum numbers of the hydrogen atom—and we shall later see that this is true to a high degree of accuracy (neglecting spin effects). Thus $|l\,m\rangle$ is a basis system in $\mathscr{R}(E)$. Since $\mathbf{A}$ is a proper vector operator with respect to the angular momentum, as stated by (3.11) and $U_P A_i\, U_P = -A_i$, it follows from the Wigner–Eckart theorem that

$$A_\kappa|l\,m\rangle = |l-1\,m+\kappa\rangle\langle l\,m\,1\,\kappa|l-1\,m+\kappa\rangle\langle l-1\|A\|l\rangle$$
$$+ |l\,m+\kappa\rangle\langle l\,m\,1\,\kappa|l\,m+\kappa\rangle\langle l\|A\|l\rangle$$
$$+ |l+1\,m+\kappa\rangle\langle l\,m\,1\,\kappa|l+1\,m+\kappa\rangle\langle l+1\|A\|l\rangle. \quad (4.1)$$

Since $|l\,m\rangle$ has been assumed to be a complete basis system, there is no parity doubling, and since A_i is a proper vector operator, it follows as in (V.4.25) that the reduced matrix element $\langle l\|A\|l\rangle = 0$. [This can also be seen to be a direct consequence of (3.18), which is related to the U_P transformation property of $\mathbf{A} \cdot \mathbf{L}$]. If we insert into (4.1) the explicit values of the Clebsch–Gordan coefficients we obtain, as a consequence of (3.11) and (3.18), the following for A_κ:

$$A_0|l\,m\rangle = \sqrt{l^2 - m^2}\,C_l|l-1\,m\rangle - \sqrt{(l+1)^2 - m^2}\,C_{l+1}|l+1\,m\rangle,$$
$$(4.2_0)$$

$$A_{+1}|l\,m\rangle = -\sqrt{(l-m)(l-m-1)/2}\,C_l|l-1\,m+1\rangle$$
$$-\sqrt{(l+m+1)(l+m+2)/2}\,C_{l+1}|l+1\,m+1\rangle, \quad (4.2_+)$$

$$A_{-1}|l\,m\rangle = -\sqrt{(l+m)(l+m-1)/2}\,C_l|l-1\,m-1\rangle$$
$$-\sqrt{(l-m+1)(l-m+2)/2}\,C_{l+1}|l+1\,m-1\rangle, \quad (4.2_-)$$

[4] For a derivation see Appendix to Section V.3.

where the C_l are as yet undetermined functions of the discrete variable l, related to $\langle l - 1\|A\|l\rangle$ by Equation (V.3.14).

The C_l are determined by (3.12). We state the result: For every natural number $n = 1, 2, 3, \ldots$, there is a function $C_l^{(n)}$ such that (3.12) is fulfilled. This function is

$$C_l^{(n)} = i\sqrt{\frac{n^2 - l^2}{4l^2 - 1}}, \qquad C_0^{(n)} = 0. \tag{4.3}$$

From (4.2) and (4.3) one finds the following properties: For every number $n = 1, 2, 3, \ldots$, there is an irreducible representation space $\mathscr{R}(n)$. The lowest angular momentum in $\mathscr{R}(n)$ is $l = 0$, because starting from a given value l by application of A_k one can always reach the value $l - 1$, according to (4.2) and (4.3), unless $l = 0$, because then $C_0^{(n)} = 0$. The highest angular momentum is $n - 1$, because according to (4.2) one can always reach the value $l + 1$ from a given value l by applying A_κ, unless $C_{l+1}^{(n)} = 0$—which, however, is the case according to (4.3) for $l = n - 1$. As the $|l\,m\rangle$ are orthogonal for different values of l, the space $\mathscr{R}(n)$ is the orthogonal direct sum of the irreducible representation spaces $\mathscr{R}^l$ of angular momentum:

$$\mathscr{R}(n) \underset{\mathscr{E}(SO(3))}{\Longrightarrow} \sum_{l=0}^{n-1} \oplus \mathscr{R}^l, \tag{4.4}$$

where the symbol $\Rightarrow$ is to indicate that the two spaces are the same when only the algebra $\mathscr{E}(SO(3))$ acts in them.

The action of L_3, $L_\pm$ in each of the spaces $\mathscr{R}^l$ and therefore in $\mathscr{R}(n)$ has been derived in Section III.3. In particular,

$$L_3|l\,m\rangle = m|l\,m\rangle, \qquad \mathbf{L}^2|l\,m\rangle = l(l + 1)|l\,m\rangle. \tag{4.5}$$

Thus L_i and A_i, and consequently the whole algebra $\mathscr{E}(SO(4))$, is known in each $\mathscr{R}(n)(n = 1, 2, 3, \ldots)$. To indicate that the vector $|l\,m\rangle$ is an element of $\mathscr{R}(n)$, i.e., that A_κ is given by (4.2) with the particular $C_l^{(n)}$, we label the vector also by this number and write: $|n\,l\,m\rangle$. A straightforward calculation, using (4.2) and (4.3), gives that the eigenvalue of the operator C_1 in the space $\mathscr{R}(n)$ is $c_1 = n^2 - 1$:

$$C_1|n\,l\,m\rangle = (n^2 - 1)|n\,l\,m\rangle. \tag{4.6}$$

This gives the set of discrete numbers that are possible for c_1.

The space $\mathscr{R}(n)$ is not only an irreducible representation space of $\mathscr{E}(SO(4))$, but also of $\mathscr{E}(SO(4))$ extended by parity U_P:

$$U_P L_i U_P = + L_i, \tag{4.7}$$

$$U_P A_i U_P = - A_i. \tag{4.8}$$

Applying the considerations of Section V.4 to the space $\mathscr{R}(n)$, where there is no parity doubling, one has

$$U_P|n\,l\,m\rangle = (-1)^l \eta_n|n\,l\,m\rangle, \tag{4.9}$$

where $\eta_n = +1$ or $\eta_n = -1$, but may be different for different spaces $\mathscr{R}(n)$. We will study the n-dependence of η below.

VI.5 The Hydrogen Spectrum

In the previous section we described all the irreducible representation spaces $\mathscr{R}(n)$ of the algebra of angular momentum and the Lenz vector [i.e., all irreducible representations of $\mathscr{E}(SO(4))$ that fulfill (3.18)]. The relation (3.17) connects these spaces with the energy eigenspaces $\mathscr{R}(E)$ of negative energy $E < 0$:

$$c_1 = (n^2 - 1) = a^2(-2E)^{-1} - 1. \tag{5.1}$$

Thus the energy in each $\mathscr{R}(n)$ is given by

$$E = E_n = -\frac{a^2}{2n^2},$$

and the $\mathscr{R}(n)$ are all possible energy eigenspaces of the hydrogen atom.

The space of physical states of the hydrogen atom $\mathscr{R}$ is the direct sum of all these $\mathscr{R}(n)$

$$\mathscr{R} = \sum_{n=1}^{\infty} \oplus \mathscr{R}(n), \tag{5.2}$$

and the spectrum of the energy operator in $\mathscr{R}$ is

$$\text{spectrum } H = E_n = -a^2/(2n^2) \qquad (n = 1, 2, 3, \ldots). \tag{5.3}$$

n is called the principal quantum number.

⟦The meaning of infinite linear combinations of $f_n \in \mathscr{R}(n)$ (i.e., $\sum_{n=1}^{\infty} f_n$) requires a precise definition. It may mean infinite sum in the Hilbert-space sense ($\sum_{n=1}^{\infty} \|f_n\|^2 < \infty$), in the Schwartz-space sense ($\sum_{n=1}^{\infty} \|f_n\|^2 n^{2p} < \infty$), or in some other sense. Even if (5.2) is meant in the Hilbert-space sense, $\mathscr{R}$ is "different" from the Hilbert space $\mathscr{H} = \{\sum_{v=0}^{\infty} \phi_v | \frac{1}{2}(\mathbf{P}^2 + \mathbf{Q}^2)\phi_v = (v + \frac{3}{2})\phi_v$ and $\sum_{v=0}^{\infty} \|\phi_v\|^2 < \infty\}$, which is the usual Hilbert space of quantum mechanics. The isometric map between $\mathscr{R}$ and $\mathscr{H}$ is physically meaningless, as it would map elements that represent different physical states into each other.

Physical states of the hydrogen atom are in any event only vectors in the dense subspace $\sum_n \mathscr{R}(n)$ of the Hilbert space, where the sum extends over an arbitrarily large but finite number of n. Any limit $n \to \infty$ is a mathematical idealization, and physically for $n \to \infty$ the physical system ceases to be a hydrogen atom; it ionizes and becomes a system consisting of a proton and electron. One can, of course, consider the enlarged physical system, consisting of hydrogen-atom states as well as electron-and-proton states. For such a physical system $\mathcal{R}$ is not the total space of physical states, though this space may still be a dense subspace of $\mathcal{H}$. The space of physical states of this system is the direct sum of $\mathcal{R}$ and a space $\mathcal{R}'$ that is the continuous direct sum of irreducible-representation spaces of the algebra $\mathcal{E}(SO(3, 1))$ generated by L_i and $A'_i = \sqrt{2H}\, \tilde{A}_i$.

The space $\mathcal{R}$ is an irreducible-representation space of the algebra $\mathcal{E}(SO(4, 1))$ [and also of $\mathcal{E}(SO(4, 2))$ and of $\mathcal{E}(E_4)$]; the space $\mathcal{R}'$ is an irreducible-representation space of $\mathcal{E}(SO(3, 2))$. Such algebras have been called spectrum-generating algebras, the underlying groups are called spectrum generating groups.]

Transitions between different energy levels are performed by the position operator Q_i (dipole transitions) and powers thereof (multipole transitions). Thus the Q_i are observables that transform between different spaces $\mathcal{R}(n)$. Using the property that the Q_κ ($\kappa = 0, \pm 1$) are the components of a proper vector operator, one obtains

$$Q_\kappa |n\, l\, m\rangle = \sum_{n'} |n'\, l - 1\, m + \kappa\rangle\langle l\, m\, 1\, \kappa | l - 1\, m + \kappa\rangle (2l + 1)^{1/2} q_l^{n'n}$$
$$+ \sum_{n'} |n'\, l + 1\, m + \kappa\rangle\langle l\, m\, 1\, \kappa | l + 1\, m + \kappa\rangle (2l + 1)^{1/2} \tilde{q}_l^{n'n},$$

$$(5.4)$$

where the reduced matrix elements

$$q_l^{n'n} = (2l + 1)^{-1/2}\langle n'\, l - 1 \| Q \| n\, l\rangle,$$
$$\tilde{q}_l^{n'n} = (2l + 1)^{-1/2}\langle n'\, l + 1 \| Q \| n\, l\rangle,$$

depend upon n', n. They can be computed (Problem VII.1), but at present we only need the property that they are different from zero for all n' and n. The physical meaning of this statement is that there exist dipole transitions between all energy levels of the hydrogen atom.

We can determine the n-dependence of the intrinsic parity η_n using the U_P transformation property of the Q_κ. Applying U_P to (5.4) gives, according to (4.9),

$$U_P Q_\kappa |n\, l\, m\rangle = \sum_{n'} (-1)^{l-1} \eta_{n'} |n'\, l - 1\, m + \kappa\rangle\langle l\, m\, 1\, \kappa | l - 1\, m + \kappa\rangle$$
$$\times (2l + 1)^{1/2} q_l^{n'n}$$
$$+ \sum_{n'} (-1)^{l+1} \eta_{n'} |n'\, l + 1\, m + \kappa\rangle\langle l\, m\, 1\, \kappa | l + 1\, m + \kappa\rangle$$
$$\times (2l + 1)^{1/2} \tilde{q}_l^{n'n}. \quad (5.5)$$

But use of (V.4.1) gives

$$U_P Q_\kappa |n\, l\, m\rangle = -Q_\kappa U_P |n\, l\, m\rangle = -(-1)^l \eta_n Q_\kappa |n\, l\, m\rangle. \quad (5.6)$$

Inserting (5.4) into the right-hand side of (5.6) and comparing the coefficients with those of (5.5) (i.e., making use of the linear independence of the basis vectors), one finds

$$\eta_{n'} q_l^{n'n} = \eta_n q_l^{n'n}, \qquad \eta_{n'} \tilde{q}_l^{n'n} = \eta_n \tilde{q}_l^{n'n}. \quad (5.7)$$

As the reduced matrix elements are different from zero, $q_l^{n'n} \neq 0$ and $\tilde{q}_l^{n'n} \neq 0$, it follows that

$$\eta_{n'} = \eta_n = \eta, \tag{5.8}$$

i.e., η_n is independent of n, and the parity of the energy eigenstate is

$$U_P | n \, l \, m \rangle = (-1)^l \eta | n \, l \, m \rangle, \tag{5.9}$$

where η is either $+1$ or -1. Equation (5.9) means that the parity of a state with a definite angular-momentum value l is the same in all energy spaces $\mathcal{R}(n)$. By convention one chooses $\eta = +1$; only the relative parity has a physical meaning.

Each energy eigenspace $\mathcal{R}(n)$ of the hydrogen atom is, according to (4.4), the direct sum of angular-momentum eigenspaces $\mathcal{R}^l$; thus each energy level E_n contains the n angular momenta

$$l = 0, 1, 2, \ldots, n - 1.$$

Each angular-momentum eigenspace $\mathcal{R}^l$ in $\mathcal{R}(n)$ is, according to (III.3.24), the direct sum of the $2l + 1$ one-dimensional spaces $\mathcal{R}_m^l$:

$$\mathcal{R}^l \xrightarrow[\mathcal{E}(SO(2))]{} \sum_{m=-l}^{+l} \oplus \mathcal{R}_m^l, \tag{5.10}$$

which are spanned by each vector $| n \, l \, m \rangle$. $\mathcal{E}(SO(2))$ denotes the subalgebra of $\mathcal{E}(SO(3))$ that is generated by L_3. The dimension of $\mathcal{R}(n)$ is therefore

$$\sum_{l=0}^{n-1} (2l + 1) = n^2, \tag{5.11}$$

i.e., there are n^2 linearly independent eigenstates of H with eigenvalue E_n, or the eigenvalue E_n is n^2-fold degenerate.

In order to compare the calculated energy spectrum with the experimental data measured in eV, we insert $a = m_e e^2 / \hbar$ from (3.5) into (5.3):

$$E_n' = \frac{E_n}{m_e} = -\frac{m_e e^4}{\hbar^2} \frac{1}{2n^2} = -R'' \frac{1}{n^2} \tag{5.12}$$

(recall that E_n is the energy in $\text{eV} \cdot m_e$, E_n' is the energy in eV). The number

$$R'' = \frac{m_e e^4}{2\hbar^2} = \frac{c^2 m_e}{2} \left(\frac{e^2}{c\hbar} \right)^2$$

can be calculated using

$$m_e c^2 = 0.51 \times 10^6 \text{ eV},$$

$$e^2 = 23 \times 10^{-20} \text{ esu}^2 = 23 \times 10^{-20} \text{ dyn cm}^2,$$

$$\alpha = \frac{e^2}{c\hbar} = \frac{1}{137.036}.$$

$R'' = m_e e^4/(2\hbar^2)$ is the Rydberg constant in energy units. Its numerical value is

$$R'' = 13.6 \text{ eV}. \qquad (5.13)$$

The energy levels of the hydrogen atom are plotted in Figure 5.1, where details in the experimental energy spectrum have not been taken into account. The numbers on the left give $E_n' - E_1'$, which is measured in eV. As one sees, (5.12) is experimentally very well fulfilled. The energy levels are spaced closer and closer to one another as n increases, and they finally converge to $E_\infty = 0$ for $n \to \infty$. $E > 0$ was excluded by our restriction to the subspace where $-2H$ is positive definite; $E > 0$ is the region in which the quantum-mechanical Kepler problem does not describe the hydrogen atom, but describes an electron that is scattered by a proton (scattering states). Thus if we choose as our physical system the hydrogen atom only, $\mathscr{R}$ of (5.2) is the space of physical states. If we consider as our physical system the system that consists of the

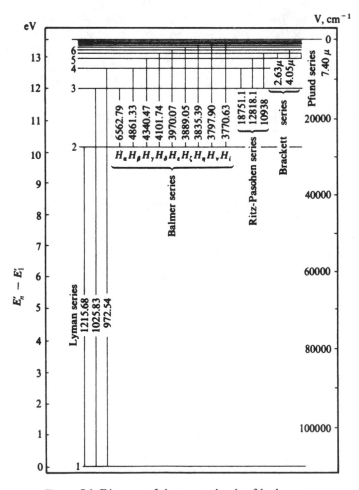

Figure 5.1 Diagram of the energy levels of hydrogen.

hydrogen atom or electron and proton, then the space of physical states is larger than $\mathscr{R}$, and $\mathscr{R}$ is the subspace of discrete energy eigenstates.[5] $E = E_\infty = 0$ is the energy value at which this physical system ceases to be a hydrogen atom; beyond this point the electron energy can take any value and one can also prepare approximate momentum eigenstates of the electron. This means that if one starts with hydrogen atoms in the ground state, one can prepare the physical system in an approximate momentum eigenstate of the electron if one transfers to it an energy of at least $-E_1$, and that the system is no longer a hydrogen atom. The continuous energy states correspond to the hyperbolic orbits of the classical Kepler problem.

The numbers on the right-hand side of the diagram give the (negative) energy in units of cm^{-1}, i.e., the wave numbers v_n that are connected with E'_n:

$$v_n = \frac{E'_n}{2\pi\hbar c}. \tag{5.14}$$

Inserting (5.12) gives for the wave numbers

$$v_n = -\frac{e^4 m_e}{\hbar^3 4\pi c} \frac{1}{n^2}. \tag{5.15}$$

The quantities

$$R' = \frac{1}{2\pi\hbar} R'' = \frac{e^4 m_e}{4\pi\hbar^3} = 3.29 \times 10^{15} \ sec^{-1}, \tag{5.16}$$

$$R = \frac{1}{2\pi\hbar c} R'' = \frac{e^4 m_e}{4\pi\hbar^3 c} = 1.097 \times 10^5 \ cm^{-1} \tag{5.16'}$$

are the Rydberg constant in frequency and wave-number units respectively. (Originally it was R that was called the Rydberg constant.)

The lines connecting the different energy levels depict the transitions in the emission or absorption of electromagnetic radiation. According to (III.1.1), the frequency and wave number of the absorbed or emitted light quanta are

$$v'_{nm} = \left| R'\left(\frac{1}{n^2} - \frac{1}{m^2}\right) \right| \quad (\text{in } sec^{-1}), \tag{5.17}$$

$$v_{nm} = \left| R\left(\frac{1}{n^2} - \frac{1}{m^2}\right) \right| \quad (\text{in } cm^{-1}). \tag{5.17'}$$

---

[5] It should be remarked that we are discussing the electron states only, i.e., describing the electron under the external influence of the nucleus. If we wanted to describe the combined system of electron and nucleus (proton), we would have to take as the space of physical states the direct-product space of the electron space $\mathscr{R}$ and the proton space $\mathscr{H}_P$: $\mathscr{R} \otimes \mathscr{H}_P$ (axiom IV), or $\mathscr{R} \otimes \mathscr{H}_{C.M.}$, where $\mathscr{H}_{C.M.}$ is the space for the center-of-mass motion. Neglecting the proton structure (e.g., proton spin), $\mathscr{H}_P$ is one-dimensional for protons at rest.

The number on each transition line in Figure 5.1 gives the wavelength of the light in angstroms:

$$\lambda = 1/v_{nm} = c/v'_{nm},$$

with $1 \text{ Å} = 10^{-8} \text{ cm} = 10^{-4} \mu$.

All transitions terminating in the same lower state form a spectral series and each series of the hydrogen atom has a historical name. The series of lines corresponding to transitions to the level $n = 1$ is called the Lyman series. The frequencies of the Lyman series are

$$v'_{1m} = R'\left(\frac{1}{1} - \frac{1}{m^2}\right) \qquad (m = 2, 3, 4, \ldots). \tag{5.18}$$

The line with the longest wavelength is $c/v'_{12} = 1215.68 \text{ Å}$. This line lies in the ultraviolet region (cf. Figure III.1.2). The series with $n = 2$,

$$v'_{2m} = R'\left(\frac{1}{2^2} - \frac{1}{m^2}\right) \qquad (m = 3, 4, 5, \ldots), \tag{5.19}$$

is called the Balmer series. It lies in the visible region of the spectrum, and (5.19) was first obtained empirically by Balmer in 1885. It played an exceptionally important role in the conjecture of the quantum theory of atoms. The other series in Figure 5.1 with $n = 3$ (Paschen), $n = 4$ (Brackett), and $n = 5$ (Pfund) lie in the infrared region.

To each energy level with energy value E_n corresponds the energy eigenspace $\mathscr{R}(n)$ of dimension n^2. The energy value E_n is n^2-fold degenerate. Thus if the energy of the hydrogen atom is measured as E_n, then the state is in general a mixture (except for $n = 1$) and the (normalized) statistical operator of this state is

$$W = (\text{Tr } \Lambda_n)^{-1}\Lambda_n, \tag{5.20}$$

where Λ_n is the projection operator on the n^2-dimensional subspace $\mathscr{R}(n)$. Energy measurements on the hydrogen atom do not reveal the value of the angular momentum; they just give an upper limit to the possible angular momentum values $l = 0, 1, \ldots, n - 1$.

The n^2-fold degeneracy of each energy level is a peculiarity of the hydrogen atom, or the $1/r$ potential. This is connected with the existence of the constant of motion A_i in addition to the angular momenta L_i. In general one has only a $(2l + 1)$-fold degeneracy of the energy values as a consequence of $[L_i, H] = 0$.

However, the degeneracy for the hydrogen atom is only approximate, and Figure 5.1 is only a rough picture of the experimental situation. The above theoretical description, like every theoretical model, gives a description of the situation in nature only up to a certain accuracy. Finer details of the hydrogen spectrum are shown in Figure 5.2 for the low-lying energy values.

Figure 5.2 Low-lying energy levels of atomic hydrogen. (Not to scale. The levels with different principal quantum number n are much further separated as is shown in Figure 5.1. This is indicated here by the broken lines separating energy levels that are far apart.)

The first number by each energy level gives the value of n. The symbol S, P, D stands for the angular-momentum value $l = 0$, 1, 2, respectively (e.g., $2P$ means $n = 2$, $l = 1$, and m can be anything such that $-l \leq m \leq +l$). According to (5.12) all $2S$ and $2P$ states should have the same energy, as shown in Figure 5.1. This is only approximately the case. The $2P$ states split into two subsets $2P^{3/2}$ and $2P^{1/2}$ with

$$E(2P^{3/2}) - E(2P^{1/2}) = 0.453 \times 10^{-4} \text{ eV} = 1095 \times 10^7 \text{ sec}^{-1}. \quad (5.21)$$

Comparing this with the numbers of (5.12), (5.13), we see that our model indeed gives an excellent approximation to the hydrogen spectrum; the deviation (5.21) is only a fraction of one percent. The splitting (5.21) (fine structure) and the $3P^{3/2}$ to $3P^{1/2}$ and $3D^{5/2}$ to $3D^{3/2}$ splittings can be explained if a new observable, the electron spin, is introduced. The superscripts $^{3/2}$, $^{1/2}$, etc. are connected with this new quantum number, the spin (see Section IX.4).

The splitting between the levels $2S^{1/2}$ and $2P^{1/2}$ cannot be explained even after the spin has been introduced. It is again an order of magnitude smaller than (5.21):

$$E(2S^{1/2}) - E(2P^{1/2}) = 0.0353 \text{ cm}^{-1}$$
$$= 0.437 \times 10^{-5} \text{ eV}$$
$$= 1058 \times 10^6 \text{ sec}^{-1}$$

and is called the *Lamb shift*. The Lamb shift is a consequence of the inter-action of the electron with the fluctuations of the radiation field, and its explanation requires a new theory called *quantum electrodynamics*.

The splitting of each term $1S^{1/2}$, $2S^{1/2}$, $2P^{1/2}$, etc. into two levels, the *hyperfine splitting*, is another order of magnitude smaller than the Lamb shift:

$$E(2S^{1/2} \text{ upper}) - E(2S^{1/2} \text{ lower}) = 0.0059 \text{ cm}^{-1},$$

$$E(2P^{1/2} \text{ upper}) - E(2P^{1/2} \text{ lower}) = 0.0020 \text{ cm}^{-1}.$$

In order to explain this splitting one needs to include observables that describe the internal structure of the nucleus (in particular, nucleon spin).

Summarizing, we have seen that our simple theoretical model describes the features of the hydrogen spectrum very well. However, to explain the fine details requires the introduction of new observables and even completely new theories (quantum electrodynamics and nuclear physics). Any given theory has only a limited domain of applicability, and this example of the hydrogen spectrum serves very well to illustrate that behind each new digit in an experimental number may be hidden a completely new domain of physics.

Problem

1.   Fill in the missing steps in the treatment of the hydrogen atom: Let L_i be the components (3.1) of the orbital-angular-momentum operator, and let H [as given by (VI.3.5)] be the energy operator of the hydrogen atom. Let $\tilde{A}_i$ denote the Lenz vector given by (3.6).

(a)   Show that as a consequence of the canonical commutation relations the angular momentum and the Lenz vector are constants of the motion, i.e., satisfy (3.8):

$$[L_i, H] = 0 \quad \text{and} \quad [\tilde{A}_i, H] = 0.$$

(b)   Define A_i by (3.9). Show that A_i is Hermitian and a constant of the motion:

$$A_i^\dagger = A_i$$

and

$$[A_i, H] = 0. \tag{3.10}$$

(c)   Show that L_i and A_j fulfill the commutation relations of (3.11) and (3.12):

$$[A_i, L_j] = i\epsilon_{ijk} A_k,$$

$$[A_i, A_j] = i\epsilon_{ijk} L_k.$$

(d)   Define the Casimir operators C_1 and C_2 by Eqs. (3.13) and (3.14). Derive (3.17) and (3.18):

$$C_1 = a^2(-2H)^{-1} - I,$$

$$C_2 = 0.$$

The following fact is useful in carrying out the above derivations: If two operators A and B commute, $[A, B] = 0$, then $[f(A), B] = 0$ for any well-defined function of the operator A [cf. Equation (II.4.46)].

CHAPTER VII

Alkali Atoms and the Schrödinger Equation of One-Electron Atoms

In Section VII.1 the concept of perturbation theory is explained using the example of the alkali atoms. Section VII.2 represents an algebraic calculation of the matrix elements of $Q^{-\nu}$ ($\nu = 1, 3, 4, \ldots$); the results are used for the calculation of the energy values of the alkali atoms. Section VII.3 gives a brief description of the solution of the Schrödinger equation for the hydrogen atom, which is used for an alternative computation of the matrix elements of $Q^{-\nu}$ and an evaluation of the alkali energy values. It also lists some properties of the spherical harmonics which are used in the second part of this book.

VII.1 The Alkali Hamiltonian and Perturbation Theory

The alkali spectra are very similar to the hydrogen spectrum. This is suggested by their classical model, according to which the alkali atom consists of one electron—the "valence electron"—that moves in the Coulomb field of the nucleus and in the average field of the other electrons that are in the orbits closer to the nucleus.

The classical potential $V_A(r)$ for this "outer" electron has, for large values of r, the form

$$V_A(r) \approx -e^2/r,$$

because the $Z - 1$ electrons in the inner orbits screen the charge of the nucleus Ze, so that the effective charge will be e. For very small r (smaller than the radius of the inner orbits)

$$V_A(r) \approx -Ze^2/r.$$

Therefore $V_A(r)$ can be written

$$V_A(r) = -\frac{e^2}{r} + V(r), \tag{1.1}$$

where $V(r)$ is negative and different from zero only in the neighborhood of the nucleus. The classical Hamiltonian function of the outer electron is then

$$h = h_{\text{Hydr}} + V(r), \tag{1.2}$$

where $V(r)$ is a small perturbation of the Hamiltonian h_{Hydr}.

The exact expression for $V(r)$ is not known since it depends upon the unknown charge distribution of the core, which consists of the nucleus and the inner electrons. If $V_A(r)$ is spherically symmetric, then $V_A(r)$ may be expanded in the form

$$V_A(r) = -c_0 \frac{e^2}{r} - c_1 \frac{e^2}{r^2} - c_2 \frac{e^2}{r^3} - \cdots, \tag{1.2'}$$

where $c_0 e$ is the charge, $c_1 e = e\mathbf{d} \cdot \mathbf{x}/r$ is the dipole contribution, $c_2 e = eq_{ij}x_i x_j/r^2$ is the quadrupole contribution, etc., of the core. For the alkali atoms the extension of the core is small compared to the distance of the outer electron, so that c_1 is small and contributions by the quadrupole moment q_{ij} and by higher multipole moments are negligible.

The Hamiltonian operator for the quantum-mechanical system is therefore, in correspondence to (1.2),

$$H = K + V, \tag{1.3}$$

where K is the Hamiltonian operator of the hydrogen atom,

$$K = \frac{\mathbf{P}^2}{2} - \frac{a}{Q}, \tag{1.3'}$$

and

$$V = V(Q) = -\frac{c_1 a}{Q^2} - \frac{c_2 a}{Q^3} - \cdots \tag{1.3''}$$

is the "small perturbation" of K. In (1.3), (1.3'), and (1.3'') we have used the units adopted in Section VI.3, i.e., we have made the replacements $Q_i/\hbar \to Q_i$ and $m_e H \to H$.

The problem is now to find the spectrum of the Hamiltonian operator H and its expectation values in the physical states. This can no longer be done exactly. Also, we do not know the exact form of $V(r)$ or $V(Q)$, though we know some of its properties. We can only hope, therefore, that the spectrum of H will not differ significantly from the spectrum of K, i.e., that V causes only a small perturbation of the spectrum of K.

The determination of the alkali spectrum is a problem of *perturbation theory*, which will be discussed in more general terms in the next chapter. One major factor in this kind of problem is the choice of an appropriate basis.

One starts with eigenstates of K, i.e., chooses a c.s.c.o. that contains K. If the other members of the c.s.c.o. do not commute with V, then V applied to this eigenvector will not only change the eigenvalue of K but also the eigenvalues of these other observables. Thus V will perturb not only the spectrum of K but also the spectrum of the other members of the c.s.c.o. Therefore the c.s.c.o. should be chosen so that as many of their members as possible commute with V.

For the alkali atoms we have

$$[V, L_k] = 0, \tag{1.4}$$

because

$$[Q, L_k] = 0.$$

Consequently, we have the ideal situation where H as well as K can be diagonalized together with $\mathbf{L}^2$ and L_3. Thus the basis $|n\,l\,m\rangle$ of the hydrogen-atom problem is a very appropriate basis to start with.

We again define

$$A_i = (-2K)^{-1/2}(\tfrac{1}{2}\epsilon_{ikl}\{P_l, L_k\} + Q_i Q^{-1}), \tag{1.5}$$

and we again have

$$C_1|n\,l\,m\rangle = (\mathbf{A}^2 + \mathbf{L}^2)|n\,l\,m\rangle = (n^2 - 1)|n\,l\,m\rangle. \tag{1.6}$$

Unlike the case of the hydrogen atom, however, it will not be true that the full Hamiltonian operator H commutes with A_i, $\mathbf{A}^2$, and C_1:

$$[H, A_i] = [V, A_i] \neq 0,$$
$$[H, C_1] = [V, C_1] \neq 0. \tag{1.7}$$

Thus H and C_1 (and consequently H and K) cannot be diagonalized together, i.e., there are no vectors in $\mathscr{R}$ that are eigenstates of H and also eigenstates of C_1. Let us denote the eigenstates of H by $|\lambda\,l\,m\rangle$ and the eigenvalues by E_λ:

$$H|\lambda\,l\,m\rangle = E_\lambda|\lambda\,l\,m\rangle. \tag{1.8}$$

H can be diagonalized together with $\mathbf{L}^2$, L_3 because of (1.4).]

Now the question arises: What are the physical states, i.e., what are the states that the alkali atoms can be prepared in? According to the basic assumption III, the alkali atoms should be in an energy eigenstate (or a mixture of energy eigenstates) if an energy measurement has been performed on the system, i.e., the statistical operator should be

$$W_H = \Lambda_\lambda, \qquad W_H^{(l)} = \Lambda_{\lambda l}, \quad \text{or} \quad W_H^{(lm)} = \Lambda_{\lambda lm} \tag{1.9}$$

if only the energy, or only the energy and angular momentum, or the energy, angular momentum, and z-component of the angular momentum have been measured, respectively. Here Λ_λ is the projector on the space of eigenvectors of H with eigenvalue E_λ, $\Lambda_{\lambda l}$ projects on the space spanned by $|\lambda\,l\,m\rangle$

$(m = -l, -l + 1, \ldots, +l)$, and $\Lambda_{\lambda l m}$ projects on the space spanned by $|\lambda\, l\, m\rangle$.

If the state of the alkali atom has not been prepared by an energy measurement but by a measurement of the observable C_1 or K, then, according to the basic assumption III, this physical system should be in an eigenstate of C_1, i.e., the statistical operator should be

$$W_K = \Lambda_n, \qquad W_K^{(l)} = \Lambda_{nl}, \qquad \text{or} \qquad W_K^{(lm)} = \Lambda_{nlm}, \qquad (1.10)$$

where Λ_n is the projection operator on the space of eigenvectors of C_1 with eigenvalue $n^2 - 1$.

Although at this stage we cannot exclude the possibility that states of atoms are prepared by a C_1 measurement, this possibility appears very unlikely, as C_1 and K have only an auxiliary meaning (except for the hydrogen atom) and the physical observable is H. (Also, in Chapter XII we shall see that states that do not change in time—and the states that correspond to the energy levels of the atoms are likely to be well described by this property— have to be eigenstates of the energy operator.) Thus, the occurrence of physical states given by (1.10) would make the formulation of the theory very unsatisfactory and would cause doubts about the appropriateness of III. Therefore it will be interesting to see how the predictions of (1.9) and (1.10) differ. In this chapter we shall perform the calculation for the alkali atoms and see that an answer to the above question cannot be obtained from the alkali spectra. In the later chapter on two-electron atoms (helium atom) we shall, however, see that experimental data require (1.9), which reassures us of the correctness of our basic assumption III—and the appropriateness of the meaning of stationary states (cf. Chapter XII).

The value predicted for the measurement of the energy operator for the state $W_K^{(lm)}$ of (1.10) is, according to the basic assumption II,

$$
\begin{aligned}
\langle H \rangle_{W_K^{(lm)}} &= \mathrm{Tr}(W_K^{(lm)} H) = \mathrm{Tr}(\Lambda_{nlm} H) = \mathrm{Tr}(\Lambda_{nlm} K) + \mathrm{Tr}(\Lambda_{nlm} V(Q)) \\
&= \langle n\, l\, m | K | n\, l\, m\rangle + \langle n\, l\, m | V(Q) | n\, l\, m\rangle \\
&= -\frac{a^2}{2n^2} + \langle n\, l\, m | V(Q) | n\, l\, m\rangle \\
&= -\frac{a^2}{2n^2} + \epsilon(n, l) \equiv E(n, l).
\end{aligned}
\qquad (1.11)
$$

For the case where the state is $W_H^{(lm)}$ of (1.9), we have

$$\langle H \rangle_{W_H^{(lm)}} = \mathrm{Tr}(W_H^{(lm)} H) = \mathrm{Tr}(\Lambda_{\lambda l m} H) = \langle \lambda\, l\, m | H | \lambda\, l\, m\rangle = E_{\lambda l}. \qquad (1.12)$$

The term $\epsilon(n, l)$ in (1.11) is the matrix element of the small perturbing Hamiltonian V:

$$\epsilon(n, l) = \langle n\, l\, m | V(Q) | n\, l\, m\rangle. \qquad (1.13)$$

As $|n\,l\,m\rangle$ is known, this matrix element can be computed if $V(Q)$ is known. The value $\epsilon(n, l)$ does not depend upon m, because of (1.4). Therefore, for the expectation value in the state Λ_{nl} one obtains

$$\langle H\rangle_{W_K^{(l)}} = \frac{\text{Tr}(W_K^{(l)}H)}{\text{Tr}\,W_K^{(l)}} = \frac{1}{2l+1} \sum_{m=-1}^{+l} \langle n\,l\,m|H|n\,l\,m\rangle = E(n, l). \quad (1.14)$$

In contrast to the expectation value $E(n, l)$, $E_{\lambda l}$ is an element of the spectrum of H. The spectrum of H and the states $|\lambda\,l\,m\rangle$ need to be determined. The spectrum of K and the states $|n\,l\,m\rangle$ are known from the hydrogen-atom problem, and in order to calculate $E(n, l)$ it only remains to calculate the "small" matrix elements $\epsilon(n, l)$.

The eigenvalues of H and the eigenvectors are determined by a perturbation-theoretical calculation. It will turn out that $E(n, l)$ of (1.11) will be the first-order approximation $E_{\lambda l}^{(1)}$ for the perturbation-theoretical calculations of the eigenvalue $E_{\lambda l}$. This is the reason we cannot distinguish between (1.9) and (1.10). As $E(n, l)$ cannot be calculated exactly [because $V(Q)$ is not known precisely], we can never tell from comparison with the experimental energy spectrum of the alkali atoms whether $E(n, l)$ is the exact energy value or only a first approximation.

VII.2 Calculation of the Matrix Elements of the Operator $Q^{-\nu}$

To obtain the correction term $\epsilon(n, l) = \langle n\,l\,m|V(Q)|n\,l\,m\rangle$, we calculate the matrix elements of the operators $Q^{-\nu}$ for any integer ν. These matrix elements will also be used for the evaluation of the fine structure splitting in Section IX.4 below.

We proceed in the following way. We first derive some relations between the operators $Q^{-\nu}$, $\mathbf{L}^2$, and

$$K = \frac{\mathbf{P}^2}{2} - \frac{a}{Q}. \quad (2.1)$$

The matrix element of one of these relations will give a recursion relation for the matrix elements $\langle n\,l\,m|Q^{-\nu}|n\,l\,m\rangle$.

We introduce the operator

$$P_r \equiv \frac{1}{2}\left\{\frac{Q_i}{Q}, P_i\right\} = \frac{Q_i}{Q}P_i - \frac{i}{Q}. \quad (2.2)$$

As a consequence of the Heisenberg commutation relation it follows that

$$[Q, P_r] = iI. \quad (2.3)$$

[Equation (2.3) follows from

$$\left[Q, \frac{1}{2}\left\{\frac{Q_i}{Q}, P_i\right\}\right] = \frac{1}{2}\left\{\frac{Q_i}{Q}, [Q, P_i]\right\} - \frac{1}{2}\left\{P_i, \left[\frac{Q_i}{Q}, Q\right]\right\}$$

by use of

$$[Q, P_i] = iQ_i/Q. \qquad (2.4)$$

The proof of (2.4) proceeds as follows:

$$[Q^2, P_i] = [Q_j Q_j, P_i] = Q_j[Q_j, P_i] + [Q_j, P_i]Q_j = 2iQ_i.$$

On the other hand

$$[Q^2, P_i] = [QQ, P_i] = Q[Q, P_i] + [Q, P_i]Q = \{Q, [Q, P_i]\},$$

and consequently

$$\{Q, [Q, P_i]\} = 2iQ_i. \qquad (2.5)$$

Furthermore,

$$\begin{aligned} [Q^2, [Q, P_i]] &= (QQ[Q, P_i] + Q[Q, P_i]Q) \\ &\quad - (Q[Q, P_i]Q + [Q, P_i]Q^2) \\ &= [Q, Q[Q, P_i]] + [Q, [Q, P_i]Q] \\ &= [Q, \{Q, [Q, P_i]\}] = [Q, 2iQ_i] = 0, \end{aligned}$$

where (2.5) has been used. However, if Q^2 commutes with $[Q, P_i]$, then $Q = (Q^2)^{1/2}$ commutes with $[Q, P_i]$ also (by the definition of the square root of an operator). Therefore it follows from (2.5) that

$$2Q[Q, P_i] = 2iQ_i,$$

which in turn gives (2.4).]

As a consequence of the definition of **L**, $L_i = \epsilon_{ijk}Q_j P_k$, and the definition (2.2), it follows by a straightforward calculation that

$$\mathbf{P}^2 = P_r^2 + \frac{1}{Q^2}\mathbf{L}^2. \qquad (2.6)$$

With this the energy operator K can be written

$$K = \frac{P_r^2}{2} + \frac{\mathbf{L}^2}{2Q^2} - \frac{a}{Q}. \qquad (2.7)$$

$P_r^2/2$ is the operator of the kinetic energy of radial motion, and P_r, the conjugate to the radius operator $Q = \sqrt{Q_i Q_i}$, is often called the *radial momentum operator*.

The radial momentum operator fulfills the commutation relation

$$[Q^{-\nu}, P_r] = -i\nu Q^{-(\nu+1)}. \qquad (2.8)$$

[We prove (2.8) by induction on ν. First note that

$$[I, P_r] = 0 = [QQ^{-1}, P_r] = Q[Q^{-1}, P_r] + [Q, P_r]Q^{-1}.$$

Then use of (2.3) gives $[Q^{-1}, P_r] = -iQ^{-2}$. Consequently (2.8) is true for $\nu = 0, \pm 1$. Assuming (2.8) is true for $\nu = n$, one calculates

$$[Q^{-n-1}, P_r] = Q^{-1}[Q^{-n}, P_r] + [Q^{-1}, P_r]Q^{-n}$$

$$= Q^{-1}(-in)Q^{-(n+1)} - iQ^{-n-2} = -i(n+1)Q^{-(n+2)}.$$

Thus if (2.8) is true for $\nu = n$, it is true for $\nu = n + 1$. In the same way, one shows (2.8) is true for $\nu = -(n + 1)$ given that it is true for $\nu = -n$. Therefore (2.8) is generally true.]

As a consequence of (2.8) and (2.7) it follows by a straightforward calculation that

$$[K, Q^{-1(\nu-1)}] = \tfrac{1}{2}(\nu - 1)(2Q^{-\nu}iP_r - \nu Q^{-(\nu+1)}). \tag{2.9}$$

A special case of (2.9) is

$$[K, Q] = -iP_r. \tag{2.9a}$$

The relation

$$[K, iP_r] = L^2 Q^{-3} - aQ^{-2} \tag{2.10}$$

can be checked by inserting (2.7) into the left-hand side and using (2.8).

With (2.9) and (2.10) one calculates

$$[K, [K, Q^{-(\nu-1)}]] = (\nu - 1)[-\nu(\nu + 1)Q^{-(\nu+2)}iP_r - \nu Q^{-(\nu+1)}P_r^2$$

$$+ \tfrac{1}{4}\nu(\nu + 1)(\nu + 2)Q^{-(\nu+3)} + L^2 Q^{-(\nu+3)} - aQ^{-(\nu+2)}]. \tag{2.11}$$

Inserting (2.9a) into (2.9) and taking the commutator of the result with K gives

$$[Q^{-\nu}[Q, K], K] - \tfrac{1}{2}\nu[Q^{-(\nu+1)}, K] + \nu[Q^{-(\nu+1)}, K]$$

$$= \frac{1}{\nu - 1}[[K, Q^{-(\nu-1)}], K] + \nu[Q^{-(\nu+1)}, K], \tag{2.12}$$

where one term has been added to both sides.

Inserting (2.11), (2.7), and (2.9) into the right-hand side of (2.12) gives

$$[Q^{-\nu}[Q, K], K] + \tfrac{1}{2}\nu[Q^{-(\nu+1)}, K] = 2\nu Q^{-(\nu+1)}K - (\nu + 1)Q^{-(\nu+3)}L^2$$

$$+ (2\nu + 1)aQ^{-(\nu+2)} + \tfrac{1}{4}\nu(\nu + 1)(\nu + 2)Q^{-(\nu+3)}. \tag{2.13}$$

The recursion relation for the matrix elements of $Q^{-\nu}$ is obtained if one takes the matrix elements of (2.13) between the vectors $|n\,l\,m\rangle$, for which

$$K|n\,l\,m\rangle = -\frac{a^2}{2n^2}|n\,l\,m\rangle,$$

$$L^2|n\,l\,m\rangle = l(l + 1)|n\,l\,m\rangle,$$

$$L_3|n\,l\,m\rangle = m|n\,l\,m\rangle.$$

The left-hand side of (2.13) between the vectors $|n\,l\,m\rangle$ gives zero, so that we get as a recursion relation

$$0 = -v\frac{a^2}{n^2}\langle n\,l\,m|Q^{-(v+1)}|n\,l\,m\rangle + (2v+1)a\langle n\,l\,m|Q^{-(v+2)}|n\,l\,m\rangle.$$

$$+ [-(v+1)l(l+1) + \tfrac{1}{4}v(v+1)(v+2)]\langle n\,l\,m|Q^{-(v+3)}|n\,l\,m\rangle.$$

$$(2.14)$$

If one takes the matrix elements of (2.13) between states $|n\,l\,m\rangle$ and $|n'\,l'\,m\rangle$, one obtains recursion relations for the off-diagonal matrix elements.[1]
   For $v = -1$, Equation (2.14) gives

$$\langle n\,l\,m|Q^{-1}|n\,l\,m\rangle = a/n^2. \tag{2.15}$$

The matrix elements of Q^{-2} cannot be obtained from (2.14); they may, however, be calculated using the wavefunctions found in the next section and performing the integral (3.29) given there with $V(r) = 1/r^2$. The result is

$$\langle n\,l\,m|Q^{-2}|n\,l\,m\rangle = \frac{a^2}{n^3(l+\frac{1}{2})}. \tag{2.16}$$

For $v = -2$, Equation (2.14) gives

$$\langle n\,l\,m|Q|n\,l\,m\rangle = \frac{n^2}{2a}\left(3 - \frac{l(l+1)}{n^2}\right).$$

Of all the matrix elements of Q^{-v} that can be obtained from (2.14), we list the lowest ones:

$$\langle n\,l\,m|Q^{-3}|n\,l\,m\rangle = \frac{a}{l(l+1)}\langle n\,l\,m|Q^{-2}|n\,l\,m\rangle$$

$$= \frac{a^3}{n^3l(l+1)(l+\frac{1}{2})}, \tag{2.17}$$

$$\langle n\,l\,m|Q^{-4}|n\,l\,m\rangle = \frac{a}{n^2}\frac{3n^2 - l(l+1)}{2l(l+1)-\frac{3}{2}}\langle n\,l\,m|Q^{-3}|n\,l\,m\rangle. \tag{2.18}$$

The constant a has, in our units, the dimension of an inverse length. According to (VI.3.5),

$$a = m_e c\alpha. \tag{2.19}$$

The above Q is, because of our choice of units, not measured in cm but rather in $(eV/c)^{-1}$ [cf. the discussion preceding Equation (VI.3.4)], i.e., the above Q is in fact $Q/\hbar$. If Q is measured in its usual units of cm, then the above a is to be

[1] N.V.V.J. Swamy and R.G. Kulkarni and L.C. Biedenharn, *J. Math. Phys.* **11**, 1165 (1970).

replaced by $a/\hbar$. Thus for Q measured in cm the above equations hold with a replaced by

$$a' = \frac{m_e c \alpha}{\hbar} = \frac{1}{r_B} = \frac{1}{0.529 \times 10^{-8} \text{ cm}}. \tag{2.20}$$

r_B is called the *Bohr radius*.

We can now give the energy value

$$E(n, l) = -\frac{a^2}{2n^2} + \langle n\,l\,m|V(Q)|n\,l\,m\rangle \tag{2.21}$$

for any V that is a sum of positive and negative powers of Q. Equation (2.21) is—as we mentioned above and as we shall see in the next chapter— also equal to the first approximation of the perturbation expansion. We shall use for $V(Q)$ only the dipole term

$$V(Q) = -c_1 a/Q^2, \tag{2.22}$$

which—as was mentioned above—should be a sufficiently good approximation.

From (2.16) and (2.22) one obtains for $E(n, l)$

$$E(n, l) = -\frac{a^2}{2n^2}\left(1 + \frac{2c_1 a}{n(l + \frac{1}{2})}\right). \tag{2.23}$$

Thus the representation spaces $\mathscr{R}(n)$ of $\mathscr{E}(SO(4))$ no longer have only one energy value, for H does not commute with A_l. [In group-theoretical language this is expressed as "$V(Q)$ breaks the $SO(4)$ symmetry."]

It is customary to write (2.23) in the form

$$E(n, l) = -\frac{2^2}{2n^{*2}} = -\frac{a^2}{2(n - \sigma)^2}, \tag{2.24}$$

where n^* is called the *effective quantum number* and $\sigma = \sigma(n, l)$ is called the *quantum defect*. Comparing (2.24) with (2.23) and using the fact that c_1 is small compared to the Bohr radius $1/a$, one obtains that the quantum defect for the alkali atoms is given by

$$\sigma = \frac{c_1 a}{l + \frac{1}{2}}. \tag{2.25}$$

The energy levels for lithium, sodium, and cesium are given in the diagrams of Figures 2.1, 2.2, and 2.3, respectively. The energy levels are labeled in the standard notation. The first number stands for the principal quantum number n, and the letters s, p, d, f, ... denote the angular-momentum quantum-number values $l = 0, 1, 2, 3, \ldots$, respectively; e.g., $3p$ means $n = 3$, $l = 1$. The lines connecting different energy levels indicate dipole transitions. The numbers in the lines give the wavelengths in angstroms. These transitions fulfill the selection rule $\Delta l = \pm 1$ [cf. (V.4.33)]. Thick lines indicate transitions of higher intensity.

Comparing these energy diagrams with the energy diagram of Figure VI.5.1 for hydrogen, we observe the l-dependence of the correction term in (2.23), which leads to a splitting of the levels with the same principal quantum number. Thus (2.23) gives a good quantitative approximation to the experimental energy spectrum. We observe further that some of the energy levels split in contradiction to (2.23); e.g., there are two energy levels $3p_1$ and $3p_2$ of sodium, whereas according to (2.23) there should be only one. According to (2.14) through (2.18), even if one takes for $V(Q)$ the general spherically symmetric expansion (1.3″), the energy values should depend upon n and l

Figure 2.1 Energy levels of lithium atoms.

Figure 2.2 Energy levels of sodium atoms.

only. The splitting of these levels with the same value of the quantum numbers n, l will be explained by the introduction of the new observable *spin*, which will be discussed in Chapter IX.

We also observe in the experimental energy diagrams of Figures 2.1–2.3 that for lithium there is no energy level with $n = 1$, that for sodium there are no energy levels with $n = 1$ or 2, that for potassium (which is not depicted here) there are no energy levels with $n = 1, 2$, or 3, and that for cesium there are no energy levels with $n = 1, 2, 3, 4$, or 5. This means that our physical system, which is the outer electron in the electrostatic potential $V_A(r) = -e^2/r + V(r)$, cannot be in a state with principal quantum number $n = 1$ (lithium), $n = 1$ or 2 (sodium), etc. In Chapter X we shall see that this can be explained as a consequence of a new basic assumption of quantum mechanics, the *Pauli exclusion principle*.

Figure 2.3 Energy levels of cesium atoms.

VII.3 Wave Functions and Schrödinger Equation of the Hydrogen Atom and the Alkali Atoms

The solution of the Schrödinger equation for the hydrogen atom is so familiar that it need not be discussed in detail here. It does not lead to any new observable results, which have not already been discussed in the previous chapter. The new quantities that it provides are the wave functions $\psi_{nlm}(\mathbf{x}) = \langle x_1, x_2, x_3 | n\,l\,m \rangle = \langle \mathbf{x} | n\,l\,m \rangle$. The quantity $|\psi(\mathbf{x})|^2$ represents, according to the results in Chapters I and II, the probability density for the observation of the electron in the state $|n\,l\,m\rangle$ at the position $\mathbf{x}$. Such a position measurement, however, cannot be performed. Nevertheless, solving the Schrödinger equation is the most common method of calculating the energy values, and it is often a very convenient method. Furthermore, the wave functions will be of physical importance for scattering problems. We give here only a brief description of this method, since detailed calculations are given in many good standard textbooks on quantum mechanics.

The Schrödinger equation for the one-electron atom is obtained by taking the "scalar product" of the energy eigenvalue equation

$$H|n\,l\,m\rangle = E|n\,l\,m\rangle$$

with the generalized eigenvector $|\mathbf{x}\rangle = |x_1 x_2 x_3\rangle$. Using (2.7) one obtains then:

$$\langle\mathbf{x}|\left(\frac{P_r^2}{2} + \frac{\mathbf{L}^2}{2Q^2} + V_A(Q)\right)|n\,l\,m\rangle = E\langle\mathbf{x}|n\,l\,m\rangle. \tag{3.1}$$

For the hydrogen atom $V_A(Q) = V_{\text{Hydr}}(Q) = -a/Q$. Equation (3.1) also applies to any other one-particle problem in three-dimensional space if $V_A(Q)$ is replaced by the appropriate potential-energy operator, and the $|n\,l\,m\rangle$ by the appropriate basis vectors.

In the basis of generalized eigenvectors $|\mathbf{x}\rangle$ of the Q_i, the position operators Q_i act like multiplication by x_i:

$$\langle\mathbf{x}|Q_i|\psi\rangle = x_i\langle\mathbf{x}|\psi\rangle. \tag{3.2a}$$

The momentum operators P_i, according to (II.7.53), act like differentiation with respect to x_i:[2]

$$\langle\mathbf{x}|P_i|\psi\rangle = \frac{1}{i}\frac{\partial}{\partial x_i}\langle\mathbf{x}|\psi\rangle. \tag{3.2b}$$

We label the generalized position eigenvectors $|\mathbf{x}\rangle$ by the polar coordinates (r, θ, ϕ) instead of the Cartesian coordinates:

$$x_1 = r\sin\theta\cos\phi, \qquad x_2 = r\sin\theta\sin\phi, \qquad x_3 = r\cos\theta. \tag{3.3}$$

From (2.2) and (3.2) it then follows that

$$\langle\mathbf{x}|P_r|\psi\rangle = \frac{1}{i}\left(\frac{\partial}{\partial r} + \frac{1}{r}\right)\langle\mathbf{x}|\psi\rangle,$$

$$\langle\mathbf{x}|P_r^2|\psi\rangle = -\frac{1}{r^2}\frac{\partial}{\partial r}\left(r^2\frac{\partial}{\partial r}\langle\mathbf{x}|\psi\rangle\right). \tag{3.4}$$

---

[2] Note that the derivation of (II.7.53) required, in addition to the Heisenberg commutation relation, the assumption that $P^2 + Q^2$ had at least one eigenvector (or equivalently $P^2 + Q^2$ was essentially self-adjoint). Thus (3.2b) is not a consequence of (VI.3.4) only, but requires an additional assumption. This assumption is natural for the case of the three-dimensional oscillator, because then $\mathbf{P}^2 + \mathbf{Q}^2$ represents the observable energy. For the hydrogen atom the energy is given by (2.1), and $\mathbf{P}^2 + \mathbf{Q}^2$ need not correspond to any observable for the hydrogen atom. Thus (3.2b) is an additional assumption, and the description of the hydrogen atom given in Chapter VI is not completely equivalent to the description by the Schrödinger equation. $\mathcal{R}$ given by (VI.5.2), considered as a Hilbertian direct sum, is a complete Hilbert space; it is a representation space of a unitary representation of SO(4) and of an irreducible unitary representation of the groups SO(4, 1) and SO(4, 2). In $\mathcal{R}$ the operator C_1 given by (VI.3.13)— and therewith, by (VI.3.17), the energy operator of the hydrogen atom—is self-adjoint. The space of square-integrable solutions of the Schrödinger equation (3.1) with discrete eigenvalues (the space of bound states) is not a complete Hilbert space. Remarkably, these distinct topological properties do not make any difference for the observable quantities.

The matrix element of the angular momentum operator $L_i = \epsilon_{ijk}Q_jP_k$ in the basis of generalized eigenvectors of the Q_i is, according to (3.2), given by

$$\langle \mathbf{x}|L_i|\psi\rangle = \frac{1}{i}\epsilon_{ijk}x_j\frac{\partial}{\partial x_k}\langle \mathbf{x}|\psi\rangle. \tag{3.5}$$

If the generalized position eigenvectors are labeled by the polar coordinates (3.3), the differential operators on the right-hand side of (3.5) have to be transformed from Cartesian to spherical coordinates; the results of that somewhat lengthy transformation are

$$\langle \mathbf{x}|L_1|\psi\rangle = \frac{1}{i}\left(-\sin\phi\frac{\partial}{\partial\theta} - \cot\theta\cos\phi\frac{\partial}{\partial\phi}\right)\langle r\,\theta\,\phi|\psi\rangle, \tag{3.6_1}$$

$$\langle \mathbf{x}|L_2|\psi\rangle = \frac{1}{i}\left(+\cos\phi\frac{\partial}{\partial\theta} - \cot\theta\sin\phi\frac{\partial}{\partial\phi}\right)\langle r\,\theta\,\phi|\psi\rangle, \tag{3.6_2}$$

$$\langle \mathbf{x}|L_3|\psi\rangle = \frac{1}{i}\frac{\partial}{\partial\phi}\langle r\,\theta\,\phi|\psi\rangle. \tag{3.6_3}$$

From this one calculates

$$\langle \mathbf{x}|\mathbf{L}^2|\psi\rangle = -\left(\frac{1}{\sin^2\theta}\frac{\partial^2}{\partial\phi^2} + \frac{1}{\sin\theta}\frac{\partial}{\partial\theta}\left(\sin\theta\frac{\partial}{\partial\theta}\right)\right)\langle r\,\theta\,\phi|\psi\rangle. \tag{3.6}$$

For $|\psi\rangle = |n\,l\,m\rangle$ the equations (3.6_3) and (3.6) become

$$\frac{1}{i}\frac{\partial}{\partial\phi}\langle r\,\theta\,\phi|n\,l\,m\rangle = m\langle r\,\theta\,\phi|n\,l\,m\rangle \tag{$3.6'_3$}$$

and

$$-\left(\frac{1}{\sin^2\theta}\frac{\partial^2}{\partial\phi^2} + \frac{1}{\sin\theta}\frac{\partial}{\partial\theta}\left(\sin\theta\frac{\partial}{\partial\theta}\right)\right)\langle r\,\theta\,\phi|n\,l\,m\rangle = l(l+1)\langle r\,\theta\,\phi|n\,l\,m\rangle. \tag{3.6'}$$

Inserting (3.4) into (3.1) gives

$$-\frac{1}{2}\frac{1}{r^2}\frac{\partial}{\partial r}\left(r^2\frac{\partial}{\partial r}\langle r\,\theta\,\phi|n\,l\,m\rangle\right) + \frac{l(l+1)}{2r^2}\langle r\,\theta\,\phi|n\,l\,m\rangle + V(r)\langle r\,\theta\,\phi|n\,l\,m\rangle$$
$$= E\langle r\,\theta\,\phi|n\,l\,m\rangle. \tag{3.7}$$

Using the expansion

$$|n\,l\,m\rangle = \int d^3x\,|\mathbf{x}\rangle\langle \mathbf{x}|n\,l\,m\rangle$$
$$= \int r^2\sin\theta\,dr\,d\theta\,d\phi|r\,\theta\,\phi\rangle\langle r\,\theta\,\phi|n\,l\,m\rangle \tag{3.8}$$

in the orthonormality condition

$$\langle n'\,l'\,m'|n\,l\,m\rangle = \delta_{nn'}\delta_{ll'}\delta_{mm'} \tag{3.9}$$

gives the normalization condition for the transition coefficients:

$$\int r^2 \sin \theta \, dr \, d\theta \, d\phi \langle n' \, l' \, m' | r \, \theta \, \phi \rangle \langle r \, \theta \, \phi | n \, l \, m \rangle = \delta_{nn'} \delta_{ll'} \delta_{mm'}. \quad (3.10)$$

Equation (3.7) for a given value of l is a differential equation in the variable r only, while (3.6$_3$) and (3.6') are differential equations in the angular variables θ, ϕ only. Therefore the transition coefficients are written

$$\langle r \, \theta \, \phi | n \, l \, m \rangle = R_{nl}(r) \, Y_{lm}(\theta, \phi). \quad (3.11)$$

Equations (3.7), (3.6'), and (3.6$_3$) then lead to

$$\left\{ -\frac{1}{2r^2} \frac{\partial}{\partial r} \left(r^2 \frac{\partial}{\partial r} \right) + \frac{l(l+1)}{2r^2} + V_A(r) \right\} R_{nl}(r) = E(n, l) R_{nl}(r), \quad (3.12)$$

$$-\left(\frac{1}{\sin^2 \theta} \frac{\partial^2}{\partial \phi^2} + \frac{1}{\sin \theta} \frac{\partial}{\partial \theta} \left(\sin \theta \frac{\partial}{\partial \theta} \right) \right) Y_{lm}(\theta, \phi) = l(l+1) Y_{lm}(\theta, \phi),$$

$$(3.13)$$

and

$$\frac{1}{i} \frac{\partial}{\partial \phi} Y_{lm}(\theta, \phi) = m Y_{lm}(\theta, \phi). \quad (3.14)$$

For the hydrogen atom $V_A(r) = -a/r$ and $E(n, l) = -a^2/(2n^2)$. The normalization condition (3.10) gives the following normalization conditions for R_{nl} and Y_{lm}:

$$\int r^2 \, dr \, \bar{R}_{n'l}(r) R_{nl}(r) = \delta_{nn'}, \quad (3.15)$$

and

$$\int \sin \theta \, d\theta \, d\phi \, \bar{Y}_{l'm'}(\theta, \phi) Y_{lm}(\theta, \phi) = \delta_{ll'} \delta_{mm'}. \quad (3.16)$$

The normalized solutions of (3.13) and (3.14) are the *spherical harmonics*

$$Y_{lm}(\theta, \phi) = \frac{(-1)^l}{2^l l!} \sqrt{\frac{2l+1}{4\pi} \frac{(l+m)!}{(l-m)!}} \, e^{im\phi} \sin^{-m} \theta \left(\frac{d}{d \cos \theta} \right)^{l-m} \sin^{2l} \theta. \quad (3.17)$$

It should be noted that different authors use different phase conventions in defining the $Y_{lm}(\theta, \phi)$. Explicit expressions for the first few spherical harmonics are given by $Y_{lm}(\theta, \phi) = \Theta_{lm}(\theta) e^{im\phi} / \sqrt{2\pi}$ where

$$\begin{aligned}
\Theta_{00}(\theta) &= \sqrt{\tfrac{1}{2}}, & \Theta_{20}(\theta) &= \sqrt{\tfrac{5}{8}}(3 \cos^2 \theta - 1), \\
\Theta_{10}(\theta) &= \sqrt{\tfrac{3}{2}} \cos \theta, & \Theta_{2 \pm 1}(\theta) &= \mp \sqrt{\tfrac{15}{4}} \sin \theta \cos \theta, \quad (3.18) \\
\Theta_{1 \pm 1}(\theta) &= \mp \sqrt{\tfrac{3}{4}} \sin \theta, & \Theta_{2 \pm 2}(\theta) &= \sqrt{\tfrac{15}{16}} \sin^2 \theta.
\end{aligned}$$

The $Y_{lm}(\theta, \phi)$ form a complete set of single-valued functions over the unit sphere:

$$\sum_{l=0}^{\infty} \sum_{m=-l}^{l} Y_{lm}(\theta, \phi) \overline{Y}_{lm}(\theta', \phi') = \frac{\delta(\theta - \theta')\delta(\phi - \phi')}{\sin \theta}. \tag{3.19}$$

Another useful relation involving the $Y_{lm}(\theta, \phi)$ is the *coupling rule*:[3]

$$Y_{l_1 m_1}(\theta, \phi) Y_{l_2 m_2}(\theta, \phi) = \sum_{l} \left[\frac{(2l_1 + 1)(2l_2 + 1)}{4\pi(2l + 1)} \right]^{1/2}$$
$$\times \langle l_1 m_1 l_2 m_2 | l\, m \rangle \langle l_1 0\, l_2 0 | l\, 0 \rangle Y_{lm}(\theta, \phi). \tag{3.20}$$

We also give here some useful relations which involve the spherical harmonics along with the *Legendre polynomials* defined by

$$P_l(\xi) = \frac{1}{2^l l!} \left(\frac{d}{d\xi} \right)^l (\xi^2 - 1)^l \quad \text{for } l = 0, 1, 2, \ldots, \tag{3.21}$$

and the *associated Legendre polynomials* defined by

$$P_l^m(\xi) = (1 - \xi^2)^{m/2} \left(\frac{d}{d\xi} \right)^m P_l(\xi) \quad \text{for } m = 0, 1, 2, \ldots, l. \tag{3.22}$$

The spherical harmonics are related to the associated Legendre polynomials by

$$Y_{lm}(\theta, \phi) = \begin{cases} (-1)^m \sqrt{\dfrac{2l + 1}{4\pi} \dfrac{(l - m)!}{(l + m)!}}\, e^{im\phi} P_l^m(\cos \theta) & \text{for } m \geq 0 \\[2ex] \sqrt{\dfrac{2l + 1}{4\pi} \dfrac{(l + m)!}{(l - m)!}}\, e^{im\phi} P_l^{-m}(\cos \theta) & \text{for } m \leq 0, \end{cases} \tag{3.23}$$

where for $m \leq 0$ we have made use of

$$Y_{lm}(\theta, \phi) = (-1)^m \overline{Y}_{l-m}(\theta, \phi). \tag{3.24}$$

When $m = 0$, (3.23) becomes

$$Y_{l0}(\theta, \phi) = \sqrt{\frac{2l + 1}{4\pi}}\, P_l(\cos \theta). \tag{3.25}$$

Lastly, we state the *addition theorem*: Let $\mathbf{n}$ and $\mathbf{n}'$ be unit vectors with directions specified by (θ, ϕ) and (θ', ϕ'), respectively; then

$$P_l(\mathbf{n} \cdot \mathbf{n}') = \frac{4\pi}{2l + 1} \sum_{m=-l}^{l} Y_{lm}(\theta, \phi) \overline{Y}_{lm}(\theta', \phi'). \tag{3.26}$$

---

[3] See, e.g., Rose (1957), Section 14.

The normalized solutions of Equation (3.12) for $V_A(r) = -a/r$ and $E(n, l) = -a^2/(2n^2)$ are

$$R_{nl}(r) = -\left(\frac{2a}{n}\right)^{3/2} \sqrt{\frac{(n-l-1)!}{2n[(n+l)!]^3}} \left(\frac{2ar}{n}\right)^l e^{-ar/n} L_{n+l}^{2l+1}\left(\frac{2ar}{n}\right), \quad (3.27)$$

where $L_{n+l}^{2l+1}(\rho)$ are the *associated Laguerre polynomials*

$$L_{n+l}^{2l+1}(\rho) = \left(\frac{d}{d\rho}\right)^{2l+1} e^\rho \left(\frac{d}{d\rho}\right)^{n+l} \rho^{n+l} e^{-\rho}. \quad (3.28)$$

$a = 1/r_{Bohr}$ is the inverse Bohr radius in units eV/c [cf. (2.19)]; in the usual units in which r and r_{Bohr} have the dimension cm, a is to be replaced by a' of (2.20).

Figure 3.1(b) shows $R_{nl}(r)$ for some of the lower values of n and l. Figure 3.1(a) gives the probability distribution for finding the electron at r.

(a)

Figure 3.1 (a) Charge distributions of the lowest states of the hydrogen atom. The radial distance r is in units of the Bohr radius. The curves are labeled by the numbers nl.

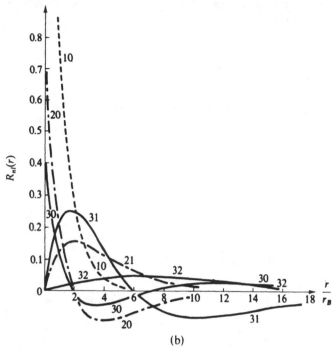

(b)

Figure 3.1 (b) The radial wave functions $R_{nl}(r)$ for hydrogenic atoms for $n = 1, 2, 3$. Each curve is labeled by nl. Note the effect of the centrifugal force in "pushing out" the wave function from the nucleus. Note also that the functions have $n - l - 1$ nodes. [Part (b) from Dicke & Wittke (1960), with permission.]

Knowledge of $\langle \mathbf{x} | n\, l\, m \rangle$ allows an alternative computation of the matrix element of $V(Q)$ to the one given in the previous section; $\epsilon(n, l) = \langle n\, l\, m | V(Q) | n\, l\, m \rangle$ can now be calculated by inserting the complete generalized basis $|\mathbf{x}\rangle$:

$$\epsilon(n, l) = \int d^3x \int d^3x' \, \langle n\, l\, m | \mathbf{x} \rangle \langle \mathbf{x} | V(Q) | \mathbf{x}' \rangle \langle \mathbf{x}' | n\, l\, m \rangle$$

$$= \int d^3 x |\langle n\, l\, m | \mathbf{x} \rangle|^2 V(r).$$

Multiplying (3.28) by $Y_{lm}(\theta, \phi)$, we obtain, according to (3.11), the transition coefficient $\langle r\, \theta\, \phi | n\, l\, m \rangle$. Substitution of $\langle r\, \theta\, \phi | n\, l\, m \rangle$ into the above equation and use of (3.16) to perform the θ, ϕ integrations then yield

$$\epsilon(n, l) = \left(\frac{2a}{n}\right)^3 \frac{(n - l - 1)!}{2n[(n + l)!]^3} \int r^2\, dr\, V(r) \left(\frac{2ar}{n}\right)^{2l} \left| L_{n+l}^{2l+1}\left(\frac{2ar}{n}\right) \right|^2 e^{-2ar/n}.$$

$$(3.29)$$

These integrals can be evaluated for the various functions $V(r) = -c/r^2$, $V(r) = -c/r^3$, $V(r) = -c/r^4$, ... and lead to the same results as given in Equations (2.14)–(2.18). From Figure 3.1(a) we can make some qualitative observations regarding the values $\epsilon(n, l)$. For a given value of n the probability distribution increases and moves towards larger values of r as l increases. Therefore, for a given value of n, $-\epsilon(n, l = 0)$ is the largest, so that $E(n, l = 0)$ will be the lowest energy value, and with increasing l the $E(n, l)$ will approach the value $-a^2/(2n^2)$. That this is indeed the case can be seen from the experimental energy diagrams of the alkali atoms in the previous section.

Problem

1. Calculate the matrix elements of the position operator Q_κ ($\kappa = -1, 0, +1$) between the SO(4) states $|n\,l\,m\rangle$ of the hydrogen atom, i.e., calculate

$$\langle n\,l'\,m'|Q_\kappa|n\,l\,m\rangle.$$

It is sufficient to do this for the third component Q_0. (*Hint*: Derive the following relation:

$$Q_i = -\tfrac{3}{4}H^{-1}\tilde{A}_i + \tfrac{3}{4}H^{-1}[H, f_i(\mathbf{Q}, \mathbf{P})],$$

and use the matrix elements for A_κ) These matrix elements are needed for the Stark effect, Problem VIII.4.

CHAPTER VIII

Perturbation Theory

In this chapter perturbation theory is developed in a general form that is easily adaptable to the discrete as well as the continuous spectrum. At the end of Section VIII.1 the Wigner–Brillouin and the Rayleigh–Schrödinger perturbation expansions are obtained as special cases. Section VIII.2 applies this general procedure to the continuous spectrum and results in the Lippmann–Schwinger equation.

VIII.1 Perturbation of the Discrete Spectrum

This section will develop the general procedure called *perturbation theory* for the evaluation of the eigenvalues E_λ and eigenvectors $|\lambda\rangle$,

$$H|\lambda\rangle = E_\lambda|\lambda\rangle, \tag{1.1}$$

of an operator

$$H = K + V, \tag{1.2}$$

which differs by a small *perturbation* V from an operator K whose spectrum ϵ_a and eigenvectors $|a\rangle$,

$$K|a\rangle = \epsilon_a|a\rangle, \tag{1.3}$$

are known. K is called the "free" Hamiltonian and $|a\rangle$ the free eigenvector. H is called the "exact" Hamiltonian and $|\lambda\rangle$ the exact eigenvector. V is called the perturbation Hamiltonian.

In this kind of perturbation-theoretical treatment it is assumed that to every eigenvector $|a\rangle$ of K there corresponds an eigenvector $|\lambda(a)\rangle$ of H such that $E_{\lambda(a)}$ is given by ϵ_a plus a small correction term such that if V is zero, then $E_{\lambda(a)} = \epsilon_a$. In the case that ϵ_a is degenerate (i.e., that there are other quantum numbers η besides a or λ), $E_{\lambda(a)}$ will in general depend upon these additional quantum numbers even though ϵ_a does not.

To have a concrete case in mind, one may assume that H, K, and V are the operators of Section VII.1; then a is the quantum number n which labels the n^2-fold degenerate eigenvalues of K: $\epsilon_{a=n} = -m_e e^4/(2\hbar^2 n^2)$.[1] Since $[L_i, H] = 0$ (due to $[L_i, V] = 0$), one can choose a basis of free and a basis of exact eigenvectors both to consist of eigenvectors of $\mathbf{L}^2$, L_3: $|a\,\eta\rangle = |a\,l\,m\rangle$, $|\lambda\,\eta\rangle = |\lambda\,l\,m\rangle$. The problem then reduces to a multitude of non-degenerate problems, one for each pair of values of the quantum numbers $\eta = (l\,m)$. If H does not commute with $\mathbf{L}^2$, L_3 and if one does not know of another set of operators which forms c.s.c.o.s with H and K separately, then the problem cannot be reduced to non-degenerate problems and degenerate case perturbation theory, which we do not discuss here, must be used. (cf. Problem 4).

Although at the moment we have in mind discrete values for ϵ_a and a, we do not want to exclude the possibility that K has a continuous spectrum;[2] in that case the summation $\sum_a$, in the equations of this section, is to be replaced by an integration over the continuous spectrum (cf. Section VIII.2.)

We define the operator

$$G(E) \equiv (EI - H)^{-1} \qquad (I = \text{unit operator}), \tag{1.4}$$

which depends upon the parameter E. $G(E)$ is called the "operator for the Green's function for the Hamiltonian" by the physicists, and is called the "resolvent of the operator H" by the mathematicians. It is (intuitively) clear that the task of finding the eigenvalues E_λ of H is equivalent to that of finding the "singularities" of $G(E)$.

We also define the operators

$$F(E) \equiv G(E) \frac{1}{g(E)}, \tag{1.5}$$

and

$$R(E) \equiv V F(E), \tag{1.6}$$

where

$$g(E) \equiv \sum_a |a\rangle\langle a| G_a(E), \tag{1.7}$$

$$G_a(E) \equiv \langle a| G(E)|a\rangle. \tag{1.8}$$

$R(E)$ is called the level shift operator, for reasons that we now explain:

---

[1] We revert to the usual cgs units in this chapter.

[2] The following treatment of perturbation theory is based on the Goldberger–Watson Collision Theory (1964), Section 8.1. It has been chosen because it displays the connection between the cases of the discrete and the continuous spectrum.

Using the above definitions and the identity[3]

$$(E - H)G(E) = I, \tag{1.9}$$

one obtains

$$(E - H)F(E)|a\rangle = \frac{1}{g(E)}|a\rangle = \frac{1}{G_a(E)}|a\rangle \tag{1.10}$$

and, for the diagonal matrix elements of $R(E)$,

$$
\begin{aligned}
R_a(E) &= \langle a|(H - K)F(E)|a\rangle \\
&= \langle a|(E - K)F(E)|a\rangle - \langle a|(E - H)F(E)|a\rangle \\
&= (E - \epsilon_a)\langle a|F(E)|a\rangle - \frac{1}{G_a(E)}.
\end{aligned} \tag{1.11}
$$

It is easily shown that

$$\langle a|F(E)|a\rangle = 1, \tag{1.12}$$

so that from (1.11) we have

$$E = \epsilon_a + R_a(E) + \frac{1}{G_a(E)}. \tag{1.13}$$

As E approaches an eigenvalue E_λ of H, $G_a(E)$ becomes infinite and (1.10) and (1.13) become

$$(E_\lambda - H)F(E_\lambda)|a\rangle = 0 \tag{1.14}$$

and

$$E_\lambda = \epsilon_a + R_a(E_\lambda). \tag{1.15}$$

Though (1.15) is valid for any E_λ and ϵ_a, it is used in practice to determine that value E_λ which is "closest" to ϵ_a, which we call $E_{\lambda(a)}$. Thus $R_a(E_\lambda)$ represents the *level shift* from the "unperturbed" energy value ϵ_a to the corresponding $E_{\lambda(a)}$. The operator $F(E)$ has been constructed such that $F(E_{\lambda(a)})$ is the operator that transforms $|a\rangle$ into the corresponding $|\lambda(a)\rangle = F(E_{\lambda(a)})|a\rangle$, which according to (1.14) is an eigenvector of H with the eigenvalue $E_{\lambda(a)}$.

As already stated above, the assumption[4] that forms the basis of this perturbation theory is that to every $|a\eta\rangle$ there corresponds a $|\lambda(a)\eta\rangle$

---

[3] E stands for $E \cdot I$; the unit operator in a product with a number will be omitted.

[4] This assumption, which underlies perturbation theory, can be shown to be fulfilled for certain V, as the following theorem states:
Let

$$H = H(\kappa) = K + \kappa K^{(1)} + \kappa^2 K^{(2)} + \cdots$$

with

$$\|K^{(k)}f\| \le \frac{M}{r^{k-1}}(\|f\| + \|Hf\|) \qquad (k = 1, 2, \ldots),$$

(i.e., there are as many $|a\,\eta\rangle$ as there are $|\lambda(a)\,\eta\rangle$), but $E_{\lambda(a)}$ may be a function of η because $R_a(E_\lambda)$ does in general depend upon η—e.g., in the example of Section VII.1

$$R_{a\eta}(E_\lambda) = \langle a\,\eta | R(E_\lambda) | a\,\eta \rangle = \langle n\,l\,m | R(E_\lambda) | n\,l\,m \rangle \qquad (1.16)$$

will in general depend upon n, l, m. In the particular case $[R(E), L_i] = 0$ it follows that $R_{a\eta}(E_\lambda)$ does not depend upon m, but it will depend upon l in addition to its dependence upon n. Thus all the $|a\,\eta\rangle$ belong to the energy level ϵ_a of the unperturbed system, and under the influence of the perturbation this level splits into sublevels $E_{\lambda(a)\eta}$ and the state $|a\,\eta\rangle$ of the unperturbed system goes into the state $|\lambda(a)\,\eta\rangle$ of the perturbed system.

The level-shift $R_a(E_{\lambda(a)})$, and thus $E_{\lambda(a)}$, are evaluated by various sequences of approximations. These various sequences are obtained in the following way: For any two invertible operators A and B it is easily shown that

$$\frac{1}{A} = \frac{1}{B} + \frac{1}{B}(B - A)\frac{1}{A}. \qquad (1.17)$$

Choosing $A = g(E)(E - H)$, $B = g(E)(E - K - O(E))$, and hence $1/A = F(E)$, $1/B = (E - K - O(E))^{-1} \cdot 1/g(E)$, one obtains

$$F(E) = \frac{1}{E - K - O(E)}\frac{1}{g} + \frac{1}{E - K - O(E)}(V - O(E))F(E). \qquad (1.18)$$

Here $O(E)$ may be any operator, but we shall only consider operators of the form

$$O(E) = \sum_{a'} O_{a'}(E)|a'\rangle\langle a'|, \qquad (1.19)$$

where $O_{a'}(E)$ are numbers (i.e., $O(E)$ is assumed to have the property $[O(E), K] - 0$).

Equation (1.18) is not defined on every vector, not even on every vector $|a\rangle$, for any value E; for example, the first term is only defined on those $|a\rangle$ and for those E for which

$$(E - \epsilon_a - O_a(E))G_a(E) \neq 0.$$

---

where the numbers M and r are positive constants. Then for $|\kappa| < r$ the eigenvalues $E(\kappa)$ of H can be written

$$E(\kappa) = E^{(0)} + \kappa E^{(1)} + \kappa^2 E^{(2)} + \cdots,$$

and the eigenvectors $|E(\kappa)\rangle$ of H can be written

$$|E(\kappa)\rangle = |E^{(0)}\rangle + \kappa|E^{(1)}\rangle + \kappa^2|E^{(2)}\rangle + \cdots.$$

For $\kappa \to 0$ it follows that

$$H = K,$$

$$E(\kappa = 0) = E^{(0)} = \epsilon_a,$$

$$|E(\kappa = 0)\rangle = |E^{(0)}\rangle = |a\rangle.$$

Applying (1.18) to $|a\rangle$ and making use of (1.7) and (1.13) gives

$$F(E)|a\rangle = \frac{1}{E - K - O(E)}(E - \epsilon_a - R_a(E))|a\rangle$$

$$+ \frac{1}{E - K - O(E)}(V - O(E))F(E)|a\rangle. \qquad (1.20)$$

$O(E)$ will be chosen such that this expression becomes the simplest possible relation for $F(E_{\lambda(a)})|a\rangle$; this can be done if $F(E_{\lambda(a)})|a\rangle$ is written as $|a\rangle$ plus a correction term. Thus $O(E)$ is to be chosen so that the first term becomes $|a\rangle$.

Two choices of $O(E)$ lead to particularly well-known perturbation expansions:

$$O(E) = R_a(E)|a\rangle\langle a|, \qquad (1.21)$$

i.e.,

$$O_{a'}(E) = R_{a'}(E)\delta_{a'a}, \qquad (1.21a)$$

and

$$O(E) = R_a(E)|a\rangle\langle a| + (E - \epsilon_a)\sum_{a' \neq a}|a'\rangle\langle a'|, \qquad (1.22)$$

i.e.,

$$O_{a'}(E) = R_{a'}(E) \quad \text{for } a' = a, \qquad O_{a'}(E) = E - \epsilon_a \quad \text{for } a' \neq a. \quad (1.22a)$$

In the case (1.21),

$$\frac{1}{E - K - O(E)} = \frac{1}{E - \epsilon_a - R_a(E)}|a\rangle\langle a| + \sum_{a' \neq a}\frac{1}{E - \epsilon_{a'}}|a'\rangle\langle a'|, \quad (1.23)$$

and in the case (1.22),

$$\frac{1}{E - K - O(E)} = \frac{1}{E - \epsilon_a - R_a(E)}|a\rangle\langle a| + \sum_{a' \neq a}\frac{1}{\epsilon_a - \epsilon_{a'}}|a'\rangle\langle a'|. \quad (1.24)$$

Thus in both cases (1.21) and (1.22) the first term on the right-hand side of (1.20) becomes $|a\rangle$. Using (1.6) and (1.12) it also follows that

$$\langle a|(V - O(E))F(E)|a\rangle = 0. \qquad (1.25)$$

Thus in both cases, (1.20) may be written as

$$F(E)|a\rangle = |a\rangle + \sum_{a' \neq a}\frac{1}{E - \epsilon_{a'} - O_{a'}(E)}|a'\rangle\langle a'|(V - O(E))F(E)|a\rangle, \quad (1.26)$$

in which it is irrelevant whether the term with $a' = a$ is included or excluded in the sum. Inserting (1.21) into (1.26) gives

$$F(E)|a\rangle = |a\rangle + \sum_{a' \neq a}\frac{1}{E - \epsilon_{a'}}|a'\rangle\langle a'|VF(E)|a\rangle. \qquad (1.27)$$

This equation can be iterated by successive substitution into $F(E)|a\rangle$ on the right-hand side:

$$F(E)|a\rangle = |a\rangle + \sum_{a' \neq a} \frac{1}{E - \epsilon_{a'}} |a'\rangle\langle a'| V|a\rangle$$

$$+ \sum_{\substack{a' \neq a \\ a'' \neq a}} \frac{1}{E - \epsilon_{a'}} \frac{1}{E - \epsilon_{a''}} |a'\rangle\langle a'|V|a''\rangle\langle a''|V|a\rangle + \cdots. \quad (1.28)$$

The level shift and therefore, by (1.15), the eigenvalue of H is obtained from (1.28) by acting on the left with V, taking the scalar product with $|a\rangle$, and evaluating at $E = E_{\lambda(a)}$:

$$E_{\lambda(a)} = \epsilon_a + \langle a|V|a\rangle + \sum_{a' \neq a} \frac{1}{E_{\lambda(a)} - \epsilon_{a'}} \langle a|V|a'\rangle\langle a'|V|a\rangle + \cdots. \quad (1.29)$$

This provides an equation for $E_{\lambda(a)}$ in terms of the unperturbed energy value ϵ_a and the matrix elements of the perturbation V between the unperturbed states $|a\rangle$. Equation (1.29), derived from the choice (1.21) for $O(E)$, constitutes the *Wigner–Brillouin* form of the perturbation expansion.

The choice (1.22) of $O(E)$ leads to the *Rayleigh–Schrödinger* perturbation series. Inserting (1.22) into (1.26) and evaluating at $E = E_{\lambda(a)}$ leads to

$$F(E_{\lambda(a)})|a\rangle = |a\rangle + \sum_{a' \neq a} \frac{1}{\epsilon_a - \epsilon_{a'}} |a'\rangle\langle a'|(V - R_a(E_{\lambda(a)})F(E_{\lambda(a)})|a\rangle, \quad (1.30)$$

and for the level shift

$$R_a(E_{\lambda(a)}) = \langle a|VF(E_{\lambda(a)})|a\rangle. \quad (1.31)$$

Approximations of $R_a(E_{\lambda(a)})$ and $F(E_{\lambda(a)})|a\rangle$ to any desired order in the perturbation V may be obtained by iterating (1.31) and (1.30). We denote by $R_a^{(n)}$ the nth-order approximation of $R_a(E_{\lambda(a)})$ and by $F^{(n)}|a\rangle$ the nth-order approximation of $F(E_{\lambda(a)})|a\rangle$. To zeroth order,

$$R_a^{(0)} = 0, \qquad F^{(0)}|a\rangle = |a\rangle. \quad (1.32)$$

To first order,

$$R_a^{(1)} = \langle a|VF^{(0)}|a\rangle = \langle a|V|a\rangle, \quad (1.33)$$

$$F^{(1)}|a\rangle = |a\rangle + \sum_{a' \neq a} \frac{1}{\epsilon_a - \epsilon_{a'}} |a'\rangle\langle a'|(V - R_a^{(0)})F^{(0)}|a\rangle$$

$$= |a\rangle + \sum_{a' \neq a} \frac{1}{\epsilon_a - \epsilon_{a'}} |a'\rangle\langle a'|V|a\rangle. \quad (1.34)$$

To second order, one may verify that

$$R_a^{(2)} = \langle a|V|a\rangle + \sum_{a' \neq a} \frac{1}{\epsilon_a - \epsilon_{a'}} \langle a|V|a'\rangle\langle a'|V|a\rangle, \qquad (1.35)$$

$$F^{(2)}|a\rangle = |a\rangle + \sum_{a' \neq a} \frac{1}{\epsilon_a - \epsilon_{a'}} |a'\rangle\langle a'|V|a\rangle$$

$$+ \sum_{\substack{a' \neq a \\ a'' \neq a}} \frac{1}{\epsilon_a - \epsilon_{a'}} \frac{1}{\epsilon_a - \epsilon_{a''}} |a'\rangle(\langle a'|V|a''\rangle - \delta_{a'a''}\langle a|V|a\rangle)\langle a''|V|a\rangle. \qquad (1.36)$$

The vectors $F^{(n)}|a\rangle$, or in general $F(E_{\lambda(a)})|a\rangle$, are once again usually labeled by the quantum numbers a. However, whereas for the unperturbed vectors $|a\rangle$ the labels are connected with eigenvalues of a complete set of operators containing K of which these vectors are eigenvectors, the "exact" vectors and the approximations $F^{(1)}|a\rangle,\ldots$ obtained by the perturbation expansion are in general not eigenvectors of the complete set of operators containing H. Therefore we will often denote them by $|(a)\rangle$, where those quantum numbers which are bracketed do not refer to eigenvalues of operators of which the $|(a)\rangle$ are eigenvectors. The meaning of the quantum numbers in the "exact" vectors is therefore only that these vectors are connected with unperturbed eigenvectors whose eigenvalues are these quantum numbers (or related to them).

VIII.2 Perturbation of the Continuous Spectrum—The Lippmann–Schwinger Equation[5]

In the previous section it was nowhere stated that ϵ_a is a discrete eigenvalue and $|a\rangle$ a proper eigenvector of K, though all equations were written as if this were the case. The assumption of a discrete spectrum is in fact unnecessary; we shall now discuss in more detail the case of the continuous spectrum, which, with the tools of Section II.7, is easily obtained by replacing the proper vectors $|a\rangle$ with the generalized vectors, and the discrete sums by integrals.

For the following discussions, we have to make some assumptions regarding the properties of the operators K and $H = K + V$. We shall assume that

$$\text{spectrum } K \subset \text{spectrum } H. \qquad (2.1)$$

This is a general enough assumption to incorporate the situations that occur for physical problems. Usually for scattering and decay processes K has only a continuous spectrum that is positive; the continuous spectrum of H agrees with the continuous spectrum of K, and in addition H has a set of negative discrete eigenvalues (and sometimes also positive discrete eigenvalues lying in the continuous spectrum) that are bounded from below.

---

[5] The results of this section will not be used until Chapter XIV.

In this case of a continuous spectrum the level shift R_a is not an observable quantity. However (1.15) still holds, for (as remarked in the previous section) it is true for any value E_λ in the spectrum of H and any value ϵ_a in the spectrum of K.

We shall consider (1.20) at the value $E_{\lambda(a)} = \epsilon_a$, i.e., we shall calculate $F(E_{\lambda(a)})|a\rangle$ for that value $E_{\lambda(a)} \in$ spectrum H which is identical to the eigenvalue $\epsilon_a \in$ spectrum K corresponding to the generalized eigenvector $|a\rangle$. Such a value $E_{\lambda(a)}$ exists in the spectrum of H because of the assumption (2.1). We call this value $E_{\lambda(a)} = E_a$. For E_a, (1.15) becomes

$$E_a - \epsilon_a = R_a(E_a) = 0. \tag{2.2}$$

If we again choose (1.21) for the operator $O(E)$ we will again arrive at (1.27), which in the continuous form is written

$$F(E_a)|a\rangle = |a\rangle + \int^P da' \, \frac{1}{E_a - E_{a'}} \, |a'\rangle\langle a'|VF(E_a)|a\rangle. \tag{2.3}$$

$^P\!\int$ means the *principal-value integral*, i.e., the integral over all a' with the omission of an "infinitesimally small interval" around $E_{a'} = E_a$. $\int da \ldots$ is a symbolic way of expressing summation over the discrete label and integration with respect to a suitable weight function over the continuous label. The label a in $|a\rangle$ stands for the energy E_a and some additional labels η_a, which may be discrete or continuous: $|a\rangle = |E_a\eta_a\rangle$. If the $|a\rangle$ are "normalized" so that

$$\langle \eta_a' E_a' | E_a \eta_a \rangle = \rho^{-1}(E_a)\delta(E_a - E_{a'}) \, \delta_{\eta_a\eta_a'},$$

then

$$\int da - \sum_{\eta_a} \int \rho(E_a) \, dE_a$$

[cf. (I.4.7c$_\mu$) and Chapter XIV]. Again (1.14) is fulfilled, i.e., $F(E_a)|a\rangle$ is a generalized eigenvector of H with eigenvalue E_a. This can also be seen directly from (2.3): Multiplying (2.3) by $(E_a - K)$ and using

$$K|a\rangle = E_a|a\rangle$$

and

$$\int^P da' \, |a'\rangle\langle a'| = I - |a\rangle\langle a|,$$

one obtains

$$(E_a - K - V)F(E_a)|a\rangle = -|a\rangle\langle a|VF(E_a)|a\rangle \tag{2.4}$$

The right-hand side of this equation is according to (1.6) and (2.2) equal to zero and one obtains:

$$(H - E_a)F(E_a)|a\rangle = 0. \tag{1.14'}$$

We now make use of a well-known relation from the theory of generalized functions,[6]

$$\frac{1}{x \pm i0} = \lim_{\eta > 0+} \frac{1}{x \pm i\eta} = P\frac{1}{x} \mp i\pi\, \delta(x).$$ (2.5)

Then (2.3) can be written

$$F(E_a)|a\rangle = |a\rangle + \int da' \frac{1}{E_a - E_{a'} \pm i0} |a'\rangle\langle a'|VF(E_a)|a\rangle$$

$$\pm i\pi \int da'\, \delta(E_a - E_{a'})|a'\rangle\langle a'|VF(E_a)|a\rangle.$$ (2.6)

Integrating and using (2.2), we see that the last term in (2.6) vanishes:

$$\pm i\pi|a\rangle\langle a|VF(E_a)|a\rangle = \pm i\pi R_a(E_a)|a\rangle = 0.$$

Because $K|a'\rangle = E_{a'}|a'\rangle$, Equation (2.6) may be written in the more usual form

$$F(E_a)|a\rangle = |a\rangle + \frac{1}{E_a - K \pm i0} V(F(E_a)|a\rangle).$$ (2.7)

This is called the *Lippmann–Schwinger equation*. It is an equation for the generalized eigenvectors

$$|\alpha\rangle = |\alpha(a)\rangle = F(E_a)|a\rangle$$ (2.8)

of the operator H with eigenvalue E_a; E_a is equal to the eigenvalue ϵ_a of K belonging to the generalized eigenvector $|a\rangle$. Thus

$$H|\alpha\rangle = H|\alpha(a)\rangle = E_a|\alpha\rangle, \qquad K|a\rangle = E_a|a\rangle.$$ (2.9)

The Lippmann–Schwinger equation is often written

$$|\alpha^{\pm}\rangle = |\alpha^{\pm}(a)\rangle = |a\rangle + \frac{1}{E_a - K \pm i0} V|\alpha^{\pm}\rangle.$$ (2.7')

Besides the solutions of the Lippmann–Schwinger equation $|\alpha^{\pm}\rangle$, there are other basis vectors of interest. These are obtained from the form (2.3) of the integral equation for $F(E_a)|a\rangle$, which in operator form is written as

$$F(E_a)|a\rangle = |a\rangle + \frac{P}{E_a - K} VF(E_a)|a\rangle,$$ (2.10)

---

[6] The proof of (2.5) may be found in Gelfand et al. (1964), Vol. 1, Chapter I, Section 2.4. Further discussion of generalized functions of the type $(x \pm i0)^{\lambda}$ may be found in Chapter I, Sections 3 and 4 of the same volume. Relations between generalized functions (distributions) like (2.5) can be understood in the conventional sense when they are multiplied with a well-behaved function $\phi(x)$ and integrated. Thus (2.5) means

$$\int dx \frac{1}{x \pm i0} \phi(x) = \overset{P}{\int} dx \frac{1}{x} \phi(x) \mp i\pi \int dx\, \delta(x)\phi(x)$$

for all well-behaved functions $\phi(x)$. Analogously, relations between generalized eigenvectors like (2.6) can be understood in the conventional sense when the scalar product is taken with a well-behaved vector $\phi \in \Phi$ (cf. Section II.8).

where P indicates that the principle-value integral is to be taken. To distinguish the solution of this equation from the two solutions of the Lippmann–Schwinger equation we denote it by $|\alpha\rangle$ and write (2.10):

$$|\alpha^P\rangle = |a\rangle + \frac{P}{E_a - K} V |\alpha^P\rangle. \tag{2.10'}$$

Problems

1. Let the energy operator be given by

$$H = K + \alpha Q^3 + \beta Q^4,$$

where

$$K = \frac{1}{2m} P^2 + \frac{m\omega^2}{2} Q^2$$

(anharmonic oscillator).

   (a) Calculate the eigenvalues to first order in the perturbation expansion.
   (b) Calculate for $\beta = 0$ the eigenvalues of H to the second order of the Rayleigh–Schrödinger perturbation series. Compare the result with that of (a).
   (c) Calculate for $\beta = 0$ the eigenvalues of H to the second order of the Wigner–Brillouin perturbation series. Compare the result with those of (a) and (b).

2. Let the exact Hamiltonian of a rotator be given by

$$H = \frac{1}{2I} L^2 + \alpha L_3 + \beta L_1$$

with $\beta \ll \alpha$, $\beta \ll \hbar/2I$. Let $\{|l\ m\rangle: m = l, l - 1, \ldots, -l\}$ be a basis of eigenvectors of L^2 and L_3 with eigenvalues $\hbar^2 l(l + 1)$ and $\hbar m$, respectively:

   (a) Using perturbation theory with $K = (1/2I)L^2 + \alpha L_3$ as the free Hamiltonian and $V = \beta L_1$ as the perturbation Hamiltonian, calculate the eigenvalues of H up to the lowest order term involving β.
   (b) H may be expressed as

$$H = \frac{1}{2I'} L'^2 + \alpha' L'_3,$$

   where the L'_i are angular momentum operators obtained by rotation from the L_i and where I' and α' are constants. Express I' and α' in terms of I, α, and β and determine the exact eigenvalues of H. Discuss whether the results obtained in part (a) are consistent with the exact results.

3. Show that [Equation (1.14')]

$$HF(E_a)|a\rangle = E_a F(E_a)|a\rangle,$$

where according to Equation (2.3)

$$F(E_a)|a\rangle = |a\rangle + \int^P da' \frac{1}{E_a - E_{a'}} |a'\rangle \langle a'|V F(E_a)|a\rangle$$

is the continuous-spectrum analogue to (1.27).

4. An external, static. homogeneous electric field $\mathscr{E}$ splits the energy levels of the hydrogen atom (Stark effect). The perturbation Hamiltonian, obtained from correspondence with the classical case, is

$$V = -\mathbf{D} \cdot \mathscr{E} = -e\mathbf{Q} \cdot \mathscr{E}$$

($\mathscr{E}$ is measured in volts/cm.)

Using the result of Problem VII.1, calculate the energy eigenvalues to first order in the perturbation V for the $n = 1$ and $n = 2$ levels of hydrogen. For $n = 1$, Equation (1.33) may be used directly with $|a\rangle = |n = 1 \; l = 0 \; m = 0\rangle$; however, because of the degeneracy of the $n = 2$ level at zeroth order, one must diagonalize V within the $n = 2$ subspace in order to obtain basis vectors for which the perturbation expansion (e.g., the second-order term) will converge. One will find that at first order the $n = 2$ level of the H-atom splits under the influence of the electric field into three sublevels. Since this splitting occurs at first order, it is called the *linear Stark effect*. Explain why there is no linear Stark effect for the alkali (and other) atoms.

CHAPTER IX

Electron Spin

In Section IX.2 the doublet fine-structure splitting of one-electron atoms is explained as a spin effect. To obtain quantitative results the interaction Hamiltonian for the spin interaction is determined in Section IX.3. In Section IX.3a the magnetic moment of the electron is determined by classical arguments; in Section IX.3b the magnetic field, which acts on the electron magnetic moment in an atom, is presented. In Section IX.4 these results are used to calculate the fine-structure splitting. In Section IX.5 selection rules for dipole transitions are derived, and the chapter closes with some general remarks concerning the visualization of quantum systems.

IX.1 Introduction

The existence of *electron spin* was suggested by the fine structure in atomic spectra. (See the description of the experimental situation at the end of Chapter VI.) The electron spin cannot be expressed in terms of the position and momentum operators of the electron. If the electron is considered as a physical object with translational and rotational degrees of freedom, then spin is an observable that corresponds to the rotational degrees of freedom in the same way that momentum corresponds to the translational degrees of freedom. Thus the electron is an elementary rotator (Section V.1) with a translational degree of freedom. In the preceding chapters we have ignored the rotational degrees of freedom of the electron, as their contributions to the energy of the electrons bound in atoms is small. We have shown in Chapter III

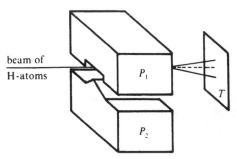

Figure 1.1 Splitting of a particle beam after traversing a Stern–Gerlach magnet.

that angular momenta of any integral and half-integral value may exist and that there is nothing peculiar about half-integral angular momenta. It turns out that the spin (*intrinsic angular momentum*) of the electron is $\frac{1}{2}$.

A direct verification of the electron-spin hypothesis came from the Stern–Gerlach experiment. The experimental arrangement is shown in Figure 1.1. A strongly inhomogeneous magnetic field was produced between pole pieces P_1 and P_2, of which P_2 had a sharp edge. A beam of hydrogen atoms[1] in the ground state was sent closely past the sharp edge of pole piece P_2, eventually hitting plate T. With no field the beam formed a narrow line on the plate (Figure 1.1 dashed line). When the magnetic field was turned on the line divided into two lines. (Figure 1.1 solid lines.)

Without going into the details for the description of this experiment (see Chapter XIII), one can already see that the model of the hydrogen atom described in Chapter VI cannot explain such a separation. According to that model the ensemble of hydrogen atoms with lowest energy is in a pure state

$$W = \Lambda_{\mathscr{R}(n=1)}, \tag{1.1}$$

because $\mathscr{R}(n=1)$ is a one-dimensional space. According to the model of Chapter VI, it would thus be impossible to separate the ensemble of the original beam into two subensembles as has been done by the magnetic field. The experimental separation into two subensembles shows that instead of $\mathscr{R}(n=1)$ we must have (at least) a two-dimensional space. If we assume that the ensembles of the separated beams are in pure states, then we must have an exactly two-dimensional space. Within a certain degree of accuracy (including only the electron structure and disregarding possible structure of the nucleus) this assumption has so far been upheld.

As the separation of the beam was caused by a magnetic field, one conjectures that the new observable, whose existence has been demonstrated by the Stern–Gerlach experiment, must be connected with the magnetic moment. As the magnetic moment is connected with rotating charges, one would suspect that this new observable must be angular momentum.

---

[1] The original experiment used silver atoms. The experiment has since been repeated with other atoms, including hydrogen.

IX.2 The Fine Structure—Qualitative Considerations

The angular momentum that causes the separation of the beam in the Stern–Gerlach experiment cannot be the orbital angular momentum $L_i = \epsilon_{ijk}Q_jP_k$, because this is zero for the ground state of the hydrogen atom. Furthermore, there are no two-dimensional spaces of states for the algebra of orbital angular momentum since $l = 0, 1, 2, \ldots$ (Problem III.1). The two-dimensional representation space of the algebra of spin angular momentum is $\mathcal{R}^{j=1/2}$ (Section III.3), containing the two one-dimensional spaces of states with opposite helicity. Denoting this two-dimensional space by $\imath^s$, we may write

$$\imath^s = \mathcal{R}^{j=1/2} \xrightarrow[\mathscr{E}(SO(2)S_3)]{} \mathcal{R}^{j=1/2}_{j_3=-1/2} \oplus \mathcal{R}^{j=1/2}_{j_3=+1/2}. \tag{2.1}$$

We thus conjecture that the electron is an elementary rotator and the electron in the hydrogen atom is a combination of the physical systems of an orbiting electron and this elementary rotator (spinning electron), the space of states of which we call $\imath^s$. Then the space of physical states of lowest energy of the hydrogen atom is

$$\mathcal{H}(n = 1) = \mathcal{R}(n = 1) \otimes \imath^s, \tag{2.2}$$

and in general the nth space of the hydrogen atom is given by

$$\mathcal{H}(n) = \mathcal{R}(n) \otimes \imath^s. \tag{2.3}$$

The orbital observables (functions of Q_i, $P_i)A_i$ and L_i, given by (VI.3.1), (VI.3.3), and (VI.3.9), act in the space $\mathcal{R}(n)$, and the new angular momenta, which are called spin and denoted by S_i, act in the space $\imath^s$.

To check whether our conjecture is in agreement with experiment, we reduce the space $\mathcal{H}(n)$ with respect to total angular-momentum. According to (VI.4.4), each space $\mathcal{R}(n)$ is the direct sum $\mathcal{R}(n) = \sum_{l=0}^{n-1} \oplus \mathcal{R}^l$. Then $\mathcal{H}(n)$ given by Equation (2.3) is the direct sum[2]

$$\mathcal{H}(n) \xrightarrow[\mathscr{E}(SO(3)_L)]{} \left(\sum_{l=0}^{n-1} \oplus \mathcal{R}^l \right) \otimes \imath^s = \sum_{l=0}^{n-1} \oplus (\mathcal{R}^l \otimes \imath^s). \tag{2.4}$$

$\mathcal{R}^l \otimes \imath^s$ is the space of physical states of the combination of two elementary rotators with angular momenta l and $s = \frac{1}{2}$. The orbital angular momenta $L_i = \epsilon_{ijk}Q_jP_k$ act in $\mathcal{R}^l$, and the spin angular momenta S_i act in $\imath^s$. The total angular momentum of the combined elementary rotators is then, according to Equation (V.2.3),

$$J_i = L_i \otimes I + I \otimes S_i. \tag{2.5}$$

---

[2] The symbol $\xrightarrow[\mathscr{E}(SO(3)_L)]{}$ means that the two spaces are equivalent under the action of operators generated by the L_i.

Using (V.2.32) we can reduce $\mathscr{R}^l \otimes \imath^s$ to the sum of total-angular-momentum eigenspaces

$$\mathscr{R}^l \otimes \imath^s = \begin{cases} \mathscr{R}^{j=1/2} & \text{if } l = 0, \\ \mathscr{R}^{j=l+1/2} \oplus \mathscr{R}^{j=l-1/2} & \text{otherwise.} \end{cases} \tag{2.6}$$

Whether the $\mathscr{R}^j$ or the $\mathscr{R}^l \otimes \imath^s$ are the spaces of physical states depends, of course, upon the preparability of these states. If physically preparable states are eigenstates of the total angular momentum, then the $\mathscr{R}^j$ are the spaces of physical states; if the physically preparable states are eigenstates of the orbital and spin angular momentum, then the $\mathscr{R}^l \otimes \imath^s$ are the spaces of physical states. The experimental data show that the physical spaces are the $\mathscr{R}^j$, for each orbital-angular-momentum level (except the $l = 0$ level) splits into two sublevels as required by (2.4) and (2.6).

From (2.4) and (2.6) it follows that

$$\mathscr{H}(n) = \mathscr{R}^{1/2}_{(l=0)} \oplus (\mathscr{R}^{1/2}_{(l=1)} \oplus \mathscr{R}^{3/2}_{(l=1)}) \oplus (\mathscr{R}^{3/2}_{(l=2)} \oplus \mathscr{R}^{5/2}_{(l=2)})$$
$$\oplus \cdots \oplus (\mathscr{R}^{n-3/2}_{(l=n-1)} \oplus \mathscr{R}^{n-1/2}_{(l=n-1)}). \tag{2.7}$$

For $n = 1$,

$$\mathscr{H}(n = 1) = \mathscr{R}^{1/2}_{(l=0)}, \tag{2.8}$$

and there is one energy level corresponding to this two-dimensional space. The two-dimensionality of $\mathscr{H}(n = 1)$ allows us to explain the separation of the beam in the Stern–Gerlach experiment into two subensembles, one corresponding to the space $\mathscr{R}^{1/2}_{(l=0)j_3 = +1/2}$ and the other to $\mathscr{R}^{1/2}_{(l=0)j_3 = -1/2}$.

For $n = 2$,

$$\mathscr{H}(n = 2) = \mathscr{R}^{1/2}_{(l=0)} \oplus \mathscr{R}^{1/2}_{(l=1)} \oplus \mathscr{R}^{3/2}_{(l=1)}, \tag{2.9}$$

and there are three energy levels, each of which corresponds to one of the spaces in (2.9):

$$\mathscr{R}^{1/2}_{(l=0)} \leftrightarrow 2s^{1/2}, \quad \mathscr{R}^{1/2}_{(l=1)} \leftrightarrow 2p^{1/2}, \quad \text{and} \quad \mathscr{R}^{3/2}_{(l=1)} \leftrightarrow 2p^{3/2}. \tag{2.10}$$

For $n = 3$,

$$\mathscr{H}(n = 3) = \mathscr{R}^{1/2}_{(l=0)} \oplus \mathscr{R}^{1/2}_{(l=1)} \oplus \mathscr{R}^{3/2}_{(l=1)} \oplus \mathscr{R}^{3/2}_{(l=2)} \oplus \mathscr{R}^{5/2}_{(l=2)}. \tag{2.11}$$

which belong to the energy levels

$$3s^{1/2}, \quad 3p^{1/2}, \quad 3p^{3/2}, \quad 3d^{3/2}, \quad 3d^{5/2},$$

respectively. For $n = 4, 5, \ldots$ the same arguments hold. These results reflect precisely (if hyperfine splitting is neglected) the experimental data shown in Figure VI.5.2 for hydrogen and Figures VII.2.1, VII.2.2, VII.2.3 for the alkali atoms. Thus to each energy level there corresponds an eigenspace of the total angular momentum; eigenspaces of $\mathbf{J}^2$ are also eigenspaces of H, and

$$[H, J_i] = 0. \tag{2.12}$$

This equation, which we obtained here from the empirical facts, expresses the rotational invariance of H and can theoretically be obtained from general symmetry arguments.

We shall denote by $\mathscr{H}^{orb}$ the space $\mathscr{R}$ given by (VI.5.2):

$$\mathscr{H}^{orb} = \sum_{n=1}^{\infty} \oplus \, \mathscr{R}(n). \tag{2.13}$$

The operator H given by (VI.3.2) or (VI.3.5) is not the total-energy operator when spin is taken into account; let us rename this approximate energy operator H_0:

$$H_0 = \frac{\mathbf{P}^2}{2} - \frac{a}{Q} \tag{2.14}$$

(or

$$H_0 = \frac{\mathbf{P}^2}{2} - \frac{a}{Q} + V(Q) \tag{2.14'}$$

for the alkali atom). The space of physical states of the hydrogen atom in a description that takes spin into account is then

$$\mathscr{H} = \mathscr{H}^{orb} \otimes \imath^s = \left(\sum_{n=1}^{\infty} \oplus \, \mathscr{R}(n) \right) \otimes \imath^s = \sum_{n-1}^{\infty} \oplus \, \mathscr{H}(n). \tag{2.15}$$

This description is the combination of the spinless hydrogen atom and an elementary rotator. A basis system in this space is thus

$$|n\,l\,m\rangle \otimes |s = \tfrac{1}{2}, s_3\rangle, \tag{2.16}$$

where the

$$|n\,l\,m\rangle \qquad (n = 1, 2, 3, \ldots; l = 0, 1, 2, \ldots, n - 1; m = -l, -l + 1, \ldots, + l)$$

form a basis system for $\mathscr{H}^{orb}$, and the $|s = \tfrac{1}{2} s_3\rangle$ $(s_3 = -\tfrac{1}{2}, +\tfrac{1}{2})$ form a basis system for $\imath^s = \mathscr{R}^{1/2}$. Each observable A in $\mathscr{H}$ is written, according to the basic postulate IV,

$$A = \sum_i A_{(i)}^{orb} \otimes A_{(i)}^{spin}, \tag{2.17}$$

where the $A_{(i)}^{orb}$ act only in $\mathscr{H}^{orb}$ and the $A_{(i)}^{spin}$ act only in $\imath^s$ [cf. (III.5.7)]. Orbital angular momentum and spin are special examples of this:

$$L_i = L_i^{orb} \otimes I^{spin} = (\epsilon_{ijk} Q_j^{orb} P_k^{orb}) \otimes I^{spin} = L_i \otimes I = (\epsilon_{ijk} Q_j P_k) \otimes I, \tag{2.18}$$

$$S_i = I^{orb} \otimes S_i^{spin} = I \otimes S_i. \tag{2.19}$$

where I^{spin}, I^{orb} are the unit operators in the spaces $\imath^s$ and $\mathscr{H}^{orb}$, respectively. (For typographical convenience we shall usually omit the superscripts "orb" and "spin" if it is clear from the context in which space an operator acts.) The total angular momentum of the combined system is the sum of these two angular momenta,

$$J_i = L_i + S_i = L_i \otimes I + I \otimes S_i. \tag{2.5}$$

The energy operator H can also be written in the form (2.17). Since $H_0 \otimes I$ already gives a very good approximation to the energy spectrum (cf. Figures VI.5.1, VII.2.1, VII.2.2, and VII.2.3), we shall write the energy operator as

$$H = H_0 + H_1 \qquad (H_0 = H_0 \otimes I), \qquad (2.20)$$

where H_1 acts on the entire space $\mathcal{H} = \mathcal{H}^{orb} \otimes \iota^s$ and is small in the sense that its contributions to the energy values are small compared with the values of H_0. H_1 cannot be of the form $H_1 \otimes I$. If it were, no fine-structure splitting between levels with the same l could occur, for H_1^{orb} could just be added to H_0^{orb} and the resultant $H = (H_0 + H_1) \otimes I$ (for a given value of l) would only lead to a shift of the energy levels. Neither can H_1 be of the form $I \otimes H_1$, for ι^s is a very simple two-dimensional space: Every operator in ι^s can be written as a linear combination of the four operators I, S_1, S_2, and S_3; thus H_1 would have to be $H_1 = \alpha^0 I + \sum_{i=1}^{3} \alpha^i S_i$ ($\alpha^i \in \mathbb{C}$). The first term would only lead to an overall shift of all energy values. The second term ($\sum_{i=1}^{3} \alpha^i S_i$), and consequently H_1, does not commute with J_k, because

$$[J_i, S_j] = i\epsilon_{ijk} S_k \qquad (2.21)$$

as a consequence of (2.5) and of

$$[S_i, S_j] = i\epsilon_{ijk} S_k. \qquad (2.21a)$$

Thus it would follow that H_1, and consequently $H = H_0 + H_1$, is not rotationally invariant, in contradiction to (2.12). Hence the most general possible form of H_1 is

$$H_1 = A \otimes I + \sum_{i=1}^{3} B_i \otimes S_i, \qquad (2.22)$$

where A and B_i are operators in $\mathcal{H}^{orb}$, i.e., functions of the operators Q_i, P_i. A is a scalar operator, while the B_i are the components of a vector operator with respect to the L_i. The former follows from the fact that every operator in $\iota^s = \mathcal{R}^{1/2}$ can be written as a linear combination of the operators I, S_1, S_2, and S_3. The latter follows from the requirement that

$$[H_1, J_i] = 0,$$

which in turn follows from (2.12), (2.20), and $[H_0, L_i] = 0$.

To see this we calculate

$$\sum_{k=1}^{3} [B_k \otimes S_k, J_i] = [B_k, L_i] \otimes S_k + i\epsilon_{kij} B_k \otimes S_j,$$

which is zero only if

$$[B_k, L_i] = i\epsilon_{kij} B_j, \qquad (2.22b)$$

i.e., if B_k is a vector operator with respect to the L_i.

As

$$[H_1, L_i] \neq 0, \qquad (2.22a)$$

which follows immediately from (2.22b), the basis vectors given by (2.16) cannot be eigenvectors of H. Physically preparable states appear to be always energy eigenstates or mixtures of energy eigenstates (cf. Chapter XII). Therefore we should use a basis of eigenvectors of the energy operator H. Because of (2.12), eigenvectors of the total-angular-momentum operators $\mathbf{J}^2$, J_3 can be eigenvectors of H. Therefore we use (V.2.12) to form the new basis vectors

$$|n\,l\,s = \tfrac{1}{2}\,j\,j_3\rangle = \sum_{l_3 s_3} |n\,l\,l_3\rangle \otimes |s = \tfrac{1}{2}\,s_3\rangle \langle l\,l_3\,s = \tfrac{1}{2}\,s_3|j\,j_3\rangle, \quad (2.23)$$

where $\langle l\,l_3\,\tfrac{1}{2}\,s_3|j\,j_3\rangle$ are the Clebsch–Gordan coefficients given in Table V.2.1. The basis vectors (2.23) are eigenvectors of the c.s.c.o.

$$H_0, \mathbf{L}^2, \mathbf{S}^2, \mathbf{J}^2, \quad J_3. \quad (2.24')$$

As $\mathbf{S}^2$ is $\tfrac{1}{2}(\tfrac{1}{2}+1)$ on the whole space $\mathscr{H}$, we can ignore it and have in

$$H_0, \mathbf{L}^2, \mathbf{J}^2, \quad J_3 \quad (2.24)$$

a c.s.c.o., of which the basis vectors of (2.23) are eigenvectors. The basis (2.23) need not be a basis of eigenvectors of H. If the basis vectors (2.23) are not eigenvectors of H, then eigenvectors of H can be obtained from (2.23) by perturbation-theoretical calculations. As we shall show below

$$H, \mathbf{L}^2, \mathbf{J}^2, \quad J_3 \quad (2.25)$$

is also a c.s.c.o., the eigenvectors of which we call

$$|E\,l\,j\,j_3\rangle. \quad (2.26)$$

The difference between (2.23) and (2.26) is that

$$H_0|n\,l\,j\,j_3\rangle = E_n^0|n\,l\,j\,j_3\rangle \quad (2.27)$$

where $E_n^0 = -a^2/2n^2$, by (VI.5.3), and

$$H|E\,l\,j\,j_3\rangle = E|E\,l\,j\,j_3\rangle \quad (2.28)$$

where E is a yet to be determined value. Thus the basis systems (2.23) and (2.26) have all but one quantum number (the energy) in common and the perturbation $(H - H_0)$ will not affect the quantum numbers l, j, j_3. Therefore after we have made the transformation (2.23) we can ignore the additional quantum numbers of the state vectors and do perturbation theory in one quantum number only, as described in Section VIII.1.

To prove that (2.25) is indeed a c.s.c.o. one needs only to show that

$$[H, \mathbf{L}^2] = 0. \quad (2.29)$$

This cannot be derived from the general form (2.22), but constitutes a condition on the A and B_i. The justification of (2.29) comes from the considerations in Section V.4 on parity, also expressed by (VI.5.9), and a new assumption about H:

$$[H, U_P] = 0 \quad \text{or} \quad U_P H U_P^{-1} = H. \quad (2.30)$$

This assumption is called *parity invariance* (of H). U_P in the spin space is given by

$$U_P|s = \tfrac{1}{2}s_3\rangle = \pi_s|s = \tfrac{1}{2}s_3\rangle, \qquad (2.31)$$

where $|\pi_s| = 1$. Equation (2.31) is (V.4.20) for the special case that $j = s = \tfrac{1}{2}$ and follows from (V.4.3) for the S_i, i.e., from

$$U_p S_i U_P^{-1} = S_i. \qquad (2.32)$$

Equations (V.4.20) and (V.4.21) with $j = l$ for the orbital angular momentum give

$$U_P|\xi\,l\,l_3\rangle = (-1)^l \eta_{\text{orb}}|\xi\,l\,l_3\rangle; \qquad (2.33)$$

here ξ may be any additional quantum number whose operator commutes with U_P, e.g., $\xi = n$ or $\xi = E$. From (2.31) and (2.32) it follows that

$$U_P(|\xi\,l\,l_3\rangle \otimes |s = \tfrac{1}{2}s_3\rangle) = (-1)^l \eta|\xi\,l\,l_3\rangle \otimes |s = \tfrac{1}{2}s_3\rangle \qquad (2.34)$$

$(\eta = \eta_{\text{orb}}\pi_s)$. It then follows from (2.23) that

$$U_P|\xi\,l\,j\,j_3\rangle = (-1)^l \eta|\xi\,l\,j\,j_3\rangle. \qquad (2.35)$$

Consequently the eigenvalue of $\mathbf{L}^2$ is connected with the eigenvalue of U_P, and the quantum number l in the basis (2.23) may be considered the parity quantum number. [The parity can have only the two values $+1$ and -1; thus for given n, j, j_3 there exist two states of opposite parity. This has already been expressed in (2.7) in the form that to every j there belong two l's, $l = j + \tfrac{1}{2}$ and $l = j - \tfrac{1}{2}$.] Thus if (2.30) holds, ξ in (2.35) may be E, the eigenvalue of H; from (2.30) it follows that E, l, j, j_3 may label the states, and consequently that (2.25) is a c.s.c.o., one which has now been shown to be identical with the c.s.c.o.

$$H, U_P, \mathbf{J}^2, \quad J_3. \qquad (2.36)$$

Now that (2.29) has been established, it can be used to obtain some new conditions on the operators A and B_i in (2.22). From (2.22) follows

$$[H_1, \mathbf{L}^2] = 0 \qquad (2.37)$$

which does not give any new restrictions on A, but, along with (2.22b), does restrict the B_i:

$$0 = [B_k, \mathbf{L}^2] = \{L_i, [B_k, L_i]\} = i\epsilon_{kij}\{L_i, B_j\}.$$

From this follows the restriction

$$B_j = f(\mathbf{Q}, \mathbf{P})L_j, \qquad (2.38)$$

where $[f(\mathbf{Q}, \mathbf{P}), L_j] = 0$, i.e., $f(\mathbf{Q}, \mathbf{P})$ is a scalar operator with respect to the L_j. We will soon find by correspondence from the classical case that $f(\mathbf{Q}, \mathbf{P}) \propto Q^{-3}$ for the spin-orbit contribution to H_1 [cf. (3.23)].

Although we have denoted the basis vectors (2.23) and (2.26) differently, it may happen that they are identical (up to a phase). We have to distinguish the two possibilities

$$[H_1, H_0] = 0, \quad \text{which implies} \quad [H, H_0] = 0, \tag{2.39}$$

and

$$[H_1, H_0] \neq 0, \quad \text{which implies} \quad [H, H_0] \neq 0. \tag{2.40}$$

If (2.39) holds, then the basis vectors (2.23) are also eigenvectors of H and

$$H|n\,l\,j\,j_3\rangle = E_{nlj}|n\,l\,j\,j_3\rangle; \tag{2.41}$$

the eigenvalue of H may in general depend upon n, l, and j, and is given by

$$E_{nlj} = E_n^0 + \epsilon_{nlj}, \qquad E_n^0 = -\frac{a^2}{2n^2}, \tag{2.42}$$

where ϵ_{nlj} is the eigenvalue of the operator H_1 in the state $|n\,l\,j\,j_3\rangle$:

$$H_1|n\,l\,j\,j_3\rangle = \epsilon_{nlj}|n\,l\,j\,j_3\rangle. \tag{2.43}$$

If (2.40) holds, then (2.23) and (2.26) are different vectors, and (2.26) can be determined from (2.23) by perturbation theory. The first-order perturbation expression $E_{nlj}^{(1)}$ for the eigenvalue of H is then given, according to (VIII.1.29) or (VIII.1.33), by

$$E_{nlj}^{(1)} = E_n^0 + \epsilon_{nlj}^{(1)}, \tag{2.44}$$

where $\epsilon_{nlj}^{(1)}$ is the expectation value of H_1 in the states (2.23),

$$\epsilon_{nlj}^{(1)} = \langle n\,l\,j\,j_3|H_1|n\,l\,j\,j_3\rangle. \tag{2.45}$$

IX.3 Fine-Structure Interaction

The calculation of quantitative results requires knowledge of H_1; H_1 can be conjectured by some plausible physical arguments from the corresponding classical situation. The fine-structure term in the energy operator consists of two contributions: (1) a contribution caused by the interaction between the magnetic moment of the electron and the magnetic field in the electron rest frame due to the motion of the proton charge as seen in this frame, and (2) a contribution caused by the variation of the mass of the electron with its velocity. Both terms arise from relativistic kinematics; although the last term is of the same order of magnitude as the magnetic-moment term, it does not contribute to the splitting of the levels with the same values of n and l. Besides the fine-structure terms there are other contributions to the energy operator, which cause the Lamb shift and the hyperfine structure, but we shall not consider those here.

IX.3a The Magnetic Moment of a Spinning Particle in Classical Physics[3]

The energy of a magnetic dipole of magnetic moment **m** in a magnetic field **B** is

$$E_1^{(m)} = -\mathbf{m} \cdot \mathbf{B}. \tag{3.1}$$

So we need to know the magnetic moment **m** of the spinning electron and the magnetic field **B** in the electron's rest frame. In this subsection we shall determine **m**, and in Section IX.3b we shall discuss **B**.

Before we consider the magnetic moment of a spinning particle, let us recall the connection between the orbital angular momentum and the magnetic moment caused by the revolution of a charged spinless particle around a center. According to classical electromagnetic theory, the magnetic dipole moment of an orbiting point charge at position **x** with mass m_e and charge $(-e)$ moving with a velocity **v** is given by

$$\mathbf{m}^{\mathrm{orb}} = \frac{1}{2c}(-e)\mathbf{x} \times \mathbf{v} = -\frac{1}{2m_e c}e\mathbf{l}. \tag{3.2}$$

(Note that we consider negatively charged revolving particles like the electron, and that $e = +4.803 \times 10^{-10}$ esu.) The connection between the magnetic moment of a spinning charged particle and its internal angular momentum (spin) **s** differs from (3.2) by a factor of $g_s \approx 2$, the Landé factor, i.e., the gyromagnetic ratio is $e/(m_e c)$ and not $e/(2m_e c)$ as in (3.2). In order to derive this factor we consider the spinning point particle.

A classical spinning particle is a physical system with two different dynamical variables, the momentum **p** and the spin **s**. If **x** denotes the position, then the total angular momentum **j** is given by

$$\mathbf{j} = \mathbf{l} + \mathbf{s} = \mathbf{x} \times \mathbf{p} + \mathbf{s}. \tag{3.3}$$

The intrinsic properties of this classical particle are assumed to be such that it possesses a charge $(-e)$ and an intrinsic magnetic moment **m**. Then a given external force **F** and external torque **T** can act upon it. The external force may arise, for example, from an external electric or magnetic field (Lorentz force); the external torque may arise, for example, from a magnetic field **B** acting upon the intrinsic magnetic moment **m**:

$$\mathbf{T} = \mathbf{m} \times \mathbf{B}. \tag{3.4}$$

The equations of motion are then

$$\frac{d\mathbf{p}}{dt} = \mathbf{F}, \tag{3.5}$$

and

$$\frac{d\mathbf{j}}{dt} = \frac{d\mathbf{l}}{dt} + \frac{d\mathbf{s}}{dt} = \mathbf{T}_{\mathrm{tot}}, \tag{3.6}$$

[3] Corben (1968).

where $\mathbf{T}_{tot}$ is the total torque, i.e.,

$$\mathbf{T}_{tot} = \mathbf{x} \times \mathbf{F} + \mathbf{T}. \tag{3.7}$$

Using (3.5)–(3.7) and the definition $\mathbf{v} = d\mathbf{x}/dt$, one obtains from (3.3):

$$\frac{d\mathbf{s}}{dt} + \mathbf{v} \times \mathbf{p} = \mathbf{T}. \tag{3.8}$$

If the momentum $\mathbf{p}$ and velocity $\mathbf{v}$ are parallel, or if either of them vanishes, then

$$\frac{d\mathbf{s}}{dt} = \mathbf{T} \tag{3.9a}$$

and one obtains by using (3.3) and (3.5) that

$$\frac{d\mathbf{l}}{dt} = \mathbf{x} \times \mathbf{F}. \tag{3.9b}$$

This means that if the momentum vector $\mathbf{p}$ and the velocity vector $\mathbf{v}$ are parallel, then the spin and orbital motions are independent; the spin motion is determined by the torque $\mathbf{T}$ and the orbital motion is determined by the force $\mathbf{F}$. If, as is always assumed in nonrelativistic mechanics, $\mathbf{p} = m\mathbf{v}$, then one can describe the orbital motion in terms of (3.9b) and the spin motion in terms of (3.9a), and the spin and orbital motions become uncoupled. By inserting (3.4) in (3.9a), one then obtains for the spin motion (of a particle for which $\mathbf{p}$ and $\mathbf{v}$ are parallel)

$$\frac{d\mathbf{s}}{dt} = \mathbf{m} \times \mathbf{B}. \tag{3.10}$$

Let us now assume that we have a particle that has *no intrinsic magnetic moment*, i.e., $\mathbf{m} = 0$; consequently, by (3.4), $\mathbf{T} = 0$, where $\mathbf{T}$ is interpreted as the torque acting on any intrinsic moment the particle may possess. *But let us assume that* $\mathbf{v} \times \mathbf{p} \neq 0$,[4] i.e., that we do not have the strictly non-relativistic relation $\mathbf{p} = m\mathbf{v}$ or the relativistic relation for free spinless particles $\mathbf{p} = m\mathbf{v}/(1 - \mathbf{v}^2/c^2)^{1/2}$. Then (3.8) becomes

$$\frac{d\mathbf{s}}{dt} = -\mathbf{v} \times \mathbf{p}, \tag{3.11}$$

Suppose the particle is moving in a uniform constant magnetic field $\mathbf{B}$. The Lorentz force on the particle, which is equated with the time derivative of the momentum, is then

$$\mathbf{F} = \frac{d\mathbf{p}}{dt} = \frac{-e}{c}\mathbf{v} \times \mathbf{B}, \tag{3.12}$$

Consequently

$$\frac{1}{2}\frac{d}{dt}(\mathbf{p}^2) = \mathbf{p} \cdot \frac{d\mathbf{p}}{dt} = -\frac{e}{c}\mathbf{p} \cdot \mathbf{v} \times \mathbf{B}. \tag{3.13}$$

---

[4] $\mathbf{v} \times \mathbf{p} \neq 0$ will occur if the time components s_{01}, s_{02}, s_{03}, which together with $s_{ij} = \epsilon_{ijk}s_k$ make up the relativistic spin tensor $s_{\mu\nu}$, are not constant in time. This is always the case for an observer for whom the particle is moving with a velocity $\mathbf{v}$.

But, using (3.11) together with the supposition that **B** is a constant field ($d\mathbf{B}/dt = 0$),

$$\frac{e}{c}\frac{d}{dt}(\mathbf{s}\cdot\mathbf{B}) = \frac{e}{c}\frac{d\mathbf{s}}{dt}\cdot\mathbf{B} = -\frac{e}{c}\mathbf{v}\times\mathbf{p}\cdot\mathbf{B}$$

$$= \frac{e}{c}\mathbf{p}\times\mathbf{v}\cdot\mathbf{B} = \frac{e}{c}\mathbf{p}\cdot\mathbf{v}\times\mathbf{B}. \tag{3.14}$$

Adding (3.13) and (3.14), we conclude

$$\tfrac{1}{2}\mathbf{p}^2 + \frac{e}{c}\,\mathbf{s}\cdot\mathbf{B} = \text{const.} \tag{3.15}$$

If we suppose that the particle possesses a constant mass m_e and define the quantity $\boldsymbol{\mu}$ as

$$\boldsymbol{\mu} \equiv -\frac{e}{m_e c}\,\mathbf{s}, \tag{3.16}$$

then we can write (3.15) as

$$\frac{\mathbf{p}^2}{2m_e} - \boldsymbol{\mu}\cdot\mathbf{B} = \text{const.} \tag{3.17}$$

This means that it is not the kinetic energy $\mathbf{p}^2/2m_e$ of the particle but rather the left-hand side of (3.17) which is a constant of motion (unless two of the vectors **v**, **p**, and **B** are parallel, in which case it would follow from (3.13) that the kinetic energy is a constant of motion). The term $-\boldsymbol{\mu}\cdot\mathbf{B}$ therefore represents an extra energy which the spinning particle, for which **v** is not parallel to **p**, acquires in the magnetic field **B**. This extra term is written in the usual form (3.1) for the energy of a magnetic moment $\boldsymbol{\mu}$ in the magnetic field **B**. Thus, even though the particle (with $\mathbf{p}\parallel\mathbf{v}$) has no intrinsic magnetic moment upon which the magnetic field can act to give a torque, it "behaves" such that there is a contribution to the energy in the magnetic field as if it had a magnetic moment $\boldsymbol{\mu}$ given by (3.16) and a gyromagnetic ratio of $e/m_e c$. Equation (3.16) is similar to (3.2), but with an additional factor of 2 (the value of the Landé factor g_s) on the right-hand side. We are thus led to the conclusion that all charged spinning particles automatically possess a magnetic moment, given by (3.16), which is of relativistic origin. [The sign in Equations (3.12)–(3.15) and (3.16) reverses for positively charged particles.]

If the electron is a quantum-mechanical particle with charge $-e$ and *without intrinsic magnetic moment*, then its magnetic-moment operator should be given by the quantum-mechanical analogue of (3.16), i.e., by

$$\mathbf{M}_s = -\frac{e}{m_e c}\mathbf{S} = -g_s\frac{e}{2m_e c}\,\mathbf{S} \qquad (g_s = 2), \tag{3.18}$$

where **S** is the spin operator. It turns out that (3.18) indeed gives a very accurate description of the electron magnetic moment. But like every theoretical description this is only an approximate description and there are deviations. For electrons these deviations are small; for other particles, called hadrons, the deviations from (3.16) are of the same order as (3.16) and are interpreted as intrinsic magnetic moments due to an intrinsic structure. The deviation from the $g_s = 2$ value for the electron comes from the radiative corrections of quantum electrodynamics and is of the same order as, and of analogous origin to, the Lamb shift. Including these corrections up to the second order gives

$$g_s = 2\left(1 + \frac{\alpha}{2\pi} - 0.328\left(\frac{\alpha}{\pi}\right)^2\right),$$

where $\alpha = e^2/(\hbar c)$ is the fine-structure constant; this value for g_s agrees with experimental values up to the eighth decimal place. These correction terms are small, due to the smallness of α, and the electrons are thus to a very good approximation particles without an intrinsic structure.

The value $g_s = 2$ was first established as far back as 1915 by an experiment of Einstein and de Haas, and was incorporated in the spin hypothesis put forward around 1926. The existence of the radiative correction terms was first discovered experimentally by Rabi and collaborators in 1947, and was calculated within the framework of quantum electrodynamics by Schwinger in 1948. The value $g_s = 2$ is also obtainable from the assumption of "minimal coupling to the electromagnetic field," first used in the Dirac relativistic wave equation for the electron and considered one of the great achievements of the Dirac equation. The above argument shows that it is already a consequence of classical considerations.

There are elementary particles with intrinsic structure; e.g., the proton has an intrinsic magnetic moment

$$\mathbf{M}_p = 5.59\,\frac{e}{2m_p c}\,\mathbf{S} = \frac{e}{2m_p c}\,(2 + 3.59)\mathbf{S}, \tag{3.18p}$$

i.e., a g-value of $g_p = 5.59$. The value of the magnetic moment in excess of the value $2(e/(2m_p c))\mathbf{S}$ given by (3.16) is called the anomalous magnetic moment; it is $1.79e/(m_p c)$ for the proton. The neutron has no charge, and therefore the magnetic moment given by (3.16) is also zero; however it has an anomalous magnetic moment

$$\mathbf{M}_n = \frac{e}{2m_n c}\,(0 - 3.83)\mathbf{S}. \tag{3.18n}$$

IX.3b The Spin–Orbit Interaction Term

The determination of the magnitude of the magnetic field $\mathbf{B}$ must take into account relativistic kinematical effects. We go into a coordinate system that moves with the electron around the proton. In this system the electron is at rest, and the proton charge moves with a velocity $\mathbf{v}$ that is equal in magnitude but opposite in direction to the electron velocity. This movement constitutes a current. The magnetic field caused by a current of a single charge $+e$ moving with a velocity $\mathbf{v}$ is, according to the Biot–Savart law, given by

$$\mathbf{B}(x) = +e\,\frac{\mathbf{v} \times \mathbf{x}}{cr^3} \tag{3.19}$$

(e is measured in esu, B in gauss), where $\mathbf{x}$ is the vector of magnitude r from the moving charge to the observation point. The angular momentum of the electron is

$$\mathbf{l} = \mathbf{x} \times (-m_e \mathbf{v}).$$

Consequently the magnetic field at the position of the electron caused by the rotating proton is

$$\mathbf{B} = + \frac{e}{m_e c r^3} \mathbf{l}. \tag{3.20}$$

Both (3.19) and (3.20) neglect relativistic effects. From (3.1) we thus obtain

$$E_1^{(m)}\big|_{\text{r.f.}} = - \frac{e}{m_e c} \frac{1}{r^3} \mathbf{l} \cdot \mathbf{m}_s \tag{3.21}$$

for the energy of the magnetic moment in this field. The subscript "r.f.," denotes that we have used a rotating frame. If the frame rotates, there is an extra contribution to the energy that reduces (3.21) by a factor of $\frac{1}{2}$. [This factor is known as the "Thomas factor," and is caused by the "Thomas precession"; a detailed calculation may be found in Jackson (1975), p. 364.] Thus the energy of the moving spin magnetic moment in the magnetic field of the proton is given by

$$E_1^{(m)} = - \frac{e}{2m_e c} \frac{1}{r^3} \mathbf{l} \cdot \mathbf{m}_s. \tag{3.22}$$

We obtain the quantum-mechanical expression corresponding to (3.22) by the usual procedure of replacing the classical quantities $\mathbf{l}$ and $\mathbf{m}_s$ by the quantum-mechanical observables $\mathbf{L}$ and $\mathbf{M}_s$ given by (VI.3.1) and by (3.18):

$$H_1^{(m)} = + g_s \frac{e^2}{4m_e^2 c^2} \frac{1}{Q^3} \mathbf{L} \cdot \mathbf{S} = \frac{1}{2} \frac{e^2}{m_e^2 c^2} \frac{1}{Q^3} L_i S_i \quad \text{(usual cgs units)}$$

$$= g_s \mu_B^2 \frac{1}{Q^3} L_i S_i \quad (\hbar = 1). \tag{3.23}$$

The quantity

$$\mu_B = \frac{e\hbar}{2m_e c} = \frac{(4.8 \times 10^{-10}\ \text{esu})(6.6 \times 10^{-16}\ \text{eV sec})}{2(3.0 \times 10^{10}\ \text{cm/sec})(9.1 \times 10^{-28}\ \text{g})}$$

$$= 0.58 \times 10^{-8} \frac{\text{esu eV sec}^2}{\text{g cm}} = 0.58 \times 10^{-8}\ \text{eV} \frac{\text{cm}^{1/2}\ \text{sec}}{\text{g}^{1/2}}$$

$$= 0.58 \times 10^{-8}\ \text{eV/gauss} = 9.3 \times 10^{-21}\ \text{erg/gauss}$$

is called the Bohr magneton. The operator $\mathbf{L} \cdot \mathbf{S}$ of (3.23) is easily calculated from (2.5),

$$\mathbf{J}^2 = (\mathbf{L} + \mathbf{S})^2 = \mathbf{L}^2 + \mathbf{S}^2 + 2\mathbf{L} \cdot \mathbf{S},$$

and the fact that

$$\mathbf{S}^2 |n\,l\,j\,j_3\rangle = s(s+1)|n\,l\,j\,j_3\rangle = \tfrac{3}{4}|n\,l\,j\,j_3\rangle.$$

We then obtain for (3.23) that (in the usual cgs units)

$$H_1^{(m)} = \frac{1}{2} \frac{e^2}{m_e^2 c^2} Q^{-3} \tfrac{1}{2}(\mathbf{J}^2 - \mathbf{L}^2 - \tfrac{3}{4}I).$$ (3.24)

IX.3c The Kinematical Correction Term

We obtain the contribution arising from the relativistic mass effect by expanding the relativistic expression for the kinetic energy

$$p_0 = \sqrt{(m_e c^2)^2 + c^2 p^2}$$

of the free electron in powers of $p/(m_e c)$ $[p = (\mathbf{p}^2)^{1/2}]$:

$$p_0 = m_e c^2 \sqrt{1 + \left(\frac{p}{m_e c}\right)^2} = m_e c^2 \left(1 + \frac{1}{2}\left(\frac{p}{m_e c}\right)^2 - \frac{1}{8}\left(\frac{p}{m_e c}\right)^4 + \cdots \right)$$

$$= m_e c^2 + \frac{p^2}{2m_e} - \frac{1}{2}\left(\frac{p^2}{2m_e}\right)^2 \frac{1}{m_e c^2} + \cdots .$$

The rest energy $m_e c^2$ of the electron is ignored because nonrelativistically the energy is determined only up to an additive constant (the rest energy would shift the energy levels by the same fixed amount). Thus classically the kinetic energy with the first-order relativistic correction is

$$E_{kin} = \frac{p^2}{2m_e} - \frac{1}{2}\left(\frac{p^2}{2m_e}\right)^2 \frac{1}{m_e c^2},$$

in contrast to the usual expression

$$E_{kin} = \frac{p^2}{2m_e}.$$

Going to quantum mechanics, we replace the numbers p_i by the operators P_i and obtain

$$H_1^{(k)} = -\frac{1}{2m_e c^2}\left(\frac{\mathbf{P}^2}{2m_e}\right)^2 = -\frac{1}{2m_e c^2}\left(H_0 + \frac{e^2}{Q}\right)^2$$ (3.25)

for the kinematical correction to

$$H_0 = \frac{\mathbf{P}^2}{2m_e} - \frac{e^2}{Q}.$$ (2.14')

[Both this equation for H_0 and (3.25) are in the usual cgs units, as opposed to the units described above (VI.3.4). The total-energy operator with the two corrections is then

$$H = H_0 + H_1^{(m)} + H_1^{(k)},$$ (3.26)

where $H_1^{(m)}$ and $H_1^{(k)}$ are given by (3.24) and (3.25), respectively.

IX.4 Fine Structure of Atomic Spectra

As $H_1 = H_1^{(m)} + H_1^{(k)}$ contains the operators Q^k ($k = -1, -2, -3$), H_1 and therefore H do not commute with H_0. It therefore appears that we have the situation described by (2.40), (2.44), (2.45), that $|n\,l\,j\,j_3\rangle$ are not the physical eigenvectors and that $E_{nlj}^{(1)}$ given by (2.44) and (2.45) is only a first approximation.

In order to calculate the matrix elements (2.45) we make use of (VII.2.15), (VII.2.16), (VII.2.17) with (VII.2.20). [Note that for $l = 0$ Equation (VII.2.17) is singular; however, for $l = 0$ the matrix element $\langle n\,l\,j\,j_3|\mathbf{L}\cdot\mathbf{S}|n\,l\,j\,j_3\rangle$ is zero—cf. (4.2) below—so that the matrix element of $\mathbf{L}\cdot\mathbf{S}/Q^3$ is always well defined.] We calculate the matrix element of the spin-orbit term $H_1^{(m)}$ and find

$$
\langle n\,l\,j\,j_3|H_1^{(m)}|n\,l\,j\,j_3\rangle = \frac{e^2}{2m_e^2 c^2}\langle n\,l\,j\,j_3|Q^{-3}|n\,l\,j\,j_3\rangle
$$
$$
\times \begin{cases} l/2 & \text{for } j = l + \tfrac{1}{2}, \\ -(l+1)/2 & \text{for } j = l - \tfrac{1}{2}, \end{cases} \tag{4.1}
$$

because

$$
\tfrac{1}{2}(\mathbf{J}^2 - \mathbf{L}^2 - \tfrac{3}{4}I)|n\,l\,j\,j_3\rangle = \tfrac{1}{2}(j(j+1) - l(l+1) - \tfrac{3}{4})|n\,l\,j\,j_3\rangle
$$
$$
= \tfrac{1}{2}|n\,l\,j\,j_3\rangle \times \begin{cases} l & \text{for } j = l + \tfrac{1}{2}, \\ -(l+1) & \text{for } j = l - \tfrac{1}{2}. \end{cases} \tag{4.2}
$$

According to (2.23),

$$
\langle n\,l\,j\,j_3|Q^{-3}|n\,l\,j\,j_3\rangle = \sum_{l_3's_3'}\sum_{l_3 s_3} \langle j\,j_3|l\,l_3\,s = \tfrac{1}{2}s_3\rangle\langle l\,l_3'\,s = \tfrac{1}{2}s_3'|j\,j_3\rangle
$$
$$
\times \langle s_3|s_3'\rangle\langle n\,l\,l_3|Q^{-3}|n\,l\,l_3'\rangle.
$$

Using the orthogonality relation (V.2.15) of the Clebsch–Gordan coefficients, and the fact that the last matrix element is zero unless $l_3 = l_3'$, which we write as

$$
\langle n\,l\,l_3'|Q^{-3}|n\,l\,l_3\rangle = \delta_{l_3 l_3'}\langle n\,l|Q^{-3}|n\,l\rangle,
$$

we obtain

$$
\langle n\,l\,j\,j_3|Q^{-3}|n\,l\,j\,j_3\rangle = \langle n\,l|Q^{-3}|n\,l\rangle.
$$

Thus from (4.1) and (VII.2.17) together with (VII.2.20) it follows that

$$
\langle n\,l\,j\,j_3|H_1^{(m)}|n\,l\,j\,j_3\rangle = \frac{1}{2}\left(\frac{1}{m_e c}\right)^2 e^2\,\frac{m_e^3 e^6}{1}\,\frac{1}{n^3(l+1)(l+\tfrac{1}{2})l}
$$
$$
\times \frac{1}{2}\begin{cases} l & \text{for } j = l + \tfrac{1}{2}, \\ -(l+1) & \text{for } j = l - \tfrac{1}{2}. \end{cases} \tag{4.3}
$$

Recalling that, according to Equation (VI.5.12) with $\hbar = 1$,

$$
E_n^0 = -\frac{m_e e^4}{1}\frac{1}{2n^2} = -\left(\frac{e^2}{c}\right)^2 m_e c^2 \frac{1}{2n^2}, \tag{4.4}
$$

(4.3) can be written as

$$\langle n \, l \, j \, j_3 | H_1^{(m)} | n \, l \, j \, j_3 \rangle = - E_n^0 \left(\frac{e^2}{c}\right)^2 \frac{1}{n(2l+1)} \begin{cases} 1/(l+1) & \text{for } j = l + \frac{1}{2}, \\ -1/l & \text{for } j = l - \frac{1}{2}. \end{cases}$$

(4.5)

The matrix element of the kinetic-energy term $H_1^{(k)}$ of (3.25) is

$$\langle n \, l \, j \, j_3 | H_1^{(k)} | n \, l \, j \, j_3 \rangle$$

$$= - \frac{1}{2m_e c^2} \{ E_n^{02} + 2E_n^0 e^2 \langle n \, l | Q^{-1} | n \, l \rangle + e^4 \langle n \, l | Q^{-2} | n \, l \rangle \}$$

$$= - E_n^0 \left\{ - \left(\frac{e^2}{c}\right)^2 \frac{1}{4n^2} + \frac{2e^2}{2m_e c^2} \frac{m_e e^2}{1} \frac{1}{n^2} + \frac{e^4 m_e^2 e^4}{2m_e c^2 E_n^0} \frac{1}{(l+\frac{1}{2})n^3} \right\}$$

$$= - E_n^0 \left(\frac{e^2}{c}\right)^2 \left\{ - \frac{1}{4n^2} + \frac{1}{n^2} - \frac{1}{n(l+\frac{1}{2})} \right\}$$

$$= - E_n^0 \left(\frac{e^2}{c}\right)^2 \frac{1}{n^2} \left(\frac{3}{4} - \frac{n}{l+\frac{1}{2}}\right).$$

(4.6)

In this calculation (VII.2.15) and (VII.2.16) together with (VII.2.20) have been used. Adding (4.5) and (4.6), we obtain for the fine-structure interaction term

$$\langle n \, l \, j \, j_3 | H_1 | n \, l \, j \, j_3 \rangle = - E_n^0 \left(\frac{e^2}{c}\right)^2 \frac{1}{n^2} \left(\frac{3}{4} - \frac{n}{j+\frac{1}{2}}\right).$$

(4.7)

Therefore the matrix element of $H = H_0 + H_1$ is (with $\hbar$ restored)

$$E_{nlj} = \langle n \, l \, j \, j_3 | H | n \, l \, j \, j_3 \rangle = E_n^0 \left(1 + \left(\frac{e^2}{\hbar c}\right)^2 \frac{1}{n^2} \left(\frac{n}{j+\frac{1}{2}} - \frac{3}{4}\right)\right).$$

(4.8)

The term in (4.8) that gives rise to the fine-structure splitting is the one proportional to $1/(j+\frac{1}{2})$. Recalling that $\alpha^2 = (e^2/(\hbar c))^2 \cong (1/137)^2$, we see that this fine-structure term is four orders of magnitude smaller than E_n^0 and gives the experimentally correct splitting between the energy levels of different values for the quantum number j. This is shown in Figure VI.5.2, where the superscripts stand for the total angular momentum value j. Thus the energy levels are labeled by $n(l)^j$ where for (l) with $l = 0, 1, 2, 3, 4, 5, \ldots$ one uses the letters $S, P, D, F, G, H, \ldots$, respectively. For example, $2P^{1/2}$ means $n = 2, l = 1, j = \frac{1}{2}$, and $2P^{3/2}$ means $n = 2, l = 1, j = \frac{3}{2}$. Equation (4.8) is independent of l and therefore cannot describe the splittings between the levels $2S^{1/2}$ and $2P^{1/2}$ or between the levels $3P^{3/2}$ and $3D^{3/2}$—i.e., between states with the same value of j but different values of l—which are also depicted in Figure VI.5.2. As mentioned before, this splitting, the Lamb shift, has a different origin and is one order of magnitude smaller than the splitting due to the spin.

Spin is not a fundamental notion of quantum theory. As discussed in Chapter III, the angular momentum value of $\frac{1}{2}\hbar$ naturally appears as a solution of the angular momentum commutation relations (III.2.16), which in turn are a consequence of rotational symmetry. The existence of angular momentum with half-integer values can thus be derived from the fundamental postulates of quantum mechanics and the properties of symmetry transformations (the Wigner theorem).

The spin value of $\frac{1}{2}\hbar$ is a property of the electron in particular, and not a peculiarity of the general postulates of quantum mechanics. In the early days of quantum theory, however, the effects of the spin were misconstrued as being difficulties inherent in the general structure of quantum physics, and this resulted in one of the most fascinating chapters in the history of quantum mechanics.

That the angular momentum quantum number can have half-integer values was first suggested by A. Landé (1921) in a systematics of the anomalous Zeeman-effect (the splitting of the energy levels of atoms in an applied magnetic field, see Problem 3). Thereafter (also in 1921) Heisenberg had the idea that if an electron is added to an atom or ion, it imparts to the original body of the atom a fraction $\frac{1}{2}\hbar$ of its angular momentum $l\hbar$, and retains the amount $(l - \frac{1}{2})\hbar$. With this hypothesis he was able to explain the doublet term splitting of the alkali atoms as it was later described by the above formula (4.7).

The suggestion that this splitting was caused by a spin of the electron was probably first made by R. Kronig (spring, 1925). He discussed it with Pauli who rejected it as untenable. The main reason for refusing the spin hypothesis was the missing factor of $\frac{1}{2}$ (the Thomas factor) in (3.22) which led to a discrepancy by a factor of 2 between the calculated doublet splitting (2 × the right-hand side of Equation (4.7)) and the observed splitting (correctly given by (4.7)). The fact that $g_s = 2$ had by then already been well established.

Then in the fall of 1925, S. Goudsmit and G. E. Uhlenbeck, ignorant about all this, published the first paper on the spin hypothesis and later, after Heisenberg had drawn their attention to it, derived the doublet splitting with the excess factor of 2. This convinced Bohr, but Pauli still objected to the electron spin; he did not believe that there existed a mechanical explanation of these effects. Finally, in 1926, L. H. Thomas showed that the additional factor of $\frac{1}{2}$ in the observed doublet splitting was a forgotten relativistic effect. That a factor of $\frac{1}{2}$ and not v/c should be a relativistic effect came as a great surprise even to the experts of relativity theory (Einstein included), but even Pauli was at last convinced of the existence of the electron spin.

IX.5 Selection Rules

Selection rules for dipole transitions have been discussed previously, in particular in Section V.4 for general angular momentum states and also briefly in Section VI.5 for the hydrogen atom without spin. The selection rules, i.e., the rules that tell when the matrix elements $\langle n'\, l'\, j'\, j'_3 | Q_i | n\, l\, j\, j_3 \rangle$ are zero, follow from the property that Q_i is a proper vector operator, i.e., that it fulfills the relations

$$[L_i, Q_j] = i\epsilon_{ijk}Q_k, \tag{5.1a}$$

$$[J_i, Q_j] = i\epsilon_{ijk}Q_k, \tag{5.1b}$$

and

$$U_P Q_i U_P = -Q_i. \tag{5.2}$$

Equation (5.1b) expresses the fact that Q_j is not only a vector operator with respect to the orbital angular momentum (an **L**-vector operator) but also a vector operator with respect to total angular momentum (a **J**-vector operator). Equation (5.1b) follows from $J_i = L_i + S_i$ together with (5.1a).

As a consequence of (5.1b), the selection rules for dipole transitions between the physical states $|n\,l\,j\,j_3\rangle$ for the hydrogen atom (and all other one-electron atoms) are:

$$\langle n'\,l'\,j'\,j_3'\,|Q_i|n\,l\,j\,j_3\rangle = 0 \quad \text{unless} \quad j' = j+1, j, \text{ or } j-1. \tag{5.3}$$

From (5.2) it follows that

$$\langle n'\,l'\,j'\,j_3'|Q_i|n\,l\,j\,j_3\rangle = 0 \quad \text{unless} \quad \pi(l)\pi(l') = -1 \tag{5.4}$$

[i.e., unless $(-1)^{l'+l} = -1$, as $\pi(l) = (-1)^l$]. From (5.3) and the fact that $j' = l' \pm \frac{1}{2}$ and $j = l \pm \frac{1}{2}$, it follows that the matrix element is zero unless $l' - l + 2, l + 1, l, l - 1, l - 2$. Thus (5.4) can be written

$$\langle n'\,l'\,j'\,j_3'|Q_i|n\,l\,j\,j_3\rangle = 0 \quad \text{unless} \quad l' = l \pm 1. \tag{5.5}$$

Equations (5.3), (5.4) or (5.5) give the selection rules for dipole transitions in the hydrogen atom and all other one-electron atoms.

IX.6 Remarks on the State of an Electron in Atoms

We should close this part on the hydrogen atom with a remark. To conjecture the algebraic structure that is the mathematical image of the hydrogen atom, we made use of the classical particle picture of the Kepler problem, i.e., we had as the point of departure for our conjecture the picture of an electron as a particle that moves in closed orbits around a center. And when we discussed spin, we even implied some analogy of the spin with the rotation of this particle around its own axis [in the comparison of (3.16) with (3.18), for instance]. This was the classical picture, which is described by the mathematical relations between the observables when they are not represented by operators but by numbers. In quantum mechanics these mathematical relations are operator relations, and are not the mathematical image of a point particle revolving around a center and rotating around its own axis. The electron in the hydrogen atom is not a particle, and the spin is not a rotation around the particle's axis. A quantum-mechanical object was called a particle when it was approximately a generalized eigenstate of the position operator, because localization is the characteristic property of a particle. The electron in the hydrogen atom is not in a generalized position eigenstate. Equally wrong is the view (which originates from the misinterpretation of the solution of the Schrödinger wave equation) that the electron in the hydrogen atom is a standing plane wave, i.e., the classical picture complementary to the particle picture. A quantum-mechanical object was called a wave when it was a generalized eigenstate of the momentum operator (Section II.9), because this state has the characteristic property of a wave (wave motion over all space). The electron in the hydrogen atom is not in a generalized momentum eigenstate. Thus neither of the complementary classical pictures of the electron is applicable to the electron in the hydrogen atom. In the hydrogen atom the electron does not appear as a wave or as a particle, but in a form different from both, namely, as an angular-momentum and energy eigenstate, in which it has neither a definite position (particle) nor a definite momentum (wave), but does have a definite angular momentum (rotator). The classical picture that comes closest to this is that of a standing spherical wave.

Problems

1. (a)   Show that every operator on $\mathfrak{t}^s$ is a linear combination of the operators $I, S_1, S_2,$ and S_3; as stated in the paragraph preceding Equation (2.21).
    (b)   Check to see whether there are other solutions of

$$i\varepsilon_{ijk}\{L_k, B_j\} = 0$$

[the equation immediately preceding (2.38)] besides (2.38).

$$B_j = f(\mathbf{Q}, \mathbf{P})L_j.$$

2. The proton possess a magnetic moment

$$\mathbf{M}_p = \frac{e}{2m_p c} (2 + 3.59)\, \mathbf{S}.$$

    (a)   Determine the Hamiltonian that describes the interaction of the electron's spin with the proton's spin in the hydrogen atom (hyperfine interaction).
    (b)   Treating the spin-spin interaction Hamiltonian in part (a) as a further perturbation term on the hydrogen-atom Hamiltonian, determine the matrix elements of this spin-spin term between states $|n\,l\,j\,j_3\rangle$ of $l = 0$ and $l = 1$. See how this correction compares with the fine-structure correction.

3. An external magnetic field breaks the rotational symmetry so that the Hamiltonian H for an atom in the external field no longer commutes with the angular momentum operators J_i.
    (a)   How does such an external symmetry breaking field affect the energy spectrum?
    The Hamiltonian for the hydrogen-atom or the Na-atom in a uniform external magnetic field $\mathbf{B}$ is given by

$$H = H_0 + H_1 + H'$$

where $H_0 + H_1$ is the rotationally invariant Hamiltonian for $\mathbf{B} = 0$: $H_0 = (1/2m)\mathbf{P}^2 + V$ and H_1 gives the fine structure. H' describes the effect of $\mathbf{B}$ and is given by

$$H' = \mu_0 \mathbf{B} \cdot (\mathbf{L} + 2\mathbf{S})$$

where $\mu_0 = eh/2mc$ (Bohr magneton).

   (b) Calculate the spectrum of H for the case that the magnetic field is very weak so that H' can be treated as a small perturbation of $H_0 + H_1$ and H_1.

   (c) List the dipole transitions between all energy levels of the 3^2P band and the 3^2S band of the Na-atom in a weak magnetic field. (Zeeman effect)

CHAPTER X

Indistinguishable Particles

In this chapter the basic assumption about the combination of identical physical systems is conjectured from the indistinguishability of these systems.

X.1 Introduction

The quantum-mechanical systems that we have considered so far have consisted of only one constituent of the same kind. The hydrogen atom was considered as one electron in an electric field. The vibrating diatomic molecule was described as the problem of one oscillator. The rotating diatomic molecule was reduced to the problem of one system rotating around a center. When the vibrating and rotating diatomic molecule was considered, it was described as the combination of one rotator and one oscillator, i.e., as one rotating oscillator. These systems are called *one-particle systems*. Thus one-particle systems are systems that consist of only one constituent of the "same kind."

We now want to consider *many-particle systems*, i.e., systems that are the combination of many ($N = 2, 3, \ldots$) one-particle systems of the same kind. Let $\mathcal{H}_1, \mathcal{H}_2, \ldots, \mathcal{H}_N$ be the spaces of physical states of the first, second, $\ldots$, Nth one-particle system. (All of the spaces $\mathcal{H}_i$ are identical; the subscript serves only to identify a space with a particular particle.) We would expect from the basic postulate IVa that the N-particle system, which is the combination of these N one-particle systems, will have as its space of states the direct product space

$$\mathfrak{H} = \mathcal{H}_1 \otimes \mathcal{H}_2 \otimes \cdots \otimes \mathcal{H}_N. \tag{1.1}$$

The algebra of the N-particle system would be the direct product of the algebras of observables of the one-particle systems, i.e., the algebra is the set of all operators

$$A = \sum A_1 \otimes A_2 \otimes \cdots \otimes A_N, \tag{1.2}$$

where A_i is an element of the algebra of observables in $\mathcal{H}_i$. (For example, if these N one-particle systems do not interact with each other, then all elements of the algebra of observables on $\mathfrak{H}$ are of the kind $A = A_1 \otimes I \otimes I \otimes \cdots \otimes I + I \otimes A_2 \otimes I \otimes \cdots \otimes I + \cdots + I \otimes I \otimes \cdots \otimes I \otimes A_N$.)

That the N-particle system would be described by (1.1) we would expect from classical considerations. Classical particles can, in principle, be numbered; i.e., we can imagine that each particle is given at some instant a label, and we can then follow the subsequent motion of each particle in its path and identify a given particle at any subsequent instant. This is not possible, however, for quantum-mechanical systems. Following the motion of each particle of a quantum-mechanical system means performing a series of position measurements. Each measurement—according to the axiom III— changes the state of the quantum-mechanical system in an uncontrollable fashion: If we have localized the particle in the neighborhood of a certain point, then we do not know what its momentum is and we do not know where it will go. The concept of path does not exist for a quantum-mechanical system (uncertainty principle). Thus in quantum mechanics it is, even in principle, impossible to follow each of a number of *identical particles* and thereby distinguish them. (By identical particles we mean particles that have the same observable values.) Thus: Identical quantum-mechanical particles are indistinguishable.

We shall now give a precise, mathematical formulation of indistinguishability and then formulate the consequences as another basic postulate of quantum mechanics. Let $|\xi_i\rangle_i$ denote a basis in the space $\mathcal{H}_i$, i.e., the one symbol ξ_i stands for the full set of quantum numbers (eigenvalues of a c.s.c.o.) necessary to label the basis system of $\mathcal{H}_i$. [For the sake of definiteness, one may assume that the N particles are electrons in a Coulomb field; then each $\mathcal{H}_i$ is the space $\mathcal{H}$ of (IX.2.15), and $|\xi_i\rangle_i = |n_i l_i j_i j_{i3}\rangle$.] The basis in $\mathfrak{H}$ of (1.1) is then given by

$$|\xi_1 \xi_2 \cdots \xi_N\rangle = |\xi_1\rangle_1 \otimes |\xi_2\rangle_2 \otimes \cdots \otimes |\xi_N\rangle_N. \tag{1.3}$$

Suppose we are given N objects (elements) in a certain order,

$$(\xi_1, \xi_2, \ldots, \xi_N);$$

such an arrangement is called a *permutation*. These N elements can be written in a different order, $(\eta_1, \eta_2, \ldots, \eta_N)$; this is called a permutation of the N objects. There are $N!$ different permutations of N objects. One particular permutation can be considered to be the "original" or "natural" or "standard" one. All other permutations can be obtained from this original one by changing the order in which the objects appear. It is clear that the *operation* of changing the order is specified by the resulting permutation;

therefore this operation is also called a permutation. Thus we may consider the permutation operation (or just "permutation") P that changes $(\xi_1, \xi_2, \ldots, \xi_N)$ into $(\eta_1, \eta_2, \ldots, \eta_N)$. For example, we can consider the permutation P_{12} that changes $(\xi_1, \xi_2, \ldots, \xi_N)$ into $(\eta_1 = \xi_2, \eta_2 = \xi_1, \eta_3 = \xi_3, \ldots, \eta_N = \xi_N)$. Or we can consider the permutation P_{ij} that changes $(\xi_1, \ldots, \xi_i, \ldots, \xi_j, \ldots, \xi_N)$ into $(\xi_1, \ldots, \xi_j, \ldots, \xi_i, \ldots, \xi_N)$ by interchanging the ith and jth elements; such permutations, which consist of the exchange of the position of two elements, are called *transpositions*. Every permutation can be obtained by a finite number of transpositions; e.g., the permutation $(\xi_2, \xi_3, \xi_1, \xi_4, \ldots, \xi_N)$ may be obtained by the transposition P_{13} followed by the transposition P_{12}: $(\xi_1, \xi_2, \xi_3, \xi_4, \ldots) \to (\xi_3, \xi_2, \xi_1, \xi_4, \ldots) \to (\xi_2, \xi_3, \xi_1, \xi_4, \ldots)$. While the decomposition of a permutation into successive transpositions is not unique, the number of transpositions will always be even or odd, depending upon the particular permutation. A permutation is *odd* (with respect to the original permutation) if it is obtained from the original permutation by an odd number of transpositions; it is *even* if it is obtained by an even number of transpositions.

Now let the N objects be the N sets of quantum numbers $(\xi_1, \xi_2, \ldots) = (n_1 l_1 j_1 j_{13}, n_2 l_2 j_2 j_{23}, \ldots)$. Each permutation $(\eta_1, \eta_2, \ldots, \eta_N)$, or each permutation operation $P: (\xi_1, \xi_2, \ldots, \xi_N) \mapsto (\eta_1, \eta_2, \ldots, \eta_N)$ can then be represented in the space $\mathfrak{H}$ by a linear operator $\mathbb{P}$ defined by

$$\mathbb{P}|\xi_1 \xi_2 \cdots \xi_N\rangle = |\eta_1 \eta_2 \cdots \eta_N\rangle. \tag{1.4}$$

For example, the transposition P_{12} is represented by the operator $\mathbb{P}_{12}$ with

$$\mathbb{P}_{12}|\xi_1 \xi_2 \cdots \xi_N\rangle = |\xi_2 \xi_1 \cdots \xi_N\rangle. \tag{1.5}$$

The operators $\mathbb{P}$ will be chosen to be unitary.[1]

[The set $\{P\}$ of all permutations of N objects forms a group known as the *symmetric* (or *permutation*) *group*, while the set $\{\mathbb{P}\}$ of all representing operators forms a *representation* of the permutation group. If all operators $\mathbb{P}$ are unitary, then the representation is called a *unitary representation* of the permutation group. Because the permutation group has a finite number of elements (is finite and consequently compact), every representation of the permutation group, according to a theorem, may be considered unitary. We shall make use of only one property of the permutation group, to be stated below, and shall not require any group theory.]

Let $\Lambda_{|\xi\rangle}$ denote the projection operator onto the one-dimension subspace spanned by $|\xi\rangle = |\xi_1 \xi_2 \cdots \xi_N\rangle$, and let $\Lambda_{|\eta\rangle}$ denote the projector onto the one-dimensional subspace spanned by $|\eta\rangle = |\eta_1 \eta_2 \cdots \eta_N\rangle$. Because of (1.4), we have the connection

$$\Lambda_{|\eta\rangle} = |\eta\rangle\langle\eta| = \mathbb{P}|\xi\rangle\langle\xi|\mathbb{P}^\dagger = \mathbb{P}\,\Lambda_{|\xi\rangle}\,\mathbb{P}^\dagger. \tag{1.6}$$

---

[1] The reason for the requirement that $\mathbb{P}$ be unitary follows from the fact that $\mathbb{P}$ is a symmetry transformation. A brief justification of this will be given in Appendix to Section XIX.2.

Now, due to *indistinguishability*, $\Lambda_{|\xi\rangle}$ and $\Lambda_{|\eta\rangle}$ can represent a (pure) physical state only if

$$\Lambda_{|\eta\rangle} = \Lambda_{|\xi\rangle}, \tag{1.7}$$

or, using (1.6),

$$\Lambda_{|\xi\rangle} = \mathbb{P} \, \Lambda_{|\xi\rangle} \, \mathbb{P}^\dagger. \tag{1.8}$$

Thus of the one-dimensional subspaces of the direct product $\mathfrak{H}$ of (1.1), only those for which (1.8) is fulfilled can represent physical states. From this we conclude that, because of indistinguishability, it is not the whole direct-product space $\mathfrak{H}$ of (1.1) but only a subspace of it that is the space of physical states. Also, it is not the whole direct-product algebra of operators given by (1.2) but only a subalgebra of it that is the algebra of observables. We now want to determine the physical subspace of (1.1). It is clear that, in general, the direct-product basis (1.3) is not a suitable basis [(1.8) means $\mathbb{P}|\eta\rangle \propto |\eta\rangle$, which cannot be fulfilled if all the quantum numbers are different]; and that, in general, the physical states are represented by linear combinations of (1.3).

Let ψ be a vector of the physical subspace [in general, a linear combination of (1.3)], and let $\mathbb{P}$ be a permutation operator. Then, if the particles are indistinguishable, ψ and $\chi = \mathbb{P}\psi$ [or Λ_ψ and $\Lambda_\chi = \Lambda_{\mathbb{P}\psi}$] represent the same physical state. The expectation value of every observable A must therefore be the same for ψ and for χ, i.e.,

$$\langle\psi|A|\psi\rangle = \langle\chi|A|\chi\rangle = \langle\psi|\mathbb{P}^\dagger A\mathbb{P}|\psi\rangle \tag{1.9}$$

for every observable A. As ψ is an arbitrary vector of the physical subspace, we conclude that

$$A = \mathbb{P}^\dagger A\mathbb{P} \tag{1.10}$$

for every observable A. Hence for any $A \in \mathscr{A}$ and for any permutation operator $\mathbb{P}$,

$$[\mathbb{P}, A] = 0. \tag{1.11}$$

[Equation (1.11) follows immediately from (1.10) for $\mathbb{P}$ unitary, which, as mentioned above, can be assumed for the permutation group. For $\mathbb{P}$ nonunitary, one must take linear combinations $\psi_1 + i\alpha\psi_2, \psi_1 + \alpha\psi_2$ in order to deduce (1.11).]

Equation (1.11) *is* the mathematical formulation of the *statement that identical particles are indistinguishable*. From this mathematical formulation of indistinguishability, one can deduce that the vectors of the physical subspace of $\mathfrak{H}$ must fulfill either

$$\mathbb{P}\psi = +\psi \quad \text{for all } \mathbb{P} \tag{1.12}$$

or

$$\mathbb{P}\psi = (-1)^p\psi \quad \text{where} \begin{cases} p \text{ is even if } \mathbb{P} \text{ is even,} \\ p \text{ is odd if } \mathbb{P} \text{ is odd.} \end{cases} \tag{1.13}$$

The ψ fulfilling (1.12) are called *symmetric*, while those fulfilling (1.13) are called *antisymmetric*.

To deduce this, we need the mathematical formulation of a physically obvious property of *the algebra $\mathscr{A} = \{A\}$ of observables A.* This can be formulated in the following way:[2] *$\mathscr{A} = \{A\}$ contains a complete set of commuting operators.* To justify this condition physically, we recall that we called a state "*pure*" (up to a certain accuracy) if there is no observable whose measurement allowed the separation of an ensemble in this state into two or more subensembles. If such a separation is possible (as in the case of the hydrogen atoms in the ground state), it leads to the introduction of a new quantum number and therewith to the introduction of a new observable (in the case of the hydrogen atom, spin). Consequently one must enlarge the algebra of observables to accommodate this new observable. What is a pure state with respect to a certain accuracy need not be a pure state with respect to a higher accuracy. But up to every desired (and observable) accuracy, every pure state is completely specified by a set of quantum numbers that are connected—with observables. (In fact, the algebra of observables is conjectured from these quantum numbers and observables.) Each label of a vector is connected with an observable, and the observables whose eigenvalues label the vectors form a complete set. [As we have seen, the statement that a certain set of commuting operators is a complete system is a *physical* statement, conjectured from the physical properties of the system, and not a mathematical statement (cf. Chapter IV)]. Therefore the apparent occurrence of pure states not fulfilling (1.12) or (1.13) is always an indication that the available set of quantum numbers is not complete and that there exist other quantum numbers completing this set that have not yet been uncovered.

Let us denote by $A_1, A_2, \ldots, A_n$ a complete system of commuting operators of the algebra of observables, by $|a\rangle = |a_1 a_2 \cdots a_n\rangle$ the corresponding eigenvectors, and by $\Lambda_{|a\rangle}$ the projectors onto the subspaces spanned by the $|a\rangle$. $\Lambda_{|a\rangle}$ is the observable whose expectation value gives the probability of obtaining $a_1, a_2, \ldots, a_n$ in a measurement of $A_1, A_2, \ldots, A_n$. As observables, the $\Lambda_{|a\rangle}$ must commute with all permutations $\mathbb{P}$ in accord with (1.11):

$$[\mathbb{P}, \Lambda_{|a\rangle}] = 0 \quad \text{for all } \mathbb{P}. \tag{1.14}$$

From (1.14) it follows that $|a\rangle$ must be an eigenvector of all $\mathbb{P}$.

PROOF. Equation (1.14) applied to a vector $|\psi\rangle$ gives $\mathbb{P} \, \Lambda_{|a\rangle}|\psi\rangle = \Lambda_{|a\rangle}\mathbb{P}|\psi\rangle$. Using $\Lambda_{|a\rangle} = |a\rangle\langle a|$, we see that $\mathbb{P}|a\rangle\langle a|\psi\rangle = |a\rangle\langle a|\mathbb{P}|\psi\rangle$, and hence that

$$\mathbb{P}|a\rangle = |a\rangle \frac{\langle a|\mathbb{P}|\psi\rangle}{\langle a|\psi\rangle}. \qquad \square \quad (1.15)$$

The possible eigenvalues, $\langle a|\mathbb{P}|\psi\rangle/\langle a|\psi\rangle$, follow from the following property (which we shall not derive here) of the permutation group: The permuta-

---

[2] It is this condition that is relaxed when parastatistics are allowed.

tion group has two one-dimensional representations: (1) the symmetrical representation in which all permutations are represented by the unit operator I, i.e.,

$$\mathbb{P}|\psi\rangle = +|\psi\rangle \quad \text{for all permutations } \mathbb{P}; \tag{1.16}$$

and (2) the antisymmetrical representation in which all even permutations are represented by the operator I and all odd permutations are represented by the operator $-I$, i.e.,

$$\mathbb{P}|\psi\rangle = (-1)^p|\psi\rangle \quad \text{where } \begin{cases} p \text{ is even if } \mathbb{P} \text{ is even,} \\ p \text{ is odd if } \mathbb{P} \text{ is odd.} \end{cases} \tag{1.17}$$

We recall that (Section III.3) an irreducible-representation space (ladder representation) of the algebra of $\{\mathbb{P}\}$ is a representation space that is obtained by applying all $\mathbb{P}$ to one element of the space. The fact that $|a\rangle$ is an eigenvector of all $\mathbb{P}$ means that $|a\rangle$ spans a one-dimensional irreducible-representation space. Therefore, according to the above property of the permutation group, we must have either

$$\mathbb{P}|a\rangle = +|a\rangle \quad \text{for all } \mathbb{P} \tag{1.18}$$

or

$$\mathbb{P}|a\rangle = (-1)^p|a\rangle \quad \text{for all } \mathbb{P}. \tag{1.19}$$

Since this is true for every basis vector $|a\rangle$, and as every vector $|\psi\rangle$ of the representation space of the algebra of observables (i.e., of the physical subspace of $\mathfrak{H}$) can be written as a linear combination of the basis vectors, we conclude that either (1.12) or (1.13) must be fulfilled.

Let us denote by $\mathscr{H}^N_+$ that subspace of $\mathfrak{H}$ which consists of the symmetric vectors,

$$\mathscr{H}^N_+ = \{|\psi\rangle \in \mathfrak{H}: \mathbb{P}|\psi\rangle = |\psi\rangle\}, \tag{1.20}$$

and let us denote by $\mathscr{H}^N_-$ that subspace of $\mathfrak{H}$ which consists of the antisymmetric vectors,

$$\mathscr{H}^N_- = \{|\psi\rangle \in \mathfrak{H}: \mathbb{P}|\psi\rangle = (-1)^p|\psi\rangle\}. \tag{1.21}$$

We can then formulate the statement, which we have derived from indistinguishability (1.11), in the following way: The space of physical states of N identical quantum-mechanical systems is either the antisymmetric space $\mathscr{H}^N_-$ or the symmetric space $\mathscr{H}^N_+$.

A basis of nonnormalized vectors in $\mathscr{H}^N_+$ is given by

$$|\xi\rangle_+ = |\xi_1 \xi_2 \cdots \xi_N\rangle_+ = \sum_{\mathbb{P}} \mathbb{P}|\xi_1 \xi_2 \cdots \xi_N\rangle, \tag{1.22}$$

where $|\xi_1 \xi_2 \cdots \xi_N\rangle$ is given by (1.3) and $\sum_{\mathbb{P}}$ is the sum over all permutations $\mathbb{P}$ of the N objects $\xi_1, \xi_2, \ldots, \xi_N$. It is clear that the order of the quantum numbers $\xi_1, \xi_2, \ldots, \xi_N$ in $|\xi\rangle_+$ is irrelevant. To prove that the $|\xi\rangle_+$ are symmetric, we calculate for an arbitrary permutation $\mathbb{P}_1$ that

$$\mathbb{P}_1|\xi\rangle_+ = \sum_{\mathbb{P}} \mathbb{P}_1 \mathbb{P}|\xi_1\,\xi_2\cdots\xi_N\rangle = \sum_{\mathbb{P}'} \mathbb{P}'|\xi_1\,\xi_2\cdots\xi_N\rangle = |\xi\rangle_+,$$

where we have set $\mathbb{P}' = \mathbb{P}_1\mathbb{P}$ and where the sum over all $\mathbb{P}$ becomes a sum over all $\mathbb{P}'$. The latter follows from the fact that if $\mathbb{P}_1$ is fixed and $\mathbb{P}$ runs over all permutations, then $\mathbb{P}'$ runs over all permutations. Thus for any permutation $\mathbb{P}_1$,

$$\mathbb{P}_1|\xi\rangle_+ = |\xi\rangle_+. \tag{1.23}$$

A basis of nonnormalized vectors in $\mathscr{H}_-^N$ is given by

$$|\xi\rangle_- = |\xi_1\,\xi_2\cdots\xi_N\rangle_- = \sum_{\mathbb{P}} (-1)^p \mathbb{P}|\xi_1\,\xi_2\cdots\xi_N\rangle. \tag{1.24}$$

To prove that they are antisymmetric, we first note that $(-1)^{p'} = (-1)^{p+p_1} = (-1)^p(-1)^{p_1}$ and hence that $(-1)^p = (-1)^{p_1}(-1)^{p'}$. Then

$$\begin{aligned}
\mathbb{P}_1|\xi\rangle_- &= \sum_{\mathbb{P}} (-1)^p \mathbb{P}_1 \mathbb{P}|\xi_1\,\xi_2\cdots\xi_N\rangle \\
&= \sum_{\mathbb{P}'} (-1)^{p_1}(-1)^{p'}\mathbb{P}'|\xi_1\,\xi_2\cdots\xi_N\rangle = (-1)^{p_1}|\xi\rangle_- \tag{1.25}
\end{aligned}$$

for any permutation $\mathbb{P}_1$. If the $|\xi_i\rangle_i$ are normalized vectors in $\mathscr{H}_i$, then the $|\xi_1\,\xi_2\cdots\xi_N\rangle$ are normalized in $\mathfrak{H}$, i.e., $\langle\xi_1\,\xi_2\cdots\xi_N|\xi_1\,\xi_2\cdots\xi_N\rangle = 1$. Consequently the $|\xi\rangle_+$ and the $|\xi\rangle_-$ are not normalized; the normalizing factors are calculated in Problem 1.

We state the result of the preceding considerations as a new basic postulate (axiom) of quantum mechanics:

IVb. The space of physical states of N identical quantum-mechanical systems (particles) is $\mathscr{H}_+^N$ if their angular momentum (spin) has an integral value, and is $\mathscr{H}_-^N$ if their angular momentum has a half-integral value.

Particles whose space of physical states is $\mathscr{H}_+^N$ are called *bosons*, and particles having $\mathscr{H}_-^N$ as their space of physical states are called *fermions*. Bosons are said to obey *Bose statistics*, while fermions are said to obey *Fermi statistics*. The above axiom IVb then states that half-integral-spin particles are fermions, that integral-spin particles are bosons, and that there are no other particles obeying some other "parastatistics" (which would have to belong to higher-dimensional representations of the permutation group). This axiom has been confirmed in all cases where it has been investigated (electrons, protons, neutrons are fermions; pions, photons, phonons, alpha particles are bosons). We remark that the axiom IVb has been described, to a large extent, from indistinguishability (1.11), which in turn was deduced from previously formulated basic postulates of quantum mechanics. The part that has not been deduced is the connection between spin and statistics.

The Pauli principle in its original form follows immediately from IVb. The Pauli exclusion principle states: The quantum numbers of two or more electrons can never entirely agree.

Problem

1. Calculate the length of the vectors

$$|\xi_1 \xi_2 \cdots \xi_N\rangle_+ = \sum_{\mathbb{P}} \mathbb{P}|\xi_1\rangle \otimes |\xi_2\rangle \otimes \cdots \otimes |\xi_N\rangle$$

and

$$|\xi_1 \xi_2 \cdots \xi_N\rangle_- = \sum_{\mathbb{P}} (-1)^P \mathbb{P}|\xi_1\rangle \otimes |\xi_2\rangle \otimes \cdots \otimes |\xi_N\rangle,$$

where the $|\xi_i\rangle$ are normalized basis vectors in the space $\mathcal{H}$ and the sum runs over all permutations $\mathbb{P}$. The factor $(-1)^P$ is $+1$ or -1 according as $\mathbb{P}$ is an even or odd permutation.

CHAPTER XI

Two-Electron Systems— The Helium Atom

The system with two electrons is studied in this chapter. Section XI.1 shows that the space of physical states of the helium atom is the sum of the para-helium and the orthohelium spaces. In Section XI.2 the ionization thresholds (i.e., the energy values $E_{n\infty}$ at which one electron is in the nth level and the other is just dissociating from the atom) are determined, and the energy levels below the first ionization threshold are discussed. Section XI.4 discusses the energy levels above the first ionization threshold without considering the interaction between these levels and the energy continuum of the (He$^+$, e) system.

XI.1 The Two Antisymmetric Subspaces of the Helium Atom

We shall illustrate the consequences of the basic assumption IVb with the example of two electrons in a Coulomb field. This is the simplest nontrivial case; although it does not demonstrate the full extent of IVb, it is mathematically simple and does not require the introduction of further properties of the representations of the permutation group.

The energy of two classical spinless particles of mass m_e with charge $-e$ that move in the field of a central charge Ze is given by

$$E = \frac{1}{2m_e}(\mathbf{p}_1^2 + \mathbf{p}_2^2) - \frac{Ze^2}{r_1} - \frac{Ze^2}{r_2} + \frac{e^2}{r_{12}}, \qquad (1.1)$$

where r_1 and r_2 are the distances of the first and second charges $-e$ from the charge Ze, and where r_{12} is the distance between these two charges. This system is the classical analogue of the helium atom (or of any two-electron ion if $Z \neq 2$). We obtain the energy operator of the helium atom by the usual procedure of replacing the numbers $p_{\alpha i}$, $x_{\alpha i}$, $r_{\alpha} = (x_{\alpha}^2)^{1/2}$, and $r_{12} = ((x_1 - x_2)^2)^{1/2}$ by the operators $P_{\alpha i}$, $Q_{\alpha i}$, $Q_{\alpha} = (Q_{\alpha}^2)^{1/2}$, and $Q_{12} = ((Q_1 - Q_2)^2)^{1/2}$, respectively. In addition we have to add a term H_1 that describes the influence of the electron spin. Thus we have

$$H = \frac{1}{2m_e}(\mathbf{P}_1^2 + \mathbf{P}_2^2) - \frac{Ze^2}{Q_1} - \frac{Ze^2}{Q_2} + \frac{e^2}{Q_{12}} + H_1 \tag{1.2}$$

for the energy operator of the helium atom ($Z = 2$), which we write in the form

$$H = H_0 + H_1. \tag{1.3a}$$

H_0 is the Hamiltonian operator corresponding to the classical Hamiltonian given by (1.1):

$$H_0 = \frac{1}{m_e}(h_1 \otimes I + I \otimes h_2) + W = H_{00} + W, \tag{1.3b}$$

where

$$H_{00} = \frac{1}{m_e}(h_1 \otimes I + I \otimes h_2), \tag{1.3c}$$

$$h_\alpha = \frac{\mathbf{P}_\alpha^2}{2} - \frac{a_Z}{Q_\alpha} \quad (a_Z \equiv m_e e^2 Z/\hbar, \quad \hbar = 1), \tag{1.3d}$$

and

$$W = \frac{e^2}{Q_{12}} = e^2 \left(\sum_{i=1}^{3} (Q_{1i} \otimes I - I \otimes Q_{2i})^2 \right)^{-1/2}. \tag{1.3e}$$

The operator H and all the other operators act in the space

$$\mathfrak{H} = \mathscr{H}_1 \otimes \mathscr{H}_2, \tag{1.4}$$

where $\mathscr{H}_\alpha$ is the space of the system that consists of one electron in the Coulomb field of the charge Ze.

For two objects ξ_1 and ξ_2 there are only $2! = 2$ permutations, (ξ_1, ξ_2) and (ξ_2, ξ_1); therefore for a fixed set of quantum numbers ξ_1 and ξ_2 with $\xi_1 \neq \xi_2$ there are only two basis vectors in $\mathfrak{H}$: $|\xi_1 \xi_2\rangle$ and $|\xi_2 \xi_1\rangle$. The normalized symmetric and antisymmetric vectors for this fixed set of quantum numbers are, according to (X.1.22) and (X.1.24),

$$|\xi_1 \xi_2\rangle_+ = \frac{1}{\sqrt{2}}(|\xi_1 \xi_2\rangle + |\xi_2 \xi_1\rangle) \tag{1.5}$$

and

$$|\xi_1 \xi_2\rangle_- = \frac{1}{\sqrt{2}}(|\xi_1 \xi_2\rangle - |\xi_2 \xi_1\rangle). \tag{1.6}$$

Thus for a fixed set of quantum numbers ξ_1 and ξ_2 with $\xi_1 \neq \xi_2$ we have a two-dimensional space spanned alternatively by $|\xi_1 \xi_2\rangle$, $|\xi_2 \xi_1\rangle$ or by $|\xi_1 \xi_2\rangle_+$, $|\xi_1 \xi_2\rangle_-$. If ξ_1 and ξ_2 are fixed with $\xi_1 = \xi_2$, we have a one-dimensional space spanned by $|\xi_1 \xi_2\rangle = |\xi_1 \xi_2\rangle_+$. The space $\mathfrak{H}$ is spanned by $|\xi_1 \xi_2\rangle$, where ξ_1 and ξ_2 independently can take any of the possible sets of values $(nljj_3)$ $(n = 1, 2, \ldots; j = 0, 1, \ldots, n - 1; \ j_3 = -j, -j + 1, \ldots, j; l = j \pm \frac{1}{2})$.[1] The space $\mathscr{H}^2_-$ is spanned by all the vectors $|\xi_1 \xi_2\rangle_-$, and the space $\mathscr{H}^2_+$ is spanned by all the vectors $|\xi_1 \xi_2\rangle_+$. Consequently

$$\mathfrak{H} = \mathscr{H}^2_- \oplus \mathscr{H}^2_+, \tag{1.7}$$

i.e., the product space is the direct sum of the symmetric and antisymmetric subspaces.

⟦Equation (1.7) is a particular feature of the case $N = 2$; for $N > 2$, $\mathfrak{H}$ of (X.1.1) is not the direct sum of the symmetric and antisymmetric subspaces (X.1.20) and (X.1.21). Rather

$$\mathfrak{H} = \mathscr{H}^N_+ \oplus \mathscr{H}^N_- \oplus \mathscr{H}^N_{\sigma_1} \oplus \mathscr{H}^N_{\sigma_2} \oplus \cdots.$$

where there are a finite number of terms, as many as the ways in which one can write N as a sum $\sum_i N_i$ of positive integers N_i. For $N = 2$ there are two ways possible (2 and $1 + 1$), and consequently there are only two different terms, as given by (1.7).⟧

According to the axiom IVb, or the Pauli principle, only the subspace $\mathscr{H}^2_-$ of (1.7) is the space of physical states for the two-electron system. (Were we considering a two-boson system, $\mathscr{H}^2_+$ would be the space of physical states.) To construct $\mathscr{H}^2_-$ and to find the properties of the algebra of observables in $\mathscr{H}^2_-$, we proceed in the following way: Each $\mathscr{H}_\alpha$ ($\alpha = 1, 2$) is, according to (IX.2.15), written as

$$\mathscr{H}_\alpha = \mathscr{H}^{\text{orb}}_\alpha \otimes \imath^s_\alpha, \tag{1.8}$$

where $\mathscr{H}^{\text{orb}}_\alpha$ is the space in which the orbital observables (i.e., the observables that are obtained as functions of the $P_{\alpha i}$ and $Q_{\alpha i}$) of the αth electron act, and where $\imath^s_\alpha$ is the space in which the spin observables of the αth electron act. We now combine the orbital and spin spaces of the two electrons separately, i.e., we form

$$\mathscr{H}^{\text{orb}^2} = \mathscr{H}^{\text{orb}}_1 \otimes \mathscr{H}^{\text{orb}}_2 \tag{1.9}$$

and

$$\imath^{s^2} = \imath^s_1 \otimes \imath^s_2. \tag{1.10}$$

We then find the symmetric and antisymmetric subspaces of $\mathscr{H}^{\text{orb}^2}$ and $\imath^{s^2}$ separately by the same procedure as described above; the ξ_α in (1.5) and

---

[1] $\xi_1 \neq \xi_2$ then means not all $(n_1 \, l_1 \, j_1 \, j_{13})$ agree with $(n_2 \, l_2 \, j_2 \, j_{23})$.

(1.6) stand for $\zeta_\alpha^{orb} = (n_\alpha \, l_\alpha \, l_{\alpha 3})$ when considering $\mathscr{H}^{orb^2}$ and for $\zeta_\alpha^s = s_{\alpha 3}$ when considering $\imath^{s^2}$. In this way we arrive at

$$\mathscr{H}^{orb^2} = \mathscr{H}_+^{orb^2} \oplus \mathscr{H}_-^{orb^2} \tag{1.11}$$

and

$$\imath^{s^2} = \imath_+^{s^2} \oplus \imath_-^{s^2}. \tag{1.12}$$

The total space $\mathfrak{H}$ is then given by[2]

$$\mathfrak{H} = \mathscr{H}^{orb^2} \otimes \imath^{s^2}$$
$$= (\mathscr{H}_+^{orb^2} \otimes \imath_+^{s^2}) \oplus (\mathscr{H}_+^{orb^2} \otimes \imath_-^{s^2}) \oplus (\mathscr{H}_-^{orb^2} \otimes \imath_+^{s^2}) \oplus (\mathscr{H}_-^{orb^2} \otimes \imath_-^{s^2}). \tag{1.13}$$

In Problem 1 it is shown that the symmetric subspace is

$$\mathscr{H}_+^2 = (\mathscr{H}_+^{orb^2} \otimes \imath_+^{s^2}) \oplus (\mathscr{H}_-^{orb^2} \otimes \imath_-^{s^2}) \tag{1.14}$$

while the antisymmetric subspace is shown to be

$$\mathscr{H}_-^2 = (\mathscr{H}_+^{orb^2} \otimes \imath_-^{s^2}) \oplus (\mathscr{H}_-^{orb^2} \otimes \imath_+^{s^2}). \tag{1.15}$$

Thus *the space of physical states $\mathscr{H}_-^2$ is the direct sum of two spaces: one is the space of symmetric orbital states and antisymmetric spin states, and the other is the space of antisymmetric orbital states and symmetric spin states.*

As has already been discussed (Section IX.2) in the case of the hydrogen atom, the basis vectors that are eigenvectors of the (total) spin and of the (total) orbital angular momentum are not a physical basis, because it is the (total) angular momentum that is the physical observable, and not the spin or orbital angular momentum. Thus to obtain the physical states one has to form those linear combinations of the direct product states $|\zeta_1^{orb} \, \zeta_2^{orb}\rangle_+ \otimes |s_{13} \, s_{23}\rangle_-$ (in $\mathscr{H}_+^{orb^2} \otimes \imath_-^{s^2}$) and $|\zeta_1^{orb} \, \zeta_2^{orb}\rangle_- \otimes |s_{13} \, s_{23}\rangle_+$ (in $\mathscr{H}_-^{orb^2} \otimes \imath_+^{s^2}$) that are eigenstates of the total angular momentum. Furthermore, it will turn out that the spaces $\mathscr{H}_+^{orb^2} \otimes \imath_-^{s^2}$ and $\mathscr{H}_-^{orb^2} \otimes \imath_+^{s^2}$ are not eigenspaces of the energy operator H of (1.3a); the reason is that H_1 does not commute with the operator of total spin $S_i = S_{1i} + S_{2i}$, but $\imath_+^{s^2}$ and $\imath_-^{s^2}$ are eigenspaces of S^2. If the physical states are eigenstates of H (as all experimental data confirm) then the physical state vectors are elements of neither $\mathscr{H}_+^{orb^2} \otimes \imath_-^{s^2}$ nor $\mathscr{H}_-^{orb^2} \otimes \imath_+^{s^2}$, but are linear combinations with a small component in one of the spaces and a large component in the other. Thus the reduction of $\mathscr{H}_-^2$ into the direct sum given by (1.15) is only approximately physical; the subspaces $\mathscr{H}_+^{orb^2} \otimes \imath_-^{s^2}$ and $\mathscr{H}_-^{orb^2} \otimes \imath_+^{s^2}$ are spaces of physical states only to the extent that the contribution of H_1 (the spin-orbit interaction) to H can be neglected. As in the case of the hydrogen atom, this will turn out to be a very good approximation.

---

[2] We wish to stress again that the appearance of only symmetric or antisymmetric subspaces is a particular feature of $N = 2$; for $N > 2$, higher-dimensional representations σ_i of the permutation group in $\mathscr{H}^{orb^N}$ and in $\imath^{s^N}$ also have to be considered. The antisymmetric space $\mathscr{H}_-^N$ then contains not only the spaces $\mathscr{H}_\mp^{orb^N} \otimes \imath_\pm^{s^N}$ as in (1.15), but also contains all direct-product spaces of the form $\mathscr{H}_\sigma^{orb^N} \otimes \imath_{\sigma'}^{s^N}$, where σ' is the irreducible representation of the permutation group that is "associated" with the irreducible representation σ in such a way that $\mathscr{H}_\sigma^{orb^N} \otimes \imath_{\sigma'}^{s^N}$ is an antisymmetric subspace of $\mathfrak{H}$.

We shall now neglect H_1 and undertake a detailed construction of all four spaces on the right-hand side of (1.15). We start with the spin spaces $\imath_+^{s^2}$ and $\imath_-^{s^2}$, as these are much simpler than the orbital spaces. The space $\imath^{s^2} = \imath_1^s \otimes \imath_2^s$ is the direct product of 2 two-dimensional spaces and is thus four-dimensional. Its direct-product basis is given by the four vectors

$$|s_{13}\rangle_1 \otimes |s_{23}\rangle_2 \qquad (s_{13} = \pm\tfrac{1}{2}, s_{23} = \pm\tfrac{1}{2}). \qquad (1.16)$$

One easily finds the antisymmetric and symmetric combinations of these four vectors. The symmetric ones are

$$|\tfrac{1}{2}\rangle_1 \otimes |\tfrac{1}{2}\rangle_2, \quad |-\tfrac{1}{2}\rangle_1 \otimes |-\tfrac{1}{2}\rangle_2,$$

$$\text{and} \quad \frac{1}{\sqrt{2}}(|\tfrac{1}{2}\rangle_1 \otimes |-\tfrac{1}{2}\rangle_2 + |-\tfrac{1}{2}\rangle_1 \otimes |\tfrac{1}{2}\rangle_2), \quad (1.17)$$

while the single antisymmetric vector is

$$\frac{1}{\sqrt{2}}(|\tfrac{1}{2}\rangle_1 \otimes |-\tfrac{1}{2}\rangle_2 - |-\tfrac{1}{2}\rangle_1 \otimes |\tfrac{1}{2}\rangle_2). \qquad (1.18)$$

The four vectors of (1.17) and (1.18) are orthonormal and therefore constitute a basis in $\imath^{s^2}$; consequently the three symmetric vectors span the symmectric space $\imath_+^{s^2}$, which is therefore three-dimensional, and the vector (1.18) spans the antisymmetric space $\imath_-^{s^2}$, which is therefore one-dimensional. $\imath^{s^2} = \mathcal{R}^{s_1 = 1/2} \otimes \mathcal{R}^{s_2 = 1/2}$ is the space of the combination of two elementary rotators. We can thus apply the results of Section V.2 and define the operator of total spin,

$$S_i = S_{1i} \otimes I + I \otimes S_{2i} \qquad (i = 1, 2, 3). \qquad (1.19)$$

Equation (V.2.32) then tells us that

$$\imath^{s^2} = \mathcal{R}^{s=1} \oplus \mathcal{R}^{s=0}, \qquad (1.20)$$

i.e., the total spin is $s = 1$ or $s = 0$, and we can introduce in $\imath^{s^2}$ the basis $|s\, s_3\rangle$ with

$$\left.\begin{array}{l} S^2|s\, s_3\rangle = s(s + 1)|s\, s_3\rangle, \\ S_3|s\, s_3\rangle = s_3|s\, s_3\rangle \end{array}\right\} \quad \left\{\begin{array}{l} s = 0, s_3 = 0; \\ s = 1, s_3 = -1, 0, 1. \end{array}\right. \qquad (1.21)$$

The observables S_i commute with the permutation operators (in this case the transposition operator $\mathbb{P}_{12}$), since

$$\mathbb{P}_{12}S_i = \mathbb{P}_{12}(S_{1i} \otimes I + I \otimes S_{2i}) = (I \otimes S_{2i} + S_{1i} \otimes I)\mathbb{P}_{12} = S_i\mathbb{P}_{12}. \qquad (1.22)$$

The two subspaces $\imath_+^{s^2}$ and $\imath_-^{s^2}$ are eigenspaces of $\mathbb{P}_{12}$ corresponding to the eigenvalues $+1$ and -1, respectively. Because of (1.22), S_i cannot transform out of either $\imath_+^{s^2}$ or $\imath_-^{s^2}$, i.e., S_i leaves $\imath_+^{s^2}$ and $\imath_-^{s^2}$ invariant. Thus S_i leaves invariant $\mathcal{R}^{s=1}$ and $\mathcal{R}^{s=0}$ on the one hand, and leaves invariant $\imath_+^{s^2}$ and $\imath_-^{s^2}$ on the other hand. Also, $\mathcal{R}^{s=1}$ is three-dimensional, as is $\imath_+^{s^2}$, and $\mathcal{R}^{s=0}$ is one-dimensional, as is $\imath_-^{s^2}$; consequently $\mathscr{R}^{s=1} = \imath_+^{s^2}$ and $\mathscr{R}^{s=0} = \imath_-^{s^2}$. We can therefore write (1.15) as

$$\mathscr{H}_-^2 = (\mathscr{H}_+^{\text{orb}^2} \otimes \mathscr{R}^{s=0}) \oplus (\mathscr{H}_-^{\text{orb}^2} \otimes \mathscr{R}^{s=1}). \tag{1.23}$$

The space of physical states (neglecting the spin-orbit interaction) is thus the direct sum of a space in which the total spin is zero (space of singlet states) and a space in which the total spin is one (space of triplet states). This, as we shall discuss below, is the explanation for para- and orthohelium, first given by Heisenberg in 1926.

XI.2 Discrete Energy Levels of Helium

We will now investigate the structure of the orbital spaces in (1.11). In each $\mathscr{H}_\alpha^{\text{orb}}$ ($\alpha = 1, 2$) we have a reducible representation of the algebra $\mathscr{E}(\text{SO}(4))$ of the orbital angular momentum

$$L_{\alpha i} = \epsilon_{ijk} Q_{\alpha j} P_{\alpha k} \tag{2.1}$$

and the Lenz vector

$$A_{\alpha i} = (-2h_\alpha)^{-1/2}\left(\tfrac{1}{2}\epsilon_{ikl}\{P_{\alpha l}, L_{\alpha k}\} + \frac{a_z Q_{\alpha i}}{Q_\alpha}\right), \tag{2.2}$$

which, according to (IX.2.13), is given by

$$\mathscr{H}_\alpha^{\text{orb}} = \sum_{n=1}^{\infty} \oplus \mathscr{R}_\alpha(n). \tag{2.3}$$

In the direct-product space

$$\mathscr{H}^{\text{orb}^2} = \mathscr{H}_1^{\text{orb}} \otimes \mathscr{H}_2^{\text{orb}} . \tag{2.4}$$

we have a representation of the algebra of orbital angular momentum and the Lenz vector given by

$$\begin{aligned} L_i &= L_{1i} \otimes I + I \otimes L_{2i}, \\ A_i &= A_{1i} \otimes I + I \otimes A_{2i}. \end{aligned} \tag{2.5}$$

The operators of (2.5) are defined in analogy to the definition (V.2.3) of the total angular momentum of the combined system of two elementary rotators. It is easy to see that L_i and A_i obey the same commutation relations as $L_{\alpha i}$ and $A_{\alpha i}$, i.e., the commutation relations of $\mathscr{E}(\text{SO}(4))$:

$$[L_i, L_j] = i\epsilon_{ijk} L_k, \qquad [L_i, A_j] = i\epsilon_{ijk} A_k, \qquad [A_i, A_j] = i\epsilon_{ijk} L_k. \tag{2.6}$$

We can define the operator

$$C_1 \equiv C_{11} \otimes I + I \otimes C_{21}, \tag{2.7}$$

where, similarly to (VI.3.13) and (VI.3.17),

$$C_{\alpha 1} = A_\alpha^2 + L_\alpha^2 = a_z^2(-2h_\alpha)^{-1} - I, \qquad a_z = \frac{Zm_e e^2}{\hbar} \tag{2.8}$$

We can further define the operator

$$C_2 \equiv A_i L_i \tag{2.9}$$

in analogy to (VI.3.14) for the hydrogen atom. The operators C_1 and C_2 commute with A_i and L_i. However, these operators do not fulfill the same relations as the operators $C_{\alpha 1}$ and $C_{\alpha 2}$. In particular, C_1 is not related to the energy operator H_0 by as simple a relation as (VI.3.17), and C_2 is no longer zero. To see the latter, insert (2.5) into (2.9); one then calculates that

$$C_2 = L_{1i} A_{1i} \otimes I + I \otimes L_{2i} A_{2i} + A_{1i} \otimes L_{2i} + L_{1i} \otimes A_{2i},$$

which gives [making use of (VI.3.18)]

$$C_2 = A_{1i} \otimes L_{2i} + L_{1i} \otimes A_{2i}. \tag{2.10}$$

This is not, in general, identically zero.

The energy operator of the helium atom in the approximation in which the influence of the spin is neglected is given by (1.3b):

$$H_0 = H_{00} + W. \tag{2.11}$$

It is customary to consider first the term (1.3c):

$$H_{00} = \frac{1}{m_e} (h_1 \otimes I + I \otimes h_2). \tag{2.12}$$

H_{00} is the energy operator for a system of two noninteracting electrons in the (nuclear) Coulomb field. As it is very unrealistic to neglect the Coulomb interaction between the two electrons (which is of the same "strength" as the Coulomb interaction between each electron and the nucleus), H_{00} is a very poor approximation to the energy operator of the helium atom. Thus we cannot expect that the spectrum of H_{00} will give a good approximation of the energy spectrum of the helium atom. As we will see later, it happens that the qualitative features of H_{00} agree with those of the energy operator $H_0 = H_{00} + W$; this justifies the usual treatment of first considering H_{00} separately.

Using (2.8) and (2.12), H_{00} can be written

$$H_{00} = -\frac{a_z^2}{2m_e} \left(\frac{1}{C_{11} + I} \otimes I + I \otimes \frac{1}{C_{21} + I} \right). \tag{2.13}$$

The spectrum of H_{00} is easily found. We introduce into $\mathcal{H}^{\mathrm{orb}^2} = \mathcal{H}_1^{\mathrm{orb}} \otimes \mathcal{H}_2^{\mathrm{orb}}$ the direct product basis

$$|n\, l\, l_3\rangle_1 \otimes |n'\, l'\, l_3'\rangle_2. \tag{2.14}$$

As, according to (1.23), we want to know the spectrum of H_{00} in the symmetric subspace $\mathcal{H}_+^{\mathrm{orb}^2}$ and in the antisymmetric subspace $\mathcal{H}_-^{\mathrm{orb}^2}$, we introduce in $\mathcal{H}^{\mathrm{orb}^2} = \mathcal{H}_+^{\mathrm{orb}^2} \oplus \mathcal{H}_-^{\mathrm{orb}^2}$ the basis system of symmetric and antisymmetric vectors

$$\frac{1}{\sqrt{2}} (|n\, l\, l_3\rangle_1 \otimes |n'\, l'\, l_3'\rangle_2 \pm |n'\, l'\, l_3'\rangle_1 \otimes |n\, l\, l_3\rangle_2). \tag{2.15\pm}$$

Equation (2.15+) gives the basis system in $\mathscr{H}_{+}^{\text{orb}}$, while (2.15−) gives the basis system in $\mathscr{H}_{-}^{\text{orb}}$. The vectors (2.15±) are eigenvectors of H_{00} and together form a complete basis system in $\mathscr{H}^{\text{orb}^2}$. The spectrum of H_{00} is therefore obtained by applying (2.13) to (2.15±); the result is

$$\text{spectrum}\left(\frac{H_{00}}{m_e}\right) = E_{nn'}^{00} = -R''_{\text{He}}\left(\frac{1}{n^2} + \frac{1}{n'^2}\right), \qquad (2.16)$$

where

$$R''_{\text{He}} = \frac{a_{Z=2}^2}{2m_e} = 4\left(\frac{m_e e^4}{2\hbar^2}\right) = 4R'' = 54.4 \text{ eV}. \qquad (2.17)$$

R''_{He} differs from the Rydberg constant for the hydrogen atom by a factor of 4. It is the Rydberg constant for a one-electron system in the Coulomb field of a charge $Ze = 2e$, i.e., for the He^+ ion. Its value in cm^{-1} (wave-number units or inverse wavelength units) is

$$R_{\text{He}} = \frac{1}{2\pi\hbar c} R''_{\text{He}} = \frac{54.4 \text{ eV}}{12.40 \times 10^{-5} \text{ eV cm}} = 4.39 \times 10^5 \text{ cm}^{-1}.$$

The basis vectors (2.15±) are not eigenvectors of the total orbital angular momentum $\mathbf{L}^2$ and L_3, and are not eigenvectors of the energy operator H_0 (recall that H_0 is the energy operator if the contribution of the spin H_1 is neglected). If the physical states are eigenstates of the energy operator H, they are very closely eigenstates of H_0. Eigenstates of H_0 can be eigenstates of the total orbital angular momentum $\mathbf{L}^2$ and L_3, but can be neither the direct-product states (2.14) of angular momentum l and l' nor the particular linear combinations (2.15±) of those direct product states. [The states (2.15±) are eigenstates of $\mathbf{L}_1^2 + \mathbf{L}_2^2$. But

$$[\mathbf{L}_1^2 + \mathbf{L}_2^2, Q_{12}^2] = -2[L_{1k}L_{2k}, (Q_{1i} - Q_{2i})(Q_{1i} - Q_{2i})]$$
$$= -4i\epsilon_{kij}(I_{\cdot 1k}Q_{2j}Q_{1i} - Q_{1i}Q_{2j}I_{\cdot 2k}) \neq 0.$$

Consequently the vectors (2.15±) are eigenvectors of an operator that does not commute with H_0.] One therefore has to couple the angular momenta l and l' in (2.15±), according to the rules of Section V.2, to obtain eigenvectors of the total orbital angular momentum. These eigenvectors, which are formed as linear combinations of (2.15±), are not yet H_0 eigenstates but still H_{00} eigenstates; H_0 eigenstates can then be formed as linear combinations with the same value of total orbital angular momentum.

Instead of approximating the helium atom by a model system that consists of two noninteracting electrons that move in the Coulomb field of a doubly charged nucleus, one can try to approximate it by a model system that consists of one electron moving in the electric field that is formed by the doubly charged nucleus and the other electron. This model is certainly much more realistic if the one electron is—in the classical picture—far away from the nucleus and from the other electron, which are close together. To obtain this approximation we write

$$H_0 = \frac{1}{m_e}h_1 + H_2^{\text{el}} = \frac{1}{m_e}h_2 + H_1^{\text{el}}, \qquad (2.18)$$

where

$$H_\alpha^{el} = \frac{1}{m_e} h_\alpha + \frac{e^2}{Q_{12}}. \tag{2.19}$$

More precisely,

$$H_1^{el} = \frac{1}{m_e} h_1 \otimes I + \frac{e^2}{Q_{12}}$$

and

$$H_2^{el} = \frac{1}{m_e} I \otimes h_2 + \frac{e^2}{Q_{12}}.$$

Equation (2.18) is exact, but H_α^{el} is not an operator in $\mathscr{H}_\alpha^{orb}$; it would be an operator in $\mathscr{H}_\alpha^{orb}$ if we made the replacement

$$Q_{12} \to Q_\alpha. \tag{2.20}$$

In our classical picture this would mean that the electron that is close to the nucleus is really at the position of the nucleus. With the asymptotic replacement (2.20), the energy operator for the electron far away from the nucleus has the asymptotic form

$$H_\alpha^{el} = \frac{\mathbf{P}_\alpha^2}{2m_e} + \left(-\frac{2e^2}{Q_\alpha} + \frac{e^2}{Q_{12}} \right) \sim \frac{\mathbf{P}_\alpha^2}{2m_e} - \frac{e^2}{Q_\alpha} = \frac{1}{m_e} H_{\alpha\,\text{Hydr}}, \tag{2.21}$$

i.e., the energy operator for the distant electron is the same as the energy operator of the electron in the hydrogen atom. In the approximation (2.21) the energy operator H_0 is approximated by

$$H_0 \to \tilde{H}_0 \equiv \frac{1}{m_e} h_1 + \frac{1}{m_e} H_{2\,\text{Hydr}} = \frac{1}{m_e} h_2 + \frac{1}{m_e} H_{1\,\text{Hydr}}, \tag{2.22}$$

where $H_{\alpha\,\text{Hydr}}$ is the hydrogen-atom Hamiltonian corresponding to (VI.3.5). Letting $n' \geq n$, the spectrum of $\tilde{H}_0$ is then

$$\text{spectrum } \tilde{H}_0 = \tilde{E}_{nn'} = -R_{\text{He}}'' \frac{1}{n^2} - R'' \frac{1}{n'^2}$$

$$= -R'' \left(\frac{4}{n^2} + \frac{1}{n'^2} \right) \geq E_{nn'}^{00} = -R'' \left(\frac{4}{n^2} + \frac{4}{n'^2} \right). \tag{2.23}$$

A more realistic approximation for H_0 of the form (2.18) is to make the replacement

$$H_\alpha^{el} \to H_{\alpha(Z_{\text{eff}})}^{el} = \frac{\mathbf{P}^2}{2m_e} - \frac{Z_{\text{eff}}^{(\alpha)} e^2}{Q_\alpha}, \tag{2.24}$$

where $Z_{\text{eff}}^{(\alpha)}$ is a number between one and two that expresses the screening of the nuclear Coulomb field by the nearby electron. A possible approximation for H_0 is thus

$$H_0 \to H_{0(Z_{\rm eff})} = H^{\rm el}_{1(Z_{\rm eff})} + H^{\rm el}_{2(Z_{\rm eff})} \qquad (2.25)$$

The value of $Z^{(\alpha)}_{\rm eff}$ should be different for different states of the helium atom. For the ground state one would expect $Z^{(1)}_{\rm eff} = Z^{(2)}_{\rm eff}$; for the state with one electron in the ground state and the other in a very high state one would expect, according to (2.22), $Z^{(1)}_{\rm eff} \approx 2$ and $Z^{(2)}_{\rm eff} \approx 1$. Thus for different subspaces we have different operators $H_{0(Z_{\rm eff})}$.

Let us now consider the subspace of states of the helium atom that can be characterized—in the classical picture—in the following way: One of the electrons has just dissociated itself from the helium atom, i.e., the system consists of a He$^+$ ion together with an electron of zero relative energy. Let us call the subspace of these states $\mathscr{H}_\infty$; on this subspace H_0 can be approximated very well by

$$H_0 = \frac{1}{m_e} h_1 + H^{\rm el}_2 = \frac{2}{m_e} h_2 + H^{\rm el}_1, \qquad (2.26)$$

where $H^{\rm el}_\alpha$ is approximated by (2.21). The subspace $\mathscr{H}_\infty$ is then defined as that subspace of $\mathscr{H}^{\rm orb2}$ on which

$$H^{\rm el}_\alpha = 0. \qquad (2.27)$$

Let us denote the eigenvectors of $H^{\rm el}_\alpha$ in $\mathscr{H}^{\rm orb}_\alpha$ by

$$|n\, l\, l_3\}_\alpha, \qquad (2.28)$$

in distinction to the eigenvectors $|n\, l\, l_3\rangle_\alpha$ of h_α. These eigenvectors have the property [under the approximation (2.21)]

$$H^{\rm el}_\alpha |n\, l\, l_3\}_\alpha = -R'' \frac{1}{n^2} |n\, l\, l_3\}_\alpha \qquad (2.29)$$

and are eigenstates of $\mathbf{L}^2_\alpha$ and $L_{\alpha 3}$. The eigenvectors on which (2.27) is fulfilled are denoted by

$$|\infty\, l\, l_3\}_\alpha. \qquad (2.30)$$

The direct-product basis in $\mathscr{H}_\infty$ is therefore given by

$$|\infty\, l'\, l'_3\}_1 \otimes |n\, l\, l_3\rangle_2, \qquad |n\, l\, l_3\rangle_1 \otimes |\infty\, l'\, l'_3\}_2. \qquad (2.31)$$

We introduce in $\mathscr{H}_\infty$ the basis of symmetric and antisymmetric vectors

$$\frac{1}{\sqrt{2}} (|\infty\, l'l'_3\}_1 \otimes |n\, l\, l_3\rangle_2 \pm |n\, l\, l_3\rangle_1 \otimes |\infty\, l'\, l'_3\}_2). \qquad (2.32\pm)$$

On $\mathscr{H}_\infty$ the energy operator (2.26) has, according to (2.27), the following spectrum:

$$E_{n\infty} = \text{spectrum } H_0|_{\mathscr{H}_\infty} = -\frac{4m_e e^4}{2\hbar^2} \frac{1}{n^2} = -R''_{\rm He} \frac{1}{n^2}. \qquad (2.33)$$

For H_0 of (2.26) applied to (2.32$\pm$) gives

$$\frac{1}{\sqrt{2}} \Bigg(|\infty\, l'\, l_3'\}_1 \otimes \frac{h_2}{m_{\mathrm{e}}} |n\, l\, l_3\rangle_2 + \underbrace{H_1^{\mathrm{el}} |\infty\, l'\, l_3'\}_1 \otimes |n\, l\, l_3\rangle_2}_{=0}$$

$$\pm \frac{h_1}{m_{\mathrm{e}}} |n\, l\, l_3\rangle_1 \otimes |\infty\, l'\, l_3'\}_2 \pm \underbrace{|n\, l\, l_3\rangle_1 \otimes H_2^{\mathrm{el}} |\infty\, l'\, l_3'\}_2}_{=0} \Bigg)$$

$$= -\frac{a_z^2}{2m_{\mathrm{e}} n^2} \frac{1}{\sqrt{2}} (|\infty\, l'\, l_3'\}_1 \otimes |n\, l\, l_3\rangle_2 \pm |n\, l\, l_3\rangle_1 \otimes |\infty\, l'\, l_3'\}_2).$$

The $E_{n\infty}$ are the energy values of the helium atom when one of the electrons has zero energy and are also the energy values at which one of the electrons dissociates from the atom (ionization thresholds). We expect that these are the highest energy values for a given quantum number n, i.e., for states in which one of the electrons has the principal quantum number n, because if the other electron has the principle quantum number n', its relative energy with the He$^+$ ion will be negative. $E_{n\infty}$ is called the nth ionization threshold; it is the value at which one of the electrons is just dissociating itself from the atom while the other is in the nth energy level.

Comparing (2.33) with the spectrum of H_{00} given by (2.16), we observe that

$$E_{nn'}^{00} \to E_{n\infty} \quad \text{from below as} \quad n' \to \infty. \tag{2.34}$$

Comparing it with (2.23), we observe

$$\tilde{E}_{nn'} \to E_{n\infty} \quad \text{from below as} \quad n' \to \infty. \tag{2.35}$$

From our above considerations we expect the eigenvalue $E_{nn'}$ of H_0 to lie between the eigenvalues of H_{00} and $\tilde{H}_0$:

$$E_{nn'}^{00} \leq E_{nn'} \leq \tilde{E}_{nn'} < E_{n\infty}. \tag{2.36}$$

In the energy diagram of helium we first draw the energy levels $E_{n\infty}$ ($n = 1, 2, 3, \ldots$). The energy levels $E_{nn'}$ lie, according to (2.34), (2.35), and (2.36), below $E_{n\infty}$, and lie closer together as n' becomes higher.

The larger the values of n', the closer are the eigenvalues of H_{00} to the energy values of the helium atom. Consequently for large n' and a given n, the space spanned by the vectors (2.15+) and (2.15−) represents to a good approximation the space of physical states. In the classical picture of the helium atom, increasing values of n' mean that the "second" electron is further away from the nucleus and from the "first" electron, so that the influence of the interaction term e^2/r_{12} is small; thus the classical picture supports the result described above. For decreasing values of n' the interaction term e^2/r_{12} becomes more important, and we expect larger deviations of the energy value of the helium atom from the value $E_{nn'}^{00}$. We expect the largest deviation for the states in which both electrons have the principal quantum number $n = n' = 1$. In this case the interaction term is expected to be largest (both electrons are closest to the nucleus and to each other), and E_{11}^{00} is a very poor approximation to the eigenvalue of H_0. In this case, therefore, the space spanned by (2.15+) with $n = n' = 1$ is far from being the space of (orbital) physical states. Thus for low values of n and n' the helium atom cannot be considered as consisting of two independent electrons (independent-particle approximation) in a Coulomb field. The physical states are not states in which one electron is in one particular state and the other electron is in another particular state; rather it is only the helium atom considered as a whole that is in the physical states.

We now consider the space of states (which we call $\hbar_1$) of the helium atom that have energy values below the value $E_{1\infty}$. The space of eigenstates of H_0 with eigenvalue $E_{1\infty}$ describes the state of the helium atom in which one electron is in the lowest state ($n' = 1$, $l' = 0$, $l'_3 = 0$) and the other electron is just dissociating from the helium atom. The eigenvectors of H_{00} in this subspace $\hbar_1$ can be obtained from the independent-particle states (2.15 ±):

$$\psi^0_{nll_3} = \frac{1}{\sqrt{2}} (|n' = 1 \, l' = 0 \, l'_3 = 0\rangle_1 \otimes |n \, l \, l_3\rangle_2$$

$$+ |n \, l \, l_3\rangle_1 \otimes |n' = 1 \, l' = 0 \, l'_3 = 0\rangle_2), \qquad (2.37)$$

$$\gamma^0_{nll_3} = \frac{1}{\sqrt{2}} (|n' = 1 \, l' = 0 \, l'_3 = 0\rangle_1 \otimes |n \, l \, l_3\rangle_2$$

$$- |n \, l \, l_3\rangle_1 \otimes |n' - 1 \, l' - 0 \, l'_3 - 0\rangle_2). \qquad (2.38)$$

The $\psi^0_{nll_3}$ span the space of symmetric states of $\hbar_1$, which we call $\hbar_{1+}$; and the $\gamma^0_{nll_3}$ span the space of antisymmetric states of $\hbar_1$, which we call $\hbar_{1-}$. That is, $\psi^0_{nll_3}$ and $\gamma^0_{nll_3}$ are eigenstates of H_{00} with eigenvalue $E^{00}_{nn'=1}$:

$$H_{00}\psi^0_{nll_3} = -R''_{He}\left(1 + \frac{1}{n^2}\right)\psi^0_{nll_3}, \qquad (2.39)$$

$$H_{00}\gamma^0_{nll_3} = -R''_{He}\left(1 + \frac{1}{n^2}\right)\gamma^0_{nll_3}. \qquad (2.40)$$

As $l' = 0$ and $l'_3 = 0$, $\psi^0_{nll_3}$ and $\gamma^0_{nll_3}$ are also eigenvectors of L^2 and L_3 with eigenvalues $l(l + 1)$ and l_3, respectively. Let us denote by ψ_{nll_3} and γ_{nll_3} the eigenvectors of H_0, L^2, and L_3 in $\hbar_{1+}$ and in $\hbar_{1-}$, respectively. The eigenvectors ψ_{nll_3} and γ_{nll_3} of H_0 can be obtained from the eigenvectors $\psi^0_{nll_3}$ and $\gamma^0_{nll_3}$ of H_{00} by acting on them with a (unitary) operator that does not change the eigenvalues of L^2, L_3, and $\mathbb{P}_{12}$:

$$\psi_{nll_3} = U^+(nl)\psi^0_{nll_3}, \qquad (2.41)$$

$$\gamma_{nll_3} = U^-(nl)\gamma^0_{nll_3}. \qquad (2.42)$$

This transformation $U^\pm(nl)$ may be different for different values of n and l; $U^\pm(nl)$ does not depend upon l_3, as $[H_0, L_i] = 0$. Equations (2.41) and (2.42) show that the eigenvectors of H_0 can still be characterized by the principal quantum number n, though they might be quite different from the eigenvectors of H_{00} with eigenvalues $E^{00}_{nn'=1} = -R''_{He}(1 + 1/n^2)$. For large

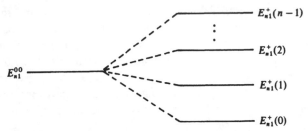

Figure 2.1 Splitting of the eigenvalues of H_{00} under the influence of W.

values of n, $U^\pm(nl)$ is very close to the unit operator. The eigenvalues of H_0 in $\hbar_1$ we denote by $E_{n1}^+(l)$ and $E_{n1}^-(l)$; in general they will depend upon l:

$$H_0\psi_{nll_3} = E_{n1}^+(l)\psi_{nll_3}, \tag{2.43}$$

$$H_0\gamma_{nll_3} = E_{n1}^-(l)\gamma_{nll_3}. \tag{2.44}$$

We are not that much interested in the calculation of the exact values of $E_{n1}^\pm(l)$, as we want only to obtain a qualitative understanding of the energy spectrum of the helium atom. Therefore it is not necessary to know the operators $U^\pm(nl)$. (They are connected with the interaction term W and can be calculated by approximation methods.)

The important conclusion for the qualitative understanding of the energy spectrum that we draw from (2.41) and (2.42) is that to each n^2-fold degenerate eigenvalue E_{n1}^{00} of H_{00} on $\hbar_{1+}$ there correspond the n energy values $E_{n1}^+(l)$, the interaction term W slpits E_{n1}^{00} into n sublevels as shown in Figure 2.1. And to each n^2-fold degenerate eigenvalue E_{n1}^{00} of H_{00} on $\hbar_{1-}$ there correspond the n energy values $E_{n1}^-(l)$. As the result of our above consideration we obtain the following energy spectrum of the helium atom below $E_{1\infty}$ (Figures 2.2 and 2.3): The lowest energy value is $E_{11}^+(l = 0)$ in the symmetric subspace $\mathscr{H}_+^{orb2}$,

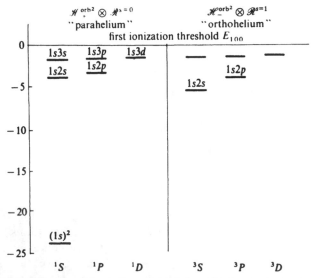

Figure 2.2 Energy levels of the helium atom below the lowest ionization threshold.

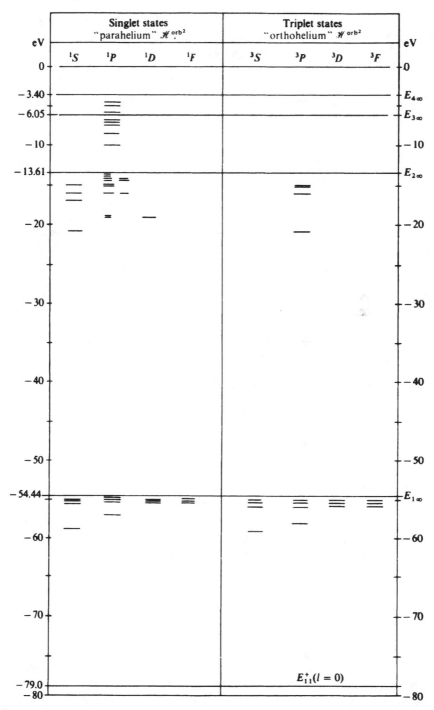

Figure 2.3 Term diagram for energy levels of helium. (The part below the first ionization threshold is again given in Figure 2.2 which is the term diagram of the helium that one usually finds.) The energy levels above the first ionization threshold $E_{1\infty}$ lie in the continuous spectrum.

which is connected with $E_{11}^{00} = -R_{He}''(1 + 1)$. There is no corresponding energy value in the antisymmetric subspace $\mathscr{H}_-^{orb^2}$, because $\gamma_{1ll_3} = 0$. The values of $\tilde{E}_{11}$ and E_{11}^{00} as calculated from (2.23) and (2.16) are

$$\tilde{E}_{11} = -68.05 \text{ eV} \quad \text{and} \quad E_{11}^{00} = -108.8 \text{ eV}.$$

The experimentally measured value of $E_{11}^+(l = 0)$, the double ionization energy of helium (minimum energy required to free both electrons), is:

$$E_{11}^+(l = 0) = -79.0 \text{ eV},$$

which shows that H_{00} is really a very poor approximation to the energy operator H_0. The energy difference

$$E_{1\infty} - E_{11}^+(0) = -54.4 \text{ eV} + 79.0 \text{ eV} = 24.6 \text{ eV}$$

is the energy that is necessary to dissociate one electron from the helium atom when it is in the ground state; it is called the "ionization energy" or "ionization potential." For $n = 2$ there are energy values $E_{12}^+(l)$ in $\mathscr{H}_+^{orb^2}$ and $E_{12}^-(l)$ in $\mathscr{H}_-^{orb^2}$. They correspond to the eigenvalues

$$E_{12}^{00} = -R_{He}''(1 + \tfrac{1}{4}) = -68.0 \text{ eV}$$

of H_{00} and

$$\tilde{E}_{12} = -R_{He}'' \cdot 1 - R'' \cdot \tfrac{1}{4} = -57.8 \text{ eV}$$

of $\tilde{H}_0$. The experimentally measured values are

$$E_{12}^-(0) - E_{11}^+(0) = 19.8 \text{ eV} \quad \text{or} \quad E_{12}^-(0) = -59.2 \text{ eV},$$

$$E_{12}^+(0) - E_{11}^+(0) = 20.6 \text{ eV} \quad \text{or} \quad E_{12}^+(0) = -58.4 \text{ eV}$$

for $l = 0$, and

$$E_{12}^-(1) - E_{11}^+(0) = 20.9 \text{ eV} \quad \text{or} \quad E_{12}^-(1) = -58.1 \text{ eV},$$

$$E_{12}^+(1) - E_{11}^+(0) = 21.2 \text{ eV} \quad \text{or} \quad E_{12}^+(1) = -57.8 \text{ eV}$$

for $l = 1$. Thus for $n = 2$ the energy values are already closer to the eigenvalues of H_{00}, as we expect from our general discussion above. For $n = 3, 4, 5, \ldots$ the agreement improves further.

Figure 2.3 gives all known (1971) discrete energy levels of the helium atom. This energy spectrum is in agreement with our discussion above. To summarize our findings for the energy spectrum below $E_{1\infty}$, there is one distorted, shifted, and split hydrogenlike spectrum of states in the space $\mathscr{H}_+^{orb^2}$, and there is one distorted, shifted, and split hydrogenlike spectrum—without the lowest level—of states in the space $\mathscr{H}_-^{orb^2}$. According to (1.23) the states in $\mathscr{H}_+^{orb^2}$ are states with total spin zero, called singlet states; the states in $\mathscr{H}_-^{orb^2}$ are states with total spin one, called triplet states. As we shall explain below, there are practically no transitions between the singlet and triplet states of the helium atom. (The 2^3S state is a quasistable state with a very long lifetime.) Therefore the splitting of the energy diagram into two parts had in earlier days lead to the hypothesis that helium was a mixture of two elements, "parahelium," described by $\mathscr{H}_+^{orb^2} \otimes \mathscr{R}^{s=0}$, and "orthohelium," described by $\mathscr{H}_-^{orb^2} \otimes \mathscr{R}^{s=1}$.

For parahelium the total angular momentum is equal to the total orbital angular momentum $L = l$ because the total spin $S = 0$. Therefore the interaction caused by the spin H_1 cannot split the energy level $E_{1n}^+(l)$, but can only cause a small shift of the same order of magnitude as the fine structure in the hydrogen atom, which has been calculated by approximation methods (singlet).

For orthohelium the total spin $S = 1$. Therefore the direct-product vectors

$$\gamma_{nll_3} \otimes |s = 1\, s_3\rangle \tag{2.45}$$

in $\hbar_{1-} \otimes \mathscr{R}^{s=1}$ are not eigenvectors of the total-angular-momentum operators $\mathbf{J}^2$ and J_3, where

$$J_i = L_i \otimes I + I \otimes S_i. \tag{2.46}$$

To obtain eigenstates of the total angular momentum we have to couple the orbital angular momentum and the spin in $\hbar_{1-} \otimes \mathscr{R}^{s=1}$ according to the rules of Section V.2. Let $\mathscr{M}^l$ denote the subspace of $\hbar_{1-} \subset \mathscr{H}^{\mathrm{orb}'}$ with orbital angular momentum l; it is the space spanned by γ_{nll_3} with a fixed value of l. The eigenspaces $\mathscr{M}^j$ of total angular momentum j are then given by

$$\mathscr{M}^l \otimes \mathscr{R}^{s=1} = \begin{cases} \mathscr{M}^{j=l+1} \oplus \mathscr{M}^{j=l} \oplus \mathscr{M}^{j=l-1} & \text{if } l \neq 0, \\ \mathscr{M}^{j=1} & \text{if } l = 0. \end{cases} \tag{2.47}$$

So we see that for orthohelium, except for the S states, all orbital-angular-momentum spaces are actually a triplet of total-angular-momentum spaces. Therefore the spin perturbation H_1 of the Hamiltonian $H = H_0 + H_1$ splits the P ($l = 1$), D ($l = 2$), and F ($l = 3$) energy levels into triplets of fine-structure levels, each belonging to an eigenspace of total angular momentum $j = l + 1, l, l - 1$.

These results agree with the experimental energy spectrum: parahelium consists only of singlet terms, whereas the orthohelium consists (except for $l = 0$) only of triplet terms (these are so close to each other that they are shown by one line in Figures 2.2 and 2.3). To avoid labeling the states by their symmetry properties under permutation of the orbital part (ortho- and para-), one also calls the S term of the orthohelium a "triplet" term, $^3S_1$, though it consists in fact of only a singlet.

XI.3 Selection Rules and Singlet–Triplet Mixing for the Helium Atom

The operator of the dipole moment for the helium atom is

$$\mathbf{d} = e(\mathbf{Q}_1 \otimes I + I \otimes \mathbf{Q}_2), \tag{3.1}$$

where $\mathbf{d}$ is a vector operator with respect to the total angular momentum operator

$$J_i = J_{1i} \otimes I + I \otimes J_{2i} = L_i \otimes I + I \otimes S_i \tag{3.2}$$

and also with respect to the total orbital angular momentum

$$L_i = L_{1i} \otimes I + I \otimes L_{2i}. \tag{2.5}$$

Thus it obeys the commutation relations

$$[J_i, d_j] = i\epsilon_{ijk} d_k \tag{3.3}$$

$$[L_i, d_j] = i\epsilon_{ijk} d_k \tag{3.3'}$$

As a consequence, one obtains by (V.3.7) for dipole radiation the selection rules

$$J \to J - 1, \quad J, \quad J + 1 \tag{3.4}$$

$$L \to L - 1, \quad L, \quad L + 1 \tag{3.5'}$$

The physical states are total angular momentum eigenstates, therefore the selection rules (3.4) hold strictly for physical states. The selection rule (3.5') holds to a very good approximation, precisely to the extent to which H_1 commutes with $\mathbf{L}^2$.

The parity operator for the two-electron system is the direct product of the parity operators in the one-particle subspaces:

$$U_p = U_{p_1} \otimes U_{p_2}. \tag{3.6}$$

Therefore, for the vectors (2.41) and (2.42), one obtains

$$U_p \psi_{nll_3} = (-1)^L \eta_0 \psi_{nll_3}, \quad U_p \gamma_{nll_3} = (-1)^L \eta_1 \gamma_{nll_3} \tag{3.7}$$

because $L = l$ below the first ionization threshold. As (3.1) is a proper vector

$$U_p \mathbf{d} U_p^{-1} = -\mathbf{d},$$

it follows by the arguments of Section V.4 that the transition $L \to L$ is forbidden, so that with (3.5') one has the selection rule

$$L \to L + 1, \quad L - 1. \tag{3.5}$$

A further selection rule follows from the fact that

$$[\mathbf{d}, \mathbb{P}_{12}] = 0. \tag{3.8}$$

For every vector

$$\gamma \in \mathcal{H}_-^{\mathrm{orb}^2} \otimes \imath_+^{s^2} \tag{3.9}$$

and

$$\psi \in \mathcal{H}_+^{\mathrm{orb}^2} \otimes \imath_-^{s^2} \tag{3.10}$$

it follows as a consequence of (3.8) that

$$\mathbf{d}\gamma = \gamma' \in \mathcal{H}_-^{\mathrm{orb}^2} \otimes \imath_+^{s^2}, \tag{3.11}$$

$$\mathbf{d}\psi = \psi' \in \mathcal{H}_+^{\mathrm{orb}^2} \otimes \imath_-^{s^2}. \tag{3.12}$$

Therefore,

$$(\psi, \mathbf{d}\gamma) = 0. \tag{3.13}$$

Consequently, there are no transitions between singlet and triplet states (the same arguments as above hold for the quadrupole and higher multipole transitions, too). As already mentioned in Section XI.2, this had led to the hypothesis that helium is a mixture of two elements—parahelium, possessing only singlet states with an energy diagram given by the left part of Figure 2.2, and orthohelium, possessing only triplet states with an energy diagram given by the right part of Figure 2.2. However, one then observed transitions of very weak intensity between energy levels of para- and orthohelium. These inter-multiplet transitions are due to "singlet–triplet mixing"[3] caused by H_1. We shall describe the eigenstates of total angular momentum in the spaces $\mathscr{H}_+^{\mathrm{orb}2} \otimes \mathscr{R}^{s=0}$, $\mathscr{H}_-^{\mathrm{orb}2} \otimes \mathscr{R}^{s=1}$, which are obtained by coupling the spin and orbital angular momentum, according to the rules of Chapter V. For $\mathscr{H}_+^{\mathrm{orb}2} \otimes \mathscr{R}^{s=0}$, we obtain:

$$\psi_{nl=J}^{JJ_3} = \psi_{nll_3=J_3} \otimes \sum_{s_3' s_3''} |s_3'\rangle_1 |s_3''\rangle_2 \langle \tfrac{1}{2}s_3', \tfrac{1}{2}s_3''|0\,0\rangle \tag{3.14}$$

where ψ_{nll_3} is given by (2.41) with (2.37) and the second factor on the right-hand side of (3.14) is the same vector as the one given by (1.18).

For $\mathscr{H}_-^{\mathrm{orb}2} \otimes \mathscr{R}^{s=1}$ the total angular momentum eigenstates are given by

$$\gamma_{nl}^{JJ_3} = \sum_{l_3 s_3} \gamma_{nll_3} \otimes \left(\sum_{s_3' s_3''} |\tfrac{1}{2}s_3'\rangle_1 |\tfrac{1}{2}s_3''\rangle_2 \langle \tfrac{1}{2}s_3', \tfrac{1}{2}s_3''|1s_3\rangle \right) \langle l\, l_3\, 1s_3|J\, J_3\rangle, \tag{3.15}$$

where γ_{nll_3} is given by (2.42) with (2.38). For a fixed n and l, $\gamma_{nl}^{JJ_3}$ is a basis in the spaces $\mathscr{M}^J$ on the right-hand side of (2.47).

According to the considerations in Chapter VIII on perturbation theory, the eigenstates of $H = H_0 + H_1$ are in first order given by

$$\tilde{\psi}_{nl}^{JJ_3} = \psi_{nl}^{JJ_3} + \sum_{\substack{n'l' \\ n' \neq n \\ l' \neq l}} \left(\frac{1}{E_{nl}^+ - E_{n'l'}^+} \psi_{n'l'}^{JJ_3} \langle \psi_{n'l'}^{JJ_3}|H_1|\psi_{nl}^{JJ_3}\rangle \right.$$

$$\left. + \frac{1}{E_{nl}^+ - E_{n'l'}^-} \gamma_{n'l'}^{JJ_3} \langle \gamma_{n'l'}^{JJ_3}|H_1|\psi_{nl}^{JJ_3}\rangle \right) \tag{3.16}$$

and by

$$\tilde{\gamma}_{nl}^{JJ_3} = \gamma_{nl}^{JJ_3} + \sum_{\substack{n'l' \\ n' \neq n \\ l' \neq l}} \left(\frac{1}{E_{nl}^- - E_{n'l'}^-} \gamma_{n'l'}^{JJ_3} \langle \gamma_{n'l'}^{JJ_3}|H_1|\gamma_{nl}^{JJ_3}\rangle \right.$$

$$\left. + \frac{1}{E_{nl}^- - E_{n'l'}^+} \psi_{n'l'}^{JJ_3} \langle \psi_{n'l'}^{JJ_3}|H_1|\gamma_{nl}^{JJ_3}\rangle \right), \tag{3.17}$$

---

[3] Mixing is the common term used for this kind of phenomenon, which is a somewhat misleading nomenclature as compared with the meaning of the word mixture of states in quantum mechanics. The singlet–triplet mixed states are not mixtures of states but rather linear combinations of state vectors.

where $E_{nl}^+ = E_{n1}^+(l)$ and $E_{nl}^- = E_{n1}^-(l)$ are the eigenvalues (2.43), (2.44) of H_0 in the singlet and triplet states, respectively.

In (3.16) and (3.17) we have already taken into account that $[H, J_i] = [H_1, J_i] = 0$. From parity conservation,

$$[H, U_p] = 0 \tag{3.18}$$

and, consequently, $[H_1, U_p] = 0$ then follows according to (3.7):

$$\langle \psi_{n'l'}^{JJ_3} | H_1 | \psi_{nl}^{JJ_3} \rangle = 0 \quad \text{unless} \quad (-1)^{l'} = (-1)^l,$$
$$\langle \gamma_{n'l'}^{JJ_3} | H_1 | \gamma_{nl}^{JJ_3} \rangle = 0 \quad \text{unless} \quad (-1)^{l'} = (-1)^l, \tag{3.19}$$

and

$$\langle \psi_{n'l'}^{JJ_3} | H_1 | \gamma_{nl}^{JJ_3} \rangle = 0 \quad \text{unless} \quad (-1)^{l'} \eta_0 = (-1)^l \eta_1. \tag{3.20}$$

In the perturbation series of (3.16), (3.17), only states of $\hbar_{1+}$ or $\hbar_{1-}$ with energy below the first ionization threshold have been included. This should suffice for the qualitative considerations that we wish to present here.[4]

The matrix elements $\langle \gamma_{n'l'}^{JJ_3} | H_1 | \gamma_{nl}^{JJ_3} \rangle$ and $\langle \psi_{n'l'}^{JJ_3} | H_1 | \psi_{nl}^{JJ_3} \rangle$ give the fine structure contributions, i.e., the splitting of the terms ${}^3(L)_J$ into three energy levels and a small shift of the energy levels of ${}^1(L)_J$. Calculations show (Bethe and Salpeter, 1957, Section 40) that the $J = 1$ and $J = 2$ levels happen to have almost the same energy.

The matrix elements $\langle \psi_{n'l'}^{JJ_3} | H_1 | \gamma_{nl}^{JJ_3} \rangle$ give the singlet–triplet mixing, which we shall discuss now.

H_1, like every observable commutes with the permutation operator $\mathbb{P}_{12}$, which represents the transposition of the two electrons. However, if we write $\mathbb{P}_{12}$ as a direct product of an orbital part $\mathbb{P}_{12}^{orb}$ and a spin part $\mathbb{P}_{12}^s$ according to the representation (1.23)

$$\mathbb{P}_{12} = \mathbb{P}_{12}^{orb} \otimes \mathbb{P}_{12}^s, \tag{3.21}$$

then

$$[H_1, \mathbb{P}_{12}^{orb}] \neq 0, \qquad [H_1, \mathbb{P}_{12}^s] \neq 0, \tag{3.22}$$

i.e., H_1 contains contributions that are not symmetrical in the orbital operators and spin operators separately. An example of such a contribution is the interaction between the spin magnetic moment and orbital magnetic moment of each electron which would, according to (IX.3.23), be given by

$$H^1 = C(Q_{(1)})\mathbf{S}_{(1)} \cdot \mathbf{L}_{(1)} + C(Q_{(2)})\mathbf{S}_{(2)} \cdot \mathbf{L}_{(2)} \tag{3.23}$$

with

$$C_{(i)} = C(Q_{(i)}) = \frac{1}{2}\left(\frac{e^2}{m_e^2 c^2}\right)\frac{1}{Q_{(i)}^3}. \tag{3.24}$$

H^1 is only one term in H_1: other terms like the relativistic mass correction term

---

[4] For more details, see Bethe and Salpeter (1957), in particular Chapter IIb.

and the interaction between the spin magnetic moment of one electron and the orbital magnetic moment of the other are of the same order of magnitude. As we want to discuss only the principle of singlet–triplet mixing, we will restrict our discussions to (3.23).

In order to calculate the matrix element $\langle \gamma_{n'L}^{JJ_3} | H^1 | \psi_{nJ}^{JJ_3} \rangle$, we use (3.14) and (3.15) and obtain

$$\langle \gamma_{n'l'}^{JJ_3} | H^1 | \psi_{nJ}^{JJ_3} \rangle = \sum_{l'_3 s_3 \, s'_3 s''_3 \atop \tilde{s}_3 \tilde{s}''_3} \left(\langle \gamma_{n'l'l'_3} | C_{(1)} \mathbf{L}_{(1)} | \psi_{nJJ_3} \rangle \langle J J_3 | l' \, l'_3 \, 1 s \rangle \langle 1 s_3 | s'_3 \, s''_3 \rangle \right.$$

$$\times \langle s'_3 | \mathbf{S}_{(1)} | \tilde{s}'_3 \rangle \langle s''_3 | \tilde{s}''_3 \rangle \langle \tilde{s}_3 \, \tilde{s}''_3 | 0 \, 0 \rangle$$

$$+ \langle \gamma_{n'l'l'_3} | C_{(2)} \mathbf{L}_{(2)} | \psi_{nJJ_3} \rangle \langle J J_3 | l' \, l'_3 \, 1 s_3 \rangle \langle 1 s_3 | s'_3 s''_3 \rangle$$

$$\left. \times \langle s'_3 | \mathbf{S}_{(2)} | \tilde{s}'_3 \rangle \langle s'_3 | \tilde{s}'_3 \rangle \langle \tilde{s}'_3 \, \tilde{s}''_3 | 0 \, 0 \rangle \right) \tag{3.25}$$

Changing in the second term the summation indices $s''_3 \rightarrow s'_3$, $\tilde{s}''_3 \rightarrow \tilde{s}_3$ and making use of the property of the Clebsch–Gordan coefficients $\langle \tfrac{1}{2} s'_3 \, \tfrac{1}{2} s''_3 | 1 s_3 \rangle = \langle \tfrac{1}{2} s''_3 \tfrac{1}{2} s'_3 | 1 s_3 \rangle$; $\langle \tfrac{1}{2} s'_3 \tfrac{1}{2} s''_3 | 0 0 \rangle = (-1) \langle \tfrac{1}{2} s''_3 \tfrac{1}{2} s'_3 | 0 0 \rangle$ one obtains

$$\sum_{l'_3 s_3 \, s'_3 s''_3 \atop \tilde{s}'_3} \langle J J_3 | l' \, l'_3 \, 1 s_3 \rangle \langle 1 s_3 | s'_3 s''_3 \rangle \langle s'_3 | \mathbf{S} | \tilde{s}'_3 \rangle \langle \tilde{s}'_3 \, s''_3 | 0 0 \rangle$$

$$\times \left(\langle \gamma_{n'l'l'_3} | C_{(1)} \mathbf{L}_{(1)} | \psi_{nJJ_3} \rangle - \langle \gamma_{n'l'l'_3} | C_{(2)} \mathbf{L}_{(2)} | \psi_{nJJ_3} \rangle \right). \tag{3.26}$$

Since ψ_{nJJ_3} is symmetric and $\gamma_{n'l'l'_3}$ is antisymmetric under $\mathbb{P}_{12}^{\text{orb}}$ one has

$$\langle \gamma_{n'l'l'_3} | C_{(2)} \mathbf{L}_{(2)} | \psi_{nJJ_3} \rangle = - \langle \gamma_{n'l'l'_3} | C_{(1)} \mathbf{L}_{(1)} | \psi_{nJJ_3} \rangle. \tag{3.27}$$

To calculate these matrix elements, one uses (2.41) and (2.42) with (2.37) and (2.38) and obtains four terms. Two of these are zero unless $l' = 0, J = 0$, and one of them is zero unless $l' = J$. In order to reduce the number of terms, we shall restrict ourselves to the case $l' \neq J = 1$ and obtain for this case

$$\langle \gamma_{n'l'l'_3} | C_{(1)} \mathbf{L}_{(1)} | \psi_{nJJ_3} \rangle = - {}_i \langle n' \, l' \, l'_3 | U^{-\dagger}(n'l') C_{(i)} \mathbf{L}_{(i)} U^+(nJ) | n \, J \, J_3 \rangle_i$$

$$= - {}_i \langle n'^- \, l' \, l'_3 | C_{(i)} \mathbf{L}_{(i)} | n^+ J \, J_3 \rangle_i \tag{3.28}$$

where i can be either 1 or 2 and $U^-(n'l')$ and $U^+(nJ)$ are here the restrictions of the unitary operators of (2.42) and (2.41) to $\mathcal{H}_i^{\text{orb}}$. All we need to know of them here is that they are scalar operators (because W is a scalar operator) and do not change the value of $l'l'_3$ and JJ_3. Thus the new vectors

$$|n^{\pm} l \, l_3 \rangle_i = U^{\pm}(nl) |n \, l \, l_3 \rangle_i \tag{3.29}$$

are again eigenvectors of $\mathbf{L}^2$ and L_3 with the same eigenvalue. Inserting (3.28) and (3.27) into (3.26), one obtains

$$\langle \gamma_{n'L}^{JJ_3} | H^1 | \psi_{nJ}^{JJ_3} \rangle$$

$$= \sum_{l_3 s_3 \, s'_3 s''_3 \atop \tilde{s}_3} \langle J J_3 | L \, l_3 \, 1 s_3 \rangle \langle 1 s_3 | \tfrac{1}{2} s'_3 , \tfrac{1}{2} s''_3 \rangle \langle s'_3 | \mathbf{S}_{(i)} | \tilde{s}_3 \rangle \langle \tfrac{1}{2} \tilde{s}_3 , \tfrac{1}{2} s''_3 | 0 0 \rangle$$

$$\times 2(-1)_i \langle n'^- L \, l_3 | C_{(i)} \mathbf{L}_{(i)} | n^+ J \, J_3 \rangle_i, \tag{3.30}$$

where $C_{(i)}$ is an $L_{(i)}$-scalar operator. $L_{(i)\kappa}, \kappa = 0 \pm 1$, are $L_{(i)}$ vector operators. Thus

$$_i\langle n'^- L\, l_3 | C_{(i)} L_{(i)\kappa} | n^+ J\, J_3 \rangle_i = \langle J\, J_3\, 1\kappa | L\, l_3 \rangle \langle n'^- L \| C_{(i)} L_{(i)} \| n^+\, J \rangle. \quad (3.31)$$

The reduced matrix element can also be further calculated using (V.3.10):

$$\langle n'^- L \| C_{(i)} L_{(i)} \| n^+\, J \rangle = \delta_{LJ} \sqrt{J(J+1)} \langle n'^- \| C_{(i)} \| n^+ \rangle_J. \quad (3.32)$$

Further evaluation of the quantity $\langle n'^- \| C_{(i)} \| n^+ \rangle_J$ requires actual calculation using perturbation theory with W as the interaction Hamiltonian or other approximation methods for the determination of $|n^{\pm} l\, l_3\rangle$ and then calculation of the matrix element $\langle n'^- \| C_{(i)} \| n^+ \rangle_J$ using (3.24). We shall not do this calculation here but just remark that these quantities are of the same order of magnitude as the fine-structure splitting terms which are given by $\langle n'^- \| C_{(i)} \| n^- \rangle$.

Inserting (3.31) and

$$_i\langle s'_3 | S_{(i)\kappa} | \tilde{s}_3 \rangle_i = \langle \tfrac{1}{2}\tilde{s}_3\, 1\, \kappa | \tfrac{1}{2} s'_3 \rangle \sqrt{\tfrac{1}{2}\tfrac{3}{2}} \quad (3.33)$$

into (3.30), one obtains

$$\langle \gamma^{JJ_3}_{n'L} | H^1 | \psi^{JJ_3}_{nl} \rangle = \sum_\kappa \sum_{l_3 s_3} \sum_{s'_3 s''_3} \sum_{\tilde{s}_3} \langle J\, J_3\, 1\, \kappa | L\, l_3 \rangle \langle \tfrac{1}{2} s''_3, \tfrac{1}{2}\tilde{s}_3 | 0\, 0 \rangle$$

$$\times \langle \tfrac{1}{2} s''_3, \tfrac{1}{2} s'_3 | 1 s_3 \rangle \langle \tfrac{1}{2}\tilde{s}_3\, 1\, \kappa | \tfrac{1}{2} s'_3 \rangle \langle L\, l_3\, 1 s_3 | J\, J_3 \rangle$$

$$\times \sqrt{3} \langle n'^- L \| C_{(i)} L_{(i)} \| n^+ J \rangle. \quad (3.34)$$

It is now a straightforward though slightly tedious calculation using the symmetry properties (V.2.42)–(V.2.47) and orthogonality relations (V.2.14), (V.2.15) of the Clebsch–Gordan coefficients to calculate (cf. Problem 3) that

$$\langle \gamma^{JJ_3}_{n'L} | H^1 | \psi^{JJ_3}_{nl} \rangle = (-1)^{J-L+1} \sqrt{2} \delta_{LJ} \sqrt{J(J+1)} \langle n'^- \| C_{(i)} \| n^+ \rangle_J. \quad (3.35)$$

Therewith we have expressed the matrix element of the spin orbit interaction between singlet and triplet states in terms of a quantity which (can be only calculated numerically) is different from zero and of the same order as the fine structure splitting term.

Inserting this quantity into (3.16) and (3.17) we see that the eigenvectors of the operator $H = H_0 + H_1$ in first order, $\tilde{\gamma}^{JJ_3}_{nl}$ and $\tilde{\psi}^{JJ_3}_{nl}$, are not pure triplet and singlet states, but have a small component in the singlet and triplet space, respectively. Thus, an orthohelium state $\tilde{\gamma}$ is mainly triplet γ with a small singlet component

$$\tilde{\gamma} = \gamma + \epsilon\psi, \qquad |\epsilon| \ll 1, \quad (3.17')$$

and the parahelium state $\tilde{\psi}$ is mainly singlet ψ with a small triplet component

$$\tilde{\psi}' = \psi' + \epsilon\gamma', \qquad |\epsilon| \ll 1. \quad (3.16')$$

The dipole matrix element between para- and orthohelium is therefore

$$(\tilde{\psi}', \mathbf{d}\tilde{\gamma}) = (\psi', \mathbf{d}\gamma) + \epsilon(\psi', \mathbf{d}\psi) + \epsilon(\gamma', \mathbf{d}\gamma) \quad (3.36)$$

and even though the first term is zero according to (3.13), this matrix element is different from zero due to the "singlet–triplet mixing."

We had mentioned above in Section VII.1 that it was impossible to verify in the spectra of one-electron atoms whether the energy values were the expectation values of the energy operator in a state which is not an energy eigenstate or the eigenvalues of the energy operator. From the helium spectrum it follows that the latter is the case; the physical para- and orthohelium states are eigenstates of $H = H_0 + H_1$ and *not* pure singlet and triplet states with the energy values being the expectation values of H between these pure singlet and triplet states. The existence of the transitions between para- and orthohelium states is an experimental verification that the physical states are eigenstates of the energy operator. In Chapter XII we will see that for theoretical reasons all stationary states have to be eigenstates of the energy operator.

XI.4 Doubly Excited States of Helium

The energy levels $E_{1n'}(l)$ of the helium atom are experimentally very well known. They have been measured by light spectroscopy, i.e., by measuring the wave number (or frequency) of the emitted or absorbed light in a transition between two energy levels according to

$$v_{n'l,\tilde{n}'\tilde{l}} = \frac{1}{2\pi\hbar c}\,(E_{1n'}(l) - E_{1\tilde{n}'}(\tilde{l})).$$

The energy levels have also been measured in energy-loss experiments as described in Section II.4, and as depicted schematically in Figure II.4.1. The collision chamber was filled with helium gas, and the intensity of the electron current was measured as a function of the energy lost in collisions with the helium atoms. The results of this experiment are shown in Figure 4.1, where the intensity I has been plotted as a function of the energy loss E. According to the interpretation of this experiment in Section II.4, the energy lost by the electrons is used to excite the atom and cause the transition from the ground state $1\,^1S$ to an excited state; therefore the intensity shows maxima at values of energy loss that agree with the energy difference between the ground state and one of the states excited by this experiment. The excitation energies that are observed in this experiment are tabulated in Table 4.1, column 3, and are compared with the energy of the $n\,^1P$ levels measured by light spectroscopy (column 2); the first column gives the energy levels that participate in the transition. Not given in the table are the bumps between 42-eV and 46-eV energy loss. These bumps are the result of two successive excitations (double scattering); the electrons in the collision chamber scatter inelastically twice, losing energy to the transition $1\,^1S \rightarrow 2\,^1P$ and to the transition $1\,^1S \rightarrow n\,^1P$, or two times to the same transition $1\,^1S \rightarrow 3\,^1P$. From Figure 4.1 we see that there is a very weak transition $1\,^1S \rightarrow 2\,^1S$, and that the transitions of high intensity are the $^1S \rightarrow\,^1P$ transitions (dipole

Table 4.1 Excitation energies of helium.

| | Excitation energy (eV) measured in | |
Transition	light spectros-copy experiment	energy-loss experiment
1^1S–2^1P	21.212	21.213
3^1P	23.081	23.084
4^1P	23.736	23.739
5^1P	24.039	24.034
6^1P	24.205	24.209
7^1P	24.304	24.308
8^1P	24.369	24.374
9^1P	24.413	24.419
10^1P	24.445	24.451
11^1P	24.469	24.474
12^1P	24.487	(24.493)
2^1S	20.610	20.612
$2s2p^1P$	60.123	60.120
$2s3p^1P$	63.651	63.651
$2s4p^1P$	64.462	(64.450)

transitions). No transition between the 1 ^1S ground state and the ^1D, ^1F, ... states or between the 1 ^1S state and the triplet states are observed, which agrees with what one would expect for dipole excitations.

Figure 4.1 shows that there are further bumps above 60 eV; these bumps look different from the lower-lying bumps and have a typical asymmetric profile. They correspond to energy levels that are well above the ionization energy for one electron. Energy levels with such a high energy above the ground state level have also been observed as spectral lines in the optical (ultraviolet) absorption spectra. Figure 4.2 shows a photograph of the absorption spectrum obtained for helium in the region 160–215 Å. The observed discrete spectral lines are superimposed upon a continuous absorption background. The lowest-lying discrete spectral line in this region is at 206.21 Å, corresponding to an energy of 60.1 eV; this agrees well with the position of the first bump of this kind in the energy-loss spectrum of Figure 4.1 (cf. also Table 4.1). It is the lowest-lying line of a single long series of discrete spectral lines of outstanding intensity that converge to the energy value 65.4 eV. (Some of these lines have corresponding bumps in the energy-loss spectrum of Figure 4.1.) Three fainter lines indicate the existence of a second series that also converges towards this energy value. A number of fainter lines can be seen in Figure 4.2 lying to shorter wavelengths from the most prominent series. These lines are grouped into series having as limit points the energy values 72.9 eV and 75.6 eV. Thus we have a series of series of spectral lines. As these lines are observed in the absorption of electromagnetic radiation by helium atoms in the ground state, we can safely assume that (at least for the high-intensity lines and probably for the fainter

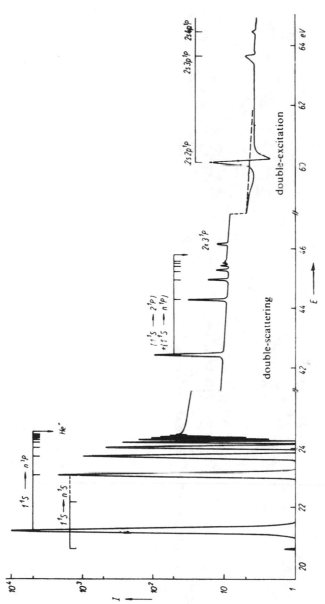

Figure 4.1 Energy-loss spectrum of helium. [From H. Boersch, J. Geiger, B. Schröder, *Abhandlungen der Deutschen Akademie der Wissenschaften*, Berlin (1967), No. 1, p. 15, with permission.]

Figure 4.2 The absorption spectrum of helium between 160 and 215 Å showing the many resonances due to the existence of two-electron excitation states in neutral helium. The photograph is a positive print (black is absorption). The white portion of each resonance indicates a reduced-adsorption zone or "window" in the continuous photoionization background absorption. [From **R. P. Madden, K. Codling**: *Astrophysical Journal* **141**, 364 (1965), with permission.]

lines too) they correspond to dipole transitions and that (according to the selection rules for dipole transitions) the corresponding excited energy levels belong to singlet states with orbital angular momentum 1, i.e., are ^1P states.

The threshold for double ionization of the helium—i.e., the energy value $E_{\infty\infty}$ at which both electrons have just dissociated from the helium nucleus and the physical system is in a state consisting of the nucleus and two electrons, both with relative energy 0—lies 79.0 eV above the ground state. We stated this experimental fact before by saying that the ground-state energy $E_{11}^+ = -79.0$ eV. [The energy scale in Figure 4.2 gives E(level) $- E_{11}^+$.]

The observations in Figure 4.2 indicate that one has a series of series of ^1P energy levels, and it appears that the limit point of this series of series lies 79.0 eV above the ground-state energy, i.e., at the threshold for double ionization. The limit points for each single series are then seen to agree with the energy values $E_{n\infty}$ as calculated from (2.33):

$$E_{n\infty} - E_{11}^+ = -R''_{He}\frac{1}{n^2} + 79.0 \text{ eV} = (-54.4 \text{ eV})\frac{1}{n^2} + 79.0 \text{ eV}; \quad (4.1)$$

hence

$$E_{1\infty} - E_{11}^1 = 24.6 \text{ eV},$$

$$E_{2\infty} - E_{11}^+ = 65.4 \text{ eV},$$

$$E_{3\infty} - E_{11}^+ = 72.9 \text{ eV}, \qquad (4.2)$$

$$E_{4\infty} - E_{11}^+ = 75.6 \text{ eV}.$$

Thus the observations in Figure 4.2 show that the discrete ^1P energy levels converge to the nth ionization threshold of the helium atom, i.e., to the energy value at which one electron has the principal quantum number n and the other has just dissociated itself from the helium atom. These experimentally measured ^1P levels above the first ionization level are depicted in Figure 2.3. Also depicted there are all the other known (1971) discrete energy levels of the helium atom. We see that the energy-level diagram below the second, third, etc. thresholds is qualitatively very similar to the energy diagram below the first threshold.

A rough explanation of the discrete energy levels above the first ionization threshold is now very simple:[5] Whereas the energy levels below $E_{1\infty}$ correspond to states in which one of the electrons is always in the ground state ($n = 1$), the energy levels below $E_{n\infty}$, $n > 1$, correspond to states in which one of the electrons is in the state with principal quantum number n and the other is in a state with $n' \geq n$. Both electrons are excited and the atom is

---

[5] Extensive theoretical calculations of these energy levels by various approximation methods have been performed and agree with the observed values; see J. Macek, *J. Phys.* B **1**, 831 (1968) and references therein. See also U. Fano: *Doubly Excited States of Atoms in Atomic Physics*, edited by V. W. Hughes *et al.*, Plenum. New York (1969).

said to be doubly excited. In the approximation where the interaction W between the two electrons is neglected, these are the states that correspond to the vectors $(2.15\pm)$ with $n' \geq n > 1$. For the case $n = 2$, $(2.15\pm)$ reads

$$\psi^0_{2l l_3, n' l' l'_3} = \frac{1}{\sqrt{2}} (|2\, l\, l_3\rangle_1 \otimes |n'\, l'\, l'_3\rangle_2 + |n'\, l'\, l'_3\rangle_1 \otimes |2\, l\, l_3\rangle_2), \quad (4.3+)$$

$$\gamma^0_{2l l_3, n' l' l'_3} = \frac{1}{\sqrt{2}} (|2\, l\, l_3\rangle_1 \otimes |n'\, l'\, l'_3\rangle_2 - |n'\, l'\, l'_3\rangle_1 \otimes |2\, l\, l_3\rangle_2), \quad (4.3-)$$

where l can take the two values $l = 0$ and $l = 1$; i.e., we have the singlet states $\psi^0_{2 1 l_3, n' l' l'_3}$ and $\psi^0_{2 0 0, n' l' l'_3}$ and the triplet states $\gamma^0_{2 1 l_3, n' l' l'_3}$ and $\gamma^0_{2 0 0, n' l' l'_3}$. If we want to explain the ^1P levels, we have to couple the angular momenta l and l' to obtain eigenvectors of total orbital angular momentum $\mathbf{L}^2 = (\mathbf{L}_1 + \mathbf{L}_2)^2$ with eigenvalue $L(L + 1) = 1(1 + 1)$. As L can be any of the values $l + l', l + l' - 1, \ldots, |l - l'|$, we can obtain $L = 1$ in the following cases:

$$
\begin{array}{lllll}
2snp & l = 0 & l' = 1 & L = l + l' & \pi = -1 \\
2pns & l = 1 & l' = 0 & L = l + l' & \pi = -1 \\
2pnp & l = 1 & l' = 1 & L = l + l' - 1 & \pi = +1 \\
2pnd & l = 1 & l' = 2 & L = l + l' - 2 & \pi = -1
\end{array}
\qquad (4.4)
$$

where π is the parity, $\pi = (-1)^l (-1)^{l'}$. Therefore one has three possibilities to obtain ^1P states with parity $\pi = -1$ as linear combinations of the states $(4.3\pm)$:

$$(n = 2, l = 0; n', l' = 1), \qquad (4.5p)$$

$$(n = 2, l = 1; n', l' = 0), \qquad (4.5s)$$

$$(n = 2, l = 1; n', l' = 2). \qquad (4.5d)$$

All of these linear combinations are eigenvectors of H_{00} with the same eigenvalue $E^{00}_{nn'}$. The energy shift due to the interaction of the two electrons depends not only on n' and L but also on l and l'; one would therefore expect three different energy values for every value of n' (except for $n' = 2$). Thus one would expect there to be three series of energy levels converging to the $E_{2\infty}$ limit. Figure 4.2 shows only two series, one with a very high transition probability from the $1\ ^1S$ ground state and the other with a very low one. This indicates the existence of an approximate selection rule that cannot be explained on the basis of the assumption that the spaces spanned by $(4.5p)$, $(4.5s)$, and by $(4.5d)$ are good approximations to spaces of physical states. It is believed that linear combinations of $(4.5p)$ and $(4.5s)$ span, to a good approximation, the spaces of energy eigenstates that belong to the two observed series,[6] and that $(4.5d)$ corresponds to a still fainter and therefore yet unobserved series.

[6] J. W. Cooper, U. Fano, and F. Prats, *Phys. Rev. Lett.* **10**, 518 (1965).

The situation becomes even more complicated for $n = 3, 4, \ldots$. With similar arguments the ^3P series can be discussed.

We conclude the discussion of the doubly excited states with a remark concerning the broad profile of the spectral lines corresponding to these states, which is shown in Figure 4.1 and is also visible in the optical spectra. These energy levels lie above the first ionization threshold. Therefore, in addition to the doubly excited states with $n = 2, n' \geq 2$ there are states with the same energy as the energy of the doubly excited states which belong to the system that consist of the He$^+$ ion in the ground state $n = 1$ and one electron (He$^+ + e$). The system He$^+ + e$ can take any possible energy value $E \geq E_{1\infty}$, because the relative energy between He$^+$ and e can take any value, i.e., the energy spectrum of He$^+$ $(n = 1) + e$ is continuous above $E_{1\infty}$. Therefore each energy value of the doubly excited states coincides with one of the energy values in the continuous spectrum of the states of the system He$^+$ $(n - 1) + e$. Thus there are two different states with the same energy: the discrete state $n = 2, n' \geq 2$, and the state of the continuum with $n = 1$ and a relative energy between He$^+$ $(n = 1)$ and e. These two states mix and interact with each other, as a result of which the doubly excited state cannot persist and easily decays into the state of a He$^+$ $(n = 1)$ and an electron. The lifetime of such doubly excited states is therefore much smaller, and the spectral line has the typical broad profile. Such states are often called resonances.

For the doubly excited states with $n = 3, n' \geq 3$, there exist two continuum states with the same energy, namely He$^+$ $(n = 1) + e$ and He$^+$ $(n = 2) + e$. For $n = 4, n' \geq 4$ there are three, and so forth.

Problems

1. Show that the symmetric and antisymmetric parts of the total space $\mathfrak{H}$ are given by

$$\mathscr{H}^2_+ = (\mathscr{H}^{\mathrm{orb}^2}_+ \otimes \iota^{s^2}_+) \oplus (\mathscr{H}^{\mathrm{orb}^2}_- \otimes \iota^{s^2}_-),$$
$$\mathscr{H}^2_- = (\mathscr{H}^{\mathrm{orb}^2}_+ \otimes \iota^{s^2}_-) \oplus (\mathscr{H}^{\mathrm{orb}^2}_- \otimes \iota^{s^2}_+),$$

respectively.

2. Check whether H^1 given by (3.23) commutes with the operator of total orbital angular momentum squared, L^2.

3. Using the properties of the Clebsch–Gordan coefficients given in Section V.2, perform the calculations that lead from (3.34) to (3.35).

CHAPTER XII

Time Evolution

In this chapter the basic assumption on time evolution is introduced. The Schrödinger picture, the Heisenberg picture, the Dirac picture, and their relations are discussed. Time-dependent external forces are only briefly mentioned at the end of the chapter.

XII.1 Time Evolution

Until now we have not mentioned time and have completely ignored the time evolution of the state of a system and the time development of an observable. We have restricted ourselves to properties of systems measured at one instant. In general, one should specify not only the physical values obtained in a measurement but also the time at which the observation is made. In this chapter we consider the time evolution—the dynamics—of physical systems. Although the notion of one instant of time makes as little sense physically as the notion of a definite position (measurements always require a finite amount of time in the same way that they always require a finite amount of space), we make the idealizing assumption that time is a continuous real parameter t that labels a sequence of states. (Note that we have called time a "parameter" as opposed to an "observable.") That is, we make the idealizing assumption that measurements can be made at a succession of arbitrarily small time intervals, and that at any instant of time the specification of a complete set of observables is possible.

Let us denote by $W(t_0)$ the state of a system that is prepared at the time $t = t_0$. We then expect that $W(t_0)$ will uniquely determine another state

$W(t_1)$ at any later time $t = t_1 > t_0$. During the evolution the system may be subject to external influences, so that the manner of evolution will depend explicitly upon the interval from t_0 to t_1—its starting time t_0 as well as its length $t_1 - t_0$. We first consider systems that evolve undisturbed by external influences (isolated systems). For such systems the evolution

$$W(t_0) \to W(t_1) \tag{1.1}$$

does not depend upon t_0 but only upon the system itself, the initial state $W(t_0)$, and the length $\tau_1 = t_1 - t_0$ of the time interval from t_0 to t_1. Our intuitive understanding of the time evolution suggests that we impose the following requirements on the map (1.1):

1. Expression (1.1) is a linear transformation of the set of states, continuous in a certain sense: If $W_1(t_0) \to W_1(t_1)$ and if $W_2(t_0) \to W_2(t_1)$, then for $a, b \in \mathbb{C}$,

    $$aW_1(t_0) + bW_2(t_0) \to aW_1(t_1) + bW_2(t_1).$$

2. Time evolution forms an additive semigroup; i.e., if $W(t_0) \to W(t_0 + \tau_1)$ for $\tau_1 = t_1 - t_0$ and if $W(t_1) \to W(t_1 + \tau_2)$ for $\tau_2 = t_2 - t_1$, then

    $$W(t_0) \to W(t_0 + \tau_1 + \tau_2).$$

Requirement 1 expresses the physical statement that first mixing the states at time t_0 and then letting the system evolve in time leads to the same results as first letting the states evolve separately in time and mixing them afterwards at time t_1; the continuity of the transformation expresses the intuitive statement that time changes continuously. Requirement 2 is intuitively obvious; it is a statement of the fact that letting a physical state evolve first for a period of length τ_1 and then letting the resultant state evolve for a period of length τ_2 leads to the same result as letting the original state evolve for a period of length $\tau_1 + \tau_2$. Requirements 1 and 2 are very general and must be satisfied by irreversible processes also.

We are here concerned with the quantum mechanics of reversible processes, and for such processes we make the additional requirement

3. The time evolution (1.1) is given by a unitary operator $U(\tau)$ satisfying

    $$W(t_0) \to W(t_1) = U^\dagger(\tau_1) W(t_0) U(\tau_1), \tag{1.2}$$

    $$U^\dagger(\tau) = U^{-1}(\tau), \tag{1.3}$$

    $$U(0) = I, \tag{1.4}$$

    $$U^{-1}(\tau) = U(-\tau), \tag{1.5}$$

    $$U(\tau_1 + \tau_2) = U(\tau_1)U(\tau_2) \qquad (-\infty < \tau_1, \tau_2 < +\infty), \tag{1.6}$$

    $U(\tau)$ is a continuous operator function[1] of the parameter τ. \hfill (1.7)

---

[1] See appendix to this chapter.

Equation (1.5) follows from (1.2) and (1.3)—specifically from

$$U(\tau_1)W(t_1)U^{-1}(\tau_1) = W(t_0) \tag{1.2'}$$

—and from the assumption that (1.2) is also valid for evolution backwards in time, i.e.,

$$W(t_0) = U^{-1}(-\tau_1)W(t_1)U(-\tau_1). \tag{1.8}$$

Equation (1.6) follows from requirement 2, which, when the time evolution is described by a unitary operator as in (1.2), is expressed as

$$U(t_2 - t_0) = U(t_1 - t_0)U(t_2 - t_1). \tag{1.6'}$$

Equation (1.6) follows for $\tau_1 = t_1 - t_0$, $\tau_2 = t_2 - t_1$. The linearity of the transformation (1.2) is obvious. The continuity of the transformation (1.1) is expressed by (1.7). We shall not give here the mathematical definition of a continuous operator function, but mention only that the formal operations, employed below, of differentiation, Taylor series expansion, etc.[2] can be rigorously defined. A continuous operator function satisfying (1.6) is called a *one-parameter group of operators*.

We differentiate $U(\tau)$ with respect to τ and define

$$A \equiv \left.\frac{dU(\tau)}{d\tau}\right|_{\tau=0}. \tag{1.9}$$

A is called the *infinitesimal operator* or *generator* of the operator $U(\tau)$ of time evolution. $U(\tau)$ is not only "differentiable," but is differentiable an arbitrary number of times. We can calculate all "derivatives" using (1.6). Differentiating (1.6) with respect to τ_1,

$$\frac{dU(\tau + \tau_1)}{d(\tau + \tau_1)}\frac{d(\tau + \tau_1)}{d\tau_1} = \frac{dU(\tau_1)}{d\tau_1}U(\tau),$$

and evaluating the resultant expression at $\tau_1 = 0$, we obtain

$$\frac{dU(\tau)}{d\tau} = AU(\tau).$$

We differentiate this again and obtain

$$\frac{d^2U(\tau)}{d\tau^2} = A\frac{dU(\tau)}{d\tau} = A \cdot AU(\tau) = A^2U(\tau).$$

We may continue to differentiate, obtaining for the pth derivative

$$\frac{d^pU(\tau)}{d\tau^p} = A^pU(\tau);$$

use of (1.4) then gives

$$\left.\frac{d^pU(\tau)}{d\tau^p}\right|_{\tau=0} = A^p.$$

---

[2] See the appendix.

The "Taylor series expansion" of $U(\tau)$ is therefore given by

$$U(\tau) = I + \tau A + \frac{\tau^2}{2!} A^2 + \cdots + \frac{\tau^n}{n!} A^n + \cdots; \qquad (1.10)$$

or, if we compare (1.10) with the series for $e^{\tau A}$, we have

$$U(\tau) = e^{\tau A}. \qquad (1.10')$$

In particular this means [according to our rule that $f(A)|a\rangle = f(a)|a\rangle$ for functions $f(A)$ of A] that

$$U(\tau)|a\rangle = e^{\tau a}|a\rangle$$

for (generalized) eigenvectors $|a\rangle$ of A corresponding to the eigenvalue a. Taking the adjoint of (1.10) gives

$$U^\dagger(\tau) = I + \tau A^\dagger + \frac{\tau^2}{2!} A^{\dagger 2} + \cdots = e^{\tau A^\dagger}. \qquad (1.11)$$

If we use (1.3) and differentiate $U^\dagger(\tau)U(\tau) = I$, we obtain

$$\frac{dU^\dagger(\tau)}{d\tau} U(\tau) + U^\dagger(\tau) \frac{dU(\tau)}{d\tau} = 0;$$

setting $\tau = 0$, we obtain

$$A^\dagger = -A. \qquad (1.12)$$

An operator that fulfills (1.12) is called *skew-Hermitian*. As τ is the time, A has the dimensions of frequency, i.e., of energy/h. The operator

$$H = -i\hbar A = -i\hbar \left.\frac{dU(\tau)}{d\tau}\right|_{\tau=0}, \qquad (1.13)$$

often also called the generator of time evolution, is therefore a Hermitian operator with the dimension of energy. H is the Hamiltonian operator or energy operator of the system, and, from (1.10'),

$$U(\tau) = e^{iH\tau/\hbar}. \qquad (1.14)$$

The precise form of H depends, of course, upon the particular physical system under consideration, and how it is found in particular cases has been discussed previously. H is an element of the algebra of observables of the particular system, and its properties are determined by the defining algebraic relations of the particular physical system. We remark that there is a one-to-one correspondence between $U(\tau)$ and H; one can therefore consider H as a quantity derived [using (1.13)] from the more fundamental time-evolution operator $U(\tau)$, or one can consider H as the more fundamental quantity and the time-evolution operator $U(\tau)$ as a quantity derived [using (1.14)] from H. It is in the spirit of our preceding formulation to consider H, an element of the algebra of observables, as the fundamental quantity, and to consider $U(\tau)$ as the derived quantity.

We formulate the result of our preceding plausibility argument as the basic assumption of quantum mechanics for time evolution—the basic dynamical law for quantum-mechanical systems:

Va (The Schrödinger Picture). A (conservative) physical system has a generator H of time translation which is a Hermitian element of the algebra of observables and which is characteristic of the physical system. The time evolution of the state of the physical system is given by

$$W(t) = U^\dagger(t)W_0 U(t), \tag{1.15}$$

where W_0 is the state of the system at some initial time $t_0 = 0$ and where $U(t)$ is given by

$$U(t) = e^{itH/\hbar}. \tag{1.16}$$

According to the axiom II, the expectation value of an observable A is given by

$$\langle A \rangle = \mathrm{Tr}(AW). \tag{1.17}$$

The observable that is represented by the operator A is defined by the prescription for its measurement, i.e., how to set up a measurement apparatus and how to take the measurement. This prescription is given for all time and does not change with time. The experimentally measured value changes in time; if we make a measurement at a time t_0, the expectation value $\langle A \rangle_{t=t_0}$ will, in general, be different from the expectation value $\langle A \rangle_{t=t_1}$ at a different time t_1. This experimental fact is represented by a description of the change of the state $W(t)$ according to (1.15). Inserting (1.15) into (1.17), we obtain

$$\langle A \rangle_t = \mathrm{Tr}(AW(t)) = \mathrm{Tr}(AU^\dagger(t)W_0 U(t)). \tag{1.18}$$

This can be written

$$\langle A \rangle_t = \mathrm{Tr}(U(t)AU^\dagger(t)W_0), \tag{1.19}$$

since the trace of a product of operators is invariant under a cyclic permutation of the order of the factor operators. We can now define a new operator $A(t)$ by

$$A(t) = U(t)AU^\dagger(t) \tag{1.20}$$

and write (1.19) as

$$\langle A \rangle_t = \langle A(t) \rangle = \mathrm{Tr}(A(t)W_0). \tag{1.21}$$

In (1.21) the change of the experimentally measured value $\langle A \rangle_t$ with time is described in a form that can be interpreted in the following way: The state of the system is described by an operator W_0; this statistical operator W_0 does not change with time, i.e., the whole history of the state of the physical system is described by the one operator W_0. The observable develops with time according to (1.20), i.e., the prescription for the measurement must include instructions as to when to take the measurement, because the observable—the experimental setup—is different for different times. Thus instead of the basic assumption Va we can formulate the time evolution by the following basic assumption:

Vb (The Heisenberg Picture). A (conservative) physical system has a generator H of time translation which is a Hermitian element of the algebra of observables and which is characteristic of the physical system. The time development of every observable $A(t)$ of the physical system is given by

$$A(t) = U(t)AU^\dagger(t), \tag{1.22}$$

where A describes the observable at some initial time $t_0 = 0$ and where $U(t)$ is given by

$$U(t) = e^{itH/\hbar}. \tag{1.23}$$

These two descriptions of the change in time of the physically measured value $\langle A \rangle_t$ are, of course, completely equivalent. In the form Va the change in time is described as a change of the state of the system. The description of the change in time in this form is called the "Schrödinger picture." In the form Vb the state is held fixed and the change in time is described by the change of the observable. The description in this form is called the "Heisenberg picture."

The time development of an observable as given by (1.22) is often written in a different form, which is obtained from (1.22) by differentiating:

$$\frac{dA(t)}{dt} = \frac{i}{\hbar} H e^{itH/\hbar} A e^{-itH/\hbar} + e^{itH/\hbar} A \left(-\frac{i}{\hbar} H \right) e^{-itH/\hbar}$$

$$= \frac{i}{\hbar} (HA(t) - A(t)H),$$

or

$$i\hbar \frac{dA(t)}{dt} = [A(t), H]. \tag{1.24}$$

This is the "Heisenberg equation of motion." An observable is a *constant of the motion* if it does not depend upon t. From (1.24) it follows that all constants of the motion commute with H. In particular, H itself is a constant of the motion. Furthermore, every member of any c.s.c.o. that contains H is a constant of the motion.

Consider the particular observables momentum P_i and position Q_i for a physical system in which neither of them is a constant of the motion. It is easily seen that $[P_i(t), Q_j(t)]$ has its usual form for all times $t > 0$ given that $[P_i, Q_j] = (\hbar/i)\delta_{ij}I$ at $t = 0$:

$$P_i(t)Q_j(t) - Q_j(t)P_i(t) = U(t)P_iU^\dagger(t)U(t)Q_jU^\dagger(t) - U(t)Q_jU^\dagger(t)U(t)P_iU^\dagger(t)$$

$$= U(t)(P_iQ_j - Q_jP_i)U^\dagger(t)$$

$$= U(t)U^\dagger(t)\frac{\hbar}{i}\delta_{ij}I = \frac{\hbar}{i}\delta_{ij}I,$$

i.e.,

$$[P_i(t), Q_j(t)] = \frac{\hbar}{i} \delta_{ij} I. \tag{1.25}$$

We are now in a position to calculate the equations of motion for the momentum and for the position and to show that these equations have the same form as those for the classical quantities. For a system whose Hamiltonian is given by

$$H = \frac{1}{2m} P_i P_i + V(\mathbf{Q}),$$

where

$$Q_i = Q_i(t) \quad \text{and} \quad P_i = P_i(t),$$

we calculate that

$$[Q_k, H] = \frac{1}{2m} [Q_k, P_i P_i] = \frac{1}{2m} \{[Q_k, P_i], P_i\}$$

$$= \frac{i\hbar}{2m} \delta_{ki} \{I, P_i\} = \frac{i\hbar}{m} P_k.$$

Inserting this into (1.24), we obtain

$$\frac{dQ_k(t)}{dt} = \frac{P_k(t)}{m}. \tag{1.26}$$

This is the quantum-mechanical analogue of the classical relation between momentum and the time derivative of position.

In classical physics the force is obtained from the potential function $V(\mathbf{x})$ by

$$f_k = - \frac{\partial V(\mathbf{x})}{\partial x_k}.$$

In analogy to this we define the "force operator" by

$$F_k \equiv - \frac{i}{\hbar} [P_k, V(\mathbf{Q})] = - \frac{\text{``} \partial V(\mathbf{Q})\text{''}}{\partial Q_k} \tag{1.27}$$

where the last symbol is defined below. Inserting this into (1.24) for $A(t) = P_k(t)$, we obtain

$$\frac{dP_k(t)}{dt} = \frac{1}{i\hbar} [P_k, H] = - \frac{i}{\hbar} [P_k, V(\mathbf{Q})] = F_k(\mathbf{Q}(t)). \tag{1.28}$$

This is the quantum-mechanical analogue of the classical relation between the time derivative of the momentum and the force. Combining (1.26) and (1.28) gives the quantum-mechanical Newton's equation,

$$m \frac{d^2 Q_k(t)}{dt^2} = F_k(\mathbf{Q}(t)). \tag{1.29}$$

["$\partial V(\mathbf{Q})/\partial Q_k$" is a symbolic way of indicating the following procedure: Replace Q_i by the numbers x_i in $V(\mathbf{Q})$ [recall that this is just the inverse to the way $V(\mathbf{Q})$ was obtained from the corresponding classical expression], differentiate $V(\mathbf{x})$ with respect to x_k, and replace x_i by Q_i in $\partial V(\mathbf{x})/\partial x_k$. That "$\partial V(\mathbf{Q})/\partial Q_k$" is in fact $(i/\hbar)$ $[P_k, V(\mathbf{Q})]$ can be seen in the following ways:

1. Write

$$V(\mathbf{Q}) = C + C_i Q_i + C_{ij} Q_i Q_j + \cdots$$
$$+ C_{i_1 \cdots i_n} Q_{i_1} \cdots Q_{i_n} + \cdots, \qquad (1.30)$$

where the $C_{i_1 \cdots i_n}$ are symmetric, i.e.,

$$C_{i_1 \cdots i_p \cdots i_q \cdots i_n} = C_{i_1 \cdots i_q \cdots i_p \cdots i_n}, \qquad (1.31)$$

because the Q_i commute with each other. Using the canonical commutation relation (1.25), calculate that

$$\frac{i}{\hbar} [P_k, V(\mathbf{Q})] = C_k + 2 C_{ki} Q_i + \cdots$$
$$+ n C_{k i_2 i_3 \cdots i_n} Q_{i_2} Q_{i_3} \cdots Q_{i_n} + \cdots, \qquad (1.32)$$

because

$$C_{i_1 \cdots i_n}[P_k, Q_{i_1} \cdots Q_{i_n}]$$
$$= C_{i_1 \cdots i_n} \sum_{p=1}^{n} [P_k, Q_{i_p}] Q_{i_1} \cdots Q_{i_{p-1}} Q_{i_{p+1}} \cdots Q_{i_n}$$
$$= C_{i_1 \cdots i_n} \sum_{p=1}^{n} \frac{\hbar}{i} \delta_{k i_p} Q_{i_1} \cdots Q_{i_{p-1}} Q_{i_{p+1}} \cdots Q_{i_n}$$
$$= n \frac{\hbar}{i} C_{k i_2 i_3 \cdots i_n} Q_{i_2} Q_{i_3} \cdots Q_{i_n},$$

where (1.31) has been used. The right-hand side of (1.32) is identical to "$\partial V(\mathbf{Q})/\partial Q_k$."

2. Calculate $[P_k, V(\mathbf{Q})]$ in the position representation:

$$\langle \mathbf{x} | [P_k, V(\mathbf{Q})] | \psi \rangle = \langle \mathbf{x} | P_k(V(\mathbf{Q}) | \psi \rangle) - \langle \mathbf{x} | V(\mathbf{Q}) P_k | \psi \rangle$$
$$= \frac{\hbar}{i} \frac{\partial}{\partial x_k} \langle \mathbf{x} | V(\mathbf{Q}) | \psi \rangle - V(\mathbf{x}) \frac{\hbar}{i} \frac{\partial}{\partial x_k} \langle \mathbf{x} | \psi \rangle$$
$$= \frac{\hbar}{i} \frac{\partial}{\partial x_k} (V(\mathbf{x}) \langle \mathbf{x} | \psi \rangle) - V(\mathbf{x}) \frac{\hbar}{i} \frac{\partial}{\partial x_k} \langle \mathbf{x} | \psi \rangle$$
$$= \frac{\hbar}{i} \frac{\partial V(\mathbf{x})}{\partial x_k} \langle \mathbf{x} | \psi \rangle. \qquad]$$

As the Heisenberg commutation relations do not change with time, all algebraic relations that are derived from them are also valid at any time, even if the operators in these algebraic relations do not commute with H. That is, any algebra of observables whose elements are functions of P_i, Q_j does not change in time, even though some of the elements may not commute with H.

A *symmetry transformation* of a set of elements is a transformation of the elements that leaves the relations among them unchanged. Thus, the time evolution of a physical system whose algebra of observables is derived from the Heisenberg commutation relations is a symmetry transformation of the algebra of observables. If time evolution is a symmetry transformation, then the mathematical structure (in particular the algebraic relations) of the algebra of observables does not change in time; this means that the physical structure is indistinguishable at two different points in time. Our experience shows that there are physical systems that have this property, and in fact it is this property that defines the isolated physical systems. Thus isolated physical systems do not age, an absolute value of time has no meaning for these systems, and only time differences are accessible to measurement. Irreversible processes do not take place in isolated physical systems defined as above. These considerations lead us to another formulation of the basic assumption V:

Vc. The time evolution of a conservative physical system is given by a continuous symmetry transformation of its algebra of observables.

Let us return to the Schrödinger picture. The time evolution (1.15) can also be expressed in differential form. Differentiating (1.15), one obtains

$$\frac{dW(t)}{dt} = \frac{i}{\hbar}(W(t)H - HW(t)) = \frac{i}{\hbar}[W(t), H]. \tag{1.33}$$

We consider now the special case where the state W_0 is a pure state, i.e.,

$$W_0 = \Lambda_{|\psi_0\rangle} = |\psi_0\rangle\langle\psi_0|, \tag{1.34}$$

where $\Lambda_{|\psi_0\rangle}$ is the projection operator onto the one-dimensional space spanned by the vector $|\psi_0\rangle$. For this case (1.15) reads

$$W(t) = U^\dagger(t)\Lambda_{|\psi_0\rangle}U(t) = U^\dagger(t)|\psi_0\rangle\langle\psi_0|U(t).$$

We define the vector $|\psi(t)\rangle$ by

$$|\psi(t)\rangle \equiv U^\dagger(t)|\psi_0\rangle = e^{-iHt/\hbar}|\psi_0\rangle \tag{1.35}$$

and have

$$W(t) = U^\dagger(t)\Lambda_{|\psi_0\rangle}U(t) = |\psi(t)\rangle\langle\psi(t)| = \Lambda_{|\psi(t)\rangle}. \tag{1.36}$$

We obtain the differential form of (1.36) by differentiating (1.35):

$$\frac{d}{dt}|\psi(t)\rangle = \frac{dU^\dagger(t)}{dt}|\psi_0\rangle = -\frac{i}{\hbar}HU^\dagger(t)|\psi_0\rangle = -\frac{i}{\hbar}H|\psi(t)\rangle.$$

This is the *Schrödinger equation* for the state vector $|\psi(t)\rangle$:

$$i\hbar\frac{d}{dt}|\psi(t)\rangle = H|\psi(t)\rangle. \tag{1.37}$$

For the Hamiltonian given by

$$H = \frac{1}{2m} P_i P_i + V(\mathbf{Q}),$$

the Schrödinger equation in the position representation is obtained from this by taking the scalar product with the position eigenvector $|\mathbf{x}\rangle$:

$$i\hbar \frac{d}{dt} \langle \mathbf{x}|\psi(t)\rangle = \frac{1}{2m} \langle \mathbf{x}|P_i P_i|\psi(t)\rangle + V(\mathbf{x})\langle \mathbf{x}|\psi(t)\rangle.$$

Using

$$\langle \mathbf{x}|P_i P_i|\psi(t)\rangle = \frac{\hbar}{i}\frac{\hbar}{i} \frac{\partial}{\partial x_i} \frac{\partial}{\partial x_i} \langle \mathbf{x}|\psi(t)\rangle$$

(see Equation (II.7.58)), the Schrödinger equation may be written in the conventional form

$$i\hbar \frac{d}{dt} \psi(\mathbf{x}, t) = -\frac{\hbar^2}{2m} \nabla^2 \psi(\mathbf{x}, t) + V(\mathbf{x})\psi(\mathbf{x}, t), \qquad (1.38)$$

where we have defined the time-dependent wave function

$$\psi(\mathbf{x}, t) \equiv \langle \mathbf{x}|\psi(t)\rangle. \qquad (1.39)$$

Summarizing, the distinction between the Heisenberg picture and the Schrödinger picture is the following: In the Heisenberg picture the observables are transformed by $U(t)$ and the states are kept fixed, while in the Schrödinger picture the observables are kept fixed and the states are transformed in the "opposite direction," i.e., by $U^\dagger(t) = U^{-1}(t)$. The time dependence of $\mathrm{Tr}(AW)$ is the same in both cases. Expressed in terms of vectors: In the Heisenberg picture the eigenvectors $|a\,t\rangle$ of the observables $A(t)$,

$$A(t)|a\,t\rangle = a|a\,t\rangle, \qquad (1.40)$$

are "rotated" by $U(t)$:

$$U(t)|a\,0\rangle = |a\,t\rangle = e^{itH/\hbar}|a\,0\rangle$$

whereas the state vectors $|\psi_0\rangle$ are kept constant. [Note that in (1.40) no time dependence occurs on the part of the eigenvalue a. This is easily shown: $A(t)|a\,t\rangle = A(t)U(t)|a\,0\rangle = U(t)A(0)|a\,0\rangle = U(t)a(0)|a\,0\rangle = a(0)|a\,t\rangle$.] In the Schrödinger picture the eigenvectors $|a\,0\rangle$ are kept constant and the state vectors are "rotated" in the opposite direction:

$$U^\dagger(t)|\psi_0\rangle = |\psi(t)\rangle = e^{-itH/\hbar}|\psi_0\rangle$$

The time dependence of the physically measurable quantities representing the probabilities is in both cases the same:

$$|\langle a\,t|\psi_0\rangle|^2 = |(U(t)|a\,0\rangle, |\psi_0\rangle)|^2$$

$$= |\langle a\,0|U^\dagger(t)|\psi_0\rangle|^2 = |\langle a\,0|\psi(t)\rangle|^2.$$

A state is called a *stationary state* if it does not change with time:

$$W(t) = U^{\dagger}(t)W_0 U(t) = W_0 = W. \tag{1.41}$$

This terminology is justified by the fact that every observable measured in this state gives a value independent of the time of the measurement; for from (1.41) it follows that the expectation value of any observable A is

$$\langle A \rangle_t = \mathrm{Tr}(AW(t)) = \mathrm{Tr}(AU^{\dagger}(t)WU(t)) = \mathrm{Tr}(AW) = \langle A \rangle_{t=0}.$$

Stationary states are always mixtures of eigenstates of the energy operator H. This follows from (1.41) in the form

$$We^{itH/\hbar} = e^{itH/\hbar}W,$$

or equivalently,

$$WH - HW = 0. \tag{1.42}$$

In particular, for a pure stationary state,

$$W = \Lambda_{|\psi_0\rangle} = |\psi_0\rangle\langle\psi_0|,$$

(1.42) reads

$$|\psi_0\rangle\langle\psi_0|H = H|\psi_0\rangle\langle\psi_0|,$$

or

$$H|\psi_0\rangle = |\psi_0\rangle \frac{\langle\psi_0|H|\psi_0\rangle}{\langle\psi_0|\psi_0\rangle}. \tag{1.43}$$

Thus pure stationary states are represented by eigenvectors of the energy operator. The time dependence of a stationary state vector is given by

$$|\psi(t)\rangle = U^{\dagger}(t)|\psi_0\rangle = e^{-itE/\hbar}|\psi_0\rangle. \tag{1.44}$$

Stationary systems are the systems that have only stationary physical states. Stationary systems can therefore only be prepared in states that are mixtures of eigenstates of the energy operator, i.e., in states that fulfill (1.42). And pure physical states of stationary systems must always be represented by energy eigenvectors. In the preceding chapters all physical systems considered were stationary physical systems, and we have now found an "explanation" for the fact that these systems always appeared in eigenstates of the energy operator.

To what extent a given physical system may be considered stationary depends upon the desired accuracy of the description. If one ignores details, atoms and molecules are stationary physical systems. If a measurement process has prepared them to be in a particular state, which has to be an energy eigenstate, then they will for "all time" remain in that state. "All time" means fairly long on atomic scales: 10^{-2} sec (for infrared transitions) to 10^{-9} sec (for ultraviolet transitions) for the excited states of atoms and molecules. If one wants to describe not only energy levels but also details of the decay process of excited states, then one cannot consider the atom as a stationary system. This will be discussed in later chapters.

The Heisenberg picture and the Schrödinger picture are the limiting cases of a more general picture, which is called the "Dirac picture" or the "interaction picture." One writes the Hamiltonian operator H as the sum of two parts,

$$H = H_0 + H_1. \tag{1.45}$$

How this splitting is performed, i.e., how one makes the choice as to which part of H is to be H_0 and which part is to be H_1, depends upon the particular situation. In principle any part of H can be called H_0 and the rest then called H_1. In practice one chooses for H_0 an operator whose eigenvalues are known or can easily be obtained. Often H_0 will represent the energy of two or more combined systems when the interaction between the systems is ignored; H_1 then represents the interaction energy.

Assuming that H is not explicitly time-dependent (a conservative system), we define, in accordance with (1.16),

$$U(t) = e^{itH/\hbar}. \tag{1.46}$$

The expectation value of an observable A at time t is then given by

$$\langle A \rangle_t = \text{Tr}(U(t)AU^\dagger(t)W_0), \tag{1.47}$$

where A represents the observable at some initial time $t_0 = 0$, and where W_0 is the state of the system at that initial time. If we define operators

$$U_0(t) \equiv e^{itH_0/\hbar} \tag{1.48}$$

and

$$U_1(t) \equiv U(t)U_0^\dagger(t) = U(t)e^{-itH_0\hbar}, \tag{1.49}$$

then (1.47) may be rewritten as

$$\begin{aligned}
\langle A \rangle_t &= \text{Tr}(U_1(t)U_0(t)AU_0^\dagger(t)U_1^\dagger(t)W_0) \\
&= \text{Tr}(U_0(t)AU_0^\dagger(t)U_1^\dagger(t)W_0 U_1(t)) \\
&= \text{Tr}(A^D(t)W^D(t)),
\end{aligned} \tag{1.50}$$

where

$$A^D(t) \equiv U_0(t)AU_0^\dagger(t), \tag{1.51}$$

and where

$$W^D(t) \equiv U_1^\dagger(t)W_0 U_1(t). \tag{1.52}$$

Note the similarity between the time development (1.22) of A in the Heisenberg picture and the time development (1.51) in the Dirac picture. In the former the evolution operator $U(t) = e^{itH/\hbar}$ is generated by the full Hamiltonian H, while in the latter $U_0(t) = e^{itH_0/\hbar}$ is generated by only part of the Hamiltonian, namely H_0. In the extreme case that $H_0 = H$ and $H_1 = 0$ we have $U_0(t) = U(t)$ and $U_1(t) = I$, so that (1.50), (1.51), and (1.52) reduce to the Heisenberg picture. Alternatively, there is a similarity between the time development (1.15) of the statistical operator W in the Schrödinger picture and the time development (1.52) in the Dirac picture. This similarity is even more pronounced if

$$[H_0, H_1] = 0, \tag{1.53}$$

for then (1.49) becomes

$$U_1(t) = e^{itH/\hbar}e^{-itH_0/\hbar} = e^{itH_1/\hbar}. \tag{1.54}$$

In the case (1.15), $U(t) = e^{itH/\hbar}$ is then generated by the full Hamiltonian H, while $U_1(t) = e^{itH_1/\hbar}$ is generated by the interaction part of the Hamiltonian, namely H_1. In this extreme case $H_0 = 0$ and $H_1 = H$, we have $U_0(t) = I$ and $U_1(t) = U(t)$, and we thereby regain the Schrödinger picture.

Suppose W_0 describes a pure state,

$$W_0 = |\psi_0\rangle\langle\psi_0|; \tag{1.55}$$

then we can define a vector

$$|\psi(t)\rangle_D \equiv U_1^\dagger(t)|\psi_0\rangle \tag{1.56}$$

such that, according to (1.52), $W^D(t)$ is expressed as

$$W^D(t) = |\psi(t)\rangle_{DD}\langle\psi(t)|. \tag{1.57}$$

Differentiating (1.56), one obtains for the most common case that (1.53) holds:

$$ih\frac{d}{dt}|\psi(t)\rangle_D = H_1|\psi(t)\rangle_D. \tag{1.58}$$

Thus the vectors representing the state in the Dirac picture obey the "Schrödinger" equation with the interaction Hamiltonian H_1.

In the interaction picture the time dependence of the physically measurable quantity $\langle A\rangle_t$ is described by letting both observable and state vary with time. The observables are rotated in one direction with a transformation generated by part of the Hamiltonian, and the states are rotated in the opposite direction by a transformation generated by the other part (the interaction part) of the Hamiltonian. Expressed in terms of vectors: The eigenvectors $|a\,t\rangle_D$ of the observable A,

$$A^D(t)|a\,t\rangle_D = a(t)|a\,t\rangle_D,$$

are given by a rotation of $|a\,0\rangle_D$ by $e^{itH_0/\hbar}$,

$$|a\,t\rangle_D = e^{itH_0/\hbar}|a\,0\rangle_D;$$

one obtains the state vectors $|\psi(t)\rangle_D$ by rotation of $|\psi_0\rangle$ in the opposite direction by $U_1^\dagger(t)$ $(= e^{-itH_1/\hbar}$ if $[H_0, H_1] = 0)$:

$$|\psi(t)\rangle_D = U_1^\dagger(t)|\psi_0\rangle.$$

The time dependence of the probabilities is the same as in the case of the Heisenberg and Schrödinger pictures:

$$|\langle a\, t|\psi_0\rangle|^2 = |\langle a\, 0|U^{\dagger}(t)|\psi_0\rangle|^2$$
$$= |\langle a\, 0|U_0^{\dagger}(t)U_1^{\dagger}(t)|\psi_0\rangle|^2$$
$$= |{}_D\langle a\, t|\psi(t)\rangle_D|^2.$$

Isolated systems are an idealization, because a physical system cannot be completely isolated from the rest of the world. However, in many cases these idealizations give a sufficiently good approximation to the physical situation. If the isolation between a physical system and some part of the rest of the world is not negligible, one can enlarge the physical system by this part of the rest of the world and then consider the combined system. If that is not sufficiently isolated, one can continue to enlarge the physical system and will ultimately end up with an isolated system to which the previous considerations about time development apply. In many situations, however, this procedure is not practical. The enlarged system might be too complicated to be described by a sufficiently simple mathematical structure (if such a structure exists at all). In such a case it is useful to consider nonconservative systems, in particular systems with time-dependent external forces.

For such systems the physical structure changes with time, and so does the mathematical structure that describes it. The observables, in particular the Hamiltonian, will depend explicitly upon time; and thus the dynamical laws will change with time. For such systems it is not possible to give the time evolution by a unitary operator as in (1.15) and (1.16), or as in (1.22) and (1.23). But one can still try to express it in a differential form in analogy to (1.24) or (1.33). The Hamiltonian operator, which is no longer the energy operator of the system, depends explicitly upon time. In this case the time dependence may be formulated in the following form, which is suggested from correspondence to classical mechanics:

$$\frac{dA(t)}{dt} = \frac{1}{i\hbar}[A(t), H(t)] + \frac{\partial A(t)}{\partial t}, \tag{1.59}$$

where $\partial A(t)/\partial t$ is the derivative of $A(t)$ with respect to its explicit time dependence. If one assumes that every observable A is a function $A = A(Q_i(t), P_j(t), t)$ of $Q_i(t)$ and $P_j(t)$ with an additional explicit dependence upon t, then (1.59) is obtained from the corresponding equation in non-relativistic classical mechanics by replacing the Poisson bracket by $1/i\hbar$ times the commutator. Equation (1.59) is a generalization of (1.24) and, in the case of an isolated physical system, goes over to (1.24), which corresponds to the classical expression for conservative systems. For systems that have quantum-mechanical observables that cannot be expressed as functions of $Q_i(t)$ and $P_j(t)$, it is not clear whether a Hamiltonian operator can be found such that the time development can be expressed by (1.59). Equation (1.59) can no longer be integrated into a form analogous to (1.22); even for the case that A does not depend explicitly upon time $[\partial A(t)/\partial t = 0]$, there does not exist a general integration theory for the solution of

$$\frac{dA(t)}{dt} = \frac{1}{i\hbar} [A(t), H(t)]. \tag{1.60}$$

XII.A Mathematical Appendix: Definitions and Properties of Operators that Depend upon a Parameter

Let $\mathfrak{S}$ be a linear space in which a limiting process is defined; such a space is called a *linear topological space*. For example, in the *Hilbert space* $\mathfrak{S} = \mathcal{H}$ a limiting process is defined in the following way: A sequence $\phi_1, \phi_2, \phi_3, \ldots$ is said to converge to $\phi \in \mathcal{H}$, denoted

$$\phi_n \xrightarrow{\mathcal{H}} \phi,$$

iff

$$\|\phi_n - \phi\| \to 0 \quad \text{as} \quad n \to \infty. \tag{A.1}$$

The Hilbert space $\mathcal{H}$ is not the only linear topological space; if one has a linear space, one can equip it with all kinds of topologies, i.e., one can give all kinds of meanings to the convergence of a sequence. For example, instead of one norm—as in the Hilbert space—one can define in a linear space $\mathfrak{S}$ a countable sequence of norms $\| \ \|_p$,

$$\|\phi\|_0 \le \|\phi\|_1 \le \|\phi\|_2 \le \cdots \quad \text{for all} \quad \phi \in \mathfrak{S},$$

and define convergence in the following way: A sequence of elements $\{\phi_n\}$ converges to ϕ with respect to the topology given by the countable sequence of norms, denoted

$$\phi_n \xrightarrow{\mathfrak{S}} \phi,$$

iff for every $p = 0, 1, 2, \ldots$

$$\|\phi_n - \phi\|_p \to 0 \quad \text{as} \quad n \to \infty. \tag{A.2}$$

Such a linear topological space $\mathfrak{S}$ is called a *countably normed space*. In quantum mechanics one usually uses the definition of convergence (A.1), i.e., one uses the Hilbert space rather than any other linear topological space. No physical justification for this choice (made by J. von Neumann) exists.

The notions of "continuity," "differentiability," etc. are defined with respect to a certain topology or meaning of convergence. Hence continuity in a Hilbert space has a different meaning from continuity in a countably normed space or continuity in any space with another kind of topology. For a given linear topological space $\mathfrak{S}$, then, we say that an operator function $A(t)$ (i.e., an operator which depends upon a parameter t) is *continuous at* $t = t_0$ iff for any $\phi \in \mathfrak{S}$

$$A(t)\phi \xrightarrow{\mathfrak{S}} A(t_0)\phi \quad \text{as} \quad t \to t_0. \tag{A.3}$$

$A(t)$ is said to be *continuous* iff (A.3) holds for all values of t_0. If one assumes that in quantum mechanics the space of physical states is the Hilbert space, then the vague statement made about the continuity of the time-evolution operator $U(t)$ has the precise meaning given by (A.3), where

$$\text{``} \overset{\mathfrak{S}}{\to} \text{''} \quad \text{means} \quad \text{``} \overset{\mathcal{H}}{\to} \text{''},$$

i.e., $U(t)$ has the property

$$\| U(t)\phi - U(t_0)\phi \| \to 0 \quad \text{as} \quad t \to t_0. \tag{A.4}$$

As mentioned in several places, in particular in Section II.8, there is no physical reason for which the Hilbert-space convergence is preferred over any other definition of convergence, and therefore there is also no reason to give the vague physical meaning of continuity the precise mathematical definition (A.4). One could as well have chosen any of the other topologies, i.e., used the definition (A.3) with any other mathematical definition of convergence in $\mathfrak{S}$. In the numerous derivations of the Hilbert space for quantum mechanics, it is usually at this point (i.e., the choice of (A.4) for the vague physical meaning of continuity) that the Hilbert space is sneaked in.

An operator function $A(t)$ is called *differentiable at* $t = t_0$ (with respect to the topology on $\mathfrak{S}$) *in a subspace* $\mathscr{D} \subseteq \mathfrak{S}$ if for any $\phi \in \mathscr{D}$

$$\frac{A(t_0 + \Delta t) - A(t_0)}{\Delta t} \phi \overset{\mathfrak{S}}{\to} \psi \in \mathfrak{S} \quad \text{as} \quad \Delta t \to 0, \tag{A.5}$$

i.e., if the limit of $((A(t_0 + \Delta t) - A(t_0))/\Delta t)\phi$ exists in $\mathfrak{S}$. The vector ψ defines a linear operator which we shall denote by $dA(t)/dt |_{t=t_0}$,

$$\psi = \frac{dA(t)}{dt}\bigg|_{t=t_0} \phi,$$

and which is called the *derivative* (with respect to the topology on $\mathfrak{S}$) *of* $A(t)$ *at* $t = t_0$. Thus the derivative $dA(t)/dt|_{t=t_0}$ is defined by

$$\frac{dA(t)}{dt}\bigg|_{t=t_0} \phi = \lim_{\Delta t \to 0}{}^{(\mathfrak{S})} \frac{A(t_0 + \Delta t) - A(t_0)}{\Delta t} \phi. \tag{A.6}$$

If one chooses the Hilbert-space convergence, then one writes

$$\frac{dA(t)}{dt}\bigg|_{t=t_0} \phi = \lim_{\Delta t \to 0}{}^{(\mathcal{H})} \frac{A(t_0 + \Delta t) - A(t_0)}{\Delta t} \phi. \tag{A.7}$$

The action of the derivative $dA(t)/dt|_{t=t_0}$ is defined on all vectors $\phi \in \mathscr{D} \subseteq \mathfrak{S}$, i.e., on all vectors for which the limit (A.5) exists.

An operator function $A(t)$ that satisfies

$$A(t_1 + t_2) = A(t_1)A(t_2) \tag{A.8}$$

and is continuous with respect to the Hilbert-space topology is called a *one-parameter group of operators* with respect to the Hilbert-space topology.

For such a one-parameter group of operators the following properties can be proved:

1. $A(t)$ is differentiable on a "dense" subspace[3] $\mathscr{D} \subseteq \mathscr{H}$. This justifies (1.9), not for all vectors, but for all vectors in $\mathscr{D}$.
2. On $\mathscr{D}$, $A(t)$ can be differentiated an arbitrary number of times.
3. The sequence $\{\phi_n\}$, where

$$\phi_n = \sum_{k=0}^{n} \frac{t^k}{k!} \frac{d^k A(t)}{dt^k}\bigg|_{t=0} \phi,$$

converges (with respect to the Hilbert-space topology) for all elements ϕ of a dense subspace $\tilde{\mathscr{D}} \subseteq \mathscr{D} \subseteq \mathscr{H}$. This justifies (1.10).

Problems

1. H, a, and $a^\dagger$ are the operators as defined by (II.3.1) and (II.2.1b). Using the axiom of time development (V), calculate da/dt and $da^\dagger/dt$.

2. The Hamiltonian H for a harmonic oscillator subject to an external force $K(t)$ is given by

$$H = \frac{P^2}{2m} + \frac{m\omega^2}{2} Q^2 - QK(t) = H_0 - QK(t).$$

Here H_0, not the Hamiltonian H, is the total energy.
   (a) Express H in terms of the operators a, $a^\dagger$ and the identity operator.
   (b) Using the axiom on time development, calculate da/dt.
   (c) Solve the differential equation obtained in part (b) and find the explicit expression for $a(t)$.

3. For the harmonic oscillator under the external force $K(t)$ in Problem 2, the orthonormal eigenvectors ψ_n of $N = a^\dagger a$ can be constructed in a manner analogous to that used in the force-free case. One finds

$$\psi_n(t) = \frac{1}{\sqrt{n!}} [a^\dagger(t)]^n \psi_0(t),$$

where $\psi_0(t)$ represents the ground state. At the time $t = 0$, the energy of the harmonic oscillator is measured to be $\hbar\omega/2$. Denote the ground state at $t = 0$ by $\phi_0 = \psi_0(t = 0)$.
   (a) What is the statistical operator for the harmonic oscillator after the energy measurement?
   (b) Show that $w_n = |(\psi_n, \phi_0)|^2$ is the probability for measuring a value of the energy $E = \hbar\omega(n + \frac{1}{2})$ at some later time t.
   (c) Show

$$w_n = \frac{|F(t)|^{2n}}{n!} e^{-|F(t)|^2},$$

---

[3] A subspace $\mathscr{D}$ is called dense if it differs from the space $\mathscr{H}$ only by limit elements. Whereas every infinite sequence with the property $\phi_m - \phi_n \to 0$ for $n, m \to \infty$ (Cauchy sequence) has a limit element ϕ in $\mathscr{H}$: $\phi_n \to \phi \in \mathscr{H}$, ($\mathscr{H}$ is complete); such a ϕ will in general not exist in $\mathscr{D}$.

where

$$F(t) = \frac{i}{\sqrt{2m\hbar\omega}} \int_0^t K(\tau)e^{i\omega\tau}\, d\tau.$$

(d)   Show that the expectation value for the position operator immediately after the energy measurement at $t = 0$ is

$$\langle Q \rangle = \frac{1}{m\omega} \int_0^t K(\tau) \sin \omega(t - \tau)\, d\tau.$$

4.   The harmonic oscillator under the influence of an external force

$$K(t) = \begin{cases} K(t) & \text{if } t > 0, \\ 0 & \text{if } t < 0 \end{cases}$$

is at the time $t = 0$ in a state in which the energy is $\hbar\omega/2$. The property that the energy at a time t is equal to or smaller than $\hbar\omega(m + \frac{1}{2})$ is characterized by the operator $\Lambda = \sum_{n=0}^{m} \Lambda_{\psi_n(t)}$, where $\Lambda_{\psi_n(t)}$ are the projection operators on the one-dimensional subspaces spanned by the $\psi_n(t)$.

(a)   What is the probability for finding the energy to be equal to or smaller than $\hbar\omega(m + \frac{1}{2})$?

(b)   What is the statistical operator W' after the measurement of Λ with a positive result?

(c)   Show that

$$W'\psi_{n'} = \begin{cases} = 0 \quad \text{for } n' > n, \\[2mm] = \dfrac{1}{\sum_{n=0}^{m} 1/n! \,|F(t)|^{2n}} \cdot \displaystyle\sum_{n=0}^{m} |\psi_n(t)\rangle e^{i(n'-n)\omega t} \\[4mm] \quad \times \dfrac{1}{\sqrt{n!n'!}} [F(t)]^n [F^*(t)]^{n'} \quad \text{for } n' \le n. \end{cases}$$

CHAPTER XIII

Some Fundamental Properties of Quantum Mechanics

This chapter illustrates some of the characteristic features by which quantum mechanics differs from classical theories. In Section XIII.1, using a *gedanken* experiment with the Stern–Gerlach apparatus it is shown that a polarized beam (a pure state) cannot be split by the magnetic field, and that the splitting of such a beam is a consequence of the measurement. In Section XIII.2 we derive the predictions of quantum mechanics for spin correlation measurements in the singlet state of pairs of spin-$\frac{1}{2}$ particles. In Section XIII.3 these predictions are confronted with Bell's inequalities, which follow from a hypothesis first proposed by Einstein, Podolsky, and Rosen. We conclude with a short discussion of hidden-variables theories.

XIII.1 Change of the State by the Dynamical Law and by the Measuring Process—The Stern–Gerlach Experiment

In the preceding chapter we have formulated how the state of a system changes in time or how an observable changes in time as a consequence of the dynamical law. This change is totally deterministic. This means that once the observable (or the state) is known at the particular time, it can be exactly predicted for all other times. In Section II.4 we have described another kind of change in time, the change of the state by a measurement. These measurement processes are not deterministic and for their outcome one can only make probabilistic predictions. Thus there are two completely different processes of time change. In this chapter we shall illustrate these

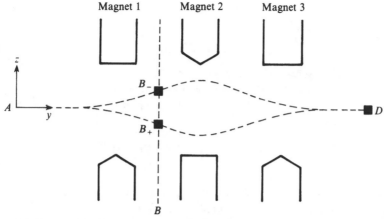

Figure 1.1 Stern–Gerlach experiment. Magnet 1 is used in a simple beam splitting experiment, Magnet 2 and Magnet 3 are added in a *gedanken* experiment to reverse the splitting of the beam.

processes by considering in detail a *gedanken* experiment with the Stern–Gerlach apparatus, which was briefly described in Section IX.1.

In the Stern–Gerlach experiment (Figure 1.1), a beam of hydrogen atoms in the ground state passes through a strongly inhomogeneous magnetic field **B**. Under the conditions of this experiment the hydrogen atoms may be considered to be the combination of two elementary physical systems (cf. the basic postulate IVa of Section III.5): Physical system I is the elementary rotator with angular momentum $\frac{1}{2}$, and describes the spinning electron; physical system II is the elementary particle that describes the motion of the structureless hydrogen atom in the experimental setup.

The space of physical states of system I is the two-dimensional spin space $\mathscr{H}_I = \imath^s = \mathscr{R}^{1/2}$ (cf. Section III.3). As a basis for $\mathscr{H}_I$ we choose the eigenvectors $|+\rangle$ and $|-\rangle$ of the spin component S_3 in the direction of the magnetic field **B**, with[1]

$$S_3|+\rangle = +\tfrac{1}{2}|+\rangle \quad \text{and} \quad S_3|-\rangle = -\tfrac{1}{2}|-\rangle. \tag{1.1}$$

The magnetic moment of this system is [Equation (IX.3.18)]

$$\mathbf{M} = -2\frac{e}{2m_e c}\mathbf{S} = -2\mu_B \mathbf{S}. \tag{1.2}$$

The energy operator of system I is a constant; because of our freedom of choice in defining the zero of energy, we may set this constant equal to zero: $H_1 = 0$.

The space of physical states of system II is the space $\mathscr{H}_{II}$ spanned by the generalized eigenvectors $|\mathbf{p}\rangle$ of the momentum operator **P** or by the generalized eigenvectors $|\mathbf{x}\rangle$ of the position operator **Q**; **Q** and **P** are the position

---

[1] We are again using units where $\hbar \equiv 1$.

and momentum of the hydrogen atom considered as a whole. The energy operator for system II is just the kinetic-energy operator

$$H_{\text{II}} = \mathbf{P}^2/2M, \tag{1.3}$$

where M is the mass of the hydrogen atom.

System I and system II are coupled by the external magnetic field $\mathbf{B}$; according to Equations (IX.3.1) and (IX.3.18) the interaction energy operator that represents this coupling is

$$H_{\text{int}} = -\mathbf{M} \cdot \mathbf{B} = -(\mathbf{M} \otimes I) \cdot (I \otimes \mathbf{B}) = 2\mu_B S_3 B_3(\mathbf{Q}) = 2\mu_B S_3 B_3(Q_3). \tag{1.4}$$

In (1.4) we have approximated the magnetic field $\mathbf{B}$ as having only a z-component and as being a function of $z = x_3$ alone.[2] Consequently the energy operator of the physical combination of these two systems, i.e., of the hydrogen atom in the ground state with electron spin in the magnetic field, is

$$H = H_{\text{I}} + H_{\text{II}} + H_{\text{int}} = \frac{1}{2M} \mathbf{P}^2 + 2\mu_B \mathbf{S} \cdot \mathbf{B}(\mathbf{Q})$$

$$= \frac{1}{2M} \mathbf{P}^2 + 2\mu_B S_3 B_3(Q_3). \tag{1.5}$$

In choosing this combination of I and II for the description of the hydrogen beam in the magnetic field we have neglected (1) the magnetic moment of the proton, (2) the influence of the magnetic field $\mathbf{B}$ upon the electron-proton interaction in the hydrogen atom, and (3) the difference between the electron position and the position of the hydrogen atom as a whole. The justification of (1) and (2) may be seen by an examination of the orders of magnitude involved: (1) The magnetic moment of the proton is [cf. (IX.3.18p)] $\mathbf{M}_p = (2 + 3.59)(e/2m_p c)\mathbf{S} = 5.59(m_e/m_p)\mu_B \mathbf{S}$. Since the mass of the proton is much greater than that of the electron ($m_p = 1836m_e$), it follows that the magnetic moment of the proton is almost three orders of magnitude smaller than the magnetic moment of the electron; the interaction energy of the proton magnetic moment in the magnetic field is therefore negligible when compared to (1.4). (2) The Bohr magneton $\mu_B = 5.795 \times 10^{-9}$ eV/gauss is sufficiently small that even in a fairly strong field of 10^3 gauss the magnetic interaction energy $\mu_B B$ is six orders of magnitude smaller than the difference between the lower energy levels of the hydrogen atom. Thus the internal states of the hydrogen atoms do not change significantly and the operator for the internal energy of the hydrogen atom [essentially the energy operator (VI.3.2) of the Kepler problem] may be considered a constant for the present problem, which we set equal to zero. Instead of the above described space $\mathcal{H}_{\text{II}}$, the space of states of system II should be $\mathcal{H}'_{\text{II}} = \mathcal{H}_{\text{II}} \otimes \mathcal{R}$, where $\mathcal{R}$ is the space of states for the Kepler problem [cf. (VI.5.2)]. But as we are keeping the hydrogen atom in the ground state [i.e., in the one-dimensional subspace

---

[2] The Maxwell equation $\nabla \cdot \mathbf{B} = \partial B_i / \partial x_i = 0$ tells us that the variation of B_1 with x is just as great as the variation of B_3 with z. We shall return to the subject of this approximation in the appendix to this section.

$\mathcal{R}(n = 1) \subset \mathcal{R}$], the space of states for system II is in fact $\mathcal{H}_{II} \otimes \mathcal{R}\,(n = 1)$, i.e., is $\mathcal{H}_{II}$.

The experimental arrangement is such that at $t = 0$ a pulse of hydrogen atoms enters the region of the magnetic field at the point A (cf. Figure 1.1) and moves with an average momentum $\tilde{p}$ in the y direction. Idealizing, we assume that in this experimental arrangement a pure state of system II has been prepared. This state is described by the state vector (statistical operator)

$$|\psi_0\rangle \quad (W_{II} = |\psi_0\rangle\langle\psi_0|) \tag{1.6}$$

in the Heisenberg picture for all times $t \geq 0$, or by

$$|\psi_0\rangle \quad (W_{II}(0) = |\psi_0\rangle\langle\psi_0|) \tag{1.7}$$

in the Schrödinger picture at time $t = 0$. System I may be in a pure state with statistical operator

$$W_I = |\phi\rangle\langle\phi| \quad \text{where} \quad |\phi\rangle = \alpha|+\rangle + \beta|-\rangle \quad (\alpha, \beta \in \mathbb{C}) \tag{1.8}$$

if the beam has been polarized before it reaches point A. (The state $|\phi\rangle$ will be an eigenstate of $\hat{\mathbf{n}} \cdot \mathbf{S}$ for some direction $\hat{\mathbf{n}}$ determined by the choice of α and β.) Alternately the system I state may be a mixture, e.g.,

$$W_I = \tfrac{1}{2}(|+\rangle\langle+| \quad + \quad |-\rangle\langle-|) = \tfrac{1}{2}I, \tag{1.9}$$

where I here is the identity operator in the spin space $\mathcal{H}_I = \mathbb{C}^s$. The state of the combined physical system is described by

$$W = W_I \otimes W_{II} \quad \text{or} \quad |\chi\rangle = |\phi\rangle \otimes |\psi_0\rangle, \tag{1.10}$$

the second description in terms of the state vector $|\chi\rangle$ being an alternative to the description in terms of the statistical operator W in the case that system I is in a pure state.[3]

In considering the time development of the pulse of hydrogen atoms we shall use both the Heisenberg and the Schrödinger pictures. By use of (XII.1.59), (1.5), and the canonical commutation relations, we see that the Heisenberg equations of motion for the position and momentum operators are

$$\frac{d\mathbf{Q}(t)}{dt} = \frac{1}{i}[\mathbf{Q}(t), H(t)] = \frac{1}{M}\mathbf{P}(t), \tag{1.11}$$

$$\frac{d\mathbf{P}(t)}{dt} = \frac{1}{i}[\mathbf{P}(t), H(t)] = \frac{1}{i}2\mu_B S_3[\mathbf{P}(t), B_3(Q_3(t))] = \left(0, 0, -2\mu_B S_3 \frac{\partial B_3}{\partial Q_3(t)}\right). \tag{1.12}$$

[3] Remark on the state of a combination of physical systems: Recall that the operators in the direct-product space are given by Equation (III.5.7). A general statistical operator in $\mathcal{H}_I \otimes \mathcal{H}_{II}$ is therefore not given simply by the direct product of statistical operators in $\mathcal{H}_I$ and in $\mathcal{H}_{II}$. However, if either of the states W_I or W_{II} is a pure state, then W_I and W_{II} uniquely determine the state W of the combined system as the product $W = W_I \otimes W_{II}$. This is the situation in the problem under investigation. W is also given by $W = W_I \otimes W_{II}$ if the state of the combined system is determined by measurements that have been performed upon systems I and II separately. If, however, measurements are performed that measure correlated properties of systems I and II, then in general W is not factorable into the form $W_I \otimes W_{II}$.

[The meaning of $\partial/\partial Q_i$ was given in the paragraph following Equation (XII.1.29).] Let us first investigate the case that system I is in the spin state $|+\rangle$ or $|-\rangle$, so that the state of the combined system is

$$|\chi \pm\rangle \equiv |\pm\rangle \otimes |\psi_0\rangle. \tag{1.13}$$

The motion of the expectation value of the momentum operator in these states is then given by

$$\frac{d}{dt}\langle\chi\pm|\mathbf{P}(t)|\chi\pm\rangle = \left(0, 0, -2\mu_B\langle\pm|S_3|\pm\rangle\langle\psi_0|\frac{\partial B_3}{\partial Q_3(t)}|\psi_0\rangle\right)$$

$$= \left(0, 0, \mp\mu_B\langle\psi_0|\frac{\partial B_3}{\partial Q_3(t)}|\psi_0\rangle\right). \tag{1.14}$$

Thus if the state has spin up, $s_3 = +\frac{1}{2}$ (spin down, $s_3 = -\frac{1}{2}$), i.e., if the state is $|+\rangle$ ($|-\rangle$), then the time development of the expectation value of the momentum's z-component fulfills

$$\frac{d}{dt}\langle\chi\pm|P_3(t)|\chi\pm\rangle = \mp\mu_B\langle\psi_0|\frac{\partial B_3}{\partial Q_3(t)}|\psi_0\rangle$$

$$= \mp\mu_B\int d^3x\,\frac{\partial B_3}{\partial x_3}\langle\psi_0|\mathbf{x}\,t\rangle\langle\mathbf{x}\,t|\psi_0\rangle, \quad (1.15\pm)$$

where the $|\mathbf{x}\,t\rangle$ are the time-dependent eigenvectors of $\mathbf{Q}$ in the Heisenberg picture. [The upper (lower) sign in $(1.15\pm)$ refers to the spin-up (spin-down) case.] The right-hand side of $(1.15\pm)$ shows that the change of the momentum depends upon the particular properties of the magnetic field; let us assume for simplicity that

$$\frac{\partial B_3}{\partial x_3} = \text{const} > 0. \tag{1.16}$$

Then

$$\frac{d}{dt}\langle\chi\pm|P_3(t)|\chi\pm\rangle = \mp\kappa\langle\psi_0|\psi_0\rangle = \mp\kappa, \tag{1.17\pm}$$

where

$$\kappa = \mu_B\frac{\partial B_3}{\partial x_3}. \tag{1.18}$$

Using the initial condition that the particle has momentum

$$\langle\chi\pm|\mathbf{P}(t=0)|\chi\pm\rangle = (0, \tilde{p}, 0) \quad \text{at} \quad t = 0,$$

we may integrate $(1.17\pm)$; the expectation value of the momentum operator at time t is then

$$\langle\chi\pm|\mathbf{P}(t)|\chi\pm\rangle = (0, \tilde{p}, \mp\kappa t). \tag{1.19\pm}$$

If we take the expectation value between $|\chi \pm\rangle$ of (1.11) and use the initial condition $\langle \chi \pm | Q(t = 0) | \chi \pm \rangle = 0$ together with (1.19±), then the expectation value of the position operator is

$$\langle \chi \pm | Q(t) | \chi \pm \rangle = \frac{1}{M} (0, \tilde{p}t, \mp \kappa t^2/2). \qquad (1.20\pm)$$

Thus the expectation value of the position operator, i.e., the position of the pulse of hydrogen atoms, moves on a parabolic orbit that bends downward (upward) in the case of a pure spin-up (spin-down) initial state, as illustrated in Figure 1.1. [This presumes that $\kappa > 0$, i.e., presumes (1.16).]

Let us now describe the same situation in the Schrödinger picture, where the state changes in time. At the time $t = 0$ the state is given by

$$|\chi \pm (t = 0)\rangle = |\chi \pm\rangle = |\pm\rangle \otimes |\psi_0\rangle, \qquad (1.21\pm)$$

which describes the pulse of hydrogen atoms with spin up (spin down) at the position A. According to (XII.1.35), the time development of this state is

$$|\chi \pm (t)\rangle = U^\dagger(t)(|\pm\rangle \otimes |\psi_0\rangle) = e^{-it(H_{II} + 2\mu_B S_3 B_3)}(|\pm\rangle \otimes |\psi_0\rangle)$$
$$= e^{-it(H_{II} \pm \mu_B B_3)}(|\pm\rangle \otimes |\psi_0\rangle) = |\pm\rangle \otimes e^{-it(H_{II} \pm \mu_B B_3)}|\psi_0\rangle. \qquad (1.22\pm)$$

Thus the initial pure state with spin up will move through the magnetic field according to

$$|\chi + (t)\rangle = |+\rangle \otimes e^{-it(H_{II} + \mu_B B_3)}|\psi_0\rangle = |+\rangle \otimes |\psi + (t)\rangle, \quad (1.23+)$$

in contrast to the motion

$$|\chi - (t)\rangle = |-\rangle \otimes e^{-it(H_{II} - \mu_B B_3)}|\psi_0\rangle = |-\rangle \otimes |\psi - (t)\rangle \quad (1.23-)$$

of the pure state with spin down. In (1.23±) we have defined

$$|\psi \pm (t)\rangle \equiv e^{-it(H_{II} \pm \mu_B B_3)}|\psi_0\rangle. \qquad (1.24\pm)$$

Now the relations connecting the momentum operator $\mathbf{P}$, the position operator $\mathbf{Q}$, and the state vectors $|\chi \pm (t)\rangle$ of the Schrödinger picture with $\mathbf{P}(t)$, $\mathbf{Q}(t)$, and $|\chi \pm\rangle$ of the Heisenberg picture are

$$\mathbf{P} = \mathbf{P}(t = 0) = U^\dagger(t)\mathbf{P}(t)U(t), \qquad (1.25a)$$

$$\mathbf{Q} = \mathbf{Q}(t = 0) = U^\dagger(t)\mathbf{Q}(t)U(t), \qquad (1.25b)$$

$$|\chi \pm (t)\rangle = U^\dagger(t)|\chi \pm (t = 0)\rangle = U^\dagger(t)|\chi \pm\rangle, \qquad (1.25c)$$

where $U(t) = e^{itH} = \exp(it(H_{II} + 2\mu_B S_3 B_3(Q_3)))$. We may write the Schrödinger-picture analogue to (1.14) by taking the expectation value of

XIII Some Fundamental Properties of Quantum Mechanics

(1.12) between the state vectors $|\chi \pm\rangle$ and shifting the time dependence from the observables to the state vectors:

$$\frac{d}{dt}\langle\chi \pm (t)|P_3|\chi \pm (t)\rangle = \frac{d}{dt}\langle\chi \pm |P_3(t)|\chi \pm\rangle$$

$$= \langle\chi \pm |\frac{dP_3(t)}{dt}|\chi \pm\rangle = \langle\chi \pm | - 2\mu_B S_3 \frac{\partial B_3}{\partial Q_3(t)}|\chi \pm\rangle$$

$$= -2\mu_B\langle\chi \pm |e^{it(H_{II} + 2\mu_B S_3 B_3)}S_3 \frac{\partial B_3}{\partial Q_3} e^{-it(H_{II} + 2\mu_B S_3 B_3)}|\chi \pm\rangle$$

$$= -2\mu_B\langle\chi \pm |e^{it(H_{II} \pm \mu_B B_3)}S_3 \frac{\partial B_3}{\partial Q_3} e^{-it(H_{II} \pm \mu_B B_3)}|\chi \pm\rangle$$

$$= -2\mu_B\langle\pm |S_3|\pm\rangle\langle\psi_0|e^{it(H_{II} \pm \mu_B B_3)} \frac{\partial B_3}{\partial Q_3} e^{-it(H_{II} \pm \mu_B B_3)}|\psi_0\rangle$$

$$= \mp\mu_B\langle\psi \pm (t)|\frac{\partial B_3}{\partial Q_3}|\psi \pm (t)\rangle.$$

(Note that $[S_3, H] = 0$ for the Hamiltonian H of (1.5); the matrix element $\langle\pm |S_3|\pm\rangle$ thus has no time dependence.) If we again assume (1.16), then

$$\langle\psi \pm (t)|\frac{\partial B_3}{\partial Q_3}|\psi \pm (t)\rangle = \int d^3x \frac{\partial B_3}{\partial x_3}\langle\psi \pm (t)|\mathbf{x}\rangle\langle\mathbf{x}|\psi \pm (t)\rangle$$

$$= \frac{\partial B_3}{\partial x_3}\langle\psi \pm (t)|\psi \pm (t)\rangle = \frac{\partial B_3}{\partial x_3}.$$

Consequently

$$\frac{d}{dt}\langle\chi \pm (t)|\mathbf{P}|\chi \pm (t)\rangle = \left(0, 0, \mp\mu_B\frac{\partial B_3}{\partial x_3}\right) = (0, 0, \mp\kappa). \qquad (1.26)$$

The expectation value

$$\frac{d}{dt}\langle\chi \pm |\mathbf{Q}(t)|\chi \pm\rangle = \frac{1}{M}\langle\mathbf{x} \pm |\mathbf{P}(t)|\chi \pm\rangle$$

of (1.11) easily transforms to the Schrödinger-picture analogue

$$\frac{d}{dt}\langle\chi \pm (t)|\mathbf{Q}|\chi \pm (t)\rangle = \frac{1}{M}\langle\chi \pm (t)|\mathbf{P}|\chi \pm (t)\rangle. \qquad (1.27)$$

Equations (1.26) and (1.27) together with the initial conditions

$$\langle\chi \pm (t = 0)|\mathbf{P}|\chi \pm (t = 0)\rangle = (0, \tilde{p}, 0)$$

and

$$\langle\chi \pm (t = 0)|\mathbf{Q}|\chi \pm (t = 0)\rangle = (0, 0, 0)$$

lead to the Schrödinger-picture analogue of (1.20±):

$$\mathbf{x}_\pm(t) \equiv \langle \chi \pm (t) | \mathbf{Q} | \chi \pm (t) \rangle = \frac{1}{M} \left(0, \tilde{p}t, \mp \frac{\kappa t^2}{2} \right). \qquad (1.28\pm)$$

As we would expect, the result is the same as in the Heisenberg picture.

The probability distribution for the position is the "expectation value" of the operator $I \otimes |\mathbf{x}\rangle\langle\mathbf{x}|$. More precisely, the probability of obtaining a value in the volume $(\mathbf{x} - \epsilon, \mathbf{x} + \epsilon)$ in a position measurement is the expectation value of the projection operator

$$I \otimes \Lambda(\mathbf{x}, \epsilon) = I \otimes \int_{\mathbf{x}-\epsilon}^{\mathbf{x}+\epsilon} d^3x' |\mathbf{x}'\rangle\langle\mathbf{x}'| \qquad (1.29)$$

[cf. Equation (II.8.8)]. Thus the position probability distribution in the state $|\chi + (t)\rangle = |+\rangle \otimes |\psi + (t)\rangle$ is given by

$$w_+(\mathbf{x}, t) = \langle +|+\rangle\langle\psi + (t)|\mathbf{x}\rangle\langle\mathbf{x}|\psi + (t)\rangle = \langle\psi + (t)|\mathbf{x}\rangle\langle\mathbf{x}|\psi + (t)\rangle. \qquad (1.30+)$$

In particular, the probability of obtaining a value in the volume $(\mathbf{x} - \epsilon, \mathbf{x} + \epsilon)$ at time t is

$$w_+(\mathbf{x} - \epsilon, \mathbf{x} + \epsilon, t) = \int_{\mathbf{x}-\epsilon}^{\mathbf{x}+\epsilon} d^3x' \langle\psi + (t)|\mathbf{x}'\rangle\langle\mathbf{x}'|\psi + (t)\rangle. \qquad (1.31+)$$

The probability distribution moves with time. The exact motion of $w_+(\mathbf{x}, t)$ may be obtained by solving the Schrödinger equation for $\langle\mathbf{x}|\psi + (t)\rangle$. From our considerations concerning the expectation value $(1.28+)$, however, we have a rough idea of how $w_+(\mathbf{x}, t)$ moves: At $t = 0$, $w_+(\mathbf{x} - \epsilon, \mathbf{x} + \epsilon, t)$ and $w_+(\mathbf{x}, t)$ will be essentially zero everywhere except in a small volume around the point $A = (0, 0, 0)$. From A the pulse moves along the lower parabola, given by $(1.28+)$; at the time $t_B = My_B/\tilde{p}$ (i.e., at the time t_B at which the pulse has the y-coordinate y_B), $w_+(\mathbf{x}, t)$ will be essentially zero everywhere except in a small volume around the point $B_+ = (0, \tilde{p}t_B/M, -\kappa t_B^2/2M)$.

Similarly, the probability distribution

$$w_-(\mathbf{x}, t) = \langle\psi - (t)|\mathbf{x}\rangle\langle\mathbf{x}|\psi - (t)\rangle \qquad (1.30-)$$

and the probability

$$w_-(\mathbf{x} - \epsilon, \mathbf{x} + \epsilon, t) = \int_{\mathbf{x}-\epsilon}^{\mathbf{x}+\epsilon} d^3x' \langle\psi - (t)|\mathbf{x}'\rangle\langle\mathbf{x}'|\psi - (t)\rangle \qquad (1.31-)$$

of finding the system in the volume $(\mathbf{x} - \epsilon, \mathbf{x} + \epsilon)$ are the ones for the state $|\chi - (t)\rangle = |-\rangle \otimes |\psi - (t)\rangle$, which was a pure state with spin down at A. Then $w_-(\mathbf{x}, t)$ and $w_-(\mathbf{x} - \epsilon, \mathbf{x} + \epsilon, t)$ move along the upper parabola $(1.28-)$, and will be significantly nonzero at time t_B only around the point $B_- = (0, \tilde{p}t_B/M, +\kappa t_B^2/2M)$.

These facts enable us to perform a measurement of the polarization—a measurement of the system-I observable S_3—by measuring the system-II observable $\mathbf{Q}$. If we put a detector plate at B (at a distance $y_B = \tilde{p}t_B/M$ from A), then it will record the presence of hydrogen atoms at $B_+ = (0, \tilde{p}t_B/M, -\kappa t_B^2/2M)$ or at $B_- = (0, \tilde{p}t_B/M, +\kappa t_B^2 2M)$ only if one started at $A = (0, 0, 0)$ with the pure spin-up state $|\chi + (t = 0)\rangle = |+\rangle \otimes |\psi_0\rangle$ or with the pure spin-down state $|\chi - (t = 0)\rangle = |-\rangle \otimes |\psi_0\rangle$, respectively. Thus a measurement of Q_3 on system II with result $z = z_+(t_B) = -\kappa t_B^2/2M$ constitutes a measurement of S_3 on system I with result $s_3 = +\frac{1}{2}$ and a measurement of Q_3 on system II with result $z = z_-(t_B) = +\kappa t_B^2/2M$ constitutes a measurement of S_3 on system I with result $s_3 = -\frac{1}{2}$.

We now investigate the case where the system-I state is the mixture (1.9), where the coefficients $\frac{1}{2}$ before each of the projectors $|+\rangle\langle+|$ and $|-\rangle\langle-|$ represent the equal *a priori* probabilities of the system being in the states $|+\rangle$ or $|-\rangle$. At $t = 0$ the beam is described by the statistical operator (1.10) with $W_I(t = 0)$ given by (1.9) and $W_{II}(t = 0)$ given by (1.7):

$$W(0) = W_I(0) \otimes W_{II}(0) = (\tfrac{1}{2}|+\rangle\langle+| + \tfrac{1}{2}|-\rangle\langle-|) \otimes (|\psi_0\rangle\langle\psi_0|)$$
$$= \tfrac{1}{2}|+\rangle \otimes |\psi_0\rangle\langle+| \otimes \langle\psi_0| + \tfrac{1}{2}|-\rangle \otimes |\psi_0\rangle\langle-| \otimes \langle\psi_0|$$
$$= \tfrac{1}{2}|\chi + (0)\rangle\langle\chi + (0)| + \tfrac{1}{2}|\chi - (0)\rangle\langle\chi - (0)|. \qquad (1.32)$$

The time development of W is

$$W(t) = \tfrac{1}{2}|+\rangle \otimes |\psi + (t)\rangle\langle+| \otimes \langle\psi + (t)|$$
$$+ \tfrac{1}{2}|-\rangle \otimes |\psi - (t)\rangle\langle-| \otimes \langle\psi - (t)|$$
$$= \tfrac{1}{2}|\chi + (t)\rangle\langle\chi + (t)| + \tfrac{1}{2}|\chi - (t)\rangle\langle\chi - (t)|. \qquad (1.33)$$

The probability distribution $w(\mathbf{x}, t) = \mathrm{Tr}((I \otimes |\mathbf{x}\rangle\langle\mathbf{x}|)W(t))$ for the position is then

$$w(\mathbf{x}, t) = \tfrac{1}{2}\langle\mathbf{x}|\psi + (t)\rangle\langle\psi + (t)|\mathbf{x}\rangle + \tfrac{1}{2}\langle\mathbf{x}|\psi - (t)\rangle\langle\psi - (t)|\mathbf{x}\rangle \quad (1.34)$$
$$= \tfrac{1}{2}w_+(\mathbf{x}, t) + \tfrac{1}{2}w_-(\mathbf{x}, t), \qquad (1.35)$$

where $w_+(\mathbf{x}, t)$ and $w_-(\mathbf{x}, t)$ are the probability distributions of $(1.30\pm)$, and their (significantly) nonzero parts move downward and upward along the parabolic trajectories $(1.28+)$ and $(1.28-)$, respectively. Thus *the unpolarized beam is split into two beams.*

This was in fact what Stern and Gerlach observed, in contradiction to the predictions of classical physics.[4] Classically an unpolarized beam contains atoms with magnetic moments equally distributed in all directions. The deviation would be proportional to the magnitude of the z-component of the magnetic moment, and hence one would expect the narrow beam to just smear out in the z-direction.

We can now add (at least in a *gedanken* experiment) second and third magnets, or another suitable arrangement of magnets, that reverse the splitting effect of the first magnet. The two beams are then brought together

---

[4] The original Stern–Gerlach experiment (1922) used silver atoms rather than hydrogen atoms.

again; and at D at some later time t_D we again have the same state as at A at time $t = 0$, i.e., a mixture described by (1.32). If this pulse is now passed through another Stern-Gerlach apparatus with the magnetic field in the x-direction, it will split into two beams. This is easily seen by representing W_1 as

$$W_1 = \tfrac{1}{2}I = \tfrac{1}{2}|s_1 = +\tfrac{1}{2}\rangle\langle s_1 = +\tfrac{1}{2}| + \tfrac{1}{2}|s_1 = -\tfrac{1}{2}\rangle\langle s_1 = -\tfrac{1}{2}|, \quad (1.36)$$

where $|s_1 = +\tfrac{1}{2}\rangle$ and $|s_1 = -\tfrac{1}{2}\rangle$ are the basis vectors of $\mathscr{H}_1$ satisfying

$$S_1|s_1 = \pm\tfrac{1}{2}\rangle = \pm\tfrac{1}{2}|s_1 = \pm\tfrac{1}{2}\rangle; \quad (1.37)$$

the above arguments are then repeated with the z-component exchanged with the x-component.

We will now analyze the Stern-Gerlach experiment for the case that the state at A is a pure state, polarized in any direction $\mathbf{n}$ other than the z-direction, e.g., the x-direction.[5] At time $t = 0$ the state is then described by the statistical operator

$$W(0) = |\chi(0)\rangle\langle\chi(0)| = |\phi\rangle\langle\phi| \otimes |\psi_0\rangle\langle\psi_0| \quad (1.38)$$

or by the vector

$$|\chi(0)\rangle = |\phi\rangle \otimes |\psi_0\rangle = \alpha|+\rangle \otimes |\psi_0\rangle + \beta|-\rangle \otimes |\psi_0\rangle, \quad (1.38')$$

where $|\phi\rangle$ is given by (1.8). This state develops in time according to (XII.1.35); using the above results (1.22$\pm$), (1.23$\pm$), and (1.24$\pm$) one obtains

$$|\chi(t)\rangle = \alpha|+\rangle \otimes |\psi + (t)\rangle + \beta|-\rangle \otimes |\psi - (t)\rangle. \quad (1.39)$$

Now in a unitary time development a pure state goes into a pure state. The system will therefore remain in a pure state while it passes through the magnetic field; *there is no splitting into two subensembles*. If the beam reaches D after having passed through the reverse splitting magnets, the pulse is again in the pure state with the same polarization that it had at A. For example, if we start at A with a beam polarized in the x-direction, it will again be polarized in the x-direction at D at the time t_D; if it is then sent through a subsequent Stern-Gerlach apparatus with the field in the x-direction, *it will not split*, but will just bend.

Suppose now that we perform a position measurement on this pure state at B, i.e., we place at B a measuring device (screen) that lets the beam pass through but that registers the presence of hydrogen atoms. Recall that the probability of finding the hydrogen atom in the volume $(\mathbf{x} - \boldsymbol{\epsilon}, \mathbf{x} + \boldsymbol{\epsilon})$ around $\mathbf{x}$ is given by the expectation value of the operator $I \otimes \Lambda(\mathbf{x}, \boldsymbol{\epsilon})$ of (1.29) with the statistical operator

$$
\begin{aligned}
W(t) = |\chi(t)\rangle\langle\chi(t)| = {} & |\alpha|^2|+\rangle \otimes |\psi + (t)\rangle\langle +| \otimes \langle\psi + (t)| \\
& + |\beta|^2|-\rangle \otimes |\psi - (t)\rangle\langle -| \otimes \langle\psi - (t)| \\
& + \alpha\bar{\beta}|+\rangle \otimes |\psi + (t)\rangle\langle -| \otimes \langle\psi - (t)| \\
& + \bar{\alpha}\beta|-\rangle \otimes |\psi - (t)\rangle\langle +| \otimes \langle\psi + (t)| \quad (1.40)
\end{aligned}
$$

---

[5] Use of the Pauli spin matrices easily gives $|s_1 = \pm\tfrac{1}{2}\rangle = (1/\sqrt{2})(|+\rangle \pm |-\rangle)$ (up to an arbitrary phase factor). For the Pauli spin matrices see Problem III.5.

corresponding to the pure state (1.39); thus

$$
\begin{aligned}
w_{|\chi(t)\rangle}(\mathbf{x} - \boldsymbol{\epsilon}, \mathbf{x} + \boldsymbol{\epsilon}) &= \mathrm{Tr}(W(t)(I \otimes \Lambda(\mathbf{x}, \epsilon))) \\
&= |\alpha|^2 \, \mathrm{Tr}_\mathrm{I}(|+\rangle\langle+|) \, \mathrm{Tr}_\mathrm{II}(|\psi + (t)\rangle\langle\psi + (t)|\Lambda(\mathbf{x}, \epsilon)) \\
&\quad + |\beta|^2 \, \mathrm{Tr}_\mathrm{I}(|-\rangle\langle-|) \, \mathrm{Tr}_\mathrm{II}(|\psi - (t)\rangle\langle\psi - (t)|\Lambda(\mathbf{x}, \epsilon)) \\
&= |\alpha|^2 w_+(\mathbf{x} - \boldsymbol{\epsilon}, \mathbf{x} + \boldsymbol{\epsilon}, t) + |\beta|^2 w_-(\mathbf{x} - \boldsymbol{\epsilon}, \mathbf{x} + \boldsymbol{\epsilon}, t),
\end{aligned}
$$

$$(1.41)$$

where $w_\pm(\mathbf{x} - \boldsymbol{\epsilon}, \mathbf{x} + \boldsymbol{\epsilon}, t)$ is given by (1.31 $\pm$). The probability distribution for the position, i.e., the "expectation value" of the "operator" $I \otimes |\mathbf{x}\rangle\langle\mathbf{x}|$, is obtained in the same manner, and is given by

$$
w_{|\chi(t)\rangle}(\mathbf{x}) = |\alpha|^2 w_+(\mathbf{x}, t) + |\beta|^2 w_-(\mathbf{x}, t). \tag{1.42}
$$

For the case that the hydrogen atoms were originally polarized in the x-direction (i.e., $|\alpha|^2 = \tfrac{1}{2}$ and $|\beta|^2 = \tfrac{1}{2}$), we have

$$
w_{|\chi(t)\rangle}(\mathbf{x}) = \tfrac{1}{2} w_+(\mathbf{x}, t) + \tfrac{1}{2} w_-(\mathbf{x}, t).
$$

Thus the position probability distribution for the pure state (1.8) is the same as the probability for a position measurement in the mixture with the system-I statistical operator

$$
W_\mathrm{I} = |\alpha|^2 |+\rangle\langle+| \;\; + \;\; |\beta|^2 |-\rangle\langle-|. \tag{1.43}
$$

From our discussion of the motions of $w_+(\mathbf{x}, t)$ and $w_-(\mathbf{x}, t)$ we conclude that the hydrogen atoms pass through the screen at B_+ and at B_- with relative probabilities $|\alpha|^2$ and $|\beta|^2$, respectively. Now Equations (1.40), (1.41), and (1.42) depend upon the assumption that the system is in the pure state $|\chi(t)\rangle$ given by (1.39). This assumption is valid only for $t < t_B$, for after the position has been measured at B the state is no longer given by $|\chi(t)\rangle$; it is changed to a new state by the measurement process. If the screen has openings only around B_+ and B_-, then, according to the basic postulate IIIb [Equation (II.4.51)], the state is given at time $t = t_B$ by

$$
\begin{aligned}
W^B(t_B) &= \Lambda(\mathbf{x}_{B+}, \epsilon)|\chi(t_B)\rangle\langle\chi(t_B)|\Lambda(\mathbf{x}_{B+}, \epsilon) \\
&\quad + \Lambda(\mathbf{x}_{B-}, \epsilon)|\chi(t_B)\rangle\langle\chi(t_B)|\Lambda(\mathbf{x}_{B-}, \epsilon).
\end{aligned} \tag{1.44}
$$

Now

$$
\Lambda(\mathbf{x}_{B\pm}, \epsilon)|\psi \pm (t_B)\rangle\langle\psi \pm (t_B)|\Lambda(\mathbf{x}_{B\pm}, \epsilon)
$$

$$
= \int_{\mathbf{x}_{B\pm} - \epsilon}^{\mathbf{x}_{B\pm} + \epsilon} d^3x' \, d^3x'' \, |\mathbf{x}'\rangle\langle\mathbf{x}'|\psi \pm (t_B)\rangle\langle\psi \pm (t_B)|\mathbf{x}''\rangle\langle\mathbf{x}''|
$$

Since $\langle\mathbf{x}'|\psi - (t_B)\rangle = 0$ for $\mathbf{x}'$ near $\mathbf{x}_{B+}$ and since $\langle\psi - (t_B)|\mathbf{x}''\rangle = 0$ for $\mathbf{x}''$ near $\mathbf{x}_{B+}$, it follows that

$$
\Lambda(\mathbf{x}_{B+}, \epsilon)|\psi - (t_B)\rangle\langle\psi - t_B|\Lambda(\mathbf{x}_{B+}, \epsilon) = 0.
$$

Similarly

$$\Lambda(\mathbf{x}_{B-}, \epsilon)|\psi + (t_B)\rangle\langle\psi + (t_B)|\Lambda(\mathbf{x}_{B-}, \epsilon) = 0.$$

Expansion of (1.44) by means of (1.39) and use of the last three equations allows us to rewrite (1.44) as

$$\begin{aligned} W^B(t_B) &= |\alpha|^2 |+\rangle\langle+| \otimes \Lambda(\mathbf{x}_{B+}, \epsilon)|\psi + (t_B)\rangle\langle\psi + (t_B)|\Lambda(\mathbf{x}_{B+}, \epsilon) \\ &+ |\beta|^2 |-\rangle\langle-| \otimes \Lambda(\mathbf{x}_{B-}, \epsilon)|\psi - (t_B)\rangle\langle\psi - (t_B)|\Lambda(\mathbf{x}_{B-}, \epsilon). \end{aligned}$$

(1.45)

We thus see that the performance of a measurement changes the situation drastically. If the beam in the pure state (1.39) passes B without a measurement being performed, then the system remains in this pure state $|\chi(t)\rangle$. If *a measurement is performed* as the beam passes B, i.e., if it is registered that the hydrogen atom has passed through the opening at B_+ or at B_-, *then the system becomes a mixture* of the electron spin-up and spin-down states. For example, if the openings in the screen are symmetrical with respect to the xy plane,

$$w_+(\mathbf{x}_{B+} - \epsilon, \mathbf{x}_{B+} + \epsilon, t_B) = w_-(\mathbf{x}_{B-} - \epsilon, \mathbf{x}_{B-} + \epsilon, t_B), \qquad (1.46)$$

and are big enough to let all of the particles in the beam pass through, then

$$\Lambda(\mathbf{x}_{B\pm}, \epsilon)|\psi \pm (t_B)\rangle\langle\psi \pm (t_B)|\Lambda(\mathbf{x}_{B\pm}, \epsilon) = |\psi \pm (t_B)\rangle\langle\psi \pm (t_B)|. \quad (1.47)$$

The state after the measurement then becomes

$$\begin{aligned} W^B(t) &= |\alpha|^2 |+\rangle\langle+| \otimes |\psi + (t)\rangle\langle\psi + (t)| \\ &+ |\beta|^2 |-\rangle\langle-| \otimes |\psi - (t)\rangle\langle\psi - (t)| \\ &= |\alpha|^2 |+\rangle \otimes |\psi + (t)\rangle\langle+| \otimes \langle\psi + (t)| \\ &+ |\beta|^2 |-\rangle \otimes |\psi - (t)\rangle\langle-| \otimes \langle\psi - (t)| \\ &= |\alpha|^2 |\chi + (t)\rangle\langle\chi + (t)| + |\beta|^2 |\chi - (t)\rangle\langle\chi - (t)| = W_M(t) \\ &\quad\text{for} \quad t > t_B. \end{aligned}$$

(1.48)

If the state (1.48) now passes through the reverse splitting magnets, it will arrive at D in a mixture of spin-up with spin-down states. A subsequent Stern–Gerlach apparatus with magnetic field in the x-direction will then split the beam.

In summary, we have found that the pure state (1.38) *is not split by the inhomogeneous magnetic fields* of the first and second magnets unless a position measurement is made at the plane B. *If such a measurement is made*, however, *the pure state splits at B into two subensembles of spin up and spin down*, and consequently into two subensembles of spin right ($s_1 = +\frac{1}{2}$) and spin left ($s_1 = -\frac{1}{2}$), by a subsequent Stern–Gerlach apparatus, whose field is in the x-direction. This concrete example illustrates the content of the basic postulate IIIb.

Appendix to Section XIII.1

In (1.4) and (1.12) we have neglected the force upon the particle except for the force in the z-direction. This requires a justification, because due to the Maxwell equation, $\nabla \cdot \mathbf{B} = 0$, a large $\partial B_3 / \partial x_3$ will also require a large

$$\frac{\partial B_1}{\partial x_1} + \frac{\partial B_2}{\partial x_2} = - \frac{\partial B_3}{\partial x_3}, \tag{1.49}$$

and therewith a large force perpendicular to the z-direction. We shall describe in the following the simplest theoretical arrangement in which the time average of the force on the particle perpendicular to the z-direction never becomes large. This can be achieved if one chooses a large magnetic field in the z-direction. This field causes a Larmor precession of the magnetic moment, as a consequence of which the time average of the force perpendicular to the z-direction is small—not the force itself.

We simplify matters by assuming the magnetic field does not vary with x_2 (no edge effects); then (1.49) reduces to

$$\frac{\partial B_1}{\partial x_1} = - \frac{\partial B_3}{\partial x_3} \tag{1.50}$$

and the Maxwell equation $\nabla \times \mathbf{B} = 0$ reduces to

$$\frac{\partial B_2}{\partial x_3} = \frac{\partial B_3}{\partial x_2} = 0, \qquad \frac{\partial B_2}{\partial x_1} = \frac{\partial B_1}{\partial x_2} = 0, \tag{1.51}$$

and

$$\frac{\partial B_3}{\partial x_1} = \frac{\partial B_1}{\partial x_3}. \tag{1.52}$$

A magnetic field compatible with this assumption and which also leads to (1.16) must be of the form

$$\mathbf{B} = (-bx_1 + \psi(x_3))\mathbf{e}_1 + B_2\mathbf{e}_2 + (bx_3 + \phi(x_1))\mathbf{e}_3, \tag{1.53}$$

where $b > 0$ and B_2 are constants and $\psi(x_3)$ and $\phi(x_1)$ are arbitrary functions except that they must satisfy

$$\frac{\partial \psi(x_3)}{\partial x_3} = \frac{\partial \phi(x_1)}{\partial x_1}, \tag{1.54}$$

which follows from (1.52). We choose $B_2 = 0$, $\phi(x_1) = 0$, and $\psi(x_1) = B_0$ with

$$B_0 \gg |bx_1| \approx |bx_3|. \tag{1.55}$$

Therewith we have a magnetic field which has a strong component in the z-direction and is inhomogeneous in the x- and the z-directions:

$$\mathbf{B} = -bx_1\mathbf{e}_1 + (B_0 + bx_3)\mathbf{e}_3. \tag{1.56}$$

In an actual experimental set-up the field will, of course, be more complicated.

For the electron, spin and magnetic moment are connected by (1.2) and the Hamiltonian is given by (1.5). Using these along with the Heisenberg equation of motion (XII.1.24), one obtains

$$\frac{dM_j}{dt} = \frac{1}{i\hbar}(-2\mu_B)(2\mu_B)[S_j, S_i]B_i. \tag{1.57}$$

From this and from the commutation relations of the S_j, there follows

$$\frac{d\mathbf{M}}{dt} = -2\mu_B \mathbf{M} \times \mathbf{B}. \tag{1.58}$$

For a static magnetic field, (1.58) has the solution

$$\mathbf{M}(t) = \mathbf{M}(0)\cos \omega t + \hat{\mathbf{B}} \times \mathbf{M}(0)\sin \omega t$$
$$+ \hat{\mathbf{B}}[\hat{\mathbf{B}} \cdot \mathbf{M}(0)](1 - \cos \omega t), \tag{1.59}$$

where $\hat{\mathbf{B}} = \mathbf{B}/B$, $B = \sqrt{\mathbf{B}^2}$, and $\omega = 2\mu_B B$ is the Larmor frequency. This solution describes the precession of $\mathbf{M}$ around the direction of $\mathbf{B}$; the component of $\mathbf{M}$ along $\mathbf{B}$ does not change in time.

For the magnetic field (1.56) with (1.55), the solution of (1.58) is given to good approximation by

$$\mathbf{M}(t) = \mathbf{M}(0)\cos \omega t + \mathbf{e}_3 \times \mathbf{M}(0)\sin \omega t$$
$$+ \mathbf{e}_3[\mathbf{e}_3 \cdot \mathbf{M}(0)](1 - \cos \omega t) \tag{1.60}$$

with the frequency

$$\omega = 2\mu_B B_0. \tag{1.61}$$

In component form, (1.60) reads

$$M_1(t) = M_1(0)\cos \omega t - M_2(0)\sin \omega t,$$
$$M_2(t) = M_1(0)\sin \omega t + M_2(0)\cos \omega t, \tag{1.60'}$$
$$M_3(t) = M_3(0),$$

which describes the precession of the magnetic moment operator $\mathbf{M}$ around the z-axis $\mathbf{e}_3$ with the frequency (1.61); the component M_3 does not change in time.

Instead of (1.4) one has now

$$H_{\mathrm{int}} = -M_1(t)bQ_1 - M_3 B_0 + M_3 bQ_3, \tag{1.62}$$

and instead of (1.12) one has for the force operator

$$\frac{d\mathbf{P}(t)}{dt} = -\frac{1}{i}[\mathbf{P}(t), Q_1]M_1(t)b + \frac{1}{i}[\mathbf{P}(t), Q_3]M_3 b. \tag{1.63}$$

The second term on the right-hand side is the same as the term in (1.12), except that we have here the special case

$$\frac{dB_3}{dQ_3(t)} = b. \tag{1.64}$$

As a consequence the expectation value of $P_3(t)$ is given by (1.19).

The first term in (1.63) results in a component of the force operator in the x-direction:

$$\frac{dP_1(t)}{dt} = M_1(t)b = [M_1(0)\cos \omega t - M_2(0)\sin \omega t]b,$$

which oscillates with the frequency (1.61). Though its expectation value may be of the same order as (1.17), it oscillates rapidly under the condition (1.55), and so does the expectation value of $P_1(t)$. The time average of the force as well as the momentum in the x-direction is zero. In an experiment, which always measures the average value of the expectation value over a finite time interval, a deviation from zero in the x-direction will not be observed.

In the conventional Stern–Gerlach experiment[6] the magnetic field is not given by the idealization (1.56) but is usually of the form

$$\mathbf{B} = B_1(x_1, x_3)\mathbf{e}_1 + (B_0 + B_3(x_1, x_3))\mathbf{e}_3$$

with $B_1(x_1, x_3)$ and $B_3(x_1, x_3)$ fulfilling (1.50) and (1.52) and $|B_0| \gg |B_1| \approx |B_2|$. Clearly, the same arguments as above apply, and (1.4) suffices to describe the observable effects.

XIII.2 Spin Correlations in a Singlet State

In the previous section we discussed the use of the Stern–Gerlach device for the measurement of spin components of a single spin-$\frac{1}{2}$ system. In particular we have described how the spin component of the physical system I (the spin degrees of freedom of an electron, an elementary rotator) can be measured by a position measurement on the physical system II (the center-of-mass motion of a hydrogen atom) to which it is attached. System II is used as an auxiliary device which permits us to use the position measurement on II as a measurement of the spin component of system I along the field. We can, therefore, consider system II together with the Stern–Gerlach device as an apparatus for the measurement of the spin component of system I.

Let $\mathbf{a}$ denote the unit vector in the direction of the inhomogeneous magnetic field. In Figure 1.1 $\mathbf{a} = \mathbf{e}_3$, the z-direction, but one may choose any direction orthogonal to the particle beam. Then the Stern–Gerlach device measures the spin component of system I along $\mathbf{a}$, i.e., the observable $\mathbf{a} \cdot \mathbf{S}$. The measured value is $+\frac{1}{2}$ $(-\frac{1}{2})$ if system II is deflected "down" ("up") and recorded by a counter in position B_+ (B_-), cf. Figure 1.1. It is more convenient now to consider instead of $\mathbf{a} \cdot \mathbf{S}$ the observable

$$\phi = 2\mathbf{a} \cdot \mathbf{S}, \tag{2.1}$$

[6] For other arrangements of a Stern–Gerlach experiment, see Myer Bloom and Karl Erdman, *Can. J. Phys.* **40**, 179 (1962).

which has the eigenvalues ± 1 instead of $\pm\frac{1}{2}$. We denote the eigenvectors of ϕ by $|\mathbf{a}\pm\rangle$, with

$$\phi|\mathbf{a}\pm\rangle = \pm 1|\mathbf{a}\pm\rangle. \tag{2.2}$$

In particular, the eigenvectors $|+\rangle$ and $|-\rangle$ of S_3 in (1.1) can be more explicitly denoted by $|\mathbf{e}_3+\rangle$ and $|\mathbf{e}_3-\rangle$, so that (1.1) now reads

$$\phi_3|\mathbf{e}_3\pm\rangle = 2\mathbf{e}_3 \cdot \mathbf{S}|\mathbf{e}_3\pm\rangle = 2S_z|\mathbf{e}_3\pm\rangle = \pm 1|\mathbf{e}_3\pm\rangle. \tag{1.1'}$$

The vector $\mathbf{a}$ can be obtained from the vector $\mathbf{e}_3$ by a rotation specified by a rotation angle θ and a unit vector $\hat{\boldsymbol{\theta}}$ parallel to $\mathbf{e}_3 \times \mathbf{a}$, which determines the direction of the rotation axis. The vectors $|\mathbf{a}\pm\rangle$ can then be obtained —up to an arbitrary phase factor—from the vectors $|\pm\rangle \equiv |\mathbf{e}_3\pm\rangle$ by the transformation which represents this rotation:

$$|\mathbf{a}\pm\rangle = e^{-i\boldsymbol{\theta}\cdot\mathbf{S}}|\pm\rangle,$$

where $\boldsymbol{\theta} = \theta\hat{\boldsymbol{\theta}}$, and where

$$e^{-i\boldsymbol{\theta}\cdot\mathbf{S}}S_3 e^{i\boldsymbol{\theta}\cdot\mathbf{S}} = \mathbf{a}\cdot\mathbf{S}.$$

Since (see Problem 1)

$$e^{-i\boldsymbol{\theta}\cdot\mathbf{S}} = \cos\frac{\theta}{2} - i2\hat{\boldsymbol{\theta}}\cdot\mathbf{S}\sin\frac{\theta}{2},$$

one has

$$|\mathbf{a}\pm\rangle = \left(\cos\frac{\theta}{2} - i2\hat{\boldsymbol{\theta}}\cdot\mathbf{S}\sin\frac{\theta}{2}\right)|\pm\rangle.$$

Thus $|\mathbf{a}\pm\rangle$ is just another basis system of the space $\mathcal{R}^{s=1/2}$; whereas the basis $|\pm\rangle$ was adapted to the Stern–Gerlach apparatus with its inhomogeneous magnetic field in the $\mathbf{e}_3$ direction, $|\mathbf{a}\pm\rangle$ is the basis adopted to a Stern–Gerlach apparatus with an inhomogeneous magnetic field in the direction of $\mathbf{a}$.

Until now we have only discussed the measurement of spin components of a one-particle system. We now consider the combination of two different spin-$\frac{1}{2}$ systems each described by $\mathcal{R}^{s=1/2}$. The state space for this combination of the spin degrees of freedom of both particles (of *two* different elementary rotators of spin $\frac{1}{2}$) is—according to the basic postulate IVa—$\mathcal{R}^{s=1/2}_{(1)} \otimes \mathcal{R}^{s=1/2}_{(2)}$. We will discuss the measurement of *spin correlations* on this system, i.e., of the correlations between certain spin components of these two particles. A system of basis vectors in this space is

$$|(1)+\rangle \otimes |(2)+\rangle, \qquad |(1)-\rangle \otimes |(2)-\rangle,$$

$$|(1)+\rangle \otimes |(2)-\rangle, \qquad |(1)-\rangle \otimes |(2)+\rangle, \tag{2.3}$$

where (1) and (2) refer to the first and second particle, respectively, and $+$ or $-$ specifies the z component of the spin as in (1.1). Instead of the basis system (2.3) one can also take—in the notation of (2.2)—the basis system

$$|(1)\mathbf{a}+\rangle \otimes |(2)\mathbf{b}+\rangle, \qquad |(1)\mathbf{a}-\rangle \otimes |(2)\mathbf{b}-\rangle,$$

$$|(1)\mathbf{a}+\rangle \otimes |(2)\mathbf{b}-\rangle, \qquad |(1)\mathbf{a}-\rangle \otimes |(2)\mathbf{b}+\rangle,$$

$$(2.3')$$

with two arbitrarily chosen but fixed unit vectors $\mathbf{a}$ and $\mathbf{b}$.

Let $S_i^{(\alpha)}$ denote the spin operators of the particle (α) acting in $\mathscr{R}_{(\alpha)}^{s=1/2}$, $\alpha = 1$ or 2. The observable "$2 \times$ spin component of particle (1) along $\mathbf{a}$" is then given by

$$2\mathbf{a} \cdot \mathbf{S}^{(1)} \otimes I^{(2)} = \phi^{(1)} \otimes I^{(2)}, \qquad (2.4)$$

and the observable "$2 \times$ spin component of particle (2) along $\mathbf{b}$" is given by

$$I^{(1)} \otimes 2\mathbf{b} \cdot \mathbf{S}^{(2)} = I^{(1)} \otimes \phi^{(2)}, \qquad (2.5)$$

where $I^{(\alpha)}$ is the unit operator in $\mathscr{R}_{(\alpha)}^{s=1/2}$. The observables (2.4) and (2.5) commute,[7] and the basis vectors $(2.3')$ are simultaneous eigenvectors of both of them, with eigenvalues $+1$ or -1. Because the observables (2.4) and (2.5) commute, they can be measured simultaneously. (We will discuss below how such simultaneous measurements can actually be performed.) The possible measured values for each of them are $+1$ or -1; the measurement of $\phi^{(1)} \otimes I^{(2)}$ in the four states $(2.3')$, for instance, will always yield $+1$, -1, $+1$, and -1, and the simultaneous measurement of $I^{(1)} \otimes \phi^{(2)}$ in the same states will yield $+1$, -1, -1, and $+1$, respectively.[8] For such simultaneous spin measurements we may define, in addition, a *spin correlation observable*. The measured value of this observable in a "single" measurement of *both* $\phi^{(1)} \otimes I^{(2)}$ and $I^{(1)} \otimes \phi^{(2)}$ is, by definition, the product of the value obtained for $\phi^{(1)} \otimes I^{(2)}$ and the value obtained for $I^{(1)} \otimes \phi^{(2)}$, and is therefore either $+1$ or -1. This spin correlation observable is described mathematically by the operator

$$\phi^{(1)} \otimes \phi^{(2)} = 2\mathbf{a} \cdot \mathbf{S}^{(1)} \otimes 2\mathbf{b} \cdot \mathbf{S}^{(2)}. \qquad (2.6)$$

The basis vectors $(2.3')$ are also eigenvectors of the operator (2.6); the eigenvalues are $+1$ for the first two and -1 for the last two basis vectors of $(2.3')$. These eigenvalues coincide, according to the above definition, with the measured values of the spin correlation observable in the four states $(2.3')$.

Such simultaneous spin measurements on two-particle systems are performed with two Stern–Gerlach devices. This is possible, however, only if the two particles of each pair are spatially separated, and each particle moves along a certain fixed axis, as shown in Figure 2.1. Only in this case may two separate Stern–Gerlach devices be applied to the two particles

---

[7] In fact, they are a complete system of commuting observables in $\mathscr{R}_{(1)}^{s=1/2} \otimes \mathscr{R}_{(2)}^{s=1/2}$.

[8] Remember that, according to Section II.4, the value obtained from a single measurement of an observable on a system in *any* state is always one of the eigenvalues of the corresponding operator. If the observable is measured on a system described by an *eigenstate* of that operator, the measured value is always the corresponding particular eigenvalue.

Figure 2.1 Simultaneous spin measurements on particle pairs (1) + (2). S is the particle source, $B_{\pm}$ are particle detectors, and **a** and **b** are field directions of the Stern–Gerlach magnets.

(1) and (2) in such a way that the orbital motion (system II, in the terminology of Section XIII.1) of each particle is used for a spin measurement on that particle (as discussed in Section XIII.1). Imagine, therefore, a particle source emitting pairs of particles, one pair at a time, such that particle (1) is always emitted to the left, and particle (2) is always emitted to the right.[9] Then a Stern–Gerlach device with inhomogeneous magnetic field along some direction **a** (perpendicular to the beam) may be applied to the left beam, consisting of particles (1), and another Stern–Gerlach device with field direction **b** may be applied to the right beam of particles (2), cf. Figure 2.1. Each Stern–Gerlach device has two counters, one at a position corresponding to B_{+} in Figure 1.1 that detects all particles deflected against the field direction **a** or **b** ("down"), and another one at B_{-} that detects all particles deflected in the field direction ("up"). Then, for instance, a click in the "up" ("down") counter of the left Stern–Gerlach device means that the measured value of the "spin component" $\not{a}^{(1)} \otimes I^{(2)}$ of the single particle (1) is -1 ($+1$).[10] Similarly, the right Stern–Gerlach device detects the spin component $I^{(1)} \otimes \not{b}^{(2)}$ of the particles (2). For *simultaneous* measurements of both $\not{a}^{(1)} \otimes I^{(2)}$ and $I^{(1)} \otimes \not{b}^{(2)}$ on single pairs, the four counters have to be used in coincidence. Since the particles (1) and (2) are emitted pairwise by the source, the two particles of a single pair pass the two Stern–Gerlach magnets and arrive at two of the four counters (almost) simultaneously. Therefore, e.g., a simultaneous response of the left "down" and the right "up" counter means that a simultaneous measurement of $\not{a}^{(1)} \otimes I^{(2)}$ with the result $+1$ and of $I^{(1)} \otimes \not{b}^{(2)}$ with the result -1 has been performed on a single particle pair. The value of the spin correlation observable $\not{a}^{(1)} \otimes \not{b}^{(2)}$ for this single pair is $(+1)(-1) = -1$. This kind of measurement is repeated N times ($N \gg 1$), and the following numbers are recorded: the number N_{++} of simultaneous clicks of the left "down" and the right "down" counter, and the analogous numbers N_{+-}, N_{-+}, N_{--} counting the simultaneous clicks left "down" and right "up," left "up" and right "down," left "up" and right "up," respectively. The measured *average values* for the

[9] We assume here that the particles are not identical. If the particles are identical one arrives at the same results by a somewhat more complicated calculation.

[10] This is only true if the spin and the magnetic moment are antiparallel as for the electron; otherwise the roles of the two counters must be interchanged. A factor 2 is included here in the definition of "spin components."

observables $\phi^{(1)} \otimes I^{(2)}$, $I^{(1)} \otimes \psi^{(2)}$ and $\phi^{(1)} \otimes \psi^{(2)}$ in this run of N measurements (here denoted by $E_1(\mathbf{a})$, $E_2(\mathbf{b})$ and $E(\mathbf{a}, \mathbf{b})$) are then given by

$$E_1(\mathbf{a}) = \frac{1}{N}(N_{++} + N_{+-} - N_{-+} - N_{--}), \tag{2.7a}$$

$$E_2(\mathbf{b}) = \frac{1}{N}(N_{++} - N_{+-} + N_{-+} - N_{--}), \tag{2.7b}$$

$$E(\mathbf{a}, \mathbf{b}) = \frac{1}{N}(N_{++} - N_{+-} - N_{-+} + N_{--}); \tag{2.7c}$$

of course, $N = N_{++} + N_{+-} + N_{-+} + N_{--}$.

According to quantum mechanics, these measured average values should coincide with the *expectation values* of the corresponding observables in the common spin state of the particle pairs emitted by the source. This spin state depends on the nature of the source. Since $\mathcal{R}^{1/2} \otimes \mathcal{R}^{1/2} = \mathcal{R}^1 \oplus \mathcal{R}^0$ (cf. Section V.2), the combination of two spin-$\frac{1}{2}$ systems may lead to total spin 1 or total spin 0. We will assume here that the particle pairs emitted by the source have total spin zero, and are therefore in the singlet state

$$\phi = \frac{1}{\sqrt{2}}(|(1)+\rangle \otimes |(2)-\rangle - |(1)-\rangle \otimes |(2)+\rangle), \tag{2.8}$$

cf. Equation (XI.1.18). A source emitting particle pairs in this spin state ϕ might contain, for example, a large number of unstable compounds (decaying states) of particles (1) and (2) with total spin zero, which are approximately at rest. Every decay of such a compound produces two particles, (1) and (2), which move in opposite directions away from the original position of the compound; however, they are still in the common spin state (2.8). Using slits and filters, then, one can select those particles (1) which move in a fixed direction (left) and the accompanying particles (2) moving in the opposite direction, in order to obtain a source of pairs with the desired properties.

A straightforward calculation (Problems 2 and 3) now yields the expectation values of the three observables considered above in the particular spin state (2.8). The result is:

$$\langle \phi^{(1)} \rangle \equiv \langle \phi | \phi^{(1)} \otimes I^{(2)} | \phi \rangle = 0, \tag{2.9a}$$

$$\langle \psi^{(2)} \rangle \equiv \langle \phi | I^{(1)} \otimes \psi^{(2)} | \phi \rangle = 0, \tag{2.9b}$$

$$\langle \phi^{(1)} \otimes \psi^{(2)} \rangle \equiv \langle \phi | \phi^{(1)} \otimes \psi^{(2)} | \phi \rangle = -\mathbf{a} \cdot \mathbf{b}. \tag{2.9c}$$

These are—according to basic postulate II—the theoretical predictions of quantum mechanics for the averages over many single measurements, (2.7a)–(2.7c).

Comparing (2.9a) with (2.7a), we see that quantum mechanics predicts that $N_{++} + N_{+-}$ and $N_{-+} + N_{--}$ (the numbers of cases in which the spin of particle (1) was found to be parallel and antiparallel to $\mathbf{a}$, respectively) will be equal for any choice of **a**. Since the spin state ϕ is rotationally invariant, we could have anticipated this result. The same argument applies to the comparison of (2.9b) and (2.7b).

At first glance, the result (2.9c) does not surprise us either. Consider, e.g., the particular choice $\mathbf{a} = \mathbf{b}$ for the orientation of the two Stern–Gerlach instruments; then (2.9c) becomes

$$\langle \phi^{(1)} \otimes \phi^{(2)} \rangle = -1. \tag{2.10}$$

Comparing this with (2.7c), and remembering that $N = N_{++} + N_{+-} + N_{-+} + N_{--}$, we see that the quantum mechanical prediction $E(\mathbf{a}, \mathbf{a}) = -1$ implies $N_{++} = N_{--} = 0$; i.e., if the spin of particle (1) is found to be parallel to **a**, the spin of the accompanying particle (2) of the same pair cannot be parallel to **a** and is thus always found to be antiparallel to **a**, and vice versa; or, more briefly: the spin components of the two particles along an arbitrarily chosen direction **a** are always *opposite* to each other. This we would also have intuitively expected: since each pair has total spin zero, the spins of particle (1) and particle (2) should indeed always "be antiparallel" to each other. A closer look at the quantum mechanical prediction (2.9c), however, will soon lead us into some quite puzzling problems.

XIII.3 Bell's Inequalities, Hidden Variables, and the Einstein–Podolsky–Rosen Paradox

According to (2.10), quantum mechanics predicts that the spin components of the two particles (1) and (2) along a fixed direction **a** are always opposite to each other. Instead of directly measuring $\phi^{(1)} \otimes I^{(2)}$ on particle (1) itself, one may therefore determine the spin component *of particle* (1) *along* **a** equally well by measuring $I^{(1)} \otimes \phi^{(2)}$ *on particle* (2) and multiplying the result by -1. This indirect "measurement" does not act on particle (1), which in fact may be very far away from the applied apparatus; nevertheless, it always yields the same value as a simultaneous direct measurement of $\phi^{(1)} \otimes I^{(2)}$ would. In view of this it appears quite natural to imagine that a single particle (1) does not somehow "get" a definite **a** component of its spin during a measurement of $\phi^{(1)} \otimes I^{(2)}$, but rather already "has" a definite value of it, either $+1$ or -1, prior to and independent of any measurement. Moreover, the direction **a** may be chosen arbitrarily except that it must be orthogonal to the beam direction. Therefore one is led to the "natural" assumption that prior to and independent of any measurement every single particle (1) possesses definite values $v(\mathbf{a})$, either $+1$ or -1, for the components of its spin, at least along all possible directions **a** orthogonal to the beam.[11] These values are just "uncovered," rather than "produced," if

---

[11] The cautious reader might notice already, however, that from the point of view of quantum mechanics this hypothesis looks not so "natural" after all, since it ascribes simultaneously fixed values to noncommuting observables.

actual spin measurements are performed. They may be visualized as hidden labels attached to every single particle (1), one (either $+1$ or -1) for every possible direction **a**. The same argument applies, of course, to all particles (2).

Without an assumption of this kind it indeed appears quite difficult to understand the perfect anticorrelation predicted by (2.10) for simultaneous measurements of $q^{(1)} \otimes I^{(2)}$ and $I^{(1)} \otimes q^{(2)}$. For if the value of $q^{(1)} \otimes I^{(2)}$ was really undetermined until it is actually measured on particle (1), it would appear impossible for particle (2)—which may be very far away—to "get informed" about this value, in order to be able to "choose" just the opposite value when $I^{(1)} \otimes q^{(2)}$ is measured on it. We will thus accept the hypothesis formulated above, and use it to derive a simple inequality. It will turn out, however, that this inequality is violated in practice!

Consider a very large number N of pairs of particles in the spin singlet state (2.8) and four arbitrarily chosen directions **a**, **b**, **c**, and **d** in the plane orthogonal to the two beams of particles produced by the source. Denote by $v_i(\mathbf{a})$ and $v_i(\mathbf{d})$ the "hidden" predetermined values of the spin components along **a** and **d**, respectively, of particle (1) in the ith pair, and by $w_i(\mathbf{b})$ and $w_i(\mathbf{c})$ the "hidden" values of the spin components along **b** and **c**, respectively, of particle (2) in the *same* pair. Then, if the spin component of particle (1) along **a** and the spin component of particle (2) along **b** are measured simultaneously for all N pairs, the results are $v_i(\mathbf{a})$ and $w_i(\mathbf{b})$ for the ith pair, and the average value (2.7c) becomes

$$E(\mathbf{a}, \mathbf{b}) = \frac{1}{N} \sum_{i=1}^{N} v_i(\mathbf{a}) w_i(\mathbf{b}). \tag{3.1}$$

But one could as well have chosen other directions—e.g., **d** in the left and **c** in the right beam—for the orientations of the two Stern–Gerlach devices. If this experiment had been performed instead with *the same* N particle pairs, it would have "uncovered" the spin components $v_i(\mathbf{d})$ and $w_i(\mathbf{c})$, instead of $v_i(\mathbf{a})$ and $w_i(\mathbf{b})$, and the observed correlation average (2.7c) would have been

$$E(\mathbf{d}, \mathbf{c}) = \frac{1}{N} \sum_{i=1}^{N} v_i(\mathbf{d}) w_i(\mathbf{c}). \tag{3.1'}$$

Analogous expressions may be written down for the results $E(\mathbf{a}, \mathbf{c})$ and $E(\mathbf{d}, \mathbf{b})$ of two other *conceivable* experiments with the same N particle pairs.[12] Clearly at most *one* of these four alternative experiments can be *actually* performed on the particle pairs $i = 1, \ldots, N$ considered. Any actual experimental determination of the four correlation averages $E(\mathbf{a}, \mathbf{b})$, $E(\mathbf{a}, \mathbf{c})$, $E(\mathbf{d}, \mathbf{b})$, and $E(\mathbf{d}, \mathbf{c})$ thus requires (at least) four *different* sets of particle pairs. However, which experiment is actually done on a given set of particle pairs may be decided arbitrarily.

---

[12] This is the "reproducibility assumption" of science: If an experiment is repeated it must reproduce the same result. (We neglect statistical fluctuations, which is justified for sufficiently large N.)

It seems reasonable to assume that an experimental determination of $E(\mathbf{d}, \mathbf{c})$ *would have produced the same result* if it had been performed on any one of the four sets of particle pairs.[12] In this sense $E(\mathbf{a}, \mathbf{b}) \cdots E(\mathbf{d}, \mathbf{c})$ are quantities characteristic of the N particle pairs and their preparation, and not dependent on which particular experiment is actually performed. In particular, if quantum mechanics is true, both $E(\mathbf{a}, \mathbf{b})$ and $E(\mathbf{d}, \mathbf{c})$ as given by the right-hand side of (3.1) and (3.1'), and also $E(\mathbf{a}, \mathbf{c})$ and $E(\mathbf{d}, \mathbf{b})$ when expressed similarly, must agree with the corresponding quantum mechanical predictions (2.9c), i.e., with $-\mathbf{a} \cdot \mathbf{b}$, $-\mathbf{d} \cdot \mathbf{c}$, $-\mathbf{a} \cdot \mathbf{c}$, and $-\mathbf{d} \cdot \mathbf{b}$, respectively. At this moment, however, we will not yet assume this, but investigate directly the "experimental" results given by equations (3.1), (3.1') and two analogous formulas.

We want to derive an estimate for $E(\mathbf{a}, \mathbf{b}) + E(\mathbf{a}, \mathbf{c}) + E(\mathbf{d}, \mathbf{b}) - E(\mathbf{d}, \mathbf{c})$. For this purpose let us first show that

$$v_i(\mathbf{a})(w_i(\mathbf{b}) + w_i(\mathbf{c})) + v_i(\mathbf{d})(w_i(\mathbf{b}) - w_i(\mathbf{c})) = \pm 2. \tag{3.2}$$

As w_i is $+1$ or -1, the first bracket is either $+2, 0$ or -2. If the first bracket is $+2$ or -2, then the second bracket is 0, and if the first bracket is 0, the second bracket is $+2$ or -2. As v_i is also $+1$ or -1, the left-hand side of (3.2) is indeed either $+2$ or -2. Summing all expressions (3.2) over $i = 1 \cdots N$ therefore yields a number between $-2N$ and $+2N$, and dividing this by N we obtain

$$\left| \frac{1}{N} \left(\sum_i v_i(\mathbf{a})w_i(\mathbf{b}) + \sum_i v_i(\mathbf{a})w_i(\mathbf{c}) + \sum_i v_i(\mathbf{d})w_i(\mathbf{b}) - \sum_i v_i(\mathbf{d})w_i(\mathbf{c}) \right) \right| \leq 2. \tag{3.3}$$

By (3.1) and its analogs, the expression on the left is easily identified in terms of $E(\mathbf{a}, \mathbf{b}) \ldots E(\mathbf{d}, \mathbf{c})$, and (3.3) becomes

$$|E(\mathbf{a}, \mathbf{b}) + E(\mathbf{a}, \mathbf{c}) + E(\mathbf{d}, \mathbf{b}) - E(\mathbf{d}, \mathbf{c})| \leq 2. \tag{3.4}$$

This inequality is the most famous and experimentally most useful of a series of similar estimates known as *Bell's inequalities*.[13]

Let us now check whether the quantum mechanical prediction (2.9c) for the spin zero state (2.8),

$$E(\mathbf{a}, \mathbf{b}) = \langle \phi^{(1)} \otimes \phi^{(2)} \rangle = -\mathbf{a} \cdot \mathbf{b}, \tag{3.5}$$

satisfies Bell's inequality (3.4). Calculating the left-hand side of (3.4) using (3.5) we obtain

$$|\mathbf{a} \cdot \mathbf{b} + \mathbf{a} \cdot \mathbf{c} + \mathbf{d} \cdot \mathbf{b} - \mathbf{d} \cdot \mathbf{c}| = |\mathbf{a} \cdot (\mathbf{b} + \mathbf{c}) + \mathbf{d} \cdot (\mathbf{b} - \mathbf{c})|$$

$$\leq |\mathbf{a}||\mathbf{b} + \mathbf{c}| + |\mathbf{d}||\mathbf{b} - \mathbf{c}|$$

$$= \sqrt{2 + 2\cos\phi} + \sqrt{2 - 2\cos\phi},$$

---

[13] J. S. Bell, *Physics* 1, 195 (1964). Various forms of Bell's inequalities are discussed in the review by J. F. Clauser and A. Shimony, *Rep. Prog. Phys.* 41, 1881 (1978). See also A. Peres, *Am. J. Phys.* 46, 745 (1978).

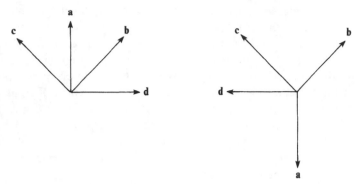

Figure 3.1 Field configurations for which Bell's inequality is maximally violated.

ϕ being the angle between $\mathbf{b}$ and $\mathbf{c}$ ($\mathbf{b} \cdot \mathbf{c} = \cos \phi$). The last expression has a maximum value $2\sqrt{2}$ for $\phi = \pi/2$ ($\cos \phi = 0$). Moreover, the inequality sign in the above example becomes an equality for the following two situations:

1. $\mathbf{a}$ and $\mathbf{b} + \mathbf{c}$, and $\mathbf{d}$ and $\mathbf{b} - \mathbf{c}$ are parallel,
2. $\mathbf{a}$ and $\mathbf{b} + \mathbf{c}$, and $\mathbf{d}$ and $\mathbf{b} - \mathbf{c}$ are antiparallel.

These configurations of directions are depicted in Figure 3.1.

Therefore we obtain for the results $E(\mathbf{a}, \mathbf{b})$ of correlation measurements as predicted by *quantum mechanics* the inequality

$$|E(\mathbf{a}, \mathbf{b}) + E(\mathbf{a}, \mathbf{c}) + E(\mathbf{d}, \mathbf{b}) - E(\mathbf{d}, \mathbf{c})| \leq 2\sqrt{2}, \qquad (3.6)$$

and for the two configurations in Figure 3.1 this becomes an equality:

$$|E(\mathbf{a}, \mathbf{b}) + E(\mathbf{a}, \mathbf{c}) + E(\mathbf{d}, \mathbf{b}) - E(\mathbf{d}, \mathbf{c})| = 2\sqrt{2}. \qquad (3.7)$$

The latter, clearly, contradicts Bell's inequality (3.4), and there are indeed infinitely many configurations of directions $\mathbf{a}$, $\mathbf{b}$, $\mathbf{c}$, and $\mathbf{d}$ for which the quantum mechanical predictions do *not* satisfy (3.4). (Any configuration sufficiently "near" to the ones depicted in Figure 3.1 will do, the latter being the configurations for which Bell's inequality is *maximally* violated.) Bell's inequality (3.4) on the one hand, and the prediction (3.5) of quantum mechanics on the other hand, are thus incompatible with each other, and at least one of them must be wrong.

In this situation the best thing one can try is to decide the conflict *empirically*. One simply has to do the experiment described in Section XIII.2 four times, with Stern–Gerlach devices oriented according to one of the configurations in Figure 3.1, to determine the four averages $E(\mathbf{a}, \mathbf{b}) \ldots E(\mathbf{d}, \mathbf{c})$ experimentally via (2.7c) (note that each of the four experiments consists of many coincidence measurements on single particle pairs), to insert them into the left-hand side of (3.7), and to look whether the result is ≤ 2 in accordance with (3.4), or equal to $2\sqrt{2}$ (within experimental errors, of course) as predicted by quantum mechanics. This is not as easy in practice as one might imagine, however. Pairs of spin-$\frac{1}{2}$ particles in the pure singlet state (2.8) are difficult to prepare. In addition to that it has been shown that, because of the uncertainty relations, Stern–Gerlach devices do not work for charged particles,[14] so that different (perhaps less precise) methods for spin measurements have to be used for pairs of charged particles. Therefore only one test of Bell's inequalities has been performed so far with spin-$\frac{1}{2}$ particles (protons).[15] The results agreed with quantum mechanics and disagreed with one of Bell's inequalities, but the evaluation of the measurements required additional hypotheses, and the results were not precise enough to remove any doubts.

Much more precise experiments can be done with correlated *photon pairs*, as produced by two-step ("cascade") transitions of atoms from a suitable excited state to the ground state. In this case, the simultaneous spin measurements considered above are to be replaced by simultaneous measurements of the transverse linear polarizations of the two emitted photons along arbitrarily chosen directions **a** and **b**. For the photon pairs emitted in cascade transitions, these polarizations are correlated in much the same way as the spin components of spin-$\frac{1}{2}$ particle pairs in the singlet state (2.8). A "natural" hypothesis analogous to the one made above again leads to Bell's inequality (3.4) for suitably defined correlation averages $E(\mathbf{a}, \mathbf{b})$, whereas quantum mechanics again yields the less stringent estimate (3.6). The quantum mechanical prediction for $E(\mathbf{a}, \mathbf{b})$ is different from (3.5), however, and therefore the configurations of directions **a**, **b**, **c**, and **d** for which (3.7) is true, and thus Bell's inequality (3.4) is maximally violated, are different from those depicted in Figure 3.1. The results of recent experiments of this kind[16] are in excellent agreement with quantum mechanics, thus proving beyond any doubt that Bell's inequality (3.4) is not respected by nature. Forced by this unequivocal verdict against (3.4), we have to renounce the apparently "natural" hypothesis made above from which this wrong result (3.4) was derived. We must realize that it was inconsistent to assume that a particle can simultaneously "have" fixed values of its spin components along different directions, regardless of whether or not they are really measured. The value of a spin component that has not been measured is not just unknown—it does not even exist. Imaginary experiments do not have results at all, not even imagined ones.[17]

Looking back at Equation (3.2), which was the crucial step in proving (3.4), we see that it indeed contains values of spin components $v_i(\mathbf{a})$ and $v_i(\mathbf{d})$ (or $w_i(\mathbf{b})$ and $w_i(\mathbf{c})$) which have not been measured simultaneously on the same (the ith) particle pair, but were only imagined to be simultaneously present on the basis of our "natural" hypothesis. The only way to avoid the

---

[14] N. F. Mott and H. S. W. Massey. *The Theory of Atomic Collisions*, 3rd ed., Section IX.2. Clarendon Press, Oxford, 1965.

[15] M. Lamehi-Rachti and W. Mittig, *Phys. Rev.* **D14**, 2543 (1976).

[16] A. Aspect, P. Grangier, and G. Roger, *Phys. Rev. Letters* **49**, 91 (1982); A. Aspect, J. Dalibard, and G. Roger, *Phys. Rev. Letters* **49**, 1804 (1982).

[17] "No elementary quantum phenomenon is a phenomenon until it is an observed phenomenon." (J. A. Wheeler.)

wrong but inevitable consequence (3.4) of (3.2) is indeed to admit that some of the quantities in (3.2) must be meaningless. If $v_i(\mathbf{a})$ and $w_i(\mathbf{b})$ are *really* measured values, Equation (3.1) is the correct formula for the correlation average measured in a real experiment. But then $v_i(\mathbf{d})$ and $w_i(\mathbf{c})$ simply do not exist, and Equation (3.1′) is completely meaningless. A real experimental value of $E(\mathbf{d}, \mathbf{c})$ used in the test of (3.4) involves particle pairs from a different set of pairs, and (3.1′) should be replaced by

$$E(\mathbf{d}, \mathbf{c}) = \frac{1}{M} \sum_{\alpha=1}^{M} v_\alpha(\mathbf{d}) w_\alpha(\mathbf{c}) \qquad (3.1'')$$

with the *measured* values $v_\alpha(\mathbf{d})$ and $w_\alpha(\mathbf{c})$, where the index α enumerates this different set of pairs. But there is no way to derive Bell's inequality (3.4) from (3.1), (3.1″) and two similar equations for $E(\mathbf{a}, \mathbf{c})$ and $E(\mathbf{d}, \mathbf{b})$.

The hypothesis that the two spin-$\frac{1}{2}$ particles "have" definite spin components $v_i(\mathbf{a})$ and $w_i(\mathbf{b})$ seems natural only from the classical point of view, in which one visualizes the particle pair to consist of two separate particles. However, the quantum mechanical state vector ϕ of (2.8) does not describe a state with separate single-particle properties. Such states would be described by the basis vectors (2.3) or (2.3′), which however are not eigenvectors of $(\mathbf{S}^{(1)} + \mathbf{S}^{(2)})^2$, i.e., have no well-defined total spin. The state ϕ describes a new entity, an *indivisible whole*, a single object whose constituent particles (1) and (2) are not definable until a measurement is made that prepares the direct product states (2.3′) (or mixtures thereof). As a state with total spin zero, this state ϕ does not have single-particle properties. And as a *pure* state it cannot be subdivided further, not even in our minds.

In quantum mechanics, due to the superposition principle, one can obtain new states for the combination of two subsystems (basic postulate IIa) which no longer have single-constituent properties like the direct product states (2.3′). These direct product states exist too, but in addition there is the possibility of a multitude of linear combinations, of which ϕ is one example.

In classical physics the building blocks of a composite system are its constituents. In quantum physics the building blocks are the subspaces of the space of physical states ("Hilbert space"). They can be any kinds of subspaces, not only the one-dimensional subspaces spanned by the direct product vectors of constituent states. The state ϕ is one of this multitude of possible states.

Historical Remark

The invalidity of the consequences of apparently "natural" ideas, like the inequality (3.4), is one of the numerous "counterintuitive" features of quantum mechanics. Such features made it difficult for many physicists, who grew up in the classical tradition, to accept the interpretation of quantum mechanics that evolved in the late 1920s and early '30s. Particularly famous are the arguments of the 1935 paper by Einstein, Podolsky, and Rosen (EPR) who describe essentially the same "paradox" as the one discussed above. They considered a pair of point particles with coordinates (one-dimensional, for simplicity) x_1 and x_2 in the (improper) state ϕ described by the wave function

$$\langle x_1 x_2 | \phi \rangle = \delta(x_1 - x_2 - a). \tag{3.8}$$

The momentum-space representation of this state ϕ is the Fourier transform of (3.8), cf. (II.7.49),

$$\langle p_1 p_2 | \phi \rangle = \frac{1}{2\pi} \int e^{-i(p_1 x_1 + p_2 x_2)} \langle x_1 x_2 | \phi \rangle \, dx_1 \, dx_2$$

$$= e^{(i/2)(p_2 - p_1)a} \, \delta(p_1 + p_2). \tag{3.9}$$

According to (3.8) simultaneous measurements of the positions Q_1 and Q_2 of the two particles always yield two values x_1 and x_2 related by $x_1 = x_2 + a$, and according to (3.9) simultaneous measurements of the momenta P_1 and P_2 always yield two values p_1 and p_2 with $p_1 = -p_2$. Therefore either the position x_1 or the momentum p_1 of particle (1) can be determined by a corresponding position or momentum measurement on particle (2).

Although these two measurements on particle (2) cannot be performed simultaneously, neither of them acts directly on particle (1). From this EPR conclude—as we did above in the analogous case of spin measurements—that such measurements cannot "produce" the measured values x_1 and p_1 but merely "uncover" them. Therefore both the position and the momentum of particle (1) should have some kind of "physical reality," independent of whether or not they are actually measured. However, according to the uncertainty principle, quantum mechanics cannot simultaneously ascribe to a particle definite values of both position and momentum! EPR conclude from this example that quantum mechanics does not provide a complete description of physical reality.

Bohr, in a reply published later in 1935, tried to defend quantum mechanics against this criticism, but apparently did not convince everyone. Since then and until quite recently some physicists have looked for theories which are "more complete" than quantum mechanics in the following sense: One assumes that there exists a more precise specification of the states of a microsystem in terms of additional "hidden" variables, such that in the new "microstates"—as specified by fixed values of the hidden variables—all observables simultaneously possess fixed values. The uncertainty relations of quantum mechanics are therefore not valid in these "microstates." On the other hand, the uncertainty relations of quantum mechanics are fulfilled even in a *pure* quantum state (cf. Chapter II). A pure state of quantum mechanics is therefore interpreted in the new theory as a *mixture* of the more precisely defined microstates of the system. Any uncertainties in the outcomes of measurements on a single system in a given pure quantum state then merely result from our ignorance of the "real" microstate of the system considered, which could be any of the microstates which are mixed up in the pure quantum state. This is the uncertainty of classical physics, which comes from not knowing enough and can in principle be overcome; it is familiar from classical statistical mechanics. Thus the novel quantum mechanical uncertainty of "not being able to know" would be converted by a hidden-variables theory into the more familiar uncertainty of "not yet knowing."

While preparing a review article on hidden-variables theories, Bell discovered in 1964 that such theories almost inevitably lead to certain estimates—now known as Bell's inequalities—which are *not* satisfied by the predictions of quantum mechanics. Indeed the above derivation of the particular example (3.4) of Bell's inequalities used almost nothing but the hypothesis that "in reality" certain spin components of a single spin-$\frac{1}{2}$ particle—whether or not they are actually measured—have fixed values. This hypothesis follows directly from the assumption that hidden variables in the sense explained above *exist*, independently of any additional details of the theory.[18]

Bell's inequalities were a decisive step towards resolving the controversy. They moved the question of hidden variables from the realm of pure theory into the realm of experimentally decidable propositions. They also provided an opportunity to falsify quantum mechanics. If the experimental data would have fulfilled Bell's inequality rather than the predictions of quantum mechanics, then quantum mechanics would have turned out to be not only incomplete (as suggested by Einstein and collaborators) but even wrong. By confirming the quantum mechanical predictions, however, the crucial experiments have not only supported quantum mechanics but also ruled out once and for all the whole class of local hidden-variables theories.

Problems

1. Let **a** be the vector which is obtained from the vector e_3 (z-direction) by a rotation θ. Let $|a\pm\rangle$ be the eigenvectors of $\mathbf{a}\cdot\mathbf{S}$ with eigenvalues $\pm\frac{1}{2}$, and $|\pm\rangle$ the eigenvectors of $S_3 = e_3\cdot\mathbf{S}$. Prove the relation

$$|a\pm\rangle = \left(\cos\frac{\theta}{2} - i\frac{\mathbf{0}}{\theta}\cdot\mathbf{S}\,2\sin\frac{\theta}{2}\right)|\pm\rangle.$$

Hint: Use that in $\mathscr{R}^{s=1/2}$ the operators S_i fulfill the anticommutation relation $\{S_i, S_j\} = \frac{1}{2}\delta_{ij}I$.

2. Using the equations (3.23), (3.22), and (3.8) of Chapter III or the Pauli matrices of Problem 5, Chapter III, show that the matrix elements of the operator $d = 2\mathbf{a}\cdot\mathbf{S}$

---

[18] We did assume in addition, however, that the "hidden" spin components of particle (1) are not influenced by spin measurements on particle (2), and vice versa. Hidden-variables theories with this property are called *local*. Bell's inequalities can be derived for local hidden-variables theories only.

between the eigenvectors $|s = \frac{1}{2} s_3 = \pm\frac{1}{2}\rangle = |\pm\rangle$ of S_3 are given by

$$\langle \pm|\hat{a}|\pm\rangle = \pm a_3,$$

$$\langle \pm|\hat{a}|\mp\rangle = a_1 \mp ia_2,$$

where $a_i = \mathbf{a} \cdot \mathbf{e}_i$ are the components of the vector $\mathbf{a}$.

3. Using the results of Problem 2 show that the expectation values of the operators $\hat{a}^{(1)} \otimes I^{(2)}$, $I^{(1)} \otimes \hat{b}^{(2)}$ and $\hat{a}^{(1)} \otimes \hat{b}^{(2)}$, defined in equations (2.4)–(2.6), in the state ϕ of (2.8) are given by $\langle \phi|\hat{a}^{(1)} \otimes I^{(2)}|\phi\rangle = 0$; $\langle \phi|I^{(1)} \otimes \hat{b}^{(2)}|\phi\rangle = 0$; $E(\mathbf{a}, \mathbf{b}) \equiv \langle \phi|\hat{a}^{(1)} \otimes \hat{b}^{(2)}|\phi\rangle = -\mathbf{a} \cdot \mathbf{b}$.

Transitions in Quantum Physical Systems—Cross Section

This is the fundamental chapter on scattering theory. Here the concept of transition probabilities is introduced and general formulas for the cross section are derived from the basic assumptions of quantum mechanics. In Section XIV.2, a general formula for the transition rate is obtained, which will be used in Section XIV.5 for the derivation of the cross-section formula and in Chapter XXI for the derivation of the decay rate. Section XIV.3 introduces the concept of cross section, and Section XIV.4 gives a brief description of the different ways to relate the cross section to fundamental physical observables. The main section of this chapter is Section XIV.5, where the cross-section formulas for very general physical situations are derived. At the end of section XIV.5 the physical state of the scattering system is further specified to obtain the well-known expressions for the cross section. For a superficial understanding of the material of this chapter one may omit parts of Section XIV.2 and Section XIV.5. For this purpose we have discussed special results of Section XIV.5 at the end of Section XIV.4.

XIV.1 Introduction

In the present chapter we collect some material that will be used in subsequent chapters to describe collision and decay processes. We derive the transition probability, introduce the notion of transition rate, and then define the cross section and derive the formula that relates it to the transition probability.

The general situation that we shall describe, and of which collision and decay processes are just two particular cases, is the change in time of a nonstationary state $W(t)$. We shall use the basic assumptions of quantum mechanics, in particular the axiom V. We shall make the further simplifying assumption that the generator H of time translation can be split into two parts:

$$H = K + V, \tag{1.1}$$

where K is the energy operator of the isolated physical system with stationary states and is assumed to be well defined. This means that it is assumed to make sense to consider an approximate description of the physical system by a stationary physical system. There are many examples for which this is possible. Thus K may be the energy operator of an atom or of a molecule. In the approximate description the states belonging to an energy level are stationary, but in reality they are only quasistationary and decay into the ground state. This transition is then caused by $V = H - K$. Another example is the combination of two physical systems that can be spatially separated far apart from each other, a situation which one usually encounters in collision processes. If they are far apart, the energy operator of the combined system is K; and $V = H - K$, which describes the interaction between the two subsystems, has a finite range.[1]

The observables that are measured in these processes are assumed to commute with the operator K. For example, a detector may register all the ground states of atoms that were initially in excited states (by registering, for example, all emitted light of a particular frequency). The observable measured is thus the projection operator on that subspace of the space of physical states which contains the ground states of the atoms, i.e., the observable measured is the property of being in the atomic ground state (cf. Section II.4). In the other example of collision experiments the detector is placed far away from the target and therefore detects eigenstates or mixtures of eigenstates of the operator K. The property B measured by the detector is described by a projection operator Λ_B, which projects onto a (in general continuous) direct sum of energy eigenspaces of K. Usually B is a more specific property. For example, the detectors may not be placed all around the target, but only at a particular angle. Λ_B is then the projection operator onto that particular subspace of the above direct sum of energy subspaces that contains states whose momentum vectors are directed into the particular solid angle.

Thus the problem that we shall discuss (in the Schrödinger picture) is the following: The state $W(t)$ changes in time according to

$$W(t) = e^{-iHt/h}We^{+iHt/h} \qquad (W \equiv W(0)). \tag{1.2}$$

---

[1] It should be remarked that the assumption (1.1) is not really necessary for the description of collision processes; it suffices to assume only that asymptotic direct-product states do exist.

What is the expectation value of an observable Λ (i.e., what is the probability of measuring the property Λ) for which

$$[\Lambda, K] = 0? \tag{1.3}$$

This problem is often interpreted as the transition between different stationary states of a physical system caused by an interaction V. This is based on the assumed existence of a K eigenstate $W^{\text{in}}(t = -\infty)$ in the remote past and of a K eigenstate $W^{\text{out}}(t = +\infty)$ in the far future that equals or contains Λ.

XIV.2 Transition Probabilities and Transition Rates

Denote the state of the physical system by W, and the observable to be measured by Λ. Then, according to the basic postulate II, the expectation value $\langle\Lambda\rangle$ of Λ is given by

$$\langle\Lambda\rangle = \text{Tr}(\Lambda W). \tag{2.1}$$

Since Λ is a projector, $\langle\Lambda\rangle$ is the probability that the system is in the subspace $\Lambda\mathscr{H}$. $\langle\Lambda\rangle$ is called the *transition probability* into the state Λ.

The transition probability is a function of time. At the beginning of the process $\langle\Lambda\rangle$ will in general be zero, and will then increase with time. According to the results of Chapter XII, this time dependence can be described in various pictures. We shall use the Schrödinger picture first, because it appeals best to our intuitive understanding of a scattering experiment with a pulsed beam or of a decay process. In the Schrödinger picture the ensemble of physical systems is described by the time-dependent statistical operator $W(t)$, which describes the beam moving towards the detector, or the excited state of an atom decaying into the ground state.

The observable Λ, which represents the apparatus, does not change in time. [We assume that Λ does not depend explicitly upon time ($\partial\Lambda/\partial t = 0$) i.e., that the apparatus is not influenced from the outside after the scattering experiment has begun.] Thus the transition probability is given by

$$\langle\Lambda\rangle_t = \text{Tr}(\Lambda W(t)), \tag{2.2}$$

where

$$W(t) = U^\dagger(t)WU(t) \qquad (W \equiv W(0));$$

W is the state of the system at some conveniently chosen time, which we call $t = 0$, and $U(t) = e^{iHt/\hbar}$ is the time-evolution operator of Chapter XII.

The *transition rate* is defined to be the time rate of change of the transition probability: $d\langle\Lambda\rangle_t/dt$. Differentiation of (2.2) and use of (XII.1.33) give

$$\frac{d}{dt}\langle\Lambda\rangle_t = \text{Tr}\left(\Lambda\frac{dW(t)}{dt}\right) = -i\,\text{Tr}(\Lambda[H, W(t)]). \tag{2.3}$$

Substitution of $H = K + V$ into (2.3) yields

$$\frac{d}{dt}\langle\Lambda\rangle_t = -i\,\text{Tr}(\Lambda[V, W(t)]) - i\,\text{Tr}(\Lambda[K, W(t)]). \qquad (2.4)$$

With the use of (1.3) the second term in (2.4) is easily shown to vanish:

$$\begin{aligned}
\text{Tr}(\Lambda[K, W(t)]) &= \text{Tr}(\Lambda[K, W(t)] + [K, \Lambda]W(t)) \\
&= \text{Tr}(\Lambda K W(t) - \Lambda W(t)K + K\Lambda W(t) - \Lambda K W(t)) \\
&= -\text{Tr}(\Lambda W(t)K) + \text{Tr}(K\Lambda W(t)) = 0.
\end{aligned}$$

Let $\{|b\rangle\}$ be a basis for the subspace $\Lambda\mathcal{H}$ onto which Λ projects. Then

$$\text{Tr}(\Lambda W(t)V) = \sum_b \langle b|W(t)V|b\rangle = \sum_b \langle b|VW(t)|b\rangle^*$$

$$= (\text{Tr}(\Lambda VW(t)))^*.$$

The right-hand side of (2.4) may then be written

$$\frac{d}{dt}\langle\Lambda\rangle_t = -i\,\text{Tr}(\Lambda VW(t)) + i(\text{Tr}(\Lambda VW(t)))^*. \qquad (2.5)$$

But $i(\alpha^* - \alpha) = 2\,\text{Im}\,\alpha$; consequently[2]

$$\hbar\frac{d}{dt}\langle\Lambda\rangle_t = 2\,\text{Im}\,\text{Tr}(\Lambda VW(t)). \qquad (2.6)$$

In order to express the transition rate in a form more easily used, we use a basis of generalized eigenvectors of K and of H:

$$K|a\rangle = E_a|a\rangle, \qquad (2.7a)$$

and

$$H|a^\pm\rangle = E_a|a^\pm\rangle. \qquad (2.7b)$$

Recall that the generalized eigenvectors $|a^\pm\rangle$ of H are related to the generalized eigenvectors $|a\rangle$ of K by means of the Lippmann–Schwinger equation [Equation (VIII.2.7′)]:

$$|a^\pm\rangle = |a\rangle + \frac{1}{E_a - K \pm i0}\,V|a^\pm\rangle. \qquad (2.8)$$

On occasion we shall make the labeling on $|a\rangle$ and $|a^\pm\rangle$ more explicit by writing

(a) $|a\rangle = |E_a\hat{a}\rangle$ and (b) $|a^\pm\rangle = |E_a\hat{a}^\pm\rangle$,

where $\hat{a} = (a_1, a_2, \ldots, a_k)$ are labels needed in addition to the energy to specify the generalized vectors. These labels (quantum numbers) are eigenvalues of additional operators $A_1, A_2, \ldots, A_k$, which together with K form a c.s.c.o. For example, $\hat{a}$ may consist of the angular-momentum quantum

---

[2] We use the units with $\hbar = 1$ and introduce only in a few important formulas the correct factors of $\hbar$.

numbers l and l_3 together with other, internal quantum numbers η; or $\hat{a}$ may consist of the direction $\mathbf{p}/|\mathbf{p}|$ of the momentum $\mathbf{p}$ together with other quantum numbers η. If the operators A_i $(i = 1, 2, \ldots, k)$ all commute with H, then the $|E_a \hat{a}^{\pm}\rangle$ are generalized eigenvectors of the c.s.c.o. $\{H, A_1, \ldots, A_k\}$; otherwise the additional labels $\hat{a}$ just serve to indicate that $|E_a \hat{a}^{\pm}\rangle$ is obtained from $|E_a \hat{a}\rangle$ by way of the Lippmann–Schwinger equation, and do not indicate that $|E_a \hat{a}^{\pm}\rangle$ is an eigenvector of all the A_i's (for more details cf. Section XV.1).

We shall assume that the K eigenvectors $|a\rangle$ are normalized according to

$$\langle a|a'\rangle = \langle E_a \hat{a}|E_{a'} \hat{a}'\rangle = \rho(E_a)^{-1}\delta_{\hat{a}\hat{a}'}\delta(E_a - E_{a'}). \tag{2.9a}$$

$[\rho(E_a)$ is a weight function that is arbitrary but fixed[3]; a convenient choice for $\rho(E_a)$ will be made later.] With such a normalization the summation $\sum_a$ is really an abbreviation for the more explicit

$$\sum_a = \int \rho(E_a)\, dE_a \sum_{\hat{a}}. \tag{2.10}$$

The generalized eigenvectors $|a^{\pm}\rangle$ of H have the same normalization

$$\langle a^{\pm}|a'^{\pm}\rangle = \langle E_a \hat{a}^{\pm}|E_{a'} \hat{a}'^{\pm}\rangle = \rho(E_a)^{-1}\delta_{\hat{a}\hat{a}'}\delta(E_a - E_{a'}) \tag{2.9b}$$

as the corresponding generalized eigenvectors $|a\rangle$ of K. The derivation of (2.9b) from (2.9a) is given in Appendix XV.A.

In general the spectrum of K and the spectrum of H are not the same. Usually K has only a continuous spectrum $\{E_a\}$ starting at a particular value, say $E = 0$, and going to infinity. But the spectrum $\{E_a\}$ of H is the combination of a continuous spectrum $\{E_a\}$, which is usually the same as the continuous spectrum of K, and a discrete spectrum $\{E_n\}$, which may be negative but is bounded from below. Physically the continuous spectrum corresponds to the scattering states and the discrete spectrum to the bound states of the projectile and target. There may be discrete eigenvalues in the continuous spectrum, but for physical reasons the energy eigenvalues cannot be arbitrarily large—in particular, not arbitrarily large and negative. We may thus choose the generalized eigenvectors of K:

$$\{|a\rangle = |E_a \hat{a}\rangle\}$$

as a basis for the space of physical states, or we may choose the set

$$\{|\alpha\rangle\} = \{|a^{\pm}\rangle = |E_a \hat{a}^{\pm}\rangle\} \cup \{|\alpha_n\rangle\}$$

of discrete eigenvectors $|\alpha_n\rangle$ and generalized eigenvectors $|a^{\pm}\rangle$ of H as a basis. The completeness property of these bases may be expressed as

$$I = \sum_a |a\rangle\langle a| = \int \rho(E_a)\, dE_a \sum_{\hat{a}} |E_a \hat{a}\rangle\langle E_a \hat{a}| \tag{2.11a}$$

---

[3] It is for the generalized energy eigenvectors $|E\rangle$ what $\rho(x)^{-1}$ is for the generalized eigenvectors $|x\rangle_\rho$ in (I.4.14) and (I.4.7c_ρ).

and as

$$I = \sum_{\alpha} |\alpha\rangle\langle\alpha| = \int \rho(E_a)\,dE_a \sum_{\hat{a}} |E_a \hat{a}^{\pm}\rangle\langle E_a \hat{a}^{\pm}| + \sum_{n} |\alpha_n\rangle\langle\alpha_n|. \quad (2.11b)$$

The trace that appears on the right-hand side of (2.6) may now be written

$$\mathrm{Tr}(\Lambda V W(t)) = \sum_{b} \langle b|V W(t)|b\rangle$$

$$= \sum_{b}\sum_{\alpha} \langle b|V|\alpha\rangle\langle\alpha|W(t)|b\rangle.$$

Use of the Schrödinger-picture time development (1.2) of $W(t)$ and insertion of another complete set of H eigenvectors $|\alpha'\rangle$ then gives

$$\mathrm{Tr}(\Lambda V W(t)) = \sum_{b}\sum_{\alpha\alpha'} \langle b|V|\alpha\rangle\langle\alpha|e^{-iHt}We^{+iHt}|\alpha'\rangle\langle\alpha'|b\rangle$$

$$= \sum_{b}\sum_{\alpha\alpha'} \langle b|V|\alpha\rangle\langle\alpha|W|\alpha'\rangle\langle\alpha'|b\rangle e^{-i(E_\alpha - E_{\alpha'})t}. \quad (2.12)$$

Suppose H does have discrete eigenvalues, corresponding to bound states $|\alpha_n\rangle$. Since bound states cannot evolve from free states, we assume that the state W contains no contributions from bound states; mathematically this assumption is expressed by

$$\langle\alpha|W|\alpha'_n\rangle = 0. \quad (2.13)$$

We may therefore restrict the summations over α and α' to the continuous spectrum of H.

Since $[\Lambda, K] = 0$, the basis $\{|b\rangle\}$ of the subspace $\Lambda\mathcal{H}$ may be chosen to consist of generalized eigenvectors of K:

$$K|b\rangle = E_b|b\rangle \qquad (|b\rangle = |E_b\hat{b}\rangle). \quad (2.14)$$

If we take the inner product of the Lippmann–Schwinger equation (2.8) with $|b\rangle$, we obtain with (2.14)

$$\langle a'^{\pm}|b\rangle = \langle a'|b\rangle + \lim_{\epsilon \to 0\pm} \langle a'^{\pm}|V \frac{1}{E_{a'} - K \mp i\epsilon}|b\rangle$$

$$= \langle a'|b\rangle + \lim_{\epsilon \to 0\pm} \langle a'^{\pm}|V|b\rangle \frac{1}{E_{a'} - E_b \mp i\epsilon}. \quad (2.15_{\pm})$$

Use of (2.15_{+})[4] in (2.12) yields

$$\mathrm{Tr}(\Lambda V W(t)) = \sum_{b}\sum_{aa'} e^{-i(E_a - E_{a'})t} \frac{\langle b|V|a^{+}\rangle\langle a'^{+}|V|b\rangle}{E_{a'} - E_b - i0} \langle a^{+}|W|a'^{+}\rangle + \mathrm{II},$$

$$(2.16)$$

---

[4] That it is more natural to choose $\epsilon \to 0+$ rather than $\epsilon \to 0-$ can be seen from Section XV.3, where we discuss in detail the significance of the choice $\epsilon \to 0+$. However the derivation could have as well been continued with (2.15_{-}) and would lead to the same results.

where

$$\text{II} = \sum_b \sum_{aa'} e^{-i(E_a - E_{a'})t} \langle b|V|a^+ \rangle \langle a'|b \rangle \langle a^+|W|a'^+ \rangle. \qquad (2.17)$$

In Appendix XV.A we shall show that the second term II vanishes; we content ourselves here with an intuitive argument: The factor $\langle a'|b \rangle$ in the second term does not describe a transition, but describes the probability of observing the configuration of the initial state. But in a scattering experiment (except for absorption measurements, which do not measure $\langle \Lambda \rangle$ directly), the detectors are placed at an angle to the direction of the incident beam so that they are not flooded by the incident beam. Thus if $|E_A \hat{A} \rangle = |A \rangle$ denotes any of the vectors that appear in the initial state $W = \sum w_A |A \rangle \langle A|$, then $|A \rangle \notin \Lambda \mathcal{H}$ and the term II does not contribute to the probability of observing Λ.[5]

The transition rate is then obtained by inserting the first term of (2.16) into (2.5). By using the Hermiticity properties

$$\langle a^+|W|a'^+ \rangle^* = \langle a'^+|W|a^+ \rangle \quad \text{and} \quad \langle b|V|a^+ \rangle^* = \langle a^+|V|b \rangle$$

of W and V, and by exchanging the summation indices a and a' in the second term arising from (2.5), one obtains

$$\frac{d}{dt} \langle \Lambda \rangle_t = -i \sum_b \sum_{aa'} e^{-i(E_a - E_{a'})t} \langle b|V|a^+ \rangle \langle a'^+|V|b \rangle \langle a^+|W|a'^+ \rangle$$

$$\times \left(\frac{1}{E_{a'} - E_b - i0} - \frac{1}{E_a - E_b + i0} \right). \qquad (2.18)$$

We will make use of this last result both in Section XIV.5 when we calculate the cross section and in Section XXI.2 when we calculate the decay rate.

XIV.3   Cross Sections

In a scattering experiment a beam is directed towards a small target. A detector is located at a large distance from the target and at a (nonzero) angle $\Omega = (\theta, \phi)$ with respect to the incident beam.

In classical physics this beam can be a beam of particles or a beam of radiation. For the case of particle scattering, for example, a beam of N_B particles of type B with mass m_B and velocity $\mathbf{v}_0$ is, during an interval of time Δt, sent towards a target consisting of N_T particles of mass m_T. The number of particles per second that are scattered and reach the detector, $N/\Delta t$—the counting rate, is proportional to both the number N_T of particles in the target and to the incoming flux (i.e., to the number of particles per second per unit area perpendicular to $\mathbf{v}_0$). The incoming flux is given by

$$\text{incident flux} = \rho_B v_0, \qquad (3.1)$$

---

[5] Note that II is the term that comes from the first term in the Lippmann–Schwinger equation.

Figure 3.1 Schematic diagram of a scattering experiment.

where ρ_B is the density of particles in the beam. Thus the counting rate $N/\Delta t$ is

$$N/\Delta t = \sigma N_T \rho_B v_0, \tag{3.2}$$

where the proportionality constant σ is called the *cross section*. The cross section for the scattering of particles by particles is thus defined by

$$\sigma = \frac{N/N_T}{\rho_B v_0 \Delta t}. \tag{3.3}$$

It has the dimensions of an area. [If the target is not at rest, then the relative speed $v = |v_0 - v_T|$ should be used in place of v_0 in (3.2).]

Cross sections are usually measured in barns (1 b $= 10^{-24}$ cm^2), millibarns (1 mb $= 10^{-27}$ cm^2) or microbarns (1 μb $= 10^{-30}$ cm^2). One can obtain an intuitive understanding of the definition (3.3) for the cross section σ in the following way: The number of particles per unit time $N/\Delta t$ which reach the detector is equal to the number of particles per unit time $\rho_B(\Delta z/\Delta t)\sigma$, which pass through the volume $\Delta z\sigma$ times the number of target particles N_T:

$$\frac{N}{\Delta t} = \rho_B \frac{\Delta z}{\Delta t} \sigma \cdot N_T.$$

σ is thus the surface area perpendicular to the incident beam which the target particle presents to the beam particle. Over how large a surface the target particle affects the beam particle depends of course upon the nature of these two particles, in particular upon their interaction. But the order of magnitude of σ is roughly defined by the "size" of the interacting particle or the range of the interaction. Typical cross sections for strongly interacting particles are therefore of the order of millibarns and microbarns, and cross sections in atomic physics (elastic scattering) are of the order of 10^{-16} cm^2.

Different kinds of cross section are given different names. If the detector detects all particles of a third kind that leave the target area in all directions, then σ is called the *production cross section*. If the detector detects particles of type B, the incident type, after they leave the target area, then σ is called the *scattering cross section*. In particular σ is called the *total scattering cross section* if the detector detects all particles B leaving the target area without regard to the direction in which the particles leave. The *differential cross section $d\sigma$* is obtained if only those dN particles which are directed into a specific cone of (infinitesimal) solid angle $d\Omega = \sin\theta \, d\theta \, d\phi$ are detected cf. Figure 3.1):

$$d\sigma = \frac{dN/\Delta t}{\rho_B N_T v_0} \quad \text{or} \quad \frac{d\sigma}{d\Omega} = \frac{dN}{d\Omega} \frac{1/\Delta t}{\rho_B N_T v_0}. \tag{3.4}$$

The other example of a scattering experiment in classical physics is the scattering of a beam of radiation incident on an obstacle (target). The detector, which is located outside the incident beam at a distance large compared with the wavelength of the radiation and with the size of the obstacle, measures the flux scattered in a given direction. The ratio of the differential scattered power per unit area dI_{scat} to the incident power per unit area (flux) I_{incid} is the differential cross section

$$d\sigma = \frac{dI_{scat}}{I_{incid}}. \tag{3.5}$$

In quantum physics the beam is a beam of quantum physical systems, and the measured quantities are probabilities. The detector measures the probability of a transition from the incident state. This transition probability is proportional to the probability that a particle before scattering is passing through a surface of unit area perpendicular to the incident velocity. The proportionality constant is the quantum-mechanical cross section:

$$\sigma = \frac{\text{transition probability}}{\text{incident probability per unit area}}. \tag{3.6}$$

As in the classical case, there are different kinds of cross section. If the detector registers all states that emerge from the target, then σ is the total cross section. If the detector registers only a subset of states (e.g., only those states which are in a state B), then $\sigma = \sigma_{BA}$ is called the partial cross section. Thus

$$\sigma_{BA} = \frac{\text{transition probability from a state } A \text{ into a state } B}{\text{incident probability per unit area}}. \tag{3.7}$$

Suppose B is the property that the system's momentum vector is directed into a specific cone of solid angle $d\Omega$ around the direction $\Omega = (\theta, \phi)$; then

$$d\sigma_{\Omega A} = \frac{\begin{array}{c}(\text{transition probability from a state } A \text{ into}\\ \text{all states with momentum vector directed into } d\Omega)\end{array}}{\text{incident probability per unit area}}, \tag{3.8}$$

or more often $d\sigma_{\Omega A}/d\Omega$ is called the differential cross section.

In typical scattering experiments the projectile is chosen to be as structureless as possible (e.g., electrons), whereas the target consists of more complicated objects (e.g., atoms). The experimental cross sections then reveal information about the structure of the target particles. The collision of the projectile with the target can be elastic, as is the case when the energy of the projectile is smaller than the difference between the internal energy levels of the target; or the collision can be inelastic, as is the case when the internal energy level of the target is changed by the collision. To fix the nomenclature we shall use "total collision cross section" for the cross section σ of a collision that involves any elastic or inelastic process. The cross section

σ_{elas} for all elastic collisions is called the *total elastic cross section* while the cross section σ_η for inelastic collisions into the ηth internal energy level will be called the *total inelastic partial cross section for the ηth level*. The *total inelastic cross section* is $\sum_\eta \sigma_\eta$. The total cross section σ is then the sum of the total elastic cross section with the total inelastic cross sections:

$$\sigma = \sigma_{\text{elas}} + \sum_\eta \sigma_\eta. \tag{3.9}$$

Corresponding notation is used for the differential cross section.

XIV.4 The Relation of Cross Sections to the Fundamental Physical Observables

Cross sections are the observable quantities. Their measurement follows in principle from their definition, given in the previous section, in terms of counting rates and flux.[6] The cross sections are related to the fundamental observables which describe the structure of the physical system consisting of the combination of projectile and target. We shall now establish the connection of the cross sections to these quantum-mechanical observables.

If the theory of the physical system (i.e., the mathematical structure that describes the system) is known, then one can calculate the cross sections in terms of the known fundamental observables and predict the outcome of a collision experiment. More often a physicist meets the reverse situation: The cross sections have been measured experimentally; from this information conjectures are made as to what the fundamental observables are, how they behave, and what mathematical structure best describes such behavior. This task is least complicated if one of the colliding subsystems (called the projectile) has a known structure that is as simple as possible. Such a projectile serves to test the structure of the target system and to provide information for a theoretical model of the target system. Some examples are electrons as projectiles with atoms or molecules as targets, protons as projectiles with nucleons as targets, and electrons or other leptons as projectiles with hadrons as targets.

To relate the cross sections to fundamental observables one may proceed in two different ways. The first and conventional way is based upon the time-development axiom V of Chapter XII. It is assumed that there exists an energy operator H in the system's algebra of operators that develops the system in a continuous fashion. It is usually further assumed that the energy operator H can be split into parts,

$$H = K + V, \qquad K = K_B \otimes I + I \otimes K_T, \tag{4.1}$$

---

[6] The actual measurement of cross sections is described in books on experimental physics. See, for example, Massey *et al.* (1969), Vol. 1.

where K is the energy operator of the combination of the physical systems in the absence of the interaction V between them, K_B is the energy operator for the beam system, and K_T is the energy operator for the target system. K is thus the energy operator when the projectile and target are far apart.[7]

The existence of a continuous unitary time development (1.2), generated by an operator H according to the basic assumption V, is a rather questionable assumption in particle physics, where a fundamental length is believed to exist. Therefore a second way was suggested by Heisenberg (1943). It is based on the realization that $W(t)$, which describes the systems while they are interacting, is really not a measurable quantity. Measurable quantities are the initial state $W^{in}(t \to -\infty)$ that describes the systems before the interaction when they are prepared for the collision process, and the final state $W^{out}(t \to +\infty)$ that describes the systems after the interaction when they are detected. Each initial state is transformed into a final state by the interaction. The operator that describes this transformation

$$W^{in}(t) \to W^{out}(t) = SW^{in}(t)S^{-1},$$

or, for pure states,

$$\Psi^{in}(t) \to \Psi^{out}(t) = S\Psi^{in}(t),$$

is called the S-operator ("S-matrix").

S is to have the following properties: (1) It transforms superpositions $a\Phi^{in} + b\Psi^{in}$ into superpositions $a\Phi^{out} + b\Psi^{out}$; therefore it must be a linear operator. (2) It transforms every normalized initial state uniquely into a normalized final state ("conservation of probability"); this, together with the assumption that the set of initial states as well as the set of final states spans the space of physical states, requires that S be unitary: $SS^\dagger = S^\dagger S = 1$.

The cross section can then be related to the matrix elements of the operator S (S-matrix). Thus in this second approach the existence of a unitary S-operator is taken as the fundamental postulate in place of the basic assumption V of a continuous unitary time development.

If the Hamiltonian H exists and is the generator of a continuous time development, then the S-matrix may be expressed in terms of H.[8] The S-operator is still a meaningful concept, however, even if this is not the case. Heisenberg's assumption that it is the S-operator (and not the Hamiltonian) that is the fundamental physical quantity has become the basis for an attempt at a new formulation of elementary-particle theory known as *S-matrix theory*.

The first way thus assumes more and allows one to relate the observable quantities, such as cross sections, to fundamental observables that reveal more of the physical system's structure. The second way denies the pos-

---

[7] The assumption (4.1) that H can be split is not really essential. One can derive expressions for the cross sections from the assumption of the existence of a time-development generator H alone. See, for example, Goldberger and Watson (1964), Chapter 5.

[8] In this context the S-matrix was first introduced by J. A. Wheeler in 1937.

sibility of this deeper insight into the structure of the physical system. For atomic scattering experiments in nonrelativistic quantum mechanics, the existence of an elementary length of less than 10^{-13} cm ($=10^{-5}$ Å) can be ignored. We can therefore connect the cross section to our basic assumptions for quantum mechanics (in particular to the axiom V) by proceeding in the first way, along which we shall meet the S operator as a derived quantity.

In the next chapter we shall examine in more detail the concept of incoming and outgoing states and their relationship to the S-matrix.

In Section XIV.5 we shall give a derivation of the cross section from the basic assumptions of quantum mechanics under very general conditions. Here we list a few special results for the benefit of the reader who does not want to go through the tedious derivations of the following section.

The cross section is expressed in terms of the matrix elements of the interaction Hamiltonian V or the T-matrix. The matrix elements of V and T are related by

$$\langle E\,\Omega\,\eta_b|\,T\,|E\,\Omega'\,\eta_A\rangle = \langle E\,\Omega\,\eta_b|\,V\,|E\,\Omega'\,\eta_A^+\rangle \qquad (5.41')$$

Here $E = p^2/2m + E_\eta$ denotes the energy of the target-projectile system, E_η denotes the internal energy (which may be a function of the internal quantum numbers η), and $\Omega = (\theta\phi)$ denotes the direction of the projectile momentum, $\mathbf{p} = p\Omega$. $|E\,\Omega\,\eta^+\rangle$ is the eigenvector of the exact operator H, and $|E\,\Omega\,\eta\rangle$ is the eigenvector of the free-energy operator $K = H - V$. If a Hamiltonian time development does not exist, then the T-matrix is connected with the S-matrix by

$$\langle E\,\Omega\,\eta|\,S\,|E'\,\Omega'\,\eta'\rangle = \langle E\,\Omega\,\eta|E'\,\Omega'\,\eta'\rangle - 2\pi i \delta(E - E')\langle E\,\Omega\,\eta|\,T\,|E\,\Omega'\,\eta'\rangle.$$

$$(\text{XV.3.36}')$$

If the generalized eigenvectors are normalized according to

$$\langle E\,\Omega\,\eta|E'\,\Omega'\,\eta'\rangle = \langle \mathbf{p}\,\eta|\mathbf{p}'\,\eta'\rangle = \delta_{\eta\eta'}\,\delta^3(\mathbf{p} - \mathbf{p}'),$$

then the differential cross section for the scattering of a projectile with mass m_A and momentum $\mathbf{p}_A$ on a target in a state η_A into a particle with mass m_b going into the direction $\Omega_b = \mathbf{p}_b/p_b$ and leaving the target in the new state η_b, is given by

$$\frac{d\sigma}{d\Omega}(E_A\,\Omega_b\,\eta_b \leftarrow \mathbf{p}_A\eta_A) = \frac{(2\pi)^4\hbar^2 m_A m_b p_b(E_A)}{p_A}|\langle E_A\,\Omega_b\,\eta_b|\,T\,|\mathbf{p}_A\,\eta_A\rangle|^2, \qquad (5.61')$$

where

$$p_b(E_A) = \sqrt{2m_b(E_A - E_{\eta_b})}, \qquad E_A = \frac{p_A^2}{2m_A} + E_{\eta_A}.$$

For elastic scattering, $m_A = m_b$, $\eta_A = \eta_b$, this goes into (5.62) at the end of Section XIV.5, which can be expressed in terms of the elastic-scattering amplitude (5.63) given by (5.64).

XIV.5[9] Derivation of Cross-Section Formulas for the Scattering of a Beam off a Fixed Target

We shall now give the precise expression for the quantum-mechanical definition of the cross section in terms of the state and the transition probability. We shall then proceed to derive various cross-section formulas.

The transition probability per unit time, corresponding to $d(N/N_T N_B)/dt$ for classical particle scattering, is in quantum mechanics given by $d\langle\Lambda\rangle_t/dt$. The probability that the system will make a transition during the time period Δt taken by the experiment is $(d\langle\Lambda\rangle_t/dt)\,\Delta t$.

The probability density for the beam position, corresponding to ρ_B/N_B for classical particle scattering, is given in quantum mechanics by $\langle\mathbf{x}|W_B^{in}(t)|\mathbf{x}\rangle$, where $W_B^{in}(t)$ is the statistical operator describing the state of the incident beam system. If we assume a small spread in the velocities of the beam's constituent particles, the incident probability per unit area per unit time for a beam that is incident in the z-direction with an average velocity v_0 is $\langle\mathbf{x}|W_B^{in}(t)|\mathbf{x}\rangle v_0$.

According to its definition (3.7) and corresponding to its classical expression (3.3), the cross section for an experiment that runs over the period Δt is then given as the proportionality factor σ in the equation

$$\frac{d\langle\Lambda\rangle_t}{dt}\,\Delta t = \sigma\langle\mathbf{x}|W_B^{in}(t)|\mathbf{x}\rangle v_0\,\Delta t. \tag{5.1}$$

If the experiment extends over a long time period, the defining equation (5.1) goes over into the equation

$$\int_{-\infty}^{+\infty} dt\,\frac{d\langle\Lambda\rangle_t}{dt} = \int_{-\infty}^{+\infty} dt\,\sigma\langle\mathbf{x}|W_B^{in}(t)|\mathbf{x}\rangle v_0. \tag{5.2}$$

If the beam does not consist of structureless quantum physical systems but has internal degrees of freedom described by a set of quantum numbers λ, then the beam's position probability density is $\sum_\lambda \langle\mathbf{x}\,\lambda|W_B^{in}(t)|\mathbf{x}\,\lambda\rangle$, where $|\mathbf{x}\,\lambda\rangle$ is a basis of generalized position eigenvectors for the space of physical states of the beam system. The quantum numbers λ usually describe the polarization (spin state), and we shall therefore generally refer to λ as the polarization, even though in the case of more complicated projectiles (e.g., atoms, molecules, or ions), λ could stand for a whole set of quantum numbers upon which the energy eigenvalues of the beam system may depend.

If there are polarizations, then one has

$$\int_{-\infty}^{+\infty} dt\,\frac{d\langle\Lambda\rangle_t}{dt} = \int_{-\infty}^{+\infty} dt\,\sigma v_0 \sum_\lambda \langle\mathbf{x}\,\lambda|W_B^{in}(t)|\mathbf{x}\,\lambda\rangle \tag{5.3}$$

instead of (5.2).

---

[9] To follow these derivations may require special effort. We have therefore presented a few special results needed for the following chapters at the end of Section XIV.4 and this section may be omitted in first reading.

We assume in this section that the target is fixed in position; v_0 is thus the average relative speed between projectile (beam) and target. We further assume that in the absence of the interaction V between the component systems, the statistical operator $W^{in}(t)$ for the combined projectile-target system would be factorable into the product

$$W^{in}(t) = W_B^{in}(t) \otimes W_T^{in}(t) \qquad (5.4)$$

of a statistical operator $W_B^{in}(t)$ describing the state of the projectile with a statistical operator $W_T^{in}(t)$ describing the state of the target. $W_B^{in}(t)$ acts in a space of states $\mathscr{H}_B$, $W_T^{in}(t)$ acts in a space $\mathscr{H}_T$, and $W^{in}(t)$ acts in the total space $\mathscr{H} = \mathscr{H}_B \otimes \mathscr{H}_T$.

The justification for these assumptions lies in the initial conditions: $W^{in}(t)$ can be thought of as the operator which describes the system before the interaction becomes effective (i.e., for $t \to -\infty$), for it is then that the beam and target system are prepared. Developing in time according to

$$W^{in}(t) = e^{-iKt} W^{in} e^{iKt},$$

where K is the interaction free energy operator, it describes a fictitious state in which the beam passes through the target without interaction. The preparation of the component systems is by way of measurements of properties which are uncorrelated at the time of preparation, with the consequence that at the time of preparation $W^{in}(t)$ is factorable into the form (5.4).[10] But then it remains so at all times, as follows from:

$$
\begin{aligned}
W^{in}(t) &= e^{-i(K_B + K_T)t} (W_B^{in} \otimes W_T^{in}) e^{+i(K_B + K_T)t} \\
&= e^{-iK_B t} W_B^{in} e^{iK_B t} \otimes e^{-iK_T t} W_T^{in} e^{iK_T t} \\
&= W_B^{in}(t) \otimes W_T^{in}(t).
\end{aligned}
\qquad (5.5a)
$$

The actual state of the combined system is $W(t)$, which, while it is essentially the same as $W^{in}(t)$ in the distant past, has a different time development [Equation (1.2)]:

$$W(t) = e^{-iHt} W e^{+iHt}. \qquad (5.5b)$$

Unlike $W^{in}(t)$, $W(t)$ is generally not factorable, because the interaction part V of its time-development generator $H = K + V$ acts nontrivially in both $\mathscr{H}_B$ and $\mathscr{H}_T$.

Rather than interrupt the calculations to follow with many explanations, we shall first make some remarks about the basis vectors which will be used and their normalizations, and shall then list some needed mathematical relations.

For the basis $\{|E\,\hat{a}\rangle\}$ of $\mathscr{H} = \mathscr{H}_B \otimes \mathscr{H}_T$ we shall use the direct-product basis consisting of the generalized K eigenvectors

$$|E\,\hat{a}\rangle = |\mathbf{p}\,\lambda\rangle \otimes |E_T\,\eta\rangle. \qquad (5.6)$$

Here $|\mathbf{p}\,\lambda\rangle$ is a generalized eigenvector with polarization λ of the momentum

---

[10] Cf. remarks in Section XIII.1 concerning the state of a combination of physical systems.

operator $\mathbf{P}$ for the beam system, and $|E_T \eta\rangle$ is a (possibly generalized) eigen-vector with polarization η of the operator $K_T = K_T^{\text{int}}$ for the internal energy of the target system.[11] The generalized eigenvectors $|\mathbf{p}\, \lambda\rangle$ have the usual δ-function normalization,

$$\langle \mathbf{p}\, \lambda | \mathbf{p}'\, \lambda'\rangle = \delta^3(\mathbf{p} - \mathbf{p}')\delta_{\lambda\lambda'} \tag{5.7}$$

while the $|E_T \eta\rangle$'s may be either proper eigenvectors $|E_T^d \eta\rangle$ corresponding to a discrete eigenvalue E_T^d and having normalization

$$\langle E_T^d \eta | E_T^{d'} \eta'\rangle = \delta_{E_T^d E_T^{d'}}\delta_{\eta\eta'}, \tag{5.8a}$$

or generalized eigenvectors $|E_T^c \eta\rangle$ corresponding to the continuous eigenvalue E_T^c and having normalization

$$\langle E_T^c \eta | E_T^{c'} \eta'\rangle = \tilde{\rho}(E_T^c)^{-1}\delta(E_T^c - E_T^{c'})\delta_{\eta\eta'}. \tag{5.8b}$$

Here $\tilde{\rho}(E)$ denotes the weight function (measure) with respect to which the generalized eigenvectors of the continuous spectrum are normalized.[3] These weight functions $\rho(E)$ of the generalized energy eigenvectors $|E\rangle$ are a convention which will be fixed for each particular case. For example, for the energy eigenvectors of the beam system, $|\mathbf{p}\, \lambda\rangle = |E_B\Omega, \lambda\rangle$, the weight function $\rho(E)$ is fixed by the normalization (5.7) and the usual convention for the solid angle δ-function $\delta^2(\Omega - \Omega')$ given in (5.17) below. Since the $|\mathbf{p}\, \lambda\rangle$'s are also the generalized eigenvectors of the beam system's energy operator

$$K_B = K_B^{\text{kin}}(\mathbf{P}) + K_B^{\text{int}}(\lambda^{\text{op}}), \tag{5.9}$$

the product vectors $|\mathbf{p}\, \lambda\rangle \otimes |E_T \eta\rangle$ are in fact the generalized eigenvectors of the energy operator $K = K_B + K_T$. In (5.9) $K_B^{\text{kin}}(\mathbf{P})$ is the kinetic energy operator for the projectile, the functional form of which depends on the nature of the projectile; for a massive nonrelativistic projectile

$$K_B^{\text{kin}}(\mathbf{P}) = \mathbf{P}^2/2m, \tag{5.10a}$$

while for a photon

$$K_B^{\text{kin}}(\mathbf{P}) = c(\mathbf{P}^2)^{1/2} = cP, \tag{5.10b}$$

where c denotes the speed of light. $K_B^{\text{int}}(\eta^{\text{op}})$ is the operator for the internal energy of the projectile. The eigenvalues of K on the basis vectors (5.6) are therefore

$$E = E_B + E_T \tag{5.11}$$

where

$$E_B = \frac{p^2}{2m} + E_B^{\text{int}}(\lambda), \qquad E_B = \frac{p^2}{2m}, \quad \text{or} \quad E_B = pc$$

---

[11] Though K_T^{int} is the operator for the internal energy of the target (i.e., has no kinetic-energy term) we shall not exclude the possibility that it also has a continuous spectrum, as may occur, for example, in collision-induced scattering.

for a nonrelativistic projectile with internal energy levels $E_B^{int}(\lambda)$, a structure-less[12] nonrelativistic projectile, or a photon, respectively.

For the sake of definiteness we shall assume that the projectile is non-relativistic and has no internal energy levels, so $E_B = p^2/2m$. For the other cases in (5.11) or for the massive relativistic case $E_B = (p^2c^2 + m^2c^4)^{1/2}$, minor modifications will be necessary. We shall also use the abbreviated notation

$$\sum_{E_T} = \sum_{E_T^d} + \int \rho(E_T^c) \, dE_T^c \qquad (5.12a)$$

and

$$\langle E_T \, \eta \, | \, E_T' \, \eta' \rangle = \delta_{E_T E_T'} \, \delta_{\eta \eta'} \qquad (5.12b)$$

for Equations (5.8a) and (5.8b). We may then express the product basis vectors $|p\,\lambda\rangle \otimes |E_T\,\eta\rangle$ in terms of the total energy E and the spherical coordinates of p:

$$|E\,\hat{a}\rangle = |E\,E_T \Omega\,\lambda\,\eta\rangle = |p\,\lambda\rangle \otimes |E_T\,\eta\rangle$$
$$= |E_B \Omega\,\lambda\rangle \otimes |E_T\,\eta\rangle, \qquad (5.13)$$

where

$$E = E_B + E_T = \frac{p^2}{2m} + E_T, \qquad (5.14a)$$

$$p = p\Omega = p(\sin\theta\cos\phi, \sin\theta\sin\phi, \cos\theta), \qquad (5.14b)$$

and

$$\Omega = (\theta, \phi). \qquad (5.14c)$$

The δ-function normalization of the product vectors $|E\,\hat{a}\rangle$ follows from the normalization (5.7) and (5.12a). We now determine the weight function $\rho(E)$ in the normalization (2.9a) and (2.10). Since from (5.7) and (5.12a) we have

$$\sum_{\hat{a}} \int \rho_{\hat{a}}(E) \, dE = \sum_{E_T \lambda \eta} \int d^3p = \sum_{E_T \lambda \eta} \int p^2 \sin\theta \, dp \, d\theta \, d\phi$$

$$= \sum_{E_T \lambda \eta} \int m\sqrt{2m(E - E_T)} \sin\theta \, d\theta \, d\phi \, dE, \qquad (5.15)$$

the basis vectors (5.13) must have—in terms of the quantum numbers E, E_T, Ω, λ and η—the following normalization:

$$\langle E\hat{a} | E'\hat{a}' \rangle = \langle EE_T \Omega\,\lambda\,\eta \, | \, E'\,E_T' \Omega'\,\lambda'\,\eta' \rangle$$

$$= \frac{1}{m\sqrt{2m(E - E_T)}} \, \delta(E - E')\delta_{E_T E_T'}\delta^2(\Omega - \Omega')\delta_{\lambda\lambda'}\,\delta_{\eta\eta'} \qquad (5.16)$$

---

[12] More precisely, a projectile with only one energy level, which still may have different polarization states.

where

$$\delta^2(\Omega - \Omega') = \delta(\cos\theta - \cos\theta')\delta(\phi - \phi'). \tag{5.17}$$

The weight function (measure) $\rho_{\hat{a}}(E)$ is therefore fixed to be

$$\rho_{E_T}(E) = \rho(E - E_T) = m\sqrt{2m(E - E_T)} = mp, \tag{5.18a}$$

and $\delta_{\hat{a}\hat{a}'}$ of (2.9) is

$$\delta_{E_T E_T'}\delta^2(\Omega - \Omega')\delta_{\lambda\lambda'}\delta_{\eta\eta'}. \tag{5.18b}$$

The mathematical relations that we need in the calculations are the identities[13]

$$\int_{-\infty}^{+\infty} dt\, e^{-ist} = 2\pi\delta(s) \tag{5.19}$$

and

$$\frac{1}{s - i0} - \frac{1}{s + i0} = 2\pi i\delta(s) \tag{5.20}$$

for the generalized function $\delta(s)$.

We shall also need the relationship

$$\langle a^+|W(t)|a'^+\rangle = \langle a|W^{\text{in}}(t)|a'\rangle, \tag{5.21}$$

which can be derived immediately (Appendix XV.A) using the results of Section XV.3 below. Equation (5.21) is plausible for the following reason: $W(t)$ is identical with $W^{\text{in}}(t)$ in the distant past before the interaction V becomes effective, and $|a^+\rangle$ is the same as $|a\rangle$ in the absence of V, so (5.21) holds at some time in the distant past. But the time developments of $W(t)$ and $W^{\text{in}}(t)$ are generated by H and by K, respectively, which have the same effect when applied to $|a^+\rangle$ and $|a\rangle$, respectively; consequently, (5.21) holds for all times t, not just in the distant past.

We now start the derivation of explicit formulas for the cross section.[14] Inserting (2.18) into (5.3), one obtains

$$-i\sum_b\sum_{aa'}\int_{-\infty}^{+\infty} dt\, e^{-i(E_a - E_{a'})t}\langle b|V|a^+\rangle\langle a'^+|V|b\rangle\langle a^+|W|a'^+\rangle$$

$$\times\left(\frac{1}{E_{a'} - E_b - i0} - \frac{1}{E_a - E_b + i0}\right)$$

$$= \int_{-\infty}^{+\infty} dt\, v_0\sigma\sum_\lambda\langle\mathbf{x}\lambda|W_B^{\text{in}}(t)|\mathbf{x}\lambda\rangle. \tag{5.22}$$

The right-hand side may be written by inserting the basis vectors $|\mathbf{p}\lambda\rangle$, using the three-dimensional version

$$\langle\mathbf{x}\,\lambda|\mathbf{p}\,\lambda'\rangle = (2\pi)^{-3/2}e^{i\mathbf{p}\cdot\mathbf{x}}\delta_{\lambda\lambda'} \tag{5.23}$$

---

[13] Gel'fand and Shilov (Vol. 1, pp. 168, 94).

[14] The result of this tedious calculation is Equation (5.38).

of (II.7.51), the time development (5.5b) of $W_B^{in}(t)$, and the identity (5.19):

$$\text{RHS of (5.22)} = i \int_{-\infty}^{+\infty} dt\, v_0 \sigma \sum_\lambda \int d^3p\, d^3p' \langle \mathbf{x}\, \lambda | \mathbf{p}\, \lambda \rangle$$

$$\times \langle \mathbf{p}\, \lambda | e^{-iK_{B}t} W_B^{in} e^{+iK_{B}t} | \mathbf{p}'\, \lambda \rangle \langle \mathbf{p}'\, \lambda | \mathbf{x}\, \lambda \rangle$$

$$= \int_{-\infty}^{+\infty} dt\, v_0 \sigma \sum_\lambda \int d^3p\, d^3p' (2\pi)^{-3} e^{i(\mathbf{p}-\mathbf{p}')\cdot\mathbf{x}} e^{-i(E_B - E_B')t}$$

$$\times \langle \mathbf{p}\, \lambda | W_B^{in} | \mathbf{p}'\, \lambda \rangle$$

$$= (2\pi)^{-2} v_0 \sigma \sum_{\lambda\lambda'} \int d^3p\, d^3p'\, e^{i(\mathbf{p}-\mathbf{p}')\cdot\mathbf{x}} \delta(E_B - E_B')$$

$$\times \langle \mathbf{p}\, \lambda | W_B^{in} | \mathbf{p}'\, \lambda' \rangle \delta_{\lambda\lambda'}$$

$$= (2\pi)^{-2} v_0 \sigma \sum_{\lambda\lambda'} \int \rho(E_B)\, dE_B\, \rho(E_B')\, dE_B'\, d\Omega\, d\Omega'$$

$$\times e^{i(\mathbf{p}-\mathbf{p}')\cdot\mathbf{x}} \delta(E_B - E_B') \langle E_B \Omega\, \lambda | W_B^{in} | E_B' \Omega'\, \lambda' \rangle \delta_{\lambda\lambda'}, \quad (5.24)$$

where

$$\mathbf{p} = p\Omega = \sqrt{2mE_B}\,\Omega, \qquad \mathbf{p}' = p'\Omega' = \sqrt{2mE_B'}\,\Omega', \quad \text{and} \quad \rho(E) = m\sqrt{2mE}$$

The calculation of the left-hand side of (5.22) goes as follows: We use (5.19) to do the t-integration, (5.20) to replace the factor in parentheses, and (5.21) to express things in terms of the state $W^{in}(t)$, which may be broken up by use of (5.4). The basis system $\{|a\rangle\}$ used in the calculation is that consisting of the generalized eigenvectors $|E\, E_T\, \Omega\, \lambda\, \eta\rangle = |E_B \Omega\, \lambda\rangle \otimes |E_T \eta\rangle$ with $E = E_a$. So

$$\text{LHS of (5.22)} = -i \sum_b \sum_{aa'} 2\pi\delta(E_a - E_{a'}) \langle b|V|a^+\rangle \langle a'^+|V|b\rangle$$

$$\times \langle a|W^{in}|a'\rangle 2\pi i\delta(E_a - E_b)$$

$$= (2\pi)^2 \sum_b \sum_{\substack{E_T\lambda\eta \\ E_T'\lambda'\eta'}} \int \rho(E - E_T)\, dE\, \rho(E' - E_T')\, dE'\, d\Omega\, d\Omega'$$

$$\times \delta(E - E')\delta(E - E_b)\langle b|V|EE_T\Omega\, \lambda\, \eta^+\rangle$$

$$\times \langle E'\, E_T'\, \Omega'\, \lambda'\, \eta'^+|V|b\rangle$$

$$\times \langle E\, E_T\, \Omega\, \lambda\, \eta|W^{in}|E'\, E_T'\, \Omega'\, \lambda'\, \eta'\rangle$$

$$= (2\pi)^2 \sum_b \sum_{\substack{E_T\lambda\eta \\ E_T'\lambda'\eta'}} \int \rho(E_B)\, dE_B\, \rho(E_B')\, dE_B'\, d\Omega\, d\Omega'$$

$$\times \delta(E_B + E_T - E_B' - E_T')\delta(E_B + E_T - E_b)$$

$$\times \langle b|V|E\, E_T\, \Omega\, \lambda\, \eta^+\rangle \langle E'\, E_T'\, \Omega'\, \lambda'\, \eta'^+|V|b\rangle$$

$$\times \langle E_B\Omega\, \lambda|W_B^{in}|E_B'\, \Omega'\, \lambda'\rangle \langle E_T\eta|W_T^{in}|E_T'\, \eta'\rangle \quad (5.25)$$

where

$$E = E_B + E_T \quad \text{and} \quad E' = E'_B + E'_T.$$

From (5.24) and (5.25) we then have

$$0 = (2\pi)^2(\text{RHS of (5.22)} - \text{LHS of (5.22)})$$

$$= \sum_{\lambda\lambda'} \int \rho(E_B) \, dE_B \, \rho(E'_B) \, dE'_B \, d\Omega \, d\Omega' \langle E_B \Omega \lambda | W_B^{\text{in}} | E'_B \Omega' \lambda' \rangle$$

$$\times \left\{ v_0 \sigma \delta(E_B - E'_B) e^{i(\mathbf{p} - \mathbf{p}')\cdot\mathbf{x}} \delta_{\lambda\lambda'} \right.$$

$$- (2\pi)^4 \sum_b \sum_{\substack{E_T\eta \\ E'_T\eta'}} \delta(E_B + E_T - E'_B - E'_T)\delta(E_B + E_T - E_b)$$

$$\left. \times \langle b | V | E E_T \Omega \lambda \eta^+ \rangle \langle E' E'_T \Omega' \lambda' \eta'^+ | V | b \rangle \langle E_T \eta | W_T^{\text{in}} | E'_T \eta' \rangle \right\}.$$

$$(5.26)$$

So far we have not specified any properties of the state W_B^{in}, but we know already that the direction of the incident momentum is usually fixed in these scattering experiments to be along the z-axis of Figure 3.1; i.e., Ω is fixed to be $\Omega_0 = (0, 0)$. The other properties of the incident beam—i.e. its polarization and its energy distribution—can vary from experiment to experiment. We shall continue the calculation under the additional assumption that the beam is completely unpolarized, in which case the density matrix is given by:

$$\langle E_B \Omega \lambda | W_B^{\text{in}} | E'_B \Omega' \lambda' \rangle = \frac{1}{g} \delta_{\lambda\lambda'} \langle E_B \Omega \| W_B^{\text{in}} \| E'_B \Omega' \rangle, \qquad (5.27)$$

where g is the number of different polarization states λ. The reduced matrix element $\langle E_B \Omega \| W_B^{\text{in}} \| E_B \Omega' \rangle$ describes the momentum distribution of the incident beam. Were the beam completely polarized with some definite value λ_0 for λ, one would have to use

$$\langle E_B \Omega \lambda | W_B^{\text{in}} | E'_B \Omega' \lambda' \rangle = \delta_{\lambda\lambda'} \delta_{\lambda\lambda_0} \langle E_B \Omega \| W_B^{\text{in}} \| E'_B \Omega' \rangle \qquad (5.28)$$

instead of (5.27) and would have to replace $\sum_\lambda 1/g$ by $\sum_\lambda \delta_{\lambda\lambda_0}$ in the calculations below.

As in the typical scattering experiment, the incident beam is always prepared so that it has a well-defined (momentum) direction Ω_0 and the density matrix of W_B^{in} has the property[15]

$$\int d\Omega \, d\Omega' \langle E_B \Omega \| W_B^{in} \| E_B' \Omega' \rangle F(\Omega, \Omega') = \langle E_B \| W_B^{in} \| E_B' \rangle F(\Omega_0, \Omega_0) \quad (5.29a)$$

for any smooth function $F(\Omega, \Omega')$. The doubly reduced matrix element

$$\langle E_B \| W_B^{in} \| E_B' \rangle = \int d\Omega \, d\Omega' \langle E_B \Omega \| W_B^{in} \| E_B' \Omega' \rangle$$

$$= \int d\Omega \langle E_B \Omega \| W_B^{in} \| E_B' \Omega \rangle \quad (5.29b)$$

of the state W_B^{in} with well-defined direction Ω_0 describes the energy distribution in the beam, which will be discussed in more detail below. The normalization of W_B^{in} requires that

$$1 = \sum_\lambda \int \rho(E_B) \, dE_B \, d\Omega \langle E_B \Omega \lambda | W_B^{in} | E_B \Omega \lambda \rangle$$

$$= \int \rho(E_B) \, dE_B \, d\Omega \langle E_B \Omega \| W_B^{in} \| E_B \Omega \rangle$$

$$= \int \rho(E_B) \, dE_B \langle E_B \| W_B^{in} \| E_B \rangle. \quad (5.29c)$$

For an unpolarized beam (5.27) with well-defined direction (5.29a), Equation (5.26) becomes

$$0 = \int \rho(E_B) \, dE_B \, \rho(E_B') \, dE_B' \langle E_B \| W_B^{in} \| E_B' \rangle \Big\{ v_0 \, \sigma \delta(E_B - E_B') e^{i(p - p')\mathbf{\Omega}_0 \cdot \mathbf{x}}$$

$$- \frac{(2\pi)^4}{g} \sum_b \sum_\lambda \sum_{\substack{E_T \eta \\ E_T' \eta'}} \delta(E_B + E_T - E_B' - E_T') \delta(E_B + E_T - E_b)$$

$$\times \langle b | V | E \, E_T \, \Omega_0 \, \lambda \, \eta^+ \rangle \langle E' \, E_T' \, \Omega_0 \, \lambda \, \eta'^+ | V | b \rangle \langle E_T \eta | W_T^{in} | E_T' \eta' \rangle \Big\}.$$

---

[15] One can convince oneself that (5.29a) is the continuous-spectrum analogue of (5.28) by calculating for an arbitrary $F_{\lambda\lambda'}$

$$\sum_{\lambda\lambda'} \langle \cdots \lambda | W_B^{in} | \cdots \lambda' \rangle F_{\lambda\lambda'} = \sum_{\lambda\lambda'} \delta_{\lambda\lambda'} \delta_{\lambda\lambda_0} \langle \cdots \| W_B^{in} \| \cdots \rangle F_{\lambda\lambda'} = \langle \cdots \| W_B^{in} \| \cdots \rangle F_{\lambda_0\lambda_0}.$$

Intuitively one may want to write as the direct analogue of (5.28)

$$\langle E_B \Omega \| W_B^{in} \| E_B' \Omega' \rangle = \delta(\Omega - \Omega') \delta(\Omega - \Omega_0) \langle E_B \| W_B^{in} \| E_B' \rangle,$$

which, however, would be mathematically incorrect. Equation (5.29a) is an idealization; according to the discussions in Section II.8 every beam must have a finite spread in momentum. Cf. also the discussion following (5.35) below.

As $p = \sqrt{2mE_B}$, the exponential under the integral contributes only unity because of the $\delta(E_B - E'_B)$. Hence

$$
0 = \int \rho(E_B)\, dE_B\, \rho(E'_B)\, dE'_B \langle E_B \|\| W^{\text{in}}_B \|\| E'_B\rangle \{v_0\,\sigma\delta(E_B - E'_B)
$$

$$
- \frac{(2\pi)^4}{g} \sum_b \sum_\lambda \sum_{\substack{E_T\eta \\ E_T\eta'}} \delta(E_B + E_T - E'_B - E'_T)\delta(E_B + E_T - E_b)
$$

$$
\times \langle b|V|E\, E_T\, \Omega_0\, \lambda\, \eta^+\rangle \langle E'\, E'_T\, \Omega_0\, \lambda\, \eta'^+|V|b\rangle \langle E_T\, \eta|W^{\text{in}}_T|E'_T\, \eta'\rangle\}.
$$

$$(5.30)$$

Recall that we have left open the possibility that the target has continuous as well as discrete energy levels E_T; although we have written (5.30) as if the E_T were discrete, it is also valid in the case (5.12) where there is a continuous part to the spectrum of K_T. In most experiments (although not for processes like collision-induced scattering) the state of the target system is in a mixture of discrete energy eigenstates (stationary states) before interacting with the beam:

$$
[K, W^{\text{in}}_T] = [K_T, W^{\text{in}}_T] = 0. \tag{5.31}
$$

Thus

$$
W^{\text{in}}_T = \sum_{E^d_T\eta\eta'} \langle \eta\| W^{\text{in}}_T(E^d_T)\|\eta'\rangle |E^d_T\, \eta\rangle \langle E^d_T\, \eta'| \tag{5.31'}
$$

or

$$
\langle E^d_T\, \eta|W^{\text{in}}_T|E^{d'}_T\, \eta'\rangle = \delta_{E^d_T E^d_{T'}}\langle \eta\| W^{\text{in}}_T(E^d_T)\|\eta'\rangle. \tag{5.32}
$$

Then (5.30) becomes

$$
0 = \int \rho(E_B)\, dE_B\, \rho(E'_B)\, dE'_B \langle E_B \|\| W^{\text{in}}_B \|\| E'_B\rangle
$$

$$
\times \left\{ v_0\,\sigma\delta(E_B - E'_B) - \frac{(2\pi)^4}{g} \sum_b \sum_\lambda \sum_{E^d_T\eta\eta'} \delta(E_B - E'_B)\delta(E_B + E^d_T - E_b) \right.
$$

$$
\left. \times \langle b|V|E\, E^d_T\, \Omega_0\, \lambda\, \eta^+\rangle \langle E'\, E^d_T\, \Omega_0\, \lambda\, \eta'^+|V|b\rangle \langle \eta\| W^{\text{in}}_T(E^d_T)\|\eta'\rangle \right\}
$$

$$
= \int \rho(E_B)\, dE_B\, \rho(E_B)\langle E_B \|\| W^{\text{in}}_B \|\| E_B\rangle
$$

$$
\times \left\{ v_0\,\sigma - \frac{(2\pi)^4}{g} \sum_b \sum_\lambda \sum_{E^d_T\eta\eta'} \delta(E_B + E^d_T - E_b) \right.
$$

$$
\left. \times \langle b|V|E\, E^d_T\, \Omega_0\, \lambda\, \eta^+\rangle \langle E\, E^d_T\, \Omega_0\, \lambda\, \eta'^+|V|b\rangle \langle \eta\| W^{\text{in}}_T(E^d_T)\|\eta'\rangle \right\}, \tag{5.33}
$$

where $E = E_B + E^d_T$ and $E' = E'_B + E^d_T$. Before we can proceed with our calculation we have to specify the energy distribution of the beam. Taking the expectation value of the beam energy K_B in the beam state W_B^{in} of (5.27) and (5.29),

$$
\begin{aligned}
\mathrm{Tr}(K_B W_B^{in}) &= \sum_\lambda \int \rho(E_B)\, dE_B\, d\Omega \langle E_B \Omega \lambda | W_B^{in} K_B | E_B \Omega \lambda \rangle \\
&= \int \rho(E_B)\, dE_B\, d\Omega\, E_B \langle E_B \Omega \| W_B^{in} \| E_B \Omega \rangle \\
&= \int \rho(E_B)\, dE_B\, E_B \langle E_B \| W_B^{in} \| E_B \rangle,
\end{aligned}
\tag{5.34}
$$

we see that

$$
F(E_B - E_{B0}) = \rho(E_B)\langle E_B \| W_B^{in} \| E_B \rangle \tag{5.35}
$$

represents the probability distribution for obtaining the value E_B when K_B is measured.

In the usual setup of scattering experiments the energy distribution is peaked around some value E_{B0}, which is under the control of the experimenter. Then $F(E_B - E_{B0})$ is one of the functions $F(x)$ depicted in Figure II.8.1. It describes the energy resolution of the apparatus which prepares the beam. The better the energy resolution, the narrower is the peak of $F(E_B - E_{B0})$. In the ideal but unphysical limit case of an exact beam energy, $F(E_B - E_{B0})$ will go into $\delta(E_B - E_{B0})$.

Even in the realistic case of a finite energy resolution one can use as the energy distribution function $F(E_B - E_{B0})$ the generalized function $\delta(E_B - E_{B0})$ if the T-matrix elements $\langle b | V | E\, E_T^d\, \Omega_0\, \lambda \eta^+ \rangle$ are slowly varying functions of E_B. Let F be different from zero only in an interval of width $2\Delta E$ around E_{B0}—describing a beam with a small energy spread $E_{B0} - \Delta E < E_B < E_{B0} + \Delta E$—then the function $F(E_B - E_{B0})$ will act like $\delta(E_B - E_{B0})$ in an integral

$$
\int dE_B F(E_B - E_{B0}) g(E_B)
$$

if the function $g(E_B)$ varies slowly over the range $E_{B0} - \Delta E_B < E_B < E_{B0} + \Delta E_B$. In (5.33) this will be the case if the matrix elements $\langle b | V | E\, E_T^d\, \Omega_0\, \lambda \eta \rangle$, where $E = E_B + E_T^d$, vary slowly as functions of E_B over the interval $2\Delta E$.[16]

---

[16] Thus if the experimental resolution can be made narrower than the interval over which the T-matrix varies significantly, the energy distribution of the state can be described by a generalized function and the state by a generalized eigenvector whose energy wave function is a generalized function. This generalized eigenvector is given by

$$
|E_0 + i\,0\rangle = \lim_{\epsilon \to 0} \frac{1}{\rho^{1/2}(E_0)} \int \rho(E')\, dE' |E'\rangle \frac{\sqrt{\epsilon/\pi}}{E' - E_0 + i\epsilon}
$$

and has as energy wave function the generalized function

$$
\langle E | E_0 + i\,0 \rangle = \frac{1}{\rho^{1/2}(E_0)} \lim_{\epsilon \to 0} \frac{\sqrt{\epsilon/\pi}}{E - E_0 - i\epsilon}.
$$

It does not describe a physically preparable state and is not an element of Φ (cf. Section II.8).

This is not always true. Near some values of E_{B0} the matrix elements may change very quickly; such a value E_{B0} is called a *resonance*.

We shall discuss the effect of the finite resolution below and continue the calculation here under the assumption that the beam is ideally monochromatic, i.e., that

$$F(E_B - E_{B0}) = \rho(E_B)\langle E_B \|W_B^{in}\| E_B\rangle \to \delta(E_B - E_{B0}). \qquad (5.36)$$

In this case the E_B integration in (5.33) can be performed to get

$$0 = \rho(E_{B0})\left\{v_0\sigma - \frac{(2\pi)^4}{g}\sum_b\sum_\lambda\sum_{E_T^d\eta\eta'}\delta(E_{B0} + E_T^d - E_b)\langle b|V|E\,E_T^d\,\Omega_0\,\lambda\,\eta^+\rangle\right.$$

$$\left. \times\,\langle E\,E_T^d\,\Omega_0\,\lambda\,\eta'^+|V|b\rangle\langle\eta\|W_T^{in}(E_T^d)\|\eta'\rangle\right\}, \qquad (5.37)$$

where $E = E_{B0} + E_T^d$. Solving for σ and reverting to rectangular coordinates $\mathbf{p}_0 = \sqrt{2mE_{B0}}\,\Omega_0$, we at long last have our basic cross-section formula:

$$\sigma(\Lambda \leftarrow \mathbf{p}_0) = \frac{(2\pi)^4\hbar^2}{v_0}\sum_b\overline{\sum_\lambda}\sum_{E_T^d\eta\eta'}\langle b|V|\mathbf{p}_0\,\lambda E_T^d\,\eta^+\rangle\langle\mathbf{p}_0\,\lambda E_T^d\,\eta'^+|V|b\rangle$$

$$\times\,\langle\eta\|W_T^{in}(E_T^d)\|\eta'\rangle\delta(E_{B0} + E_T^d - E_b), \qquad (5.38)$$

where $E_{B0} = p_0^2/2m$, $v_0 = p_0/m$, and where we have introduced the notation

$$\overline{\sum_\lambda} = \frac{1}{g}\sum_\lambda$$

for the averaging over the polarizations in the initial state. We have also restored the $\hbar$'s in order to express this important result in the usual units.

Equation (5.38) is the cross section for the scattering of an unpolarized beam with a well-defined momentum $\mathbf{p}_0$ off a fixed target in a mixture W_T^{in} of discrete energy eigenstates and into any final configuration whose quantum numbers $b = (E_b, \hat{b})$ appear in the summation $\sum_b$. [To get the differential or partial cross section $d\sigma(b \leftarrow \mathbf{p}_0)$ for scattering into the "state" $\Lambda_b = |b\rangle\langle b|$, one merely omits the summation over b and replaces $\sigma(\Lambda \leftarrow \mathbf{p}_0)$ by $d\sigma(b \leftarrow \mathbf{p}_0)$.]

Suppose that the energy spread ΔE_B is not negligible, as will be the case if $\langle b|V|EE_T^d\Omega_0\lambda\eta^+\rangle$ varies considerably when $E_B = E - E_T^d$ varies within the interval ΔE_B (this might happen in the neighborhood of a resonance unless ΔE_B is much smaller than the width of the resonance, as will be discussed in Chapter XVIII). Then the substitution (5.36) is not possible, and consequently one cannot calculate $\sigma(\Lambda \leftarrow \mathbf{p}_0)$ as in (5.37). The quantity one can calculate then from (5.33) is[17]

$$\int \rho(E_B)\,dE_B\,F(E_B - E_{B0})v\sigma(E_B) \equiv (\rho v\sigma) * F, \qquad (5.39a)$$

where $v = v(E_B) = p(E_B)/m = \sqrt{2mE_B}/m$.

[17] If the spread in velocity is not negligible, then the incident probability per unit area is $\int dz\,\langle x|W_B^{in}(t)|x\rangle = \int dt\,\langle x|\mathcal{V}W_B^{in}(t)|x\rangle$ instead of $\int dt\,v_0\langle x|W_B^{in}(t)|x\rangle$, where $\mathcal{V}$ is the velocity operator. This results in a replacement of v_0 by $v(E_B)$ in (5.33).

This, according to (5.33), is then equal to

$$(\rho v \sigma) * F = (2\pi)^4 \int \rho(E_B)\, dE_B\, F(E_B - E_{B0}) \sum_b \overline{\sum_\lambda} \sum_{E_T^d \eta \eta'} \langle b | V | \mathbf{p}\, \lambda\, E_T^d \eta^+ \rangle$$

$$\times \langle \mathbf{p}\, \lambda\, E_T^d \eta'^+ | V | b \rangle \langle \eta \| W_T^{\text{in}}(E_T^d) \| \eta' \rangle \delta(E_B + E_T^d - E_b). \qquad (5.39b)$$

As $\rho(E_B)$ and $v(E_B)$ are slowly varying functions of E_B compared to the rapidly varying functions $F(E_B - E_{B0})$ and $|\langle b | V | E E_T^d \Omega_0\, \lambda \eta^+ \rangle|^2$, one can take them out of the integrals on the left-hand side of (5.39a) and on the right-hand side of (5.39b) and replace them by their average values, $\rho(E_{B0})$ and $v(E_{B0})$. Then one obtains the cross section formula for coarse-resolution experiments:

$$\sigma * F = \int dE_B F(E_B - E_{B0})\sigma(E_B) = (2\pi)^4 \hbar^2 \frac{1}{v(E_{B0})} \int dE_B F(E_B - E_{B0})$$

$$\times \sum_b \overline{\sum_\lambda} \sum_{E_T^d \eta \eta'} \delta(E_B + E_T^d - E_b)\langle b | V | \mathbf{p}\, \lambda\, E_T^d \eta^+ \rangle$$

$$\times \langle \mathbf{p}\, \lambda\, E_T^d \eta'^+ | V | b \rangle \langle \eta \| W_T^{\text{in}}(E_T^d) \| \eta' \rangle. \qquad (5.40)$$

We continue now with the expression (5.38) for the idealized energy resolution, but will return to the case (5.40) of a limited energy resolution in Section XVIII.8.

The cross section $\sigma(\Lambda \leftarrow \mathbf{p}_0)$ is often expressed in terms of the *T-matrix*. If the interaction Hamiltonian V is given, the T-matrix may be defined by

$$\langle E\, \hat{a} | T | E\, \hat{a}' \rangle \equiv \langle E\, \hat{a} | V | E\, \hat{a}'^+ \rangle. \qquad (5.41)$$

This does not fully define a transition operator T, because not all matrix *elements* $\langle E\, \hat{a} | T | E'\, \hat{a}' \rangle$ of T are defined by (5.41), but only those matrix elements "on the energy shell" $E = E'$. The matrix elements $\langle E\, \hat{a} | T | E'\, \hat{a}' \rangle$ that are "off the energy shell," i.e., those for which $E \neq E'$, may be defined in various ways; and one gets differing transition operators depending on how these off-the-energy-shell matrix elements are taken. One such operator is T^+, defined by

$$\langle E\, \hat{a} | T^+ | E'\, \hat{a}' \rangle \equiv \langle E\, \hat{a} | V | E'\, \hat{a}'^+ \rangle. \qquad (5.42)$$

Obviously the restriction of (5.42) to the energy shell gives the quantities $\langle E\, \hat{a} | T | E\, \hat{a}' \rangle$ of (5.41). Another transition operator T^- is defined by

$$\langle E\, \hat{a} | T^- | E'\, \hat{a}' \rangle \equiv \langle E\, \hat{a} | V | E'\, \hat{a}'^- \rangle. \qquad (5.43)$$

Although not obviously, this too agrees with (5.41) when restricted to the energy shell. Any useful definition of a transition operator T^ϵ must agree with (5.41) when on the energy shell:

$$\langle E\, \hat{a} | T^\epsilon | E\, \hat{a}' \rangle = \langle E\, \hat{a} | T | E\, \hat{a}' \rangle. \qquad (5.44)$$

Because of the presence of the factor $\delta(E_{B0} + E_T - E_b)$ (which expresses energy conservation) in (5.38), it is only the on-the-energy-shell T-matrix elements that will contribute, a situation one always encounters when dealing with physically observable quantities. Since all transition operators agree on the energy shell, it makes no difference which T^ϵ is used.

In the S-matrix approach mentioned in the previous section, it is not the existence of an interaction Hamiltonian V that is assumed, but rather that of the T-matrix $\langle E \, \hat{a} \,|\, T \,|\, E \, \hat{a}' \rangle$, which is considered the fundamental quantity. The expression of the cross section formula in terms of the T-matrix,

$$\sigma(\Lambda \leftarrow \mathbf{p}_0) = \frac{(2\pi)^4 \hbar^2}{v_0} \sum_b \sum_\lambda \sum_{E_T^d \cdot \eta \eta'} \delta(E_{B0} + E_T^d - E_b)$$

$$\times \langle b \,|\, T \,|\, \mathbf{p}_0 \, \lambda \, E_T^d \eta \rangle \langle \mathbf{p}_0 \, \lambda \, E_T^d \eta' \,|\, T \,|\, b \rangle \langle \eta \| W_T^{\mathrm{in}}(E_T^d) \| \eta' \rangle \quad (5.38')$$

may therefore be considered valid in either approach.[18] If a V exists, then $\langle E \, \hat{b} \,|\, T \,|\, E \, \hat{a}' \rangle$ is defined in terms of (5.41) and

$$\langle E \, \hat{b} \,|\, T \,|\, E \, \hat{a}' \rangle = \sum_{\hat{a}} \langle E \, \hat{b} \,|\, E \, \hat{a} \rangle \langle E \, \hat{a} \,|\, T \,|\, E \, \hat{a}' \rangle; \quad (5.45)$$

otherwise it is the T matrix $\langle E \, \hat{a} \,|\, T \,|\, E \, \hat{a}' \rangle$ that is fundamental and which cannot be further determined.

In order to calculate the cross section from (5.38) or from (5.38'), one has to know both V and $|\mathbf{p}_0 \, \lambda \, E_T \eta^+ \rangle$ or the matrix $\langle E \, \hat{b} \,|\, T \,|\, \mathbf{p}_0 \, \lambda \, E_T^d \lambda \rangle$, respectively. Even if V is considered to be known one still requires knowledge of the $|\mathbf{p}_0 \, \lambda \, E_T^d \eta^+ \rangle = |E_0 \, E_T^d \Omega_0 \, \lambda \, \eta^+ \rangle \equiv |E_0 \hat{a}^+ \rangle$ $(E_0 = p_0^2/2m + E_T^d)$, which are solutions of the Lippmann–Schwinger integral equation. In principle the $|E_0 \hat{a}^+ \rangle$ may be obtained by an iteration process similar to the one described in Section VIII.1. The approximation that one obtains for the lowest order of the iteration process, i.e., the approximation

$$\langle b \,|\, T^+ \,|\, E_0 \, \hat{a} \rangle = \langle b \,|\, V \,|\, E_0 \, \hat{a}^+ \rangle \approx \langle b \,|\, V \,|\, E_0 \, \hat{a} \rangle, \quad (5.46)$$

is called the *Born approximation*, and is usually sufficient in the case of a high initial beam energy E_{B0} and a weak interaction V.

Equation (5.38) [or (5.38')] gives the cross section for the rather general situation in which the target state is left unspecified except for the requirement that it be a mixture of discrete energy eigenstates. We shall now give the cross section for more specific target states. We first assume that W_T is diagonal in some of the quantum numbers $\tilde{\eta}$ of $\eta = (\tilde{\eta}, \bar{\eta})$ and that no measurement has been made with respect to the other quantum numbers $\bar{\eta}$. For example, if the target consists of hydrogen or alkali atoms, the internal quantum numbers η are $\eta = (n, j, j_3, \pi)$, and $E_T = E_T(n, j, \pi)$ is a function of n, j, and π. Usually the atom is in a specified mixture of energy and angular-momentum and parity eigenstates, while the z-component of angular momentum has not been measured. Then

$$\langle n \, j \, j_3 \, \pi \| W_T^{\mathrm{in}}(E_T) \| n' \, j' \, j_3' \, \pi' \rangle = \frac{1}{2j + 1} \delta_{\pi\pi'} \delta_{j_3 j_3'} \delta_{jj'} \, W_T^{\mathrm{in}}(E_T(n, j, \pi)) \delta_{nn'}.$$

---

[18] Strictly speaking, the matrix elements of an operator are the numbers that result from placing the operator between two vectors from the *same* basis. Since the $\hat{b}$ and $\hat{a}$ may refer to eigenvalues of different operators, $\langle E \, \hat{b} \,|\, T \,|\, E \, \hat{a} \rangle$ need not be a matrix element, but it is a linear combination of matrix elements. For convenience, however, we shall hereafter refer to $\langle E \, \hat{b} \,|\, T \,|\, E \, \hat{a} \rangle$ as a T-matrix element.

Similarly, in the general situation

$$\langle \tilde{\eta}\tilde{\eta} \| W_T^{in}(E_T^d) \| \tilde{\eta}'\tilde{\eta}' \rangle = \frac{1}{g_T} \delta_{\tilde{\eta}\tilde{\eta}'} \delta_{\tilde{\eta}\tilde{\eta}'} W_T^{in}(E_T^d, \tilde{\eta}), \qquad (5.47)$$

where g_T is the number of values $\tilde{\eta}$ can take on. The cross section obtained in the case (5.47) is

$$\sigma(\Lambda \leftarrow \mathbf{p}_0, \tilde{\eta}) = \frac{(2\pi)^4 \hbar^2 m}{p_0} \sum_b \overline{\sum_\lambda} \sum_{E_T^d \tilde{\eta}} \overline{\sum_{\tilde{\eta}}} |\langle b | T | \mathbf{p}_0 \, \lambda \, E_T^d \, \tilde{\eta} \, \tilde{\eta} \rangle|^2$$

$$\times W_T^{in}(E_T^d, \tilde{\eta}) \delta(E_{B0} + E_T^d - E_b), \qquad (5.48)$$

where we use the notation $\overline{\sum_\lambda} = (1/g) \sum_\lambda$ and $\overline{\sum_\eta} = (1/g_T) \sum_\eta$ (averaging).

The weight $W(E_T^d \tilde{\eta})$ of each energy level may depend upon all the quantum numbers $\tilde{\eta} E_T^d = (\eta_1 \eta_2 \cdots \eta_\kappa, E_T^d)$ or only upon a subset of them, e.g., η_1, E_T^d. Very often it depends only upon E_T^d: $W(E_T^d \tilde{\eta}) = W(E_T^d)$. For example, if the target system is in thermal equilibrium, then W_T is given by (II.4.50) and $W(E_T^d)$ is given by a Gibbs distribution

$$W(E_n) = \frac{e^{-E_n/kT}}{\sum\limits_{\tilde{\eta} E_n} e^{-E_n/kT}} = \frac{e^{-E_n/kT}}{\sum\limits_{E_n} (\dim \mathcal{H}_n) e^{-E_n/kT}}$$

where $\dim \mathcal{H}_n$ is the dimension of the energy eigenspace with eigenvalue E_n.

A very special but rather common situation is the case in which the target is in a pure energy eigenstate, e.g., the ground state of an atom specified by the quantum numbers (E^0, η_0):

$$\langle \eta \| W_T^{in}(E_T^d) \| \eta' \rangle = \delta_{\eta\eta_0} \delta_{\eta_0\eta'} \delta_{E_T^d E_T^0}, \qquad (5.49)$$

or

$$W_T^{in} = |E_T^0 \, \eta_0 \rangle \langle E_T^0 \, \eta_0|. \qquad (5.49')$$

The cross section (5.38) for scattering of an unpolarized beam off such targets then becomes

$$\sigma(\Lambda \leftarrow \mathbf{p}_0 E_T^0 \eta_0) = \frac{(2\pi)^4 \hbar^2 m}{p_0} \sum_b \overline{\sum_\lambda} |\langle b | V | \mathbf{p}_0 \, \lambda \, E_T^0 \, \eta_0^+ \rangle|^2 \delta(E_0 - E_b)$$

$$= \frac{(2\pi)^4 \hbar^2 m}{p_0} \sum_b \overline{\sum_\lambda} |\langle b | T | \mathbf{p}_0 \, \lambda \, E_T^0 \, \eta_0 \rangle|^2 \delta(E_0 - E_b), \qquad (5.50)$$

where $E_0 = E_{B0} + E_T^0$. In the Born approximation and for structureless projectiles, this reduces to the well-known formula

$$\sigma(\Lambda \leftarrow \mathbf{p}_0 E_T^0 \eta_0) = \frac{(2\pi)^4 \hbar^2 m}{p_0} \sum_b |\langle b | V | \mathbf{p}_0 \, E_T^0 \, \eta_0 \rangle|^2 \delta(E_{B0} + E_T^0 - E_b). \qquad (5.51)$$

Usually, the transition matrix is independent of the polarization λ. Then the averaging over the polarizations in (5.50) is trivial, and the quantum numbers λ could just as well be omitted. For observables that do not depend upon the polarization, an unpolarized beam can be treated like a beam of structureless projectiles; e.g., for an unpolarized electron beam one can neglect the spin of the electron.

The most common experiments are those in which the differential cross section for scattering into a particular solid angle $\Delta\Omega_{D0}$ is measured by placing a counter at a particular angle Ω_{D0} to the incident direction. $\Lambda\mathcal{H}$ is then the subspace of states with momentum pointing in any of the directions $\Omega_{D0} \pm \Delta\Omega_{D0}$. If the detector detects only those states within a certain energy range, then $\Lambda\mathcal{H}$ is further restricted.

So far we have not specified the basis system $|b\rangle = |E\,\hat{b}\rangle$ in the space of final states. In order to obtain the differential cross section for scattering into a particular direction specified by the angles $\Omega = (\theta, \phi)$, one conveniently chooses a basis system of generalized eigenvectors labeled by Ω or by the momentum $\mathbf{k}$ of the detected (scattered) particle. Thus we choose

$$|b\rangle = |E\,\hat{b}\rangle = |\mathbf{k}\,\xi\rangle \otimes |\epsilon\,\zeta\rangle = |E_D\,\Omega_D\,\xi\rangle \otimes |\epsilon\,\zeta\rangle$$
$$= |E_D\,\epsilon\,\Omega_D\,\xi\,\zeta\rangle. \tag{5.52}$$

where $E = E_D + \epsilon$ and $\mathbf{k} = k\Omega_D$. ξ are the internal quantum numbers of the detected particle (polarizations) and ϵ, ζ are the internal energy and other internal quantum numbers of the target. Thus our description is general enough to include the case where the detected particle may be of a different kind than that of the incoming beam, and the particle left behind (post-collision target) may be different from those the target originally consisted of. For example, a photon γ might be incident on an atom A and ionize it, leaving behind an ion A^+ with an electron coming off: $\gamma + A \rightarrow e + A^+$. A simpler example for the reader to keep in mind is the elastic or inelastic scattering of an electron or photon beam by an atom: $e + A \rightarrow e' + A^*$.

If the particles detected are nonrelativistic, then

$$E_D = k^2/2m_D + E_D^{\text{int}}(\xi), \tag{5.53a}$$

while if the particles detected are photons,

$$E_D = kc. \tag{5.53b}$$

We shall impose the same normalizations as in (5.7) and (5.8) on the $|\mathbf{k}\,\xi\rangle$ and $|\epsilon\,\zeta\rangle$:

$$\langle \mathbf{k}\,\xi | \mathbf{k}'\,\xi'\rangle = \delta^3(\mathbf{k} - \mathbf{k}')\delta_{\xi\xi'} \tag{5.54}$$

and

$$\langle \epsilon\,\zeta | \epsilon'\,\zeta'\rangle = \delta_{\epsilon\epsilon'}\delta_{\zeta\zeta'}. \tag{5.55}$$

Then in the case (5.53a) we have by derivations similar to those of (5.15) and (5.16) that

$$\sum_b = \sum_b \int \bar{\rho}_{\hat{b}}(E)\,dE = \sum_{\xi\epsilon\zeta} \int m_D\sqrt{2m_D(E - E_D^{\text{int}} - \epsilon)}\,dE\,d\Omega_D$$

$$\equiv \sum_{\xi\epsilon\zeta} \int \bar{\rho}_{\xi\epsilon}(E)\,dE\,d\Omega_D \tag{5.56}$$

and

$$\langle E \epsilon \Omega_D \xi \zeta | E' \epsilon' \Omega'_D \xi' \zeta' \rangle$$
$$= m_D^{-1} \sqrt{2m_D(E - E_D^{int}(\xi) - \epsilon)} \, \delta(E - E') \delta_{\epsilon \epsilon'} \delta^2(\Omega_D - \Omega'_D) \delta_{\xi \xi'} \delta_{\zeta \zeta'}.$$

(5.57)

Equation (5.50) is then written in detail as

$$\sigma(\Lambda \leftarrow \mathbf{p}_0 E_T^0 \eta_0) = \frac{(2\pi)^4 \hbar^2 m_B m_D}{p_0} \sum_\lambda \sum_{\xi' \epsilon' \zeta'} \int dE' \, d\Omega'_D \, k'(E', \epsilon', \xi')$$
$$\times |\langle E' \, \epsilon' \, \Omega'_D \, \xi' \, \zeta' | T | \mathbf{p}_0 \, \lambda \, E_T^0 \, \eta_0 \rangle|^2 \delta(E_0 - E'), \quad (5.58)$$

where

$$E^0 = p_0^2/2m_B + E_T^0 \quad \text{(initial energy)}$$

$$k' = \sqrt{2m_D(E' - E_D^{int}(\xi') - \epsilon')} \quad \text{(momentum of the detected particle)}$$

The ranges of the summation over ξ', ϵ', ζ' and the integration over E', Ω'_D depend on the nature of the detection apparatus specified by Λ. From (5.58) we see that the differential cross section per unit solid angle for scattering in the direction Ω_D is

$$\frac{d\sigma}{d\Omega_D}(\Omega_D \leftarrow \mathbf{p}_0 E_T^0 \eta_0) = \frac{(2\pi)^4 \hbar^2 m_B m_D}{p_0} \sum_\lambda \sum_{\xi' \epsilon' \zeta'} \int dE' k'(E', \epsilon', \xi')$$
$$\times |\langle E' \, \epsilon' \, \Omega_D \, \xi' \, \zeta' | T | \mathbf{p}_0 \, \lambda \, E_T^0 \, \eta_0 \rangle|^2 \delta(E_0 - E').$$

(5.59)

We shall now assume that the projectile and the detected particle are the same ($m_B = m_D = m$) and that it is structureless or that the polarization quantum numbers λ and ζ can be ignored.[19] η_0 and ζ are then eigenvalues of the same set of operators η^{op}. The internal energies E_T^0 or ϵ', which are eigenvalues of the same operator K_T, are assumed to be already determined by the internal quantum numbers, i.e., $K_T = K_T(\eta^{op})$ as is the usual convention for targets with discrete energy spectrum.[20] Then the expression for (5.59) simplifies to

$$\frac{d\sigma}{d\Omega}(\Omega \leftarrow \mathbf{p}_0 \eta_0) = \frac{(2\pi)^4 \hbar^2 m^2}{p_0} \sum_{\zeta'} \int dE' \, k'(E', \zeta') \delta(E_0 - E')$$
$$\times |\langle E' \, \Omega, \zeta' | T | \mathbf{p}_0 \eta_0 \rangle|^2, \quad (5.60)$$

where

$$k' = \sqrt{2m(E' - E_T(\zeta'))} \quad \text{and} \quad E_0 = p_0^2/2m + E_T(\eta_0).$$

If not only the initial total energy is fixed (by fixing η_0 and the beam momentum p_0) but also the final total energy (by choosing a detector that

[19] One could also imagine that the polarization λ is incorporated in the quantum numbers η.

[20] If the target is the hydrogen atom, then $\eta = (n, j, j_3, \pi)$ and $E_T(\eta) = E'_n$ given by (VI.5.12).

registers only a particular momentum k), and if also the final internal quantum numbers are fixed (by triggering the detector only if the final internal quantum numbers have the value ζ), then the differential cross section is obtained from (5.60) as

$$\frac{d\sigma}{d\Omega}(E\,\Omega\,\zeta \leftarrow \mathbf{p}_0\eta_0) = \frac{(2\pi)^4\hbar^2m^2k}{p_0}|\langle E\,\Omega\,\zeta|T|\mathbf{p}_0\,\eta_0\rangle|^2. \qquad (5.61)$$

where

$$k = \sqrt{2m(E - E_T(\zeta))},$$

which must be equal to

$$\sqrt{2m\left(\frac{p_0^2}{2m} + E_T(\eta_0) - E_T(\zeta)\right)}$$

because of the enery-conservation δ-function $\delta(E_0 - E)$. For the case of elastic scattering, i.e., when the internal quantum numbers are not changed ($\eta_0 = \zeta$), one obtains the well-known expression

$$\frac{d\sigma}{d\Omega}(E_0\,\Omega\,\eta_0 \leftarrow \mathbf{p}_0\,\eta_0) = (2\pi)^4\hbar^2m^2|\langle E_0\,\Omega\,\eta_0|T|\mathbf{p}_0\,\eta_0\rangle|^2. \qquad (5.62)$$

The generalized eigenvectors in (5.62) [and also in the preceding expressions (5.61) and (5.60)] are normalized by (5.54) and (5.7) according to

$$\langle E\,\Omega\,\eta|E'\Omega'\eta'\rangle = \langle \mathbf{k}\,\eta|\mathbf{k}'\,\eta'\rangle = \delta_{\eta\eta'}\delta^3(\mathbf{k} - \mathbf{k}').$$

If one uses generalized eigenvectors with a different normalization, then this expression changes correspondingly.

For a spherically symmetric interaction, the T-matrix (and therefore the differential cross section) will not depend upon the angle ϕ around the incident direction $\mathbf{p}_0/p_0$, but only upon the angle θ between the incident direction and the final direction $\Omega = \mathbf{k}/k$ (cf. Figure 3.1). (A derivation of this statement will be given in Chapter XVI.) θ is called the scattering angle. In this case one often uses instead of the T-matrix element the elastic-scattering amplitude $T(p_0, \theta)$ defined by

$$T(p_0, \theta) = -\hbar 4\pi^2 m\langle E_0\Omega\eta_0|T|\mathbf{p}_0\eta_0\rangle. \qquad (5.63)$$

The differential cross section (5.62) is then written as

$$\frac{d\sigma}{d\Omega}(E_0, \theta) = |T(p_0, \theta)|^2. \qquad (5.64)$$

Problems

1. The quantity $\mathbf{q} = \mathbf{k} - \mathbf{p}_0$, where $\mathbf{p}_0$ is the projectile momentum before scattering and $\mathbf{k}$ is the projectile momentum after scattering, is called momentum transfer. (It represents the momentum that is transferred to the projectile by the target.)

(a) Show that the T-matrix $\langle \mathbf{k}, \eta_0 | T | \mathbf{p}_0, \eta_0 \rangle$ in the Born approximation is a function of the momentum transfer only and does not depend upon the momenta $\mathbf{p}_0$ and $\mathbf{k}$ separately if the interaction Hamiltonian V is a function of the projectile position.

(b) Show that in the Born approximation the forward T-matrix element

$$\langle \mathbf{p}_0, \eta_0 | T | \mathbf{p}_0, \eta_0 \rangle$$

is independent of scattering energy.

(c) Assume that the interaction Hamiltonian V is spherically symmetric, $[V, L_i] = 0$, and express the momentum transfer in the Born approximation of the T-matrix in terms of *scattering momentum* and *scattering angle* (for nonrelativistic particles). Show that at high energies (when the Born approximation is usually good), the scattering falls off as the scattering angle θ increases from 0 to π.

2. Let the generalized eigenvectors $|p, \lambda\rangle = |E, \Omega, \lambda\rangle$ of the energy operator $K = \mathbf{P}^2/2m$ be normalized according to

$$\langle \mathbf{p}' \lambda' | \mathbf{p} \lambda \rangle = \delta^3(\mathbf{p} - \mathbf{p}')\delta_{\lambda' \lambda}.$$

Show that this fixes the weight function $\rho(E)$ for the energy normalization

$$\langle E' \Omega' \lambda' | E \Omega \lambda \rangle = \rho^{-1}(E)\delta(E - E')\delta^2(\Omega - \Omega')\delta_{\lambda' \lambda}.$$

Here Ω denotes the direction of the momentum

$$\mathbf{p} = p\Omega = p(\sin \theta \cos \phi, \sin \theta \sin \phi, \cos \theta),$$

and the solid angle δ-function is defined by

$$\delta^2(\Omega - \Omega') = \delta(\cos \theta - \cos \theta') \, \delta(\phi - \phi').$$

Calculate $\rho(E)$.

3. Obtain the expression for the differential cross section (5.62) in terms of the T-matrix

$$\langle E \Omega | T | E \Omega' \rangle$$

defined with the generalized eigenvectors $|E \Omega\rangle$, which are normalized according to

$$\langle E \Omega | E' \Omega' \rangle = \delta(E - E')\delta^2(\Omega - \Omega'),$$

where the solid angle δ-function $\delta(\Omega - \Omega')$ is defined by

$$\int d\Omega f(\Omega)\delta^2(\Omega - \Omega_0) = f(\Omega_0)$$

with $d\Omega = \sin \theta \, d\theta \, d\phi$.

4. The Yukawa interaction was introduced to describe nuclear interactions and led to the prediction of the meson. It is described by the potential

$$V(r) = ge^{-\mu r}/r$$

and can also serve as a simple model for the screened Coulomb field of an atom.

(a) Calculate the scattering amplitude T and the differential cross section in the Born approximation.

(b) The second-order term in the Born series of the scattering amplitude is given by [cf. (XV.3.38)]

$$T^{(2)}(\mathbf{k} \leftarrow \mathbf{p}_0) = -4\pi^2 m \langle \mathbf{k}\, \eta_0 | T^{(2)} | \mathbf{p}_0 \eta_0 \rangle$$

$$= -4\pi^2 m \int d^3 p' \, \frac{\langle \mathbf{k} | V | \mathbf{p}' \rangle \langle \mathbf{p}' | V | \mathbf{p}_0 \rangle}{E_{p_0} - E_{p'} + i0}$$

Calculate the forward-scattering amplitude, i.e., the scattering amplitude for $\mathbf{p}_0 = \mathbf{k}$, up to the second order of the Born series, and find the values of the parameters m, μ, and g, and the energies for which the necessary condition for the validity of the Born approximation, $|T^{(2)}| \ll |T^{(1)}|$, is fulfilled.

(c) Obtain the total cross section for the Yukawa potential in the Born approximation.

(d) The parameter $r_0 = 1/\mu$ in the Yukawa potential characterizes the range of the interaction ($V(r)$ is practically zero for $r > 2r_0$) and g its strength. Assume that $r_0 = 10^{-13}$ cm and $\frac{g}{\hbar c} = 0.5$, as would be typical if $V(r)$ is to describe the strong interaction of protons scattering on nuclei. Let the projectile have the mass $m = m_{\text{proton}}$ and the scattering energy $E = 100$ MeV. Calculate the value of the total cross section.

CHAPTER XV

Formal Scattering Theory and Other Theoretical Considerations

Notions used in Chapter XIV for the derivation of the cross-section formula will be further discussed in this chapter to provide a deeper understanding of the material. In Section XV.1 the Lippmann–Schwinger equation is discussed again. Sections XV.2 and XV.3 are on formal scattering theory. In the Appendix a derivation of (XIV.2.9b), (XIV.2.17), and (XIV.5.21) is given, using the material developed in this chapter.

XV.1 The Lippmann–Schwinger Equation

The Lippmann–Schwinger equation

$$|a^{\pm}\rangle = |a\rangle + \frac{1}{E_a - K \pm i0} V|a^{\pm}\rangle \qquad (1.1\pm)$$

was introduced in Section VIII.2. To each generalized eigenvector $|a\rangle$ of K with eigenvalue E_a the Lippmann–Schwinger equation relates a generalized eigenvector $|a^+\rangle$ or $|a^-\rangle$ of $H = K + V$ with the same eigenvalue E_a. (In Section VIII.2 it was assumed that the continuous parts of the spectra of H and of K were identical and that E_a was an element of these continuous spectra.) The generalized eigenvectors $|a\rangle$ of K and $|a^{\pm}\rangle$ of H are labeled by the energy E_a together with some additional labels $\hat{a} = (a_1, a_2, \ldots, a_k)$:

$$|a\rangle = |E_a\hat{a}\rangle, \qquad (1.2a)$$

$$|a^{\pm}\rangle = |E_a\hat{a}^{\pm}\rangle. \qquad (1.2b)$$

In Section VIII.1 it was assumed that the $\hat{a}$ were additional quantum numbers for the K basis, i.e., that

$$\{K, A_1, A_2, \ldots, A_k\} \tag{1.3}$$

was a complete system of commuting observables (c.s.c.o) with the sets of eigenvalues $(E, a_1, a_2, \ldots, a_k)$. Suppose that

$$\{H, A_1, A_2, \ldots, A_k\} \tag{1.4}$$

is also a c.s.c.o. Both the generalized eigenvectors $|a\rangle$ of (1.3) and the proper and generalized eigenvectors $|a_n\rangle$ and $|a^+\rangle$ of (1.4) are a basis of the space of physical states.

   The assumption that (1.4) is a c.s.c.o. was not used in the derivation of the Lippmann–Schwinger equation and may be relaxed accordingly. Suppose that

$$\{K, B_1, B_2, \ldots, B_k\} \tag{1.5}$$

is a c.s.c.o. but that $\{H, B_1, B_2, \ldots, B_k\}$ is not, i.e., that

$$[B_i, H] = [B_i, V] \neq 0 \tag{1.6}$$

for some or all of the B_i's $(i = 1, 2, \ldots, k)$. By repeating the arguments of Section VIII.2, the basis vectors

$$|b\rangle = |E_b \hat{b}\rangle \tag{1.7}$$

of K may be related by way of the Lippmann–Schwinger equation to the generalized eigenvectors

$$|b^\pm\rangle \equiv |E_b(\hat{b})^\pm\rangle = |E_b \hat{b}\rangle + \frac{1}{E_b - K \pm i0} V |E_b(\hat{b})^\pm\rangle \tag{1.8}$$

of H. However, because of (1.6), the vectors $|E_b(\hat{b})\rangle$ are not eigenvectors of all the operators B_i. In this case the b_i's are defined not by a c.s.c.o. containing H, but only through the Lippmann–Schwinger equation and the c.s.c.o. containing K. The parentheses around the label $\hat{b}$ are a temporary notation used to indicate that the $\hat{b} = b_1, b_2, \ldots, b_k$ are labels only, whereas the E_b are eigenvalues of the operator H.

   As an example of these possibilities consider the scattering of structureless particles off a fixed target. Let $\mathbf{Q}$ and $\mathbf{P}$ be the (canonically conjugate) position and momentum operators for the projectile, and assume the interaction of the projectile with the fixed target is given by a spherically symmetric potential $V = V(Q)$ $[Q = (\mathbf{Q}^2)^{1/2}]$; the Hamiltonian for the system is then $H = K + V$, where $K = \mathbf{P}^2/2m$ is the kinetic-energy operator for the projectile. Since $[L_i, V] = [L_i, K] = 0$, it follows that

$$\{K, \mathbf{L}^2, L_3\} \tag{1.9a}$$

and

$$\{H, \mathbf{L}^2, L_3\} \tag{1.9b}$$

are both c.s.c.o's and that we have the case (1.4). The $|Ell_3\rangle$ and the $|Ell_3^{\pm}\rangle$ are corresponding generalized eigenvectors. $\{|Ell_3\rangle\}$ is a basis for the entire space of physical states, but $\{|Ell_3^{\pm}\rangle\}$ is a basis for the "subspace of scattering states" only. The projectiles in scattering experiments are usually prepared as collimated beams going in a specific direction, and it is therefore more convenient to use for the K-basis the generalized eigenvectors

$$|b\rangle = |\mathbf{p}\rangle = |E\,p_1 p_2\rangle = |E\,\theta\,\phi\rangle \qquad (1.10)$$

of the momentum operators P_i. The c.s.c.o. of the vectors (1.10) is

$$\{P_1, P_2, P_3\} \quad \text{or} \quad \{K, P_1, P_2\} \quad \text{or} \quad \{K, \text{directions of } \mathbf{P}\}; \qquad (1.11)$$

we then have the case (1.5) with (1.6), because

$$[P_i, H] = [P_i, V] \neq 0.$$

The label $\mathbf{p}$ for the H eigenvectors

$$|(b)^{\pm}\rangle = |(\mathbf{p})^{+}\rangle \qquad (1.12)$$

does not then mean that the $|(\mathbf{p})^{\pm}\rangle$ are eigenvectors of the P_i but only means that the $|(\mathbf{p})^{\pm}\rangle$ are connected to the momentum eigenvectors $|\mathbf{p}\rangle$ by the Lippmann–Schwinger equation,

$$|(\mathbf{p})^{\pm}\rangle = |\mathbf{p}\rangle + \frac{1}{E - K \pm i0}\,V\,|(\mathbf{p})^{\pm}\rangle \qquad (E = \mathbf{p}^2/2m). \qquad (1.13)$$

The assumption that (1.3) is a c.s.c.o. for a particular physical system can only be justified by physical motivation. It is equivalent to the assumption that the set

$$\{|E\,\hat{a}\rangle : (E, \hat{a}) \in \text{spectrum } (K, A_1, \ldots, A_k)\} \qquad (1.14)$$

of generalized eigenvectors of (1.3) forms a basis for the space of physical states $\mathscr{H}$ of the physical system. Although the set

$$\{|E\,\hat{a}\rangle_H : (E, \hat{a}) \in \text{spectrum } (H, A_1, \ldots, A_k)\} \qquad (1.15)$$

of generalized eigenvectors of (1.4) is also a basis for $\mathscr{H}$, the sets

$$\{|E\,\hat{a}^{+}\rangle : (E, \hat{a}) \in \text{spectrum } (K, A_1, \ldots, A_k)\} \qquad (1.16+)$$

and

$$\{|E\,\hat{a}^{-}\rangle : (E, \hat{a}) \in \text{spectrum } (K, A_1, \ldots, A_k)\} \qquad (1.16-)$$

of generalized eigenvectors of (1.4), which are obtained through the Lippmann–Schwinger equations (1.1+) and (1.1−), are in general not a complete basis system for $\mathscr{H}$. In the case that (1.4) is not a c.s.c.o., the sets

$$\{|E(\hat{a})^{\pm}\rangle : (E, \hat{a}) \in \text{spectrum } (K, A_1, \ldots, A_k)\} \qquad (1.17\pm)$$

are in general also not a complete basis system. Usually the operator K has only a continuous spectrum, but H has in addition to the continuous spectrum a discrete spectrum $\{E_n\}$ as well. The proper eigenvectors $|E_n\hat{a}\rangle$ with energy eigenvalues E_n in the discrete spectrum of H describe pure physical states corresponding to the bound states of the projectile-target system, and span a *subspace of bound states* $\mathcal{H}_{bnd}$. The orthogonal complement of $\mathcal{H}_{bnd}$, i.e., the subspace of all vectors of $\mathcal{H}$ which are orthogonal to $\mathcal{H}_{bnd}$, is called the *space of scattering states* $\mathcal{H}_{scat}$. Thus[1]

$$\mathcal{H} = \mathcal{H}_{bnd} \oplus \mathcal{H}_{scat}. \qquad (1.18)$$

The subspaces $\mathcal{H}_+$ and $\mathcal{H}_-$ that are spanned by the sets $(1.16+)$ and $(1.16-)$ need not necessarily agree with $\mathcal{H}_{scat}$ or with each other. However we will consider only physical scattering systems, for which the subspaces $\mathcal{H}_+$, $\mathcal{H}_-$, and $\mathcal{H}_{scat}$ agree:

$$\mathcal{H}_+ = \mathcal{H}_- = \mathcal{H}_{scat} \qquad (1.19)$$

The system for which (1.19) is fulfilled is said to be *asymptotically complete.*[2]

Under very general assumptions we thus have the following situation: The space of physical states $\mathcal{H}$ is spanned by the set $\{|E\,\hat{a}\rangle\}$ of generalized eigenvectors of K, and is the direct sum $\mathcal{H} = \mathcal{H}_{bnd} \oplus \mathcal{H}_{scat}$ of the space $\mathcal{H}_{bnd}$ spanned by the set $\{|E_n\hat{a}\rangle\}$ of proper eigenvectors of H and the space $\mathcal{H}_{scat}$ spanned by either of the sets $\{|E\hat{a}^+\rangle\}$ or $\{|E\hat{a}^-\rangle\}$ of generalized eigenvectors of H related to the $|E\hat{a}\rangle$ by the Lippmann–Schwinger equation $(1.1\pm)$.

The Lippmann–Schwinger equation is an equation for the $|a^\pm\rangle$ and can be iterated as described in Section VIII.1. It can also be solved formally. For any two invertable operators A and B the following easily established relation holds:

$$\frac{1}{A} - \frac{1}{B} = \frac{1}{A}(B - A)\frac{1}{B} = \frac{1}{B}(B - A)\frac{1}{A}. \qquad (1.20)$$

A special case of this is

$$\frac{1}{E - H \pm i0} - \frac{1}{E - K \pm i0} = \frac{1}{E - H \pm i0} V \frac{1}{E - K \pm i0}. \qquad (1.21\pm)$$

By applying this to $V|a^\pm\rangle$ and then using $(1.1\pm)$ twice, we obtain the "solution"

$$|a^\pm\rangle = |a\rangle + \frac{1}{E_a - H \pm i0} V|a\rangle. \qquad (1.22\pm)$$

---

[1] This statement is not difficult to envisage if the discrete and continuous spectra of H are disjoint. It may happen, however, that some of the discrete eigenvalues coincide with values of the continuous spectrum (e.g., the doubly excited states of the He atom; see Sections XI.2 and XI.3). To such eigenvalues correspond both proper eigenvectors in $\mathcal{H}_{bnd}$ and generalized eigenvectors in the set of generalized eigenvectors that spans $\mathcal{H}_{scat}$. Such proper and generalized eigenvectors are orthogonal, even though they belong to the same eigenvalue of H and the same eigenvalues $\hat{a} = (a_1, a_2, \ldots, a_k)$ of $\{A_1, A_2, \ldots, A_k\}$.

[2] If the interaction V is described by a potential $V(Q)$ which falls off more rapidly than r^{-3} at infinity, is less singular than r^{-2} at the origin and is sufficiently smooth in between, then asymptotic completeness can be proven.

The operator analogue of Equation (XIV.5.20) is

$$\frac{1}{E - H + i0} - \frac{1}{E - H - i0} = -2\pi i\delta(E - H), \qquad (1.23)$$

which together with (1.22 +) and (1.22 −) allows us to relate the H eigenvectors $|a^+\rangle$ and $|a^-\rangle$ to each other:

$$|a^+\rangle - |a^-\rangle = -2\pi i\delta (E_a - H)V|a\rangle. \qquad (1.24)$$

The formal solution (1.22 ±) is not of much practical use, but it does serve as a way of introducing the S-matrix, which we shall meet in Section XV.3.

XV.2 In-States and Out-States

In this section we discuss the significance of the labels + and − that appear in the generalized eigenvectors $|E\hat{a}^+\rangle$ and $|E\hat{a}^-\rangle$ of the exact Hamiltonian $H = K + V$. In these discussions we shall consider pure physical states as described by state vectors; the results are then easily carried over to statistical mixtures as described by statistical operators.

Let us first describe the fictitious case where there is no interaction between beam and target. This means that the beam passes through the target without being affected at all. In the simplest case of potential scattering this means that the potential is turned off. The state vector $\Phi(t)$ that describes this fictitious situation develops in time according to the free Hamiltonian K:

$$\Phi(t) = e^{-iKt}\Phi \qquad (\Phi = \Phi(0)). \qquad (2.1+)$$

It is called the *free state vector*, and the state it describes is called the *free state*. To simplify the discussion we shall assume that Φ is an eigenvector corresponding to the eigenvalues $\hat{a} = (a_1, a_2, \ldots, a_k)$ of those operators A_1, $A_2, \ldots, A_k$ that together with K form a c.s.c.o. (We assume the eigenvalues of $A_1, A_2, \ldots, A_k$ to be discrete.) Sometimes we will write $\Phi(\hat{a}, t)$ instead of $\Phi(t)$ in order to emphasize $\Phi(t)$'s preparation as an eigenstate of the A_i's. The probability distribution for obtaining the energy eigenvalue E when K is measured in the state Φ is

$$\langle E\,\hat{a}|\Phi\rangle\langle\Phi|E\,\hat{a}\rangle = |\langle E\,\hat{a}|\Phi\rangle|^2. \qquad (2.2+)$$

Let us assume that the energy distribution is described by the function $\phi(E)$, which is related to $\langle E\,\hat{a}|\Phi\rangle$ by

$$\phi(E) = 2\pi\rho_a(E)\langle E\,\hat{a}|\Phi\rangle, \qquad (2.3+)$$

where $\rho_a(E)$ is the normalization function for the generalized eigenvectors of K and the generalized eigenvectors of H as given by (XIV.2.9a) and (XIV.2.9b). That is, we assume that the state $\Phi(t)$ has been prepared at some time such that at $t = 0$, and therefore at any other time t, it has the probability distribution

$$\langle E\,\hat{a}|\Phi(t)\rangle\langle\Phi(t)|E\,\hat{a}\rangle = \langle E\,\hat{a}|\Phi\rangle\langle\Phi|E\,\hat{a}\rangle$$
$$= (2\pi\rho_a(E))^{-2}|\phi(E)|^2 \qquad (2.4+)$$

for the eigenvalue E of the energy operator K. Φ may then be written in terms of the generalized eigenvectors $|E\,\hat{a}\rangle$ of K as

$$\Phi = \Phi(0) = \int \rho_a(E)\,dE\,\langle E\,\hat{a}|\Phi\rangle|E\,\hat{a}\rangle = \frac{1}{2\pi}\int dE\,\phi(E)|E\,\hat{a}\rangle.$$
$$(2.5+)$$

The free state vector at any other time t is given by

$$\Phi(t) = e^{-iKt}\Phi = \frac{1}{2\pi}\int dE\,e^{-iEt}\phi(E)|E\,\hat{a}\rangle. \qquad (2.6+)$$

Let us now consider the state vector

$$\Phi^+ = \Phi^+(0) = \frac{1}{2\pi}\int dE\,\phi(E)|E\,\hat{a}^+\rangle = \int \rho_a(E)\,dE\,\langle E\,\hat{a}|\Phi\rangle|E\,\hat{a}^+\rangle,$$
$$(2.7+)$$

where $|E\,\hat{a}^+\rangle$ are the generalized eigenvectors of H that are connected with the generalized eigenvectors $|E\,\hat{a}\rangle$ of K by $(1.1+)$. The H energy distribution of Φ^+ is the expectation value of the "operator" $\sum_{a'}|E\,\hat{a}'^+\rangle\langle E\,\hat{a}'^+|$:

$$\langle\Phi^+|\left(\sum_{a'}|E\,\hat{a}'^+\rangle\langle E\,\hat{a}'^+|\right)|\Phi^+\rangle = \sum_{a'}|\langle E\,\hat{a}'^+|\Phi^+\rangle|^2 = |\langle E\,\hat{a}|\Phi\rangle|^2.$$
$$(2.8+)$$

[Here we have used both the expansion $(2.7+)$ of Φ^+ and the normalization (XIV.2.9b) of the $|E\,\hat{a}^+\rangle$'s.] Thus, the probability distribution for obtaining the value E when H is measured in the state Φ^+ is the same as the probability distribution for obtaining the value E when the free energy K is measured in the (fictitious) state Φ [or $\Phi(t)$]. For any other time t, earlier or later, the state that develops to or from the state under the influence of the interaction is given by

$$\Phi^+(t) = e^{-iHt}\Phi^+ = \frac{1}{2\pi}\int dE\,e^{-iEt}|E\,\hat{a}^+\rangle\phi(E). \qquad (2.9+)$$

$\Phi^+(t)$ is called the *exact state vector*. $\Phi^+(t)$ and $\Phi(t)$ are states with the same quantum numbers $\hat{a}$ and the same energy distribution $\phi(E)$; their difference is that $\Phi^+(t)$ develops according to the exact Hamiltonian and describes the development of an actual state, while $\Phi(t)$ develops according to the free Hamiltonian and describes the development of the same state as if there were no interaction.

The connection between $\Phi^+(t)$ and $\Phi(t)$ is obtained if one inserts the Lippmann–Schwinger equation (1.1+) into the integral of (2.9+):

$$\Phi^+(t) = \frac{1}{2\pi}\int dE\, e^{-iEt}\phi(E)|E\,\hat{a}\rangle + \frac{1}{2\pi}\int dE\, e^{-iEt}\phi(E)\frac{1}{E - K + i0}V|E\,\hat{a}^+\rangle$$

$$= \Phi(t) + \frac{1}{2\pi}\int dE\, e^{-iEt}\phi(E)\frac{1}{E - K + i0}V|E\,\hat{a}^+\rangle \qquad (2.10+)$$

The first term in (2.10+) is given by (2.6+); the calculation of the second term is a purely mathematical task and is given below, after (2.20−). Using the result of that calculation one obtains

$$\Phi^+(t) = \Phi(t) + \int_{-\infty}^{\infty} dt'\, G_0^+(t - t')V\Phi^+(t') \qquad (2.11+)$$

where

$$G_0^+(t) = -i\theta(+t)e^{-iKt} = \begin{cases} -ie^{iKt} & \text{if } t > 0, \\ 0 & \text{if } t < 0. \end{cases} \qquad (2.12+)$$

$\theta(t)$ is the unit step function

$$\theta(t) = \begin{cases} 1 & \text{if } t > 0, \\ 0 & \text{if } t < 0. \end{cases} \qquad (2.13)$$

Suppose we now take the limit $t \to -\infty$ in (2.11+). As the limit is taken, we eventually get $t < t'$ for any value of t', so that the integrand vanishes. Thus

$$\Phi^+(t) \to \Phi(t) \quad \text{as} \quad t \to -\infty. \qquad (2.14+)$$

This means that if in the remote past when the interaction was not effective a state had been prepared so as to have the energy distribution $\phi(E)$ (and the quantum numbers $\hat{a}$), then this state would develop so that at time t it is given by $\Phi^+(t)$ of (2.9+). Thus $\Phi^+(t)$ describes a state that develops from a state $\Phi(t)$ prepared in the remote past when the interaction V was not effective. The prepared state is called an *in-state*:

$$\Phi(t) = \Phi^{in}(t)[\equiv \Phi^{in}(\hat{a}, t)] \qquad (2.15+)$$

Describing the behavior of $\Phi^+(t)$ in the distant future when the interaction V ceases to be effective is another free state, the *out-state*:

$$\Phi^{out}(t)[\equiv \Phi^{out}(\hat{a}, t)] = e^{-iKt}\Phi^{out} \qquad [\Phi^{out} = \Phi^{out}(0)]. \qquad (2.16+)$$

In the present situation, where the exact state $\Phi^+(t)$ is prescribed by its behavior in the distant past as $\Phi^{in}(t)$, the outstate is unknown and un-controlled, except through our knowledge of V and our control over $\Phi^{in}(t)$. As will be discussed in the next section, the connection between the in-state $\Phi^{in}(t)$ and the out-state $\Phi^{out}(t) = S\Phi^{in}(t)$ is the scattering operator S.

We shall now repeat the same arguments used above but using the generalized eigenvectors $|E \hat{b}^-\rangle$ of H that appear in the Lippmann–Schwinger equation

$$|E \hat{b}^-\rangle = |E \hat{b}\rangle + \frac{1}{E - K - i0} V|E \hat{b}^-\rangle \qquad (2.17-)$$

instead of the $|E \hat{a}^+\rangle$ appearing in (1.1+). Here $\hat{b} = (b_1, b_2, \ldots, b_k)$ are the eigenvalues of a set of operators $B_1, B_2, \ldots, B_k$, which are possibly different from the operators $A_1, A_2, \ldots, A_k$, but which together with K form a c.s.c.o. (The spectra of $B_1, B_2, \ldots, B_k$ are assumed for simplicity to be discrete.) We thus consider the free state

$$\Psi(t)[\equiv \Psi(\hat{b}, t)] = e^{-iKt}\Psi = \frac{1}{2\pi} \int dE \, e^{-iEt} \psi(E)|E \hat{b}\rangle, \qquad (2.6-)$$

where the function $\psi(E)$ describes the energy distribution of $\Psi(t)$ as it will be prepared at some time. $\psi(E)$ is connected with the probability distribution for the energy operator K by

$$\langle E \hat{b}|\Psi(t)\rangle\langle\Psi(t)|E \hat{b}\rangle = (2\pi\tilde{\rho}_{\hat{b}}(E))^{-2}|\psi(E)|^2. \qquad (2.4-)$$

The function $\tilde{\rho}_{\hat{b}}(E)$ is a weight function for the normalization of the $|E \hat{b}\rangle$'s, just as $\rho_{\hat{a}}(E)$ was a weight function for the normalization (XIV.2.9a) of the $|E \hat{a}\rangle$'s. From its expansion (2.6−), $\Psi(t)$ is obviously an eigenvector of $B_1, B_2, \ldots, B_k$ corresponding to the eigenvalues $\hat{b}$.

The exact state vector $\Psi^-(t)$ corresponding to $\Psi(t)$ is defined to be

$$\Psi^-(t) = e^{-iHt}\Psi^- = \frac{1}{2\pi} \int dE \, e^{-iEt}\psi(E)|E \hat{b}^-\rangle, \qquad (2.9-)$$

where $|E \hat{b}^-\rangle$ are the generalized eigenvectors of H that are connected with the generalized eigenvectors $|E \hat{b}\rangle$ of K by (2.17−). Again $\Psi^-(t)$ and $\Psi(t)$ are states with the same quantum numbers $\hat{b}$ and the same energy distribution $\psi(E)$, but $\Psi^-(t)$ develops according to the exact Hamiltonian H, while $\Psi(t)$ develops according to the free Hamiltonian K. To obtain the connection between $\Psi^-(t)$ and $\Psi(t)$ we insert (2.17−) into (2.9−) and obtain

$$\Psi^-(t) = \frac{1}{2\pi} \int dE \, e^{-iEt}\psi(E)|E \hat{b}\rangle + \frac{1}{2\pi} \int dE \, e^{-iEt}\psi(E)\frac{1}{E - K - i0} V|E \hat{b}^-\rangle$$

$$= \Psi(t) + \frac{1}{2\pi} \int dE \, e^{-iEt}\psi(E)\frac{1}{E - K - i0} V|E \hat{b}^-\rangle. \qquad (2.10-)$$

The result of a calculation similar to the calculation to be given for the second term in (2.10+) gives

$$\Psi^-(t) = \Psi(t) + \int_{-\infty}^{\infty} dt' \, G_0^-(t - t')V\Psi^-(t'), \qquad (2.11-)$$

where

$$G_0^-(t) = +i\theta(-t)e^{-iKt} = \begin{cases} +ie^{-iKt} & \text{if } t < 0, \\ 0 & \text{if } t > 0. \end{cases} \qquad (2.12-)$$

If we now take the limit $t \to +\infty$ in (2.11−), we obtain

$$\Psi^-(t) \to \Psi(t) \quad \text{as } t \to +\infty \qquad (2.14-)$$

This means that if in the distant future a state is measured that has the energy distribution $\psi(E)$ and quantum numbers $\hat{b}$, then this state was given at time t by $\Psi^-(t)$ of (2.9−). Thus $\Psi^-(t)$ describes a state that will develop into a known state in the distant future when the interaction V is no longer effective. This state will be called an *out-state*:

$$\Psi(t) = \Psi^{\text{out}}(t)[\equiv \Psi^{\text{out}}(\hat{b}, t)] \qquad (2.15-)$$

Describing the behavior of $\Psi^-(t)$ in the remote past before the interaction was effective is another free state, the *in-state*:

$$\Psi^{\text{in}}(t)[\equiv \Psi^{\text{in}}(\hat{b}, t)] = e^{-iKt}\Psi^{\text{in}} \qquad [\Psi^{\text{in}} \equiv \Psi^{\text{in}}(0)]. \qquad (2.16-)$$

Now the exact state $\Psi^-(t)$ is prescribed by what its behavior will be in the distant future as $\Psi^{\text{out}}(\hat{b}, t)$, and it is the in-state that is unknown and uncontrolled. The connection between the in- and out-states is again $\Psi^{\text{out}}(t) = S\Psi^{\text{in}}(t)$—or, since it is our control over $\Psi^{\text{out}}(t)$ that should be emphasized, $\Psi^{\text{in}}(t) = S^{-1}\Psi^{\text{out}}(t) = S^\dagger\Psi^{\text{out}}(t)$.

We have thus found the meaning of the labels $+$ and $-$: $\Phi^+(\hat{a}, t)$ describes a state that in the remote past, before the interaction V became effective, was prepared as $\Phi^{\text{in}}(\hat{a}, t)$ with well-defined quantum numbers $\hat{a}$ and a certain energy distribution $\phi(E)$. In the distant future it will again become a free state, an out-state $\Phi^{\text{out}}(t)$; however, this state is not simple, since it is determined not only by the preparation but also by the scattering process. $\Psi^-(\hat{b}, t)$ describes a state that in the distant future, after the interaction V has ceased, will be given by the free state $\Psi_b^{\text{out}}(t)$ with a simple energy distribution $\psi(E)$ and well-determined values $\hat{b}$ for the other quantum numbers. In the remote past $\Psi^-(\hat{b}, t)$ was also a free state $\Psi^{\text{in}}(t)$, but its properties must have been more complicated. Since in scattering experiments it is the behavior of the system in the distant past over which we exercise control, it is the $+$ states $\Phi^+(\hat{a}, t)$ that are natural to use when describing scattering experiments.

Our results may be summarized as integral equations for the exact states $\Phi^+(\hat{a}, t)$ and $\Psi^-(\hat{b}, t)$ in terms of the controlled free states $\Phi^{\text{in}}(\hat{a}, t)$ and $\Psi^{\text{out}}(\hat{b}, t)$:

$$\Phi^+(\hat{a}, t) = \Phi^{\text{in}}(\hat{a}, t) + \int_{-\infty}^{\infty} dt' \, G_0^+(t - t')V\Phi^+(\hat{a}, t'), \qquad (2.18+)$$

$$\Psi^-(\hat{b}, t) = \Psi^{\text{out}}(\hat{b}, t) + \int_{-\infty}^{\infty} dt' \, G_0^-(t - t')V\Psi^-(\hat{b}, t'). \qquad (2.18-)$$

(There are also corresponding integral equations for $\Phi^+(\hat{a}, t)$ in terms of $\Phi^{\text{out}}(\hat{a}, t)$ and for $\Psi^-(\hat{b}, t)$ in terms of $\Psi^{\text{in}}(\hat{b}, t)$, but in view of the uncontrolled nature of $\Phi^{\text{out}}(\hat{a}, t)$ and $\Psi^{\text{in}}(\hat{b}, t)$, such equations have little meaning.)

The preceding results are easily extended by linearity to the case of a mixture. The exact state that was described in the distant past by the statistical operator

$$W^{\text{in}}(t) = \sum_{mn} w_{mn} |\Phi_m^{\text{in}}(\hat{a}_m, t)\rangle \langle \Phi_n^{\text{in}}(\hat{a}_n, t)| \qquad (2.19+)$$

is

$$W^+(t) = \sum_{mn} w_{mn} |\Phi_m^+(\hat{a}_m, t)\rangle \langle \Phi_n^+(\hat{a}_n, t)|, \qquad (2.20+)$$

and the exact state that will be described in the distant future by

$$\tilde{W}^{\text{out}}(t) = \sum_{mn} w_{mn} |\Psi_m^{\text{out}}(\hat{b}_m, t)\rangle \langle \Psi_n^{\text{out}}(\hat{b}_n, t)| \qquad (2.19-)$$

is

$$\tilde{W}^-(t) = \sum_{mn} w_{mn} |\Psi_m^-(\hat{b}_m, t)\rangle \langle \Psi_n^-(\hat{b}_n, t)|. \qquad (2.20-)$$

We shall now take up the purely mathematical task of showing that Equations $(2.11\pm)$ follow from $(2.10\pm)$, using results from distribution theory, in particular results concerning the Fourier transforms of generalized functions.

⟦The *Fourier transform* $\tilde{g} = F[g]$ of a function $g(t)$ is defined by[3]

$$\tilde{g}(E) = F_E[g(t)] = \int_{-\infty}^{\infty} dt \, e^{iEt} g(t). \qquad (2.21)$$

The *inverse Fourier transformation* F^{-1} is given by

$$F_t^{-1}[\tilde{g}(E)] = \frac{1}{2\pi} \int_{-\infty}^{\infty} dE \, e^{-iEt} \tilde{g}(E) \qquad (2.22)$$

Strictly speaking, (2.21) is the definition of a well-behaved function. The Fourier tranform of a generalized function f, defined by the linear functional $f(\Psi) = (f, \Psi)$, is defined by $(F[f], F[\Psi]) = 2\pi(f, \Psi)$. But as the rules for Fourier transforms are retained for generalized functions, when interpreted properly one may ignore this mathematical precision and use (2.21) and (2.22) also for generalized functions.[4] In particular, one always has

$$g = F^{-1}[\tilde{g}] = F^{-1}[F[g]] = F[F^{-1}[g]]. \qquad (2.23)$$

---

[3] Note that only in this mathematical insert do we use the mathematician's convention for the Fourier transform which is used in Gel'fand and Shilov (1964), in order to facilitate comparisons.

[4] Gel'fand and Shilov (1964), Vol. 2, Chapter III; Vol. 1, Chapter II.

Translation in the space of Fourier-transformed functions is obviously given by

$$\tilde{g}(E - E') = F_E[e^{-iE't}g(t)]. \qquad (2.24)$$

If we apply (2.24) to the Fourier transforms

$$F_E[\theta(\pm t)] = \pm i \frac{1}{E \pm i0} \qquad (2.25\pm)$$

of the generalized function $\theta(+t)$ and $\theta(-t)$, we get

$$\frac{1}{E - E' \pm i0} = F_E[\mp ie^{-iE't}\theta(\pm t)]. \qquad (2.26\pm)$$

The *convolution* $f * g$ of two functions f and g is defined by

$$(f * g)(t) = \int_{-\infty}^{\infty} dt'\, f(t - t')g(t') \qquad (2.27)$$

and has the particularly simple Fourier transform[5]

$$F[f * g] = F[f] \cdot F[g], \qquad (2.28)$$

or equivalently,

$$F^{-1}[\tilde{f} \cdot \tilde{g}] = f * g. \qquad (2.29)$$

With the above mathematical facts we shall show the equivalence of the second terms in (2.10+) and (2.11+). If we take the scalar product of $\Phi^+(t)$ as given by (2.10+) with an arbitrarily chosen K basis vector $|a'\rangle = |E'\,\hat{a}'\rangle$ [more precisely, if we consider the value of the functional $|\Phi^\pm(t)\rangle \in \Phi^{\times\times} = \Phi$ at the point $\langle a'| \in \Phi^\times$], we obtain

$$\langle a'|\Phi^+(t)\rangle = \langle a'|\Phi(t)\rangle + \frac{1}{2\pi} \int dE\, e^{-iEt}\phi(E)\frac{1}{E - E' + i0}$$
$$\times \langle a'|V|E\,\hat{a}^+\rangle. \qquad (2.30+)$$

Equation (2.9+) implies

$$\langle a'|V|\Phi^+(t)\rangle = \frac{1}{2\pi} \int dE\, e^{-iEt}\phi(E)\langle a'|V|E\,\hat{a}^+\rangle$$
$$= F_t^{-1}[\phi(E)\langle a'|V|E\,\hat{a}^+\rangle], \qquad (2.31+)$$

or equivalently,

$$\phi(E)\langle a'|V|E\,\hat{a}^+\rangle = F_E[\langle a'|V|\Phi^+(t)\rangle]. \qquad (2.32+)$$

The second term on the right-hand side of (2.30+) is now recognized as the inverse Fourier transform of the product of two Fourier

---

[5] This statement has to be qualified somewhat in the case of generalized functions. See Gel'fand and Shilov (1964), Vol. 2, Chapter III.

transforms [cf. (2.26+) and (2.32+)]. By using (2.29) we may rewrite (2.30+) as

$$\langle a'|\Phi^+(t)\rangle = \langle a'|\Phi(t)\rangle + (-i\theta(+t)e^{-iE't}) * (\langle a'|V|\Phi^+(t)\rangle)$$

$$= \langle a'|\Phi(t)\rangle - i\int_{-\infty}^{\infty} dt'\, \theta(+(t-t'))e^{-iE'(t-t')}$$

$$\times \langle a'|V|\Phi^+(t')\rangle. \qquad (2.33+)$$

Since $|a'\rangle$ was an arbitrary K basis vector, we have shown

$$|\Phi^+(t)\rangle = |\Phi(t)\rangle + \int_{-\infty}^{\infty} dt'\, G_0^+(t-t')V|\Phi^+(t')\rangle, \quad (2.11+)$$

as we set out to do. The equivalence of the second terms in (2.10−) and (2.11−) may be shown by a similar calculation.

The operators

$$G_0^\pm(t) = \mp i\theta(\pm t)e^{-iKt}$$

are seen by (2.26±) to be the inverse Fourier transforms of the resolvents

$$\tilde{G}_0^\pm(E) = (E - K \pm i0)^{-1} \qquad (2.34)$$

of the operator K. The operator $G_0^+(t)$ is known as the *retarded Green's function*, and $G_0^-(t)$ is known as the *advanced Green's function*. (In physics the distinction between a function $g(t)$ and its Fourier transform $\tilde{g}(E) = F_E[g(t)]$ is not always reflected by the terminology, and so the resolvents $\tilde{G}_0^\pm(E)$ are sometimes also called "Green's functions," as mentioned in Section VIII.1. Physicists frequently blur the notation as well, and write $G_0^\pm(E)$ for $\tilde{G}_0^\pm(E)$; the argument of the function is then used to tell whether the function or its Fourier transform is meant.)]

The operators $G_0^\pm(t)$ are both *free* Green's functions, as opposed to the *exact* Green's functions given by

$$G^+(t) = -i\theta(+t)e^{-iHt} = \begin{cases} 0 & \text{if } t < 0, \\ -ie^{-iHt} & \text{if } t > 0, \end{cases} \qquad (2.35+)$$

$$G^-(t) = +i\theta(-t)e^{-iHt} = \begin{cases} +ie^{-iHt} & \text{if } t < 0, \\ 0 & \text{if } t > 0. \end{cases} \qquad (2.35-)$$

These latter Green's functions may be used to express formal solutions of (2.18+) and (2.18−):

$$\Phi^+(\hat{a}, t) = \Phi^{in}(\hat{a}, t) + \int_{-\infty}^{+\infty} dt'\, G^+(t-t')V\Phi^{in}(\hat{a}, t'), \qquad (2.36+)$$

$$\Psi^-(\hat{b}, t) = \Psi^{out}(\hat{b}, t) + \int_{-\infty}^{+\infty} dt'\, G^-(t-t')V\Psi^{out}(\hat{b}, t'). \quad (2.36-)$$

The derivation of these solutions, which is left as an exercise, parallels the derivations of (2.11±) [or equivalently of (2.18±)] except that where the derivations of the latter use the Lippmann–Schwinger equation (1.1+) and (2.17−), the derivations of the former use the formal solution (1.22+) and the formal solution for the $|E\,\hat{b}^{-}\rangle$'s corresponding to (1.22).

One final remark should be made: None of our arguments have depended on the assumptions that $\Phi^{in}(t)$ is an eigenvector $\Phi^{in}(\hat{a}, t)$ of the operators $A_1, A_2, \ldots, A_k$ and that Ψ^{out} is an eigenvector $\Psi^{out}(\hat{b}, t)$ of the operators $B_1, B_2, \ldots, B_k$. All equations for pure states (as opposed to statistical mixtures) are therefore equally valid, by linearity, if we remove the restrictions that $\Phi^{in}(t)$ and $\Psi^{out}(t)$ must be eigenvectors of the operators $A_1, A_2, \ldots, A_k$ and of the operators $B_1, B_2, \ldots, B_k$, respectively.

XV.3 The *S*-Operator and the Møller Wave Operators

In a scattering experiment a state is prepared before the projectile and target start interacting with each other, and a state is detected after they stop interacting with each other. Thus in scattering experiments in-states are transformed into out-states. The operator that describes this transformation is called the *scattering operator* or *S-operator*. For every physical system undergoing collisions we postulate the existence of a unitary operator S that transforms in-states into out-states. The *S*-operator is so immediately related to the directly observable quantities (like cross sections) that knowledge of it leads at once to the prediction of these quantities.

As remarked in Section XIV.4, the concept of the *S*-operator makes sense even if a generator H of time development is not defined and the time-development axiom (V in Chapter XII) does not hold. If H is not defined, then neither are its generalized eigenvectors $|a^{\pm}\rangle$ nor the exact states $\Phi^{+}(t)$ and $\Psi^{-}(t)$. However, the states Φ^{in} and Ψ^{out}, which describe the prepared and detected states, and the operator

$$S: \Phi^{in} \rightarrow \Phi^{out} = S\Phi^{in}, \tag{3.1}$$

which transforms the prepared (and hence controlled) in-state Φ^{in} into the uncontrolled out-state Φ^{out}, are still valid concepts. If the time development axiom does hold (and there has been no evidence to the contrary in non-relativistic quantum physics), then the *S*-operator and its matrix, the *S*-matrix

$$(\Psi^{out}, S\Phi^{in}) \tag{3.2}$$

may be expressed in terms of more fundamental and less directly observable quantities.

We shall now do this by giving a precise definition of S in terms of previously introduced quantities. The modulus squared of the matrix element

$$(\Psi^{-}(\hat{b}, t), \Phi^{+}(\hat{a}, t)) \tag{3.3}$$

gives the probability of finding the state $\Psi^-(\hat{b}, t)$, which is observed after the interaction V has ceased to be effective as the state $\Psi^{out}(\hat{b}, t)$, if the state of the system is $\Phi^+(\hat{a}, t)$, which was prepared before V became effective as the state $\Phi^{in}(\hat{a}, t)$. That is to say, (3.3) describes the probability for a transition from an initial configuration, described by the quantum numbers $\hat{a}$ and the energy distribution $\phi(E)$, into a final configuration, described by the quantum numbers $\hat{b}$ and the energy distribution $\psi(E)$. We first develop an expression for (3.3) in terms of the in-state Φ^{in} and the out-state Ψ^{out}. To do this we shall use the formal solutions $(2.36+)$ and $(2.36-)$,

$$\Phi^+(t) = \Phi^{in}(t) + \int_{-\infty}^{+\infty} dt'\, G^+(t - t')V\Phi^{in}(t') \tag{3.4+}$$

and

$$\Psi^-(t) = \Psi^{out}(t) + \int_{-\infty}^{+\infty} dt'\, G^-(t - t')V\Psi^{out}(t'), \tag{3.4-}$$

of $(2.18+)$ and $(2.18-)$. Equations $(2.6\pm)$ and $(2.15\pm)$ imply

$$\Phi^{in}(t') = e^{-iK(t'-t)}\Phi^{in}(t) \tag{3.5+}$$

and

$$\Psi^{out}(t') = e^{-K(t'-t)}\Psi^{out}(t). \tag{3.5-}$$

These expressions may be substituted into the integrands of $(3.4+)$ and $(3.4-)$ to get

$$\Phi^+(t) = \Omega^+\Phi^{in}(t) \tag{3.6+}$$

and

$$\Psi^-(t) = \Omega^-\Psi^{out}(t), \tag{3.6-}$$

where we have defined the *Møller wave operators*

$$\begin{aligned}
\Omega^\pm &\equiv I + \int_{-\infty}^{+\infty} dt'\, G^\pm(t - t')Ve^{-iK(t'-t)} \\
&= I \mp i \int_{-\infty}^{+\infty} dt'\, G^\pm(t - t')VG_0^\mp(t' - t) \\
&= I \mp i \int_{-\infty}^{+\infty} dt''\, G^\pm(-t'')VG_0^\pm(t'').
\end{aligned} \tag{3.7$\pm$}$$

An immediate consequence of $(3.6\pm)$ is that the mixtures $W^+(t)$ of $(2.20+)$ and $\tilde{W}^-(t)$ of $(2.20-)$ are related to the mixtures $W^{in}(t)$ of $(2.19+)$ and $\tilde{W}^{out}(t)$ of $(2.19-)$ by

$$W^+(t) = \Omega^+W^{in}(t)\Omega^{+\dagger} \tag{3.8+}$$

and by

$$\tilde{W}^-(t) = \Omega^-\tilde{W}^{out}(t)\Omega^{-\dagger}. \tag{3.8-}$$

Use of $(3.6\pm)$ in (3.3) gives

$$(\Psi^-(\hat{b}, t), \Phi^+(\hat{a}, t)) = (\Psi^{\text{out}}(\hat{b}, t), \Omega^{-\dagger}\Omega^+\Phi^{\text{in}}(\hat{a}, t)) \qquad (3.9)$$

This last equation motivates

$$S \equiv \Omega^{-\dagger}\Omega^+ \qquad (3.10)$$

as the definition of the scattering operator.

The last expression for $\Omega^\pm$ in $(3.7\pm)$ shows $\Omega^\pm$ to be time-independent, which implies that S is also time-independent. So is the matrix element (3.9), for $\Phi^+(t)$ is given both by

$$\Phi^+(t) = e^{-iHt}\Phi^+ = e^{-iHt}\Omega^+\Phi^{\text{in}}(0)$$

and by

$$\Phi^+(t) = \Omega^+\Phi^{\text{in}}(t) = \Omega^+ e^{-iKt}\Phi^{\text{in}}.$$

Since $\{\Phi^{\text{in}}\}$ spans the space of physical states, we conclude that

$$e^{-iHt}\Omega^\pm = \Omega^\pm e^{-iKt}. \qquad (3.11\pm)$$

[The upper equations follow from the stated argument; a similar argument gives $(3.11-)$.] Equivalently, one has the so-called *intertwining relations*

$$H\Omega^\pm = \Omega^\pm K. \qquad (3.12\pm)$$

It is then easily shown that

$$[S, K] = 0, \qquad (3.13)$$

a consequence of which is the time independence of (3.9):

$$(\Psi^-(\hat{b}, t), \Phi^+(\hat{a}, t)) = (\Psi^{\text{out}}(\hat{b}, 0), e^{+iKt}Se^{-iKt}\Phi^{\text{in}}(\hat{a}, 0))$$

$$= (\Psi^{\text{out}}(\hat{b}, 0), S\Phi^{\text{in}}(\hat{a}, 0)). \qquad (3.14)$$

Equation (3.13), it should be noted, may be interpreted as a statement of energy conservation between the in-state Φ^{in} and the out-state Ψ^{out}.

If we substitute $(2.6+)$ and $(2.9+)$ into the left-hand side and right-hand side of $(3.6+)$, respectively, we obtain

$$\frac{1}{2\pi}\int dE\, e^{-iEt}\phi(E)|E\,\hat{a}^+\rangle = \frac{1}{2\pi}\int dE\, e^{-iEt}\phi(E)\Omega^+|E\,\hat{a}\rangle.$$

Since $\phi(E)$ may be any well-behaved function, it follows that

$$\Omega^+|E\,\hat{a}\rangle = |E\,\hat{a}^+\rangle. \qquad (3.15+)$$

Equations $(2.6-)$, $(2.9-)$, and $(3.6-)$ may be used in a similar manner to conclude

$$\Omega^-|E\,\hat{a}\rangle = |E\,\hat{a}^-\rangle. \qquad (3.15-)$$

Since we have assumed the system is asymptotically complete [Equation

(1.19)], we may conclude that the Møller operators Ω^+ and Ω^- map the space of physical states $\mathcal{H}$ onto the space of scattering states $\mathcal{H}_{\text{scat}}$:

$$\Omega^+ \mathcal{H} = \Omega^- \mathcal{H} = \mathcal{H}_{\text{scat}}. \tag{3.16}$$

In other words, the domain of definition of both Ω^+ and Ω^- is the entire space $\mathcal{H}$, but the range of Ω^+ and Ω^- is the (proper if bound states exist) subspace $\mathcal{H}_{\text{scat}}$.

⟦An operator A is said to be *isometric* if it preserves the norm of vectors, i.e., if

$$(\psi, \psi) = (A\psi, A\psi) = (\psi, A^\dagger A\psi) \tag{3.17}$$

for all vectors ψ. An equivalent definition is that A satisfies the condition

$$A^\dagger A = I. \tag{3.18}$$

If A satisfies the stronger condition

$$A^\dagger A = I \quad \text{and} \quad AA^\dagger = I, \tag{3.19}$$

then A is *unitary*. A unitary operator A is defined on all of $\mathcal{H}$ and has all of $\mathcal{H}$ as its range, i.e., maps $\mathcal{H}$ onto $\mathcal{H}$ in a one-to-one fashion. A unitary operator is necessarily isometric, but not vice versa.⟧

Because $\Omega^\pm \mathcal{H} = \mathcal{H}_{\text{scat}}$ is usually not all of $\mathcal{H}$, we should not expect the Møller operators to be unitary; they are, however, isometric, a fact we shall now show.

By inserting

$$\Phi^+(t') = e^{-iH(t'-t)}\Phi^+(t)$$

[which follows from (2.9+)] into (2.18+), we obtain

$$\Phi^{\text{in}}(t) = \Phi^+(t) - \int_{-\infty}^{+\infty} dt' \, G_0^+(t - t')Ve^{-iH(t'-t)}\Phi^+(t)$$

$$= \left(I + i\int_{-\infty}^{+\infty} dt' \, G_0^+(t - t')VG^-(t' - t)\right)\Phi^+(t) \tag{3.20}$$

$$= \left(I + i\int_{\infty}^{\infty} dt'' \, G_0^+(-t'')VG^-(+t'')\right)\Phi^+(t).$$

The second line follows from the definition (2.35−) of $G^-(t)$ and the fact that $G_0^+(t - t')$ is nonzero only for $t - t' > 0$. From the definitions (2.12±) and (2.35±) of the Green's functions and from the Hermiticity of K and H it easily follows that

$$G_0^\pm(-t)^\dagger = G_0^\mp(t) \tag{3.21a\pm}$$

and

$$G^{\pm}(t)^{\dagger} = G^{\mp}(-t). \qquad (3.21b\pm)$$

Upon use of $(3.21\pm)$ and the definition $(3.7+)$ of Ω^+, Equation (3.20) becomes

$$\Phi^{\text{in}}(t) = \left(I - i \int_{-\infty}^{\infty} dt''\, G^+(-t'')VG_0^-(t'')\right)^{\dagger} \Phi^+(t) = \Omega^{+\dagger}\Phi^+(t). \qquad (3.22+)$$

But $\Phi^+(t) = \Omega^+ \Phi^{\text{in}}(t)$, so

$$\Phi^{\text{in}}(t) = \Omega^{+\dagger}\Omega^+\Phi^{\text{in}}(t).$$

We have not made use of any property of $\Phi^{\text{in}}(t)$ except that it can be expanded in terms of the generalized eigenvectors $|E\,\hat{a}\rangle$ of K, so $\Phi^{\text{in}}(t)$ is an arbitrary element of $\mathscr{H}$, Consequently,

$$\Omega^{+\dagger}\Omega^+ = I, \qquad (3.23+)$$

i.e., Ω^+ is isometric. A similar derivation gives

$$\Omega^{-\dagger}\Omega^- = I. \qquad (3.23-)$$

The failure of $\Omega^{\pm}$ to be unitary is now easily shown: By use of $(3.15\pm)$ and $(\text{XIV}.2.11)$ one obtains

$$\Omega^{\pm}\Omega^{\pm\dagger} = \left(\sum_a \Omega^{\pm}|a\rangle\langle a|\right)\left(\sum_{a'} \Omega^{\pm}|a'\rangle\langle a'|\right)^{\dagger}$$

$$= \sum_{aa'} |a^{\pm}\rangle\langle a|a'\rangle\langle a'^{\pm}| = \sum_a |a^{\pm}\rangle\langle a^{\pm}| \qquad (3.24\pm)$$

$$= \Pi_{\text{scat}} = I - \Pi_{\text{bnd}}$$

where

$$\Pi_{\text{scat}} = \sum_a |a^{\pm}\rangle\langle a^{\pm}| \qquad (3.25a)$$

and

$$\Pi_{\text{bnd}} = \sum_n |\alpha_n\rangle\langle \alpha_n| \qquad (3.25b)$$

are the projectors onto the subspaces of scattering states and bound states, $\mathscr{H}_{\text{scat}}$ and $\mathscr{H}_{\text{bnd}}$, respectively. Unless the *unitary deficiency* Π_{bnd} is zero, i.e., unless there are no bound states $|\alpha_n\rangle$, the operators $\Omega^{\pm}$ can only be isometric and not unitary.

The adjoint Møller operators $\Omega^{\pm\dagger}$ map $\mathscr{H}_{\text{scat}}$ back onto $\mathscr{H}$; more precisely, they annihilate the $\mathscr{H}_{\text{bnd}}$ part of $\mathscr{H} = \mathscr{H}_{\text{bnd}} \oplus \mathscr{H}_{\text{scat}}$ and map the rest, $\mathscr{H}_{\text{scat}}$, back onto $\mathscr{H}$:

$$\Omega^{\pm\dagger}|\alpha_n\rangle = \left(\sum_{a'} \Omega^{\pm}|a'\rangle\langle a'|\right)^{\dagger}|\alpha_n\rangle = \sum_{a'} |a'\rangle\langle a'^{\pm}|\alpha_n\rangle = 0, \qquad (3.26\pm)$$

or equivalently,

$$\Omega^{\pm\dagger}\Pi_{\text{bnd}} = 0; \qquad (3.26'\pm)$$

and

$$\Omega^{\pm\dagger}|a^{\pm}\rangle = \Omega^{\pm\dagger}\Omega^{\pm}|a\rangle = I|a\rangle = |a\rangle. \qquad (3.27\pm)$$

We are now in a position to show that S is unitary:

$$SS^{\dagger} = (\Omega^{-\dagger}\Omega^{+})(\Omega^{+\dagger}\Omega^{-}) = \Omega^{-\dagger}(I - \Pi_{\mathrm{bnd}})\Omega^{-} = I \qquad (3.28a)$$

and

$$S^{\dagger}S = (\Omega^{+\dagger}\Omega^{-})(\Omega^{-\dagger}\Omega^{+}) = \Omega^{+\dagger}(I - \Pi_{\mathrm{bnd}})\Omega^{+} = I. \qquad (3.28b)$$

In writing Equation (3.28) we have used the definition (3.10) of S, the iso-metricity (3.23$\pm$) of $\Omega^{\pm}$, and Equations (3.24$\pm$) and (3.26$\pm$); in order for Equation (3.28) to make sense—indeed, for the definition (3.10) of S to make sense—we had to assume domain $\Omega^{\pm\dagger} = $ range $\Omega^{\mp\dagger} = \Omega^{\mp}\mathscr{H}$, or, in other words, asymptotic completeness in the form (3.16). In this way we have proven that the S-operator defined by (3.10) and (3.7$\pm$) in terms of V and K is unitary.

We define the S-*matrix* to be the matrix $\{S_{aa'}\}$ of the scattering operator with respect to the K basis:

$$S_{aa'} \equiv \langle a|S|a'\rangle = \langle a|\Omega^{-\dagger}\Omega^{+}|a'\rangle = \langle a^{-}|a'^{+}\rangle. \qquad (3.10')$$

The S-matrix is therefore the transformation matrix between the two basis systems (1.16$+$) and (1.16$-$) of the space of scattering states $\mathscr{H}_{\mathrm{scat}}$:

$$|a^{+}\rangle = \sum_{E'\hat{a}'}|a'^{-}\rangle\langle a'^{-}|a^{+}\rangle = \sum_{E'\hat{a}'}|E'\hat{a}'^{-}\rangle\langle E'\hat{a}'|S|E\hat{a}\rangle. \qquad (3.10'')$$

By using the relation (1.24) we may write $S_{aa'}$ either as

$$S_{aa'} = [\langle a^{+}| - 2\pi i\langle a|V\delta(E_{a} - H)]|a'^{+}\rangle$$
$$= \langle a^{+}|a'^{+}\rangle - 2\pi i\delta(E_{a} - E_{a'})\langle a|V|a'^{+}\rangle \qquad (3.29)$$

or as

$$S_{aa'} = \langle a^{-}|(|a'^{-}\rangle - 2\pi i\delta(E_{a'} - H)V|a'\rangle)$$
$$= \langle a^{-}|a'^{-}\rangle - 2\pi i\delta(E_{a'} - E_{a})\langle a^{-}|V|a'\rangle. \qquad (3.30)$$

If we use (XIV.2.9b) (which we shall finally prove at the end of this section), we see that the first terms on the right-hand sides of (3.29) and (3.30) are equal. Thus

$$\langle a|V|a'^{+}\rangle = \langle a^{-}|V|a'\rangle \quad \text{when} \quad E_{a} = E_{a'} = E,$$

i.e.,

$$\langle E\hat{a}|V|E\hat{a}'^{+}\rangle = \langle E\hat{a}^{-}|V|E\hat{a}'\rangle. \qquad (3.31)$$

In Section XIV.5 we introduced the on-the-energy-shell T-matrix

$$T_{aa'}(E) \equiv \langle E\hat{a}|V|E\hat{a}'^{+}\rangle \qquad (3.32)$$

(which does not define an operator) and its off-the-energy-shell extension

$$T_{aa'}^{+} = \langle a|T^{+}|a'\rangle \equiv \langle a|V|a'^{+}\rangle = \langle E_{a}\hat{a}|V|E_{a'}\hat{a}'^{+}\rangle, \qquad (3.33)$$

which defines an operator T^+. Equation (3.31) suggests a different extension of the T-matrix, namely

$$T^-_{aa'} = \langle a | T^- | a' \rangle \equiv \langle a^- | V | a' \rangle = \langle E_a \, \hat{a}^- | V | E_{a'} \, \hat{a}' \rangle, \qquad (3.34)$$

which defines the operator T^-. $\{T^+_{aa'}\}$ and $\{T^-_{aa'}\}$ are extensions of $\{T_{aa'}(E)\}$ in the sense that

$$T^+_{aa'} = T^-_{aa'} = T_{aa'}(E) \quad \text{when} \quad E_a = E_{a'} = E. \qquad (3.35)$$

The S-matrix element $S_{aa'}$ may then be expressed as

$$S_{aa'} = \langle a | a' \rangle - 2\pi i \delta(E_a - E_{a'}) T_{aa'}(E_a)$$

$$= \rho(E_a)^{-1} \delta_{\hat{a}\hat{a}'} \, \delta(E_a - E_{a'}) - 2\pi i \delta(E_a - E_{a'}) T_{aa'}(E_a). \qquad (3.36)$$

Note that it is the on-the-energy-shell T-matrix elements that contribute to the S-matrix; matrix elements of S between generalized eigenvectors $|E_a \, \hat{a}\rangle$ and $|E_{a'} \, \hat{a}'\rangle$ vanish for $E_a \neq E_{a'}$.

An integral equation for the T^+-matrix may be obtained by inserting the Lippmann–Schwinger equation (1.1+) into (3.33):

$$T^+_{aa'} = \langle a | V \left(| a' \rangle + \frac{1}{E_{a'} - K + i0} V | a'^+ \rangle \right)$$

$$= \langle a | V | a' \rangle + \sum_{a''} \frac{\langle a | V | a'' \rangle \langle a'' | V | a'^+ \rangle}{E_{a'} - E_{a''} + i0}$$

$$= \langle a | V | a' \rangle + \sum_{a''} \frac{\langle a | V | a'' \rangle T^+_{a''a'}}{E_{a'} - E_{a''} + i0}. \qquad (3.37)$$

Equation (3.37) may be iterated by repeated substitution of the right-hand side into itself:

$$T^+_{aa'} = \langle a | V | a' \rangle + \sum_{a''} \frac{\langle a | V | a'' \rangle \langle a'' | V | a' \rangle}{E_{a'} - E_{a''} + i0} + \cdots. \qquad (3.38)$$

The *Born approximation*

$$T^+_{aa'} \approx \langle a | V | a' \rangle = \langle E_a \, \hat{a} | V | E_{a'} \, \hat{a}' \rangle, \qquad (3.39)$$

in which only the first term of the iterative solution (3.38) is kept, is usually adequate in the case of high energies $E_a = E_{a'} = E$ and a weak interaction V.

From the unitarity of the operator S it follows that the S-matrix is also unitary:

$$\sum_{a''} S_{aa''} S^*_{a'a''} = \sum_{a''} S^*_{a''a} S_{a''a'} = \delta_{aa'} \qquad (3.40)$$

or in more detail,

$$\sum_{\hat{a}''} \int dE_{a''} \rho_{\hat{a}''}(E_{a''}) \langle E_a \, \hat{a} | S | E_{a''} \, \hat{a}'' \rangle \langle E_{a'} \, \hat{a}' | S | E_{a''} \, \hat{a}'' \rangle^*$$

$$= \sum_{\hat{a}''} \int dE_{a''} \rho_{a''}(E_{a''}) \langle E_{a''} \, \hat{a}'' | S | E_a \, \hat{a} \rangle^* \, \langle E_{a''} \, \hat{a}'' | S | E_{a'} \, \hat{a}' \rangle$$

$$= \rho(E_a)^{-1} \delta_{\hat{a}\hat{a}'} \delta(E_a - E_{a'}). \qquad (3.40')$$

By inserting (3.40') into (3.36) we may express the unitarity of $\{S_{aa'}\}$ in terms of the matrix elements $T_{\hat{a}\hat{a}'}(E)$:

$$T_{\hat{a}\hat{a}'}(E) - T_{\hat{a}'\hat{a}}^{*}(E) = -2\pi i \sum_{\hat{a}''} \rho_{\hat{a}''}(E) T_{\hat{a}\hat{a}''}(E) T_{\hat{a}\hat{a}''}^{*}(E). \tag{3.41}$$

The unitarity condition as expressed by (3.41) is sometimes called the *generalized optical theorem.*

In summary, we have obtained both the S-matrix and the T-matrix by assuming the axiom V of continuous time development. We have expressed the S-matrix and T-matrix in terms of quantities, like V and $|a^{+}\rangle$, that depend on the existence of the time-development generator H [cf. (3.32) and (3.36)]. The generalized eigenvectors $|E\,\hat{a}^{+}\rangle$ of H and the corresponding states $|\Phi^{+}(\hat{a}, t)\rangle$ describe the situation at the time of collision; they are therefore quantities that are never accessible to experimental determination. In contrast, the generalized eigenvectors $|E\,\hat{a}\rangle$ of K and the corresponding states $|\Phi^{\text{in}}(\hat{a}, t)\rangle$ and $|\Phi^{\text{out}}(\hat{a}, t)\rangle$ are experimentally accessible by means of the initial preparation and final observation of a collision experiment. They, together with the operator that describes the transition between them, are indispensable in the theoretical description of a collision experiment; whereas experimentally nonaccessible quantities like $|\Phi^{+}(\hat{a}, t)\rangle$ are theoretical tools and rest on additional theoretical assumptions (e.g., the axiom V).

The theoretical description which is based only on the experimentally accessible quantities is called *S-matrix theory.* The S-matrix and the on-energy-shell T-matrix, which are connected by way of (3.36), are then the fundamental physical quantities; furthermore, the only assumption entering into (3.36) is the conservation of energy between the in-state and the out-state ($[S, K] = 0$). Unitarity, which implies the condition (3.41) on the T-matrix, is a separate fundamental assumption in S-matrix theory and is justified by the requirement of conservation of overall probability. Another fundamental assumption in S-matrix theory is the analyticity of the T-matrix as a function of energy and other physical parameters. This is argued to be connected with causality. (We shall discuss this connection in Section XVIII.4.) In the conventional formulation based on the assumption of a continuous time development, properties such as unitarity and analyticity (in a certain domain) are derived from the properties of H and of V.

Symmetry considerations place the most significant conditions on the S-matrix. In the conventional formulation symmetries are formulated as transformation properties of V and H under various symmetry transformations, and these transformation properties then lead to derived symmetry properties of the S-matrix. (We shall discuss the consequences of rotational symmetry in Chapter XVI.) In S-matrix theory the symmetries are formulated directly in terms of symmetries of the S-matrix or in terms of transformation properties of the T-operator.

If V and H are known, then the conventional approach to the description of collision processes is more "fundamental," although based on more assumptions. After Equation (3.37) has been solved for the T-matrix, one may predict experimental results. As mentioned in previous chapters, in physics we most often have the reverse situation: One wants to conjecture from experimental data (e.g., cross sections) the properties of the fundamental observables and then use these properties to make predictions about other experimental quantities. In this situation a formulation of the properties in terms of the T-operator is just as valid as a formulation in terms of V, and is often more practical.

XV.A Appendix

With the help of the isometricity $(3.23\pm)$ of the Møller operators, $\Omega^{\pm\dagger}\Omega^{\pm} = I$, it is now trivial to prove two equalities that were used in Chapter XIV. (The proof of isometricity did not depend upon these equalities.) Equation (XIV.2.9b) follows from $(3.15+)$ and isometricity:

$$\langle a^{\pm}|a'^{\pm}\rangle = (\langle a|\Omega^{\pm\dagger})(\Omega^{\pm}|a'\rangle)$$
$$= \langle a|(\Omega^{\pm\dagger}\Omega^{\pm})|a'\rangle = \langle a|a'\rangle. \qquad (A.1)$$

To prove (XIV.5.21) we use $(3.15+)$ and $(3.8+)$ together with isometricity:

$$\langle a^{+}|W^{+}(t)|a'^{+}\rangle = (\langle a|\Omega^{+\dagger})(\Omega^{+}W^{\mathrm{in}}(t)\Omega^{+\dagger})(\Omega^{+}|a'\rangle)$$
$$= \langle a|(\Omega^{+\dagger}\Omega^{+})W^{\mathrm{in}}(t)|(\Omega^{+\dagger}\Omega^{+})|a'\rangle$$
$$= \langle a|W^{\mathrm{in}}(t)|a'\rangle; \qquad (A.2+)$$

in a similar fashion one can prove

$$\langle b^{-}|\tilde{W}^{-}(t)|b'^{-}\rangle = \langle b|\tilde{W}^{\mathrm{out}}(t)|b'\rangle. \qquad (A.2-)$$

Lastly we show that the term II given by (XIV.2.17) vanishes. The key to doing so is Equation $(A.2+)$. If we substitute (XIV.2.9b) and $(A.2+)$ at $t = 0$ into (XIV.2.17), then

$$\mathrm{II} = \sum_{b}\sum_{aa'} e^{-i(E_a - E_{a'})t}\langle b|V|a\rangle\rho(E_b)^{-1}\delta_{a'\hat{b}}\delta(E_{a'} - E_b)\langle a|W^{\mathrm{in}}|a'\rangle$$

$$= \int\rho(E_b)\,dE_b\sum_{\hat{b}}\int\rho(E_a)\,dE_a\,\rho(E_{a'})\,dE_{a'}\sum_{\hat{a}\hat{a}'} e^{-i(E_a - E_{a'})t}$$
$$\times \langle E_b\hat{b}|V|E_a\hat{a}\rangle\rho(E_b)^{-1}\delta_{\hat{a}'\hat{b}}\delta(E_{a'} - E_b)\langle E_a\hat{a}|W^{\mathrm{in}}|E_{a'}\,\hat{a}'\rangle$$

$$= \int\rho(E_b)\,dE_b\sum_{\hat{b}}\int\rho(E_a)\,dE_a\sum_{\hat{a}} e^{-i(E_a - E_b)t}\langle E_b\hat{b}|V|E_a\hat{a}\rangle$$
$$\times \langle E_a\hat{a}|W^{\mathrm{in}}|E_b\hat{b}\rangle$$

$$= \sum_{b}\sum_{a} e^{-i(E_a - E_b)t}\langle b|V|a\rangle\langle a|W^{\mathrm{in}}|b\rangle. \qquad (A.3)$$

But the summation $\sum_b$ is only over those values of b for which $|b\rangle \in \Lambda\mathcal{H}$, and by hypothesis the initial state W^{in} is so prepared that $|A\rangle \notin \Lambda \mathcal{H}$ for all states $|A\rangle$ that give a nonzero contribution ($w_A \neq 0$) to

$$W^{\text{in}} = \sum_A w_A |A\rangle \langle A|.$$

Therefore

$$W^{\text{in}}|b\rangle = 0$$

for all b in the summation, so that

$$\text{II} = 0. \tag{A.4}$$

The contribution of II both to the transition rate $d\langle \Lambda \rangle_t/dt$ of (XIV.2.5) and to the transition probability $\langle \Lambda \rangle_\infty$ of (XIV.5.2) is then, of course, zero.

CHAPTER XVI

Elastic and Inelastic Scattering for Spherically Symmetric Interactions

In this chapter we derive general results for spherically symmetric situations, i.e., for situations when in addition to the free Hamiltonian K, the interaction Hamiltonian V and the transition operator T also commute with the operator of angular momentum. For the sake of simplicity we consider the spin-zero case. In Section XVI.1, the partial-wave expansion is discussed and the partial cross sections are derived. In Section XVI.2, the phase shift is introduced as a consequence of the unitarity of the S-matrix. In Section XVI.3, a graphical representation of the partial-wave amplitude, the Argand diagram, is introduced.

XVI.1 Partial-Wave Expansion

Many scattering problems are spherically symmetric. This means that both $H = K + V$ and V commute with the (orbital) angular-momentum operators L_i or, if the scattering is described by a transition operator T, that T (and hence the scattering operator S) commutes with the L_i's:

$$[V, L_i] = 0, \qquad [T, L_i] = 0. \qquad (1.1)$$

Then

$$\{K, \mathbf{L}^2, L_3, \eta^{\mathrm{op}}\} \qquad (1.2a)$$

and

$$\{H, \mathbf{L}^2, L_3, \kappa^{\mathrm{op}}\}, \qquad (1.2b)$$

where η^{op} and κ^{op} stand for the internal observables (their eigenvalues η and κ being the internal quantum numbers), are both c.s.c.o.'s, and one may choose their generalized eigenvectors

$$|E_a l\, l_3 \eta\rangle \tag{1.3a}$$

and

$$|E_\alpha l\, l_3 \kappa\rangle \tag{1.3b}$$

as basis vectors of the space of physical states.

The vectors

$$|E_\alpha l,\, l_3 \eta^+\rangle \tag{1.3c}$$

which are obtained from (1.3a) by the Lippmann–Schwinger equation (XV.1.1+) or by (XV.1.22+) are in general not eigenvectors of (1.2b). However, if $[V, \eta^{op}] = 0$ and therefore $[H, \eta^{op}] = 0$, then the collection of operators $\kappa^{op} = \kappa_1^{op} \cdots \kappa_k^{op}$ in (1.2b) can be chosen to be η^{op}. Then the generalized eigenvectors (1.3b)—which are determined only up to an arbitrary phase factor that can still depend upon the quantum numbers—can be chosen to be identical with (1.3c). In general $[V, \eta^{op}] \neq 0$, and then (1.3c) are not eigenvectors of the c.s.c.o. (1.2b) and the meaning of the label η is defined through the Lippmann–Schwinger equation as explained in Section XV.1.

Still another c.s.c.o. consists of the projectile momentum $\mathbf{P}$ (or the momentum $\mathbf{P}$ in the center-of-mass system when the target is not fixed) and the operators for the internal quantum numbers η^{op}.

$$\{\mathbf{P}_1, \mathbf{P}_2, \mathbf{P}_3, \eta^{op}\}. \tag{1.4}$$

The corresponding basis consists of the generalized momentum eigenvectors

$$|\mathbf{p}\, \eta\rangle. \tag{1.5}$$

The additional quantum numbers $\eta = (\eta_1 \eta_2 \cdots \eta_k)$ may refer to the internal structure of the target, of the projectile, or of both. For example, if the target is a hydrogen atom, then H will contain the quantum numbers n, j, j_3, and π that characterize the state of the electron in the hydrogen atom (cf. Chapters VI and IX). According to (1.2), $[L_i, \eta^{op}] = 0$; thus we exclude the possibility that η includes the quantum numbers of an internal angular momentum. Then the total angular momentum is equal to the orbital angular momentum L_i, and we avoid the complication of coupling the internal angular momentum with the orbital angular momentum to get the total angular momentum. Most processes encountered in experiment involve at least one internal angular momentum; however, to understand the principal problems of this chapter it is easier if one omits at the beginning the inessential complications introduced by the coupling of angular momenta.

The simplest case we can discuss is the case of a structureless projectile incident on a structureless target. Whether the target and/or projectile may be considered to be structureless depends on the energies involved. If, for example, the target is a helium atom, then it may be considered structureless for projectile energies less than the energy of the first excited state, and the quantum numbers η are unnecessary. If the projectile energy is comparable in size to the difference between the energy levels of the helium atom, then the target's structure must be considered, and η will contain the quantum numbers n, j, j_3, π (singlet or triplet) of the helium atom.

We would like to express the matrix element[1]

$$\langle \mathbf{p}_b \eta_b | T | \mathbf{p}_A \eta_A \rangle \equiv \langle \mathbf{p}_b \eta_b | V | (\mathbf{p}_A)(\eta_A)^+ \rangle \tag{1.6}$$

of the transition operator T with respect to the momentum basis (1.5) in terms of the matrix elements of T with respect to the angular-momentum basis (1.3a). To do this we expand $|\mathbf{p}\,\eta\rangle$ in terms of the angular-momentum basis,

$$|\mathbf{p}\,\eta\rangle = \sum_{ll_3} \int \rho_\eta(E)\, dE\, |E\, l\, l_3\, \eta\rangle \langle E\, l\, l_3\, \eta | \mathbf{p}\,\eta\rangle. \tag{1.7}$$

The determination of the transition coefficients $\langle \mathbf{p}\,\eta | E\, l\, l_3\, \eta \rangle = \langle E\, l\, l_3\, \eta | \mathbf{p}\,\eta \rangle^*$ is a problem similar to the determination of the wave functions $\langle \mathbf{x} | n\, l\, m \rangle$ of the hydrogen atom, which was discussed in Section VII.3. Instead of (VII.3.5) we have the analogue

$$\langle \mathbf{p}\,\eta | L_i | \psi \rangle = -\frac{1}{i} \epsilon_{ijk}\, p_j \frac{\partial}{\partial p_k} \langle p\,\theta_p\,\phi_p\,\eta | \psi \rangle, \tag{1.8}$$

from which follow the analogues of (VII.3.6$'_3$) and (VII.3.6$'$). Here $(p\,\theta_p,\,\phi_p)$ are the spherical coordinates of the momentum vector $\mathbf{p}$. Instead of the "radial equation" (VII3.7) or (VII.3.1) we have

$$\langle p\,\theta_p\,\phi_p\,\eta | \left(\frac{\mathbf{P}^2}{2M} + K^{int} \right) | E\, l\, l_3\, \eta \rangle = E \langle p\,\theta_p\,\phi_p\,\eta | E\, l\, l_3\, \eta \rangle, \tag{1.9}$$

which is a consequence of the (nonrelativistic) relation (cf. XIV.5.31)

$$K = \frac{\mathbf{P}^2}{2M(\eta^{op})} + K^{int}(\eta^{op}) \tag{1.10}$$

between the energy operator K, the momentum $\mathbf{P}$, and the internal energy K^{int}. Although the point has not been raised previously, the mass of the particle may depend upon the internal quantum numbers η of the system; this dependence allows us to consider cases where the outgoing particle is different from the incoming particle. From (1.9) and the analogues of (VII.3.6$'$) and (VII.3.6$'_3$) it follows that the transition coefficients have the form [cf. (VII.3.11)]

$$\langle p\,\theta_p\,\phi_p\,\eta | E\, l\, l_3\, \eta \rangle = f_\eta(p)\delta\left(E - \frac{p^2}{2m_\eta} - E_\eta^{int} \right) Y_{ll_3}(\theta_p, \phi_p) \tag{1.11}$$

---

[1] The meaning of putting the quantum numbers in parentheses has been explained in Section XV.1.

The function $f_\eta(p)$ is determined (up to a phase factor) by the normalization (XIV.2.9a) of the generalized K-eigenvectors $|E\,l\,l_3\,\eta\rangle$, for

$$\rho_\eta(E)^{-1}\delta(E-E')\delta_{ll'}\delta_{l_3l_3'} = \langle E\,l\,l_3\,\eta|E'\,l'\,l_3'\,\eta\rangle$$

$$= \int d^3p\,\langle E\,l\,l_3\,\eta|\mathbf{p}\,\eta\rangle\langle\mathbf{p}\,\eta|E'\,l'\,l_3'\,\eta\rangle$$

$$= \int p^2\sin\theta_p\,dp\,d\theta_p\,d\phi_p\,f_\eta^*(p)\,\delta\left(E-\frac{p^2}{2m_\eta}-E_\eta^{\text{int}}\right)$$

$$\times\,Y_{ll_3}^*(\theta_p,\phi_p)f_\eta(p)\delta\left(E'-\frac{p^2}{2m_\eta}-E_\eta^{\text{int}}\right)Y_{l'l_3'}(\theta_p,\phi_p)$$

$$= m_\eta\rho_\eta(E)|\,f_\eta(p_\eta(E))|^2\delta(E-E')\delta_{ll'}\delta_{l_3l_3'}. \tag{1.12}$$

In this calculation use has been made of (VII.3.16) and the integration with respect to p has been evaluated in the following manner[2] using the substitution

$$\epsilon = p^2/2m_\eta - E_\eta^{\text{int}}, \qquad p = p_\eta(\epsilon) = \sqrt{2m_\eta(\epsilon+E_\eta^{\text{int}})}:$$

$$\int p^2|\,f_\eta(p)|^2\delta\left(E-\frac{p^2}{2m_\eta}-E_\eta^{\text{int}}\right)\delta\left(E'-\frac{p^2}{2m_\eta}-E_\eta^{\text{int}}\right)dp$$

$$= \int d\epsilon\,m_\eta\sqrt{2m_\eta(\epsilon-E_\eta^{\text{int}})}|\,f_\eta(p(\epsilon))|^2\,\delta(\epsilon-E)\delta(E'-\epsilon)$$

$$= m_\eta\sqrt{2m_\eta(E-E_\eta^{\text{int}})}|\,f_\eta(p(E))|^2\delta(E'-E). \tag{1.13}$$

For the positive square root in (1.13), which represents the absolute value of the momentum as a function of energy, the following notation has been used:

$$p_\eta(E) = \sqrt{2m_\eta(E-E_\eta^{\text{int}})}. \tag{1.14}$$

Equation (1.12), together with

$$\rho_\eta(E) = m_\eta p_\eta(E) \tag{1.15}$$

of (XIV.5.18a), gives

$$|\,f_\eta(p_\eta(E))| = \rho_\eta(E)^{-1}, \tag{1.16}$$

so that with an appropriate choice of phase for the generalized eigenvectors $|\mathbf{p}\,\eta\rangle$, the transition coefficients (1.11) are given by

$$\langle p\,\theta_p\,\phi_p\,\eta|E\,l\,l_3\,\eta\rangle = (\rho_\eta(E))^{-1}\delta\left(E-\frac{p^2}{2m_\eta}-E_\eta^{\text{int}}\right)Y_{ll_3}(\theta_p,\phi_p). \tag{1.17}$$

Equation (1.1) tells us that T is a scalar operator; its matrix elements with respect to the angular-momentum basis are then [using the Wigner–Eckart theorem (V.3.6)]

$$\langle E\,l\,l_3\,\eta_b|T|E'\,l'\,l_3'\,\eta_A\rangle = \langle l'\,l_3'\,0\,0|l\,l_3\rangle\langle E\,l\,\eta_b\|T\|E'\,l'\,\eta_A\rangle$$

$$= \delta_{ll'}\delta_{l_3l_3'}\langle E\,l\,\eta_b\|T\|E'\,l\,\eta_A\rangle. \tag{1.18}$$

---

[2] To evaluate the integral one can also use the mathematical formalism for $\delta(f(x))$ developed in Gelfand and Shilov (1964), Vol. 1, Section II.2.5.

Use of (1.7), (1.17), and (1.18) in (1.6) gives

$$\langle \mathbf{p}_b \eta_b | T | \mathbf{p}_A \eta_A \rangle = \sum_{ll_3} \int dE \, dE' \, \delta\left(E - \frac{p_b^2}{2m_b} - E_b^{\text{int}}\right) \delta\left(E' - \frac{p_A^2}{2m_A} - E_A^{\text{int}}\right)$$

$$\times \langle E \, l \, \eta_b \| T \| E' \, l \, \eta_A \rangle \, Y_{ll_3}(\Omega_b) Y_{ll_3}^*(\Omega_A), \tag{1.19}$$

where $\Omega_A = \mathbf{p}_A/|\mathbf{p}_A| = (\sin\theta_A \cos\phi_A, \sin\theta_A \sin\phi_A, \cos\theta_A)$ is the direction of the incident particle and $\Omega_b = \mathbf{p}_b/|\mathbf{p}_b| = (\sin\theta_b \cos\phi_b, \sin\theta_b \sin\phi_b, \cos\theta_b)$ is the direction of the detected particle. If we perform the summation $\sum_{l_3}$ by using the addition theorem (VII.3.26)

$$\sum_{l_3=-l}^{+l} Y_{ll_3}(\Omega_b) Y_{ll_3}^*(\Omega_A) = \frac{2l+1}{4\pi} P_l(\Omega_b \cdot \Omega_A) \tag{1.20}$$

for spherical harmonics, then (1.19) becomes

$$\langle \mathbf{p}_b \, \eta_b | T | \mathbf{p}_A \, \eta_A \rangle = \frac{1}{4\pi} \sum_l (2l+1) P_l(\cos\theta) \langle E_b \, l \, \eta_b \| T \| E_A \, l \, \eta_A \rangle, \tag{1.21}$$

where θ is the angle between Ω_A and Ω_b (cf. Figure XIV.3.1), and where[3]

$$E_A = E(p_A, \eta_A) = \frac{p_A^2}{2m_A} + E_{\eta_A}^{\text{int}} \tag{1.22a}$$

$$E_b = E(p_b, \eta_b) = \frac{p_b^2}{2m_b} + E_{\eta_b}^{\text{int}} \tag{1.22b}$$

The reduced matrix element $\langle E_b \, l \, \eta_b \| T \| E_A \, l \, \eta_A \rangle$ depends upon the energy and the internal quantum numbers η_b and η_A, but does not depend upon the direction. As one would have expected for a spherically symmetric problem in which the initial conditions distinguish only the direction Ω_A, the T-matrix element (1.21) does not depend upon the angle ϕ around the direction Ω_A.

Equation (1.21) proves the statement preceding and justifying the definition (XIV.5.63). Written in terms of the scattering amplitude, (1.21) becomes, for $\eta_b = \eta_A$ and $E_b = E_A$,

$$T(p_A, \theta) = \sum_l (2l+1) P_l(\cos\theta)(-\pi m)\hbar \langle E_A \, l \, \eta_A \| T \| E_A \, l \, \eta_A \rangle \tag{1.21'}$$

In the expressions for the cross section [e.g., (XIV.5.38), (XIV.5.50), and (XIV.5.60)] and in any other physical quantity, the factor $\delta(E_b - E_A)$ appears as a consequence of energy conservation [cf. (XV.3.13)]. Consequently only the reduced matrix elements

$$\langle \eta_b \| T_l(E_A) \| \eta_A \rangle \equiv \langle E_b = E_A \, l \, \eta_b \| T \| E_A \, l \, \eta_A \rangle \tag{1.23}$$

on the energy shell, i.e., for $E_b = E_A$, have physical meaning. Energy conservation for the nonrelativistic case is given by

$$p_A^2/2m_A + E_{\eta_A}^{\text{int}} = E_A = E_b = p_b^2/2m_b + E_{\eta_b}^{\text{int}}. \tag{1.24}$$

In nonrelativistic collision experiments one usually has $m_b = m_A$, but does not have $E_{\eta_A}^{\text{int}} = E_{\eta_b}^{\text{int}}$ except in the case of elastic scattering. In the following

---

[3] Note the change in notation: E_A here denotes the total energy, whereas in Section XIV.5 E_A denoted the projectile (kinetic) energy.

we shall examine in detail the case in which the detected particle is the same as the incident particle, so that $m_b = m_A$; first, however, we give the expressions for the cross section in the general (spinless) case.

The differential cross section per unit solid angle for scattering from an initial state with momentum $\mathbf{p}_A$ and internal quantum numbers η_A into a final state with internal quantum numbers η_b is [cf. (XIV.5.60) or (XIV.5.61') at the end of Section XIV.4]

$$\frac{d\sigma^{\eta_b}}{d\Omega_b} = (2\pi)^4 \frac{m_A m_b}{p_A} \hbar^2 \int p_b(E_b)\, dE_b |\langle E_b\, \theta_b\, \phi_b\, \eta_b | T | p_A\, \theta_A\, \phi_A\, \eta_A \rangle|^2$$

$$\times\ \delta(E_b - E_A). \tag{1.25}$$

By using (1.21) we may rewrite $d\sigma^{\eta_b}/d\Omega_b$ in terms of the reduced matrix elements (1.23):

$$\frac{d\sigma^{\eta_b}}{d\Omega_b} = \pi^2 \frac{m_A m_b}{p_A} \hbar^2 p_b(E_A) | \sum_l (2l+1) P_l(\cos\theta) \langle \eta_b \| T_l(E_A) \| \eta_A \rangle|^2. \tag{1.26}$$

The dependence of the momentum $p_b = p_b(E_b, \eta_b)$ of the detected particle on $E_b(=E_A)$ and η_b is

$$p_b(E_A) = \sqrt{2m_b(E_A - E_{\eta_b}^{\text{int}})}, \tag{1.27}$$

as determined by (1.24) or (1.14).

The differential cross section given by (1.26) may be expressed as a series in the Legendre polynomials P_l, but the result is complicated (this will be done in Section XVIII.8, where it is applied to the phase shift analysis). We therefore give here only the expression for the total cross section, which may be expressed simply in terms of the reduced matrix elements. By using the orthogonality property

$$\int_{-1}^{+1} d\xi\, P_l(\xi) P_{l'}(\xi) = \frac{2}{2l+1}\delta_{ll'} \tag{1.28}$$

of the Legendre polynomials one obtains

$$\sigma^{\eta_b} = \int d\Omega_b \frac{d\sigma^{\eta_b}}{d\Omega_b}$$

$$= 4\pi^3 \frac{m_A m_b}{p_A} \hbar^2 p_b(E_A) \sum_l (2l+1) |\langle \eta_b \| T_l(E_A) \| \eta_A \rangle|^2 \tag{1.29}$$

for the total cross section. This suggests that we define the lth *partial cross section*

$$\sigma_l^{\eta_b} \equiv 4\pi^3 \frac{m_A m_b}{p_A} \hbar^2\, p_b(E_A)(2l+1) |\langle \eta_b \| T_l(E_A) \| \eta_A \rangle|^2 \tag{1.30}$$

and that we write the total cross section as

$$\sigma^{\eta_b} = \sum_l \sigma_l^{\eta_b}. \tag{1.31}$$

$\sigma_l^{\eta_b}$ and σ^{η_b} in (1.30) and (1.31) are the partial and total cross sections, respectively, for scattering from a state with internal quantum numbers $\eta = \eta_A$ into a state with internal quantum numbers $\eta = \eta_b$. Often the detector registers not just those states with a well-defined value η_b of η but all states with values of η within some range. This happens, for example, if η are the internal quantum numbers of the target and the detector registers scattered projectiles with any possible energy. The projectile may excite the target into any internal state η by losing the corresponding amount of kinetic energy, and the detector cannot distinguish between elastically scattered projectiles and inelastically scattered projectiles. In such a case the lth partial cross section is given by

$$\sigma_l = \sum_{\eta_b} \sigma_l^{\eta_b} = 4\pi^3 \frac{m_A m_b}{p_A} \hbar^2 (2l+1) \sum_{\eta_b} p_b(E_A) |\langle \eta_b \| T_l(E_A) \| \eta_A \rangle|^2, \quad (1.32)$$

and the total cross section is given by

$$\sigma = \sum_{\eta_b} \sigma^{\eta_b} = \sum_l \sigma_l = \sum_{\eta_b l} \sigma_l^{\eta_b}. \quad (1.33)$$

Instead of the reduced matrix elements $\langle \eta_b \| T_l(E_A) \| \eta_A \rangle$ it is convenient to use the "partial-wave amplitudes." They are commonly used in nonrelativistic scattering theory (particularly in the case of elastic scattering), and formulas written in terms of them can be immediately taken over to the relativistic theory. One proceeds from the assumption that the initial state is fixed by the internal quantum numbers η_A, which usually describes the state of lowest energy. From the reduced matrix elements one defines the *partial-wave amplitudes* by[4]

$$T_{l\eta_A}^{\eta_b}(E_A) = T_l^{\eta_b}(p_A) \equiv -\pi \sqrt{m_A m_b} \hbar \langle \eta_b \| T_l(E_A) \| \eta_A \rangle, \quad (1.34)$$

where p_A is given in terms of E_A and η_A by (1.22a) and in terms of p_b, $E_{\eta_b}^{int}$, and $E_{\eta_A}^{int}$ by (1.24). No additional quantum numbers η_b are required to describe elastic scattering, for which $\eta_b = \eta_A$, and one simply writes[5]

$$T_l(p_A) \equiv T_{l\eta_A}^{\eta_A}(p_A) = -\pi \sqrt{m_A m_A} \hbar \langle \eta_A \| T_l(E_A) \| \eta_A \rangle. \quad (1.35)$$

The partial-wave amplitude $T_l^{\eta_b}(p_A)$ for the inelastic case ($\eta_b \neq \eta_A$) is often called the *partial-wave reaction amplitude*, in distinction to the *partial-wave amplitude* $T_l(p_A)$ for elastic scattering. In terms of the partial-wave amplitudes the differential cross section (1.26) is written

$$\frac{d\sigma^{\eta_b}}{d\Omega_b} = \frac{p_b(E_A)}{p_A} \left| \sum_l (2l+1) P_l(\cos\theta) T_l^{\eta_b}(p_A) \right|^2, \quad (1.36)$$

---

[4] We have restored here the factor $\hbar$ which had been set equal to unity in the preceding expressions for the cross sections. In (1.21') the factor on the right-hand side is $\hbar$. With $\hbar = 1$ the length units, cm, are identical to the inverse of the momentum units, erg sec/cm or eV/c, and the cross section will have the units of inverse momentum squared instead of area.

[5] We shall suppress the label η_A designating the internal properties of the initial state, whenever this does not obfuscate the notation.

while the lth partial cross section (1.30) is written

$$\sigma_l^{\eta b} = 4\pi \frac{p_b(E_A)}{p_A} (2l + 1) |T_l^{\eta b}(p_A)|^2. \tag{1.37}$$

The lth partial cross section (1.32) for elastic scattering is then

$$\sigma_l(\text{elastic}) = \sigma_l^{\eta A} = 4\pi(2l + 1)|T_l(p_A)|^2, \tag{1.38}$$

and the lth partial cross section for inelastic scattering is

$$\sigma_l(\text{inelastic}) = \sum_{\eta_b \neq \eta_A} \sigma_l^{\eta b} = \frac{4\pi}{p_A}(2l + 1) \sum_{\eta_b \neq \eta_A} p_b(E_A)|T_l^{\eta b}(p_A)|^2. \tag{1.39}$$

The final momentum $p_b(E_A)$ that appears in the above equations may be expressed in terms of the initial momentum and the difference between the initial and final internal energies by

$$p_b(E_A)^2/2m_b = p_A^2/2m_A - (E_b^{int} - E_A^{int}). \tag{1.40}$$

The S-matrix is given in terms of the (on-the-energy-shell) T-matrix by[6]

$$\langle E_b \, l \, l_3 \, \eta_b | S | E_A \, l' \, l_3' \, \eta_A \rangle = \langle E_b \, l \, l_3 \, \eta_b | E_A \, l' \, l_3' \, \eta_A \rangle$$
$$- 2\pi i \delta(E_b - E_A) \langle E_A \, l \, l_3 \, \eta_b | T | E_A \, l' \, l_3' \, \eta_A \rangle. \tag{1.41}$$

Spherical symmetry, which was initially stated by (1.1), is equivalently stated by

$$[S, L_i] = 0. \tag{1.42}$$

S is thus a scalar operator and by the Wigner–Eckart theorem has matrix elements of the form

$$\langle E_b \, l \, l_3 \, \eta_b | S | E_A \, l' \, l_3' \, \eta_A \rangle = \delta_{ll'} \delta_{l_3 l_3'} \langle E_b \, l \, \eta_b \| S \| E_A \, l \, \eta_A \rangle. \tag{1.43}$$

From energy conservation [Equation (XV.3.13)],

$$[S, K] = 0, \tag{1.44}$$

it follows that

$$\langle E_b \, l \, \eta_b \| S \| E_A \, l \, \eta_A \rangle = \delta(E_b - E_A) \rho_b(E_A)^{-1/2} \rho_A(E_A)^{-1/2} \langle \eta_b \| S_l(E_A) \| \eta_A \rangle, \tag{1.45}$$

where $\rho_\eta(E)$ is given by (1.14) and (1.15). The factoring off of the weight functions $\rho_b^{-1/2}$ and $\rho_A^{-1/2}$ is a convention made in the definition of $\langle \eta_b \| S_l(E_A) \| \eta_A \rangle$ [but not in definition of $\langle \eta_b \| T_l(E_A) \| \eta_A \rangle$] in order to simplify later expressions (e.g., the unitarity condition in terms of the S- and T-matrices, which then can be carried over directly to the relativistic case).

---

[6] The S- and T-operators were discussed in more detail in Section XV.3, and (1.41) is identical with (XV.3.36). If Chapter XV was skipped in the first reading, the reader may take (1.41) as the definition of the S-matrix.

After substitution of (1.18) and (1.23) and of (1.43) and (1.45), the connection (1.41) between the reduced matrix elements of S and T becomes

$$\langle \eta_b \| S_l(E_A) \| \eta_A \rangle = \delta_{\eta_b \eta_A} - 2\pi i \sqrt{\rho_b(E_A)\rho_A(E_A)} \langle \eta_b \| T_l(E_A) \| \eta_A \rangle. \quad (1.46)$$

In terms of the partial-wave amplitudes this connection is given by

$$S_l(E_A) \equiv \langle \eta_A \| S_l(E_A) \| \eta_A \rangle = 1 + 2i p_A T_l(p_A) \quad (1.47)$$

for elastic scattering and by

$$S_l^{\eta_b}(E_A) \equiv \langle \eta_b \| S_l(E_A) \| \eta_A \rangle = 2i \sqrt{\rho_b(E_A)p_A} T_l^{\eta_b}(p_A) \quad (\eta_b \neq \eta_A) \quad (1.48)$$

for inelastic reaction processes.

XVI.2 Unitarity and Phase Shifts

In the previous section we expressed the cross section in terms of the transition matrix and also in terms of the partial-wave amplitudes [Equations (1.26) and (1.36)]. The only condition used was the assumption (1.1) of spherical symmetry. This allowed us to write the cross section in terms of some known functions of the scattering angle and the partial-wave amplitudes. For a given scattering problem the partial-wave amplitudes can be determined once the potential is known, in the same way that the scattering amplitude can be determined from the potential, by solving the integral equation (XV.3.37). There is, however, a general condition that permits us to obtain some general properties of the scattering amplitude. This is the unitarity of the S-operator, which is an expression of the conservation of overall probability and which is a consequence (for the case that time development is described by a Hamiltonian, as specified by the axiom V) of the Hermiticity of the Hamiltonian.

The unitarity of the scattering operator [cf. (XV.3.28)][7]

$$S^\dagger S = I \quad \text{(and } SS^\dagger = I), \quad (2.1)$$

is written in the angular-momentum basis as

$$\sum_{l l_3 \eta_b} \int \rho_b(E_b) \, dE_b \langle \tilde{E} \tilde{l} \tilde{l}_3 \tilde{\eta} | S^\dagger | E_b \, l \, l_3 \, \eta_b \rangle \langle E_b \, l \, l_3 \, \eta_b | S | E_A \, l' \, l'_3 \, \eta_A \rangle$$

$$= \langle \tilde{E} \tilde{l} \tilde{l}_3 \tilde{\eta} | E_A \, l' \, l'_3 \, \eta_A \rangle = \rho_A(E_A)^{-1} \delta(\tilde{E} - E_A) \delta_{\tilde{l} l'} \, \delta_{\tilde{l}_3 l'_3} \delta_{\tilde{\eta} \eta_A}. \quad (2.2)$$

Using (1.43) and (1.45), this leads to the unitarity condition for the reduced S-matrix elements:

$$\sum_{\eta_b} \langle \eta_b \| S_l(E_A) \| \tilde{\eta} \rangle^* \langle \eta_b \| S_l(E_A) \| \eta_A \rangle = \delta_{\tilde{\eta} \eta_A}. \quad (2.3)$$

For the special case $\tilde{\eta} = \eta_A$ Equation (2.3) may be written as

$$|S_l(E_A)|^2 + \sum_{\eta_b \neq \eta_A} |\langle \eta_b \| S_l(E_A) \| \eta_A \rangle|^2 = 1, \quad (2.4)$$

---

[7] For the case that one does not assume a Hamiltonian time development (2.1) is taken as one of the basic postulates for the S-operator.

where $\sum_{\eta_b}$ extends over all possible values of η_b except $\eta_b = \eta_A$. Inserting (1.47) and (1.48) into (2.4) leads to the unitarity condition expressed in terms of the partial-wave amplitudes:

$$\text{Im } T_l(p_A) = \sum_{\eta_b} p_b(E_A)|T_l^{\eta_b}(p_A)|^2$$

$$= p_A |T_l(p_A)|^2 + \sum_{\eta_b \neq \eta_A} p_b(E_A)|T_l^{\eta_b}(p_A)|^2. \qquad (2.5)$$

Equation (2.5) is the partial-wave amplitude form of the "generalized optical theorem" (XV.3.41). The first term on the right-hand side corresponds to elastic scattering; the second term corresponds to inelastic reaction processes.

For a particular collision experiment the values of η_b over which the sum in (2.3) and (2.5) extends may be limited to those permitted by the initial energy. If the kinetic energy of the projectile is lower than the energy necessary to excite the target, i.e., if

$$p_A^2/2m_A < E_{\eta_b}^{\text{int}} - E_{\eta_A}^{\text{int}} \qquad (\eta_b \neq \eta_A)$$

(for an "infinitely heavy" target that remains always at rest), then the projectile can only be scattered elastically, and all $\langle \eta_b \| S_l(E_A) \| \eta_A \rangle$ vanish for $\eta_b \neq \eta_A$. One says that only the "elastic channel" is open. For instance, if in e-H scattering the kinetic energy $p^2/2m$ of the incident electron is lower than the energy difference between the ground state and the first excited state, $E(n = 2) - E(n = 1)$, then only elastic scattering

$$e + H \rightarrow e + H$$

is energetically possible, so only the elastic channel is open. As the kinetic energy of the incident beam is increased, one reaches a point above which it is energetically possible to raise the target into the first excited state η_1 with the projectile losing a corresponding amount of kinetic energy. This point is the *threshold* of an inelastic process. The threshold momentum for an infinitely heavy target is given by

$$p_{\text{thresh}}^2/2m_A = E_{\eta_1}^{\text{int}} - E_A^{\text{int}}.$$

Above this value for p_A, both $\langle \eta_A \| S_l(E_A) \| \eta_A \rangle$ and $\langle \eta_1 \| S_l(E_A) \| \eta_A \rangle$ are different from zero. One says that an "inelastic channel" has opened up. As the incident energy is further increased, other inelastic processes may become possible, and an increasing number of reduced S-matrix elements in (2.4) become nonzero.

There can be different kinds of inelastic channels. The simplest are the excitation channels, in which the internal state of the projectile (target) remains unchanged whereas the target (projectile) performs a transition from the initial state into an excited state. η_A and η_b then refer to the different internal quantum numbers of the target (projectile). (If the projectile is also a system with internal structure and its internal quantum numbers are contained in the set η, then these quantum numbers will remain fixed.) In

the example of e-H scattering the excitation channels are the final states of
the process

$$e + \text{H}(n = 1) \rightarrow e + \text{H*}(n \geq 2).$$

The threshold energy for this process is $|E(n = 1) - E(n = 2)|$. If the incident
energy is increased further, the target may ionize, i.e., the target may be
raised to the energy at which it breaks up into two particles; one then says
that the "ionization channel" has opened up. For e-H scattering the ionization channel is the final state of the process

$$e + \text{H}(n = 1) \rightarrow e + e + \text{H}^+$$

and the ionization threshold is $|E(n = 1) - E(n \rightarrow \infty)| = 13.6 \text{ eV}$.

The internal quantum numbers η_b, or at least a subset of them, now also
extend over a continuous set of values. In these cases it is often more practical
to use for the postcollision system a different set of quantum numbers than
for the precollision system. This becomes even more necessary for more
complicated collision processes like rearrangement collision, in which the
projectile is absorbed by the target and another particle is emitted. A simple
example of such a process in atomic physics is

$$\gamma + \text{He} \rightarrow \text{He}^+ + e.$$

For more complicated precollision systems the number of possible scattering
channels increases—e.g., in the collision of He atoms and protons the
following channels are possible:

$$
\begin{aligned}
\text{H}^+ + \text{He} &\rightarrow \text{H}^+ + \text{He} &&\text{(elastic channel)} \\
&\rightarrow \text{H}^+ + \text{He*} &&\text{(excitation channels)} \\
&\rightarrow \text{H}^+ + \text{He}^+ + e &&\text{(ionization channel)} \\
&\rightarrow \text{H}^+ + \text{He}^{+*} + e &&\text{(ionization-excitation channels)} \\
&\rightarrow \text{H} + \text{He}^+ &&\text{(rearrangement channel)} \\
&\rightarrow \text{H*} + \text{He}^+ &&\text{(rearrangement-excitation channels)} \\
&\vdots
\end{aligned}
$$

Many-channel processes *also* occur in nuclear physics and in particle
physics, and it is in these areas that many-channel problems become particularly important. For instance in pion-proton (π-p) scattering one has (among
others) the following channels:

$$
\begin{aligned}
\pi^- + p &\rightarrow \pi^- + p &&\text{(elastic channel)} \\
&\rightarrow \pi^- + \pi^0 + p \\
&\rightarrow \eta + n \\
&\rightarrow K^0 + \Lambda \\
&\rightarrow \pi^- + p^*
\end{aligned}
$$

(η and K^0 denote mesons heavier than π; n and Λ denote the neutron and the "strange" baryon Λ, which is heavier than the proton and the neutron; p^* denotes various excited states of the proton.)

All scattering channels (i.e., the elastic channel and all inelastic channels) are simultaneously open if the initial energy is high enough. These scattering channels are various final states, differing in the internal quantum numbers η_b, and the sums in (2.4) and (2.5) must be extended over all channels that are open, i.e., energetically possible. The interaction can, of course, be such that an open channel, say $\eta_{\bar{b}}$, does not contribute much to a particular partial-wave amplitude, i.e., that $|\langle \eta_{\bar{b}} \| S_l(E_A) \| \eta_A \rangle|$ is small or zero even for energies above the threshold for the channel $\eta_{\bar{b}}$. These are detailed properties of the interaction ("the dynamics") which require further information.

We initially assume that all inelastic channels are closed. Equation (2.4) then becomes

$$|S_l(E_A)|^2 = 1, \tag{2.6}$$

For elastic scattering it thus follows that $S_l(E_A)$ is a function of E_A (or, equivalently, of p_A) of modulus 1 and consequently may be written

$$S_l(E_A) = e^{2i\delta_l(E_A)}; \tag{2.7}$$

where $\delta_l(E_A)$ is a real-valued function defined up to an integral multiple of π by (2.7). $\delta_l(E_A)$ is called the *scattering phase shift* for reasons we will discuss later.

The elastic partial-wave amplitude that appears in (1.47) may also be expressed in terms of the phase shift,

$$T_l(p_A) = \frac{S_l(p_A) - 1}{2ip_A} = \frac{1}{2ip_A}(e^{2i\delta_l(p_A)} - 1)$$

$$= \frac{1}{p_A} e^{i\delta_l(p_A)} \sin \delta_l(p_A) = \frac{1}{p_A} \frac{\tan \delta_l(p_A)}{1 - i \tan \delta_l(p_A)}$$

$$= \frac{1}{p_A} \frac{1}{\cot \delta_l(p_A) - i}, \tag{2.8}$$

which in turn allows us to express the lth partial elastic cross section of (1.38) in terms of the phase shift:

$$\sigma_l(\text{elastic}) = \frac{4\pi}{p_A^2}(2l + 1)\sin^2 \delta_l(p_A) = \frac{2\pi}{p_A^2}(2l + 1)(1 - \cos 2\delta_l) \tag{2.9}$$

We would like to extend the phase-shift formalism to include the inelastic or reaction channels. If the sum over the reaction channels in (2.4) is nonzero (i.e., if E_A is sufficiently high that some of the reaction channels are open), then $S_l(E_A)$ can no longer be written as in (2.7) with real $\delta_l(E_A)$; but it may be parametrized in the form[8]

$$S_l(E_A) = \eta_l(E_A)e^{2i\delta_l(E_A)} \tag{2.10}$$

---

[8] The η_l that appears in (2.10) is not to be confused with our use of η to denote the internal quantum numbers. We use it here because this notation of δ_l and η_l is often used in particle physics; in other areas of physics the notation may be different—e.g., in atomic physics η_l is often the symbol for the phase shift denoted by δ_l here.

with real $\delta_l(E_A)$ and with real $\eta_l(E_A)$ satisfying

$$0 \leq \eta_l(E_A) \leq 1. \tag{2.11}$$

That η_l satisfies (2.11) may be seen by inserting (2.10) into (2.4),

$$\eta_l(E_A)^2 = 1 - \sum_{\eta_b \neq \eta_A} |\langle \eta_b \| S_l(E_A) \| \eta_A \rangle|^2, \tag{2.12}$$

and observing that

$$\sum_{\eta_b \neq \eta_A} |\langle \eta_b \| S_l(E_A) \| \eta_A \rangle|^2$$

cannot be larger than 1. Equation (2.12) can be written in terms of the partial-wave reaction amplitudes by using (1.48):

$$\eta_l(E_A)^2 = 1 - 4p_A \sum_{\eta_b \neq \eta_A} p_b(E_A) |T_l^{\eta_b}(p_A)|^2. \tag{2.13}$$

The partial-wave amplitude for the elastic channel may be expressed in terms of $\delta_l(p_A)$ and $\eta_l(p_A)$ as [cf. (1.47)]

$$\begin{aligned}
T_l(p_A) &= \frac{S_l(p_A) - 1}{2ip_A} = \frac{\eta_l(p_A)e^{2i\delta_l(p_A)} - 1}{2ip_A} \\
&= \frac{\eta_l(p_A) \sin 2\delta_l(p_A)}{2p_A} + i\frac{1 - \eta_l(p_A) \cos 2\delta_l(p_A)}{2p_A}.
\end{aligned} \tag{2.14}$$

From (1.38) and (2.14) it then follows that the lth partial elastic cross section is

$$\sigma_l(\text{elastic}) = \frac{\pi}{p_A^2}(2l + 1)(1 + \eta_l(p_A)^2 - 2\eta_l(p_A) \cos 2\delta_l(p_A)), \tag{2.15}$$

while from (1.39) and (2.13) we have

$$\sigma_l(\text{inelastic}) = \frac{\pi}{p_A^2}(2l + 1)(1 - \eta_l(p_A)^2) \tag{2.16}$$

for the lth partial inelastic cross section. The total lth partial cross section is the sum of these:

$$\sigma_l = \sigma_l(\text{total}) = \frac{2\pi}{p_A^2}(2l + 1)(1 - \eta_l(p_A) \cos 2\delta_l(p_A)). \tag{2.17}$$

For $\eta_l = 1$ both (2.15) and (2.17) are identical with (2.9), which gives σ_l in the case that all inelastic channels are closed, and (2.16) vanishes. From this and from (2.10) and (2.11) one sees that η_l is a quantitative characterization of the amount of inelasticity involved in the scattering; η_l is therefore called the *inelasticity coefficient*.

The total lth partial cross section may also be expressed in terms of the lth elastic partial-wave amplitude by adding (1.38) and (1.39) and using the generalized optical theorem (2.5) to get

$$\sigma_l = \frac{4\pi}{p_A}(2l + 1) \operatorname{Im} T_l(p_A). \tag{2.18}$$

If we sum (2.18) over all l and use the imaginary part at $\theta = 0$ of the *elastic-scattering amplitude* (in the form obtained from (1.21') and (1.35)),

$$T(p_A, \theta) \equiv \sum_l (2l + 1) P_l(\cos \theta) T_l(p_A), \tag{2.19}$$

we then obtain

$$\sigma = \sum_l \sigma_l = \frac{4\pi}{p_A} \sum_l (2l + 1) \, \mathrm{Im} \, T_l(p_A) = \frac{4\pi}{p_A} \mathrm{Im} \, T(p_A, \theta = 0). \tag{2.20}$$

[Also used was the property $P_l(1) = 1$.] Equation (2.20), which connects the total cross section and the imaginary part of the forward-scattering amplitude, is usually called the *optical theorem*.

XVI.3 Argand Diagrams

In order to familiarize ourselves with the properties of the scattering amplitude and its behavior under certain conditions we consider a graphical representation in which $p_A T_l(p_A)$ is regarded as a vector in the complex plane. These graphical representations are called *Argand diagrams* and play a particularly important role in the description and detection of resonances.

We start from the expression (2.14) for the elastic partial wave amplitude, which is written in the form

$$p_A T_l(p_A) - \frac{i}{2} = -\frac{i}{2} \eta_l(p_A) e^{2i\delta_l(p_A)}. \tag{3.1}$$

For the case of purely elastic scattering (e.g., in the region of low momentum where no inelastic channels are open) $\eta_l(p_A) = 1$, so $p_A T_l(p_A) - i/2$ is a complex number that lies on a circle of radius $1/2$. If $\delta_l(p_A)$ varies between $0 \le \delta_l \le \pi$ (modulo π) when p_A is varied over the range in which only elastic scattering occurs, then all points of this circle are covered. In general, however, $\delta_l(p_A)$ will not vary over all of the interval $0 \le \delta_l \le \pi$, and hence $p_A T_l(p_A) - i/2$ will not vary over the full circle as p_A is varied over that range in which only elastic scattering is possible.

For elastic scattering the values of

$$p_A T_l(p_A) = \frac{i}{2} - \frac{i}{2} e^{2i\delta_l(p_A)} \qquad [\eta_l(p_A) = 1] \tag{3.2}$$

lie on a circle with center at $i/2$ and radius $1/2$. This is in agreement with the optical theorem (2.5), according to which $\mathrm{Im} \, T_l(p_A) \ge 0$. The circle around $i/2$ with radius $1/2$ is called the *unitary circle*.

If there is absorption of energy by inelastic processes, then

$$p_A T_l(p_A) = \frac{i}{2} - \frac{i}{2} \eta_l(p_A) e^{2i\delta_l(p_A)} \qquad [\eta_l(p_A) < 1], \tag{3.3}$$

and the value of $p_A T_l(p_A)$ moves on a path inside the unitary circle as p_A is varied. Figure 3.1 shows the graphical representation (Argand diagram) of

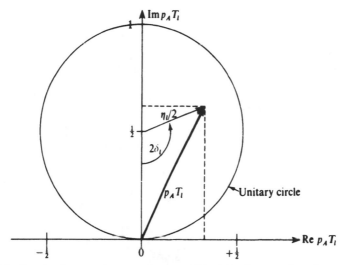

Figure 3.1 The elastic-scattering amplitude $p_A T_l(p_A)$ in the complex plane. (Argand diagram.)

$p_A T_l(p_A)$. One calls the circle the "unitary circle" because, as a consequence of the unitary condition, which is expressed by (3.3), it follows that the path of $p_A T_l(p_A)$ must lie inside this circle. The exact path followed by $p_A T_l(p_A)$ depends upon the particular interaction. The most important case of these diagrams is the case in which the interaction between projectile and target is resonant. We will consider these cases below in Section XVIII.8.

Argand diagrams may also be drawn for the partial-wave amplitude $T_l^{\eta_b}(p_A)$ of an inelastic channel ($\eta_b \neq \eta_A$). One expresses the partial-wave amplitude $T_l^{\eta_b}(p_A)$ [or the reduced S-matrix element $\langle \eta_b \| S_l(E_A) \| \eta_A \rangle$ that corresponds to it by (1.48)] in a form analogous to (2.10):[9]

$$\sqrt{p_b(E_A) p_A} T_l^{\eta_b}(p_A) = \frac{1}{2i} \langle \eta_b \| S_l(E_A) \| \eta_A \rangle = \frac{1}{2i} \eta_l^b(p_A) e^{2i\delta_l^b(p_A)}. \quad (3.4)$$

Unitarity of $S((2.3)$ or $(2.4))$ imposes the restrictions

$$0 \leq \eta_l^b(p_A) \leq 1) \quad (3.5a)$$

and from (2.12)

$$\eta_l(p_A)^2 + \sum_{\eta_b \neq \eta_A} \eta_l^b(p_A)^2 = 1. \quad (3.5b)$$

In the Argand diagram, $\alpha \equiv \sqrt{p_b(E_A) p_A} \, T_l^{\eta_b}(p_A)$ then moves as a function of p_A on a path that lies inside the circle with center at the origin and radius $1/2$ (cf. Figure 3.2). (The path would lie on the circle if there were neither elastic scattering nor inelastic scattering into any channel other than the one with quantum numbers η_b.)

---

[9] η_b, η_A denote the internal quantum numbers (channels). η_l^b denotes the inelasticity coefficient for channel η_b as in Equation (3.4).

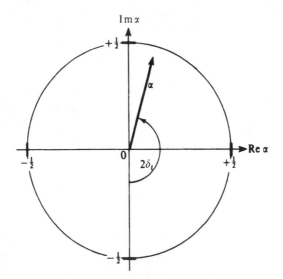

Figure 3.2 The amplitude $\alpha = \sqrt{p_b(E_A)p_A}\,T_l^{\eta b}(p_A)$ for scattering from channel η_A to an inelastic channel η_b. In this case the center of the circle is at the origin and the amplitude is restricted to $|\alpha| \leq \frac{1}{2}$.

Problems

1. (a)   Show that the Born approximation violates unitarity.

   (b)   Where in the Argand diagram for elastic scattering can the Born approximation be considered a good approximation?

2.   Obtain the inelastic and the elastic total cross section for scattering (of a structureless projectile) by a black disc (hard sphere) of radius a. A black disc is described by the following properties: It is totally absorbing, i.e.,

   (a) $\eta_l = 0$ (zero inelasticity coefficient), and

   (b)   only partial amplitudes with $l \leq L$ contribute to the cross section where $L = p_A a$.

((a) specifies that the disc is black, (b) specifies that it has a well-defined edge.) Discuss the result, in particular the value for the elastic total cross section.

CHAPTER XVII

Free and Exact Radial Wave Functions

The present chapter treats the spatial properties of the scattering process as described by the wave function. After an introduction, in Section XVII.2 the differential equation for the radial wave function (the Schrödinger equation) is derived. Section XVII.3 presents the solutions of the free radial wave equation, and lists some of their properties. The properties of the exact radial wave functions, in particular their asymptotic forms and their connection to the phase shifts and the S-matrix, are discussed in Section XVII.4. In Section XVII.5 the connection between bound states and poles of the S-matrix on the positive imaginary axis is established. Some material about functions of a complex variable, which is needed in this chapter and in Chapter XVIII, is reviewed in a mathematical appendix.

XVII.1 Introduction

In the preceding sections we discussed those properties of the scattering amplitude and the cross section that were consequences of very general assumptions like unitarity and spherical symmetry. We derived relations between the phase shifts, the inelasticity coefficients, the partial-wave amplitudes, and the lth partial S-matrix elements. The cross section was expressed in terms of these quantities. Nowhere in our discussion did we consider the spatial properties of the scattering process; in particular, we did not have to make use of position-dependent quantities such as the probability wave function.

The probability wave function, however, serves to provide a pictorial representation of the scattering process. In fact, it is in scattering theory that the wave function has a direct relation to observation, since it represents the probability for position measurements that can actually be performed. (For stationary states such measurements are not performed in practice, as has been discussed in the first half of the book; the wave function then is merely an auxiliary mathematical quantity.)

In the present chapter we introduce the wave function. As in the previous chapters, we consider two different points of view: Firstly, we assume time development of the system as given by a Hamiltonian $H = K + V$. This will provide a "deeper" foundation for the relations between the more directly accessible quantities. We then see, however, that these relations have a more general validity.

In addition to providing us with the wave function, this chapter will also establish the connection between our more general coordinate-free formulation of scattering theory and the usual formulation in terms of solutions of the Schrödinger wave equation. This latter formulation is directly obtained if the internal quantum numbers η are ignored.

Though the Schrödinger differential equation constitutes a powerful tool in many cases (described by an interaction potential), it will turn out that the integral equation for the wave function (which is, in fact, just the Lippmann–Schwinger equation in coordinate form) is of more immediate use.

XVII.2 The Radial Wave Equation

We start with the Lippmann–Schwinger equation in the angular momentum basis

$$|E\, l\, l_3\, \eta^+\rangle = |E\, l\, l_3\, \eta\rangle + \frac{1}{E - K + i0}\, V |E\, l\, l_3\, \eta^+\rangle.$$

It is assumed that the position operator for projectile relative to target, $\mathbf{Q}$, commutes with the internal observables η^{op} of the system:

$$[Q_i, \eta^{\mathrm{op}}] = 0.$$

By taking the scalar product of the Lippmann–Schwinger equation with the generalized position eigenvector

$$|\mathbf{x}\, \eta'\rangle \equiv |x_1\, x_2\, x_3\, \eta'\rangle \equiv |r\, \theta\, \phi\, \eta'\rangle,$$

we then get

$$\langle r\, \theta\, \phi\, \eta' | E\, l\, l_3\, \eta^+\rangle = \langle r\, \theta\, \phi\, \eta' | E\, l\, l_3\, \eta\rangle$$

$$+ \langle r\, \theta\, \phi\, \eta' | \frac{1}{E - K + i0}\, V |E\, l\, l_3\, \eta^+\rangle, \quad (2.1)$$

where $\langle r\,\theta\,\phi\,\eta'|E\,l\,l_3\,\eta\rangle$ and $\langle r\,\theta\,\phi\,\eta'|E\,l\,l_3\,\eta^+\rangle$ are called the *free* and *exact wave functions*, respectively.

After scattering, the projectile's position $\mathbf{x}$ and momentum $\mathbf{p}$ have the same direction (see Figure XIV.3.1):

$$\mathbf{x}/|\mathbf{x}| = \mathbf{p}/|\mathbf{p}|. \qquad (2.2)$$

Thus (θ,ϕ) are the angular coordinates of $\mathbf{x}$ as well as of $\mathbf{p}$. On account of spherical symmetry one can write, using the same arguments that led to Equation (VII.3.11),

$$\langle r\,\theta\,\phi\,\eta'|E\,l\,l_3\,\eta\rangle = \langle\theta\,\phi|l\,l_3\rangle\langle r\,\eta'|E\,\eta\rangle_l$$

$$= Y_{l l_3}(\theta,\phi)\delta_{\eta\eta'}\langle r|E\rangle_l^\eta, \qquad (2.3a)$$

$$\langle r\,\theta\,\phi\,\eta'|E\,l\,l_3\,\eta^+\rangle = \langle\theta\,\phi|l\,l_3\rangle\langle r\,\eta'|E\,\eta^+\rangle_l$$

$$= Y_{l l_3}(\theta,\phi)\langle r\,\eta'|E\,\eta^+\rangle_l. \qquad (2.3b)$$

The reduced matrix elements $\sqrt{\pi/2}\langle r|E\rangle_l^\eta$ and $\sqrt{\pi/2}\langle r\,\eta'|E\,\eta^+\rangle_l$ are called the *free* and *exact radial wave functions*, respectively.[1]

Using the Wigner–Eckart theorem for the scalar operator V, one obtains

$$\langle r\,\theta\,\phi\,\eta'|V|E\,l\,l_3\,\eta^+\rangle = \sum_{\tilde{E}\tilde{l}\tilde{l}_3\tilde{\eta}}\langle r\,\theta\,\phi\,\eta'|\tilde{E}\,\tilde{l}\,\tilde{l}_3\,\tilde{\eta}^+\rangle\langle\tilde{E}\,\tilde{l}\,\tilde{l}_3\,\tilde{\eta}^+|V|E\,l\,l_3\,\eta^+\rangle$$

$$= \sum_{\tilde{E}\tilde{l}\tilde{l}_3\tilde{\eta}}\langle r\,\theta\,\phi\,\eta'|\tilde{E}\,\tilde{l}\,\tilde{l}_3\,\tilde{\eta}^+\rangle\delta_{l\tilde{l}}\delta_{l_3\tilde{l}_3}\langle\tilde{E}\,\tilde{\eta}^+\|V\|E\,\eta^+\rangle_l$$

$$= \sum_{\tilde{E}\tilde{\eta}}Y_{l l_3}(\theta,\phi)\langle r\,\eta'|\tilde{E}\,\tilde{\eta}^+\rangle_l\langle\tilde{E}\,\tilde{\eta}^+\|V\|E\,\eta^+\rangle_l$$

where $\langle\tilde{E}\,\tilde{\eta}^+\|V\|E\,\eta^+\rangle_l$ is the reduced matrix element of V. Thus

$$\langle r\,\theta\,\phi\,\eta'|V|E\,l\,l_3\,\eta^+\rangle = Y_{l l_3}(\theta,\phi)\langle r\,\eta'\|V\|E\,\eta^+\rangle_l, \qquad (2.3c')$$

where we have defined the reduced transition matrix element $\langle r\,\eta'\|V\|E\,\eta^+\rangle_l$ in terms of the reduced matrix elements of V by

$$\langle r\,\eta'\|V\|E\,\eta^+\rangle_l = \sum_{\tilde{E}\tilde{\eta}}\langle r\,\eta'|\tilde{E}\,\tilde{\eta}^+\rangle_l\langle\tilde{E}\,\tilde{\eta}^+\|V\|E\,\eta^+\rangle_l.$$

The reduced transition matrix element is a function of l, r, E, η', and η only. For spherically symmetric problems, V is an operator function of the radius operator $Q \equiv (\mathbf{Q}^2)^{1/2}$. In general it also depends upon the internal observables, i.e., upon the operators that change the internal quantum numbers η.

We use the fact that $[Q, V] = 0$ to write

$$\langle r\,\eta'\|V\|\tilde{r}\,\tilde{\eta}\rangle = \langle r|\tilde{r}\rangle\langle\eta'\|\,V(r)\,\|\tilde{\eta}\rangle,$$

where $\langle r|\tilde{r}\rangle$ is the δ-function with respect to the continuous summation

---

[1] $\langle r\,\eta'|E\,\eta^+\rangle_l$ cannot, in general, be written as $\langle r|E^+\rangle^\eta\delta_{\eta\eta'}$, because $[H,\eta^{\mathrm{op}}] = 0$ might not be true. Thus in general $|E\,\eta^+\rangle$ is not necessarily an eigenvector of η^{op}. Only if $[H,\eta^{\mathrm{op}}] = [V,\eta^{\mathrm{op}}] = 0$ is this true. Cf. the remark following Equation (XVI.1.3).

over $\tilde{r}$ and $\langle \eta' \|\| V(r) \|\| \tilde{\eta} \rangle$ is a doubly reduced matrix element depending only on r, $\tilde{\eta}$, and η'.[2] Therefore

$$\langle r \, \eta' \| V \| E \, \eta^+ \rangle_l = \sum_{\tilde{r} \tilde{\eta}} \langle r \, \eta' \| V \| \tilde{r} \, \tilde{\eta} \rangle \langle \tilde{r} \, \tilde{\eta} | E \, \eta^+ \rangle_l$$

$$= \sum_{\tilde{r} \tilde{\eta}} \langle r | \tilde{r} \rangle \langle \eta' \|\| V(r) \|\| \tilde{\eta} \rangle \langle \tilde{r} \, \tilde{\eta} | E \, \eta^+ \rangle_l.$$

Performing the summation over $\tilde{r}$ and inserting the result,

$$\langle r \, \eta' \| V \| E \, \eta^+ \rangle_l = \sum_{\tilde{\eta}} \langle \eta' \|\| V(r) \|\| \tilde{\eta} \rangle \langle r \, \tilde{\eta} | E \, \eta^+ \rangle_l,$$

into (2.3c'), one obtains

$$\langle r \, \theta \, \phi \, \eta' | V | E \, l \, l_3 \, \eta^+ \rangle = Y_{l l_3}(\theta, \phi) \sum_{\tilde{\eta}} \langle \eta' \|\| V(r) \|\| \tilde{\eta} \rangle \langle r \, \tilde{\eta} | E \, \eta^+ \rangle_l.$$

$$(2.3c)$$

Using the same arguments that led to (2.3c'), one can also show that

$$\langle r \, \theta \, \phi \, \eta' | \frac{1}{E - K + i0} V | E \, l \, l_3 \, \eta^+ \rangle = Y_{l l_3}(\theta, \phi) \langle r \, \eta' | \frac{1}{E - K + i0} V | E \, \eta^+ \rangle_l.$$

$$(2.3d)$$

We shall now use the results (2.3) to obtain the Schrödinger differential equation for the radial wave function. In Section XVII.3 we shall consider the differential equation for the free radial wave function. Then, from the Lippmann–Schwinger equation (2.1) for the wave function, we shall obtain, in Section XVII.4, the integral equation for the radial wave function. The advantage of the integral equation over the differential equation is that the former incorporates the boundary conditions; in this particular case the boundary conditions appropriate to the Lippmann–Schwinger equation for $|E \, \hat{a}^+ \rangle$ are those for a free incident state, as discussed in Section XV.2.

We assume that the incident beam has projectiles of one kind, whose mass m is not changed by the scattering process. The differential equation is then obtained by taking the transition matrix element of the operator $H = K + V = \mathbf{P}^2/2m + K^{int}(\eta^{op}) + V$, where $\mathbf{P}$ is the projectile momentum operator:

$$E \langle r \, \theta \, \phi \, \eta' | E \, l \, l_3 \, \eta^+ \rangle = \langle r \, \theta \, \phi \, \eta' | H | E \, l \, l_3 \, \eta^+ \rangle$$

$$= \langle r \, \theta \, \phi \, \eta' | (\mathbf{P}^2/2m + K^{int} + V) | E \, l \, l_3 \, \eta^+ \rangle$$

$$= \frac{1}{2m} \left(\frac{-1}{r^2} \frac{d}{dr} \left(r^2 \frac{d}{dr} \right) + \frac{l(l+1)}{r^2} + 2mE^{int} \right) \langle r \, \theta \, \phi \, \eta' | E \, l \, l_3 \, \eta^+ \rangle$$

$$+ \langle r \, \theta \, \phi \, \eta' | V | E \, l \, l_3 \, \eta^+ \rangle$$

$$(2.4)$$

---

[2] This statement can be viewed as the Wigner–Eckart theorem for the group generated by Q applied to the operator V, which is a scalar operator with respect to this group.

[The second equality follows from Equation (VII.2.6) and (VII.3.4) and is the analogue of (VII.3.7).]

Using (2.3a, b, c) in (2.4), we obtain a system of coupled differential equations:

$$\sum_{\tilde{\eta}} \left[\left\{ \frac{1}{r^2} \frac{d}{dr} \left(r^2 \frac{d}{dr} \right) - \frac{l(l+1)}{r^2} + p_\eta(E)^2 \right\} \delta_{\eta'\tilde{\eta}} \right.$$

$$\left. - 2m \langle \eta' ||| V(r) ||| \tilde{\eta} \rangle \right] \langle r \, \tilde{\eta} | E \, \eta^+ \rangle_l = 0, \quad (2.5)$$

where

$$p_\eta(E) = \sqrt{2m(E - E_\eta^{int})}. \quad (2.6)$$

In order to solve this system of equations one has to know the reduced matrix elements of the potential $\langle \eta' ||| V(r) ||| \eta \rangle$. These depend, of course, upon the particular process under investigation. For instance, in the many-channel problem of the elastic and inelastic scattering of electrons by atoms these reduced matrix elements can be determined from the Coulomb interaction and the atomic wave function.[3]

We shall not discuss the many-channel problem here any further[4] but restrict ourselves to the case

$$[V, \eta^{op}] = 0. \quad (2.7)$$

Then the problem reduces to a one-channel problem for each value of η (i.e., for each set of values of the internal quantum numbers). To perform the reduction, we use the above arguments to write

$$\langle \eta' ||| V(r) ||| \tilde{\eta} \rangle = \delta_{\eta'\tilde{\eta}} V_{\tilde{\eta}}(r). \quad (2.8)$$

Since we now have $[H, \eta^{op}] = 0$, we can also write

$$\langle r \, \eta' | E \, \eta^+ \rangle_l = \delta_{\eta'\eta} \langle r | E^+ \rangle_l^\eta. \quad (2.9)$$

Under these conditions (2.5) goes over into a set of uncoupled equations, one for each channel η:

$$\left[\frac{1}{r^2} \frac{d}{dr} \left(r^2 \frac{d}{dr} \right) - \frac{l(l+1)}{r^2} - 2m V_\eta(r) + p_\eta(E)^2 \right] \langle r | E^+ \rangle_l^\eta = 0. \quad (2.10)$$

In the following sections we shall discuss the properties of the radial wave functions $\langle r | E^+ \rangle_l^\eta$. Though $V_\eta(r)$ may vary with η, and p_η^2 may differ from $2mE = p^2$ by different constants $2mE_\eta^{int}$, the label η is inessential, as it does not change in the scattering process. We shall therefore suppress it in most of what follows, reinstating it at the end to emphasize the fact that further quantum numbers may be present.

---

[3] For the scattering of electrons by hydrogen atoms this is discussed in Massey *et al.* (1969), Vol. 1, Section 7.1. A more general case is treated in Smith (1971), Section 2.1.

[4] We shall return to the many-channel problem in Chapter XX.

XVII.3 The Free Radial Wave Function

The equation for the free radial wave function $\langle r|E\rangle_l$ is obtained from the eigenvalue equation of K:

$$E\langle r\,\theta\,\phi\,\eta'|E\,l\,l_3\,\eta\rangle = \langle r\,\theta\,\phi\,\eta'|K|E\,l\,l_3\,\eta\rangle$$

in the same way as (2.5) is obtained from (2.4). It differs from (2.5) only in that the term in V is missing and that since $[K, \eta^{op}] = 0$,

$$\langle r\,\tilde{\eta}|E\,\eta\rangle_l = \delta_{\tilde{\eta}\eta}\langle r|E\rangle_l^{\eta}. \tag{3.1}$$

We thus have

$$\left\{\frac{1}{r^2}\frac{d}{dr}\left(r^2\frac{d}{dr}\right) - \frac{l(l+1)}{r^2} + p^2\right\}\langle r|E\rangle_l^{\eta} = 0, \tag{3.2}$$

where $p = p_\eta(E) \equiv \sqrt{2m(E - E_\eta^{int})}$. We shall write $\langle r|E\rangle_l^{\eta} = \langle r|p\rangle_l^{\eta}$. Equation (3.2) is a well-known equation, which has as its linearly independent solutions the spherical Bessel and Neumann functions. The solution that is regular at $r = 0$, to which $\langle r|p\rangle_l$ is proportional, is the (proper) *spherical Bessel function* (spherical Bessel function of the first kind, or Riccati–Bessel function):[5]

$$j_l(pr) = j_l(z) = \left(\frac{\pi}{2z}\right)^{1/2} J_{l+1/2}(z)$$

$$= (-z)^l\left(\frac{1}{z}\frac{d}{dz}\right)^l\left(\frac{\sin z}{z}\right)$$

$$= z^l\sum_{n=0}^{\infty}\frac{(-z^2/2)^2}{n!(2l + 2n + 1)!!}, \qquad pr = z, \tag{3.3}$$

and the solution irregular at $r = 0$ is the *spherical Neumann function* (spherical Bessel function of the second kind or Riccati–Neumann function):

$$n_l(z) = -(-1)^l\left(\frac{\pi}{2z}\right)^{1/2} J_{-l-1/2}(z) = (-1)^{l+1}j_{-l-1}(z)$$

$$= -(-z)^l\left(\frac{1}{z}\frac{d}{dz}\right)^l\left(\frac{\cos z}{z}\right)$$

$$= -z^{-l-1}\sum_{n=0}^{\infty}\frac{(z^2/2)^n(2l - 2n - 1)!!}{n!}. \tag{3.4}$$

Here $J_{l+1/2}(z)$ and $J_{-l-1/2}(z)$ are ordinary Bessel functions of the first kind.

---

[5] The double factorial is defined by

$$n!! \equiv \begin{cases} n(n-2)\cdots(5)(3)(1) & \text{if } n \text{ is odd,} \\ n(n-2)\cdots(4)(2) & \text{if } n \text{ is even.} \end{cases}$$

For further properties and of the spherical Bessel functions and their proofs see Abramowitz et al. (1965), Chapter 10, and Watson (1958).

The *spherical Hankel functions*, defined by

$$h_l(z) = j_l(z) + in_l(z),$$ (3.5a)

$$h_l^*(z) = j_l(z) - in_l(z),$$ (3.5b)

are then also independent solutions of (3.2). We note for later use that all the above relations also hold for complex values of z.

The asymptotic behavior of $j_l(z)$, $n_l(z)$, and $h_l(z)$ for real $z \to \infty$ is

$$j_l(z) \underset{z \to \infty}{\sim} \frac{1}{z} \sin\left(z - \frac{l\pi}{2}\right),$$ (3.6a)

$$n_l(z) \underset{z \to \infty}{\sim} -\frac{1}{z} \cos\left(z - \frac{l\pi}{2}\right),$$ (3.6b)

$$h_l(z) \underset{z \to \infty}{\sim} \frac{(-i)^{l+1}}{z} e^{iz}.$$ (3.6c)

For future reference we note the bounds for complex values of z:

$$|j_l(z)| \leq \text{const} \left(\frac{1}{1 + |z|}\right)^{l+1} e^{|\text{Im}\, z|},$$ (3.6d)

$$|h_l(z)| \leq \text{const} \left(\frac{1}{1 + |z|}\right)^{-l} e^{-\text{Im}\, z}.$$ (3.6e)

Near the origin we have

$$j_l(z) \underset{z \to 0}{\to} \frac{z^l}{(2l + 1)!!},$$ (3.7a)

$$n_l(z) \underset{z \to 0}{\to} [-(2l - 1)!!]z^{-l-1}.$$ (3.7b)

The normalization of the j_l's is

$$\int_0^\infty r^2\, dr\, j_l(pr)j_l(p'r) = \frac{\pi}{2p^2} \delta(p - p'),$$ (3.8a)

$$\int_0^\infty p^2\, dp\, j_l(pr)j_l(pr') = \frac{\pi}{2r^2} \delta(r - r').$$ (3.8b)

The spherical Neumann function cannot be delta-function normalized in this way. All spherical functions $j_l(z)$, $n_l(z)$, $h_l(z)$, and $h_l^*(z)$ fulfill the functional equations

$$\frac{d}{dz} f_l(z) = \frac{1}{2l + 1}[lf_{l-1}(z) - (l + 1)f_{l+1}(z)]$$ (3.9a)

and the following equality for the indefinite integral:

$$\int r^2\, dr\, f_l(pr)f_l^*(p'r) = \frac{r^2}{p^2 - p'^2}[p' f_l(pr)f_{l-1}^*(p'r) - pf_{l-1}(pr)f_l^*(p'r)],$$ (3.9b)

of which (3.8) is a consequence.

To find the proportionality factor between $\langle r|p\rangle_l$ and $j_l(pr)$, we calculate the normalization integral of the free radial wave function $\langle r|p\rangle_l$. In spherical coordinates the δ-function normalization of the generalized position eigenvectors reads

$$\langle \mathbf{x}|\mathbf{x}'\rangle = \delta^3(\mathbf{x} - \mathbf{x}') = \frac{1}{r^2 \sin \theta}\delta(r - r')\delta(\theta - \theta')\delta(\phi - \phi'); \quad (3.10)$$

Inserting the sum over the complete system of basis vectors gives:

$$\langle \mathbf{x}|\mathbf{x}'\rangle = \sum_{l l_3} \int \rho(E)\, dE\, \langle r\,\theta\,\phi|E\,l\,l_3\rangle \langle E\,l\,l_3|r'\,\theta'\,\phi'\rangle$$

$$= \sum_{l l_3} \int \rho(E)\, dE\, Y_{l l_3}(\theta, \phi)\langle r|E\rangle_l\, Y_{l l_3}^*(\theta', \phi')\langle r'|E\rangle_l^*$$

$$= \frac{1}{\sin \theta}\delta(\theta - \theta')\delta(\phi - \phi')\int \rho(E)\, dE\, \langle r|E\rangle_l\langle r'|E\rangle_l^*,$$

where we have used

$$\sum_{l_3} Y_{l l_3}(\theta, \phi)Y_{l l_3}^*(\theta', \phi') = \frac{1}{\sin \theta}\delta(\theta - \theta')\delta(\phi - \phi'). \quad \text{(VII.3.19)}$$

Comparing this with (3.10) shows that the normalization of the radial wave function is

$$\int \rho(E)\, dE\, \langle r|E\rangle_l\langle r'|E\rangle_l^* = \frac{1}{r^2}\delta(r - r'), \quad (3.11)$$

or, in terms of p,

$$\int p^2\, dp\, \langle r|p\rangle_l\langle r'|p\rangle_l^* = \frac{1}{r^2}\delta(r - r'). \quad (3.11')$$

As $\langle r|p\rangle_l$ is a solution of (3.2) regular at $r = 0$, and therefore proportional to $j_l(pr)$, one obtains by comparison of (3.11') with (3.8b) that

$$\langle r|p\rangle_l = \sqrt{2/\pi}j_l(pr). \quad (3.12)$$

XVII.4 The Exact Radial Wave Function

The precise properties of the exact radial wave function $\langle r|E^+\rangle_l$ depend upon the interaction Hamiltonian V, through the interaction potentials $V_\eta(r) = \langle \mathbf{x}\,\eta|V|\mathbf{x}\,\eta\rangle$. However, provided V fulfills some very general conditions, one can make some general statements about $\langle r|E^+\rangle_l$ without knowing the detailed form of V. The content of these conditions is roughly as follows:[6]

1. The interaction is not effective for large projectile-target separation, i.e., $V(r) \rightarrow 0$ as $r \rightarrow \infty$. Which precise assumption one has to make

[6] See, e.g., Taylor (1972).

for the rate of falling off of V depends on what one wants to achieve: Usually one assumes that $V(r)$ falls off at least as fast as $1/r^3$. For the derivation of some properties one has to assume that $V(r)$ decreases faster than any power of $1/r$, and often one assumes a finite range for V.

2. The interaction does not change too abruptly as the projectile approaches the target: $V(r)$ is a continuous function of r except at a finite number of finite discontinuities.

3. There exists a lowest energy value (since the binding energy must be finite) or, in other words, the spectrum of H has a lower bound. In terms of the potential this means that $V(r)$ is not too singular at the origin $r = 0$ (less singular than $r^{-3/2}$ if $V(r) < 0$).

Almost all physical potentials fulfill these conditions—potentials for electrons scattering off an atom, potentials for atom-atom scattering, the Yukawa potential, the square-well potential, etc. The Coulomb potential does not fulfill condition 1 and therefore requires a separate mathematical treatment. In practice, however, this is not of much consequence, because electromagnetic screening is always present: One never really has the exact Coulomb potential $1/r$, but a screened potential $\psi(r)/r$, with $\psi(r) = 0$ for large r.[7] For the following discussion we assume that the above three conditions are fulfilled in whatever precise form we need.

If

$$V(r) > 0 \quad \text{for all } r, \tag{4.1}$$

then the potential is said to be *repulsive*, while if

$$V(r) < 0 \quad \text{for all } r, \tag{4.2}$$

then it is said to be *attractive*.[8] If $V(r) < 0$ in some interval $r_1 < r < r_2$, then $V(r)$ is said to be attractive in that interval. If $V(r)$ is attractive in some interval, there may be a certain number of solutions of the differential equation (2.10) for a certain number of discrete, negative values of $p_n^2/2m = E_n$. These solutions correspond to bound states of the projectile-target system with binding energies $|E_n|$, as we shall discuss in the next section. They must fall off sufficiently rapidly as $r \to \infty$, in addition to being regular at $r = 0$.

The exact radial wave function $\langle r|E^+\rangle_l = \langle r|p^+\rangle_l$ for $p^2 \geq 0$ is the solution of the differential equation (2.10) with certain boundary conditions: In addition to being regular (i.e., $\langle r|p^+\rangle < \infty$ at $r = 0$), one would want, as $r \to \infty$, $\langle r|p^+\rangle_l$ to represent the lth spherical component of an incident plane wave and an outgoing scattered wave, and that for $V \to 0$, $\langle r|p^+\rangle_l \to \langle r|p\rangle_l = \sqrt{2/\pi} j_l(pr)$, the acceptable solution of the free wave equation.

Instead of solving (2.10) under these boundary conditions (see Problem 1), we shall obtain the exact radial wave function $\langle r|p^+\rangle_l$ from the Lippmann–Schwinger equation (2.1), which has the boundary conditions

---

[7] Note also that in general any assumed $V(r)$ is a mathematical idealization of the actual situation.

[8] The forces between projectile and target are then also called repulsive and attractive, respectively.

already built in. Inserting (2.3a), (2.3b), (2.3d), (2.8), and (2.9) into (2.1), one obtains the integral equation for the radial wave function $\langle r|E^+\rangle_l$:

$$\langle r|E^+\rangle_l$$

$$= \langle r|E\rangle_l + \int \rho(E')\,dE'\,r'^2\,dr'\,\langle r|E'\rangle_l \frac{1}{E - E' + i0}\langle r'|E'\rangle_l^* V(r')\langle r'|E^+\rangle_l. \tag{4.3}$$

Now define the (lth-partial-wave) Green's function by

$$G_l^p(r;r') = \frac{1}{2m}\int_0^\infty \rho(E')\,dE'\frac{\langle r|E'\rangle_l\langle r'|E'\rangle_l^*}{E - E' + i0} \tag{4.4}$$

$$= \int_0^\infty p'^2\,dp'\frac{\langle r|p'\rangle_l\langle r'|p'\rangle_l^*}{p^2 - p'^2 + i0} \tag{4.5}$$

$$= \frac{2}{\pi}\int_0^\infty p'^2\,dp'\frac{j_l(p'\,r)j_l(p'\,r')}{p^2 - p'^2 + i0} \tag{4.6}$$

$$= \frac{1}{\pi}\int_{-\infty}^{+\infty} p'^2\,dp'\frac{j_l(p'r)j_l(p'r')}{p^2 - p'^2 + i0}, \tag{4.7}$$

where we have used the fact that the integrand is even. The integral equation (4.3) then becomes

$$\langle r|p^+\rangle_l = \langle r|p\rangle_l + 2m\int_0^\infty r'^2\,dr'\,G_l^p(r;r')V(r')\langle r'|p^+\rangle_l$$

$$= \sqrt{2/\pi}j_l(pr) + 2m\int_0^\infty r'^2\,dr'\,G_l^p(r;r')V(r')\langle r'|p^+\rangle_l. \tag{4.8}$$

Note that if the potential has a finite range, then the integral is over a finite interval.

The Green's function is evaluated using Equation (3.5) to write

$$G_l^p(r;r') = \frac{1}{2\pi}\int_{-\infty}^{+\infty} dq\,q^2\frac{j_l(qr)h_l(qr')}{p^2 - q^2 + i\epsilon} + \frac{1}{2\pi}\int_{-\infty}^{+\infty} dk\,k^2\frac{j_l(kr)h_l^*(kr')}{p^2 - k^2 + i\epsilon}. \tag{4.9}$$

Suppose $r' > r$. Then because of the asymptotic behavior (3.6c) of h_l, and (3.6d), the product $j_l(qr)h_l(qr')$ decreases exponentially in the upper half complex q-plane, and one may extend the integration path of the first integral to include a large semicircle in the upper half plane (cf. Figure 4.1). From Equations (3.3), (3.4), and (3.5) it is seen that

$$j_l(kr) = (-1)^l j_l(-kr) \quad \text{and} \quad h_l^*(kr') = (-1)^l h_l(-kr')$$

and hence with the substitution $q = -k$, the second integral of (4.9) is the same as the first integral. We may now rewrite (4.9) as

$$G_l^p(r;r') = \lim_{R\to\infty}\frac{-2}{2\pi}\oint_{\Gamma(R)} dq\frac{q^2 j_l(qr)h_l(qr')}{(p+q)(q-p-i\epsilon')} \quad \left(\epsilon' = \frac{\epsilon}{p+q}\right),$$

where $\Gamma(R)$ is the closed contour indicated by Figure 4.1. The spherical

Figure 4.1 The integration path $\Gamma(R)$ used to evaluate the first integral in Equation (4.9). As $R \to \infty$, the contribution from the semicircle vanishes, leaving only the integral along the Re q axis.

Bessel functions are entire functions, and the function $q^2 j_l(qr)h_l(qr')/(p + q)$ is analytic inside $\Gamma(R)$, so by the Cauchy residue theorem[9] we obtain

$$G_l^p(r; r') = -2i \frac{p^2}{p + p} j_l(pr)h_l(pr')$$

$$= -ipj_l(pr)h_l(pr') \qquad (r' > r). \qquad (4.10a)$$

A similar procedure, but with r and r' interchanged, gives the Green's function for $r' < r$:

$$G_l^p(r; r') = -ipj_l(pr')h_l(pr) \qquad (r' < r). \qquad (4.10b)$$

Equations (4.10) are often combined as

$$G_l^p(r; r') = -ipj_l(pr_<)h_l(pr_>) \qquad (4.11)$$

where

$$r_< = \min(r, r') \qquad (4.12a)$$

and

$$r_> = \max(r, r'). \qquad (4.12b)$$

We already know the asymptotic behavior of the free radial wave function $\langle r|p \rangle_l = \sqrt{2/\pi} j_l(pr)$ for $r \to \infty$ [cf. (3.6a)]. Let us now examine the asymptotic behavior of the exact radial wave function $\langle r|p^+ \rangle_l$. Using (3.6a), (3.6c), and (4.10b) in (4.8), we obtain

$$\sqrt{\frac{\pi}{2}} \langle r|p' \rangle_l \underset{r \to \infty}{\sim} \frac{1}{pr} \sin\left(pr - \frac{l\pi}{2}\right)$$

$$+ 2m\sqrt{\frac{\pi}{2}} \int_0^\infty r'^2 \, dr' \left[-ipj_l(pr') \frac{(-i)^{l+1}}{pr} e^{ipr} \right] V(r') \langle r'|p^+ \rangle_l$$

$$= \frac{-1}{2ipr} \left\{ e^{-i(pr - l\pi/2)} \right.$$

$$\left. - e^{i(pr - l\pi/2)} \left[1 - (2ip)(\sqrt{2\pi}m) \int_0^\infty r'^2 \, dr' \, j_l(pr') V(r') \langle r'|p^+ \rangle_l \right] \right\}$$

$$(4.13)$$

---

[9] For a review of important results in the theory of functions of a complex variable, see Appendix XVII.A.

The first term represents an incoming partial wave, whereas the second term represents an outgoing one. (We shall discuss this further at the end of this section.) The outgoing wave is modified with respect to the incoming one by the factor in square brackets. This factor may be related to the elastic scattering phase shift defined by Equation (XVI.2.7) for every value of the internal quantum number η. Note that having chosen the quantum numbers η such that (2.7) holds, $\langle \eta_b \| S_l(E) \| \eta_A \rangle$ vanishes for $\eta_b \neq \eta_A$. Thus we have here only elastic scattering for every value of η.

Recall, now, that the lth partial-wave amplitude defined by (XVI.1.35), (XVI.1.18), and (XVI.1.23) is

$$T_l(p) = -\pi m \langle E\,l\,l_3 | T | E\,l\,l_3 \rangle = -\pi m \langle E\,l\,l_3 | V | E\,l\,l_3^+ \rangle, \qquad (4.14)$$

where $E = E(p) = p^2/2m + E^{\text{int}}$, the label η having been suppressed. In terms of the radial wave function this is written

$$T_l(p) = -\pi m \int r^2 \sin\theta\, dr\, d\theta\, d\phi\, \langle E\,l\,l_3 | r\,\theta\,\phi \rangle \langle r\,\theta\,\phi | V | E\,l\,l_3^+ \rangle$$

$$= -\pi m \left[\int \sin\theta\, d\theta\, d\phi\, Y_{l l_3}^*(\theta, \phi) Y_{l l_3}(\theta, \phi) \right]$$

$$\times \left[\int r^2\, dr\, \langle r | E \rangle_l^* V(r) \langle r | E^+ \rangle_l \right]$$

$$T_l(p) = -\pi m \int_0^\infty r^2\, dr\, \sqrt{\frac{2}{\pi}} j_l(pr) V(r) \langle r | p^+ \rangle_l, \qquad (4.15)$$

where Equations (2.3a), (2.3c), (2.8), (2.9), (VII.3.16), and (3.12) have been used. Using (XVI.1.47) and (XVI.2.7), it follows from (4.15) that

$$S_l(p) = e^{2i\delta_l(p)} = 1 + 2ip T_l(p)$$

$$= 1 - (2ip)(\sqrt{2\pi m}) \int_0^\infty r^2\, dr\, j_l(pr) V(r) \langle r | p^+ \rangle_l. \qquad (4.16)$$

Although Equations (4.15) and (4.16) are important in themselves, expressing the lth partial-wave amplitude and the phase shift in terms of the potential and the exact radial wave function, they are of little practical use. This is because the differential equation (2.10) and the integral equation (4.8) are not easily solved for $\langle r | p^+ \rangle_l$, nor is $\langle r | p^+ \rangle_l$ as directly related to the experimental data as are $\delta_l(p)$ and $T_l(p)$.

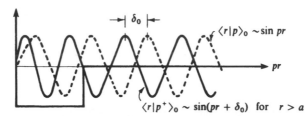

Figure 4.2 The $l = 0$ free and exact radial wave functions for an attractive square well.

We make use of (4.15) and (4.16) to rewrite (4.13) in various useful forms:

$$\sqrt{\pi/2}\langle r|p^+\rangle_l \underset{r\to\infty}{\sim} \frac{1}{pr}\sin\left(pr - \frac{l\pi}{2}\right) + pT_l(p)\frac{e^{i(pr - l\pi/2)}}{pr} \tag{4.17a}$$

$$\underset{r\to\infty}{\sim} \frac{1}{2ipr}\{e^{i(pr - l\pi/2)}e^{2i\delta_l(p)} - e^{-i(pr - l\pi/2)}\} \tag{4.17b}$$

$$\underset{r\to\infty}{\sim} \frac{e^{i\delta_l(p)}}{pr}\sin\left(pr - \frac{l\pi}{2} + \delta_l(p)\right) \tag{4.17c}$$

$$\underset{r\to\infty}{\sim} \frac{e^{-il\pi/2}}{2ipr}\{e^{ipr}S_l(p) - (-1)^l e^{-ipr}\}. \tag{4.17d}$$

Comparison of (4.17c) with the asymptotic form of the free wave function obtained from (3.12) and (3.6a),

$$\sqrt{\frac{\pi}{2}}\langle r|p\rangle_l = j_l(pr) \underset{r\to\infty}{\sim} \frac{1}{pr}\sin\left(pr - \frac{l\pi}{2}\right), \tag{4.18}$$

explains why $\delta_l(p)$ is called the "phase shift": The interaction shifts the phase of the asymptotic exact radial wave function $\langle r|p^+\rangle_l$ by $\delta_l(p)$ as compared with the asymptotic free radial wave function $\langle r|p\rangle_l$. This property also serves as an alternative definition of δ_l in place of (XVI.2.7). The phase shift for the $l = 0$ wave in a square-well potential is shown in Figure 4.2.

For an *attractive potential* (like the square well in Figure 4.2) the wave function is "pulled" into the interaction region and *the phase shift is positive*. The exact wave function emerges from the interaction region with its phase in advance of the free wave function. For a *repulsive potential* the wave function is "pushed" out of the interaction region and *the phase shift is negative*.

That this connection between the sign of the phase shift and the sign of the potential is generally fulfilled is best established in the small-phase-shift approximation:

$$e^{i\delta_l(p)}\sin\delta_l(p) \approx \delta_l(p).$$

From (4.15), using (XVI.2.8), one then obtains

$$\delta_l(p) \approx -\pi mp \int_0^\infty r^2\, dr\, \sqrt{2/\pi} j_l(pr)V(r)\langle r|p^+\rangle_l,$$

which in the Born approximation is given by

$$\delta_l(p) \approx -2mp \int_0^\infty r^2\, dr\, \{j_l(pr)\}^2\, V(r).$$

To fully justify the above statement that (4.13) and (4.17) represent an incoming and an outgoing partial wave, we have to consider the time-dependent wave function of a state $\phi^+(t)$ that develops in time from a prepared in-state.

The exact radial wave function $\langle r | p^+ \rangle_l$ is the pth component of the radial wave function of an angular-momentum eigenstate $\phi_{ll_3}^+(t)$ that develops in time according to the equation

$$\phi_{ll_3}^+(t) = \frac{1}{2\pi} \int dE \; \tilde{\phi}(E) | E \, l \, l_3^+ \rangle e^{-iEt}, \qquad (4.19)$$

where $\tilde{\phi}(E) \equiv \tilde{\phi}(p)$, $E = p^2/2m$, is the energy (momentum) distribution of this state in the remote past.[10] The wave function of this state is

$$\langle r \, \theta \, \phi | \phi_{ll_3}^+(t) \rangle = Y_{ll_3}(\theta \, \phi) \langle r | \phi^+(t) \rangle_l, \qquad (4.20)$$

where the radial wave function $\langle r | \phi^+(t) \rangle_l$ of the state $\phi^+(t)$ is given by

$$\langle r | \phi^+(t) \rangle_l = \frac{1}{2\pi} \int_0^\infty dE \; \tilde{\phi}(E) \langle r | E^+ \rangle_l e^{-iEt}. \qquad (4.21)$$

Multiplying (4.17d) by $\sqrt{2/\pi}(1/2\pi)\tilde{\phi}(E)e^{-iEt}$ and integrating over E then gives

$$\langle r | \phi^+(t) \rangle_l \underset{r \to \infty}{\sim} \frac{\sqrt{2\pi} e^{-i(\pi/2)(l-1)}}{4\pi^2 m} \int_0^\infty dp \; \tilde{\phi}(p) e^{-iEt} \frac{1}{r} [e^{-ipr}(-1)^l - S_l(p)e^{ipr}]$$

$$\langle r | \phi^+(t) \rangle_l \underset{r \to \infty}{\sim} \frac{\sqrt{2\pi} e^{-i(\pi/2)(l-1)}}{4\pi^2 m} \left[\int_0^\infty dp \; \tilde{\phi}(p) \frac{e^{-i(pr+Et)}(-1)^l}{r} \right.$$

$$\left. - \int_0^\infty dp \; \tilde{\phi}(p) S_l(p) \frac{e^{i(pr-Et)}}{r} \right]. \qquad (4.22)$$

Thus the radial wave function of the angular-momentum eigenstate $\phi_{ll_3}^+(t)$ consists asymptotically of an incoming wave with the original momentum distribution and an outgoing wave in which the original momentum distribution has been modified by $S_l(p)$ due to the interaction. Equations (4.13) and (4.17) merely state this result in the time-independent form for the pth component of the radial wave function, i.e., for an (unphysical) exact momentum eigenstate.

The asymptotic forms of the radial wave function (4.22) and (4.17) have been derived from the Lippmann–Schwinger equation under the assumption that the interaction is described by an interaction Hamiltonian V. Then $S_l(p)$ is given in terms of $V(r)$ by (4.16). Even if the time-development axiom is not assumed, there exists a unitary operator S that transforms states before the interaction into states after the interaction. If this interaction is of finite range, then outside the interaction region the wave function should still be a superposition of incoming and outgoing waves. Thus the radial wave function for large distances r should still be given by (4.22), where $S_l(p)$, the lth S-matrix element, is now the fundamental quantity. Consequently, the

---

[10] That this $\phi^+(t)$, connected with the basis vectors $|E \, l \, l_3^+ \rangle$, is a state that develops from a prepared in-state has been shown in Section XV.2 and is here of no further consequence.

pth component of the radial wave function, $\langle r | p^+ \rangle_l$, should have the asymptotic form (4.17).

Thus in the framework of a Hamiltonian time development, the asymptotic forms (4.17) and (4.22) are derived, while in the framework of S-matrix theory the asymptotic forms (4.17) and (4.22) are assumed and are justified as a consequence of the superposition principle for the wave function outside the interaction region.

XVII.5 Poles and Bound States[11]

In scattering processes, the variable p that we have been considering in the preceding section is the magnitude of the incident momentum, and therefore a real positive quantity. Accordingly, the variable $E = p^2/2m$ representing the kinetic energy (or $E \quad E_n^{int} = p^2/2m$ if the internal energy has to be taken into account) is a real positive quantity. We know, however, that other values of E also have physical significance. For example, the solutions of (2.10) for negative values of $p^2 = 2mE$ are the bound-state radial wave functions, and the discrete negative energy values E_n, for which (2.10) has normalizable solutions $\langle r | E_n \rangle_l$, are the energy levels of the projectile-target bound system (as has been discussed in Section VII.3).

One should then also be able to extend the integral equation (4.8) [with G_l^p given by (4.11)], which follows from (2.10), to negative-energy solutions. Such solutions can be obtained by the following line of reasoning: Replace p by the complex variable z. For Im $z \geq 0$, the arguments that lead to the Green's function (4.11) are still valid. (For Im $z < 0$, similar arguments, wherein one closes the contour in the lower half of the complex z-plane, hold, and one gets analogous expressions with h_l replaced by h_l^*.)

The fact that (4.11) holds in this general case does not necessarily mean that (4.8) can also be continued to Im $z > 0$. Let us assume, however, that the potential $V(r)$ is such that it can be so continued, and let us assume that H has a bound state with energy $E = -\alpha^2/2m$. Then, at the point $z = i\alpha$, the radial wave function $\langle r | i\alpha^+ \rangle_l$ (being, as a wave function of a bound state, an element of the space of infinitely differentiable, rapidly decreasing functions that is a realization of the Schwartz space) must asymptotically decrease faster than any power of r^{-1}, e.g., it must decrease exponentially. Now according to (3.6d), $j_l(i\alpha r)$ diverges exponentially. The second term in the right-hand side of (4.8) is bounded as $r \to \infty$, because $h_l(i\alpha r)$ decreases exponentially according to (3.6e). Thus $\langle r | i\alpha^+ \rangle_l$ can decrease for $r \to \infty$ only if the term $\langle r | p \rangle_l = \sqrt{2/\pi} j_l(pr)$, originating according to (4.13) or (4.17a) from the incident state, drops out of Equation (4.8). When can this be the

---

[11] For this section, as well as for some sections of Chapter XVIII, we shall require some basic results from the theory of functions of a complex variable. These results are outlined in a mathematical appendix to this chapter, Appendix XVII.A. The reader who is unfamiliar with complex variable theory, or who merely wishes to refresh his memory, may consult the appendix before reading this section.

case? In order to check this we consider, instead of (4.8), the equation obtained by multiplying Equation (4.8) by $(z - i\alpha)$:

$$(z - i\alpha)\langle r|z^+\rangle_l = (z - i\alpha)\langle r|z\rangle_l$$

$$- 2izm \int_0^\infty r'^2 \, dr' \, j_l(zr_<) h_l(zr_>) V(r')(z - i\alpha)\langle r'|z^+\rangle_l.$$

(5.1)

Equation (5.1) is identically fulfilled at $z = i\alpha$ unless $\langle r|z^+\rangle_l$ has a pole at $z = i\alpha$, i.e., unless in the neighborhood of $i\alpha$ it is of the form[12]

$$\langle r|z^+\rangle_l = \frac{1}{z - i\alpha} \langle r|E = -\alpha^2/2m\rangle_l$$

$$+ \text{(function of } z \text{ analytic around } z = i\alpha\text{).} \quad (5.2)$$

With the ansatz (5.2) for $\langle r|z^+\rangle_l$ one obtains from (5.1) at $z = i\alpha$

$$\langle r|E = -\alpha^2/2m\rangle_l = 2m\alpha \int_0^\infty r'^2 \, dr' \, j_l(i\alpha r_<) h_l(i\alpha r_>) V(r')\langle r'|E = -\alpha^2/2m\rangle_l.$$

(5.3)

In (5.3) the troublesome term $\langle r|p\rangle_l$ has disappeared and we have obtained a homogeneous integral equation for the discrete number of normalizable solutions of the Schrödinger equation (2.10) with eigenvalues $E = -\alpha^2/2m$. Thus we see that a bound-state solution is obtained if we assume that $\langle r|z^+\rangle$ is of the form (5.2), i.e., that it has a pole on the positive imaginary momentum axis at the point $z = i\alpha = i\sqrt{2m|E|}$, where E is the bound-state energy eigenvalue for angular momentum l. The bound-state radial wave function $\langle r|E = -\alpha^2/2m\rangle_l$ is then the residue at this pole.

From (4.15) and (4.16) one then concludes that the lth partial-wave amplitude and the lth S-matrix element have poles on the positive imaginary momentum axis at the values $p = i\sqrt{2m|E_n^l|}$, where E_n^l ($n = 1, 2, \ldots$) are the discrete eigenvalues of the Hamiltonian H. In a theory with Hamiltonian time development, these discrete eigenvalues correspond to the energy levels of the projectile-target bound states of angular momentum l.

Though the above arguments[13] show that a bound state of the Hamiltonian with energy E_n^l corresponds to a pole of the lth partial S-matrix element $S_l(p)$ on the positive imaginary axis, the reverse need not hold: $T_l(p)$ and $S_l(p)$ may have singularities that are not connected with the bound states.[14]

Nevertheless, the hypothesis that bound states of angular momentum l correspond to poles of $S_l(p)$ on the positive imaginary axis (and vice versa)

---

[12] The above arguments will also hold for a pole of higher order, but one can prove that these poles are simple (which we shall not, however, do here).

[13] For potential scattering, the connection between simple poles and bound-state eigenvalues of H can be made more precise. Cf. Taylor (1972), Chapter 12.

[14] These "redundant" poles may occur if the asymptotic expressions in the complex p-plane differ from (4.17). For interactions that are cut off at a finite distance and also for potentials that fall off at infinity faster than any exponential (e.g. $e^{-\mu r}$), "redundant" poles are absent.

has become quite generally accepted. Especially in those cases where a Hamiltonian time development is not assumed to exist (relativistic S-matrix theory) and there is no operator whose spectral properties characterize the bound states, one usually assumes a one-to-one correspondence between bound states and poles on the positive imaginary axis.

XVII.6 Survey of Some General Properties of Scattering Amplitudes and Phase Shifts

The precise properties of the scattering amplitudes and the phase shifts depend upon the interaction. If the interaction is described by a potential, one can calculate phase shifts and partial-wave scattering amplitudes as a function of scattering energy, as has been done in some of the problems. There are, however, general properties of these functions that do not depend upon the particular form of the interaction, and we want to list them here.

The integral equation for the radial wave function can be iterated, thereby yielding:

$$\langle r|p^+\rangle_l = \langle r|p\rangle_l + 2m \int r'^2 \, dr' \, G_l^p(r,r')V(r')\langle r'|p\rangle_l$$

$$+ (2m)^2 \int r'^2 \, dr' \, G_l^p(r, r')V(r') \int r''^2 \, dr'' \, G_l^p(r', r'')V(r'')\langle r''|p\rangle_l$$

$$+ \cdots$$

Inserting this into (4.15) one obtains a series which contain powers of the potential and the spherical Bessel functions. For weak potentials and for high energies the higher-order terms in the potential and in the spherical Bessel functions are small (due to the property (3.6a) for the spherical Bessel functions). For this reason the Born approximation of the partial-wave amplitude, obtained from (4.15) by restricting the above expansion to the lowest-order term:

$$T_l^{(\text{Born})}(p) = -\pi m \frac{2}{\pi} \int_0^\infty r^2 \, dr \, j_l(pr)V(r)j_l(pr), \tag{6.1}$$

is believed to be a good approximation for high energies and weak potentials.

The asymptotic behavior of the lth partial S-matrix element is then

$$S_l(p) \to 1 \quad \text{as} \quad p \to \infty. \tag{6.2}$$

This follows immediately from (4.16) using the Born approximation (6.1) and the asymptotic behavior (3.6a) of $j_l(pr)$. An intuitive argument for (6.2) is that with increasing energy of the projectile the effect of a given interaction will become less important.

The phase shift $\delta_l(p)$, therefore, tends to a multiple of π as $p \to \infty$. As the phase shift is defined by (XVI.2.7) only up to a multiple of π, one can remove this modulo π ambiguity by defining

$$\delta_l(p) \to 0 \quad \text{as} \quad p \to \infty. \tag{6.3}$$

Requiring then that $\delta_l(p)$ be a continuous function of p (which is possible because $S_l(p)$ is continuous) makes $\delta_l(p)$ unique.

Clearly $S_l(p) \to 1$ also if the interaction goes to zero. Then with the convention (6.3) it follows that also

$$\delta_l(p) \to 0 \quad \text{for interaction going to zero} \quad [V(r) \to 0]. \tag{6.4}$$

For a given potential and given energy

$$T_l \to 0 \quad \text{and} \quad S_l \to 1 \quad \text{as} \quad l \to \infty. \tag{6.5}$$

Intuitively, this can be understood from (2.10) by regarding the term $l(l+1)/2mr^2$ as a repulsive centrifugal potential. The larger l is, the more repulsive is this centrifugal barrier and the less effective does the actual potential $V(r)$ become. Let R be the range of the interaction. Then for values l such that $l(l+1)/2mR^2$ is much larger than the kinetic energy $E = p^2/2m$ the projectile is unlikely to penetrate into the range of the interaction. Thus for

$$l(l+1) \gg p^2 R^2, \quad \text{or} \quad l \gg pR,$$

the partial-wave amplitude $T_l(p)$ will be negligible:

$$T_l(p) \approx 0, \qquad S_l(p) \approx 1, \quad \text{for} \quad l \gg pR. \tag{6.6}$$

It follows from (6.6) that for the phase shift

$$\delta_l \approx n\pi, \qquad n \text{ integer}, \quad \text{for} \quad l \gg pR. \tag{6.7}$$

Equations (6.6) and (6.7) are important properties of which one makes use when one expresses the differential cross section in terms of the partial-wave amplitudes as in (XVI.1.36). For a given value of $E = p^2/2m$ the infinite sum in (XVI.1.36) can, because of (6.6), be approximated by a finite sum.

To obtain the phase-shift behavior at low energies, we turn to (4.15). For small values of p, $\langle r | p^+ \rangle_l$ and $j_l(pr)$ have the same p-dependence, given by (3.7a), since according to (4.11),

$$G_l^p(r, r') \to -ip \frac{p^l r_<^l}{(2l+1)!!} \left(\frac{p^l r_>^l}{(2l+1)!!} - i(2l-1)!! \, p^{-l-1} r_>^{-l-1} \right)$$

$$G_l^p(r, r') \to -\frac{r_<^l r_>^{-l-1}}{2l+1} \quad \text{for} \quad p \to 0 \tag{6.8}$$

and $\langle r | p^+ \rangle_l$ and $j_l(pr)$ are connected by (4.8). Consequently, it follows from (4.15), for potentials that vanish for r greater than some R, that

$$T_l(p) \to -\pi m \left(\int_0^R r^2 \, dr \sqrt{\frac{2}{\pi}} r^l V(r) r^l \frac{1}{[(2l+1)!!]^2} \frac{\langle r | p^+ \rangle_l}{j_l(pr)} \right) p^{2l}$$

$$T_l(p) \to -a_l p^{2l} \quad \text{for} \quad p \to 0, \tag{6.9}$$

where a_l is a constant given by the integral in parentheses, which does not depend upon p, because $\langle r|p^+ \rangle_l / j_l(pr) \to$ (function of r) for $p \to 0$.[15]

The constants a_l are called the *scattering lengths*. Only the s-wave scattering length a_0 has the dimension of a length. (6.9) shows that for $p = 0$ all partial-wave amplitudes except the s-wave vanish. For the s-wave

$$T_0(p) \to -a_0 \quad \text{as} \quad p \to 0, \tag{6.10}$$

and consequently by (XVI.1.36) and (XVI.1.38),

$$\frac{d\sigma}{d\Omega} \to a_0^2, \qquad \sigma \to 4\pi a_0^2, \quad \text{as} \quad p \to 0. \tag{6.11}$$

Using (XVI.2.8) it follows further that

$$\delta_l(p) - n\pi \to -a_l p^{2l+1} \quad \text{as} \quad p \to 0, \tag{6.12}$$

and consequently also

$$\tan \delta_l(p) \to -a_l p^{2l+1} \quad \text{as} \quad p \to 0. \tag{6.13}$$

Equation (6.13) shows that

$$p^{2l+1} \cot \delta_l(p) \to -\frac{1}{a_l}$$

or their inverses are the appropriate functions to consider near $p = 0$ (as long as the scattering lengths $a_l \neq 0$ or $a_l \neq \infty$). The power series for this quantity near $p = 0$ is:

$$p^{2l+1} \cot \delta_l(p) = -\frac{1}{a_l} + \frac{r_l}{2} p^2 + O(p^4). \tag{6.14}$$

That only even powers appear in the power-series expansion of $p^{2l+1} \cot \delta_l(p)$ follows from $\delta_l(-p) = -\delta_l(p)$, which is established in Equation (XVIII.5.7) below.

For $l = 0$, (6.14) is a very useful approximation called the *effective-range approximation*:

$$p \cot \delta_0(p) = -\frac{1}{a_0} + \frac{r_0}{2} p^2. \tag{6.15}$$

The constant r_0 is called the effective range of the potential; one can show that r_0 is roughly proportional to the range of the potential. Equation (6.15) is a good approximation for energies small compared to the potential energy and has been one of the most important parameterizations of low-energy scattering data, in particular for neutron-proton scattering.

---

[15] If the potential does not have a finite range, some modifications are necessary, but one can show that for $V(r)$ decreasing faster than any power of $1/r$ the result (6.9) still holds.

XVII.A   Mathematical Appendix on Analytic Functions

In this appendix we review some basic *results* in the theory of analytic functions of a complex variable. For proofs of these results, and for more on the theory of analytic functions, the reader is referred to, e.g., Smirnov (1964), Vol. II, Part 2.

a. Definition of Analyticity

The function $f(z) = u + iv$ of the complex variable $z = x + iy$ is said to be analytic in some region R of the z-plane when

1. df/dz is continuous and independent of the direction of dz in the region R. (The following conditions are equivalent.)
2. $\partial u/\partial x = \partial v/\partial y$, $\partial u/\partial y = -\partial v/\partial x$ (Cauchy–Riemann conditions). These derivatives are also continuous in R.
3. $\oint_C f(z)\,dz = 0$ for any closed contour C lying within R.

If $f(z)$ is analytic in R, then so is $d^n f/dz^n$ for all $n = 0, 1, 2, \ldots$.

b. Cauchy Integral Formula

If $f(z)$ is analytic within some simply connected region R, then

$$\oint \frac{f(z)\,dz}{z - a} = 2\pi i f(a), \tag{A.1a}$$

$$\oint \frac{f(z)\,dz}{(z - a)^{n+1}} = \frac{2\pi i}{n!} f^{(n)}(a) \equiv \frac{2\pi i}{n!} \left.\frac{d^n f(z)}{dz^n}\right|_{z=a} \tag{A.1b}$$

for contours entirely within R and enclosing the point a. If a is excluded, the integrals vanish. (Integration is assumed to be counterclockwise.) Thus the values of $f(z)$ within R are completely determined by the values on the boundary.

c. Taylor Series

If $f(z)$ is analytic within and on a circle C centered at $z = z_0$, then one can expand $f(z)$ about z_0 :

$$f(z) = \sum_{n=0}^{\infty} a_n (z - z_0)^n, \tag{A.2a}$$

where

$$a_n = \frac{1}{n!} f^{(n)}(z_0). \tag{A.2b}$$

This Taylor series is uniformly convergent as long as the radius of C is less than the "radius of convergence," which is the distance from z_0 to the nearest singularity of $f(z)$, i.e., the nearest point where f is nonanalytic. Conversely, any power series represents an analytic function within its radius of convergence.

d. Laurent Series

If $f(z)$ is analytic within and on the annular region between two concentric circles C_1 and C_2 centered at $z = z_0$, then for any point z within the annular region one has the uniformly convergent power series

$$f(z) = \sum_{n=-\infty}^{\infty} a_n (z - z_0)^n, \tag{A.3a}$$

where

$$a_n = \frac{1}{2\pi i} \oint_C \frac{f(z)\,dz}{(z - z_0)^{n+1}}, \tag{A.3b}$$

C being any contour within the annulus.

e. Singularities

Points of nonanalyticity of f are called singularities of f.

1. A single-valued function f can have *isolated singularities* of the following kinds:
    (a) A singularity at $z = a$ is called a *pole of order n* if as $z \to a$

    $$f(z) \to \frac{g(z)}{(z - a)^n} \qquad (n = 1, 2, \ldots), \tag{A.4}$$

    where $g(z)$ is analytic at $z = a$, and $g(a) \neq 0$. A pole of order 1 is called a *simple pole*.
    (b) If $f(z)(z - a)^n$ diverges as $z \to a$ for all finite n, then a is called an *essential singularity* of f.
    In the Laurent series for f about a pole of order n, $-n$ is the lowest power one encounters. The Laurent series about an essential singularity contains infinitely many terms of negative power.
2. A multivalued function f (e.g., $f = z^{1/2}$) has *branch points*, which always occur in pairs. At a branch point $z = a$, the function $f(a + \epsilon e^{i\phi})$, where ϵ is such that $a + \epsilon e^{i\phi}$ does not include another branch point of f for all ϕ, is not periodic in ϕ with period 2π.

A function that is analytic in a region of the complex plane, except for a finite number of poles in that region, is said to be *meromorphic* in that region. If the function is meromorphic in the entire z-plane, except at the point ∞, it is called a *meromorphic function* of z. If the function is analytic in the entire z-plane, except at ∞, it is called an *entire* function.

f. Cauchy Residue Theorem

The coefficient a_{-1} in the Laurent series is called the *residue* of $f(z)$ at $z = z_0$. If $f(z)$ is analytic within and on a closed contour C, except for a finite number of poles and essential singularities, then

$$\oint_C f(z)\,dz = 2\pi i \sum (\text{residues in } C) \qquad (A.5)$$

g. Analytic Continuation

Let two functions be analytical in a region R, and have the same values (a) in some subregion of R, or (b) on a line segment in R, or (c) at a denumerably infinite number of points having a limit point within R. Then they are identical throughout R. However, if one knows only that they differ at most by an amount ϵ in (a), (b), or (c), then, no matter how small $|\epsilon|$ is, the two functions may differ vastly within the rest of R.

The facts stated above illustrate the justification and the limitations of *analytic continuation*: If $f_1(z)$ is defined on a line segment Γ, and $f_2(z)$ is analytic in a region R containing Γ, and if $f_1(z) = f_2(z)$ on Γ, then $f_2(z)$ is the unique analytic continuation of $f_1(z)$ onto R.

In the same spirit one can also do the following: Let $f_1(z)$ be given within a circle R_1, e.g., by a power series about the center z_1. Let $f_2(z)$ be the power-series expansion of $f_1(z)$ about a point z_2 near the periphery of R_1, the circle of convergence being R_2, which extends beyond R_1. Then the values of $f_2(z)$ in R_2 are uniquely determined by the values of $f_1(z)$ in R_1, and $f_2(z)$ is the analytic continuation of $f_1(z)$ into R_2. One can repeat this process several times, obtaining a unique analytic continuation $F(z)$ of $f_1(z)$ into the region $R_1 \cup R_2 \cup R_3 \cup \ldots$ (see Figure A.1). A function cannot be analytically continued into a region containing a singularity of that function.

After repeating the process of analytic continuation several times, the nth circle of convergence may overlap the first one. Then the values of $f_n(z)$

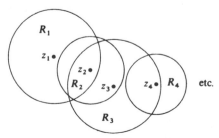

Figure A.1 Extension of a function defined in R_1 into the region $R_2 \cup R_3 \cup R_4 \cup \cdots$ by analytic continuation.

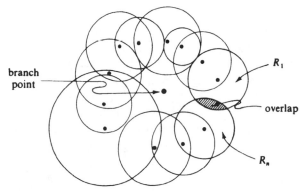

Figure A.2 Analytic continuation of a multivalued function around a branch point.

in the overlap region may or may not coincide with those of $f_1(z)$. In the latter case the function is multivalued, and the region of encirclement (see Figure A.2) contains a simple branch point of the function.

We have seen above how the values of an analytic function in a large region of the complex plane are determined by its values in an arbitrary small region of analyticity. However, the analytic continuation does not depend on these initial values in a continuous manner. Therefore, it is impossible to express $F(z)$ in R explicitly in terms of the values in a certain small region. Though $F(z)$ is uniquely prescribed by its values in the subregion, the problem of constructing $F(z)$ in R from its values in the small subregion is not a correctly formulated one.

h. Schwarz Reflection Principle

If $f(z)$ is analytic in a region R that includes a segment of the real axis and if $f(z)$ is real on this segment, then $f(z)$ can be continued onto the region $R^* = \{z^* | z \in R\}$ and satisfies $f(z) = [f(z^*)]^*$ for all $z \in R \cup R^*$.

i. Multivalued Functions

The function $w(z)$ can be considered as a map of the z-plane onto the w-plane. Consider the example $w(z) = z^2$, Figure A.3. It is clear from the figure that

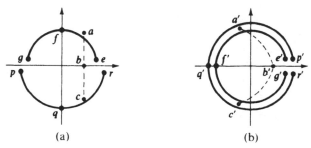

(a)   (b)

Figure A.3 The function $w(z) = z^2$ (a) z-plane; (b) w-plane.

the upper half of the z-plane, including the positive real axis and excluding the negative real axis, is mapped onto the entire w-plane. So also is the lower half of the z-plane, including the negative real axis and excluding the positive real axis. Thus one complete circuit of the z-plane (e.g., $efgpqr$) involves two complete circuits of the w-plane (e.g., $e'f'g'p'q'r'$) provided the former circuit encloses the origin.

The inverse of a function such as $w(z)$ is a *multivalued function*. To continue with the above example, consider the inverse "function"

$$f(z) = z^{1/2}.$$

Writing z and f in polar form,

$$z = \rho e^{i\phi}, \qquad f = r e^{i\theta},$$

we have

$$r = \rho^{1/2}, \qquad \theta = \phi/2.$$

In making one complete circuit around the origin of the z-plane, $0 \le \phi < 2\pi$, one has

$$f_1 = \rho^{1/2} e^{i\phi/2},$$

and the part of the f-plane so covered is the upper half including the $+\mathrm{Re}\, f$ axis and excluding the $-\mathrm{Re}\, f$ axis. Going around the z-plane once more ($2\pi \le \phi < 4\pi$), the point with polar angle ϕ now has polar angle $\phi + 2\pi$; consequently

$$f_2 = \rho^{1/2} e^{i(\phi + 2\pi)/2} = -\rho^{1/2} e^{i\phi/2}.$$

Making a third circuit would give back the value f_1 for f. Clearly f is a double-valued function of z (Figure A.4).

We have seen that it requires two circuits around the origin of the z-plane to encircle the origin of the f-plane, thus giving rise to the double-valued-ness of f. On the other hand, starting with some point z_0 ($\ne 0$) in the z-plane *with well-defined argument*[16] ϕ and following a closed contour C, one traces

---

[16] More precisely, defined up to integral multiples of 4π.

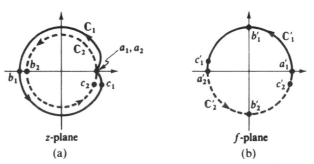

Figure A.4 The function $f(z) = z^{1/2}$ (a) z-plane; (b) f-plane.

Figure A.5 Closed contour in one branch of the function $f(z) = z^{1/2}$ (a) z-plane; (b) f-plane.

out a closed contour C' in the f-plane (Figure A.5). Thus, as long as one does not encircle the origin one can treat f as a single-valued function.

The two values of $z^{1\,2}$ that occur for each value of z form two independent sets, called *branches* of $z^{1/2}$ (e.g., upper and lower $z^{1/2}$-plane). On crossing the positive real z-axis one goes over from one branch of $z^{1/2}$ into the other. The positive real axis is called a *branch line*. If one analytically continues one branch of the function $z^{1/2}$ along a circle that enclosed the origin, one ends up with the other branch. The region of encirclement then contains a singularity of $z^{1/2}$. Since the circle can be made arbitrarily small, it is clear that the singularity is $z = 0$. This is called a *branch point* of the function $z^{1/2}$.

At a branch point, the multivalued function $f(z)$ has the same value for all branches of z. Branch points always occur in pairs, with a branch line joining a pair of branch points. In our example $z = \infty$ is the other branch point. Any curve joining these two branch points can serve as a branch line —the choice is a matter of definition and is usually dictated by convenience.

j. Riemann Surfaces

The theory of analytic functions outlined above for single-valued functions can be extended to a wide class of multivalued functions using a geometrical construct called a *Reimann surface*.

We shall describe the Riemann surface for the function $f(z) = z^{1/2}$. We take two z-planes, call them R_1 and R_2, and define the branch line to be the positive real axis. We "cut" R_1 along the branch line (a *branch cut*), or actually "a little below" it; we do the same for R_2.

We now join the lower lip of R_1 to the upper lip of R_2, and the lower lip of R_2 to the upper lip of R_1. The surface thus obtained is the Riemann surface for the function $z^{1/2}$ (Figure A.6).

We now *define* ϕ on R_1 to have the range $0 \le \phi < 2\pi$, and ϕ on R_2 to have the range $2\pi \le \phi < 4\pi$. Thus on the entire Riemann surface ϕ takes on the values $0 \le \phi < 4\pi$. Starting from $\phi = 0$ in R_1 and encircling the branch point, one reaches the branch cut. Crossing the branch cut takes us onto the upper region, R_2. Crossing the branch cut once more after encircling the origin, one returns to R_1. For z in R_1 $[R_2]$ one has $f(z) = f_1(z)$ $[f_2(z)]$,

Figure A.6 Riemann surface for $f(z) = z^{1/2}$.

i.e., each *Riemann sheet* of the composite Riemann surface corresponds to a single-valued branch of the composite multivalued function.

What we have achieved is the following: From a sequence of single-valued functions [branches of the multivalued $f(z)$] defined on *one* complex z-plane, we have obtained *one* continuous single-valued function defined on the Riemann surface. $f(z)$ is now analytic over the entire Riemann surface except for the branch points, which are now to be treated as (isolated) singularities.

A branch point is said to be of order n if the (multivalued) function is returned to its original value after encircling the branch point at least $n + 1$ times. If this cannot be done for finite n, the branch point is said to be of infinite order. Thus for the function $z^{1/n}$, the branch points $z = 0$ (and $z = \infty$)[17] are of order $n - 1$. The Riemann surface has n sheets.

An example of an infinite-valued function is the function

$$f(z) = \ln z$$

$$= \ln \rho + i(\phi + 2\pi n); \qquad (A.6)$$

each encirclement of the branch point $z = 0$ increases the value of $\ln z$ by $2\pi i$. Accordingly, $z = 0$ is a branch point of infinite order, and the Riemann surface has an infinity of sheets.

Problems

1. The square-well potential is given by

$$V(r) = \begin{cases} -V_0 & \text{if } r < a. \\ 0 & \text{if } r \leq a. \end{cases}$$

   (a) Calculate the Born approximation of the scattering amplitude.
   (b) Show that the exact expression for the lth partial-wave amplitude is given by

$$T_l(p) = \frac{1}{ip} \frac{pj_l(ka)j_l'(pa) - kj_l'(ka)j_l(pa)}{kj_l'(ka)h_l(pa) - pj_l(ka)h_l'(pa)},$$

   where $k = \sqrt{p^2 + 2mV_0}$ (momentum inside the well) and $j_l'(z) = dj_l/dz$, $h_l'(z) = dh_l/dz$.
   (c) Show that at low energies the s-wave, $l = 0$, dominates so that $T(p, \theta) \approx T_0(p)$.

---

[17] $f(z)$ is said to have a singularity at $z = \infty$ if $f(1/z)$ has a singularity at $z = 0$.

(d)   Compare the correct low-energy amplitude with the Born approximation and show that at low energies the Born approximation is good only for a shallow well.

2.   Consider the square-well potential of Problem 1 for the case of small energies.
   (a)   Obtain the scattering cross section and calculate the scattering length a_0 in terms of a and V_0 in the limiting case

$$p^2 \ll 2mV_0, \qquad pa \ll \delta_0.$$

   (b)   Obtain the scattering cross section and the scattering length for the case that V_0 is also small:

$$V^0 \ll \frac{1}{a^2}\frac{1}{2m}.$$

3.   The repulsive square-well potential is given by

$$V(r) = \begin{cases} V_0 & \text{for } r < a, \\ 0 & \text{for } r \geq a. \end{cases}$$

   (a)   Obtain an equation for the s-wave phase shift δ_0 and express the value of δ_0 in terms of scattering momentum p and a for the limit $V_0 \to \infty$.
   (b)   Calculate the scattering cross section if the energy is so small that $pa \ll 1$.
   (c)   Compare the results of (a) and (b) with the results of Problem XVI.2 for the the hard sphere.
   (d)   Study the p-wave phase shift δ_{l-1} for the repulsive square-well potential, in particular for $V_0 \to \infty$, and show that it behaves like $(pa)^3$. Compare it with δ_0 for the infinite repulsive square well $V_0 \to \infty$.
   (e)   Obtain the partial-wave amplitude for the infinite repulsive square well $V_0 \to \infty$ and for energies so small that $(pa)^n \approx 0$ for $n = 3, 4, \ldots$.
   (f)   Calculate the scattering length and the effective range for s-wave scattering for the infinite repulsive square well $V_0 \to \infty$.

Resonance Phenomena

Resonance phenomena constitute some of the most interesting and striking features of scattering experiments. This chapter discusses in detail the connection between quasistationary states and resonance phenomena, and culminates in the derivation of the Breit–Wigner formula. In Section XVIII.2 the concept of "time delay" is introduced and its relation to the phase shift derived. Various formulations of causality are given in Section XVIII.3. In Section XVIII.4 the causality condition is used to derive certain analyticity properties of the S-matrix. These properties are discussed further in Section XVIII.5. In Section XVIII.6, the central section of this chapter, the connection between quasistationary states, defined by a large time delay, and resonances, defined by characteristic structures in the cross section, is derived. Section XVIII.7 describes the observable effects of virtual states. Section XVIII.8 discusses the effect resonances have on the Argand diagram. The actual appearance of resonances in experimental data when the effects of the resonant phase shift, the nonresonant background, and the limited resolution of the apparatus are taken into account is discussed in Section XVIII.9.

XVIII.1 Introduction

In our study of the discrete energy spectra of atoms and molecules we treated the excited states as if they were infinitely long-lived. When we discussed in Section XI.4 the helium atom and its levels, which have the same energy value as the physical system consisting of an He^+ ion and an electron, we ignored the presence of the He^+-e^- system and used an approximate description in which the helium atom was considered to be an isolated stationary system, even though the helium-atom system can never really be isolated from the He^+-e^- system with which it interacts. In this approximate description of excited states the transition processes could not be described. In the more accurate description, which does not ignore transitions, excited states have a finite lifetime. This lifetime is fairly long for the ordinary discrete energy levels, but is not so long for the discrete energy levels that lie in the continuous spectrum of the He-(He^+-e^-) system; the lifetimes of both sorts of states are still, however, much longer than the time taken by the transition processes.

The helium energy-loss experiment, whose result is depicted in Figure XI.4.1, is an inelastic scattering process in which a long-lived intermediate state He* is produced, which subsequently decays either into a helium ion and electron or into a helium atom and photon:

$$e^- + \text{He} \rightarrow e^{-\prime} + \text{He*}$$
$$\phantom{e^- + \text{He} \rightarrow e^{-\prime} + \text{He*}} \begin{array}{l} \rightarrow \text{He}^+ + e^- \\ \rightarrow \text{He} + \gamma. \end{array} \qquad (1.1)$$

The lifetime of a singly excited He*(S), which has an energy level below the first ionization threshold, is orders of magnitude larger than the lifetime of the doubly excited He*(D).[1] This is due to the fact that the interaction V_{II} that causes the transition

$$\text{He*}(D) \rightarrow \text{He}^+ + e^-$$

is much stronger than the interaction V_I that causes the transitions

$$\text{He*}(D) \rightarrow \text{He} + \gamma$$

and

$$\text{He*}(S) \rightarrow \text{He} + \gamma.$$

[The transition $\text{He*}(S) \rightarrow \text{He}^+ + e^-$ is energetically impossible.] In the approximation of Section XI.4, in which both V_I and V_{II} are ignored, the energy operator is given by

$$H = K + V,$$

where $K = K_{\text{electron}} + K_{\text{He}}$ with K_{He} given by (XI.1.2), and V describes the interaction between He and e^-. Consequently both He*(S) and He*(D) have infinite lifetimes. If V_I is ignored but the stronger interaction V_{II} is not, then He*(S) lives infinitely long, but He*(D) has a finite lifetime, which is determined by V_{II}. If neither V_I nor V_{II} is ignored, then both He*(S) and He*(D) have finite lifetimes.

---

[1] The width of the bumps in Figure XI.4.1 is mainly due to the resolution of the apparatus — caused, for example, by the energy spread in the incident electron beam — and not a measure of the lifetime.

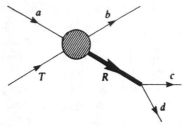

Figure 1.1 Schematic diagram of a production process.

Systems that have finite lifetimes are called *quasistationary* or *metastable states*, in distinction to the *stationary* or *stable states*, which are "infinitely" long lived. Below in Section XVIII.6 we will justify the name *resonances* for the quasistationary states. When to consider a state as a resonance and when to consider it as stable is a matter of the accuracy required of the description; the distinction between (relatively) stable states [particles like He*(S)] and unstable states [resonances like He*(D)] is in principle quantitative rather than qualitative.

There are two main types of experiments in which resonances can be obtained: *formation experiments* and *production experiments*. The process (1.1) is an example of a production experiment. These experiments are characterized by processes of the form

$$a + T \rightarrow b + R \rightarrow b + c + d, \tag{1.2}$$

and are depicted by diagrams like that of Figure 1.1 Production experiments always involve inelastic scattering processes. Formation experiments, on the other hand, consist of simpler processes

$$a + T \rightarrow R \rightarrow a' + T' \tag{1.3}$$

and are illustrated by diagrams like that of Figure 1.2.

A beautiful example of a resonance obtained in a formation experiment in atomic physics is the Schulz resonance in the elastic scattering process

$$e^- + \mathrm{He} \rightarrow \mathrm{He}^- \rightarrow \mathrm{He} + e^-. \tag{1.4}$$

If one measures the elastic scattering cross section as a function of the scattering energy (kinetic energy of the projectile e^- with respect to the target He), one observes that at the energy $E = E_R = 19.31$ eV something unusual happens. At this value the cross section changes violently, as shown in the data of various elastic-collision experiments [cf. Figures 1.3, 1.4, and 1.5,

Figure 1.2 Schematic diagram of a formation process.

Figure 1.3 Typical current-energy plots for electrons in helium as observed by C. E. Kuyatt, J. A. Simpson, and S. R. Mielczarek. [From *Physical Review* **138**, A385 (1965), with permission.]

which show the observed cross sections (Figures 1.4(a), 1.4(c), 1.5) as well as the intensity of the transmitted electron current, i.e., the intensity of the current of electrons that are not scattered (Figures 1.3, 1.4(b), 1.4(d))]. We shall show that the effect shown in these figures can be explained from the hypothesis that He and e^- form a long-lived compound that belongs to one of the energy levels of the He$^-$ ion, as indicated by Equation (1.4).

A *resonance* is usually defined as a sharp structure in the cross section together with a rapidly increasing phase shift going through $\pi/2$. In the remainder of this chapter we shall see that in general the formation of a quasistationary state is connected with these effects, and shall study the characteristic properties of the S-matrix and phase shifts for these phenomena. We shall in this chapter restrict ourselves to the case where only one channel is open. This means that we are in an energy range below the first inelastic threshold, where only the elastic channel

$$a + T \rightarrow R \rightarrow a + T \tag{1.5}$$

is open, so that the only reduced S-matrix element different from zero is

$$S_l(E_A) = \langle \eta_A \| S_l(E_A) \| \eta_A \rangle = e^{2i\delta_l(E_A)}$$

(Recall that η_A are the quantum numbers that determine the internal state of the a-T system; we omit the label η_A on S_l and δ_l.)

Resonances in a multichannel system will be discussed in Chapter XX; it will be shown there that with a few changes multichannel resonances may be treated in much the same manner as the single-channel resonance case studied in this chapter.

Figure 1.4 (a) Variation of intensity of scattering of electrons at 72° by helium atoms with electron energy, as observed by Schulz, *Phys. Rev. Lett.* **13**, 583 (1964). (b) Variation with electron energy of the transmission of electrons through helium as observed by Simpson, *Phys. Rev. Lett.* **11**, 158 (1964). (c) Variation with electron energy of the total cross section for elastic scattering of electrons by helium atoms, as observed by Golden and Bandel, *Phys. Rev.* **138**, A14 (1965). ●, Experimental points. (d) Variation with electron energy of the transmission of electrons through helium as observed by Golden and Nakano, *Phys. Rev.* **144**, 71 (1966).

Figure 1.5 Variation with incident electron energy of the intensity of electrons elastically scattered in helium at different angles of scattering. The intensity variations may be judged from the lines shown on the right-hand side, which indicate 10 percent of the total intensity near the resonance. From D. Andrick and H. Ehrhardt, Z. *Phys.* **192,** 99 (1966).

XVIII.2 Time Delay and Phase Shifts

The lifetime of a quasistationary state formed in a scattering experiment of the form (1.3)[2] is roughly measured by the time by which the projectile is delayed in the scattering region as a consequence of the interaction. Let $\mathscr{R}$, the interaction region around the target, be characterized by a radius R, which is chosen large enough such that there is no interaction outside of $\mathscr{R}$. This is possible because the interactions encountered in practice do have a finite range. We describe the state of the projectile-target system in the Schrödinger picture by the statistical operator $W(t)$. We also imagine a fictitious noninteracting system described by the operator $W^{in}(t)$. If $H = K + V$ is the total-energy operator, then the time development of $W(t)$ is generated by H, while that of $W^{in}(t)$ is generated by K [cf. Equation (XIV.5.5)]. Since we shall consider only elastic scattering, with the target remaining in the pure state $W_T = |\eta_A\rangle\langle\eta_A|$, we can ignore the factor W_T in $W^{in} = W_B^{in} \otimes W_T$ and consider W as referring to the projectile alone. $W(t)$ then describes the state as it develops when the target is present in the center of $\mathscr{R}$, while $W^{in}(t)$ describes the state when there is no target in $\mathscr{R}$.

---

[2] To keep the discussion simple we assume the projectile mass remains unchanged, i.e., $m_{a'} = m_a = m$.

The probability that the projectile is in the region $\mathscr{R}$ at the time t is given by

$$I(t) = \int_{\mathscr{R}} d^3x \, \langle \mathbf{x} | W(t) | \mathbf{x} \rangle \tag{2.1}$$

(cf. Sections II.8 and II.9), where $|\mathbf{x}\rangle$ are the generalized eigenvectors of the projectile position operator with respect to the target. The time the projectile spends in $\mathscr{R}$ is then

$$T = \int_{-\infty}^{+\infty} dt \, I(t) = \int_{-\infty}^{\infty} dt \int_{\mathscr{R}} d^3x \, \langle \mathbf{x} | W(t) | \mathbf{x} \rangle. \tag{2.2}$$

If there were no interaction between projectile and target, the time spent in $\mathscr{R}$ would be

$$T^{\text{in}} = \int_{-\infty}^{+\infty} dt \int_{\mathscr{R}} d^3x \, \langle \mathbf{x} | W^{\text{in}}(t) | \mathbf{x} \rangle. \tag{2.3}$$

The time that the projectile is delayed by the interaction with the target is therefore

$$t_D^{(\mathscr{R})} = T - T^{\text{in}} = \int_{-\infty}^{+\infty} dt \int_{\mathscr{R}} d^3x \, (\langle \mathbf{x} | W(t) | \mathbf{x} \rangle - \langle \mathbf{x} | W^{\text{in}}(t) | \mathbf{x} \rangle). \tag{2.4}$$

This "time delay"[3] may be positive or negative. If a quasistationary state is formed it will be positive and large.

We now investigate the relationship of the time delay $t_D^{(\mathscr{R})}$ with the phase shifts and the S-matrix. We insert complete systems of K and H eigenvectors into Equation (2.4) and use Equations (XIV.5.5) and (XIV.5.21) (which was proved in Appendix XV.A) to obtain

$$
\begin{aligned}
t_D^{(\mathscr{R})} &= \int_{-\infty}^{\infty} dt \int_{\mathscr{R}} d^3x \sum_{aa'} \{ \langle \mathbf{x} | a^+ \rangle \langle a^+ | e^{-itH} W e^{itH} | a'^+ \rangle \langle a'^+ | \mathbf{x} \rangle \\
&\quad - \langle \mathbf{x} | a \rangle \langle a | e^{-itK} W^{\text{in}} e^{itK} | a' \rangle \langle a' | \mathbf{x} \rangle \} \\
&= \int_{-\infty}^{\infty} dt \sum_{aa'} e^{-it(E_a - E_{a'})} \langle a | W^{\text{in}} | a' \rangle \\
&\quad \times \int_{\mathscr{R}} d^3x \{ \langle \mathbf{x} | a^+ \rangle \langle a'^+ | \mathbf{x} \rangle - \langle \mathbf{x} | a \rangle \langle a' | \mathbf{x} \rangle \}.
\end{aligned}
\tag{2.5}
$$

$\mathscr{R}$ is the region within which the interaction takes place. The wave function $\langle \mathbf{x} | a^+ \rangle$ in this region depends strongly upon the particular nature of the interaction. On the other hand, the wavefunction has a very general form outside the region $\mathscr{R}$, where the projectile moves freely, the effect of the interaction being expressed by a phase shift relative to the asymptotic form of the free wave function $\langle \mathbf{x} | a \rangle$. Since we want to obtain statements about the time

---

[3] Time delay has been introduced previously in a similar way by F. T. Smith, *Phys. Rev.* **118**, 349 (1960) based on E. P. Wigner, *Phys. Rev.* **98**, 145 (1955).

delay that are very generally true, we attempt to replace the integral over $\mathscr{R}$ by an integral outside of $\mathscr{R}$. To this end, we first show that the time delay over all space ($R \to \infty$) vanishes. For $R \to \infty$ the last integral in (2.5) is given by:

$$\int_\infty d^3x \, \{\langle \mathbf{x}|a^+\rangle\langle a'^+|\mathbf{x}\rangle - \langle \mathbf{x}|a\rangle\langle a'|\mathbf{x}\rangle\}$$

$$= \int_\infty d^3x \, \{\langle a'^+|\mathbf{x}\rangle\langle \mathbf{x}|a^+\rangle - \langle a'|\mathbf{x}\rangle\langle \mathbf{x}|a\rangle\}$$

$$= \langle a'^+|a^+\rangle - \langle a'|a\rangle, \tag{2.6}$$

where we have used the relation

$$\int_\infty d^3x \, |\mathbf{x}\rangle\langle \mathbf{x}| = I.$$

According to Equation (XIV.2.9a,b), the generalized eigenvectors of H and K have the same normalization. (This is proved in Appendix XV.A.) Thus the right-hand side of (2.6) and consequently $t_D^{(\infty)}$ vanish. We now split the integral over the whole space in (2.6) into a sum of an integral over the region $\mathscr{R}$ and an integral over the rest of the space outside the region $\mathscr{R}$, which we denote by $(\infty - \mathscr{R})$, and have

$$0 = \int_\mathscr{R} d^3x \, \{\langle \mathbf{x}|a^+\rangle\langle a'^+|\mathbf{x}\rangle - \langle \mathbf{x}|a\rangle\langle a'|\mathbf{x}\rangle\}$$

$$+ \int_{\infty - \mathscr{R}} d^3x \, \{\langle \mathbf{x}|a^+\rangle\langle a'^+|\mathbf{x}\rangle - \langle \mathbf{x}|a\rangle\langle a'|\mathbf{x}\rangle\}.$$

Inserting this into (2.5), the time delay over region $\mathscr{R}$ can then be written as

$$t_D^{(\mathscr{R})} = \int_{-\infty}^{\infty} dt \sum_{aa'} e^{-it(E_a - E_{a'})}\langle a|W^{\text{in}}|a'\rangle$$

$$\times \int_{(\infty - \mathscr{R})} d^3x \, \{\langle \mathbf{x}|a\rangle\langle a'|\mathbf{x}\rangle - \langle \mathbf{x}|a^+\rangle\langle a'^+|\mathbf{x}\rangle\}. \tag{2.7}$$

Assuming spherical symmetry, i.e., $[H, L_i] = [K, L_i] = 0$, we may choose angular-momentum bases $\{|E \, l \, l_3 \, \eta^+\rangle\}$ and $\{|E \, l \, l_3 \, \eta\rangle\}$ for $\{|a^+\rangle\}$ and $\{|a\rangle\}$.[4] Then the time delay is given by

$$t_D^{(\mathscr{R})} = \int_{-\infty}^{+\infty} dt \sum_{ll_3} \int \rho(E) \, dE \, \rho(E') \, dE' \, e^{-it(E - E')}\langle E \, l \, l_3|W^{\text{in}}|E' \, l \, l_3\rangle$$

$$\times \int_R^{\infty} r^2 \, dr \, \{\langle r|E\rangle_l\langle r|E'\rangle_l^* - \langle r|E^+\rangle_l\langle r|E'^+\rangle_l^*\}, \tag{2.8}$$

---

[4] Here η can be any set of additional quantum numbers, and we shall usually suppress the label η. If it is present then the position eigenket must also have this label: $|\mathbf{x}, \eta\rangle$.

where ρ is the normalization function of (XIV.2.9a), and where the relations

$$\langle r\,\theta\,\phi|E\,l\,l_3\rangle = Y_{ll_3}(\theta,\,\phi)\langle r|E\rangle_l, \tag{2.9a}$$

$$\langle r\,\theta\,\phi|E\,l\,l_3^+\rangle = Y_{ll_3}(\theta,\,\phi)\langle r|E^+\rangle_l, \tag{2.9b}$$

$$\int d\Omega \; Y_{ll_3}(\Omega)Y_{l'l_3'}(\Omega) = \delta_{ll'}\,\delta_{l_3l_3'} \tag{2.9c}$$

have been used. We can write (2.8) in the form

$$t_D^{(\mathscr{R})} - \int_{-\infty}^{\infty} dt \sum_{ll_3} \int \rho(E)\,dE\;\rho(E')\,dE'\;e^{-it(E-E')}\langle E\,l\,l_3|\,W^{\mathrm{in}}|E'\,l\,l_3\rangle J, \tag{2.10}$$

where

$$J \equiv \int_R^{\infty} r^2\,dr\,\{\langle r|E\rangle_l\langle r|E'\rangle_l^* - \langle r|E^+\rangle_l\langle r|E'^+\rangle_l^*\}. \tag{2.11}$$

In Equation (2.11), which holds for all R larger than the radius of the interaction region $\mathscr{R}$, we choose R to lie in the asymptotic region. The integral J can then be evaluated using the asymptotic forms (XVII.4.17) and (XVII.4.18) of the exact and free radial wave functions.[5] Though derived under the assumption of a Hamiltonian time development, these asymptotic forms exist independently of H, depending as they do only on the S-matrix through the phase shift δ_l. We have, therefore,

$$J = \frac{2}{\pi}\int_R^{\infty} r^2\,dr\,\frac{1}{r^2 pp'}\left\{\sin\left(pr - \frac{l\pi}{2}\right)\sin\left(p'r - \frac{l\pi}{2}\right)\right.$$

$$\left. - e^{i(\delta_l(p) - \delta_l(p'))}\sin\left(pr - \frac{l\pi}{2} + \delta_l(p)\right)\sin\left(p'r - \frac{l\pi}{2} + \delta_l(p')\right)\right\}. \tag{2.12}$$

Because the t-integration in (2.10) gives a factor $\delta(E - E')$, one only needs to evaluate J in the limit $E' \to E$, i.e., $p' \to p$. The evaluation of this integral is an exercise in distribution theory, with the result

$$J(p \to p') = \frac{1}{\pi}\frac{1}{p^2}\left\{\frac{d\delta_l(p)}{dp} - (-1)^l\frac{1}{p}\cos(2pR + \delta_l(p))\sin\delta_l(p)\right\}. \tag{2.13}$$

[The calculation that gives (2.13) proceeds in the following way. We rewrite the integrand in (2.12) as follows:

$$\pi pp'J = \frac{1}{2}\int_{-\infty}^{+\infty} dr\,\theta(r - R)\{e^{i(p - p')r}[1 - e^{2i(\delta_l(p) - \delta_l(p'))}]$$

$$+ (-1)^l[e^{i(p + p')r}(e^{2i\delta_l(p)} - 1) + e^{-i(p + p')r}(e^{-2i\delta_l(p')} - 1)]\}. \tag{2.14}$$

---

[5] Only for $l = 0$ are the exact and free radial wave functions given by the trigonometric function such that for $l = 0$ (2.12) holds for any value of R outside the *interaction* region and one can choose for R the smallest possible value, the effective radius of the scatterer. The larger the value of l the larger one has to choose the value of R in order to be in the asymptotic region.

The integral over the first term in the integrand is the Fourier transform of the θ-function and may be calculated using Equation (XV.2.25):

$$\text{first term} = \frac{1}{2} \int_{-\infty}^{+\infty} dr \theta(r - R) e^{i(p-p')r} [1 - e^{2i(\delta_1(p) - \delta_1(p'))}]$$

$$= \tfrac{1}{2} [1 - e^{2i(\delta_1(p) - \delta_1(p'))}] e^{i(p-p')R} \int_{-\infty}^{\infty} dr \theta(r - R) e^{i(p-p')(r-R)}$$

$$= \tfrac{1}{2} [1 - e^{2i(\delta_1(p) - \delta_1(p'))}] e^{i(p-p')R} \frac{i}{p - p' + i0} . \qquad (2.15)$$

Using the relation[6] (VIII.2.5):

$$\frac{1}{p - p' \pm i0} = \frac{1}{p - p'} \mp i\pi\delta(p - p'), \qquad (2.16)$$

we rewrite (2.15), noting that the contribution of the term containing $\delta(p - p')$ vanishes, since

$$\delta(p - p')[1 - e^{2i(\delta_1(p) - \delta_1(p'))}] = 0.$$

We are left with

$$\text{first term} = \tfrac{1}{2} [1 - e^{2i(\delta_1(p) - \delta_1(p'))}] e^{i(p-p')R} \frac{i}{p - p'} . \qquad (2.17)$$

Now let $p \to p'$. Expanding the exponential in the square brackets, we get

$$\text{first term} \to \tfrac{1}{2} [-2i(\delta_1(p) - \delta_1(p'))] e^{i(p-p')R} \frac{i}{p - p'} \to \frac{d\delta_1(p)}{dp} . \qquad (2.18)$$

The integrals over the second and third terms in (2.14) are obtained similarly, with the result

$$\text{second and third terms} = (-1)^l \left[i \frac{e^{i(p+p')R}}{p + p' + i0} \cdot \tfrac{1}{2}(e^{+2i\delta_1(p)} - 1) \right.$$

$$\left. - i \frac{e^{-i(p+p')R}}{p + p' - i0} \cdot \tfrac{1}{2}(e^{-2i\delta_1(p')} - 1) \right] . \qquad (2.19)$$

Here there is no problem with taking the limit $p' \to p$, and after elementary simplification one obtains the second term on the right-hand side of (2.13).]

---

[6] Gel'fand and Shilov (1964), Vol. 1.

If we insert J of (2.13) into (2.10), perform the integration over t [letting the bound of the integral go to infinity and using (XIV.5.19)], and then integrate over E', we obtain

$$t_D^{(\mathscr{R})} = 2\sum_{l l_3} \int \rho^2(E)\, dE\, \langle E\, l\, l_3|\, W^{\mathrm{in}}|E\, l\, l_3\rangle$$

$$\times \frac{1}{p^2}\left[\frac{d\delta_l(p)}{dp} - (-1)^l \frac{1}{p}\cos(2pR + \delta_l(p))\sin\delta_l(p)\right]. \quad (2.20)$$

Since the expression in the square brackets does not depend upon l_3, we can sum over that component of the angular momentum. Denoting

$$\sum_{l_3} \langle E\, l\, l_3|\, W^{\mathrm{in}}|E\, l\, l_3\rangle = W_l(E) = W_l(p), \quad (2.21)$$

which because of the normalization of W^{in} fulfills

$$\sum_{l l_3} \int \rho(E)\, dE\, \langle E\, l\, l_3|\, W^{\mathrm{in}}|E\, l\, l_3\rangle = \sum_l \int \rho(E)\, dE\, W_l(E) = 1, \quad (2.22)$$

one can write (2.20) as

$$t_D^{(\mathscr{R})} = \sum_l \int \rho(E)\, dE\, W_l(E) t_D^{(\mathscr{R})l}(E), \quad (2.23)$$

where for a nonrelativistic particle with mass m, the quantities p, E, and ρ are related by [cf. (XIV.5.11) and (XIV.5.18a)]

$$E = p^2/2m, \qquad \rho(E) = mp(E), \quad (2.24)$$

and where we have defined

$$t_D^{(\mathscr{R})l}(E) = 2\rho(E)\frac{1}{p^2}\left[\frac{d\delta_l(p)}{dp} - \frac{(-1)^l}{p}\cos(2pR + \delta_l(p))\sin\delta_l(p)\right]$$

$$= 2m\frac{1}{p}\left[\frac{d\delta_l(p)}{dp} - \frac{(-1)^l}{p}\cdot \tfrac{1}{2}[\sin 2(pR + \delta_l) - \sin 2pR]\right]. \quad (2.25)$$

The second term on the right-hand side of (2.25) is of limited variation and depends upon the radius R of the region $\mathscr{R}$, which can be arbitrarily chosen, the only restriction being that it be greater than the range of the interaction.[7] We therefore average over R to obtain a quantity that does not depend upon the irrelevant exact choice of the radius R and is characteristic of the scattering process only. This average of $t_D^{(\mathscr{R})l}$ is

$$t_D^l(E) = 2\rho(E)\frac{1}{p^2}\frac{d\delta_l(p)}{dp} = 2m\frac{1}{p}\frac{d\delta_l(p)}{dp} = 2\frac{d\delta_l(E)}{dE}. \quad (2.26)$$

The average time delay, which for short we again just call the time delay, is thus

$$t_D = \sum_l \int \rho(E)\, dE\, W_l(E) t_D^l(E). \quad (2.27)$$

---

[7] Note that for $l \neq 0$, R in the above derivation has to be chosen in the asymptotic region.

The incident beam W^{in} is usually not prepared with a definite value of the orbital angular momentum l, but with a momentum pointing in a well-defined direction. $W_l(E)$ then describes the weight of the lth angular-momentum component in the mixture W^{in}. However, if one arranges that the incident "beam" has a well defined angular momentum l_A, then one has

$$\rho(E)W_l(E) = \delta_{ll_A}F_l(E - E_A),\qquad(2.28)$$

where $F_{l_A}(E - E_A)$ fulfills, because of (2.22), the condition

$$\int dE\, F_{l_A}(E - E_A) = 1$$

and describes the energy distribution in the "beam" [cf. (XIV.5.35)]. The incident beam is usually not monoenergetic. However, if it can be considered monoenergetic, then [cf. (XIV.5.36)]

$$F_{l_A}(E - E_A) = \delta(E - E_A).\qquad(2.29)$$

Inserting (2.29) and (2.28) into (2.27), one obtains for the time delay of this particular "state" with angular momentum l_A and energy E_A

$$t_D = t_D^{l_A}(E_A).\qquad(2.30)$$

Thus the quantity (2.25) or (2.26) is the time delay occurring in the scattering of a state with angular momentum l and energy E. And the time delay in the state W^{in} (2.27) is the weighted average of the time delays $t_D^l(E)$ over all angular-momentum and energy values.

Restoring the dependence upon the additional quantum numbers η, the time delay of a "state" with quantum numbers $\eta l E$ is given by:

$$t_D^{\eta l}(E) = 2\frac{d\delta_l^{\eta}(E)}{dE} = \frac{2m}{p}\frac{d\delta_l^{\eta}(p)}{dp}.\qquad(2.31)$$

A monoenergetic beam describes a steady state for which time delay really has no observable meaning. For a measurement of the time delay one needs a beam of pulses of finite duration and, therefore, of nonzero energy spread.[8] Then the quantity t_D of (2.27) and also the time delay for a particular angular-momentum value,

$$t_D^{\eta l} = \int dE\, F(E - E_A)t_D^{\eta l}(E) = 2\int dE\, F(E - E_A)\frac{d\delta_l^{\eta}(E)}{dE},\qquad(2.32)$$

can—at least in a *gedanken* experiment—be measured as the difference in time that a pulse of projectiles needs to pass through a region $\mathscr{R}$ with and without the target present. Scattering experiments are, however, usually not carried out in a way that allows the measurement of such a delay, and the significance of (2.31) or (2.26) lies in its theoretical consequences, which we shall discuss now.

---

[8] Cf. the discussion in Section II.9.

If during the scattering process for a particular set of values $(\eta l E) = (\eta_R l_R E_R)$ a quasistationary state is formed, that is, if the projectile is temporarily captured by the target, then for this set of values the delay time $t_D^{\eta l}(E)$ must be large. Thus the condition for a quasistationary state is

$$\frac{d\delta_l^\eta(E)}{dE} \text{ has a sharp positive maximum at } (\eta_R l_R E_R), \qquad (2.33)$$

which in particular means that

$$\left.\frac{d^2\delta_{l_R}^{\eta_R}(E)}{dE^2}\right|_{E=E_R} = 0, \qquad \left.\left|\frac{d^3\delta_{l_R}^{\eta_R}(E)}{dE^3}\right|\right|_{E=E_R} \text{ is large.} \qquad (2.34)$$

The main contribution in the expression (2.27) for the delay time then comes from the term with these particular values $(l_R E_R)$.

It may, of course, happen that for a particular scattering system all of the $\delta_l^\eta(E)$ are such functions of E that the $d\delta_l^\eta(E)/dE$ do not have any maxima, and therefore no quasistationary state exists. However, more often than not at least some of the $d\delta_l^\eta(E)/dE$ do have such isolated maxima, and we shall investigate the consequences of such an occurrence below, in particular in Section XVIII.6.

XVIII.3 Causality Conditions

The time $T = T^{(\mathcal{R})}$ which a projectile spends within a certain region $\mathcal{R}$ of radius R around the target must always be positive. Thus the condition

$$T^{(\mathcal{R})} \geq 0$$

is obviously fulfilled. This condition is, through the definition of T in (2.2), connected with the *van Kampen causality condition*, which states:

$$\int_\mathcal{R} d^3x \, \langle \mathbf{x}| W(t)|\mathbf{x}\rangle \geq 0 \quad \text{at any time } t. \qquad (3.1)$$

(3.1) is also obviously fulfilled as it represents the probability to find the projectile in the region $\mathcal{R}$. The strongest form of the condition (3.1) is obtained if one chooses for R its smallest possible value, i.e, the effective radius of the scatterer.

The time $T^{(\mathcal{R})}$ that the projectile spends in the interaction region can be obtained according to (2.4) from

$$T^{(\mathcal{R})} = t_D^{(\mathcal{R})} + T^{\text{in}(\mathcal{R})}, \qquad (3.2)$$

where $t_D^{(\mathcal{R})}$ is given by (2.23) and (2.25) and $T^{\text{in}(\mathcal{R})}$ is given by (2.3) which can be written:

$$T^{\text{in}(\mathcal{R})} = \int_{-\infty}^{+\infty} dt \sum_{ll_3} \int \rho(E) \, dE \, \rho(E') \, dE' \, e^{-it(E-E')}\langle E \, l \, l_3| W^{\text{in}}| E' \, l \, l_3\rangle(-J^{\text{in}})$$

$$(3.3)$$

with J^{in} given by

$$-J^{in} = \int_0^R r^2 \, dr \, \langle r|E\rangle_l \langle r|E'\rangle_l^* = \frac{2}{\pi} \int_0^R r^2 \, dr \, j_l(pr) j_l(p'r). \qquad (3.4)$$

Equation (3.3) with (3.4) is obtained by the same arguments that led from (2.4) to (2.10) with (2.11), except that in (3.4) we did not have to replace the integral over the interaction region $\mathscr{R}$ by an integral over $(\infty - \mathscr{R})$.

The integral in (3.4) is calculated using (XVII.3.9b) and (XVII.3.7a), the result being

$$-J^{in} = \frac{2}{\pi} \frac{R^2}{(p + p')(p - p')} (p' j_l(pR) j_{l-1}(p'R) - p j_{l-1}(pR) j_l(p'R))$$

$$\approx \frac{2}{\pi} \frac{R^2}{(p + p')(p - p')} \left(\frac{1}{pR^2} \sin \left(pR - \frac{\pi l}{2} \right) \sin \left(p'R - \frac{\pi(l - 1)}{2} \right) \right.$$

$$\left. - \frac{1}{p'R^2} \sin \left(pR - \frac{\pi(l - 1)}{2} \right) \sin \left(p'R - \frac{\pi l}{2} \right) \right). \qquad (3.5)$$

For the second equality in (3.5) the asymptotic form (XVII.3.6a) was used, which is justified for R in the asymptotic region or in case $l = 0$ for R just outside the interaction region. Using well-known relations between trigonometric functions, this becomes

$$-J^{in} = \frac{1}{\pi} \frac{1}{pp'} \left[\frac{1}{p - p'} \sin (p - p')R - \frac{(-1)^l}{p + p'} \sin (p + p')R \right] \qquad (3.6)$$

and for $p' \to p$

$$-J^{in} = \frac{1}{\pi p^2} \left(R - \frac{(-1)^l}{2p} \sin 2p R \right). \qquad (3.7)$$

This expression is inserted into (3.3) after one has performed in (3.3) the integration over t using $\int_{-\infty}^{+\infty} e^{itx} \, dt = 2\pi\delta(x)$ and over E' which leads to

$$T^{in(\mathscr{R})} = \sum_{l_3} \int \rho^2(E) \, dE \, \langle E \, l \, l_3| \, W^{in}| E \, l \, l_3\rangle \left(\frac{2R}{p^2} - \frac{(-1)^l}{p^3} \sin 2pR \right). \qquad (3.8)$$

With the definition (2.21) and (2.24) one obtains

$$T^{in(\mathscr{R})} = \sum_l \int \rho(E) \, dE \, W_l(E) T^{in(\mathscr{R})l}(E) \qquad (3.9)$$

where

$$T^{in(\mathscr{R})l}(E) = m \left(\frac{2R}{p(E)} - \frac{(-1)^l}{p(E)^2} \sin 2p(E)R \right). \qquad (3.10)$$

Equations (3.9) and (3.10) have been written in complete analogy to (2.23) and (2.25). The quantity $T^{in(\mathscr{R})l}(E)$ is thus the time which a "state" with angular momentum l and energy E spends in the region $\mathscr{R}$ if no interaction takes place. The time $T^{(\mathscr{R})}$ which the projectile spends in the region $\mathscr{R}$ when the interaction is present is in analogy to (3.9) and (2.23) also written as

$$T^{(\mathscr{R})} = \sum_l \int \rho(E)\, dE\, W_l(E) T^{(\mathscr{R})l}(E), \qquad (3.11)$$

where according to (3.2)

$$T^{(\mathscr{R})l}(E) = t_D^{(\mathscr{R})l}(E) + T^{\text{in}(\mathscr{R})l}(E). \qquad (3.12)$$

Inserting (2.25) and (3.10) in (3.12), one obtains

$$T^{(\mathscr{R})l}(E) = 2m\frac{1}{p}\left(\frac{d\delta_l}{dp} + R\right) - (-1)^l \frac{m}{p^2} \sin 2(pR + \delta_l(p)) \qquad (3.13)$$

$T^{(\mathscr{R})l}(E)$ is the time that the "part" of the projectile with energy E and angular momentum l spends in the interaction region.

The condition (3.1) must be fulfilled for any arbitrary state of the projectile, i.e., for any set of positive functions $W_l(E)$. Consequently we conclude from (3.11) and (3.1) that

$$T^{(\mathscr{R})l}(E) \geq 0 \quad \text{for all } l \text{ and } E. \qquad (3.14)$$

This condition (3.14) can be rewritten using (3.13):

$$\frac{d\delta_l}{dp} \geq -R + (-1)^l \frac{\sin 2(\delta_l + pR)}{2p} \geq -\left(R + \frac{1}{2p}\right). \qquad (3.15)$$

In the form (3.15) the condition is called *Wigner's causality inequality*. The smaller R the stronger is the condition (3.15); but only for $l = 0$ can one choose in our above derivation for R the effective radius of the scatterer.

Wigner's causal inequality states that the phase shift cannot decrease faster than at a certain rate. Thus if the phase changes rapidly, then it must be increasing. We shall make use of this fact below.

The condition (3.1) is so obviously fulfilled that one may wonder why the conditions (3.1) and (3.15) were given the name *causality conditions*. The connection with causality becomes most apparent if one rewrites (3.14) into yet another form using (3.12) with (3.10):

$$t_D^{(\mathscr{R})l}(E) \geq -T^{\text{in}(\mathscr{R})l}(E) = -\left[\frac{m \cdot 2R}{p} - (-1)^l \frac{m}{p}\frac{\sin 2pR}{p}\right]. \qquad (3.16)$$

The first term on the right-hand side of (3.16) and of (3.10) is the free time of flight that a classical projectile with velocity p/m would need to traverse the region of diameter $2R$. The second term represents the quantum effects; it is negligible for sufficiently fast particles, $pR \gg 1$, and arises from the fact that quantum particles cannot be localized within dimensions smaller than a de Broglie wavelength.

If the region of diameter $2R$ is the interaction region in which the projectile interacts with the target, then—depending upon the kind of interaction—the projectile may be delayed ($t_D > 0$) or hastened ($t_D < 0$) in it. The projectile may be delayed an arbitrarily long time, and we shall discuss this phenomena in the subsequent sections. However, it cannot be advanced an arbitrarily large time, because according to causality outgoing projectiles cannot appear before the incoming projectiles have reached the scattering region. Classically, the maximal time advance occurs if the projectile has hardly entered the interaction region when it is repelled at the surface, and leaves the interaction region $\mathscr{R}$ at a time $m2R/p$ earlier than a particle that passes through such a region without interaction—i.e., its time delay is $-m2R/p$. A more negative time delay is not possible, because then the outgoing projectile would have to leave the surface of the interaction region before the incoming projectile had reached it. For quantum systems these arguments have to be corrected by the term $(m/p)(\sin 2pR)/p$, and then lead to (3.16) as a mathematical formulation of the above statement of causality.

XVIII.4 Causality and Analyticity

In chapter XVI we considered the lth partial T-matrix elements $\langle \eta_b \| T_l(E) \| \eta_A \rangle$ and the lth partial S-matrix elements $\langle \eta_b \| S_l(E) \| \eta_A \rangle$; in particular we considered the lth partial T- and S-matrix elements for elastic scattering, $T_l(E) = T_l(p)$ and $S_l(E) = S_l(p)$, where the momentum p and energy E of the projectile of mass m are connected by $E = p^2/2m$. If the interaction is described by a Hamiltonian $H = K + V$, then T and S are connected to the interaction Hamiltonian V by expressions like (XIV.5.41), but as mentioned above (e.g., in Section XIV.4), the notions of S- and T-matrix make sense also if there exists no Hamiltonian time development. Then the S- or T-matrix is the ultimate basis for the description of all the information required to compute observable quantities, such as cross sections [(XVI.1.26), (XVI.1.29), (XVI.1.30)]. The properties of the S-matrix then follow from general physical principles believed to be satisfied by any interaction. One *class* of general physical principles are symmetry principles, and we have, in fact, already made use of such a symmetry principle, namely the rotational symmetry (XVI.1.1), when we derived that the S-matrix elements are independent of l_3. The general physical principle that we shall utilize now is the causality condition discussed in the preceding section. The causality condition leads to analytic properties of the S-matrix elements when the energy and momentum of the projectile are extended to complex values.[9] When the interaction is described in terms of a potential function $V(r)$, then this kind of analyticity of the S-matrix can be derived from the properties of the potential function.[10] Following the spirit of the presentation in this book, we shall not start from a potential function, but establish some analyticity of $S_l(p)$ as a consequence of causality. We shall not give an extensive discussion of this subject, especially as not all questions on this subject

---

[9] One obtains dispersion relations when these analytic properties are expressed in terms of integral relations between different matrix elements for real values of the variables.

[10] Taylor (1972), Chapter 12.

have been answered yet, and content ourselves in this section with the formulation of the analyticity properties of $S_l(p)$ that will be needed in the subsequent sections in connection with resonance phenomena.

The probability of finding the projectile at an arbitrary time t somewhere in space is unity:

$$1 = \int_{-\infty}^{+\infty} d^3x \, \langle \mathbf{x}|W(t)|\mathbf{x}\rangle = \int_{\mathscr{R}} d^3x \, \langle \mathbf{x}|W(t)|\mathbf{x}\rangle + \int_{(\infty - \mathscr{R})} d^3x \, \langle \mathbf{x}|W(t)|\mathbf{x}\rangle. \tag{4.1}$$

Consequently the van Kampen causality condition (3.1) can be reformulated to read

$$\int_{(\infty - \mathscr{R})} d^3x \, \langle \mathbf{x}|W(t)|\mathbf{x}\rangle \le 1 \quad \text{at any time } t. \tag{4.2}$$

Inserting a complete system of basis vectors $|E \, l \, l_3^+\rangle$ and using (XIV.5.21) and (2.9a,b,c), one obtains for (4.2) [by the same calculation that led from (2.4) to (2.7) except for the change from the integral over $\mathscr{R}$ to the integral over $(\infty - \mathscr{R})$]

$$\sum_{l l_3} \iint \rho(E) \, dE \, \rho(E') \, dE' \, e^{-it(E-E')} \langle E \, l \, l_3| W^{\text{in}}| E' \, l \, l_3\rangle$$

$$\times \int_{r=R}^{r=\infty} \langle r|E^+\rangle_l \langle E'^+|r\rangle_l r^2 \, dr \le 1. \tag{4.3}$$

In the region $(\infty - \mathscr{R})$, where no interaction takes place and the projectiles move freely, the wave function $\langle r|E^+\rangle_l = \langle r|p^+\rangle_l$ has the general asymptotic form (XVII.4.17), in which the effect of the interaction is expressed in terms of the S-matrix $S_l(p) = e^{2i\delta_l(p)}$:

$$\langle r|p^+\rangle_l \sim \sqrt{\frac{2}{\pi}} \frac{1}{2ipr} (e^{i(pr - l\pi/2)}S_l(p) - e^{-i(pr - l\pi/2)}). \tag{4.4}$$

Inserting this into the integral over r in (4.3), one obtains for R from the asymptotic region

$$\text{integral} = \int_{-\infty}^{+\infty} r^2 \, dr \, \theta(r - R)\langle r|p^+\rangle_l \langle r|p'^+\rangle_l^*$$

$$= \frac{2}{\pi} \frac{1}{4pp'} e^{-i(p-p')R} \bigg\{ S_{lR}(p)S_{lR}^*(p') \int_{-\infty}^{+\infty} dr \, \theta(r - R)e^{i(p-p')(r-R)}$$

$$+ \int_{-\infty}^{+\infty} dr \, \theta(r - R)e^{-i(p-p')(r-R)}$$

$$- e^{-il\pi}S_{lR}(p) \int_{-\infty}^{+\infty} dr \, \theta(r - R)e^{i(p+p')(r-R)}$$

$$- e^{il\pi}S_{lR}^*(p') \int_{-\infty}^{+\infty} dr \, \theta(r - R)e^{-i(p+p')(r-R)} \bigg\},$$

where we have defined[11]

$$S_{lR}(p) = e^{2ipR}S_l(p). \tag{4.5}$$

The Fourier transform of the θ-function has been calculated in (XV.2.25+); using that result one obtains

$$\text{integral} = \frac{2}{\pi} \frac{e^{-i(p-p')R}}{4pp'} \left\{ S_{lR}(p)S_{lR}^*(p') \frac{i}{p-p'+i0} + \frac{i}{p'-p+i0} \right.$$
$$\left. + (-1)^l \left(S_{lR}^*(p') \frac{i}{p+p'-i0} - S_{lR}(p) \frac{i}{p+p'+i0} \right) \right\}.$$

Using the relation between the distributions, $(2.16) = (VIII.2.5)$, this can be rewritten

$$\text{integral} = \frac{1}{2\pi pp'} e^{-i(p-p')R} \left\{ (S_{lR}(p)S_{lR}^*(p') - 1) \frac{i}{p-p'} \right.$$

$$+ (-1)^l (S_{lR}^*(p') - S_{lR}(p)) \frac{i}{p+p'}$$

$$+ \pi \delta(p - p')(S_{lR}(p)S_{lR}^*(p') + 1)$$

$$\left. + \pi(-1)^l \delta(p + p')(S_{lR}^*(p') + S_{lR}(p)) \right\}$$

Inserting this integral into (4.3), one obtains, using $\rho \, dE = p^2 \, dp$,

$$\sum_{ll_3} \int_0^\infty \int_0^\infty p^2 \, dp \, p'^2 \, dp' \, e^{-it(E-E')} \langle p \, l \, l_3 | W^{in} | p' \, l \, l_3 \rangle \frac{1}{2\pi pp'} e^{-i(p-p')R}$$

$$\times \left\{ (1 - S_{lR}(p)S_{lR}^*(p')) \frac{1}{i(p-p')} - (-1)^l (S_{lR}^*(p') - S_{lR}(p)) \frac{1}{i(p+p')} \right.$$

$$\left. + \pi \delta(p - p')(S_{lR}(p)S_{lR}^*(p') + 1) + \pi(-1)^l \delta(p + p')(S_{lR}^*(p') + S_{lR}(p)) \right\} \leq 1. \tag{4.6}$$

Performing the integral over p and p' in the last term gives a zero contribution

---

[11] The physical meaning of the exponential factor e^{-2ipR} in

$$S_l(p) = e^{2i\delta_l(p)} = e^{2i(-pR)}S_{lR}(p)$$

is that it represents the phase advance corresponding to a path difference from the surface of the interaction region to the center and back. According to the description of Figure XVII.4.2, a repulsive potential gives a negative phase shift. An impenetrable sphere of radius R has a phase shift of $-pR$, because the wave is reflected at the surface of the sphere, as compared with the one going through the center, cf. Problem XVII.3. Correspondingly an outgoing signal can appear up to $2R/v = 2Rm/p$ earlier than would have been possible in the absence of the scatterer. $S_{lR}(p)$ then describes the deviation of the effect of the scatterer from the effect of an impenetrable sphere with the radius of the interaction region.

because of $\delta(p + p')$. After integrating the next to the last term over p' and using the unitary condition $S_{lR}(p)S_{lR}^*(p) = 1$, one obtains the contribution

$$\sum_{ll_3} \int dp\, p^2 \langle p\, l\, l_3 | W^{in} | p\, l\, l_3 \rangle = 1.$$

For $t = 0$ the above inequality (4.6) therefore becomes

$$\sum_{ll_3} \int_0^\infty \int_0^\infty p^2\, dp\, p'^2\, dp'\, \langle p\, l\, l_3 | W^{in} | p'\, l\, l_3 \rangle \frac{1}{2\pi pp'} e^{-i(p-p')R}$$

$$\times (S_{lR}(p)S_{lR}^*(p') - 1)\frac{1}{i(p - p')}$$

$$\geq \sum_{ll_3} \int_0^\infty \int_0^\infty p^2\, dp\, p'^2\, dp'\, \langle p\, l\, l_3 | W^{in} | p'\, l\, l_3 \rangle \frac{1}{2\pi pp'} e^{-i(p-p')R}(-1)^l$$

$$\times (S_{lR}(p) - S_{lR}^*(p'))\frac{1}{i(p + p')}. \tag{4.7}$$

As this has to be fulfilled for any state W^{in}, it follows that a corresponding inequality must hold for any term in the sum $\sum_{ll_3}$ separately and for any physically permitted momentum distribution function.

One can write (4.7) in a different form by introducing for W^{in} the momentum wave function $A_R^{ll_3}(p)$ [in analogy to the $f(p - p')$ of (II.9.1p)] by

$$A_R^{ll_3}(p)A_R^{ll_3*}(p') = pe^{-ipR}\langle p\, l\, l_3 | W^{in} | p'\, l\, l_3 \rangle e^{ip'R}p'. \tag{4.8}$$

Thus one obtains from (4.7)

$$\int_0^\infty \int_0^\infty dp\, dp'\, A_R^{ll_3}(p)A_R^{ll_3*}(p')(S_{lR}(p)S_{lR}^*(p') - 1)\frac{1}{i(p - p')}$$

$$\geq \int_0^\infty \int_0^\infty dp\, dp'(-1)^l A_R^{ll_3}(p)A_R^{ll_3*}(p')(S_{lR}(p) - S_{lR}^*(p'))\frac{1}{i(p + p')} \tag{4.9}$$

for every value of l and l_3 and for any well-behaved function (i.e., for any element of the Schwartz space), $A_R^{ll_3}(p)$.

From the form (4.9) of the van Kampen causal inequality it follows, by a purely technical proof that we shall not reproduce here:[12]

> $S_l(p)$ has an analytic continuation without singularities in the first
>
> quadrant $0 \leq \arg p \leq \pi/2 - \delta(\delta > 0)$ of the complex p plane. (4.10)

In fact, one can prove more about the analyticity property of $S_l(p)$ than just this, and in Section XVIII.5 below we shall summarize some results. It is, however, only (4.10) that we shall need for our derivation of the Breit–Wigner formula in Section XVIII.6. We wish to emphasize that (4.10) holds for any value that one chooses for R, as long as it is finite.

[12] N. G. van Kampen, *Phys. Rev.* **91**, 1267 (1953), Section II.

XVIII.5 Brief Description of the Analyticity Properties of the S-Matrix

Though a derivation of the following result on the analyticity properties of $S_l(p)$ exceeds the scope of this book, and we shall also not make explicit use of these properties in what follows, the knowledge of these properties will greatly enhance the picture of the S-matrix. We shall, therefore, state these analyticity properties here with a few connecting remarks.[13] These properties are independent of the specific form of the interaction and follow from very general principles—essentially from the causality condition in the form (4.9).

In order to obtain the properties of $S_l(p)$ beyond the first quadrant, we have to continue $S_l(p)$ across the imaginary axis and into the lower half plane. In order to do the first we have to know the properties of $S_l(p)$ on the positive imaginary axis. We know already from the discussions in Section XVII.5 that bound states correspond to poles on the positive imaginary axis, so that we cannot expect $S_l(p)$ to be analytic there. Figure 5.1 shows these poles and zeros, which are, of course, also possible on the real axis.

As a further consequence of the causality condition (4.9) one can prove that

$$\text{Im}\,(-1)^l S_{lR}(p) \le \frac{\text{Re}\,p}{\text{Im}\,p}, \qquad \text{Im}\,S_{lR}(p) \le 1 \quad \text{in} \quad 0 \le \arg p < \pi/2 \quad (5.1)$$

and

$$|S_{lR}(p)| \text{ is bounded in the first quadrant.}^{14} \tag{5.2}$$

This is not sufficient to obtain information about the nature of the singularities on the imaginary axis, except to conclude that if these are poles, they cannot be of higher than the first order.

In order to conclude anything more, one has to make an assumption in addition to the causality condition. If the interaction is described by a Hamiltonian H, this assumption amounts to the requirement that H be a semibounded operator, i.e., there is a lowest energy eigenvalue B and the

---

[13] For more on this subject, see Nussenzveig (1972).

[14] This is a very weak condition for large values of R, because it allows $S_l(p)$ to vary rapidly on the real axis also in the absence of any poles or zeros. But in the above derivation of the causality condition (4.9) R has to be chosen from the asymptotic region because (4.4) is the asymptotic form for the radial wave function outside the interaction region. The smaller the value of l, the smaller can one choose the value of r for which the asymptotic form (4.4) is valid. But only for the case $l = 0$ is (4.4) identical with the radial wave function outside the interaction region. Therefore, only for $l = 0$ can one choose for R the smallest possible value namely the effective radius of the scatterer. In order to obtain the strongest result, i.e. the above statements for R equal to the interaction radius of the scatterer, also for $l > 0$, one has to take instead of the asymptotic form (4.4) the exact expression for the radial wave function for r outside the interaction region:

$$\langle r|p^+\rangle_l = \sqrt{\tfrac{1}{2}\pi}(S_l(p)h_l(pr) + h_l^*(pr)) \tag{4.4'}$$

where $h_l(pr)$ and $h_l^*(pr)$ are the spherical Hankel functions (XVII.2.5). The mathematical proof that (5.1) and (5.2) follow also if one takes (4.4') instead of (4.4) has not yet been given.

binding energy cannot be infinite. In the case where one does not have a Hamiltonian time development and the bound states are described by singularities on the imaginary p-axis, one would assume this to translate into the statement that there are no singularities of $S_l(p)$ on the imaginary p-axis above the value iK, where $K = +\sqrt{2mB}$.

It has in fact been shown[15] that from a precise formulation of the finiteness of the binding energy it follows that

$$\operatorname{Im}(-1)^l S_{lR}(p) \geq -\frac{\operatorname{Re} p}{\operatorname{Im} p} \quad \text{in the first quadrant, for } K = 0 \qquad (5.3)$$

and

$$\operatorname{Im} S_{lR}(p) \to 0 \quad \text{for} \quad \operatorname{Re} p \to 0_+, \quad \text{for any} \quad K \geq 0. \qquad (5.4)$$

Consequently

$$S_l(p) \text{ is real on the positive imaginary axis except between 0 and } iK \quad (5.5)$$

For scattering states the physical values of the momentum p must be positive; however, p occurs quadratically in the Schrödinger equation and also in the Lippmann–Schwinger equation. Consequently the equations are invariant with respect to the change of sign of p, and therefore, if p is replaced by $-p$ in the radial wave function, the resulting function must again be a solution of these equations. Replacing p by $-p$ in (XVII.4.17d) gives

$$\langle r| - p^+ \rangle \sim \frac{e^{-i(\pi/2)(l+1)}}{\sqrt{2\pi}} (-1)^l S_l(-p)\frac{1}{pr}(-(-1)^l e^{-ipr} + S_l^{-1}(-p)e^{ipr}).$$
$$(5.6)$$

As $\langle r|p^+ \rangle$ and $\langle r| - p^+ \rangle$ describe the same physical content, the right-hand side of (XVII.4.17d) and (5.6) must both describe an incoming spherical wave and an outgoing spherical wave modified by the scattering matrix. We therefore require that the right-hand side of (5.6) agrees with the right-hand side of (XVII.4.17d) except for an irrelevant phase factor $((-1)S_l(-p))$. This leads to the condition that for negative values of p the S matrix element S_l be given by

$$S_l^{-1}(-p) = S_l(p) \qquad (5.7a)$$

or

$$S_l(-p) = S_l^{-1}(p). \qquad (5.7b)$$

Using unitarity,

$$S_l(p)S_l^*(p) = 1, \qquad (5.8)$$

one obtains

$$S_l(-p) = S_l^*(p) \quad \text{for real } p. \qquad (5.7c)$$

This is called the *symmetry relation* for the lth partial S-matrix element.

[15] N. G. van Kampen, *Physica* (Utrecht), **20**, 115 (1954).

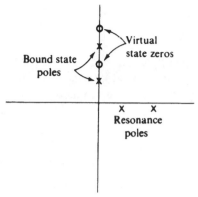

Figure 5.1 Some poles (indicated by ×) and zeros (indicated by ○) of the S-matrix element in the p-plane.

The lth partial S-matrix element is usually considered as a function of $E = p^2/2m$, $S_l(E)$. In the mapping from p to E the complex p-plane (see Figure 5.1) is mapped onto a two-sheeted Riemann surface [cf. XVII.Ai and XVII.Aj] with the branch cut from 0 to ∞. The upper half p-plane Im $p > 0$ corresponds to the first sheet; the first quadrant of the p-plane corresponds to the upper half of the first sheet. The *physically* meaningful values of p in scattering states, p real, $p \geq 0$, correspond to the upper rim of the first sheet. The first sheet is called the "physical" sheet (Figure 5.2). If one continues through the cut one comes to the second sheet, called the "unphysical" sheet, which corresponds to the lower half of the p-plane. $S_l(E)$ is a function on the two-sheeted Riemann surface. The bound-state poles of $S_l(p)$ on the positive imaginary p-axis correspond to poles on the negative real E-axis of the "physical" sheet. Statement (4.10) then means that

$$S_l(E) \text{ is analytic in the upper half plane of the physical sheet.} \qquad (5.9)$$

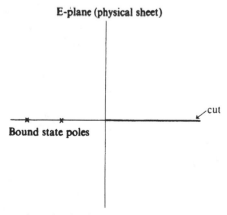

Figure 5.2 Poles and cut in the first ("physical") sheet of the energy plane.

The relation (5.7c) for S_l considered as function of E is then stated as

$$S_l(E - i\epsilon) = S_l^*(E + i\epsilon), \qquad E > 0. \tag{5.10}$$

From the Schwarz reflection principle (XVII.Ah) and condition (5.5) follows that $S_l(E)$ can be continued into the lower half energy plane across the negative real energy axis and is given there by

$$S_l(E) = S_l^*(E^*) \tag{5.11}$$

Since the lower half energy plane corresponds to the second quadrant of the momentum plane and E^* corresponds to $-p^*$, (5.11) leads to

$$S_l(p) = S_l^*(-p^*) \tag{5.12}$$

as an extension of the symmetry relation (5.7c) to complex values of p. Thus, at points symmetrically placed with respect to the imaginary axis, $S_l(p)$ takes on complex conjugate values. As it has no singularity in the first quadrant, it has also no singularity in the second quadrant. If $S_l(p)$ has a zero at p_1, as shown in Figure 5.3, then it will also have a zero at $-p_1^*$.

One can now continue $S_l(p)$ into the lower half p-plane (Figure 5.3), by extending the relation (5.7a) to complex values of p. For every value in the second quadrant (5.7a) defines a value of $S_l(p)$ in the fourth quadrant, except at zeros in the second quadrant which produce poles in the fourth quadrant. In the same way (5.7a) defines a function analytic in the third quadrant except for poles coming from zeros in the first quadrant. Thus, $S_l(p)$ is in general not analytic in the lower half plane and if it has zeros above the positive p-axis it will have poles below the p-axis, the resonance poles of Figure 5.1 and their counterparts in the third quadrant.

In this way we have seen that $S_l(p)$ is a meromorphic function in the whole p-plane that is analytic in the upper half p-plane except perhaps on the imaginary axis. The poles in the lower half plane occur in pairs, symmetrically distributed with respect to the negative imaginary axis, except for those on the negative imaginary axis, corresponding to zeros on the positive imaginary

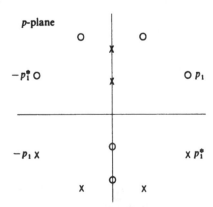

Figure 5.3 Relation between zeros (indicated by ○) and poles.

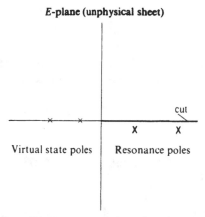

Figure 5.4 Positions of possible resonance poles, virtual-state poles and cuts (capture state poles not shown) in the second ("unphysical") sheet of the energy-plane.

axis. [The number of poles can be infinite, but they cannot have an accumulation point for finite values of p, as an analytic function (in the upper half plane) cannot have a finite accumulation point of zeros.] The singularities on the imaginary axis can be shown to lie within the interval $(-iK, +iK)$, where $B = K^2/2m$ is the maximal binding energy. Nothing can be derived from causality about the behavior of $S_l(p)$ in the neighborhood of this interval. It is, however, believed that for an interaction that is effective only in a region of finite size, one has poles, which must then be simple.[16]

If we translate these properties of $S_l(p)$ into statements about $S_l(E)$ in the energy plane, we can summarize our results, which are principally a consequence of the causality condition, in the following way (Figures 5.2 and 5.4): $S_l(E)$ is a meromorphic function on the two-sheeted Riemann surface with a branch point at $E = 0$ and a cut from 0 to ∞. The physical values of E in collision processes lie on the upper edge of the cut on the "physical sheet." Bound-state poles lie on the negative real axis of the physical sheet, and except for these bound-state poles $S_l(E)$ is an analytic function on the physical sheet. Further poles (of any order) may lie on the second, "unphysical" sheet, coming from possible zeros on the first sheet. Poles at various locations on the "unphysical" sheet have various physical interpretations. Poles on the negative real axis of the unphysical sheet [coming from zeros of $S_l(p)$ on the positive imaginary axis] are called virtual-state poles. A virtual state is one that would be bound if the interaction were more attractive and a virtual state close to threshold causes a large cross section at low energy; we will discuss such states in Section XVIII.7. Poles of $S_l(E)$ on the unphysical sheet, if they are close to the positive real axis, are of particular

---

[16] If the interaction is described by a potential $V(r)$ that decreases faster than any exponential for $r \to \infty$, then $S_l(p)$ is meromorphic in the whole p-plane and analytic in the upper half plane except for a finite number of simple poles. On the other hand, for a Yukawa-type potential $V(r) = \int_m^\infty \sigma(\mu)e^{-\mu r}/r \, d\mu, \ m > 0$, one has, in addition to possible poles of $S_l(p)$, also cuts along the imaginary axis from $im/2$ to $i\infty$ and from $-im/2$ to $-i\infty$.

importance. They are called resonance poles or Siegert poles, and we shall study them in the remainder of this book. Each pole in the second sheet below the real axis has a counterpart in the second sheet above the real axis. As we shall see in Chapter XXI, the pole below the real axis describes a decaying state and the pole above the real axis describes a forming (capture) state.

XVIII.6   Resonance Scattering—Breit–Wigner Formula for Elastic Scattering

Experimentally, resonances are usually associated with a sharp variation of the cross section as a function of energy. If the elastic or inelastic cross sections exhibit sharp maxima or minima, one says that a resonance has occurred. From (XVI.2.9) for the elastic lth partial cross section one sees that a maximum occurs either at the energies for which $\delta_l(E) = \pi/2$ (modulo π) or for which $\delta_l(E)$ has a maximum. The latter possibility, however, cannot lead to a sharp maximum, because according to the causality condition (3.15), the phase shift can never decrease with a steep slope. Thus a sharp maximum in the lth partial cross section occurs at those value of E for which $\delta_l(E) = \pi/2$ (modulo π); and the more rapidly $\delta_l(E)$ increases by π, the sharper is the maximum. Similarly, a minimum for the elastic lth partial cross section occurs for $\delta_l(E) = \pi$ (modulo π). The connection between the change in cross sections and the change in phase shifts for these two cases and for some intermediate cases is shown in Figure 6.1. Note that as a consequence of the causality condition (3.15), $\delta_l(E)$ must be mainly increasing; if it decreases, this decrease can be only very gradually. In every case a sharp structure in the cross section $\sigma_l(E)$ is connected with a sharp increase of the phase shift $\delta_l(E)$ by π.

We therefore want to take as the preliminary definition of a resonance of angular momentum l at the energy E_0 that

$$\delta_l(E) \text{ increases rapidly by approximately } \pi \text{ when } E \text{ passes through } E_0. \quad (6.1)$$

If there is a sharp increase of one δ_l by π around E_0, and if all other phase shifts are constant or slowly varying around E_0, then rapid variations will also show up in the total cross section $\sigma = \sum \sigma_l$ or the differential cross section at certain angles. However, not all structures in the cross section should be ascribed to resonances. Resonances can usually be distinguished from nonresonant phenomena in the cross sections by the fact that resonances appear only in the single partial wave, while nonresonant phenomena are the result of cooperative contributions from many partial waves.

We now want to explain the association of resonances and quasistationary states that we have already alluded to in the introduction (Section XVIII.1). According to our considerations in Section XVIII.2, a quasistationary state occurs at those values of $E = E_0$ for which $d\delta_l(E)/dE$ has a sharp

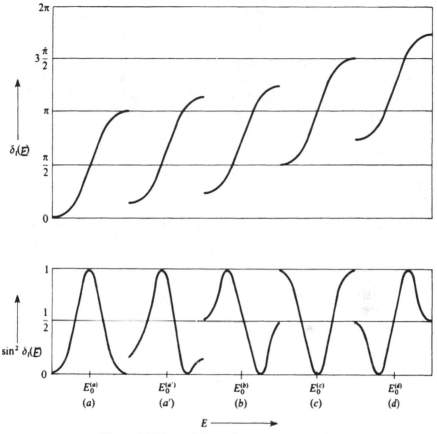

Figure 6.1 Phase shifts and resonance profiles.

maximum. We therefore take as the definition of a quasistationary state of angular momentum l at the energy E_0 that

$$\frac{d^2\delta_l(E)}{dE^2}\bigg|_{E=E_0} = 0, \quad \frac{d^3\delta_l(E)}{dE^3}\bigg|_{E=E_0} < 0; \quad \frac{1}{E_0^2}\bigg|\frac{d\delta_l(E)}{dE}\bigg|_{E_0} \ll \bigg|\frac{d^3\delta_l(E)}{dE^3}\bigg|_{E_0}. \quad (6.2)$$

Whereas the first two conditions of (6.2) state that E_0 is a maximum of $d\delta_l(E)/dE$, the last condition of (6.2) states that this maximum is to be very sharp. The first part of the conditions (6.2) is already incorporated in the drawings of the phase shifts in Figure 6.1, where the point E_0 around which $\delta_l(E)$ changes rapidly is a point of inflection. However, it is not clear that the two phenomena characterized by (6.1) and (6.2) are related. We shall show in this section that as a consequence of very general physical assumptions (causality), these two phenomena are indeed connected: that a quasistationary state leads to a resonance and that a resonance leads to a quasistationary state. We will see that (6.2) leads to a pole in the lth partial S-matrix below and near to the positive real E-axis. And we shall obtain from (6.2) a representation of $S_l(E)$, valid in the neighborhood of E_0, from which (6.1) follows immediately.[17] The notion of resonance poles was introduced in Section XVIII.5, when we described the analyticity property of the S-matrix. We argued there that $S_l(E)$ can have poles anywhere in the second sheet, which came in pairs unless they lie on the negative real axis, and we mentioned that they are called resonance poles if they are close to the positive real axis. Here we show that they appear as a consequence of the existence of a quasistationary state, (6.2), and provide therewith a physical explanation of their existence.

To start our derivation we introduce the function

$$f(E) = \left(\frac{d\delta_l(E)}{dE}\right)^{-1}, \tag{6.3}$$

which as a consequence of (6.2) has a sharp minimum at $E = E_0$, so that $f' = f'(E)_{E=E_0} = 0$ and $f'' = f''(E)_{E=E_0}$ is large. More precisely, a simple calculation shows that

$$f'' = -\left(\frac{d\delta_l}{dE}(E_0)\right)^{-2}\frac{d^3\delta_l}{dE^3}(E_0), \tag{6.4}$$

and consequently,

$$\sqrt{\frac{2f}{E_0^2 f''}} = \sqrt{-\frac{2}{E_0^2}\frac{d\delta_l}{dE}(E_0)\bigg/\frac{d^3\delta_l}{dE^3}(E_0)} \ll 1 \tag{6.5}$$

from the last two inequalities (6.2).

We want to expand $f(E)$ in a Taylor series around $E = E_0$:

$$f(E) = f + \frac{f''}{2}(E - E_0)^2 + F(E - E_0), \tag{6.6}$$

where the "remainder" term is

$$F(E - E_0) = \frac{f'''}{3!}(E - E_0)^3 + \cdots + \frac{f^{(n)}}{n!}(E - E_0)^n + \cdots \tag{6.7}$$

The Taylor series converges if for all n larger than a certain value N

$$f^{(n+1)} < f^{(n)}\frac{n+1}{(E - E_0)}, \tag{6.8}$$

certainly not too stringent a condition.

We shall first assume that $F(E - E_0)$ is negligibly small $[F(E - E_0) = 0]$ in a certain energy range around E_0; this would be the case if $\delta_l^{(4)}$ and the higher derivatives of the phase shifts at E_0 could be neglected. In a second step we will discuss the corrections that arise from a small $F(E - E_0)$.

---

[17] This derivation is based on Goldberger and Watson (1964), Section 8.5.

With $F(E - E_0) = 0$ it follows from (6.3) and (6.6) that

$$\delta_l(E) = \int \frac{dE}{f + f''/2(E - E_0)^2}$$

$$= \sqrt{\frac{2}{ff''}} \arctan\left(\frac{E - E_0}{\sqrt{2f/f''}}\right) + \gamma_l'$$

$$= -\sqrt{\frac{2}{ff''}} \operatorname{arccot}\left(\frac{E - E_0}{\sqrt{2f/f''}}\right) + \gamma_l \tag{6.9}$$

$$\delta_l(E) = -\sqrt{\frac{2}{ff''}} \arctan\left(\frac{\sqrt{2f/f''}}{E - E_0}\right) + \gamma_l, \tag{6.10}$$

where γ_l' and γ_l are arbitrary integration constants,

$$\gamma_l - \gamma_l' = \sqrt{(2/ff'')}\pi/2.$$

Equation (6.10) suggests the introduction of the two new parameters:

$$r = \sqrt{\frac{2}{ff''}} = \sqrt{-\frac{2(\delta_l^{(1)})^3}{\delta_l^{(3)}}} \tag{6.11}$$

and

$$\Gamma = 2\sqrt{\frac{2f}{f''}} = 2\sqrt{-\frac{2\delta_l^{(1)}}{\delta_l^{(3)}}}. \tag{6.12}$$

r is dimensionless, and Γ has the dimension of energy. As a consequence of (6.5) (which followed from the requirement that there exist a quasistationary state at E_0), Γ fulfills the condition

$$\Gamma \ll 2E_0. \tag{6.13}$$

Inserting (6.11) and (6.12) into (6.10), one obtains for the lth partial S-matrix element

$$S_l(E) = e^{i2\delta_l(E)} = e^{i2r \arctan((\Gamma/2)/(E_0 - E))} e^{i2\gamma_l} \tag{6.14}$$

The phase shift (6.10) is written as a sum of two terms:

$$\delta_l(E) = \delta_l^{(R)}(E) + \gamma_l, \tag{6.15}$$

where $\delta_l^{(R)}(E)$ is a rapidly varying function due to (6.13):

$$\delta_l^{(R)}(E) = r \arctan \frac{\Gamma/2}{E_0 - E}, \tag{6.16}$$

whose value changes by almost $r\pi$ when E passes through E_0. γ_l is a constant under the assumption $F(E - E_0) = 0$.

Figure 6.2 Change of phase shift at a quasistationary state.

If $F(E - E_0)$ is not zero, then one obtains instead of (6.9):

$$\delta_l(E) = \int \frac{dE}{f + (f''/2)(E - E_0)^2 + F(E - E_0)} = \int \frac{dE}{f + (f''/2)(E - E_0)^2}$$

$$- \int \frac{dE\, F(E - E_0)}{[f + (f''/2)(E - E_0)^2][f + f''/2(E - E_0)^2 + F]}. \qquad (6.17)$$

Writing $\delta_l(E)$ again in the form (6.15), we obtain for γ_l

$$\gamma_l = \gamma_l(E) = - \int \frac{dE\, F(E - E_0)}{[f + (f''/2)(E - E_0)^2][f + (f''/2)(E - E_0)^2 + F]}, \qquad (6.18)$$

which for small $F(E - E_0)$ is a slowly varying function of E.

Thus the phase shift in the neighborhood of a quasistationary state defined by the requirements (6.2) is the sum of a term $\delta_l^{(R)}(E)$, which changes rapidly by almost πr, and a slowly varying term $\gamma_l(E)$:

$$\delta_l(E) = \delta_l^{(R)}(E) + \gamma_l(E), \qquad (6.19)$$

This is depicted in Figure 6.2. The quantity $\gamma_l(E)$ is called the background phase shift, or also the potential part[18] of the phase shift, as opposed to $\delta_l^{(R)}(E)$, which is called the resonant part. Thus, if $r = 1$, we have derived the condition (6.1) as a consequence of the condition (6.2). In particular, if the background phase shift is zero, we have the situation depicted in Figure 6.1(a).

We should now, therefore, ask what values the parameter r can take. In order to answer this question, we use the analyticity property (4.10).[19] The function

$$\arctan \frac{\Gamma/2}{E_0 - E} = \frac{i}{2}[\ln(E - E_0 + i\Gamma/2) - \ln(E - E_0 - i\Gamma/2)] \qquad (6.20)$$

---

[18] If the interaction is described by a potential, then γ_l comes from the long-range component of this potential, whereas δ_l^R comes from a short-range attractive component.

[19] For the present discussion not all of the analyticity (4.10) is required; one just needs that $S_l(E)$ and, therewith, $T_l(E)$ are analytic in a domain above the real positive energy axis that contains the point $E = E_0 + i\Gamma/2$, a very weak condition indeed.

has a branch point of infinite order at $E = E_0 + i\Gamma/2$ (and also one at $E = E_0 - i\Gamma/2$).[20] If E moves from the real axis around this branch point and returns to the initial point on the real axis, always staying in the domain of analyticity, then $\ln(E - E_0 - i\Gamma/2)$ increases by $2\pi i$, and $\arctan[(\Gamma/2)/(E_0 - E)]$ by π. Thus moving along this closed contour around the point $E_0 + i\Gamma/2$ results, according to (6.14), in the change

$$S_l(E) \to S_l(E)e^{2\pi ir}. \tag{6.21}$$

On the other hand, since $S_l(E)$ is analytic in this domain, its value should not change. This is only possible if

$$r = 0, 1, 2, \ldots. \tag{6.22}$$

We shall now discuss the simplest nontrivial case, $r = 1$. From (6.14) and (6.15) one then obtains

$$S_l(E) = \frac{E_0 + i\Gamma/2 - E}{E_0 - i\Gamma/2 - E} e^{i2\gamma_l} = \left(1 + \frac{i\Gamma}{E_0 - i\Gamma/2 - E}\right) e^{i2\gamma_l}$$

$$= e^{i2\delta_l^{(R)}(E)} e^{i2\gamma_l}. \tag{6.23}$$

This shows that the S-matrix has a simple pole at the value $E = E_0 - i\Gamma/2$, which is called a Siegert pole. For $r = -1$ we obtain another representation of the S-matrix

$$S_l(E) = \frac{E_0 - i\Gamma/2 - E}{E_0 + i\Gamma/2 - E} e^{i2\gamma_l} = \left(1 + \frac{-i\Gamma}{E_0 + i\Gamma/2 - E}\right) e^{i2\gamma_l} \tag{6.23'}$$

which has a pole at $E = E_0 + i\Gamma/2$. These two poles are one pair of the resonance poles mentioned in Section XVIII.5. There we showed that resonance poles come in pairs; (6.23) is the representation of the S-matrix most suitable for energies in the neighborhood of the pole below the real energy axis and (6.23') is the representation most suitable for energies in the neighborhood of the pole above the real axis.

We shall at the moment consider the case $r = 1$. From (6.16) follows:

$$\tan \delta_l^R(E) = \frac{\Gamma/2}{E_0 - E}. \tag{6.24}$$

The S-matrix element (6.23) is then written

$$S_l(E) = 1 + 2ipT_l(E) = e^{i2\gamma_l}(1 + 2ipT_l^{(R)}(E)), \tag{6.25}$$

where one has defined the resonant partial-wave amplitude $T_l^{(R)}(E)$ in such a way that it agrees with the partial-wave amplitude $T_l(E)$ if the background can be ignored ($\gamma_l = 0$). From comparison of (6.25) and (6.23) one then obtains for the resonant partial-wave amplitude

$$T_l^{(R)}(E) = \frac{1}{p} \frac{\Gamma/2}{E_0 - i\Gamma/2 - E}. \tag{6.26}$$

[20] Cf. Appendix XVII.A.

In conclusion we have shown the following: if a quasistationary state with angular momentum l is formed at the energy E_0, then the phase shift has the resonant behavior specified by (6.1), and—if the background can be neglected—the partial-wave amplitude is given for the simplest case $r = 1$ by (6.26).

The lth partial cross section that one obtains from (6.26), which is therefore the pure resonance cross section (any background being neglected), is then given [using (XVI.1.38)] by

$$\sigma_l^R(E) = \frac{4\pi}{p^2}\,(2l + 1)\,\frac{(\Gamma/2)^2}{(E_0 - E)^2 + (\Gamma/2)^2}. \tag{6.27}$$

This function of E is shown in Figure 6.3 together with the corresponding resonance phase shift. We see that the resonance cross section has a maximum at E_0, and the full width at half maximum of this peak is Γ. E_0 is called the *resonance energy*, and Γ is called the *resonance width*.

At the resonance energy the cross section takes the value

$$\sigma_l^{(R)} = \frac{4\pi}{p^2}(2l + 1), \tag{6.28}$$

which is the maximum value for the elastic partial cross section compatible with unitarity, as seen from (XVI.2.9).

The quantity Γ that has emerged in these discussions can be calculated from the phase shifts using (6.12) if these are known, e.g., from the interaction potential in potential scattering. However, as mentioned already in several

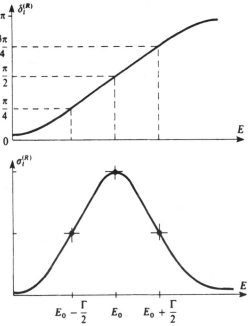

Figure 6.3 The Breit–Wigner cross section and its relation to the resonance phase shift.

places, the situation is more often than not the reverse: One knows the cross section from an experiment and notices that it shows, in the neighborhood of a certain value E_0, the features shown in the lower part of Figure 6.1. Then one tries to fit this experimental cross section with a formula

$$\sigma_l^{\text{exp}} = 4\pi(2l + 1)|T_l^R(E) + T^{\text{bg}}(E)|^2, \qquad (6.29)$$

where $T_l^R(E)$ is the function (6.26) and $T^{\text{bg}}(E)$ is a slowly varying background term. In this way one determines the values (E_0, Γ).

Equation (6.27) is the celebrated Breit-Wigner formula; (6.26) is called the Breit-Wigner amplitude. It hardly ever gives a completely accurate description of the experimental situation, which indeed one would not expect in view of the idealizations involved. It almost never fails to give a useful parametrization whenever resonance phenomena are involved. It is, perhaps, the most frequently used formula of quantum physics. Its fundamental significance comes from its introduction of the parameter Γ: On the level of accuracy at which excited states are considered stationary, they are characterized by one parameter,[21] their energy value E_0; on the level of accuracy at which they are considered *quasistationary states*, they *are characterized by two parameters, their energy and width* (E_0, Γ). From Chapter XII we know that stationary states are described by eigenvectors of the energy operator with eigenvalue E_0; in Chapter XXI we will see that, analogously, quasistationary states are described by generalized eigenvectors of the [essentially self-adjoint] energy operator with eigenvalue $E_0 + i\Gamma/2$.

Like every theoretical description, the description of the scattering amplitude by a Breit-Wigner function (6.26) is an approximate description. In a realistic scattering process there is always a background or even one or more additional resonances as we shall discuss below.

We now want to establish the connection between resonances and poles of S_l. We have already introduced, in Section XVIII.5, the name resonance pole for a pole of $S_l(E)$ in the second sheet of the energy plane immediately below the positive real axis. As seen, e.g., from (6.23), the Breit-Wigner formula furnishes such a pole in the lower half plane at $E = E_0 - i\Gamma/2$. The smaller the value of Γ, and consequently the sharper the resonance, the closer this pole is to the real axis. We have already mentioned that there is no pole on the lower half of the first sheet of the energy plane.[22] Consequently, this pole furnished by the Breit-Wigner formula must be on the second sheet of the plane, and thus there must be a pole of $S_l(p)$ in the fourth quadrant of the p-plane close to the real axis.

---

[21] In addition to their other quantum numbers like l, η.

[22] We did not give a complete proof of the statements in Section XVIII.5, but argued that causality, the finiteness of the binding energies for possible bound states, and the finite range of the interaction are sufficient, though certainly much less is necessary.

On the other hand, every simple pole of $S_l(p)$ at the point $p = p_R = k_R - iK_R$ in the fourth quadrant near the real axis ($K_R \ll k_R$) may, but does not necessarily, lead to the resonance phenomenon. This is easily seen by expanding $S_l(p)$ around this pole in a Laurent series. In a sufficiently small neighborhood of the pole, $S_l(p)$ may be approximated by the principal part of the Laurent series $\sim 1/(p - p_R)$. However, $S_l(p)$ must be unitary $[|S_l(p)| = 1]$, so the factor multiplying $1/(p - p_R)$ must be chosen to be $p - p_R^*$ if the rest of the series is to be expressed as an exponential $e^{i2\gamma_l}$. Thus in the neighborhood of the pole the expansion of $S_l(p)$ compatible with unitarity is

$$S_l(p) = e^{i2\delta_l(p)} \approx e^{i2\gamma_l(p)} \frac{p - p_R^*}{p - p_R} = e^{i2\gamma_l(p)} \frac{p - k_R + iK_R}{p - k_R - iK_R}. \quad (6.30)$$

Multiply denominator and numerator by p/m and note that

$$p(p - k_R)\frac{1}{m} = \frac{p^2}{m} - \frac{pk_R}{m} = \frac{p^2}{2m} - \frac{k_R^2}{2m} + \frac{(p - k_R)^2}{2m} \approx \frac{p^2}{2m} - \frac{k_R^2}{2m} + \frac{K_R^2}{2m}$$

$$(6.31)$$

for real values of p that differ from k_R by approximately $K_R \ll k_R$, i.e., for values of p that fulfill $|p - k_R| \approx K_R$. Then with

$$E = \frac{p^2}{2m}, \qquad E_R = \frac{1}{2m}(k_R^2 - K_R^2), \quad (6.32)$$

one obtains

$$S_l(E) = e^{i2\gamma_l(E)} \frac{E - E_R - i\Upsilon/2}{E - E_R + i\Upsilon/2} \quad \text{with} \quad \Upsilon = -\frac{2K_R p}{m} = \Upsilon(p). \quad (6.33)$$

With the same accuracy as the approximate equality (6.31),

$$\Upsilon(p) \approx -\frac{2K_R k_R}{m} \equiv \Gamma. \quad (6.34)$$

Then (6.33) is identical with the resonance formula (6.23), and consequently a pole of $S_l(p)$ at p_R describes a resonance.

The above considerations would suggest a one-to-one correspondence between poles of the S-matrix immediately below the positive real axis and a resonance phenomenon characterized by (6.1); however, they do not prove such a correspondence, because we have always assumed that the γ_l is a slowly varying background phase shift. If γ_l varies rapidly without having the Breit–Wigner form (6.24), then it may compensate the effect of the pole, and the total phase shift $\delta_l = \delta_l^{(R)} + \gamma_l$ may be slowly varying. It can, indeed, be shown [23] that phase shifts $\tilde{\gamma}_l$ can be constructed that approach the function (6.24) as close as one wishes in an interval $E_0 - \Delta \le E \le E_0 + \Delta$,

[23] L. Fonda, *Fortschr. Phys.* **20**, 135 (1972).

$\Delta \gg \Gamma$, and that therefore describe a resonance phenomenon (6.1), but that are not associated with a singularity of $e^{i2\tilde{\gamma}_l}$. Thus choosing $S_l = e^{i2\tilde{\gamma}_l}$ gives an S-matrix with a resonance phenomenon but without a pole, and choosing $\gamma_l = -\tilde{\gamma}_l$ in (6.33) gives an S-matrix with a pole but without a resonance phenomenon.

However, if the fourth and higher derivatives of δ_l are small, then, as we saw in the first part of this section, γ_l will be slowly varying. Therefore, one generally assumes the *one-to-one correspondence between poles immediately below the real axis and resonances*.

We also have seen above that a quasistationary state characterized by a sharp maximum in the time delay is a resonance. Further, as from (6.19), (6.24), and (2.26) it follows immediately that

$$\tfrac{1}{2}t_D^l(E) = \frac{d\delta_l(E)}{dE} = \frac{d\delta_l^{(R)}}{dE} + \frac{d\gamma_l}{dE} = \frac{\Gamma/2}{(E_0 - E)^2 + \Gamma^2/4} + \frac{d\gamma_l(E)}{dE}, \quad (6.35)$$

we see that a resonance leads to a sharp maximum in the time delay for the resonating partial wave. Thus *a quasistationary state, a resonance, and a pole of S_l immediately below the real axis are in fact one and the same phenomenon.*

We shall now discuss the connection between the width Γ and the time delay. According to the discussions in Section XVIII.2, (6.35) is the "time delay for a monoenergetic beam;" that means the energy spread ΔE of the beam is much smaller than the width of any structure in the cross section, $\Delta E \ll \Gamma$. If a resonance occurs, then the derivative of the background phase shift is negligible, so that the resonant time delay is well approximated by

$$\tfrac{1}{2}t_D^l(E) = \frac{d\delta_l^{(R)}}{dE} = \frac{\Gamma/2}{(E_0 - E)^2 + \Gamma^2/4}. \quad (6.35')$$

The "time delay for a monoenergetic beam" that has energy $E = E_0$ identical to the resonance energy is therefore

$$t_D^l(E)|_{E=E_0} = \frac{4}{\Gamma}. \quad (6.36)$$

This idealized time delay is not really a physical quantity, because the monoenergetic beam is a steady state and the time delay is not an observable quantity, as mentioned already in Section XVIII.2.

In the opposite extreme, $\Delta E \gg \Gamma$, the time delay is essentially determined by the experimental energy distribution of the beam, as we see by inserting (6.35) into (2.32):

$$t_D^l = 2\int dE\, F(E - E_A)\frac{\Gamma/2}{(E_0 - E)^2 + \Gamma^2/4}. \quad (6.37)$$

Noting that for $\Delta E \gg \Gamma$ one can take $\Gamma/2 \to 0$ in the integral, and using

$$\frac{\Gamma/2}{(E_0 - E)^2 + \Gamma^2/4} \to \pi\delta(E_0 - E) \quad \text{for} \quad \Gamma \to 0,$$

we obtain

$$t_D^l = 2\pi F(E_0 - E_A).$$
(6.38)

For a Gaussian energy distribution of the incident beam,[24] with $\Delta E \gg \Gamma$,

$$F(E - E_A) = \frac{1}{\sqrt{2\pi}} \frac{1}{\Delta E} e^{-(E - E_A)^2/2(\Delta E)^2},$$
(6.39)

one obtains at the resonance energy

$$t_D^l|_{E_A = E_0} = \sqrt{2\pi} \frac{1}{\Delta E}.$$
(6.40)

And for a Lorentzian energy distribution of the incident beam,[24]

$$F(E - E_A) = \frac{1}{\pi} \frac{\Delta E/2}{(E_A - E)^2 + (\Delta E/2)^2},$$
(6.41)

with $\Delta E \gg \Gamma$, one obtains

$$t_D^l|_{E_A = E_0} = \frac{1}{\Delta E}.$$
(6.42)

In general, if the energy distribution in the incident beam caused by the resolution of the apparatus has a width ΔE of the same order as the resonance width Γ, one obtains a time delay that depends on Γ and on the experimental energy distribution. A particular situation arises when the experimental energy distribution is Lorentzian [Equation (6.41)] with an energy spread that is equal to the resonance width $\Delta E = \Gamma$. Then one obtains by inserting (6.41) into (6.37)

$$t_D^l = 2/\Gamma \quad (= 2\hbar/\Gamma \text{ in conventional units of time}).$$
(6.43)

Thus we have seen that the time delay has something to do with the inverse of the resonance width.[25] For an idealized monochromatic beam this relation is given by (6.36). But for more realistic situations the effect of the energy resolution of the incident beam will result in other connections between time delay and width—e.g., the one given in (6.43).

So far we have considered for the parameter r, which we introduced in (6.10), only the value $r = 1$. However, according to (6.22) it can take any integer value.

The case $r = 1$ seems to be the only case that is realized in quantum-scattering experiments, though some time ago there was thought to exist in

---

[24] The normalization of F has to be chosen so that $F(E - E_A) \to \delta(E - E_A)$ as $\Delta E \to 0$ according to (XIV.5.36). Note that F is the energy distribution of the incident noninteracting beam at $t = 0$, i.e., the beam has been prepared in the remote past in such a way that an energy measurement at $t = 0$ with no interaction present would give the result $F(E - E_A)$ for the probability distribution to measure E.

[25] As we shall see in Chapter XXI, the inverse resonance width is the lifetime of a decaying state.

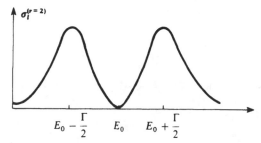

Figure 6.4 Illustration of the energy dependence of the scattering cross section (6.45) for a dipole resonance.

particle physics some evidence for the existence of a "dipole," which is a quasistationary state with $r = 2$. We shall discuss this case briefly here. With $r = 2$ we obtain from (6.14), ignoring the background ($\gamma_l = 0$),

$$S_l(E) = \left(\frac{E_0 - E + i\Gamma/2}{E_0 - E - i\Gamma/2}\right)^2 \qquad (6.44)$$

and therewith for the lth partial cross section

$$\sigma_l^{(r=2)}(E) = \frac{4\pi(2l + 1)}{p^2} \frac{\Gamma^2(E - E_0)^2}{[(E - E_0)^2 - \Gamma^2/4]^2 + \Gamma^2(E - E_0)^2}. \qquad (6.45)$$

The shape of the cross section for such a dipole resonance is depicted in Figure 6.4.

This is to be compared with Figure 6.1(a). When background scattering is present, the cross section will be modified accordingly.

XVIII.7 The Physical Effect of a Virtual State

Besides the bound-state poles and the resonance poles, we mentioned in Section XVIII.5 the virtual-state poles on the negative imaginary axis of the momentum plane. If a virtual-state pole is sufficiently close to the real axis in the momentum plane, it can have an appreciable influence on the cross section at low energies.

Denoting the momentum value of the virtual state pole by $p_v = -i\kappa$, $\kappa > 0$, we can write the $l = 0$ S-matrix element in the neighborhood of the pole as

$$e^{2i\delta_0(p)} = S_0(p) = -\frac{p - p_v^*}{p - p_v} = -\frac{p - ik}{p + ik}. \qquad (7.1)$$

This follows by arguments similar to those leading to (6.30): The principal part of the Laurent series $\sim 1/(p - p_v)$ has to be multiplied by a factor $(p - p_v^*)$ such that $S_0(p)$ is unitary, i.e., $|S_0(p)| = 1$. The remaining phase factor has to be chosen (-1) so that

$$\delta_0(p) = \frac{1}{2i} \ln \frac{ik - p}{ik + p}$$

fulfills the causality condition (3.15). From (7.1) follows

$$\frac{d\delta_0(p)}{dp} = \frac{k}{p^2 + k^2} > 0 > -\left(R + \frac{1}{2p}\right),$$

whereas the choice of the opposite sign in (7.1) would have lead to

$$\frac{d\delta_0(p)}{dp} = \frac{-k}{p^2 + k^2}$$

which can not always fulfill (3.15).

The 0th partial-wave amplitude is obtained by inserting (7.1) into (XVI.2.8):

$$T_0(p) = \frac{-1}{ip - k}. \tag{7.2}$$

The 0th partial cross section is then

$$\sigma_0 = 4\pi|T_0|^2 \approx \frac{4\pi}{\kappa^2 + p^2}. \tag{7.3}$$

In the limit of zero scattering energy, when the virtual pole is close to the real axis, this goes into

$$\sigma_0 \approx \frac{4\pi}{\kappa^2}, \tag{7.3a}$$

which is identical to the total cross section since, according to (XVII.6.6), higher partial waves can be ignored at low energies. Thus we see that for a virtual-state pole close to the real axis the scattering cross section can be extremely large. Also large is the scattering length, obtained according to (XVII.6.10) from (7.2) as:

$$a_0 = -\frac{1}{\kappa}. \tag{7.4}$$

It is easy to see that a bound state with sufficiently low binding energy will also lead to (7.3) for the cross section. Therefore, a virtual state can only be detected in the low-energy scattering cross section if low-energy bound states are absent.

Virtual states have been observed in nature. In the case of proton-neutron scattering there exists a bound state, the deuteron (isospin 0, angular momentum 1). The scattering length is $a_0^{(0, 1)} = 5.4 \times 10^{-13}$ cm. For angular momentum equal to zero (isospin 1), no bound state is present, but $a_0^{(1, 0)} = -23.7 \times 10^{-13}$ cm. Thus, there is a virtual state in the singlet (angular momentum 0) p-n system, whose position is closer to the threshold than the triple bound state and which dominates low-energy scattering.

Virtual states also cause long time delay. As we are at low energies, the phase shift can be obtained from (cf. XVII.6.15)

$$p \cot \delta_0(p) = -\frac{1}{a_0}. \tag{7.5}$$

From this we obtain

$$\frac{d\delta_0}{dp} = -\frac{a_0}{1 + p^2 a_0^2} = \frac{1}{\kappa} \frac{1}{1 + p^2/\kappa^2}.$$

Thus the average time delay (2.26) for a virtual state with scattering length (7.4),

$$t_D^{(virt)} = 2\frac{m}{p}\frac{1}{\kappa}\frac{1}{1 + p^2/\kappa^2}, \tag{7.6}$$

can reach very high values. An illustration of this effect is obtained if we compare the time delay with the time that a noninteracting particle moving with a velocity p/m spends in the region of radius R, which is given by (cf. Section XVIII.3):

$$T^{cl} = \frac{m \cdot 2R}{p} \tag{7.5}$$

Therefore the ratio of the time delay caused by a virtual state to this classical time delay is

$$\frac{t_D^{virt}}{T^{cl}} = \frac{1}{R\kappa}\frac{1}{1 + p^2/\kappa^2} = -\frac{a_0}{R}\frac{1}{1 + p^2 a_0^2}. \tag{7.7}$$

Depending upon the value of κ and R, this ratio can be very high. E.g., for the p-n virtual state, taking $R \approx 2.5 \times 10^{-13}$ cm and $a_0 \approx -24 \times 10^{-13}$ cm, we see the time delay caused by a virtual state is an order of magnitude larger than the classical time.

A virtual state is thus not simply a fictitious, mathematical state, but a physical state in which projectile and target spend a considerable length of time together.

XVIII.8 Argand Diagrams for Elastic Resonances and Phase-Shift Analysis[26]

The Argand diagram introduced in Section XVI.3 is a useful tool for the detection and display of resonances and the determination of their parameters (E_R, Γ). To understand this we investigate in detail the behavior of the scattering amplitude in the Argand diagram. We first study the idealized case of an elastic resonance without background. For this the scattering amplitude is given by (6.26), which we write in the form

$$pT_l^{(R)}(E) = \frac{1}{\epsilon - i}, \tag{8.1}$$

[26] G. Bialkowski has helped me with the writing of this section.

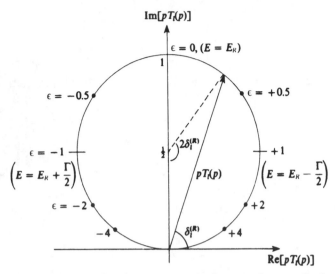

Figure 8.1 Argand diagram for the resonant elastic scattering amplitude: $pT_l(p) = (\epsilon - i)^{-1}$ and $\cot \delta_l^{(R)} = \epsilon$.

where we have introduced the abbreviation

$$\epsilon = (E_R - E)\frac{2}{\Gamma} = \cot \delta_l^{(R)}(E). \tag{8.2}$$

Figure 8.1 shows the Argand diagram for Equation (8.1). As we see, this elastic Breit–Wigner amplitude lies on the unitary circle. The top of the circle corresponds to the resonant energy $\epsilon = 0$, at which value the phase shift $\delta_l(E_R)$ is $\pi/2$ (cf. Figure XVI.3.1). With increasing energy E, pT_l^R moves in a counterclockwise sense—slowly along the lower part of the circle, and more rapidly as E gets closer to the resonance value E_R. This is indicated by the values for ϵ on the path of pT_l^R in Figure 8.1.

To take the background into account we use (6.25) and obtain

$$pT_l = e^{i\gamma_l} \sin \gamma_l + e^{2i\gamma_l} pT_l^{(R)} = pT_l^{(bg)} + e^{2i\gamma_l} pT_l^{(R)}, \tag{8.3}$$

where we have defined the background amplitude $T_l^{(bg)}$ in analogy to (XVI.2.8) by

$$T_l^{(bg)} = \frac{1}{p} e^{i\gamma_l} \sin \gamma_l = \frac{1}{2ip}(e^{i2\gamma_l} - 1). \tag{8.4}$$

From (8.3) we see that the effect of the background is to rotate the resonant amplitude by the angle $2\gamma_l$. The quantity pT_l for constant background amplitude is shown in Figure 8.2. The resonance circle starts at the point B and traverses the unitary circle, the only difference with the case of Figure 8.1 being that now the resonance energy is not at the point where the resonance amplitude is pure imaginary, but at the point diametrically opposite B. If the background changes with energy, then the point of resonance is not well defined.

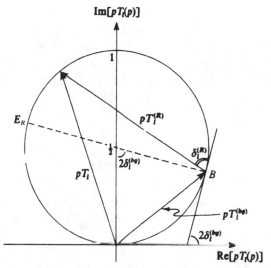

Figure 8.2 Argand diagram for the resonant elastic scattering amplitude in the presence of background: $T_l = T_l^{(R)} + T_l^{(bg)}$.

As we have seen above, the resonance amplitude has a characteristic circular path in the Argand diagram, the resonance energy being the value where the phase shift changes most rapidly. As we shall see in Section XX.3, this feature will remain even when inelasticity is taken into account, only there the circles become smaller and distorted.

The Argand diagram can be used to display the energy and width of a resonance for a particular partial wave. For this purpose one has just to draw pT_l as function of energy in the Argand diagram. If pT_l runs along a circle close to the unitary circle, one knows that this particular partial wave has an elastic resonance, and the energy value at which the amplitude has its most rapid change will be the resonance energy E_R. If there is no background, E_R will be the point at which pT_l crosses the imaginary axis. In order to draw $pT_l(E)$ one has to determine it from the experimental data, and this is the main difficulty of phase-shift analysis. In the following we will give a brief description of this procedure.

The data determined in an experiment are the cross sections: the total cross section at various momenta $\sigma(p)$ and the differential cross sections $d\sigma/d\Omega(\theta, p)$ at various angles and momenta (or the differential cross sections $d\sigma/d\cos\theta = 2\pi \, d\sigma/d\Omega$). According to (XVI.1.36) these cross sections are connected to the partial-wave amplitudes by

$$\frac{d\sigma}{d\cos\theta}(p,\cos\theta) = 2\pi \left| \sum_{l=0}^{\infty} (2l+1)P_l(\cos\theta)T_l(p) \right|^2. \tag{8.5}$$

And by the optical theorem (XVI.2.20),

$$\sigma(p) = \frac{4\pi}{p} \sum_{l=0}^{\infty} (2l+1)\operatorname{Im} T_l(p). \tag{8.6}$$

These sums extend to infinity and would be of little practical use for the determination of the partial-wave amplitude $T_l(p)$. Fortunately, however, according to (XVII.6.6) the higher partial-wave amplitudes for a particular value of the momentum p can be neglected, so that for a particular value of p the sum in (8.5) extends in fact only over a finite number of terms:

$$\frac{d\sigma}{d\cos\theta}(p,\cos\theta) = 2\pi\left|\sum_{l=0}^{L}(2l+1)P_l(\cos\theta)T_l(p)\right|^2, \qquad (8.5')$$

and so does the sum in (8.6). How high a value of L one has to take for a particular value of p depends upon the problem under investigation and has to be determined empirically by the "principle of stability." One does this by choosing some value L_0. Then one determines, by the procedure described below, $T_0(p), \ldots, T_{L_0}(p)$; chooses $L_0 + 1$; and determines from the same set of data a new set of amplitudes $\tilde{T}_0(p), \ldots, \tilde{T}_{L_0}(p), \tilde{T}_{L_0+1}(p)$. If T_l and $\tilde{T}_l$ determined in this way differ significantly, L_0 was chosen too small for that particular value of energy and one has to try a higher value.

Using the relation for the Legendre polynomials[27]

$$P_l(\cos\theta)P_{l'}(\cos\theta) = \sum_{L=|l-l'|}^{l+l'}\langle l\,0\,l'\,0|L\,0\rangle^2 P_L(\cos\theta), \qquad (8.7)$$

where $\langle l\,m\,l'\,m'|L, m+m'\rangle$ are the SO(3) Clebsch–Gordan coefficients, one can write (8.5') in the form

$$\frac{d\sigma}{d\cos\theta} = 2\pi\sum_{j=0}^{2L}P_j(\cos\theta)C_j(p). \qquad (8.8)$$

The inversion of (8.8) is[28]

$$C_l(p) = \frac{2l+1}{4\pi}\int_{-1}^{+1}d\cos\theta\,\frac{d\sigma(p,\theta)}{d\cos\theta}P_l(\cos\theta). \qquad (8.8')$$

For the sake of definiteness let us do this for the particular values $L = 1$ and $L = 2$. For $L = 1$, (8.5') becomes

$$\frac{d\sigma}{d\cos\theta} = 2\pi\{|T_0|^2 + 3(T_0 T_1^* + T_0^* T_1)P_1 + 9|T_1|^2 P_1^2\} \qquad (8.9_1)$$

For $L = 2$ it becomes

$$\frac{d\sigma}{d\cos\theta} = 2\pi\{[|T_0|^2 + 3|T_1|^2 + 5|T_2|^2]P_0(\cos\theta)$$

$$+ 6\,\mathrm{Re}(T_0^* T_1 + 2T_1^* T_2)P_1(\cos\theta)$$

$$+ [10\,\mathrm{Re}\,(T_0^* T_2) + 6|T_1|^2 + \tfrac{50}{7}|T_2|^2]P_2(\cos\theta)$$

$$+ 18\,\mathrm{Re}\,(T_1^* T_2)P_3(\cos\theta) + \tfrac{90}{7}|T_2|^2 P_4(\cos\theta)\}, \qquad (8.9_2)$$

---

[27] This is a special case of (VII.3.20).

[28] This is obtained by multiplying (8.8) with $P_l(\cos\theta)$ and using (VII.3.16) with (VII.3.25).

where the numerical factors are determined from the values of $2l + 1$ and of the Clebsch–Gordan coefficients. Thus, for example, for $L = 2$,

$$C_0(p) = |T_0|^2 + 3|T_1|^2 + 5|T_2|^2,$$

$$C_1(p) = 6\mathrm{Re}(T_0^* T_1 + 2T_1^* T_2), \qquad \text{etc.} \tag{8.10}$$

Now if $d\sigma/d\cos\theta$ $(p, \cos\theta)$ has been measured at a particular value of p, then one calculates the $C_l(p)$ from (8.8′). These "experimental" $C_l(p)$ one uses in equations like (8.10) to determine the "experimental" values of the partial-wave amplitudes. There are $2L + 1$ equations (8.10). In addition one has the optical theorem

$$\sigma(p) = \frac{4\pi}{p} \sum_{l=0}^{L} (2l + 1)\, \mathrm{Im}\, T_l(p). \tag{8.6′}$$

So one has $2L + 2$ nonlinear equations for the $2L + 2$ real quantities $\mathrm{Im}\, T_l(p)$, $\mathrm{Re}\, T_l(p)$, $l = 0, 1, \ldots, L$, or the $2L + 2$ quantities $\delta_l(p)$ and $\eta_l(p)$ of (XVI.2.10).

If the scattering at that particular energy is elastic, then one has only the $L + 1$ unknowns $\delta_l(p)$, and one need not consider the $C_j(p)$ for $j > L$. There are then $L + 1$ equations for $L + 1$ unknowns in (8.10). [The optical theorem (8.6′) is then not independent of (8.10).]

One has to go through the above procedure of determining the partial-wave amplitude for every value of p that is plotted in the Argand diagram. For small momenta, small values of L (e.g., $L = 0$) are sufficient, but as the momentum increases, higher values of L have to be chosen. As the equations are nonlinear, the solutions are not unique. Further, the experimental data have errors. The uncertainties that result can be overcome partially by considering $pT_l(p)$ at several values of p and requiring that it change smoothly with momentum. Sometimes further relations, such as dispersion relations, are used to obtain this smoothness. In this way one obtains a picture of pT_l as a function of p.

XVIII.9 Comparison with the Observed Cross Section: The Effect of Background and Finite Energy Resolution

Due to the effect of the background in the resonating partial cross section and the background of other nonresonating partial cross sections, the experimental cross section near resonance does not usually look as simple as in Figure 6.1a. We first consider in detail the interference between the resonance and the background.

If we insert (8.3) into (XVI.1.38), we obtain for the resonant partial cross section

$$\sigma_l = 4\pi(2l + 1)\left| \frac{1}{p} e^{i\gamma_l} \sin\gamma_l + e^{2i\gamma_l} T_l^R \right|^2. \tag{9.1}$$

We introduce the Fano shape parameter[29]

$$q = + \cot \gamma_l, \qquad \epsilon = (E_R - E)\frac{2}{\Gamma} = \cot \delta_l^{(R)}(E). \qquad (9.2)$$

q is called the profile index or shape profile parameter, and ϵ is defined as in (8.2). After some straightforward calculation one obtains for (9.1)

$$\sigma_l = 4\pi(2l + 1)\left[\frac{1}{p^2}\sin^2 \gamma_l + |T_l^R|^2\left(\frac{q^2 + 2\epsilon q - 1}{q^2 + 1}\right)\right]. \qquad (9.3)$$

This is rewritten as

$$\sigma_l = \sigma_l^{(bg)} + \sigma_l^{(R)}\left(\frac{q^2 + 2\epsilon q - 1}{q^2 + 1}\right), \qquad (9.4)$$

where

$$\sigma_l^{(R)} = \frac{4\pi}{p^2}(2l + 1)\frac{1}{\epsilon^2 + 1} \qquad (9.5)$$

is the Breit–Wigner cross section of (6.27), and $\sigma_l^{(bg)}$ has been defined in analogy to (XVI.2.9) by

$$\sigma_l^{(bg)} = \frac{4\pi}{p^2}(2l + 1)\sin^2 \gamma_l. \qquad (9.6)$$

Inserting (9.5) and (9.6) with (9.2) back into (9.4), one can write the resonant partial cross section also as

$$\sigma_l = \frac{4\pi}{p^2}(2l + 1)\frac{(q + \epsilon)^2}{(q^2 + 1)(\epsilon^2 + 1)} = \frac{4\pi}{p^2}(2l + 1)\sin^2\gamma_l\frac{(q + \epsilon)^2}{\epsilon^2 + 1}, \qquad (9.7)$$

or

$$\sigma_l = \sigma_l^{(bg)}\frac{(q + \epsilon)^2}{\epsilon^2 + 1}. \qquad (9.8)$$

The factor $(q + \epsilon)^2/(\epsilon^2 + 1)$, which gives the shape of the resonance, is shown for various values of the shape profile parameter q in Figure 9.1. We see that the shape of the resonance in the resonant partial cross section depends considerably upon the value of the parameter q, which is connected with the background phase shift γ_l. (If the background phase shift depends upon the energy ϵ, which is generally the case, then the shape becomes even more complicated.) For positive values of q we see that for increasing energies E, i.e., for increasing values of $-\epsilon$, the cross section increases above background, then falls sharply at the value of the resonance energy $\epsilon = 0$, is zero at $\epsilon = -q$, and increases thereafter.

For negative values of q the cross section decreases with increasing energy, reaches zero at $\epsilon = q$ (i.e., before resonance), and then increases sharply.

---

[29] The original parameters introduced by U. Fano [*Phys. Rev.* **124**, 1866 (1961)] were the negative of the parameters defined by (9.2).

We are now in a position to explain the behavior of the experimental data depicted in some of the figures of Section XVIII.1. As we have so far only discussed the effect of the background, with the present results we can only interpret those experiments in which the energy distribution in the beam and the resolution of the detector can be ignored. These are the high-resolution experiments whose results are depicted in Figure 1.3 and Figure 1.4(d). These experiments measure the transmitted current, which is proportional to the incident current minus the total cross section. As the total cross section is the sum of the resonant lth partial cross section (9.8) plus the slowly varying background cross section from the other partial waves, a dip in the resonant partial cross section should show up as a peak in the transmitted current, and a peak in the resonant partial cross section as a dip in the transmitted current. According to the results of Figure 9.1, the resonance in

Figure 9.1 The resonance profile function for different values of the profile index q. The arrow above E indicates the direction in which E increases. [From Massey & Burhop (1969), vol. 1, with permission.]

the lth partial cross section should show up in the following way:

For $q > 0$, as a dip around resonance followed by a peak right after resonance.

For $q < 0$, as a peak right before resonance followed by a dip right after resonance.

In the high-resolution experiment (very narrow spread of the electron beam) of Figure 1.4(d), and in particular in the very high-resolution experiment of Figure 1.3, one can see that the situation described for $q < 0$ is observed. (If one knew the exact value for E_R, one could even infer the value of q). Thus our hypothesis of Section XVIII.1, that in the collision process (1.4) a quasistationary state He$^-$ is formed at around 19.31 eV, is justified. We shall see that the other experiments of Figure 1.4 and Figure 1.5 lead to the same conclusion. For these experiments the resolution of the apparatus is coarse, and the quantity measured is not the cross section at a particular energy value, but the total cross section in a whole energy interval.

The total cross section (XVI.1.33) is written as

$$\sigma = \sigma_l + \sum_{l' \neq l} \sigma_{l'} = \sigma_l + \sigma_b, \tag{9.9}$$

where σ_l is the resonant partial cross section given by (9.8) (describing the resonance and the background) and it is assumed that there is no resonance of any angular momentum l' other than l. The background cross section σ_b is then given by

$$\sigma_b = \frac{4\pi}{p^2} \sum_{l' \neq l} (2l' + 1) \sin^2 \gamma_{l'}, \tag{9.10}$$

where $\gamma_{l'}(E)$ are slowly varying background phase shifts.

Inserting (9.10) and (9.8) into (9.9), we obtain for the cross section

$$\sigma = \sigma_l^{(bg)} \frac{(q + \epsilon)^2}{\epsilon^2 + 1} + \sigma_b, \tag{9.11}$$

where $\sigma_l^{(bg)}$ and σ_b vary slowly with energy.

At energies far away from the resonance energy, $\epsilon \to \infty$, we obtain from (9.11)

$$\sigma \to \sigma_l^{(bg)} + \sigma_b \equiv \sigma_{(nr)} \tag{9.12}$$

where we call $\sigma_{(nr)}$ the slowly varying nonresonant background. The maximum of the cross section (9.11) occurs at $\epsilon = +1/q$, and at this maximum the cross section is

$$\sigma_{max} = \sigma_l^{(bg)} + \sigma_b + q^2 \sigma_l^{(bg)} \tag{9.13}$$

Because of (9.12) the cross section (9.11) is conveniently written in the form

$$\sigma = \sigma_l^{(bg)} + \sigma_b + \frac{q^2 + 2q\epsilon - 1}{1 + \epsilon^2} \sigma_l^{(bg)} = \sigma_{(nr)} + \frac{q^2 + 2q\epsilon - 1}{1 + \epsilon^2} \sigma_l^{(bg)}, \tag{9.14}$$

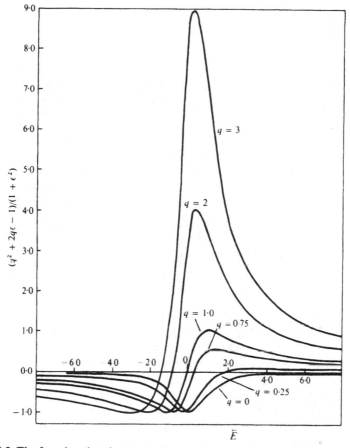

Figure 9.2 The function that determines the form of the resonance effect relative to the smooth background. [Massey & Burhop (1969), vol. 1, with permission.]

which shows that the resonance occurs above a gradually varying background $\sigma_{(\text{nr})}$ and has the characteristic shape determined by the factor $(q^2 + 2q\epsilon - 1)/(1 + \epsilon^2)$. Thus, if there is background from other partial waves, then in distinction to (9.8), the cross section no longer vanishes at $\epsilon = q$ but has its minimum value there, $\sigma_{\min} = \sigma_b$, which lies below the value for the background $\sigma_{(\text{nr})}$. The shape is now determined by the shape factor shown in Figure 9.2.

If $|q| < 1$,

$$\tfrac{1}{4}\pi < \gamma_l < \tfrac{3}{4}\pi,$$

then the resonance effect results in a reduction from the background on one side of the resonance that is greater than the increase above the background on the other side. For $|q| > 1$ the reverse is true.

So far in our discussions we have always assumed that the energy resolution of the experimental apparatus is ideal. This means that the initially prepared beam can be considered as ideally monoenergetic, so that the energy distribution function can be described by the δ-function according to (XIV.5.36):

$$F(E - E_A) = \rho(E) \langle E ||| W_B^{in} ||| E \rangle = \delta(E - E_A), \qquad (9.15)$$

where E_A is the central energy value in the beam. Such an ideal resolution can be achieved in two ways: either by an ideal preparation apparatus (a high-resolution monochromator), or by an ideal detection apparatus (a high-resolution analyzer in the experiment of Figure II.4.1). Both will lead to the same result because of energy conservation. As already mentioned in Chapter XIV, what constitutes ideal resolution depends upon the effects that are to be observed: The energy spread of the beam ΔE must be small compared to the width of the structure in the scattering amplitude in order that (9.15) may be used.

The observed quantity is not the cross section $\sigma(E)$ at a definite value E, but the energy-averaged cross section

$$\sigma * F = \int dE \, F(E - E_A) \sigma(E), \qquad (9.16)$$

where F is determined by the energy resolution of the detection (or preparation) apparatus, because every realistic detection apparatus detects particles not only with a definite energy E_A, but with an energy in a finite interval ΔE around E_A. If $\sigma(E)$ varies slowly as compared to $F(E - E_A)$, then $\sigma * F \approx \sigma(E_A)$, which is equivalent to using (9.15). But if ΔE is not small compared with the width Γ of a resonance, and consequently $\sigma(E)$ given by (6.27) does not vary slowly, then one cannot approximate $\sigma * F$ by $\sigma(E_A)$. The observed quantity is then the quantity given in terms of the T-matrix elements by (XIV.5.40). $\sigma(E)$ is just an abbreviation for the expression under the integral besides $F(E - E_A)$ on the right-hand side of (XIV.5.40) and is not an observable quantity. Thus for a resonance of the lth partial wave with background in the lth partial wave and the other partial waves, $\sigma(E) = \sigma(\epsilon)$ given by (9.14) is not the observable physical quantity if the width Γ is of the same order as the energy resolution ΔE. The observable physical quantity is

$$\langle \sigma(E_A) \rangle = \sigma * F = \int dE \, \sigma(E) F(E - E_A)$$

$$= \int dE \, (\sigma_l^{(bg)}(E) + \sigma_b(E)) F(E - E_A)$$

$$+ \int \frac{q^2 + 2q\epsilon - 1}{1 + \epsilon^2} \sigma_l^{(bg)}(E) F(E - E_A) \, dE. \qquad (9.17)$$

The precise value of this depends upon the energy distribution [the detector efficiency function $F(E - E_A)$] for the particular experiment. For the sake of definiteness let us assume that the probability of measuring the energy E

Figure 9.3 Assumed energy distribution for the beam with energy resolution $\Delta E \gg \Gamma$.

is zero except for values in an interval of ΔE around E_A. Then

$$
F(E - E_A) = \begin{cases} \dfrac{1}{\Delta E} & \text{for } E_A - \dfrac{\Delta E}{2} \leq E \leq E_A + \dfrac{\Delta E}{2}, \\ 0 & \text{otherwise}, \end{cases} \tag{9.18}
$$

depicted in Figure 9.3, will describe the apparatus resolution. Equation (9.18) is of course also an unphysical distribution like $\delta(E - E_A)$, and a Gaussian around E_A would often best reflect reality. However, for a beam with energy spread $\Delta E \gg \Gamma$, the assumption (9.18) is of sufficient accuracy. With (9.18) we obtain for the observed cross section (9.17):

$$
\langle \sigma(E_A) \rangle = \frac{1}{\Delta E} \int_{E_A - \Delta E/2}^{E_A + \Delta E/2} \sigma(E)\, dE = \frac{1}{\Delta E} \int_{E_A - \Delta E/2}^{E_A + \Delta E/2} (\sigma_l^{(bg)}(E) + \sigma_b(E))\, dE
$$

$$
+ \frac{1}{\Delta E} \int_{E_A - \Delta E/2}^{E_A + \Delta E/2} \frac{q^2 + 2q\epsilon - 1}{1 + \epsilon^2} \sigma_l^{(bg)}(E)\, dE. \tag{9.19}
$$

$\sigma_l^{(bg)}(E)$ and $\sigma_b(E)$ are slowly varying functions of E. Therefore

$$
\langle \sigma_{\mathrm{nr}}(E_A) \rangle = \frac{1}{\Delta E} \int_{E_A - \Delta E/2}^{E_A + \Delta E/2} (\sigma_l^{(bg)}(E) + \sigma_l(\epsilon))\, dE
$$

$$
\approx \sigma_l^{(bg)}(E_A) + \sigma_b(E_A) \approx \sigma_l^{(bg)}(E_R) + \sigma_b(E_R)
$$

is a slowly varying function of the average beam energy E_A whose value is well approximated by the value it takes when the beam energy E_A is equal to the resonance energy E_R.

The second integral on the right-hand side of (9.19) is given by

$$
\langle \sigma_r(E_A) \rangle = -\frac{\Gamma}{2\Delta E} \sigma_l^{(bg)}(E_A) \int_{\epsilon_A + \Delta E/\Gamma}^{\epsilon_A - \Delta E/\Gamma} \frac{q^2 + 2q\epsilon - 1}{1 + \epsilon^2}\, d\epsilon,
$$

$$
\epsilon_A = (E_R - E_A)\frac{2}{\Gamma}.
$$

For E_A far away from E_R, i.e., for ϵ_A far away from zero, this will be close to zero, as can be seen from Figure 9.2, and the averaged cross section will be given there essentially by $\langle \sigma_{nr}(E_A) \rangle$. At the value of the average beam energy $E_A = E_R$ the value of $\langle \sigma_r(E_A) \rangle$ is given by

$$\langle \sigma_r(E_R) \rangle = -\frac{\Gamma}{2\Delta E} \sigma_l^{(bg)}(E_R) \int_{\Delta E/\Gamma}^{-\Delta E/\Gamma} \frac{q^2 + 2q\epsilon - 1}{1 + \epsilon^2} d\epsilon$$

$$\approx \frac{\Gamma}{2\Delta E} \sigma_l^{(bg)} \int_{-\infty}^{+\infty} \frac{q^2 + 2q\epsilon - 1}{1 + \epsilon^2} d\epsilon$$

$$= \frac{\Gamma}{2\Delta E} \sigma_l^{(bg)} \pi (q^2 - 1).$$

In performing the integration we have taken the limits of the integral over the shape factor as $\pm\infty$, which for $\Delta E \gg \Gamma$ does not cause an appreciable error, as can be seen from Figure 9.2. Inserting this and

$$\sigma_l^{(bg)}(E_R) = \frac{4\pi}{(p(E_R))^2} (2l + 1) \sin^2 \gamma_l(E_R) \tag{9.20}$$

into (9.19), we obtain for the value of the averaged cross section when the beam energy E_A is set at the resonance energy E_R

$$\langle \sigma(E_R) \rangle = \langle \sigma_{nr}(E_R) \rangle + \langle \sigma_r(E_R) \rangle$$

$$= \langle \sigma_{nr}(E_R) \rangle + \frac{2\pi^2}{(p(E_R))^2} \frac{\Gamma}{\Delta E} (2l + 1) \cos 2\gamma_l(E_R). \tag{9.21}$$

Thus if a value of $\langle \sigma(E_R) \rangle$ is observed in a cross-section measurement with an apparatus with poor energy resolution, then it may be above or below or even equal to the slowly varying background $\langle \sigma_{nr}(E_A) \rangle$, depending upon the value of the background phase shift in the resonant partial wave, $\gamma_l(E_R) \approx \gamma_l(E_A)$. If $\cos 2\gamma_l > 0$, then *the resonance will show* in the observed cross section $\langle \sigma(E_A) \rangle$ *as a peak* above background $\langle \sigma_{nr}(E_A) \rangle$ at the value $E_A = E_R$. If $\cos 2\gamma_l < 0$, then *the resonance will show as a depression* below background. For $\cos 2\gamma_l \approx 0$ the resonance will not show up at all in the observed cross section.

With these results we can interpret the experimental data in Figures 1.4(b), 1.4(c), and 1.5, which were taken in an experiment with energy spread of the electron beam, ΔE, larger than the resonance width. Figures 1.4(b) and 1.4(c) show an increase of the transmitted current and a decrease of the total elastic cross section around the value $E_A \approx E_R \approx 19.3$ eV. This is what one expects from (9.21) if the background phase shift γ_l satisfies $\cos 2\gamma_l(E_R) < 0$, or

$$\tfrac{1}{4}\pi < \gamma_l(E_R) < \tfrac{3}{4}\pi. \tag{9.22a}$$

The same effect is observed in the averaged differential cross section at all angles depicted in Figure 1.5.

From the data of the high-resolution experiments in Figure 1.3 and Figure 1.4(d), we concluded above that $q < 0$. Using (9.2), this means that

$$\tfrac{1}{2}\pi < \gamma_l(E_R) < \pi. \tag{9.22b}$$

Thus combining (9.22a) and (9.22b), we conclude as a result of our qualitative considerations that

$$\tfrac{1}{2}\pi < \gamma_l(E_R) < \tfrac{3}{4}\pi. \tag{9.23}$$

This shows that the background phase shift in the resonant partial wave for the process $e + \text{He} \to \text{He}^- \to \text{He} + e$ is not at all small. Taking just T_l^R of (6.26) as the lth partial scattering amplitude and $\sigma_l^{(R)}$ of (6.27) as the cross section around resonance energy would certainly be unjustified.

We have now explained the experimental data for the elastic scattering process $e + \text{He} \to e + \text{He}$ as a resonance of narrow width around 19.3 eV interfering with the background phase shift fulfilling (9.23). In order to obtain more information about this He^- resonance, we turn to the experiment of Figure 1.5. This is a low-resolution experiment with $\Delta E = (50–100) \times 10^{-3}$ eV in which the intensity of the elastically scattered electrons is measured as a function of average beam energy at various values of the scattering angle, $\theta = 10°, 40°, 60°, 90°, 110°$. Thus Figure 1.5 gives us the averaged elastic differential cross section as function of scattering angle and energy:

$$\left\langle \frac{d\sigma}{d\Omega}(E_A, \theta) \right\rangle = \int dE\, \frac{d\sigma}{d\Omega}(E, \Omega) F(E - E_A),$$

where F is the energy-resolution function of the apparatus in this experiment.

With the experimental data of Figure 1.5, one can perform the phase-shift analysis as described in the second part of Section XVIII.8. Assuming that in the energy range considered the partial-wave amplitudes T_l with $l > 2$ can be ignored, one can use (8.9) for the determination of the phase shifts δ_0, δ_1, and δ_2 at various values of the momentum. As the inelasticity coefficients η_l are zero in the present case, where only the elastic channel is open, the $T_l(p)$ of (8.9) are connected with $\delta_l(p)$ by (cf. XVI.2.8)

$$p T_l(p) = e^{i\delta_l} \sin \delta_l. \tag{9.24}$$

The Legendre polynomials that occur in (8.9) are

$$P_0 = 1, \quad P_1 = \cos\theta, \quad P_2 = \tfrac{1}{2}(3\cos^2\theta - 1),$$

$$P_3 = \tfrac{1}{2}(5\cos^3\theta - 3\cos\theta), \quad P_4 = \tfrac{1}{8}(35\cos^4\theta - 30\cos^2\theta + 3).$$

$$\tag{9.25}$$

One can first ignore the partial-wave amplitudes above T_1 and use (8.9₁). At $\theta = 90°$ one has, according to (8.9₁), only the contributions from T_0. As according to Figure 1.5 there is still a resonance effect at $\theta = 90°$, one concludes that the resonance occurs in the S-wave:

$$l_R = 0. \tag{9.26}$$

One can now apply (9.21) to the cross section $d\sigma/d\Omega(E_A, \theta = 90°)$. Instead of the apparatus resolution function (9.18), one may have to take a different one, depending upon the experiment.[30] Then the expression on the right-hand side of (9.21) will be different, but the principle of determining the resonant phase shift $\delta_0 = \pi/2 + \gamma_0$ is the same. The curve in Figure 1.5 for $\theta = 90°$ respresents

$$\left\langle \frac{d\sigma}{d\Omega} (E_A, \theta = 90°) \right\rangle .$$

It shows again the decline below the nonresonant differential cross section characteristic of a resonating partial wave with background phase shift given by (9.22a). From the values of

$$\left\langle \frac{d\sigma}{d\Omega} (E_A, \theta = 90°) \right\rangle$$

at various energies E_A around the resonance energy E_R, one can determine the background phase shift γ_0 and the true resonance width Γ. The values obtained from this experiment are

$$\Gamma \approx (15\text{--}20) \times 10^{-3} \, \text{eV}, \qquad \gamma_0 \approx 100°. \tag{9.27}$$

γ_0 is in agreement with the restrictions (9.23) obtained from the other experiments above.

One may now use the data of Figure 1.5 at $\theta = 60°$. $P_2(\cos 60°)$ is close to zero, and the influence of a possible T_2 term, which would be considerable for $P_2(\cos \theta) \neq 0$ according to (8.9$_2$), is minimized. Fitting (8.9$_1$) to the experimental data for $\theta = 60°$ with $\delta_0^R = \pi/2$ and $\gamma_0 = 100°$, one obtains

$$\delta_1 = \gamma_1 \approx 25°. \tag{9.28}$$

To check that the contribution from the higher partial waves is small, one can now fit the experimental data for the other scattering angles θ with (8.9$_2$). For example, with $\gamma_0 = 100°$ and $\gamma_1 = 25°$, one obtains from the experimental curve for $\theta = 10°$

$$\delta_2 = \gamma_2 \approx 4°. \tag{9.29}$$

This is indeed small, and justifies the working hypothesis made at the beginning of our analysis that T_2 and all higher partial waves can be ignored.

The He in the collision process

$$e + \text{He} \rightarrow \text{He}^- \rightarrow \text{He} + e \tag{1.4}$$

is in the ground state $^1S_0$ with $l = 0$, $s = 0$. The e in this process has $s = \frac{1}{2}$, and as we know from (9.26), $l = 0$. Consequently, the quasistationary state He$^-$ has $l = 0$ and $s = \frac{1}{2}$, i.e., it is a $^2S_{1/2}$ state. It is very reminiscent of the

---

[30] Most frequently the apparatus resolution function is a Gaussian (6.39). The observed cross section is then according to (9.17) the convolution of a Gaussian and a Lorentzian function, which is known as the Voigt integral or Voigt profile. The Voigt integral cannot be found analytically, and has been computed and tabulated. The Lorentzian width Γ and the Gaussian width ΔE can be obtained from a fit of the experimental data with the Voigt profile. See, e.g., H. G. Kuhn, *Atomic Spectra*, Academic Press, 1969, Section VIID.

$^2S_{1/2}$ ground state in the lithium spectrum (cf. Figure VII.2.1). The only difference is that the charge of the nucleus in lithium is $+3$, whereas that in He$^-$ is $+2$, and the He$^-$ state has a width different from zero and given by (9.27).

Problems

1. For the infinite repulsive square-well potential of Problem XVII.3, obtain the time $T^{(\mathcal{R})l=0}$ which the projectile spends in the interaction region $\mathcal{R}$ of radius a and the time delay $t_D^{l=0}$ caused by this potential. Discuss and interpret this result.

2. (a) Obtain the S-wave phase shift and the $l = 0$ S-matrix element $S_0(p)$ for the square-well potential.

(b) Find the energy values of the bound states of angular momentum 0 as poles of $S_0(p)$. Discuss the relations for the mass, the depth V_0, and the radius R of the well that have to be fulfilled in order for there to be (i) no bound state, (ii) one bound state, (iii) two bound states.

(c) What are the virtual states of angular momentum zero? Under which conditions do there exist no virtual states for $l = 0$?

(d) Consider a very deep well fulfilling $2mV_0 R^2 \gg 1$. Show that there are regularly spaced s-wave resonances. Determine the energy and width of these resonances.

(e) Obtain the scattering matrix element $S_0(p)$ at energies close to a resonance energy, and obtain the background phase shifts. Calculate the elastic scattering cross section ($l = 0$), and show that it can be written as the sum of three terms: one from background scattering, one from resonance scattering, and one from the interference between resonance and background. Give a graphical representation of the elastic scattering cross section as a function of energy.

3. Calculate the time delay caused by a narrow resonance of angular momentum l_R and with resonance parameters (E_R, Γ) of a beam with an energy distribution given by

$$\sum_{l_3} \langle E \mid l\, l_3 | W^{\text{in}} | E \mid l\, l_3 \rangle = W_l(E) = \frac{1}{\rho(E)} F(E - E_R) \quad \text{for every } l,$$

where

$$F(E - E_R) = \begin{cases} \dfrac{1}{\Delta E} & \text{for } E_R - \dfrac{\Delta E}{2} \le E \le E_R + \dfrac{\Delta E}{2}, \\ 0 & \text{otherwise,} \end{cases}$$

with $\Delta E \gg \Gamma$. Give the result in the approximation of constant background phase shifts.

4. Calculate the time delay of a beam with a Lorentzian energy distribution (6.41) caused by a resonance with resonance parameters $(E_0 = E_A, \Gamma = \Delta E)$.

5. Taking for the scattering amplitude an expression of the form

$$\tilde{T}_l = \frac{1}{p} \frac{\Gamma/2}{(E - E_0) - i(\Gamma/2)}$$

will also lead to a Breit–Wigner cross section (6.27). Why can $\tilde{T}_l$ not describe the partial-wave amplitude for resonance scattering?

6. (a)   Describe the behavior of the resonance phase shift for a "dipole" resonance (6.44).

   (b)   Discuss the cross section near resonance energy as a function of energy for a dipole resonance, taking the effect of a slowly varying background $\gamma_l(E)$ into account. Draw graphs in analogy to Figure 6.1.

7.   Compare the case of a dipole resonance with the case of two resonances close to each other that are described by a scattering amplitude

$$T_l(p) = \frac{1}{p} e^{i2\gamma_l^R(E)} \left(\frac{\frac{1}{2}\gamma_R}{E_R - E - i(\Gamma_R/2)} + \frac{\frac{1}{2}\gamma_X e^{i2(\gamma_l^X(E) - \gamma_l^R(E))}}{E_X - E - i(\Gamma_X/2)} \right)$$

where $\gamma_R, \Gamma_R, \gamma_X, \Gamma_X$ are real constants: $E_i \gg \Gamma_i > \gamma_i$ ($i = R, X$) and $|E_R - E_X|$ are of the same orders as Γ_R and Γ_X; and $\gamma_l^R(E), \gamma_l^X(E)$ are slowly varying functions of E. Consider first the special case of zero background shifts $\gamma_l^R = \gamma_l^X = 0$.

8. (a)   For a resonance, $pT_l(E)$ moves counterclockwise in the Argand diagram. Show that this is related to causality.

   (b)   Discuss when $pT_l(E)$ can move clockwise in the Argand diagram.

9.   Consider the quasistationary state with energy E_0 and width Γ for which the parameter r in

$$S_l(E) = \exp \left(i2r \arctan \frac{\Gamma/2}{E_0 - E} \right)$$

has the value $r = 3$. Obtain the slope of the resonant cross section for such a "tripole" resonance. (Ignore the background.)

10.   Explain, using the shape of the function $\sigma_l(E)$ in Figure 6.1 for a background phase shift $\gamma_l = \pi/4$, why the cross section $\langle \sigma(E_A) \rangle$ observed by an apparatus with poor energy resolution $\Delta E \gg \Gamma$ need not show a structure at the energy value E_A equal to the resonance energy E_R.

11.   Show that the effect of a bound state with low binding energy in elastic scattering is the same as that of a virtual-state pole close to the real axis.

CHAPTER XIX

Time Reversal

For the discussion of multichannel resonances in Chapter XX, we require the properties of the S-matrix that follow from time-reversal invariance. In Section XIX.1 space-inversion invariance is applied to determine some properties of the S-matrix. Time reversal is introduced in Section XIX.2. The main purpose of Section XIX.1 is to introduce Section XIX.3, where the consequences of time-reversal invariance for the properties of the S-matrix are examined.

XIX.1 Space-Inversion Invariance and the Properties of the S-Matrix

In Section V.4 we introduced the observable *parity*, represented in the space of physical states by the unitary operator U_P fulfilling the relations (V.4.1)–(V.4.3) with the position, momentum, and angular-momentum observables. We mentioned in Section V.4 that not all energy operators commute with the parity operator—in particular, the interaction Hamiltonian that causes the weak transitions, and consequently the transition operator for the weak interaction, do not commute with U_P. The weak interaction violates parity invariance. However, many scattering and decaying systems are parity-invariant, i.e., in addition to

$$[K, U_P] = 0 \tag{1.1}$$

they also fulfill

$$[V, U_P] = 0 \quad \text{or} \quad [T, U_P] = 0, \tag{1.2}$$

or in terms of the S-operator

$$S = U_P^\dagger S U_P. \tag{1.3}$$

Parity invariance leads to some constraints on the T-matrix and S-matrix elements. For example, it follows from (V.4.2) and (V.4.5) that

$$U_P|\mathbf{p}\,\eta\rangle = \pi_\eta|-\mathbf{p}\,\eta^\pi\rangle, \tag{1.4}$$

where π_η is $+1$ or -1, and for the angular-momentum eigenvectors (XVI.1.3), in analogy to (V.4.39),

$$U_P|E\,l\,l_3\,\eta\rangle = (-1)^l\pi_\eta|E\,l\,l_3\,\eta^\pi\rangle. \tag{1.5}$$

The additional, internal quantum numbers η usually do not change under U_P; thus $\eta^\pi = \eta$. The internal parity π_η may be an element of the set of additional quantum numbers η.

From (1.4) it follows for the S- and T-matrix elements in the momentum basis, using (1.3) or (1.2), that

$$\begin{aligned}\langle\mathbf{p}'\eta'|S|\mathbf{p}\,\eta\rangle &= \langle\mathbf{p}'\eta'|U_P^\dagger S U_P|\mathbf{p}\,\eta\rangle \\ &= \langle-\mathbf{p}'\eta'|S|-\mathbf{p}\,\eta\rangle\pi_{\eta'}\pi_\eta,\end{aligned} \tag{1.6}$$

or

$$\langle\mathbf{p}'\eta'|T|\mathbf{p}\,\eta\rangle = \langle-\mathbf{p}'\eta'|T|-\mathbf{p}\,\eta\rangle\pi_{\eta'}\pi_\eta. \tag{1.7}$$

From (1.5) it follows for the T-matrix elements in the angular-momentum basis (assuming T is spherically symmetric) that

$$\langle E\,l'\,l_3'\,\eta'|T|E\,l\,l_3\,\eta\rangle = \langle E\,l'\,l_3'\,\eta'|T|E\,l\,l_3\,\eta\rangle(-1)^{l+l'}\pi_{\eta'}\pi_\eta. \tag{1.8}$$

The conditions (1.7) and (1.8) constitute restrictions on the properties of the T-matrix elements.

We shall also restrict ourselves to the case discussed in Chapter XVI, namely, where the internal quantum numbers η do not contain internal angular momenta. Then, using (XVI.1.18), one obtains from (1.8)

$$\langle E\,l'\,l_3'\,\eta'|T|E\,l\,l_3\,\eta\rangle = \pi_{\eta'}\pi_\eta\langle E\,l\,l_3'\,\eta'|T|E\,l\,l_3\,\eta\rangle\delta_{ll'}\,\delta_{l_3 l_3'} \tag{1.9}$$

or for the partial-wave amplitude defined by (XVI.1.34), (XVI.1.23), (XVI.1.18),

$$T_{l\eta}^{\eta'}(E) = \pi_{\eta'}\pi_\eta\, T_{l\eta}^{\eta'}(E). \tag{1.10}$$

This shows that in the case of elastic scattering of spin-zero particles, parity invariance does not impose any additional restrictions that are not already consequences of rotational invariance. For $\eta \neq \eta'$, (1.10) states that the partial-wave reaction amplitude will be different from zero only if the internal parity of the initial state is equal to the internal parity of the final state $\pi_\eta = \pi_{\eta'}$ or if the internal parity is conserved. Parity invariance (1.2) leads to parity conservation (1.7), (1.8), (1.10), in the same way as rotational invariance (XVI.1.1) led to angular-momentum conservation (XVI.1.18).

XIX.2 Time Reversal

The main purpose of the preceding discussion on parity invariance is to serve as an introduction to another kind of invariance, which superficially appears to be very similar to parity invariance but is of a completely different kind: the invariance under time reversal. Naively, time reversal is the replacement of t by $t^R = -t$; however, this is not a physical operation like taking the mirror image $x_i \rightarrow x_i^R = -x_i$, as we cannot turn time backwards. But we can compare the state of a physical system having a position x_i and a velocity v_i with the state of the physical system having the same position x_i and the velocity $-v_i$. If in a scattering process all positions and velocities undergo this transition at all times:

$$\mathcal{T} : \mathbf{x} \rightarrow +\mathbf{x}, \quad \mathcal{T} : \mathbf{v} \rightarrow -\mathbf{v} \quad \text{or} \quad \mathcal{T} : \mathbf{x} \rightarrow \mathbf{x}, \quad \mathcal{T} : \mathbf{p} \rightarrow -\mathbf{p}, \quad (2.1)$$

then one obtains a process which looks like the original process going backward in time.

The time-reversal operation $\mathcal{T}$ is the velocity reversal (2.1). For a quantum physical system this transformation $\mathcal{T}$ is represented by an operator in the space of physical states. We call this operator the time-reversal operator A_T. If A_T represents a usual observable, like those we have considered so far, then A_T should be a linear operator. We shall see instantly that A_T cannot be a linear operator.

In analogy to (V.4.1) and (V.4.2) for the parity operator U_P, one would translate (2.1) for the quantum physical observables Q_i and P_i into

$$A_T Q_i A_T^{-1} = Q_i, \quad (2.2)$$

$$A_T P_i A_T^{-1} = -P_i. \quad (2.3)$$

We shall justify this below.

For the orbital-angular-momentum operator $L_i = \epsilon_{ijk} Q_j P_k$ and thence for any angular-momentum operator J_i, it follows from (2.2) and (2.3) that

$$A_T J_i A_T^{-1} = -J_i. \quad (2.4)$$

Let us now consider the commutation relation between the momentum operator P_i and the angular-momentum operator J_i:

$$[J_i, P_k] = i\epsilon_{ikl} P_l, \quad (2.5)$$

and insert (2.3) and (2.4) into (2.5):

$$A_T [J_i, P_k] A_T^{-1} = -i\epsilon_{ikl} A_T P_l A_T^{-1}. \quad (2.6)$$

If A_T were an ordinary linear operator, then we would obtain by multiplying (2.6) with A_T^{-1} from the left and A_T from the right

$$[J_i, P_k] = -i\epsilon_{ikl} P_l.$$

This would contradict (2.5). Thus if A_T is to reverse the direction of the momenta (2.6), keeping the positions unchanged, then A_T cannot be a linear operator.

[A one-to-one mapping A of a linear space $\mathscr{H}$ over the complex numbers $\mathbb{C}$ is called a *semilinear* operator if

$$A(\psi + \phi) = A\psi + A\phi \quad \text{for all} \quad \psi, \phi \in \mathscr{H}, \qquad A\phi = 0 \Rightarrow \phi = 0,$$
$$(2.7a)$$

and, for all $\alpha \in \mathbb{C}$

$$A\alpha\phi = \alpha' A\phi, \quad \text{where} \quad \alpha' = \alpha \quad \text{or} \quad \alpha' = \alpha^*. \qquad (2.7b)$$

A semilinear operator is called *antilinear* if $\alpha' = \alpha^*$ (complex conjugate of α), i.e., if

$$A\alpha\phi = \alpha^* A\phi. \qquad (2.7c)$$

A semilinear operator is called *linear* if $\alpha' = \alpha$. Thus in distinction to a linear operator, already defined by (I.3.1) and (I.3.2), an antilinear operator also transforms the complex numbers into their complex conjugates:

$$Ai = -iA. \qquad (2.7')$$

In a scalar-product space $\mathscr{H}$ one can uniquely define for every antilinear operator A (defined everywhere in $\mathscr{H}$ or in a dense subspace thereof) an antiadjoint operator $A^\dagger$ (often simply called the adjoint operator) by

$$(Af, g) = (A^\dagger g, f) = (f, A^\dagger g)^*, \qquad f, g \in \mathscr{H}. \qquad (2.8)$$

The following relations hold for a linear operator L and antilinear operators A_1, A_2:

$$(A_1 A_2)^\dagger = A_2^\dagger A_1^\dagger, \qquad (AL)^\dagger = L^\dagger A^\dagger, \qquad (A_1 L A_2)^\dagger = A_2^\dagger L^\dagger A_1^\dagger. \qquad (2.9)$$

An antilinear operator is called antiunitary if it preserves the norm, i.e., if

$$(Ag, Af) = (A^\dagger Af, g) = (f, g) \quad \text{or} \quad A^\dagger = A^{-1}. \qquad (2.10)$$

The product of two antiunitary operators is a unitary operator.

For antilinear operators one has, in analogy to the case for linear operators,

$$\text{Tr}(AW) = \text{Tr}(WA), \qquad \text{Tr}(QAWA^\dagger) = \text{Tr}(A^\dagger QAW), \qquad (2.11)$$

where W and Q are Hermitian operators. It is always assumed that the operators are such that all traces that occur in the proof are well defined.]

If we choose for the representative of the velocity reversal or time reversal, A_T, not a unitary operator as we did for the space inversion in Section V.4, but an antiunitary operator, then (2.6) will not lead to a contradiction, but — using (2.7c) — back to (2.5).

We therefore make the following statement: The time-reversed or velocity-reversed state $W^T(t)$ of the state $W(t)$ is obtained from $W(t)$ by

$$W^T(t) = A_T W(t) A_T^\dagger, \tag{2.12}$$

where the time-reversal operator A_T is an antiunitary operator

$$A_T^\dagger A_T = I \tag{2.13}$$

that fulfills the defining relations (2.2)–(2.4).[1]
In addition A_T fulfills

$$A_T^2 = \epsilon I, \quad \text{where} \quad \epsilon = +1 \text{ or } -1. \tag{2.14}$$

Furthermore we require that the energy and transition operators commute with the time-reversal operator A_T:[2]

$$A_T K - K A_T = 0, \qquad A_T H - H A_T = 0, \qquad A_T T - T A_T = 0, \tag{2.15}$$

and that the S-operator fulfills

$$A_T^\dagger S A_T = S^\dagger. \tag{2.15'}$$

(2.15) and (2.15′) are the statement of time reversal invariance.

The justification of this statement consists of two parts. The first and more difficult part is to show that the operation of time reversal is represented by a semiunitary operator [i.e., an operator fulfilling (2.7a), (2.7b), (2.13)]. In our heuristic arguments above, we have simply assumed that — in resemblance to the basic assumption I — time reversal is represented by something like a linear operator. The justification of this is, essentially, the celebrated Wigner theorem which applies to a much wider class of transformations than that of time reversal. This part is briefly discussed in the appendix to this section.

The second part in the justification of the above statement is to establish the particular properties of the time reversal operator expressed by the relations (2.2), (2.3), (2.4), (2.15).

From the physical interpretation of the time-reversal or velocity reversal operation, the following requirement results: If the expectation value of the momentum in the state $W(t)$ is $\bar{p}$, then the expectation value of the momentum in the state $W^T(t)$ should be $-\bar{p}$. This means

$$\mathrm{Tr}(P_i W(t)) = -\mathrm{Tr}(P_i W^T(t)) = -\mathrm{Tr}(P_i A_T W(t) A_T^\dagger) = -\mathrm{Tr}(A_T^\dagger P_i A_T W(t)). \tag{2.16}$$

---

[1] If there exist additional independent observables that cannot be expressed in terms of P_i, Q_i, J_i, then their relation to A_T also has to be specified.

[2] So far, this requirement has been found to hold except for decaying K^0 mesons, and there may be a superweak part of the Hamiltonian that does not fulfill (2.15).

As this is to be fulfilled for any state $W(t)$ and its corresponding time-reversal state $W^T(t)$, it follows that

$$P_i = -A_T^\dagger P_i A_T,$$

from which (2.3) is obtained. Equation (2.2) follows in the same way. Equation (2.4) follows from (2.3) and (2.2), as mentioned above. It has also already been mentioned that from (2.6) the antilinearity of A_T follows.

Executing the operation of time reversal two times on all states,

$$W^{TT}(t) = A_T W^T(t) A_T^\dagger = A_T A_T W(t) A_T^\dagger A_T^\dagger, \tag{2.17}$$

should lead to the original state. From this requirement it follows that

$$A_T A_T = \epsilon I \quad \text{with} \quad \epsilon \epsilon^* = 1. \tag{2.18}$$

But as $A_T A_T^2 = A_T A_T A_T = A_T^2 A_T$ and consequently $A_T \epsilon = \epsilon A_T$, it follows from the antilinearity that ϵ must be real. From this (2.14) follows.

Whether one has to choose for the time-reversal operator an A_T fulfilling (2.14) with $\epsilon = +1$ or with $\epsilon = -1$ has—at least at the present time—to be decided by experience. The following choice has proved to be without contradiction to experimental observations:

$$\epsilon = (-1)^{2s}, \tag{2.19}$$

where s is the spin of the physical system.[3]

The requirement (2.15) is empirical in nature: No interaction has been found (except for the above-mentioned K^0 system, which we shall ignore here) that violates time reversal invariance as stated by (2.15). Also, if the Heisenberg equation of motion (XII.1.24) with (XII.1.26) is to be fulfilled by the position operator, then it follows from (2.2) and (2.3) and the antilinearity of A_T that H must commute with A_T.

Equation (2.15′) is a consequence of (2.15) and the antiunitarity of A_T.

The state W and its A_T-transformed state W^T may be considered as different states of the same physical system. It is, however, more practical to restrict the scope of the physical system and to consider all the A_T-transformed states of all the states of a physical system as belonging to a new system, the A_T-transformed system of the original system. To obtain an idea about the property of the A_T-transformed system, let us consider a particular state:

$$W^T(t) = A_T W(t) A_T^\dagger.$$

From the time-development axiom V—in particular, from (XII.1.15) and (XII.1.16)—it follows that

$$\begin{aligned} W^T(t) &= A_T e^{-itH} W_0 e^{itH} A_T^\dagger = e^{+itH} A_T W_0 A_T^\dagger e^{-itH} \\ &= e^{-i(-t)H} W_0^T e^{+i(-t)H}. \end{aligned} \tag{2.20}$$

---

[3] E. P. Wigner, in *Group Theoretical Concepts and Methods in Elementary Particle Physics* (Ed. F. Gursey), Gordon & Breach, New York, 1964, p. 37.

Thus the A_T-transformed system develops forward in time in the same way as the original system would develop backward: $W(-t) = e^{-i(-t)H}W_0 e^{i(-t)H}$. If an original scattering system develops from a particular in-state W^{in} into a particular out-state W^{out}, then the time-reversed system develops from an in-state $(W^T)^{in}$ agreeing with the out-state of the original system $(W^T)^{in} = W^{out}$ into the out-state $(W^T)^{out} = W^{in}$.

Appendix to Section XIX.2

In this Appendix we give a brief description of the conditions and arguments that lead to the statement that the transformation of a physical system is described by a semiunitary operator. We choose here the transformation of time reversal.

Let $W^T(t)$ denote the state that is obtained from the state $W(t)$ by the operation of time reversal $\mathcal{T}$:

$$\mathcal{T}: W(t) \rightarrow W^T(t).$$

It is natural to require that:

1. If two states have no property in common, then the $\mathcal{T}$-transform of these two states should also have no property in common.
2. If one state has a more specific property than a second state, then its $\mathcal{T}$-transform should also have a more specific property than the $\mathcal{T}$-transform of the second state.

Let us restrict ourselves to states that are described by projection operators Λ_1 and Λ_2. Then requirement 1 means:

$$\Lambda_1\Lambda_2 = 0 \Rightarrow \Lambda_1^T\Lambda_2^T = 0,$$

or

$$\Lambda_1\mathcal{H} \perp \Lambda_2\mathcal{H} \Rightarrow \Lambda_1^T\mathcal{H} \perp \Lambda_2^T\mathcal{H}$$

(orthogonal properties go into orthogonal properties). Requirement 2 means:

$$\Lambda_1\Lambda_2 = \Lambda_1 \Rightarrow \Lambda_1^T\Lambda_2^T = \Lambda_1^T.$$

Λ_1 is the state with the more specific property than Λ_2. (As an example, Λ_2 may be the state that has angular momentum $j = 2$, and Λ_1 the state that has angular momentum $j = 2$ and angular-momentum component $j_3 = +2$.)

It is further natural to require that:

3. Time reversal is a one-to-one mapping in the set of states.

From these three requirements it follows by a general mathematical theorem[4] that time reversal $\mathcal{T}$ is described by a semilinear operator A_T:

$$\mathcal{T}: W(t) \rightarrow W^T(t) = A_T W(t) A_T^\dagger.$$

---

[4] E. Artin, *Geometric Algebra*, Academic Press, New York, 1951, p. 88.

It is further natural to require that:

4.   Overall probability should be preserved by the operation of time
reversal. This means that for *all* W and their corresponding W^T

$$\text{Tr } W = \text{Tr } W^T.$$

From this follows (2.13).

Thus, the operation of time reversal is represented by a semiunitary
(unitary or antiunitary) operator.

So far no specific property of the time-reversal operation has been used,
and the above arguments apply to any transformation of the physical states
with the properties 1, 2, 3, and 4. Thus every such transformation is represented by a semiunitary operator which is in general determined up to
a phase factor. The celebrated Wigner theorem,[5] which states that a symmetry operation of a quantum physical system is induced by a unitary or
antiunitary operator, is contained in the above statement.

XIX.3   Time-Reversal Invariance and the Properties of the *S*-Matrix

Before we can apply the assumption of time-reversal invariance to find the
properties of the S-matrix and T-matrix elements, we have to determine
the action of A_T on the basis vectors. This is done in the same way as it was
done for the parity operator U_P in Section V.4. For the momentum eigen-
vector $|\mathbf{p}\,\eta\rangle$ it follows from (2.14) that

$$P_i(A_T|\mathbf{p}\,\eta\rangle) = -p_i(A_T|\mathbf{p}\,\eta\rangle). \tag{3.1}$$

Consequently, $(A_T|\mathbf{p}\,\eta\rangle)$ is also an eigenvector of P_i but with eigenvalue
$-p_i$. If there are no additional eigenvectors and $\mathbf{p}$ itself labels the basis
system completely, then it follows from (3.1) that

$$A_T|\mathbf{p}\rangle = \alpha(\mathbf{p})|-\mathbf{p}\rangle,$$

where α is a phase factor, which, because of

$$A_T^2|\mathbf{p}\rangle = \epsilon|\mathbf{p}\rangle = \alpha(\mathbf{p})A_T|-\mathbf{p}\rangle = \alpha(\mathbf{p})\alpha(-\mathbf{p})|\mathbf{p}\rangle,$$

$$\alpha^2 = \epsilon \quad (= +1 \text{ for the spinless case}) \tag{3.2}$$

and can be shown to be independent of $\mathbf{p}$.

For the angular-momentum eigenvectors $|E\,l\,l_3\rangle$ it follows in the same
way, from (2.4) and (2.14), that

$$A_T|E\,l\,l_3\rangle = \alpha|E\,l-l_3\rangle. \tag{3.3}$$

---

[5] Proofs of the Wigner theorem can be found in Ludwig, (1954), p. 101; U. Uhlhorn, *Ark.
Fys.* **23**, 307 (1963); L. O'Raifeartaigh, G. Rasche, *Ann. of Phys.* **25**, 155 (1963); V. Bargmann,
Ann. of Math. **59**, 1 (1954). See also J. M. Jauch, in *Group Theory and Its Applications* (Ed. E. M.
Loebl), Academic Press, New York, 1968, p. 131, where further references are given.

If there are additional quantum numbers η (internal quantum numbers, channel labels), then the action of A_T on the basis vectors $|\ldots\eta\rangle$ depends upon the relation between A_T and η^{op}. Let us assume that η^{op} does not contain any internal angular momenta. Let us further assume that all η^{op} commute with A_T:

$$[A_T, \eta^{op}] = 0. \tag{3.4}$$

This will be the case if η^{op} stands for the internal-energy operator of the target, for instance. Then

$$A_T|\mathbf{p}\,\eta\rangle = \alpha(\eta)|-\mathbf{p}\,\eta\rangle \quad \text{and} \quad A_T|E\,l\,l_3\,\eta\rangle = \alpha(\eta)|E\,l-l_3\,\eta\rangle, \tag{3.5}$$

where $\alpha = \alpha(\eta)$ may depend upon η and perhaps also upon l.

Let us now consider the S-matrix element in the angular-momentum basis (XVI.1.42) and the reduced S-matrix elements $\langle\eta\|S_l(E)\|\eta'\rangle$ defined by (XVI.1.46) and (XVI.1.44). One can either use (2.15) for the T-operator together with (3.5) and the definition of the S-matrix given by (XVI.1.41), or one can use (2.15) for the S-operator directly together with (3.15), to calculate, using (2.8) and (3.5),

$$\langle E\,l\,l_3\,\eta|A_T^\dagger S A_T|E'\,l'\,l_3'\,\eta'\rangle = (A_T|E\,l\,l_3\,\eta\rangle, S A_T|E'\,l'\,l_3'\,\eta'\rangle)^*$$
$$= \alpha(\eta)\alpha^*(\eta')\langle E\,l-l_3\,\eta|S|E'\,l'-l_3'\,\eta'\rangle^* \tag{3.6}$$

On the other hand, we obtain from (2.15):

$$\langle E\,l\,l_3\,\eta|A_T^\dagger S A_T|E'\,l'\,l_3'\,\eta'\rangle = \langle E\,l\,l_3\,\eta|S^\dagger|E'\,l'\,l_3'\,\eta'\rangle$$
$$= (S|E\,l\,l_3\,\eta\rangle, |E'\,l'\,l_3'\,\eta'\rangle)$$
$$= \langle E'\,l'\,l_3'\,\eta'|S|E\,l\,l_3\,\eta\rangle^*. \tag{3.7}$$

Thus

$$\langle E'\,l'\,l_3'\,\eta'|S|E\,l\,l_3\,\eta\rangle^* = \langle E\,l-l_3\,\eta|S|E'\,l'-l_3'\,\eta'\rangle^*\alpha(\eta)\alpha^*(\eta') \tag{3.8}$$

or, taking the complex conjugate of (3.8),

$$\langle E'\,l'\,l_3'\,\eta'|S|E\,l\,l_3\,\eta\rangle = \langle E\,l-l_3\,\eta|S|E'\,l'-l_3'\,\eta'\rangle\alpha(\eta')\alpha^*(\eta) \tag{3.9}$$

In order to determine the phase factors, we calculate

$$\langle E\,l\,l_3\,\eta|A_T^\dagger A_T|E'\,l'\,l_3'\,\eta'\rangle = \alpha(\eta)\alpha^*(\eta')\langle E\,l\,l_3\,\eta|E'\,l'\,l_3'\,\eta'\rangle^*. \tag{3.10}$$

On the other hand, because of (2.13),

$$\langle E\,l\,l_3\,\eta|A_T^\dagger A_T|E'\,l'\,l_3'\,\eta'\rangle = \langle E\,l\,l_3\,\eta|E'\,l'\,l_3'\,\eta'\rangle$$
$$= \rho^{-1}(E)\delta(E-E')\delta_{ll'}\delta_{l_3 l_3'}\delta_{\eta\eta'} \tag{3.11}$$

if the generalized eigenvectors are "normalized" in the usual way with real $\rho(E)$. Consequently, comparison of (3.10) and (3.11) shows that we require

$$\alpha(\eta)\alpha^*(\eta) = 1. \tag{3.12}$$

Equations (3.8) and (3.9) with (3.12) are called the *reciprocity theorem*[6] (for the spinless case). This relation is quite distinct from the corresponding relations—e.g., (1.7) or (1.8)—following from parity invariance, as it relates two different processes with each other. The left-hand side of (3.9) describes the process $\eta \to \eta'$, whereas the right-hand side describes the process $\eta' \to \eta$. The reciprocity theorem then states that these two processes have the same transition probability. The measurement of these transition probabilities provides a possibility of checking whether time-reversal invariance is fulfilled.

We now want to determine the consequences of time-reversal invariance upon the lth partial reduced S-matrix element $\langle \eta' \| S_l(E) \| \eta \rangle$ defined by (XVI.1.45) and (XVI.1.43). Inserting these into (3.8), we obtain

$$\langle \eta' \| S_l(E) \| \eta \rangle^* = \langle \eta \| S_l(E) \| \eta' \rangle = \langle \eta' \| S_l^\dagger(E) \| \eta \rangle; \tag{3.13}$$

inserting them into (3.9) gives

$$\langle \eta' \| S_l(E) \| \eta \rangle = \langle \eta \| S_l(E) \| \eta' \rangle \tag{3.14}$$

Thus the lth partial reduced S-matrix element $\langle \eta' \| S_l(E) \| \eta \rangle$ is not only a unitary [cf. (XVI.2.4)] but also a symmetric matrix. We shall make use of this fact in the following section.

For an experimental check of the reciprocity theorem we obtain the consequences of (3.14) for the partial cross section. Inserting (3.14) into (XVI.1.48), we obtain for the partial-wave amplitudes (XVI.1.34) the following relation:

$$\sqrt{p'(E)p}\,T_{l\eta}^{\eta'}(E) = \sqrt{p(E')p'}\,T_{l\eta'}^{\eta}(E'). \tag{3.15}$$

Inserting this into (XVI.1.37), one obtains the following relation between the partial cross sections $\sigma_l^{\eta' \leftarrow \eta}$ and $\sigma_l^{\eta \leftarrow \eta'}$ describing the processes $\eta \to \eta'$ and $\eta' \to \eta$, respectively:

$$\sigma_l^{\eta \leftarrow \eta'} = \sigma_l^{\eta' \leftarrow \eta} \frac{p^2}{p'^2}. \tag{3.16}$$

An analogous relation holds for the total and differential cross sections. Equation (3.16) has been derived for the case of spinless particles. If projectile and target for the system η have the spins s and s_T, and for the system η' the spins s' and s'_T, then (3.16) is replaced by

$$\sigma^{\eta \leftarrow \eta'} = \sigma^{\eta' \leftarrow \eta} \frac{p^2}{p'^2} \frac{(2s + 1)(2s_T + 1)}{(2s' + 1)(2s'_T + 1)}. \tag{3.16'}$$

For a check of the principle of detailed balance (3.16), we consider the experimental data of Figure 3.1.

---

[6] The reciprocity theorem is also called the principle of detailed balance or the principle of microreversability.

Figure 3.1. Relative cross section as function of energy for various scattering processes: A, elastic scattering of protons on aluminum. B, reaction $Al^{27} + p \rightarrow \alpha + Mg^{24}$. C, reaction $Mg^{24} + \alpha \rightarrow p + Al^{27}$. D, elastic scattering of α-particles on magnesium. [From S. G. Kaufmann *et al.*, *Phys. Rev.* **88**, 673 (1952), with permission.]

This figure depicts the counting rate (which is proportional to the cross section) as a function of the energy for various processes. Parts B and C are for the processes

$$Al^{27} + p \rightarrow \alpha + Mg^{24},$$

$$Mg^{24} + \alpha \rightarrow p + Al^{27},$$

which are related by the principle of detailed balance, and (3.16) should apply, with η characterizing the Al^{27}-proton system and η' characterizing the Mg^{24}-α system. The energy scale has been appropriately adjusted, and the curves should agree with each other (except for a factor). The data were taken from two different experiments with different resolution; still, the agreement is convincing. The various bumps correspond to resonances of the Si^{28} system, and we shall discuss this, as well as the comparison with the curves in parts A and D, in the following chapter.

Problems

1. Show that the product of two antilinear operators is a linear operator and that the product of an antilinear operator and a linear operator is antilinear.

2. Show that as a consequence of the antiunitarity the time reversal operator must fulfill $A_T^2 = \epsilon I$, where $\epsilon = +1$ or $\epsilon = -1$.

3. Show that from the definition of the Møller wave operators (XV.3.7$\pm$) it follows that $\Omega^{\pm} = A_T^{\dagger} \Omega^{\mp} A_T$ if H and K commute with A_T. Use this to show that (2.15) is a consequence of (2.14) if the S-operator is defined in terms of the energy operator.

4. Prove that the time-reversal operator transforms the exact energy eigenvectors $|\mathbf{p}\,\eta^{+}\rangle$ into the exact energy eigenvectors $|-\mathbf{p}\,\eta'^{-}\rangle$. (The label $+$ or $-$ refers to the definition according to the Lippmann–Schwinger equation.)

5. Show that the time-reversal operator transforms an out-state into an in-state and vice versa.

6. Consider a multichannel system of spinless particles. Show that the transformation property under U_P and A_T implies that the S-matrix in the momentum basis is symmetric.

7. Show that from the Heisenberg equation of motion (XII.1.24) for the position operator, $A = Q_i$, with (XII.1.26) and the antilinearity of A_T follows that H commutes with A_T.

Resonances in Multichannel Systems

In the present chapter we discuss resonances in scattering processes from an initial state into several possible final states, where the internal quantum numbers of the resonances are, in general, different from the quantum numbers of the initial state and the final states. In Section XX.2 we first discuss, in detail, the case of a single multichannel resonance and then the case of a double multichannel resonance (which occurs if there are two resonances with different internal quantum numbers in the same partial wave, with both the resonances coupling to initial and final states). In Section XX.3 the Argand diagrams for inelastic resonances are described.

XX.1 Introduction

In the discussion in Chapter XVIII we assumed that there is only one channel open in the scattering experiment and that it is this particular channel to which the resonance phenomenon is restricted. Then there is only one reduced S-matrix element $\langle \eta \| S_l(E) \| \eta \rangle$ that is different from zero, and it is this reduced matrix element that resonates. This is the case of elastic scattering.

$$a + T \rightarrow R \rightarrow a + T. \tag{1.1}$$

η are the quantum numbers of the initial state (a, T), of the final state (a, T), *and* of the resonance R.

This is, of course, a very special situation, which in general will not be fulfilled. When the energy is high enough, so that several final channels are open, a resonance can occur not only in the elastic scattering process described by $S_l(E) = \langle \eta_A \| S_l(E) \| \eta_A \rangle$ but also in the inelastic processes described by $\langle \eta \| S_l(E) \| \eta_A \rangle$ with $\eta \neq \eta_A$. In order that one particular resonance shows up in several channels this resonance state must have nonzero transition matrix elements to *all* initial and final states with quantum numbers η, η', η''. If one also wants to assign internal quantum numbers, say κ, to the resonance state, then these must be different than the quantum numbers η; the corresponding observables κ^{op} and η^{op} cannot commute.

It may be a matter of taste and convenience whether one characterizes these resonance states by internal quantum numbers $\kappa = (\kappa_1 \kappa_2 \cdots \kappa_n)$ and considers the resonance energy as a derived quantity $E_R = E(\kappa_R)$, $\kappa_R = (\kappa_{R1} \kappa_{R2} \cdots \kappa_{Rn})$ determined by the internal quantum numbers, or whether one characterizes the resonance state directly by the resonance energy E_R. The former view is prevailing in particle physics, the latter in nuclear physics. We shall follow here the former view, assuming that the resonance state is characterized in addition to the angular momentum by a set of "internal" quantum numbers κ, because this allows a very simple derivation of multichannel resonance formulas from the results of Chapter XVIII.

XX.2 Single and Double Resonances

The reduced matrix elements of the S-matrix $\langle \eta \| S_l(E) \| \eta' \rangle$ form a unitary matrix, according to (XVI.2.3). According to a well-known theorem of linear algebra,[1] for any unitary matrix there exists another unitary matrix that diagonalizes it, and the nonzero elements of the diagonal matrix (the eigenvalues) are of absolute value one. Thus for any value of E, there exists a unitary matrix $\langle \kappa \| \eta \rangle$ (with an adjoint $\langle \eta \| \kappa \rangle = \overline{\langle \kappa \| \eta \rangle}$),

$$\langle \kappa | \kappa' \rangle = \delta_{\kappa\kappa'} = \sum_\eta \langle \kappa \| \eta \rangle \langle \eta \| \kappa' \rangle, \qquad \delta_{\eta'\eta} = \langle \eta' \| \eta \rangle = \sum_\kappa \langle \eta' \| \kappa \rangle \langle \kappa \| \eta \rangle \tag{2.1}$$

such that

$$\delta_{\kappa\kappa'} e^{i2\delta_l^{(\kappa)}(E)} = \sum_{\eta\eta'} \langle \kappa \| \eta \rangle \langle \eta \| S_l(E) \| \eta' \rangle \langle \eta' \| \kappa' \rangle. \tag{2.2}$$

$\delta_l^{(\kappa)}(E)$ are called the eigenphase-shifts or eigenphases. The matrix $\langle \kappa \| \eta \rangle$ may be different for different values of E and, therefore, in general, a function of the scattering energy. From the time-reversal invariance it follows that the reduced S-matrix is symmetric [Equation (XIX.3.14)]. From the general theorems[1] of linear algebra it follows that for a symmetric unitary matrix there exists an orthogonal matrix that diagonalizes it. Thus, the $\langle \kappa \| \eta \rangle$ fulfill not only (2.1) but also

$$\langle \kappa \| \eta \rangle = \langle \eta \| \kappa \rangle = \langle \kappa \| \eta \rangle^*. \tag{2.1a}$$

[1] See, e.g., A. Lichnerowicz (1967) or V. I. Smirnov (1964), Vol. III, Part 1, Section 42.

Inverting (2.2) with the help of (2.1), we obtain

$$\langle \eta \| S_l(E) \| \eta' \rangle = \sum_\kappa e^{i2\delta_l^{(\kappa)}(E)} \langle \eta \| \kappa \rangle \langle \kappa \| \eta' \rangle. \tag{2.3}$$

To understand the physics behind this mathematical transformation of the reduced S matrix into diagonal form, let us consider the basis vectors that were used for the space of the initial states (a, T):

$$|E\, l\, l_3\, \eta \rangle = |E\, l\, l_3 \rangle \otimes |\eta \rangle.$$

The orthogonal matrix $\langle \eta \| \kappa \rangle$ transforms this basis system into a new basis system

$$|E\, l\, l_3\, \kappa \rangle = \sum_\eta |E\, l\, l_3\, \eta \rangle \langle \eta \| \kappa \rangle \tag{2.4}$$

with the quantum numbers $E\, l\, l_3$ unaffected but with new internal quantum numbers κ. The observables whose eigenvalues are κ do not commute with the observables whose eigenvalues are η:

$$[\kappa^{\mathrm{op}}, \eta^{\mathrm{op}}] \neq 0. \tag{2.5}$$

The quantum numbers η were chosen so that the initial state (a, T) is characterized by a definite value of η: $\eta = \eta_A = \eta(a, T)$. Now it may be that the quasistationary state R that is formed in the process $a + T \to R \to a + T$ does not have a definite value of the quantum numbers η but a definite value of the quantum numbers κ, say $\kappa = \kappa_R$. Then we can repeat the same considerations as in Sections XVIII.2 and XVIII.6—this time not for the internal quantum numbers η, but for the internal quantum numbers κ, in particular for that value κ_R of the quantum numbers κ that constitutes the internal quantum numbers of R. The time delay $t_D^{(\kappa_R)}$ and phase shift $\delta_l^{(\kappa_R)}$ are then characterized by this quantum number in the same way as in Chapter XVIII they are characterized by the quantum numbers η_A of the initial (and final) projectile-target system (a, T). Thus it is not for the quantum numbers η_A of the initial state (a, T) that (XVIII.6.23) holds, but for the quantum numbers κ_R of the resonance state R, which now has quantum numbers that are incompatible with the quantum numbers η_A, i.e., they fulfill (2.5). Thus instead of $S_l = \langle \eta \| S_l(E) \| \eta_A \rangle$, we now write Equation (XVIII.6.23) for $\langle \kappa_R \| S_l(E) \| \kappa_R \rangle$:

$$\langle \kappa_R \| S_l(E) \| \kappa_R \rangle = e^{i2\delta_l^R(\kappa_R)} e^{i2\gamma_l(\kappa_R)} = e^{i2\delta_l(\kappa_R)}$$

$$= \left(1 + \frac{i\Gamma_R}{E_R - E - i\Gamma_R/2} \right) e^{i2\gamma_l(\kappa_R)}. \tag{2.6}$$

$\delta_l^R(\kappa_R)$ here denotes the resonant phase shift, and $\gamma_l(\kappa_R)$ the background phase shift, so that the phase shift is again written

$$\delta_l(\kappa_R) = \delta_l^R(\kappa_R) + \gamma_l(\kappa_R).$$

The reduced S-matrix elements $\langle \eta \| S_l(E) \| \eta' \rangle$ between the (reduced) basis vectors $\| \eta \rangle$ have then to be expressed in terms of the reduced S-matrix elements $\langle\kappa\|S_l(E)\|\kappa'\rangle$ between the (reduced) basis vectors $\|\kappa\rangle$, using the transformation

$$\|\eta\rangle = \sum_{\kappa} |\kappa\rangle\langle\kappa\|\eta\rangle, \qquad (2.4')$$

which leads to (2.3) with

$$\delta_{\kappa\kappa'} e^{i2\delta_l^{(\kappa)}} = \langle\kappa\|S_l\|\kappa'\rangle.$$

The matrix element $\langle\kappa\|\eta\rangle$ is the number that characterizes the strength of the coupling between the states with the internal quantum numbers η and κ. As mentioned above, they depend in general upon the energy E.

It may of course be that a quasistationary state is formed not only for one value κ_R of the quantum numbers κ, but also for other values $\kappa = \kappa_{X_1}, \kappa_{X_2}, \ldots$. This may even happen at the same value l of the angular momentum and even at close energies E_R and E_{X_1}. We assume for the sake of definiteness that there are two values κ_R and κ_X for which quasistationary states are formed with angular momentum l, and their energies and widths are (E_R, Γ_R) and (E_X, Γ_X), respectively. Then in addition to (2.6) one has also

$$\langle\kappa_X\|S_l(E)\|\kappa_X\rangle = e^{i2\delta_l(\kappa_X)} = \left(1 + \frac{i\Gamma_X}{E_X - E - i\Gamma_X/2}\right)e^{i2\gamma_l(\kappa_X)}. \qquad (2.7)$$

We also assume that all the other S-matrix elements are nonresonant, i.e.,

$$\langle\kappa\|S_l(E)\|\kappa\rangle = e^{i2\delta_l(\kappa)} = e^{i2\gamma_l(\kappa)} \quad \text{for} \quad \kappa \neq \kappa_R, \kappa_X, \qquad (2.8)$$

where $\gamma_l(\kappa)$ are the background phase shifts, which are functions of energy E. One may be tempted to assume that the background phase shifts in the resonant and non-resonant channels are slowly varying functions of the energy. This is, however, in general not the case[2] and we shall not make use of such an assumption here.

We shall call the above-described object a double resonance or double multichannel resonance of quantum numbers (κ_R, κ_X). Thus the formation experiment in the scattering process

$$a + T \rightarrow a' + T$$

can go two ways:

$$(a + T)_{\text{with q.n. } \eta_A} \quad \overbrace{\begin{array}{l} \rightarrow R_{\text{with q.n. } \kappa_R} \\ \searrow X_{\text{with q.n. } \kappa_X} \end{array}} \quad \rightarrow (T' + a')_{\text{with q.n. } \eta}.$$

---

[2] These rapid energy variations follow from Wigner's eigenphase repulsion theorem: The eigenphases $\delta_l(\kappa)$ do not cross (modulo π). As a consequence, near the resonance energy in the channel κ_R the eigenphases in all other channels κ must also vary. This is assured by the existence of branch cuts in the individual eigenphases (and the corresponding transition matrix elements $\langle\kappa\|\eta\rangle$ which are also functions of E) which are in the complex energy plane much closer to the real axis than the resonance pole. These branch cuts do not occur in the complete S matrix and therefore do not have physical significance. The occurrence of these branch cuts in the $\langle\kappa\|\eta\rangle$ and $\gamma_l(\kappa)$ must be just such that all these branch cuts cancel out in the background term (2.16), which can be a slowly varying function of energy. A discussion of these singularities of the individual eigenphases is given in C. J. Goebel, K. W. McVoy, Phys. Rev. **164**, 1932 (1967); H. A. Weidenmuller, Phys. Letters **24B**, 441 (1967); and R. H. Dalitz, R. G. Moorhouse, Proc. Roy. Soc. **A318**, 279 (1970).

Figure 2.1

In the case of elastic scattering, $\eta = \eta_A$, $(T' = T, a' = a)$; in the case of a reaction, $\eta = r \neq \eta_A$. If one wants to illustrate this by a diagram analogous to Figure XVIII.1.2, one would draw Figure 2.1.

The S-matrix for the double multichannel resonance is obtained by inserting (2.6), (2.7) and (2.8) into (2.3):[3]

$$\langle\eta\|S_l(E)\|\eta'\rangle = \sum_\kappa \langle\eta\|\kappa\rangle\langle\kappa\|\eta'\rangle e^{i2\gamma(\kappa)} + \frac{i\Gamma_R\langle\eta\|\kappa_R\rangle\langle\kappa_R\|\eta'\rangle e^{i2\gamma(\kappa_R)}}{E_R - E - i\Gamma_R/2}$$

$$+ \frac{i\Gamma_X\langle\eta\|\kappa_X\rangle\langle\kappa_X\|\eta'\rangle e^{i2\gamma(\kappa_X)}}{E_X - E - i\Gamma_X/2}. \tag{2.9}$$

The sum over κ extends over all channels including the resonating channels κ_R and κ_X.

The amplitude for elastic scattering connected to the S matrix by (XVI.1.48), or (XVI.2.8), is obtained from (2.9) with $\eta = \eta' = \eta_A$ after a short calculation using (2.1):

$$pT_l(p) = e^{i2\gamma(\kappa_R)}\left(\frac{\frac{1}{2}\Gamma_R|\langle\eta_A\|\kappa_R\rangle|^2}{E_R - E - i\Gamma_R/2} + \frac{\frac{1}{2}\Gamma_X|\langle\eta_A\|\kappa_X\rangle|^2 e^{i2(\gamma(\kappa_X) - \gamma(\kappa_R))}}{E_X - E - i\Gamma_X/2}\right)$$

$$+ \sum_\kappa \langle\eta_A\|\kappa\rangle\langle\kappa\|\eta_A\rangle e^{i\gamma(\kappa)} \sin\gamma(\kappa). \tag{2.10}$$

In the case of inelastic scattering from the initial state with quantum numbers $\eta' = \eta_A$ into the final state with quantum numbers $\eta = r$, the scattering amplitude (XVI.1.48) is obtained from (2.9) as

$$\sqrt{p_r p}\,T_{r,l}^{(A\to r)}(p) = e^{i2\gamma(\kappa_R)}\left(\frac{\frac{1}{2}\Gamma_R\langle r\|\kappa_R\rangle\langle\kappa_R\|\eta_A\rangle}{E_R - E - i\Gamma_R/2}\right.$$

$$\left. + \frac{\frac{1}{2}\Gamma_X\langle r\|\kappa_X\rangle\langle\kappa_X\|\eta_A\rangle e^{i2(\gamma(\kappa_X) - \gamma(\kappa_R))}}{E_X - E - i\Gamma_X/2}\right)$$

$$+ \sum_\kappa \langle r\|\kappa\rangle\langle\kappa\|\eta_A\rangle e^{i2\gamma(\kappa)}/(2i) \tag{2.11}$$

Equations (2.9)–(2.11) are approximations to the same extent to which (2.6) and (2.7) and (XVIII.6.23) are approximations. They have been derived here from the ordinary Breit–Wigner approximation and the assumption that the resonance (or quasistationary state) has internal quantum numbers κ incompatible with the internal quantum numbers η of the initial and final state. Due to this derivation (2.9)–(2.11) contain the individual eigenphases $\gamma(\kappa)$ though these eigenphases belonging to the intermediate quantum

---

[3] We omit the index l in the phase shift γ_l and δ_l whenever possible. γ then always denotes the background phase of that partial wave in which the resonance occurs.

numbers κ may be unphysical quantities. The physically observable quantities are the resonance terms and the total background term (summed over all κ) and only these quantities should be compared with the experimental data.

XX.2a Single Multichannel Resonance

Before we continue the discussion of a double resonance, we treat the simpler case that there is only one resonance with quantum number κ_R. Such a resonance we shall call a single multichannel resonance. This case is obtained from the above formulas (2.10) and (2.11) by omitting the second term in the bracket.

It is customary to introduce the following notation:

$$\Gamma_\eta^{1/2} = \langle \eta \| \kappa_R \rangle \Gamma^{1/2}, \qquad \Gamma = \Gamma_R. \tag{2.12}$$

At this point we use time-reversal invariance. If time-reversal invariance holds, then the unitary matrix $\langle \eta \| \kappa \rangle$ is orthogonal, i.e., the matrix elements are real [Equation (2.1a)]. Thus for time-reversal-invariant interactions $\Gamma_\eta^{1/2}$ are real.[4]

Because of (2.1), the Γ_η fulfill

$$\sum \Gamma_\eta = \Gamma = \Gamma_R. \tag{2.13}$$

Γ_η is called the partial width for the channel η. With this notation the elastic scattering amplitude for a single resonance of quantum number κ_R is written

$$p T_l(p) = e^{i2\gamma(\kappa_R)} \frac{\Gamma_{\eta_A}/2}{E_R - E - i\Gamma/2} + b(\eta_A), \tag{2.14}$$

and the reaction amplitude is given by:

$$\sqrt{p_r p}\, T_{r,l}(p) = e^{i2\gamma(\kappa_R)} \frac{\frac{1}{2}\Gamma_r^{1/2}\Gamma_{\eta_A}^{1/2}}{E_R - E - i\Gamma/2} + b(r), \tag{2.15}$$

where

$$b(\eta_A) = \sum_\kappa \langle \eta_A \| \kappa \rangle \langle \kappa \| \eta_A \rangle e^{i\gamma(\kappa)} \sin \gamma(\kappa) \tag{2.16A}$$

and

$$b(r) = \sum_\kappa \langle r \| \kappa \rangle \langle \kappa \| \eta_A \rangle e^{i2\gamma(\kappa)}/(2i), \tag{2.16r}$$

which can be a slowly varying function of energy even if the individual $\gamma(\kappa)$ vary rapidly. The representation (2.14), (2.15) is called the generalized Breit–Wigner approximation. These expressions are very similar to the corresponding expression (XVIII.6.26) for the one-channel resonance, except that the height of the multichannel resonance is smaller by the factor

---

[4] If time-reversal invariance does not hold, $\Gamma_\eta^{1/2}$ is complex and one defines Γ_η by $\Gamma_\eta = |\Gamma_\eta^{1/2}|^2$. We shall not discuss this case here.

$\langle \eta_A \| \kappa_R \rangle^2$ or $\langle \eta_A \| \kappa_R \rangle \langle r \| \kappa_R \rangle$. The first factor represents the transition amplitude (square root of the transition probability) for a transition from a state with quantum numbers η_A to a state with quantum numbers κ_R and back to a state with quantum numbers η_A; the second factor represents the transition amplitude from η_A to κ_R and from κ_R to r. Thus the smaller height is due to the inelasticity, i.e., the fact that the resonance contributes not only to the channel η_A or r, but also to all the other channels $\eta \neq \eta_A$ or $\eta \neq r$.

The cross sections corresponding to the amplitudes (2.15) will then not look much different from the expressions for the cross sections given in Sections XVIII.6 ..nd XVIII.9. Using the abbreviations:

$$\epsilon = \frac{E_R - E}{\Gamma/2}, \tag{2.17}$$

$$R = R^\eta = \langle \eta \| \kappa_R \rangle \langle \kappa_R \| \eta_A \rangle, \tag{2.18}$$

where κ_R are the quantum numbers of the resonance R, and η_A the quantum numbers of the initial state $A = (a, T)$, one can write the lth partial cross section $\sigma_l^{(A \to \eta)}$ using (2.14) for $\eta = \eta_A$ and (2.15) for $\eta = r$ with (2.16):

$$\sigma_l^{(A \to \eta_A)} = 4\pi(2l + 1)\frac{1}{p^2}\left| \sum_\kappa \langle \eta_A \| \kappa \rangle \langle \kappa \| \eta_A \rangle e^{i\gamma(\kappa)} \sin \gamma(\kappa) + e^{i2\gamma(\kappa_R)} \frac{R^{\eta_A}}{\epsilon - i} \right|^2$$

$$\tag{2.19A}$$

For $\eta = r$ one obtains from (2.15) with (2.16r):

$$\sigma_l^{(A \to r)} = 4\pi(2l + 1)\frac{1}{p^2}\left| \sum_\kappa \langle r \| \kappa \rangle \langle \kappa \| \eta_A \rangle \frac{e^{i2\gamma(\kappa)}}{2i} + e^{2i\gamma(\kappa_R)} \frac{R^r}{\epsilon - i} \right|^2. \tag{2.19r}$$

After some calculation this can be written as

$$\sigma_l^{(A \to \eta_A)} = 4\pi(2l + 1)\frac{1}{p^2}\left[|b(\eta_A)|^2 + \frac{R^{\eta_A}}{\epsilon^2 + 1}\left(R^{\eta_A} + \sum_\kappa \langle \eta_A \| \kappa \rangle \langle \kappa \| \eta_A \rangle \right. \right.$$

$$\left. \left. \times \sin \gamma(\kappa)(2\epsilon \cos\{2(\gamma(\kappa_R) - \gamma(\kappa))\} - 2 \sin\{2(\gamma(\kappa_R) - \gamma(\kappa))\}) \right) \right],$$

$$\tag{2.20A}$$

$$\sigma_l^{(A \to r)} = 4\pi(2l + 1)\frac{1}{p^2}\left[|b(r)|^2 + \frac{R^r}{\epsilon^2 + 1}\left(R^r + \sum_\kappa \langle r \| \kappa \rangle \langle \kappa \| \eta_A \rangle \right. \right.$$

$$\left. \left. \times (-\epsilon \sin\{2(\gamma(\kappa_R) - \gamma(\kappa))\} - \cos\{2(\gamma(\kappa_R) - \gamma(\kappa))\}) \right) \right], \tag{2.20r}$$

where $b(\eta_A)$ and $b(r)$ are given by (2.16). Equation (2.20A) is the elastic cross section and (2.20r) is the reaction cross section.

For an energy E far away from the resonance energy E_R, i.e., for $|\epsilon| \gg 0$, the second term in the bracket on the right-hand side of (2.20) is very small and vanishes for $|\epsilon| \to \infty$. The cross section is then given by

$$\sigma_l^{(A \to \eta)} \to \sigma_{(nr)} = 4\pi(2l + 1)\frac{1}{p^2}|b(\eta)|^2 \quad \text{for} \quad |\epsilon| \to \infty. \tag{2.21}$$

Thus we see that (2.20) is very similar to (XVIII.9.14): a single resonance in the many-channel case will appear as a structure above a gradually varying background. The shape of this structure will in general not be the same as the one given in Figure XVIII.9.2 and will depend upon the various background phase shifts γ. However, if these background phase shifts are, in the neighborhood of the resonance energy $E \approx E_R$, approximately the same modulo π for all quantum numbers κ:

$$\gamma_l(\kappa) = \gamma_l(E) = \gamma(E), \tag{2.22}$$

then one can write (using $\langle \eta_A | \eta_A \rangle = \sum_\kappa \langle \eta_A \| \kappa \rangle \langle \kappa \| \eta_A \rangle = 1$),

$$\sigma_l^{(A \to \eta)} \to \sigma_{(nr)} + \frac{Rq^2 + 2q\epsilon - 2 + R}{\epsilon^2 + 1} R\sigma_l^{(bg)} \quad \text{for} \quad \eta = \eta_A, \tag{2.23}$$

where q is again defined by

$$q = \cot \gamma_l(E) \tag{2.24}$$

and $\sigma_l^{(bg)}$ is given by

$$\sigma_l^{(bg)} = 4\pi(2l + 1) \frac{1}{p^2} \sin^2 \gamma. \tag{2.25}$$

Equation (2.23) is, indeed, very similar to (XVIII.9.14) with (XVIII.9.6). The shape of a single resonance in the many-channel case will, therefore, be very similar though not identical to the shape depicted in Figure XVIII.9.2.

From (2.20) we see that the multichannel resonance shows up in the elastic channel, $\eta = \eta_A$, as well as in all the reaction channels $\eta = r$, at the same energy E_R and with the same width Γ. The shape of the resonance depends upon the background phase shifts $\gamma(\kappa)$. And even if (2.22) holds, the resonance with quantum number κ_R will have different though similar shapes in the various channels η, because R in (2.23) depends in general upon η according to (2.18).

However, if one has the particular case that the background for all quantum numbers κ—the resonating $\kappa = \kappa_R$ and all the nonresonating κ—is negligible, then one has

$$\sigma_l^{(R)(A \to \eta)} = 4\pi(2l + 1) \frac{1}{p^2} \frac{\frac{1}{4}\Gamma_\eta \Gamma_{\eta_A}}{(E_R - E)^2 + \Gamma^2/4}. \tag{2.26}$$

Thus, the resonance without background appears in all channels as a Breit–Wigner resonance at the same energy and with the same width, but with different heights depending upon the value of the partial width Γ_η.

Further, since η_A can really take any value, the same Breit–Wigner resonance also appears for any initial state. Thus, independent of what the initial and final states of a scattering process are, if only their transition amplitudes to the state with quantum numbers κ_R are different from zero, the resonance will appear at the same energy and with the same width. This is, of course, what one must expect of a quasistationary state, which is entirely determined by its "quantum numbers" $(E_R, \Gamma, l, \kappa_R)$ and independent of the experimental conditions.[5]

Experimental examples of single multichannel resonances are shown in Figure XIX.3.1. This figure shows the various resonances ("energy levels") of Si^{28} as they occur in the processes

$$A: \quad Al^{27} + p \to (Si^{28})^* \to p + Al^{27},$$

$$B: \quad Al^{27} + p \to (Si^{28})^* \to \alpha + Mg^{24},$$

$$C: \quad Mg^{24} + \alpha \to (Si^{28})^* \to p + Al^{27},$$

$$D: \quad Mg^{24} + \alpha \to (Si^{28})^* \to \alpha + Mg^{24}.$$

(Si^{28})-resonances are observed at the following energy values:[6] 2.75, 2.93, 2.95, 3.00, 3.14, 3.20, 3.28 MeV. Fine discrepancies between the curves—like the double peak around 3.4 MeV in process B, which is matched only by a single peak in procees C—can be explained in terms of differences in the energy resolution for the experiments.

Let η_A denote the quantum numbers of the p-Al^{27} system, r the quantum numbers of α-Mg^{24}, and $\kappa_{R_1}, \kappa_{R_2}, \ldots$ the quantum numbers of the various (Si^{28})-resonances. The energy values of these resonances are sufficiently far apart that one can consider each separately as a single resonance and need not apply the considerations for a double resonance described below.

We see that in general, each single resonance occurs in all four processes. However, whereas in the reactions B and C they usually occur as pure Breit–Wigner resonance, in the elastic process A and in the process D they have considerable background phases. For example, the resonance at 3.203 MeV, whose quantum number we may call κ_{R_6}, occurs in B and C as a bump and in A and D as a dip. This indicates that $\langle \eta_A \| \kappa_{R_6} \rangle$ as well as $\langle r \| \kappa_{R_6} \rangle$ are different from zero, and that the sum over κ in (2.20) is negligible for $\eta = r$ but not negligible for $\eta = \eta_A = $ (quantum numbers of the p-Al^{27} system) and $\eta = r = $ (quantum numbers of the α-Mg^{24} system). A possible explanation for this effect is that the background phases $\gamma(\kappa)$ are not negligible, but approximately independent of κ, so that for the elastic processes something like (2.23) holds. Under the same assumption one obtains for the inelastic processes, using $\langle \eta | \eta_A \rangle = 0$,

$$\sigma_l^{(A \to \eta)} \approx \sigma_{(nr)} + \frac{(\langle \eta \| \kappa_R \rangle \langle \kappa_R \| \eta_A \rangle)^2}{\epsilon^2 + 1}, \tag{2.23'}$$

[5] A resonance is defined by the pair of poles of the l-th partial S-matrix element $\langle \eta \| S_l(E) \| \eta_A \rangle$ at the positions $E = E_R \mp i\Gamma/2$; it has the additional ("intrinsic"- or "channel"-) quantum numbers κ_R. If the matrix elements $\langle \kappa_R \| \eta \rangle$ and $\langle \eta \| \kappa_R \rangle$ are different from zero this resonance will show up in the transition from channel η_A to channel η. The resonance is an autonomous physical entity characterised by the quantum numbers $(E_R, \Gamma, l, \kappa_R)$ in the same way as a stationary state is characterised by the quantum numbers (E_0, l, κ) because its lifetime $\hbar/\Gamma$ is infinite.

[6] These are the energy values on the α-particle energy scale. The difference in the origin of the energy axis for the α-particle energy and the proton energy is 1.61 MeV, and the energy scales of the processes A, B and C, D have been adjusted by bringing each pair of resonance peaks of the reaction curves B and C into coincidence.

so that in a reaction the resonance occurs as a pure Breit–Wigner above a possible background $\sigma_{(nr)}$.

XX.2b Double Multichannel Resonance

We now return to the case of two resonating lth partial waves: one with the internal quantum numbers κ_R and energy and width (E_R, Γ_R), and the other with internal quantum numbers κ_X and resonance parameters (E_X, Γ_X). We start from (2.10) and (2.11) and consider simultaneously the cases of elastic scattering of (2.10) $\eta = \eta_A$ and of the reaction (2.11), $\eta = r$. We use the notation

$$R = R^\eta = \langle \eta \| \kappa_R \rangle \langle \kappa_R \| \eta_A \rangle,$$

$$X = X^\eta = \langle \eta \| \kappa_X \rangle \langle \kappa_X \| \eta_A \rangle e^{i2(\gamma(\kappa_X) - \gamma(\kappa_R))}. \qquad (2.27)$$

We shall always assume time-reversal invariance (2.1b), so that R is real and the phase of X is entirely determined by the difference of the background phases. Equations (2.10) and (2.11) are then written

$$pT_l(p) = e^{i2\gamma(\kappa_R)} \left(\frac{(\Gamma_R/2)R^\eta}{E_R - E - i\Gamma_R/2} + \frac{(\Gamma_X/2)X^\eta}{E_X - E - i\Gamma_X/2} \right) + b, \qquad (2.28)$$

where b is given by (2.16).

To simplify our following discussion we ignore the background term b. It will have effects similar to those discussed above for the single resonance and in Section XVIII.9.

The cross section for the double multichannel resonance then becomes (if we assume for the sake of simplicity that X is real):

$$\sigma_l^{(R)} = 4\pi(2l + 1)\frac{1}{p^2}\left(\frac{(\Gamma_R^2/4)R^2}{(E_R - E)^2 + \Gamma_R^2/4} + \frac{(\Gamma_X^2/4)X^2}{(E_X - E)^2 + \Gamma_X^2/4} \right.$$

$$\left. + \frac{(\Gamma_R/4)\Gamma_X 2RX[(E_R - E)(E_X - E) + (\Gamma_R/4)\Gamma_X]}{[(E_R - E)^2 + (\Gamma_R^2/4)][(E_X - E)^2 + (\Gamma_X^2/4)]} \right). \qquad (2.29)$$

Thus the lth partial cross section is the sum of two Breit–Wigner cross sections, with different widths and different heights, and a third term (the interference term). A double resonance is characterized by four parameters $(E_R, \Gamma_R; E_X, \Gamma_X)$ and the quantities R and X, which depend according to their definition (2.27) upon the final-state quantum numbers η and also upon the initial-state quantum numbers η_A. Thus for one channel R and X may have the same sign, and for another channel they may have opposite signs. *The shape of the cross section of a double resonance, therefore, varies with the channel quantum number.* If in one channel it appears as a double hump, it may show up in another channel as one single broad hump. The most favorable circumstances for a double hump are $R \approx X$. Then the third term is (2.29) has two maxima, one near $E = E_R$ and the other near $E = E_X$, which enhance the humps at these values that come from the two Breit–Wigner cross sections in (2.29). Then the double resonance will appear in the cross section as two maxima with different widths, provided the background can be ignored.

To obtain an idea of the possible shapes of such a double resonance, let us consider some special cases. It is useful to introduce the new variable

$$\mathscr{E} = E - \frac{E_R + E_X}{2} \tag{2.30}$$

and denote

$$\Delta = E_R - E_X. \tag{2.31}$$

Then the cross section formula (2.29) is written as

$$\sigma_l^{(R)} = 4\pi(2l + 1)\frac{1}{p^2}\frac{1}{4}\left(\frac{R^2\Gamma_R^2}{(\Delta/2 - \mathscr{E})^2 + (\Gamma_R/2)^2} + \frac{X^2\Gamma_X^2}{(-\Delta/2 - \mathscr{E})^2 + (\Gamma_X/2)^2}\right.$$

$$\left. + \frac{2RX(\mathscr{E}^2 - (\Delta/2)^2 + \Gamma_R\Gamma_X/4)\Gamma_R\Gamma_X}{[(\Delta/2 - \mathscr{E})^2 + (\Gamma_R/2)^2][(-\Delta/2 - \mathscr{E})^2 + (\Gamma_X/2)^2]}\right). \tag{2.32}$$

The interesting situation arises when E_R is close but not identical to E_X and if the widths of the two resonances have approximately the same magnitude as $\Delta = E_R - E_X$. Therefore we shall consider the case

$$\Gamma_R = \Gamma_X = \Delta. \tag{2.33_1}$$

Note that in this case the third term in (2.32) has a zero and a minimum (if sign R = sign X) or maximum (if sign R = $-$sign X). With (2.33) one obtains from (2.32)

$$\sigma_l^{(R)} = 4\pi(2l + 1)\frac{1}{p^2}\frac{\Gamma^2}{4}\left(\frac{R^2}{(\Delta/2 - \mathscr{E})^2 + (\Delta/2)^2}\right.$$

$$\left. + \frac{X^2}{(-\Delta/2 - \mathscr{E})^2 + (\Delta/2)^2} + \frac{2RX\mathscr{E}^2}{\mathscr{E}^4 + (\Delta^2/2)^2}\right). \tag{2.34}$$

The term in the bracket of (2.34)—i.e., $\sigma_l^{(R)}$ except for the factor

$$4\pi(2l + 1)\frac{1}{p^2}\frac{\Gamma^2}{4}$$

—is depicted in Figure 2.2 for various values of X ($R = 1$). Ignoring $e^{i2(\gamma(\kappa_R) - \gamma(\kappa_X))}$, X is connected to the coupling constants according to (2.27) by

$$X = \frac{\langle\eta\|\kappa_X\rangle\langle\kappa_X\|\eta_A\rangle}{\langle\eta\|\kappa_R\rangle\langle\kappa_R\|\eta_A\rangle}.$$

The correspondence between the curves and the values of X is as given in the legend. We see that the cross section may look quite different for different couplings between the initial and final channels and the resonance channel. For $X = 1$ one obtains a double peak with the maxima slightly

Figure 2.2 Cross section for double resonance: the case (2.33_1) for various values of X:

Symbol:	$\times$	$+$	$\triangledown$	$\triangle$	$\bigcirc$
X:	1.0	0.5	0.0	-0.5	-1.0

below $E = E_X$ ($\mathscr{E} = -\Delta$) and slightly above $E = E_R$ ($\mathscr{E} = +\Delta$). For $X = -1$, the cross section has one single broad bump centered around $E = \frac{1}{2}(E_R + E_X)$. For $X = 0$ there is, of course, only one bump at $E = E_R$, representing the resonance R, as in this case the resonance X does not couple to the channels η or η_A. But also for $X = -0.5$ there is only one single bump, which however does not have the form of a Breit–Wigner resonance. Thus, for different final channels η, a double resonance may show up in the cross section as a double peak, as a broad single peak with maximum between the two resonance energies, or as a single peak around one of the resonance energies.

The cross sections in Figure 2.2 correspond to the most symmetrical case, (2.33_1). If Γ_X or Γ_R differ from Δ the picture changes slightly. Figure 2.3 depicts (2.32) [except for the factor $4\pi(2l + 1)(1/p^2)\Gamma^2$] for the case

$$\Gamma_X = \Delta, \qquad \Gamma_R = \tfrac{1}{2}\Delta. \tag{2.33_2}$$

The variation in the shape of the cross section with the coupling between the channels is similar to that in Figure 2.2.

In the above considerations we have ignored the background term b, so that the above discussions are valid only if (2.16) vanishes. Also we have ignored $\gamma(\kappa_X) - \gamma(\kappa_R)$. From our discussions in Section XVIII.9, and from the footnote on p. 520 we know already that the background phase shifts may be large, so that this idealization may well be inadmissable. Taking the interference between resonances into account will make the complicated features of a double resonance even more complicated. Taking in addition

Figure 2.3 Cross section for double resonance: the case (2.33_2). Symbols as in Figure 2.2.

the background from other partial waves and the resolution of the experimental apparatus into account, one concludes that the observation of a double multichannel resonance as a double peak in the cross section is an unlikely possibility. Thus two resonances with different quantum numbers κ_R and κ_X but with energies E_R and E_X, that differ from each other by an amount comparable with the width and that both couple to the initial and final channel, will be very hard to observe.

Nonetheless, examples of double resonances have been seen in nuclear scattering processes. The most prominent examples are the Be^8 energy levels with (angular momentum)$^{parity} = l^P = 2^+$. These resonances have been observed in the processes

$$He^4 + He^4 \rightarrow (Be^8)^* \rightarrow He^4 + He^4 \tag{2.35_1}$$

and

$$He^4 + He^4 \rightarrow (Be^8)^* \rightarrow Be^8 + \gamma, \tag{2.35_2}$$

at excitation energies of 16.6 and 16.9 MeV of the Be^8, with widths of approximately 110 and 80 keV, respectively. We do not want to describe the details of this system, and shall just establish the connection with the quantities of our formulas (2.28) and (2.29). We ignore the background phases; then (2.28) can be written

$$pT_i^{A \rightarrow \eta_A} = \frac{\frac{1}{2}\Gamma_X \langle \eta_A \| \kappa_X \rangle \langle \kappa_X \| \eta_A \rangle}{E_X - E - i\Gamma_X/2} + \frac{\frac{1}{2}\Gamma_R \langle \eta_A \| \kappa_R \rangle \langle \kappa_R \| \eta_A \rangle}{E_R - E - i\Gamma_R/2} \tag{2.36_1}$$

and

$$pT_i^{A \rightarrow \eta} = \frac{\frac{1}{2}\Gamma_X \langle \eta \| \kappa_X \rangle \langle \kappa_X \| \eta_A \rangle}{E_X - E - i\Gamma_X/2} + \frac{\frac{1}{2}\Gamma_R \langle \eta \| \kappa_E \rangle \langle \kappa_R \| \eta_A \rangle}{E_R - E - i\Gamma_R/2}. \tag{2.36_2}$$

Figure 2.4 Fits of the doublet structure in Be^8 in the He^4-He^4 channel, Note that the energy decreases from left to right. [From W. D. Callender *et al.*[8] with permission.]

Here $|\eta_A\rangle$ denotes the internal state of the (He^4, He^4) system, $|\eta\rangle$ denotes the state of the (Be^8, γ) system, and $|\kappa_X\rangle$ and $|\kappa_R\rangle$ denote the internal states of the Be^8* (16 MeV) and Be^8* (16.9 MeV), respectively, so that (2.36$_i$) is the scattering amplitude for the process (2.35$_i$). Depending upon the ratio $\langle\eta_A\|\kappa_X\rangle^2/\langle\eta_A\|\kappa_R\rangle^2$, the cross section for the process (2.35$_1$) will then have the shape of one of the curves in Figure 2.2 or 2.3. The cross sections have been measured[7,8] and the experimental data points together with a fit to the data are plotted in Figure 2.4.[8]

The curve in the upper plot shows a fit to the data obtained with an expression (2.29) for a double resonance. The lower plot illustrates the fit with the sum of two Breit–Wigner cross sections,

$$\sigma \sim \frac{A_1}{(E_X - E)^2 + (\Gamma_X/2)^2} + \frac{A_2}{(E_R - E)^2 + (\Gamma_R/2)^2}, \qquad (2.37)$$

[7] A. D. Bacher, F. G. Resmini, H. E. Conzett, R. deSwinairski, H. Meiner, J. Ernst, *Phys. Rev. Letters* **29**, 1331 (1972).

[8] W. D. Callender, C. P. Browne, *Phys. Rev.* **2**, 1 (1970). This figure does not give the cross section for the formation process (2.35$_1$) but for the production process

$$B^{10} + d \rightarrow \alpha + Be^8*$$
$$\quad\quad\quad \hookrightarrow \alpha + \alpha.$$

R and X in (2.29) are therefore not

$$R = \langle\eta_A\|\kappa_R\rangle\langle\kappa_R\|\eta_A\rangle, \quad X = \langle\eta_A\|\kappa_X\rangle\langle\kappa_X\|\eta_A\rangle$$

but

$$R = g_R\langle\kappa_R\|\eta_A\rangle, \quad X = g_X\langle\kappa_X\|\eta_A\rangle$$

with

$$g_R \approx g_X.$$

Figure 2.5 Fits of the double structure in Be^8 in the Be^8-γ channel. [From Nathan et al.[9], *Phys. Rev. Lett.* **35**, 1137 (1975), with permission.]

which would hold if the states at 16.6 and 16.9 MeV were unrelated single resonances. This fit with two noninterfering Breit–Wigner curves is clearly unacceptable, whereas the fit in the upper plot with a double resonance (two interfering Breit–Wigner curves) is excellent. Thus we conclude that the double-peak structure observed around 16.6 and 16.9 MeV of the Be^{8*} system is a double resonance. Both bumps have approximately the same height, the bump at 16.9 McV is slightly narrower, and the fitted curve resembles the curve (marked $\times$) in Figure 2.3 for $X/R \approx 1$. Both states $|\kappa_R\rangle$ and $|\kappa_X\rangle$ couple, therefore, with approximately the same strength to the incident channel:

$$\langle \kappa_X \| \eta_A \rangle \approx \langle \kappa_R \| \eta_A \rangle. \qquad (2.38)$$

Figure 2.5 shows a plot of the experimental data[10] for the reaction (2.35$_2$). The solid line is a fit with (2.29); the dashed line is a fit with (2.37). Again the sum of two noninterfering Breit–Wigner curves is clearly ruled out, and a double resonance gives a very good fit. This time, however the curve resembles more the curve (marked $+$) in Figure 2.3 for $X/R \approx 0.5$. From this and (2.38) we conclude that the coupling of the two $|\kappa_R\rangle$ and $|\kappa_X\rangle$ to the outgoing channel (Be^8, γ) fulfills the relation

$$|\langle \eta \| \kappa_X \rangle| < |\langle \eta \| \kappa_R \rangle|. \qquad (2.39)$$

The experimental data for the Be^8 system around 16.6 and 16.9 MeV give a beautiful illustration of the fact that a double resonance may show up in the cross section with a variety of shapes.

[9] A. M. Nathan, G. T. Garvey, P. Paul, E. K. Warburton, *Phys. Rev. Lett.* **35**, 1137 (1975).

XX.3 Argand Diagrams for Inelastic Resonances

In Section XVIII.8 we discussed the behavior of $pT_l(p)$ in the Argand diagram for the case of purely elastic scattering. In the second part of that section we also described how $T_l(p)$ can be obtained from the experimental data. In the present section we shall give a brief discussion of the behavior of the elastic partial-wave amplitude $pT_l(p)$ in the presence of inelastic processes and of the partial-wave reaction amplitude $\sqrt{p_r p}\, T_{r,l}(p)$. The determination of these amplitudes from the experimental data follows essentially the same procedure as described in the second part of the Section XVIII.8: for the determination of $T_l(p)$ one uses the elastic differential cross sections, and for $T_{r,l}(p)$ the differential cross section for the reaction (initial state η_A) $\rightarrow$ (final state r).

We will now assume that there is a single multichannel resonance and discuss how this single resonance will show up in the Argand diagrams for the elastic scattering amplitude for the reaction amplitude. We first treat the idealized case without background. The scattering amplitudes are then given according to (2.14) and (2.15) by

$$pT_l^{(R)}(p) = \frac{\frac{1}{2}\Gamma_{\eta_A}}{E_R - E - i\Gamma/2}, \tag{3.1}$$

$$\sqrt{p_r p}\, T_{r,l}^{(R)}(p) = \frac{\frac{1}{2}\Gamma_r^{1/2}\Gamma_{\eta_A}^{1/2}}{E_R - E - i\Gamma/2}. \tag{3.2}$$

Here p is the initial momentum and p_r is the final momentum, which is a function of the initial momentum and the difference of internal energies in the initial and final states.[10]

Γ_r and Γ_{η_A} are the partial widths (2.12) for the reaction channel r and for the initial channel η_A, respectively. E_R and $\Gamma_R = \Gamma$ are the energy and total width of the resonance. We use again the abbreviation

$$\epsilon = (E_R - E)\frac{2}{\Gamma} \tag{3.3}$$

and also define

$$\sqrt{x_\eta} = \langle \eta \| \kappa_R \rangle = \sqrt{\Gamma_\eta/\Gamma}. \tag{3.4}$$

$\sqrt{x_\eta}$ may be negative.

Equations (3.1) and (3.2) can then be written

$$pT_l^{(R)} = \frac{x_{\eta_A}}{\epsilon - i}, \tag{3.5}$$

$$\sqrt{p_r p}\, T_{r,l}^{(R)} = \frac{\sqrt{x_r x_{\eta_A}}}{\epsilon - i}. \tag{3.6}$$

---

[10] For a massive nonrelativistic projectile of mass m_A before and mass m_b after the reaction, $p_r = p_b$ is given by the formula (XVI.1.24).

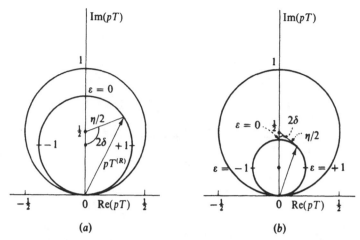

Figure 3.1 The resonant amplitude for the elastic channel is shown for two different values of the elasticity x_{η_A}: (a) $x_{\eta_A} = 0.75$ and the phase shift for the amplitude goes through 90° at the resonant energy; (b) $x_{\eta_A} = 0.4$ and $\delta = 0°$ at the resonant energy. Notice that the phase of the resonant amplitude in the resonance circle goes through 90° in both cases. [From Barbaro–Galtieri (1968), with permission.]

$x_{\eta_A} = \langle \eta_A \| \kappa_R \rangle^2$ is called the elasticity of the resonance and is the transition probability for a transition from a state with quantum numbers κ_R into a state with quantum numbers η_A at a particular value of E. Thus x_{η_A} represents the "fraction" of the resonance with internal quantum numbers κ_R that is coupled to the elastic channel η_A. For $x_{\eta_A} = 1$ the resonance couples to the elastic channel only, all x_η with $\eta \neq \eta_A$ are zero, and $pT_l^{(R)}$ describes the unitary circle for an elastic resonance as described in Section XVIII.8. The elasticity x_{η_A} is related to the inelasticity coefficient $\eta_l(E)$ defined in (XVI.2.10). At the resonance energy this relation is $\eta_l(E_R) = 2x_{\eta_A} - 1$. In contrast to $\eta_l(E)$, which has a strong energy dependence, empirical data show that the elasticity x_{η_A} is only very weakly energy-dependent and is therefore a more convenient parameter.

The Argand diagram of (3.5) is depicted in Figure 3.1.[11] It is a circle of diameter x_{η_A} lying inside the unitary circle. The resonance energy, $\epsilon = 0$, lies on the top of the circle, and the points $\epsilon = \mp 1$ correspond to $E = E_R \pm \Gamma/2$. One has to distinguish two cases:

(a) $x_{\eta_A} \geq 0.5$; then the resonance occurs at a phase-shift value

$$\delta_l^{(\eta_A)}(E_R) = \pi/2.$$

(b) $x_{\eta_A} < 0.5$; then the phase shift at resonance is $\delta_l^{(\eta_A)}(E_R) = 0$.

However, there is no qualitative difference between these two cases; in both cases the resonant eigen-phase-shift $\delta^{(\kappa_R)}(E)$ goes through $\pi/2$ at $E = E_R$.

---

[11] Figures 3.1, 3.2, and 3.3 are from Barbaro–Galtieri (1968).

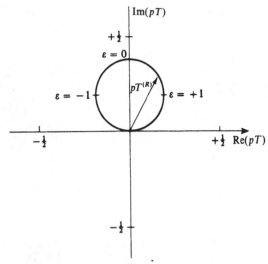

Figure 3.2 The resonant amplitude for scattering into an inelastic channel. [From Barbaro–Galtieri (1968), with permission.]

The Argand diagram for the resonant reaction amplitude (3.6) is depicted in Figure 3.2. For the diameter of the circle one takes $|\sqrt{x_r x_{\eta_A}}|$. It lies inside the circle of Figure XVI.3.2 for the reaction amplitude, and

$$x_r x_{\eta_A} < \tfrac{1}{4}.$$

If the background is included, the elastic scattering amplitude (2.14) is written

$$pT_l = e^{i2\gamma} \frac{x_{\eta_A}}{\epsilon - i} + b. \tag{3.7}$$

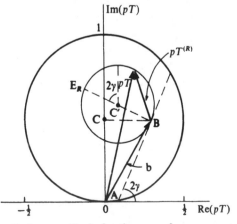

Figure 3.3 Elastic scattering amplitude for the case of a resonance and a background amplitude. If the background is assumed to be constant throughout the resonance region, the amplitude pT is on the circle with center C'. In this figure the background amplitude is not purely elastic, but represents some absorption. [From Barbaro–Galtieri (1968), with permission.]

The graphical representation of (3.7) for $\gamma = 0$ is obtained from the graphical representation of (3.5) in Figure 3.1 by displacing the circle by the vector b and starting the resonance circle at the endpoint B of b. The resonance energy is then at the top of this resonance circle. The effect of the background phase shift γ is then a rotation of the resonance circle by the angle 2γ. The resulting Argand diagram for (3.7) is depicted in Figure 3.3.

The Argand diagrams depicted in the figures of this section and also in the figures of Section XVIII.8 are idealizations. The energy dependence of the background $b(E)$, the background phase shift $\gamma(E)$, and the elasticity distorts the circular behavior of the amplitude. An example of an experimental Argand diagram, obtained from the experimental data by a procedure described in the second part of Section XVIII.8, is depicted in Figure 3.4 below.

The existence and nature of a resonance in the Argand diagram are then determined by comparison with the idealized cases depicted in the above figures. If there is an indication of a circle, one takes this as a sign for a resonance. The resonance energy E_R is then determined as that value of E

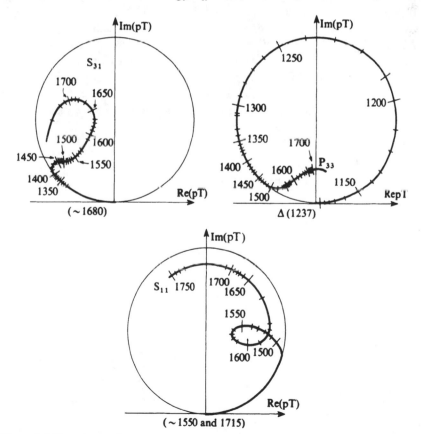

Figure 3.4 Experimental Argand diagrams for some pion–nucleon partial-wave amplitudes. [From A. H. Rosenfeld *et al., Data for Elementary Particle Physics*, 1965, with permission.]

at which $pT_l(E)$ and the phase shift $\delta_l(E)$ change most rapidly. Around E_R the path $pT_l(E)$ may show as a segment of a circle, and the elasticity x_{η_A} is then obtained as the radius of this circle. The total width Γ may then be obtained from the energy difference between the points $\frac{1}{4}$ circle to the left and $\frac{1}{4}$ circle to the right of the resonance energy on the circle segment. In this way Argand diagrams have been of great help for displaying resonance behavior and for determining the resonance parameters.

As an illustration, we depict in Figure 3.4 Argand diagrams obtained from the experimental data for some partial-wave amplitudes in π-p elastic scattering.[12] Each diagram shows the unitary circle and, inside it, $pT_l(E)$ for various partial waves. (For π-p scattering there are several partial waves for each value of l, characterized by isospin I and total angular momentum $j = l \pm \frac{1}{2}$; we cannot discuss these things here, they are also of no relevance for our present discussion.) The numbers at the curves for $pT(E)$ indicate the energy of the π-p system in MeV $= 10^6$ eV. The second diagram (labeled by P_{33}) shows the almost purely elastic resonance Δ at the energy 1236 MeV. The Argand diagram is almost exactly as described in Figure XVIII.8.1. The third diagram (labeled S_{11}) is the Argand diagram of another partial wave with two resonances, one around 1550 MeV and the other around 1715 MeV, both inelastic. The first diagram (labeled S_{31}) shows another case of an inelastic resonance around 1680 MeV. We see from these examples that the behavior of different partial-wave amplitudes is quite distinct and in general far from the idealized behavior. The partial wave in the second diagram, P_{33}, is in fact an exception; all other resonances are inelastic and have a large background term.

---

[12] π stands for π-meson and p for proton.

CHAPTER XXI

The Decay of Unstable Physical Systems

In Section XXI.1 the decaying state is introduced as a resonance for which the production process is ignored. Section XXI.2 gives a heuristic discussion of decay probability (decay rate) and its measurement. Section XXI.3 describes a decaying state by a generalized eigenvector of a Hermitian energy operator with complex eigenvalue. Using this novel concept, the calculation of the decay rate is very simple and is given in Section XXI.5. Section XXI.6 contains a discussion of the partial decay rates and the use of various basis systems for their calculation.

XXI.1 Introduction

In Chapters XVIII and XX we discussed scattering processes in which an unstable physical system was formed as a quasistationary intermediate state. The measurements were performed at the initial state, which was prepared in the distant past, and at the final state, which was observed when the interaction was no longer in effect and the quasistationary state had ceased to exist. We have already seen in Chapter XX that the property of the quasistationary state is—at least to a high degree of accuracy—independent of the process in which it is observed. Its characteristics are the energy E_R, the width Γ, and the internal quantum numbers κ_R.

The formation process

$$a + T \to R \to a' + T' \tag{1.1}$$

537

Figure 1.1 Resonance scattering.

in which the quasistationary state is detected is depicted by the diagram in Figure 1.1, and (ignoring background) is described according to (XX.2.15) by the scattering amplitude

$$\sqrt{p_r p}\, T_{r,i} = \langle r|\kappa_R\rangle \frac{\Gamma/2}{E_R - E - i\Gamma/2} \langle \kappa_R|\eta_A\rangle. \tag{1.2}$$

If Γ is sufficiently small and the delay time according to (XVIII.6.36) and (XVIII.6.43) sufficiently large, one may ignore the formation process. One then starts the description at a given time, $t = 0$, with a physical state $W(t = 0) = W^R$, and describes only that part of the process of Figure 1.1 which is depicted by Figure 1.2(a). R is then considered as a decaying physical system or a decaying state of a physical system. Figure 1.2(b) depicts the earlier part of the resonance scattering process terminating at $t = 0$. Here R is considered as a forming state.

Examples of decaying states are numerous in physics; "real" stable states are very rare. The radiative transitions of excited atoms or molecules A^* to lower states A,

$$A^* \rightarrow A + \gamma, \tag{1.3a}$$

are typical examples of decay processes. The unstable physical system had, of course, first to be formed, and this may have happened by a process like

$$\gamma + A \rightarrow A^*. \tag{1.3b}$$

But for the excited states of atoms one ignores the mode of formation and starts with an initial state W^R describing the decaying system A^*.

The crucial problem is thus the choice of W^R at the arbitrarily chosen initial time $t = 0$. From W^R, if the time development is described by a Hamiltonian H according to the basic assumption V of Chapter XII, one can then obtain the state of the decaying system at any later time t by

$$W(t) = e^{-iHt} W^R e^{+iHt} \rightarrow W^{\text{out}}(t \rightarrow +\infty). \tag{1.4}$$

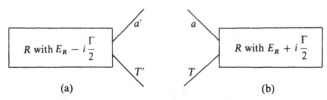

(a)   (b)

Figure 1.2 (a) Decaying state; (b) forming state.

Equation (1.4) is valid if the decaying system is isolated and evolves undisturbed. Whether for a realistic decay process the decaying system really evolves undisturbed is questionable, because it almost unavoidably interacts with its surroundings, especially when the decay process is observed.[1] Therefore the decaying system may be subjected to measurement processes occurring at random times. For example, a decaying particle in a bubble chamber leaves a track of bubbles caused by the interaction of the decaying particle with the environment. Each bubble means a measurement in which the system has been found undecayed. Such a measurement at time t_1 changes the state, according to the basic assumption IIIb, into the state

$$W'(t_1) = \Lambda_b W(t_1)\Lambda_b, \tag{1.5}$$

where Λ_b is the projection operator on the space of states of the undecayed system or a subspace of it. Thus in addition to the change of state (1.4) one has to consider the randomly occurring change of state (1.5), and the observed lifetime will not be the same as the lifetime of the unobserved decaying state developing according to (1.4).

All these problems we shall not discuss here.[2] We shall treat the idealized situation where the decaying state can be considered as an isolated physical system whose lifetime is not affected by a measurement process (1.5) caused by the environment.

XXI.2 Lifetime and Decay Rate

In a decay experiment one has an ensemble of unstable physical systems R in a volume $\mathcal{V}$, which is surrounded by counters that detect the decay products a'. The counting rate, i.e., the number of decays per second, N/T, is proportional to the number of unstable particles, $N_R = \rho_R \mathcal{V}$. The proportionality constant

$$\lambda = \frac{N/T}{N_R} \tag{2.1}$$

is called the *initial decay rate*, or often just *decay rate*.

The number of decays per second, N/T, is equal to the rate of decrease of the number of unstable particles R, dN_R/dt. Therefore, by (2.1),

$$-\frac{dN_R}{dt}(t) = \lambda N_R(t), \tag{2.2}$$

[1] A Beskow, J. S. Nilsson, *Arkiv för Fysik* **34**, 561 (1967).

[2] For a review of these subjects the reader is referred to L. Fonda, G. C. Ghirardi, and A. Rimini. Decay theory of unstable quantum systems. *Reports on Progress in Physics* **41**, 587 (1978). Further methods not discussed in this review can be found in A. P. Grecos: Decaying states in quantum systems. *Singularities and Dynamical Systems*, edited by F. N. Pnevmatikos, North Holland, Amsterdam (1985) and A. George, F. Henin, F. Mayne, I. Prigogine, New quantum rules for dissipative systems. *Hadronic Journal* **1**, 520 (1978).

at least at $t = 0$. And if there is no further source or sink of unstable particles R in $\mathscr{V}$, one is led from (2.2) to the exponential decay law

$$N_R(t) = N_R(0)e^{-\lambda t}, \tag{2.3}$$

where $N_R(0)$ is the number of unstable particles R at the time $t = 0$ when the decay process started.

As $N_R(0) = -\int_{t=0}^{\infty} dN_R(t)$, it follows that $-dN_R(t)/N_R(0)$ is the probability of a decay at time t during the time interval dt. The decay rate, i.e., the probability per unit time for a decay at t, is therefore

$$\frac{d\mathscr{P}}{dt} \equiv \dot{\mathscr{P}}(t) = -\frac{1}{N_R(0)}\frac{dN_R(t)}{dt}. \tag{2.4}$$

From (2.2) it then follows that

$$\dot{\mathscr{P}}(t)|_{t=0} = \lambda, \tag{2.5}$$

which explains the name "initial decay rate" for λ.

The average lifetime of an unstable particle R,

$$\tau = \int_0^\infty t\dot{\mathscr{P}}(t)\, dt, \tag{2.6}$$

is called the *mean life*, or just the *lifetime*. If (2.3) holds, then

$$\tau = \int_0^\infty \frac{1}{N_R(0)} N_R(0)e^{-\lambda t}\lambda t\, dt = \frac{1}{\lambda} = \frac{1}{\dot{\mathscr{P}}(t=0)}. \tag{2.7}$$

In quantum physics the decaying particles are quantum physical systems and the observed quantities are probabilities. The observables measured on the decaying state $W(t)$ may be the "property" of being in the original resonance state W^R. The expectation value of W^R in the state $W(t) = e^{-iHt}W^R e^{iHt}$ represents the probability that R has not decayed. It is called the *nondecay probability*:

$$\mathscr{P}_R(t) = \text{Tr}(W^R W(t)). \tag{2.3a}$$

It corresponds to the observed quantity $N_R(t)/N_R(0)$.

Another observable that may be measured is the projection operator Λ on the space of physical states of the decay products (a', T'), i.e., the property Λ. The expectation value of Λ in the state $W(t)$ represents the probability for observing a' and T'. This transition probability from the state $W(t)$ to Λ,

$$\mathscr{P}(t) = \langle\Lambda\rangle_t = \text{Tr}(\Lambda W(t)), \tag{2.8}$$

is called the *decay probability* of R into $(a'\, T')$. Obviously, if $R \to a' + T'$ is the only decay possible, then the sum of the probability that R has not decayed and the probability that R has decayed into $a' + T'$ must be one:

$$\mathscr{P}_R(t) + \mathscr{P}(t) = 1. \tag{2.9}$$

If R has several decay channels, i.e., if it can decay into several decay products labeled by η, whose spaces of physical states are $\Lambda_\eta \mathcal{H}, \eta = 1, 2, 3, \ldots$, with $\Lambda_\eta \Lambda_{\eta'} = 0$ for $\eta \neq \eta'$, then

$$\Lambda \equiv \Lambda_1 + \Lambda_2 + \Lambda_3 + \cdots$$

and the probability for a decay into any of the channels η is

$$\mathscr{P}(t) = \sum_\eta \mathscr{P}_\eta(t), \qquad \mathscr{P}_\eta(t) = \text{Tr}(\Lambda_\eta W(t)). \tag{2.10}$$

(2.9) is also valid for this transition probability and states that at any arbitrary time t either R has decayed or has not decayed.

The quantum-mechanical decay rate is the transition rate from $W(t)$ into Λ:

$$\frac{d\mathscr{P}}{dt} = \dot{\mathscr{P}}(t) = \frac{d}{dt} \langle \Lambda \rangle_t. \tag{2.11}$$

In (XIV.2.18) we have already calculated an expression for it, which we shall use below.

The experimentally measured quantity is the lifetime τ or the decay rate λ. For large lifetimes (from 1 sec to 1 year), one measures $N_R(t)$ as a function of t and obtains λ from (2.3). For shorter lifetimes down to 10^{-9} sec one measures the lifetime by electronic methods. The decay to be studied must be preceded by another event, e.g., the formation of the unstable system, which is used to establish the origin of time. A counter is activated by this event and remains activated for a time t. The number of times the activated counter detects a decay divided by the number of times the counter has been activated is the observed quantity which is proportional to the probability that a decay has occurred during the time interval between 0 and t. This is according to (2.4) and (2.3) given by

$$\int_0^t \dot{\mathscr{P}}(t') \, dt' = \int_0^t dt' \, \lambda e^{-\lambda t'} = (1 - e^{-\lambda t}).$$

Thus

$$k \int_0^t dt' \, \lambda e^{-\lambda t'} = k(1 - e^{-\lambda t}) = \frac{\text{Number of times the activated counter detects a decay}}{\text{Number of times the counter is activated}}$$

$$\tag{2.12}$$

One measures the ratio on the right-hand side as a function of t. As k is a constant independent of t [connected with $N_R(0)$ and the detector efficiency], one can calculate λ if one knows (2.12) for several values of t.

For lifetimes shorter than 10^{-9} sec one cannot directly measure τ, but has to resort to indirect methods. As we shall see below, the decay rate is equal to the width. One then obtains the lifetime from the measurement of the width in resonance scattering.

XXI.3   The Description of a Decaying State and the Exponential Decay Law

In Section XVIII.6 we have discussed resonances, quasistationary states, Breit–Wigner amplitudes, and resonance poles of the S-matrix. The connection is summarized in the diagram of Figure 3.1. A pair of second sheet poles of the S-matrix is the property that implies the three other properties listed above it in the diagram. Whereas a resonance is always a quasistationary state and vice versa, these two properties lead to a Breit–Wigner amplitude only if the fourth and higher derivatives of $\delta_l(E)$ are small. And a Breit–Wigner amplitude does not necessarily imply the existence of an S-matrix with a pair of second sheet poles, as the S-matrix entails further properties like analyticity (causality) and unitarity. We will, therefore, take the pair of second sheet poles of the S-matrix near the real energy axis as the definition of a quasistationary state. From this definition we will deduce in the next section the vectors that describe decaying states, which we will call *Gamow vectors* or *Gamow kets*. In the present section we will introduce the decaying states more heuristically and state some of their properties, which will be derived in the next section. These properties will be used to calculate the decay probability and in Section XXI.5, the decay rate.

The decaying state W^R can be prepared by a scattering experiment in which the time delay is large. One obtains the decaying state if one isolates the intermediate quasistationary system, ignores its mode of formation and considers it as an initial state which at time $t = 0$ starts the time development which leads to the decay. As the scattering amplitude of a resonance scattering process has the Breit–Wigner form, we expect the decaying state $W(0) = W^R$ to have a Breit–Wigner energy distribution.

For the sake of definiteness we will assume that the spectra of H and K are continuous and go from 0 to ∞. This is not a serious restriction of

Resonance (defined by a rapidly increasing $\delta_l(E)$)

$$t_D^l(E) = 2\frac{d\delta_l(E)}{dE}$$

Quasistationary State (defined by a large value of the time delay $t_D^l(E)$)

$$S_l(E) = e^{i2\delta_l(E)} = 1 + \frac{i\Gamma}{E_R - E - i\Gamma/2}$$

only if $\dfrac{d^n\delta(E)}{dE^n}$ is small for $n \geq 4$

Breit–Wigner Amplitude (Siegert pole)

only if the S-matrix is analytic (causality) and unitary

Pair of Second Sheet Poles of the S-matrix

Figure 3.1 Relations between a resonance, quasistationary state, Breit–Wigner amplitude, and pair of second sheet poles of the S-matrix.

generality, because bound states can be added without difficulty. The density matrix[3] should then be given by:

$$\langle E_a \hat{a}^- \,|\, W^-(0) \,|\, E_a \,\hat{a}'^- \rangle = \langle E_a \,\hat{a} \,|\, W^{out}(0) \,|\, E_a \,\hat{a}' \rangle$$
$$= F(E_a - E_R)\rho(E)^{-1}\langle \hat{a} \| W^R \| \hat{a}' \rangle \qquad (3.1)$$

where $F(E_a - E_R)$ is a Lorentzian energy distribution function with the resonance width as the width of the energy distribution:

$$F(E - E_R) = \frac{1}{\pi} \frac{\Gamma/2}{(E_R - E)^2 + (\Gamma/2)^2} \quad \text{for} \quad 0 \le E < \infty. \qquad (3.2)$$

Here we have normalized $F(E - E_R)$ in such a way that[4]

$$F(E - E_R) \to \delta(E - E_R) \quad \text{for} \quad \Gamma/2 \to 0. \qquad (3.3)$$

As $W(t)$ goes into an observed out-state it will have to be the state $W^-(t)$ and we shall use the basis vectors $|E\,\hat{a}^-\rangle$. We shall also normalize the reduced matrix element to one:

$$\sum_{\hat{a}} \langle \hat{a} \| W^R \| \hat{a} \rangle = 1 \qquad (3.4)$$

and discuss the normalization of $W(0) = W^R$. We calculate

$$\text{Tr } W(0) = \sum_a \langle a^- | W(0) | a^- \rangle = \sum_{\hat{a}} \int_0^\infty \rho(E)\, dE \langle E\,\hat{a}^- | W(0) | E\,\hat{a}^- \rangle$$

$$= \sum_{\hat{a}} \int_0^\infty dE\, F(E - E_R) \langle \hat{a} \| W^R \| \hat{a} \rangle$$

$$= \frac{\Gamma}{2\pi} \int_0^\infty dE\, \frac{1}{(E - E_R)^2 + (\Gamma/2)^2}$$

$$= \frac{1}{\pi} \int_{x = -E_R 2/\Gamma}^\infty dx\, \frac{1}{x^2 + 1}$$

$$= \frac{1}{\pi} \left[\frac{\pi}{2} + \arctan\left(E_R \frac{2}{\Gamma} \right) \right], \qquad (3.5)$$

where the new variable x is

$$x = (E - E_R)\frac{2}{\Gamma}. \qquad (3.6)$$

The integration over E in the summation over the complete set of basis vectors should extend over the spectrum of H, i.e., from 0 to ∞. If the integration over E would extend from $-\infty$ to $+\infty$ then the lower bound in the integral over x would also be $-\infty$ instead of $-2E_R/\Gamma$ and we would have

$$\text{Tr } W^R = \text{Tr } W(0) = 1. \qquad (3.5a)$$

---

[3] The first line of (3.1) has been proven in Appendix XV.A.

[4] See, e.g., Equation (II.8.12).

Equation (3.5a) can also be obtained from (3.5) in the limit of an infinitely narrow width, i.e., if one takes instead of (XVIII.6.13) the limiting case

$$2E_R/\Gamma \to \infty \qquad (3.7)$$

(because then arctan $2E_R/\Gamma \to \pi/2$). For sufficiently large $2E_R/\Gamma$, the extension of the integration over E to $-\infty$ would make only a very small contribution to (3.5). When we derive below the decaying state from the resonance pole of the S-matrix, we shall see that the energy integration in (3.5) and similar integrals will indeed extend from $-\infty$ to $+\infty$ and not from 0 to ∞.[5]

The expectation value of the energy operator in an energy eigenstate $W = |E_0\rangle\langle E_0|$ is $\mathrm{Tr}(H|E_0\rangle\langle E_0|) = E_0$. We shall now calculate the expectation value of H in the state $W(0) = W^R$ given by (3.1) which is to describe a resonance with energy E_R and width Γ:

$$\mathrm{Tr}(HW^R) = \sum_{\hat{a}} \langle \hat{a}\| W^R \|\hat{a}\rangle \int dE \, \frac{\Gamma}{2\pi} \frac{E}{(E - E_R)^2 + (\Gamma/2)^2}$$

$$= \frac{E_R}{\pi} \int_{x=-E_R \, 2/\Gamma}^{\infty} dx \, \frac{1}{x^2 + 1} + \frac{\Gamma}{2\pi} \int_{x=-E_R \, 2/\Gamma}^{\infty} dx \, \frac{x}{x^2 + 1}. \qquad (3.8)$$

In the limit (3.7) or if the integration over E extends to $-\infty$, the first integral, according to (3.5), goes to π, and the second integral goes to zero, so that

$$\mathrm{Tr}(HW^R) = E_R. \qquad (3.9)$$

This is the result we desire from comparison with the stationary energy eigenstate.

Equation (3.1) gives only the diagonal matrix elements of the statistical operator W^R for a decaying system. To fix W^R fully we also need to specify its matrix elements between generalized energy eigenvectors of different energy.

According to our discussion in Section II.9—in particular, according to (II.9.1)—$W(0)$ should in general be given by

$$W(0) = W^R$$

$$= \sum_{\hat{a}\hat{a}'} \int \rho(E_a) \, dE_a \, \rho(E_a') \, dE_a' \, f(E_a' - E_R) |E_a' \, \hat{a}'^-\rangle$$

$$\times \langle \hat{a}'\| W^R \|\hat{a}\rangle \langle E_a \, \hat{a}^- | f^*(E_a - E_R). \qquad (3.10)$$

Hence

$$\langle E \, \hat{a}'^- | W^R | E \, \hat{a}^-\rangle = |f(E - E_R)|^2 \langle \hat{a}'\| W^R \|\hat{a}\rangle, \qquad (3.11)$$

---

[5] We should remind the reader that the Breit–Wigner amplitude for resonance scattering must be an approximation, as remarked in Section XVIII.6. Therefore the choice, (3.2) with the restriction on E may also be only an approximation and the more appropriate energy distribution will be (3.2) with $-\infty < E < +\infty$. As remarked, for sufficiently small $2\Gamma/E_R$ there will be only very little observable difference between these two.

so that by comparison with (3.1) we must have

$$|f(E - E_R)|^2 = \rho(E)^{-1}F(E - E_R) = \frac{\Gamma}{2\pi\rho(E)} \frac{1}{(E - E_R)^2 + (\Gamma/2)^2}. \quad (3.12)$$

Except for an arbitrary factor of modulus 1, there are only two possibilities from (3.12):

$$f(E - E_R) = \sqrt{\frac{\Gamma}{2\pi\rho}} \frac{1}{E_R + i\Gamma/2 - E} \quad (3.13)$$

and

$$f(E - E_R) = \sqrt{\frac{\Gamma}{2\pi\rho}} \frac{1}{E_R - i\Gamma/2 - E}. \quad (3.14)$$

As we shall discuss in the following section, for a decaying state one has to choose (3.14). Equation (3.13) will lead to a forming or "capture" state.

To simplify the notation we shall now specify the additional quantum numbers a and then ignore them. As discussed in the preceding chapters, the resonance R will have a definite value of angular momentum $l = l_R$ and internal quantum numbers $\kappa = \kappa_R$. Therefore a choice of angular-momentum eigenvectors

$$|E \hat{a}^-\rangle = |E l l_3 \kappa^-\rangle \quad (3.15)$$

is convenient. The reduced matrix element of W^R in this basis is

$$\langle \kappa l l_3 \| W^R \| \kappa' l' l_3' \rangle = \delta_{\kappa\kappa_R} \delta_{ll_R} \delta_{\kappa'\kappa_R} \delta_{l'l_R} \delta_{l_3 l_3'} \frac{1}{2l + 1} \quad (3.16)$$

for a completely unpolarized state. W^R of (3.10) is then

$$W^R = \frac{1}{2l + 1} \sum_{l_3} \int \rho(E) \, dE \int \rho(E') \, dE' f(E' - E_R)|E' l_R l_3 \kappa_R^-\rangle$$
$$\times \langle E l_R l_3 \kappa_R^- | f^*(E - E_R), \quad (3.17)$$

which we will write with (3.14) as

$$W^R = \frac{\Gamma}{2\pi} \int_0^\infty dE \int_0^\infty dE' \frac{1}{E_R - i\Gamma/2 - E'}$$
$$\times |E'^-\rangle\langle E^-| \frac{1}{E_R + i\Gamma/2 - E}, \quad (3.18)$$

where we have defined for the sake of simplicity

$$|E^-\rangle = |E l l_3 \kappa^-\rangle \rho^{1/2}(E) \quad (3.19)$$

and ignored the additional quantum numbers. The normalization and summation for these new Dirac kets $|E^\pm\rangle$ is now fixed to be

$$\langle E'^\mp | E^\mp \rangle = \delta(E' - E); \qquad \sum_E = \int_0^\infty dE \quad (3.20)$$

instead of the general normalization given by (XIV.2.9) and (XIV.2.10).

A proper, stationary energy eigenstate (pure if we ignore the additional quantum numbers) of the observable H with eigenvalue E_R is written as

$$\Lambda_R = |E_R)(E_R|, \tag{3.21}$$

where

$$H|E_R) = E_R|E_R) \tag{3.22}$$

and the $|E_R)$ are proper eigenvectors and elements of the space of physical states:[6]

$$(E_R|E_R) = 1 \qquad (|E_R) \in \Phi \subset \mathscr{H} \subset \Phi^\times). \tag{3.23}$$

In order to write the decaying state W^R in an analogous way we define the vector:[7]

$$\left| E_R - i\frac{\Gamma^-}{2} \right) = |f^D\rangle = \frac{1}{i} \int dE |E^-\rangle f^{(\rho = 1)}(E - E_R)$$

$$= \frac{1}{i}\left(\frac{\Gamma}{2\pi}\right)^{1/2} \int dE |E^-\rangle \frac{1}{E_R - i\Gamma/2 - E}. \tag{3.24}$$

This vector has the energy distribution

$$\langle E^-|f^D\rangle = \frac{1}{i}\left(\frac{\Gamma}{2\pi}\right)^{1/2} \frac{1}{E_R - i\Gamma/2 - E} = \frac{1}{i} f^{(\rho = 1)}(E - E_R)$$

$$= \langle f^D|E^-\rangle^*. \tag{3.25}$$

The second equality in (3.25) follows if we treat $|f^D\rangle$ and $|E^-\rangle$ like well-behaved vectors and $\langle \ | \ \rangle$ as their scalar product. Therefore we define the conjugate vector $\langle f^D|$ by:

$$\left(E_R + i\frac{\Gamma^-}{2} \right| = \langle f^D| = -\frac{1}{i}\left(\frac{\Gamma}{2\pi}\right)^{1/2} \int dE \langle E^-| \frac{1}{E_R + i\Gamma/2 - E}; \tag{3.26}$$

then the second equality in (3.25) is a consequence of this definition.

With (3.24) and (3.26), W^R in (3.18) can be written as

$$W^R = |f^D\rangle\langle f^D|. \tag{3.27}$$

The integration over E in (3.18), (3.24), (3.26) is to be taken from 0 to ∞ if we take the Breit–Wigner distribution (3.2) as the exact form. As we mentioned already, we will introduce only a minor error if we take the integration from $-\infty$ to $+\infty$, extending the function $f(E - E_R)$ and therewith the generalized vectors $|E^-\rangle$ to negative values of energy. From the discussions

---

[6] As it is important to distinguish between the proper and generalized eigenvectors, we shall here use a separate notation $|E)$ for a proper eigenvector and $|E\rangle$ for a generalized eigenvector (cf. Section II.7).

[7] Since we use for the $|E^-\rangle$ the normalization (3.20), i.e. the convention $\rho(E) = 1$ rather than (XIV.2.9b), we have to make the same convention for the wave function; this is indicated by the superscript $(\rho = 1)$ on f.

in Section XVIII.5 we are already familiar with the continuation to "unphysical" values of the scattering energy. There we also learned that the "unphysical" values of the energy cover the second sheet of a two-sheeted Riemann surface cut along the positive real axis. The integral over the spectrum of H runs along the upper rim of the cut of the physical sheet; we will see below that when we extend it to negative values of E these will lie on the second sheet.

The length of the vector $|f^D\rangle$ is given by[8]

$$\langle f^D | f^D \rangle = \frac{\Gamma}{2\pi} \iint dE\, dE'\, \langle E'^- | E^- \rangle \frac{1}{E_R + i\Gamma/2 - E'} \frac{1}{E_R - i\Gamma/2 - E}$$

$$= \frac{\Gamma}{2\pi} \int_0^\infty dE \frac{1}{(E_R - E)^2 + (\Gamma/2)^2}$$

$$= \frac{1}{\pi} \int_{x=-2E_R/\Gamma}^{\infty} dx \frac{1}{x^2 + 1} = 1 - \frac{1}{\pi}\left(\frac{\Gamma}{2E_R}\right) - \frac{1}{3}\left(\frac{\Gamma}{2E_R}\right)^3 + \cdots \quad (3.28)$$

if we use integration over the spectrum of H. It is equal to 1 if we integrate from $-\infty$ to $+\infty$, which is sufficiently close to (3.28) for $\Gamma/2E_R \ll 1$.

Equation (3.27) would be precisely the same as in the stationary case, (3.21), if $|f^D\rangle$ was an eigenvector of H. We will show below that something like (3.22) is indeed fulfilled for the $|f^D\rangle$, namely

$$H|f^D\rangle = \left(E_R - i\frac{\Gamma}{2}\right)|f^D\rangle \quad (3.29a)$$

and (if $\langle f^D | \phi \rangle = \langle \phi | f^D \rangle^*$):

$$\langle f^D | H = \left(E_R + i\frac{\Gamma}{2}\right)\langle f^D |. \quad (3.29b)$$

Equation (3.29) is not to be understood as a proper eigenvector equation like (3.22) but as a generalized eigenvector equation valid if $|f^D\rangle$ is considered as a functional over roughly half of the space of physical states.[9]

The time development of $|f^D\rangle$ is a formal consequence of (3.29):

$$e^{-iHt}|f^D\rangle = e^{-iE_Rt}e^{-(\Gamma/2)t}|f^D\rangle \quad \text{for} \quad t \geq 0. \quad (3.30)$$

As the time t for the decay process starts at $t = 0$, we need to consider (3.30) only for $t \geq 0$. It will turn out that (3.30) can be proven as a generalized eigenvector equation precisely for these values of t.

[8]
$$\int_{-x_R}^\infty dx \frac{1}{x^2 + 1} = \frac{1}{\pi}\left(\frac{\pi}{2} + \arctan x_R\right)$$

$$= 1 - \frac{1}{\pi}\left(\frac{1}{x_R} - \frac{1}{3x_R^3} + \frac{1}{5x_R^5} - \cdots\right) \text{ for } x_R \gg 1.$$

[9] A self-adjoint operator cannot have complex proper eigenvalues but it can have generalized eigenvalues which are complex.

The generalized eigenvector $|f^D\rangle$ is according to (3.24) and (3.26) specified by the two real numbers E_R, Γ or by the complex number

$$Z_R = E_R - i\frac{\Gamma}{2} \tag{3.31}$$

which characterize a Breit–Wigner resonance. Following the usual convention of labelling the proper eigenvectors and the generalized eigenvector with real eigenvalue (the Dirac kets) by their eigenvalue, it is appropriate to also label $|f^D\rangle$ by its eigenvalue and write

$$|f^D\rangle = \left|E_R - i\frac{\Gamma^-}{2}\right\rangle = |Z_R\rangle, \tag{3.32a}$$

$$\langle f^D| = \left(E_R + i\frac{\Gamma^-}{2}\right| = (Z_R^{*-}|. \tag{3.32b}$$

With this notation (3.27) is written

$$W^R = \left|E_R - i\frac{\Gamma^-}{2}\right\rangle\left(E_R + i\frac{\Gamma^-}{2}\right|. \tag{3.33}$$

This is the form for the statistical operator of a decaying state with resonance parameters (E_R, Γ) that is in perfect analogy to the form (3.21) for a stationary state. W^R has also, like a projection operator describing a pure state, the property that

$$W^R W^R = W^R. \tag{3.34}$$

This follows immediately from (3.28) (with the integration extended from $-\infty$ to $+\infty$). W^R is however not exactly like Λ_R and $|f^D\rangle$ is not exactly like $|E_R\rangle$; this one notices immediately if one tries to calculate the length of the vector $H|f^D\rangle$ in the same manner as used in (3.28) (cf. Problem 1).

For later reference we now give the expression for W^R with the additional quantum numbers $\hat{a}$ reinstated. Equation (3.10) is written in the form (3.27) as

$$W^R = \sum_{a'a} \left|E_R - i\frac{\Gamma}{2}, \hat{a}'^-\right\rangle\langle \hat{a}'\|W^R\|\hat{a}\rangle\left(E_R + i\frac{\Gamma}{2}, \hat{a}^-\right|. \tag{3.35}$$

Here we have defined in analogy to (3.32) with (3.24):

$$\left|E_R - i\frac{\Gamma}{2}, \hat{a}'^-\right\rangle = \frac{1}{i}\sqrt{\frac{\Gamma}{2\pi}}\int dE'\, \rho^{1/2}(E')|E'\,\hat{a}'^-\rangle\frac{1}{E_R - i\Gamma/2 - E'}. \tag{3.24a}$$

which is "normalized" in the sense of (3.28). The occurrence of $\rho^{1/2}(E)$ in (3.24a) indicates that the $|E'\,\hat{a}^-\rangle$ have the general "normalization" (XIV.2.9b). For example, if the reduced matrix elements of W^R are given by (3.16), corresponding to a resonance with angular momentum l_R and internal quantum number κ_R, then W^R becomes

$$W^R = \sum_{l_3} \frac{1}{2l+1}\left|E_R - i\frac{\Gamma}{2}, l_R\, l_3\, \kappa_R^-\right\rangle\left(E + i\frac{\Gamma}{2} l_R\, l_3\, \kappa_R^-\right|. \tag{3.36}$$

From the form (3.33) or (3.35), (3.36) of W^R the exponential decay law for the nondecay amplitude follows immediately. Using (1.4) and (3.30), one obtains for $t > 0$:

$$W(t) = e^{-iHt} \left| E_R - i\frac{\Gamma^-}{2} \right) \left(E_R + i\frac{\Gamma^-}{2} \right| e^{iHt}$$

$$= e^{-i(E_R - i\Gamma/2)} e^{i(E_R + i\Gamma/2)} \left| E_R - i\frac{\Gamma^-}{2} \right) \left(E_R + i\frac{\Gamma^-}{2} \right|. \qquad (3.37)$$

Thus

$$W(t) = e^{-\Gamma t} W^R. \qquad (3.38)$$

For the nondecay probability (2.3a) we then obtain, with (3.34):

$$\mathscr{P}_R(t) = \mathrm{Tr}(W^R W(t)) = \mathrm{Tr}(W^R W^R) e^{-\Gamma t} = e^{-\Gamma t} \, \mathrm{Tr} \, W^R = e^{-\Gamma t}. \qquad (3.39)$$

Equations (3.38) and (3.39) show that the state W^R is an exponentially decaying state.[10]

Inserting (3.39) into (2.9) we obtain

$$\mathscr{P}(t) = 1 - e^{-\Gamma t} \qquad (3.40)$$

from which it follows for the initial decay rate

$$\dot{\mathscr{P}}(t)|_{t\to 0} = \Gamma \qquad (3.41)$$

Comparing this with (2.5) and (2.7) we obtain

$$\tau = \frac{1}{\lambda} = \frac{1}{\Gamma}. \qquad (3.42)$$

Thus the lifetime of a resonance with resonance parameters (E_R, Γ), considered as an unstable physical system, is equal to the inverse width.

XXI.4 Gamow Vectors and Their Association to the Resonance Poles of the S-Matrix

The present section will give a derivation of notions used in the previous section and a mathematically more precise explanation of the statements in that section. This section will make use of results obtained in Sections XV.2 and XV.3.

In Sections XVII.5 and XVII.6 we related the poles on the negative real axis in the physical sheet of the energy Riemann surface for the S-matrix,

---

[10] As is well known to specialists, in the usual precise Hilbert space formulation one obtains deviations from the exponential decay law [L. Fonda et al. (1978)]. Deviations for large values of t follow from the condition that the spectrum of H be bounded from below which corresponds to a finite lower limit in the integrals (3.5), (3.8), (3.24), (3.26) and which is not the case if one integrates from $-\infty$ to $+\infty$ as we did to prove (3.34). Deviations from the exponential law for small times t follow from the condition that the energy in the decaying state be finite, which corresponds to the condition that the decaying state vector be in the domain of the Hilbert space operator $\bar{H}$ (or even in Φ, the domain of H) which is also not the case for the vector $|f^D\rangle$ as shown in Problem 1.

the bound state poles, to eigenvectors of the energy operator H. These eigenvectors with discrete eigenvalues describe according to Chapter XII (pure) stationary states. For the stationary states we have, therefore, a dual description by a pole of the S-matrix or by an eigenvector of H.

The continuous set of real energy values along the cut from 0 to ∞ on the physical sheet of the Riemann surface for the S-matrix correspond—according to Section XVIII.5—to the physical values of the scattering energy. These energy values are generalized eigenvalues of H (belonging to the continuous spectrum). The generalized eigenvectors, or Dirac kets, belonging to these eigenvalues describe idealized scattering states, i.e., their continuous super-positions (e.g., (XV.2.7 +)) or continuous mixtures (e.g., (XIV.5.27) with (XIV.5.29b) and (XIV.5.36)) describe pure scattering states ("wave packets") or mixtures thereof. Thus for scattering states we have also a dual description, either by the cut along the positive real axis of the Riemann surface for the S-matrix or by generalized eigenvectors of H.

In the same way one can also describe the quasistationary states, defined by the resonance poles of the S-matrix, as generalized eigenvectors of the energy operator H.[11] These generalized eigenvectors, the Gamow vectors or Gamow kets, will be functionals over roughly half the space of physical states Φ, in distinction to the Dirac kets which are functionals over the whole of Φ (cf. Section I.7 or Section II.8). We shall now deduce the Gamow vectors from the resonance poles of the S-matrix.

As in this section we are interested only in the principles we will first reduce the problem to the essentials. For this purpose we choose the angular momentum basis and ignore internal angular momenta (polarization):

$$|a^\pm\rangle = |E\,\hat{a}^\pm\rangle = |E\,l\,l_3\,\eta^\pm\rangle.$$

Further we restrict ourselves to just one channel, the elastic channel $\eta = \eta' = \eta_A$. In this case as a consequence of conservation of angular momentum, (XVI.1.42), the S-matrix defined in (XV.3.10′) reduces to the $S_l(E)$ of (XVI.1.47). The relation that defines the S-matrix, (XV.3.10″), then simplifies using (XVI.1.43), (XVI.1.45), and (XV.2.10), to

$$|E\,l\,l_3\,\eta^+\rangle = \sum_{\eta'} \frac{\rho_{\eta'}(E)}{\rho_\eta(E)}\,|E\,l\,l_3\,\eta'^-\rangle\langle\eta'\|S_l\|\eta\rangle$$

$$= |E\,l\,l_3\,\eta^-\rangle S_l(E + i\epsilon). \tag{4.1}$$

To simplify the notation to the essentials we will again use the definition (3.19),

$$|E^\pm\rangle = |E\,l\,l_3\,\eta^\pm\rangle\rho_\eta^{1/2}(E), \tag{3.19$\pm$}$$

and write (4.1) as

$$|E^+\rangle = |E^-\rangle S(E). \tag{4.2}$$

Here we have also omitted the l from the S-matrix element.

---

[11] Virtual states, defined by the poles on the negative real axis of the second sheet, can also be described by generalized eigenvectors. We will not discuss this here but refer the interested reader to A. Bohm, M. Gadella (1989).

We start our consideration from the S-matrix element (XV.3.2), (XV.3.14) for the scattering of a pure physical state ϕ^{in} into a pure physical state ψ^{out} which we assume—for the sake of simplicity of notation—to have a well-defined value for the quantum numbers $\hat{a}$:

$$\phi^{in} = \phi^{in}(\hat{a}, 0) = \phi^{in}(l\, l_3\, \eta, 0),$$

$$\psi_{out} = \psi_{out}(\hat{b}, 0) = \psi_{out}(l\, l_3\, \eta, 0). \tag{4.3}$$

We then ignore these quantum numbers in the S-matrix element:

$$(\psi^{out}(t), S\phi^{in}(t)) = (\Omega^- \psi^{out}(t), \Omega^+ \phi^{in}(t)) = (\psi^-(t), \phi^+(t))$$

$$= (\psi^{out}, S\phi^{in}) = (\psi^-(0), \phi^+(0)) = (\psi^-, \phi^+). \tag{4.4}$$

We now expand ψ^{out} and ϕ^{in} according to (XV.2.6$\pm$) with respect to the basis system of eigenvectors of K (XV.1.14):

$$\phi^{in} = \sum_{l l_3 \eta} \int \rho_\eta(E)\, dE\, |E\, l\, l_3\, \eta\rangle\langle E\, l\, l_3\, \eta|\phi^{in}\rangle = \int dE\, |E\rangle\langle E|\phi^{in}\rangle,$$

$$\psi^{out} = \sum_{l l_3 \eta} \int \rho_\eta(E)\, dE\, |E\, l\, l_3\, \eta\rangle\langle E\, l\, l_3\, \eta|\psi^{out}\rangle = \int dE\, |E\rangle\langle E|\psi^{out}\rangle, \tag{4.5}$$

where we have used for the unperturbed basis vectors the simplifying notation

$$|E\rangle = |E\, l\, l_3\, \eta\rangle\rho_\eta^{1/2}(E). \tag{3.19°}$$

The integration in (4.5) extends over the spectrum of K which we have assumed to be identical with the continuous spectrum of H, Spect.H, which we choose (without restricting generality) to be from 0 to $+\infty$.

It is important to mention that ϕ^+ represents the pure physical state which developed from the prepared in-state ϕ^{in}; it is therefore a well-behaved vector. Similarly ψ^- represents a pure physical state that develops into the measured out-state ψ^{out}; it is therefore also a well-behaved vector $\psi^- \in \Phi \subset \mathcal{H} \subset \Phi^\times$. And so are ϕ^{in} and ψ^{out}. The energy wave functions

$$\langle E|\phi^{in}\rangle = \langle E^+|\phi^+\rangle \quad \text{(given by the incident beam)}, \tag{4.6$^+$}$$

$$\langle E|\psi^{out}\rangle = \langle E^-|\psi^-\rangle \quad \text{(measured by the detector)} \tag{4.6$^-$}$$

are therefore well-behaved functions of energy. The equality in (4.6$\pm$) is a special case of (A.2$\pm$) in the Appendix to Chapter XV.

Inserting (4.5) into (4.4) and making use of (4.6$\pm$), the S-matrix element (4.4) can be written (omitting all inessential quantum numbers):

$$(\psi^-, \phi^+) = \sum_{l l_3 \eta} \sum_{l' l_3' \eta'} \iint dE'\, \rho_{\eta'}(E')\, dE\, \rho_\eta(E)$$

$$\times \langle \psi^{out}|E'\, l'\, l_3'\, \eta'\rangle\langle E'\, l'\, l_3'\, \eta'|S|E\, l\, l_3\, \eta\rangle\langle E\, l\, l_3\, \eta|\phi^{in}\rangle$$

$$= \int_{\text{Spect.}H} dE\langle \psi^-|E + i\epsilon^-\rangle S(E + i0)\langle E + i\epsilon^+|\phi^+\rangle. \tag{4.7}$$

Here we have used (XVI.1.43), (XVI.1.45), (XVI.1.47), and the assumption that ϕ^{in} and ψ^{out} have the same value for $l\, l_3\, \eta$ and avoided the irrelevant complication in the notation by using (3.19). The integration in (4.7) is along the spectrum of H which is along the upper rim of the cut of the physical sheet of the S-matrix. We have indicated this by writing $E + i0$ as the argument of the S-matrix. Integration along the upper rim of the cut in the physical sheet is, according to the property of the S-matrix described in Section XVIII.5, the same as integrating along the lower rim of the cut in the second sheet because $S(E + i0) = S(E_{II} - i0)$, where the subscript II refers to the second sheet.

We want to deform the contour of integration of (4.7) into the lower half of the second sheet of the Riemann surface for the S-matrix, where—according to Section XVIII.5—the resonance poles are located. As in this section we are only interested in the principle that establishes the connection between resonance poles and Gamow vectors, we will make the simplest possible assumption about our resonance model. We assume that our physical system is such that the S-matrix has no bound state poles and no singularities other than just one pair of resonance poles.[12] This is depicted in Figure 4.1. Figure 4.1(a) shows the upper half of the first sheet and the lower half of the second sheet with the resonance pole at $Z_R = E_R - i\Gamma/2$. Figure 4.1(b), showing the upper half of the second sheet with the resonance pole at $Z_R^* = E_R + i(\Gamma/2)$ and the lower half of the first sheet, will be used in the latter part of this section.

The contour of integration of (4.7), which was originally along Spect. H, has been deformed into the contour $\mathscr{C}_-$ and a circle around the pole. This contour deformation is possible if the well-behaved wave functions $\langle \psi^- | E^- \rangle$ and $\langle E^+ | \phi^+ \rangle$, which were initially only defined on the spectrum of H, are boundary values of analytic functions in the lower half of the second sheet. For further deformations of the contour we also want to assume that these analytic functions vanish sufficiently fast at the lower infinite semicircle.[13] This is a condition on the set of well-behaved vectors ψ^-; not all vectors representing physical states fulfill this condition. Thus the set of vectors ψ^- for which the contour of integration in (4.7) can be deformed into the contour shown in Figure 4.1(a) cannot be the set of all pure physical states, i.e., cannot be the set of all vectors in Φ. But if we consider a decay process only, i.e., the later half of a resonance scattering process as depicted by Figure 1.2, then the $\{\psi^-\}$ need not represent all the pure physical states but only roughly half of them, namely only those which start developing at $t = 0$ and go for $t \to \infty$ into the observed out-states ψ^{out}. The remaining

---

[12] If there are more than one pair of resonance poles one obtains for each pair a pair of Gamow vectors and if there should be other singularities the background term introduced below will have a different form, but otherwise the arguments carry through in exactly the same way. If bound states exist one must carefully distinguish between $\mathscr{H}$ and $\mathscr{H}_{scat}$ (cf. Chapter XV), but otherwise our arguments will not be affected.

[13] The S-matrix in the second sheet is polynomially bounded at infinity for many interaction potentials. Assuming this to be the case also for our model, we need wave functions which vanish sufficiently fast so that the integrals over the infinite semicircle are zero.

Figure 4.1 Deformation of the path of integration into the second sheet of the energy plane. (a) for the decaying state; (b) for the growing state. [Reprinted from A. Bohm, *J. Math. Phys.* **22**, 2817 (1981).] Sp $\bar{\mathsf{H}}$ denotes the spectrum of the (closure of) the Hamiltonian.

half of the pure physical states is the set $\{\psi^+\}$ of vectors ψ^+ which come from prepared in-states ψ^{in} for $t \to -\infty$ and stop developing at $t = 0$. This set will be needed when we consider the contour deformation into the upper half of the second sheet in the discussion of the formation process depicted in Figure 1.2(b).

The physical separation of the set of states has its mathematical counterpart in the division of the space of physical states $\Phi \subset \mathscr{H} \subset \Phi^\times$. In order to show this we have to introduce some more mathematics:

⟦The Hilbert space of square integrable functions on the real line H (a possible realization of the space $\mathscr{H}$) is the direct sum

$$\mathsf{H} = \mathsf{H}_+^2 \oplus \mathsf{H}_-^2 \tag{4.8}$$

of the space of Hardy class functions from above H_+^2 and the space of Hardy class functions from below H_-^2.

A complex function $G(E)$ on the real line is a *Hardy class function*[14] from above (below) if

   (1) $G(E)$ is the boundary value of a function $G(\omega)$ of the complex variable (complex energy) $\omega = E + i\eta$ that is analytic in the half plane $\eta > 0$ ($\eta < 0$)

[14] P. L. Duren, *Theory of H^p-Spaces*, Academic Press, New York, 1970.

(2) $\displaystyle\int_{-\infty}^{\infty} |G(E + i\eta)|^2 \, dE < k < \infty$

for all η with $0 < \eta < \infty \ (-\infty < \eta < 0)$.

The function $G(E)$ is called an $H_{\pm}^p$-class function if the power 2 in the integral is replaced by the power $p = 1, 2, 3, \ldots$.

Hardy class functions have the following important properties:

(A)[15] If $G(E) \in H_+^2$ or $G(E) \in H_-^2$ then $G(E)$ (on the real axis) is uniquely determined by its values on the positive real axis.

(B)[16] If $G(E) \in H_+^p$, $1 < p < \infty$, then for all $\omega = E + i\eta$,

$$\frac{1}{2\pi i} \int_{-\infty}^{+\infty} \frac{G(E)}{E - \omega} \, dE = \begin{cases} G(\omega) & \text{for } \eta = \operatorname{Im} \omega > 0, \\ 0 & \text{for } \eta = \operatorname{Im} \omega < 0. \end{cases}$$

If $G(E) \in H_-^p$, $1 < p < \infty$, then for all $\omega = E + i\eta$,

$$-\frac{1}{2\pi i} \int_{-\infty}^{+\infty} \frac{G(E)}{E - \omega} \, dE = \begin{cases} 0 & \text{for } \eta = \operatorname{Im} \omega > 0, \\ G(\omega) & \text{for } \eta = \operatorname{Im} \omega < 0. \end{cases}$$

This is often referred to as the *Titchmarsh Theorem*.

(C)[17] If $G_{\pm}(E) \in H_{\pm}^p$, $p = 1, 2$; then the Fourier transform

$$\hat{G}_{\pm}(t) = \frac{1}{\sqrt{2\pi}} \int_{-\infty}^{+\infty} e^{-itE} G_{\pm}(E) \, dE$$

has the property

$$\hat{G}_{\pm}(t) = 0 \quad \text{for} \quad t \lessgtr 0.$$

This is called the *Paley–Wiener Theorem*.

(D) If $G(E) \in H_+^p$ then its complex conjugate $G^*(E)$ belongs to H_-^p and vice versa.

We will call a well-behaved vector $\psi^- \in \Phi$ a Hardy class vector from above, and write $\psi^- \in \Phi_+$, if its wave function $\langle E^- | \psi^- \rangle \in H_+$ (precisely if the values of $\langle E^- | \psi^- \rangle$ on the spectrum determine by property A a Hardy class function from above). Then $\langle \psi^- | E^- \rangle \in H_-$. Similarly we call $\psi^+ \in \Phi$ a Hardy class function from below, and write $\psi^+ \in \Phi_-$, if $\langle E^+ | \psi^+ \rangle \in H_-$.

In this way the space of physical states separates into two non-intersecting spaces Φ_+ and Φ_-. We mention without further discussion that they have a nuclear locally convex topology and each of them is dense in $\mathscr{H}$, where $\mathscr{H}$ is the space realized by the square-integrable functions on the spectrum, i.e., the positive real line, not by H. Thus we can construct two rigged Hilbert spaces[18]

$$\Phi_+ \subset \mathscr{H} \subset \Phi_+^{\times} \quad \text{and} \quad \Phi_- \subset \mathscr{H} \subset \Phi_-^{\times}, \tag{4.9}$$

where $\Phi_{(\pm)}^{\times}$ is the space of continuous antilinear functionals on $\Phi_{(\pm)}$.]

---

[15] C. van Winter, *J. Math. Anal.* **47**, 633 (1974); *Trans. Am. Math. Soc.* **162**, 103 (1971).
[16] P. L. Duren, *ibid.*, Theorem 11.8.
[17] P. L. Duren, *ibid.*, Theorem 11.9.
[18] For details see A. Bohm, M. Gadella (1989).

After this mathematical interlude we can continue with the discussion of the contour deformation in the integral (4.7). Now we admit to (4.7) only those ψ^- which can represent the state of decay products ψ^{out} for $t \to \infty$ that start to appear only at $t = 0$ (for $t < 0$ such decay products have not yet come into existence). This we then translate—on the basis of the discussion in the mathematical insert—into the mathematical statement

$$\psi^- \in \Phi_+ \tag{4.10a}$$

which means

$$\langle E^- | \psi^- \rangle \in H^2_+, \qquad \langle \psi^- | E^- \rangle \in H^2_- \tag{4.10b}$$

and

$$E^n \langle E^- | \psi^- \rangle \in H^2_+, \quad E^n \langle \psi^- | E^- \rangle \in H^2_- \quad \text{for every} \quad n = 1, 2, 3, \ldots. \tag{4.10c}$$

(4.10c) is fulfilled because $\langle E^- | \psi^- \rangle$ is well-behaved (i.e., $\psi^- \in \Phi$).

After the contour deformations have been performed, assuming that $\langle \psi^- | \omega^- \rangle$ and $\langle \omega^+ | \phi^+ \rangle$ have the mathematical properties required for it, (4.7) goes over into

$$(\psi^-, \phi^+) = \int_{\mathscr{C}_-} d\omega \langle \psi^- | \omega^- \rangle S_{\mathrm{II}}(\omega) \langle \omega^+ | \phi^+ \rangle$$
$$+ \int_{\mathscr{O}} d\omega \langle \psi^- | \omega^- \rangle \frac{\jmath_{-1}}{\omega - Z_R} \langle \omega^+ | \phi^+ \rangle, \tag{4.11}$$

Here $S_{\mathrm{II}}(\omega) = S(\omega_{\mathrm{II}})$ denotes the value of the S-matrix in the second sheet and $\jmath_{-1}$ is the residue of $S_{\mathrm{II}}(\omega)$ in the second sheet at Z_R:

$$S_{\mathrm{II}}(\omega) = \frac{\jmath_{-1}}{\omega - Z_R} + \jmath_0 + \jmath_1(\omega - Z_R) + \cdots. \tag{4.12}$$

The second integral in (4.11), the integral around the pole at Z_R will be considered separately below; it is this integral which will give the Gamow vector. We will first consider the integral along $\mathscr{C}_-$ which we call the background. As the integrand does not have any further singularities in the lower half of the second sheet, we can deform the contour $\mathscr{C}_-$ further and write the background as an integral from 0 to $-\infty$ plus an integral over the infinite lower semicircle in the second sheet, which however vanishes (because $S_{\mathrm{II}}(\omega)$ does not grow faster than a polynomial when $|\omega| \to \infty$). Thus for the background we obtain:

$$\text{background} = \int_0^{-\infty_{\mathrm{II}}} dE_{\mathrm{II}} \langle \psi^- | E_{\mathrm{II}}^- \rangle S(E_{\mathrm{II}} - i\epsilon) \langle E_{\mathrm{II}}^+ | \phi^+ \rangle. \tag{4.13}$$

Here we have indicated by E_{II} that the integral is running along the negative real axis in the second sheet and by $-\infty_{\mathrm{II}}$ that it goes to infinity on the second sheet.

We now consider the pole term. Using the Cauchy integral formula (A.1a) of the Appendix to Chapter XVII we obtain for the second integral in (4.11):

$$\text{pole term} = \int_{\mathscr{O}} d\omega \langle \psi^- | \omega^- \rangle \frac{\jmath_{-1}}{\omega - Z_R} \langle \omega^+ | \phi^+ \rangle$$
$$= -2\pi i \jmath_{-1} \langle \psi^- | Z_R^- \rangle \langle Z_R^+ | \phi^+ \rangle. \tag{4.14}$$

Adding (4.13) and (4.14) we can write (4.7) in the form

$$(\psi^-, \phi^+) = \int_0^{-\infty_{\text{II}}} dE_{\text{II}} \langle \psi^- | E_{\text{II}} - i\epsilon^+ \rangle \langle E_{\text{II}} - i\epsilon^+ | \phi^+ \rangle$$
$$- \langle \psi^- | Z_R^- \rangle (2\pi\Gamma) \langle Z_R^+ | \phi^+ \rangle. \tag{4.15a}$$

Here we have used (4.2) extended to "unphysical" values of E:

$$| E_{\text{II}}^+ \rangle = | E_{\text{II}}^- \rangle S(E_{\text{II}}). \tag{4.2a}$$

We have also made use of the fact that as a consequence of the unitarity the residue of the S-matrix is related to the width by

$$\jmath_{-1} = -i\Gamma \tag{4.16}$$

which can be immediately seen by comparing (4.12) with the representation (XVIII.6.23′) of the S-matrix. ψ^- was an arbitrary vector except that it had to be an element of Φ_+. We can omit this arbitrary vector in (4.15a) and write (4.15a) as

$$\phi^+ = \int_0^{-\infty} dE_{\text{II}} | E_{\text{II}}^+ \rangle \langle E_{\text{II}}^+ | \phi^+ \rangle$$
$$- | Z_R^- \rangle (2\pi\Gamma) \langle Z_R^+ | \phi^+ \rangle \quad \text{on } \Phi_+. \tag{4.15b}$$

This looks very much like the generalized basis vector expansion (II.7.16), (XV.2.7+), or (I.4.4g) except that for those basis vector expansions one could have taken the scalar product with a ψ^- which could have been from the whole space Φ, whereas in (4.15b) one can only take the scalar product with $\psi^- \in \Phi_+$. In other words, the generalized basis vectors $| E_{\text{II}}^+ \rangle$ and $| Z_R^- \rangle$ are only defined as functionals over Φ_+, whereas the Dirac kets $| E^+ \rangle$ with $E \in$ Spect. H are defined over all of Φ.

The pole term integral in (4.14) does not yet look as if it has anything to do with a resonance. A resonance has a Breit–Wigner energy distribution, so we will perform further contour deformations to bring the integral in (4.14) into a form exhibiting the typical Breit–Wigner amplitude. The circular contour in (4.14) is deformed into a straight line from $-\infty$ to $+\infty$ right below the real axis plus an infinite semicircle. If the integral over the infinite semicircle vanishes then (4.14) takes the form

$$\text{pole term} = \int_{-\infty}^{+\infty} dE_{\text{II}} \langle \psi^- | E_{\text{II}}^- \rangle \langle E_{\text{II}}^+ | \phi^+ \rangle \frac{\jmath_{-1}}{E_{\text{II}} - Z_R}$$
$$= (-2\pi i \jmath_{-1}) \langle \psi^- | Z_R^- \rangle \langle Z_R^+ | \phi^+ \rangle. \tag{4.17}$$

This integral contains the Breit–Wigner distribution $1/(E_{\text{II}} - Z_R)$ but it extends over the entire real axis.

Equation (4.17) is the same statement as Theorem B (the Titchmarsh theorem) for the function $G_-(E_{\text{II}}) = \langle \psi^- | E_{\text{II}}^- \rangle \langle E_{\text{II}}^+ | \phi^+ \rangle \jmath_{-1}$. Thus in order that the Breit–Wigner integral emerges from the pole term the function $G_-(E_{II})$ must be an element of H_-^1.

If we use the theorem B for $\langle \psi^- | E_{II}^- \rangle \in H_-^2$ then

$$-\frac{1}{2\pi i} \int_{-\infty}^{+\infty} dE_{II} \langle \psi^- | E_{II}^- \rangle \frac{1}{E_{II} - Z_R} = \langle \psi^- | Z_R^- \rangle. \tag{4.18}$$

Omitting in (4.18) the arbitrary $\psi^- \in \Phi_+$ we get a definition of a vector $|Z_R^-\rangle$ which has meaning only as a functional over Φ_+, which means that one can only take the "scalar products" of it with vectors which are $\in \Phi_+$:

$$|Z_R^-\rangle = -\frac{1}{2\pi i} \int_{-\infty_{II}}^{+\infty_{II}} dE_{II} |E_{II}^-\rangle \frac{1}{E_{II} - Z_R} \equiv \left(\frac{1}{2\pi\Gamma}\right)^{1/2} |f^D\rangle$$

$$\equiv \left(\frac{1}{2\pi\Gamma}\right)^{1/2} \left| E_R - i\frac{\Gamma^-}{2}\right\rangle. \tag{4.19}$$

This vector is—except for a "normalization" factor—identical to the vector $|f^D\rangle$ of (3.24), as indicated by the second equality in (4.19). This vector we call, because of its time development derived below, the *Gamow vector* or *Gamow ket*. We now also understand why f^D in (3.24) is to be defined by an integral over the Breit–Wigner amplitude from $-\infty$ to $+\infty$ and not from 0 to ∞ as one would have expected from the spectrum of H: The theoretically correct Breit–Wigner amplitude, i.e., the one connected with the resonance pole of the S-matrix, has its support on the whole real axis, more precisely from $-\infty$ to 0 on the second sheet and from 0 to ∞ on either the second sheet immediately below the cut or on the first sheet immediately above the cut. The "normalization" integral in (3.28) should, therefore, extend from $-\infty$ to $+\infty$ rather than from $-2E_R/\Gamma$ to ∞ so that $\langle f^D | f^D \rangle = 1.$[19] However, as already mentioned, for practical purposes the distinction is imperceivable if $\Gamma/E_R \ll 1$.

The complex conjugate of (4.18) gives

$$+\frac{1}{2\pi i} \int_{-\infty}^{\infty} dE_{II} \langle E_{II}^- | \psi^- \rangle \frac{1}{E_{II} - Z_R^*} = (\langle \psi^- | Z_R^- \rangle)^*. \tag{4.20}$$

Using for the left-hand side Theorem B this leads to

$$\langle Z_R^{*-} | \psi^- \rangle = (\langle \psi^- | Z_R^- \rangle)^*. \tag{4.21}$$

[19] To avoid misunderstanding concerning the mathematics it must be emphasized that the Breit–Wigner amplitude $\langle E^- | f^D \rangle$ is an element of H in which the scalar product is defined by integration from $-\infty$ to $+\infty$. $|f^D\rangle$ is not an element of $\mathscr{H}$, but an element of $\Phi_+^\times$ (functionals over Φ_+). $\mathscr{H}$ is realized by the space of square integrable functions on the positive real line $L^2(\mathbb{R}^+)$ (the spectrum of H), not by $H = L^2(\mathbb{R})$. The physical scalar product is given by the integral over the spectrum or also—according to (4.15b)—by the integral from 0 to $-\infty$ plus the integral from $-\infty$ to $+\infty$ which differs from the scalar product in H by the integral from 0 to $-\infty$. Therefore vectors in H become generalized vectors for $\mathscr{H}$.

Again omitting the arbitrary $|\psi^-\rangle$ we obtain

$$\langle Z_R^{*\,-}| = \frac{1}{2\pi i} \int_{-\infty}^{\infty} dE_{\mathrm{II}} \langle E_{\mathrm{II}}^- | \frac{1}{E_{\mathrm{II}} - Z_R^*}$$

$$\equiv \left(\frac{1}{2\pi\Gamma}\right)^{1/2} \langle f^D| \equiv \left(\frac{1}{2\pi\Gamma}\right)^{1/2} \left(E_R + i\frac{\Gamma^-}{2}\right| \qquad (4.22)$$

which is the same as (3.26).

Using the "normalized" vectors and (4.16) the pole term (4.14) can be written

$$\text{pole term} = -\left\langle \psi^- \left| E_R - i\frac{\Gamma^-}{2} \right)\left(E_R - i\frac{\Gamma^+}{2} \right| \phi^+ \right\rangle \qquad (4.23)$$

and the generalized basis vector expansion (4.15b) has the form

$$\phi^+ = \int_0^{-\infty} dE_{\mathrm{II}} |E_{\mathrm{II}}^+\rangle \langle E_{\mathrm{II}}^+ |\phi^+\rangle -- \left| E_R - i\frac{\Gamma^-}{2} \right)\left(E_R - i\frac{\Gamma^+}{2} \right| \phi^+ \right\rangle \qquad (4.15c)$$

which is an equality between continuous functionals in $\Phi_+^\times$. If instead of one resonance pole one has n resonances $R_1, R_2, \ldots, R_n$ then the last term will be replaced by a sum of n such terms. Thus in the new basis vector expansion (4.15a)–(4.15c) the resonances appear in very much the same way as the bound states in the ordinary spectral decomposition.

We shall now derive (3.29). To this end we consider the vector $H\psi^-$. By assumption (well-behavedness) this is again $\in \Phi_+$. We can then repeat the whole above procedure starting with (4.7) and replacing everywhere ψ^- by $H\psi^-$ on the left-hand side and $\langle \psi^- | E^-\rangle$ by $E\langle \psi^- | E^-\rangle$ under the integral. When we reach (4.18) it will now read

$$\langle H\psi^- | Z_R^-\rangle = -\frac{1}{2\pi i} \int_{-\infty}^{+\infty} dE_{\mathrm{II}} \, E_{\mathrm{II}} \langle \psi^- | E_{\mathrm{II}}^-\rangle \frac{1}{E_{\mathrm{II}} - Z_R}. \qquad (4.24)$$

Since $E\langle \psi^- | E^-\rangle$ is also a Hardy class function from below we can apply the Titchmarsh theorem with $G(E_{\mathrm{II}}) = E_{\mathrm{II}} \langle \psi^- | E_{\mathrm{II}}^-\rangle$ to the right-hand side and obtain

$$\langle H\psi^- | Z_R^-\rangle = Z_R \langle \psi^- | Z_R^-\rangle \quad \text{for every } \psi^- \in \Phi_+. \qquad (4.25)$$

This is the precise formulation of the statement that $|Z_R^-\rangle$ is a generalized eigenvector of H over Φ_+ with eigenvalue Z_R, which was written formally as (3.29a). Equation (3.29b) is then immediately obtained by taking the complex conjugate of (4.24) and using the Titchmarsh theorem for the Hardy class function from above $G(E) = E\langle E^- |\psi^-\rangle$.

Equation (3.30) may appear to be an immediate consequence of (3.29), but then one would wonder about the restriction $t \geq 0$. It is indeed not straightforward. $|f^D\rangle = |Z_R^-\rangle$ is an element of $\Phi_+^\times$, the space of continuous antilinear functionals over Φ_+. Thus the unitary time development operator $U(t) = e^{-iHt}$ in the Hilbert space has to be extended to a time development operator in $\Phi_+^\times$. This can only be done if $U^*(t) = e^{iHt}$ leaves the space Φ_+ invariant, i.e., if and only if $e^{iHt}\psi^- \in \Phi_+$ for every $\psi^- \in \Phi_+$.

We will now apply the above procedure starting with (4.7) to the vector $e^{iHt}\psi^-$ instead of ψ^-. Under the integral on the right-hand side of (4.7) we then have $e^{-iEt}\langle\psi^-|E^-\rangle$ in place of $\langle\psi^-|E^-\rangle$. Everything in the above procedure can be repeated as long as $e^{-iEt}\langle\psi^-|E^-\rangle$ is a Hardy class function from below. But if $G(E)\in H^2_-$ then $e^{-iEt}G(E)\in H^2_-$ iff $t\geq 0$. Therefore only for $t\geq 0$ can one repeat the above procedure and arrive in place of (4.18) at

$$\langle e^{iHt}\psi^-|Z_R^-\rangle = -\frac{1}{2\pi i}\int dE_{\rm II}\, e^{-iE_{\rm II}t}\langle\psi^-|E_{\rm II}^-\rangle\frac{1}{E_{\rm II}-Z_R}, \qquad \text{for } t\geq 0 \text{ only.}$$

(4.26)

To the right-hand side one can again apply the Titchmarsh theorem and obtain

$$\langle e^{iHt}\psi^-|Z_R^-\rangle = e^{-iZ_R t}\langle\psi^-|Z_R^-\rangle \quad \text{for} \quad t\geq 0 \quad \text{and every} \quad \psi^-\in\Phi_+.$$

(4.27)

Equation (4.27) is the precise form of (3.30).

If e^{iHt} is a continuous operator from Φ_+ to Φ_+ one can define the conjugate operator in $\Phi_+^\times$ by the usual definition of the conjugate operator in the rigged Hilbert space:[20]

$$\langle e^{iHt}\psi^-|Z_R^-\rangle = \langle\psi^-|(e^{iHt}_+)^\times|Z_R^-\rangle.$$

(4.28)

This operator, defined only for $t>0$ is what is meant by the operator in (3.30):

$$(e^{iHt}_+)^\times = ``e^{-iHt}", \qquad t\geq 0.$$

(4.29)

Here the subscript $+$ indicates that it is defined in $\Phi_+^\times$ only, where it is the extension of the Hilbert space operator e^{-iHt}.

We have herewith established all the results that have been used and justified only heuristically in Section XXI.3. We have given more precise definitions and statements of Section XXI.3's new entities and their properties. And we have shown that the exponentially decaying state, the Gamow vectors (4.19) or (3.24), come from the resonance pole of the S-matrix in the lower half plane.

To complete our discussions we shall now very briefly demonstrate that the resonance poles above the real axis lead to exponentially growing states.

We start from the complex conjugate of (4.7) and use the symmetry relation of the S-matrix $S^*(E+i0)=S(E-i0)$, (XVIII.5.10). Then we obtain

$$(\phi^+,\psi^-) = \int_{\text{Spect. }H} dE\langle\phi^+|E^+\rangle S(E-i0)\langle E^-|\psi^-\rangle.$$

(4.7')

If the integration in (4.7) was along the upper rim of the cut the integration is now along the lower rim of the cut in the physical sheet, which is indicated by writing $E-i0$ as the argument of S. We use the same symbols ϕ^+ and ψ^- as above but these ϕ^+ and ψ^- have now an entirely different physical interpretation and therefore different mathematical properties than those

[20] For example, A. Bohm (1978), Chapter III.

entities denoted by the same symbol above. Indeed, what we consider now is the earlier half of the resonance scattering process in which one starts in the remote past, $t \to -\infty$, with an in-state ϕ^{in} that becomes ϕ^+ developing with the Hamiltonian H into a state ϕ^+ at $t = 0$. Thus whereas in (4.7) ψ^- described the controlled state, i.e., the state determined by the experimental arrangement for the decay products ψ^{out}, in (4.7′) it is the vector ϕ^+ which is the controlled state. ϕ^+ in (4.7′) is controlled by the preparation apparatus for ϕ^{in} and ψ^- in (4.7′) is the state that results as the consequence of the dynamics.

For these controlled states ϕ^+, which will have to be gone by the time $t = 0$, and which therefore constitute roughly half of the space of physical states which we call Φ_-, we postulate that the wave functions $\langle E^+ | \phi^+ \rangle$ are Hardy class functions from below (because then $\int \langle E^+ | \phi^+(t) \rangle \, dE = 0$ for $t > 0$ by the Paley–Wiener theorem). Thus $\langle \phi^+ | E^+ \rangle$ in (4.7′) are $\in H^2_+$ and we may deform the contour of integration in (4.7′) through the cut into the upper half plane of the second sheet as shown in Figure 4.1(b). (4.7′) then goes into:

$$(\phi^+, \psi^-) = \int_{\mathscr{C}_+} d\omega \, \langle \phi^+ | \omega^+ \rangle S_{\mathrm{II}}(\omega) \langle \omega^- | \psi^- \rangle$$

$$+ \int_{\circlearrowright} d\omega \, \langle \phi^+ | \omega^+ \rangle \frac{s_{-1}}{\omega - Z_R^*} \langle \omega^- | \psi^- \rangle, \qquad (4.11')$$

where s_{-1} is the residue of $S(\omega)$ in the second sheet at Z_R^*:

$$S_{\mathrm{II}}(\omega) = \frac{s_{-1}}{\omega - Z_R^*} + s_0 + s_1(\omega - Z_R^*) + \cdots . \qquad (4.12')$$

and $\mathscr{C}_+$ is the path above the pole at Z_R^* and $\circlearrowright$ is the circular path around Z_R^* as shown in Figure 4.1(b). Again assuming sufficiently nice properties of $\langle \phi^+ | \omega^+ \rangle$, $\langle \omega^- | \psi^- \rangle$ and $S(\omega)$, the first integral, the background, can be written

$$\text{background} = \int_0^{-\infty_{\mathrm{II}}} dE_{\mathrm{II}} \, \langle \phi^+ | E_{\mathrm{II}}^+ \rangle S(E_{\mathrm{II}}) \langle E_{\mathrm{II}}^- | \psi^- \rangle. \qquad (4.13')$$

The pole term is evaluated using the Cauchy integral formula (XVII.A.1a) and we obtain

$$\text{pole term} = \int_{\circlearrowright} d\omega \langle \phi^+ | \omega^+ \rangle \frac{s_{-1}}{\omega - Z_R^*} \langle \omega^- | \psi^- \rangle$$

$$= 2\pi i s_{-1} \langle \phi^+ | Z_R^{*+} \rangle \langle Z_R^{*-} | \psi^- \rangle. \qquad (4.14')$$

The integral in (4.7′) can therefore be written as

$$(\phi^+, \psi^-) = \int_0^{-\infty_{\mathrm{II}}} dE_{\mathrm{II}} \, \langle \phi^+ | E_{\mathrm{II}}^- \rangle \langle E_{\mathrm{II}}^- | \psi^- \rangle$$

$$- \langle \phi^+ | Z_R^{*+} \rangle 2\pi \Gamma \langle Z_R^{*-} | \psi^- \rangle, \qquad (4.15'a)$$

where we have used an extension of (4.2) to the second sheet,

$$|E_{\text{II}}^+\rangle S^*(E_{\text{II}}) = |E_{\text{II}}^-\rangle, \tag{4.2'}$$

and the consequence of the unitarity of the S-matrix[21]

$$s_{-1} = i\Gamma. \tag{4.16'}$$

The integral in the pole term (4.11') is now transformed into a Breit–Wigner integral by the contour deformation using the fact that

$$\langle\phi^+|E_{\text{II}}^+\rangle\langle E_{\text{II}}^-|\psi^-\rangle s_{-1} \in H_+^1:$$

$$\text{pole term} = \int_{-\infty}^{+\infty} dE_{\text{II}}\langle\phi^+|E_{\text{II}}^+\rangle\langle E_{\text{II}}^-|\psi^-\rangle s_{-1}\,\frac{1}{E_{\text{II}} - Z_R^*}$$

$$= 2\pi i s_{-1}\langle\phi^+|Z_R^{*+}\rangle\langle Z_R^{*-}|\psi^-\rangle. \tag{4.17'}$$

$\langle\phi^+|E^+\rangle$ is now $\in H_+^2$, i.e., $\phi^+ \in \Phi_-$. The Titchmarsh theorem applied to this function then gives

$$\langle\phi^+|Z_R^{*+}\rangle = \frac{1}{2\pi i}\int_{-\infty}^{+\infty} dE_{\text{II}}\,\langle\phi^+|E_{\text{II}}^+\rangle\,\frac{1}{E_{\text{II}} - Z_R^*}. \tag{4.18'}$$

Writing this without the arbitrary vector $\phi^+ \in \Phi_-$, we obtain

$$|Z_R^{*+}\rangle = \frac{1}{2\pi i}\int_{-\infty}^{+\infty} dE_{\text{II}}\,|E_{\text{II}}^+\rangle\,\frac{1}{E_{\text{II}} - Z_R^*}$$

$$\equiv \left(\frac{1}{2\pi\Gamma}\right)^{1/2}|f^G\rangle \equiv \left(\frac{1}{2\pi\Gamma}\right)^{1/2}\left|E_R + i\frac{\Gamma^+}{2}\right). \tag{4.19'}$$

The vector $|f^G\rangle = |E_R + i\Gamma/2^+)$, G standing for growing or forming, defined by (4.19') is the Gamow vector associated with the pole of the S-matrix in the upper half of the second sheet. It has meaning only as a functional over Φ_-. Thus there are two Gamow kets associated with each resonance, the exponentially decaying functional over Φ_+, (4.19), and the functional (4.19') which is exponentially growing, as we shall derive below.

The complex conjugate of (4.18') leads to

$$\langle Z_R^+| = -\frac{1}{2\pi i}\int_{-\infty}^{+\infty} dE\langle E^+|\frac{1}{E - Z_R}$$

$$\equiv \left(\frac{1}{2\pi\Gamma}\right)^{1/2}\langle f^G| \equiv \left(\frac{1}{2\pi\Gamma}\right)^{1/2}\left(E_R - i\frac{\Gamma^+}{2}\right|. \tag{4.22'}$$

---

[21] Note that in (4.16) an arbitrary phase factor has been fixed (incorporated in the definition of the kets which also contain an arbitrary phase). The phase in (4.16') is then no longer arbitrary but is given by time-reversal invariance.

In place of (4.15c) we obtain now from (4.15′a) for the generalized basis vector expansion over Φ_-

$$\psi^- = \int_0^{-\infty} dE_{\text{II}} |E_{\text{II}}^-\rangle\langle E_{\text{II}}^-|\psi^-\rangle - \left|E_R + i\frac{\Gamma}{2}^+\right)\left(E_R + i\frac{\Gamma}{2}^-\right|\psi^-\rangle \tag{4.15′c}$$

which is an equality between continuous functionals in $\Phi_-^\times$. (Note that the symbols ϕ^+ and ψ^- have different meanings in the primed and unprimed equations.)

The proof that $|Z_R^{*+}\rangle$ is a generalized eigenvector of H with eigenvalue $Z_R^* = E_R + i\Gamma/2$ is a straightforward application of the Titchmarsh theorem to $E_{\text{II}}\langle\phi^+|E_{\text{II}}^+\rangle \in H_+^2$. In the same way as for (4.25) one shows

$$\langle H\phi^+|Z_R^{*+}\rangle = Z_R^*\langle\phi^+|Z_R^{*+}\rangle \quad \text{for every} \quad \phi^+ \in \Phi_- \tag{4.25′}$$

which one also writes formally as

$$H|f^G\rangle = \left(E_R + i\frac{\Gamma}{2}\right)|f^G\rangle, \tag{4.25′a}$$

$$\langle f^G|H = \left(E_R - i\frac{\Gamma}{2}\right)\langle f^G|. \tag{4.25′b}$$

The time development of $|f^G\rangle$ is now obtained in complete analogy to (4.27). We have to define a time-development operator in $\Phi_-^\times$ as an extension of the Hilbert space time development operator e^{-iHt}. This operator can only be defined as the conjugate operator to the operator e^{iHt} in Φ_- if for every $\phi^+ \in \Phi_-$ also $e^{iHt}\phi^+ \in \Phi_-$. This means that it can be defined only if with $\langle\phi^+|E^+\rangle \in H_+^2$ also $\langle e^{iHt}\phi^+|E^+\rangle = e^{-iEt}\langle\phi^+|E^+\rangle \in H_+^2$. This is the case only if $t \leq 0$. Therefore one obtains for $\phi^+ \in \Phi_-$ from

$$\langle e^{iHt}\phi^+|Z_R^{*+}\rangle = \frac{1}{2\pi i}\int dE_{\text{II}}e^{-iE_{\text{II}}t}\langle\phi^+|E_{\text{II}}^+\rangle\frac{1}{E_{\text{II}} - Z_R^*} \tag{4.26′}$$

by application of the Titchmarsh theorem to right-hand side:

$$\langle e^{iHt}\phi^+|Z_R^{*+}\rangle = e^{iZ_R^*t}\langle\phi^+|Z_R^{*+}\rangle \quad \text{only for } t \leq 0 \text{ and for every } \phi^+ \in \Phi_-. \tag{4.27′}$$

Thus one can define the time development operator in $\Phi_-^\times$ as the conjugate of the operator e^{iHt} in Φ_- only for $t \leq 0$:

$$(e_-^{iHt})^\times = \text{``}e^{-iHt}\text{''}, \quad t \leq 0. \tag{4.29′}$$

This is the extension of the unitary time-development operator e^{-iHt} in $\mathcal{H}$ to the space $\Phi_-^\times$. Formally one writes (4.27′) as

$$\text{``}e^{-iHt}\text{''}|f^G\rangle = e^{-iE_Rt}e^{+\Gamma/2t}|f^G\rangle \quad \text{for} \quad t \leq 0 \text{ only,} \tag{4.27′a}$$

keeping in mind that one can take the "scalar product" of this equation only with elements of Φ_-. One also writes

$$\text{``}e^{-iHt}\text{''}|f^G\rangle\langle f^G|\text{``}e^{iHt}\text{''} = e^{\Gamma t}|f^G\rangle\langle f^G| \quad \text{for} \quad t \leq 0 \text{ only.} \tag{4.27′b}$$

This is to be compared with (3.38) for the exponentially decaying state. Thus we see that the Gamow vector $|f^G\rangle$, which derives from the resonance pole above the real axis, represents an exponentially growing state.

XXI.5 The Golden Rule

The formula that expresses the initial decay rate in terms of the T-matrix or the matrix elements of the interaction Hamiltonian V is called the *Golden Rule*. The decay rate is—according to (XIV.2.3)—the time derivative of the decay probability (2.8), which is the transition probability from the decaying state $W(t) = e^{-iHt}|f^D\rangle\langle f^D|e^{iHt}$ into the states of the decay products described by the subspace $\Lambda\mathcal{H}$. Though one could use the results of Section XIV.2 and derive immediately the decay rate (XIV.2.6) or (XIV.2.18), it is instructive to derive first the expression for the decay probability

$$\mathcal{P}(t) = \text{Tr}(\Lambda W(t)) \tag{5.1}$$

and then obtain from it the conventional form of the Golden Rule for the initial decay rate $\mathcal{P}(t = 0)$ by differentiation.

In order not to obscure this simple derivation by unimportant details we will first ignore all the additional quantum numbers $\hat{a}\,(=ll_3\kappa)$ and only afterwards take into consideration that the state W^R is a mixture of $|f^D\rangle = |E_R - i\Gamma/2, \hat{a}^-\rangle$ as expressed by (3.35).

$\Lambda\mathcal{H}$ describes the final decay products after the interaction has ceased. Therefore—as in Section XIV.2—

$$\Lambda = \sum_b |b\rangle\langle b|, \quad \text{where} \quad |b\rangle = |E_b, \hat{b}\rangle$$

are eigenvectors of the free Hamiltonian K. Using these basis vectors for the calculation of the trace in (5.1) we obtain

$$\mathcal{P}(t) = \sum_b \langle b|e^{-iHt}|f^D\rangle\langle f^D|e^{iHt}|b\rangle. \tag{5.2}$$

The sum in (5.2) goes over all values of $b = (E_b, \hat{b})$ if Λ is the projection operator on the space of physical states of all decay products into which the decaying state $|f^D\rangle\langle f^D|$ can decay; if only a part of the decay products are observed, i.e., if only a partial decay rate is measured, then Λ is the projector on the subspace of these observed decay products and the sum in (5.2) is only over those basis vectors which span this subspace.

The vectors $|b\rangle$ in (5.2) are replaced by the $V|b\rangle$ and the eigenvectors of $H, |b^-\rangle$, using the solution of the Lippmann–Schwinger equation (XV.1.22). One obtains

$$\langle f^D|e^{-iHt}|b\rangle = \langle f^D|e^{-iHt}|b^-\rangle - \langle f^D|e^{-iHt}V|b\rangle\frac{1}{E_b - (E_R + i\Gamma/2) - i\epsilon}, \tag{5.3}$$

where the property (3.29b) of the decaying state vector $\langle f^D| = (E_R + i\Gamma/2^-|$ was used.

Inserting (5.3) into (5.2) gives:

$$\mathscr{P}(t) = \int_0^\infty dE_b e^{-iE_b t}\langle E_b^- |Z_R^-\rangle(Z_R^{*-}|E_b^-\rangle e^{iE_b t}$$

$$- \sum_b (Z_R^{*-}|e^{iHt}|b^-\rangle\langle b|Ve^{-iHt}|Z_R^-) \frac{1}{E_b - (E_R - i\Gamma/2) + i\epsilon}$$

$$- \sum_b \langle b^-|e^{-iHt}|Z_R^-)(Z_R^{*-}|e^{iHt}V|b\rangle \frac{1}{E_b - (E_R + i\Gamma/2) - i\epsilon}$$

$$+ \sum_b \langle b|Ve^{-iHt}|Z_R^-)(Z_R^{*-}|e^{iHt}V|b\rangle \frac{1}{(E_b - E_R)^2 + (\Gamma/2)^2}. \quad (5.4)$$

In the first term we have replaced $\sum_b$ by the integral over the spectrum of H which is identical with the spectrum of K. The second and the third term in (5.4) are the complex conjugates of each other. We now use the property (3.30) = (4.27) of the decaying Gamow vectors; (5.4) then becomes

$$\mathscr{P}(t) = \int_0^\infty dE_b\,\langle E_b^- |Z_R^-)(Z_R^{*-}|E_b^-\rangle$$

$$- e^{-\Gamma t}\sum_b ((Z_R^{*-}|b^-\rangle\langle b|V|Z_R^-) \frac{1}{E_b - (E_R - i\Gamma/2) + i\epsilon} + \text{c.c.})$$

$$+ e^{-\Gamma t}\sum_b \langle b|V|Z_R^-)(Z_R^{*-}|V|b\rangle \frac{1}{(E_b - E_R)^2 + (\Gamma/2)^2}. \quad (5.5)$$

Here c.c. denotes the complex conjugate of the preceding second term, which we denote by II. This term is now elaborated further by using again the solution of the Lippmann–Schwinger equation (XV.1.22−) for $|b^-\rangle$:

$$\text{II} = -e^{-\Gamma t}\sum_b ((Z_R^{*-}|b\rangle\langle b|V|Z_R^-) \frac{1}{E_b - (E_R - i\Gamma/2) + i\epsilon}$$

$$+ \frac{1}{(E_b - E_R)^2 + (\Gamma/2)^2}(Z_R^{*-}|V|b\rangle\langle b|V|Z_R^-)).$$

The first term in II is zero. The reason for this is analogous to the reason that led to II = 0 following equation (XIV.2.17) or preceding equation (XV.A.4):[22] $(Z_R^{*-}|$ describes the decaying state, $|b\rangle$ spans the space of the decay products when the interaction V performing the transition is switched off; thus $(Z_R^{*-}|b\rangle$ is the probability amplitude for no transition, whereas $\langle b|V|Z_R^-)$ is the probability amplitude for the transition. The second term in II is real, therefore c.c. in (5.5) is identical to II. This is also identical (except for the opposite sign) to the fourth term in (5.5). The first term in

---

[22] Instead of using (5.1) as the starting point of our derivation we could have started with (XIV.2.18) for the decay rate, in which this term is already gone. This would however have made the calculation slightly less transparent, cf. Problem 3.

(5.5) is obtained from (3.28) to be equal to unity, except for a small correction $\Gamma/2\pi E_R + \cdots$. Thus we obtain for the decay probability:

$$\mathscr{P}(t) = \left(1 - \frac{\Gamma}{2\pi E_R} + \cdots\right)$$
$$- e^{-\Gamma t} \sum_b \langle b|V|f^D\rangle\langle f^D|V|b\rangle \frac{1}{(E_b - E_R)^2 + (\Gamma/2)^2}. \quad (5.6)$$

This formula for the decay probability of a quasistationary state with energy energy E_R and width Γ does not contain any approximation. It has the form expected from (3.40) for $\Gamma/2E_R \to 0$ with

$$\mathscr{P}_R(t) = \frac{2\pi e^{-\Gamma t}}{\Gamma} \sum_b |\langle b|V|f^D\rangle|^2 \left(\frac{\Gamma/2\pi}{(E_b - E_R)^2 + (\Gamma/2)^2}\right). \quad (5.7)$$

According to (3.39), $\mathscr{P}_R(t)$ at $t = 0$ should be equal to 1 (and $\mathscr{P}(t) \to 0$). As (cf. Equation (II.8.12)):

$$\frac{\Gamma/2\pi}{(E_b - E_R)^2 + (\Gamma/2)^2} \to \delta(E_b - E_R) \quad \text{for} \quad \Gamma/2E_R \to 0 \quad (5.8)$$

we obtain from (5.7):

$$\Gamma = 2\pi \sum_b |\langle b|V|f^D\rangle|^2 \, \delta(E_b - E_R). \quad (5.9)$$

For the decay rate we obtain from (5.6) by differentiation with respect to time:

$$\dot{\mathscr{P}}(t) = e^{-\Gamma t} 2\pi \sum_b |\langle b|V|f^D\rangle|^2 \frac{\Gamma/2\pi}{(E_b - E_R)^2 + (\Gamma/2)^2}$$
$$= e^{-\Gamma t} 2\pi \int_0^\infty dE_b |\langle E_b|V|f^D\rangle|^2 \frac{\Gamma/2\pi}{(E_b - E_R)^2 + (\Gamma/2)^2}. \quad (5.10)$$

This formula for the decay rate contains no approximation. The factor

$$b(E_b) = \frac{\Gamma/2\pi}{(E_b - E_R)^2 + (\Gamma/2)^2} \quad (5.11)$$

is called the *natural line width*. It can be observed if the detector has a very good energy resolution $\Delta E_b \ll \Gamma$ so that one can measure the decay rate per unit energy:

$$\frac{d\dot{\mathscr{P}}(t)}{dE_b} = e^{-\Gamma t} 2\pi |\langle E_b|V|f^D\rangle|^2 \frac{\Gamma/2\pi}{(E_b - E_R)^2 + (\Gamma/2)^2}. \quad (5.12)$$

In the limit of narrow width $\Gamma/2E_R \to 0$ the initial decay rate, (5.10) for $t = 0$, becomes

$$\dot{\mathscr{P}}(0) = 2\pi \sum_b |\langle b|V|f^D\rangle|^2 \, \delta(E_b - E_R). \quad (5.13)$$

Comparing this with (5.9) we see, as already observed before (3.14) that the initial decay rate is equal to the width.

The formula (5.13) is the conventional form of the Golden Rule.[23] As (5.6) to (5.12) are different forms of one and the same result we will refer to all these formulas as the Golden Rule.

Though (5.6), (5.10) and (5.12) do not entail any approximation, they contain some quantities which are not easily accessible and can usually only be obtained by approximation methods—like the generalized eigenvalue of the exact Hamiltonian, $E_R - i\Gamma/2$, and its eigenvector $|f^D\rangle$. For narrow resonances, like the excited states of atoms and molecules and in any other case when the interaction Hamiltonian V describes a weak interaction, Γ is orders of magnitude smaller than E_R and $|f^D\rangle$ can be replaced by the eigenvector of K which is stationary with respect to the time development e^{-iKt} (Born approximation).

Now that the fundamental results have been derived for the simplest model, we will reincorporate the additional quantum numbers $\hat{a}$ in order to apply the Golden Rule to realistic cases. This is done by replacing $|f^D\rangle\langle f^D|$ in the above formulas with W^R of (3.35). The result for the decay rate is

$$\dot{\mathcal{P}}(0) = 2\pi \sum_b \sum_{\hat{a}\hat{a}'} \delta(E_b - E_R)\langle b|V|E_R, \hat{a}')(E_R, \hat{a}|V|b\rangle\langle \hat{a}'\|W^R\|\hat{a}\rangle \quad (5.14)$$

where we have written $|E_R, \hat{a})$ for the decaying state vector $|E_R - i\Gamma/2, \hat{a}^-)$. For the familiar case of a decaying state with angular momentum l_R and internal quantum numbers κ_R [Equation (3.16)], this goes over into

$$\dot{\mathcal{P}}(t = 0) = 2\pi \frac{1}{(2l_R + 1)} \sum_{l_3} \sum_b \delta(E_b - E_R)|\langle b|V|E_R, l_R, l_3, \kappa_R)|^2. \quad (5.15)$$

(5.14) is the general form of the Golden Rule for the initial decay rate. If the summation in b extends over all values of the quantum numbers $b = (E_b, \hat{b})$ then $\mathcal{P}$ is called the total decay rate.

Equations (5.6)–(5.15) are formulas used for the calculation of decay rates in all branches of physics. Usually the interaction is weak, so that the Born approximation can be used for the transition matrix element $\langle b|V|E_R, \hat{a})$. This means $|E_R, \hat{a})$ is replaced by the eigenvector of the free Hamiltonian

$$K = H - V. \quad (5.16)$$

The $|b\rangle$ are already eigenvectors of K [cf. (XIV.2.14)]. So the program for the calculation of the initial decay rate is the following: One finds the eigenstates and generalized eigenvectors of the free-energy operator. These are the energy eigenvectors in the approximation in which the decaying system is considered stable. For instance, in the decay (1.3a), $|E_R, l_R, l_3, \kappa_R)$ will be the proper eigenvector of the energy operator of the atom when the interaction with the radiation is ignored and all energy levels are stable. $|b\rangle$ is the direct product

$$|b\rangle = |E_R^{gr}, l_{gr}, l_3, \kappa_{gr}) \otimes |\gamma\rangle,$$

---

[23] This formula of Dirac has given so much praiseworthy service in many areas of quantum physics that Fermi called it a Golden Rule (the "Golden Rule No. 2").

where the first factor is the ground-state energy eigenvector of the atom, and $|\gamma\rangle$ is a basis vector of the one-photon space. Then one has to conjecture the interaction Hamiltonian V (or the transition operator T if the Born approximation is not satisfactory). For the case of the radiative decays of atoms this can be done to a certain extent by using the correspondence with a classical system; in other situations, e.g., for the decay of elementary particles, one has only very few general principles to limit the possibilities for this conjecture. After V and the solution of the free problem are known, one has to calculate the matrix elements of V between the free eigenvectors and insert it into (5.14) or (5.15). In this way the transition is considered as a transition from one "stable" state into another, caused by the interaction Hamiltonian V or by the transition operator T.

XXI.6 Partial Decay Rates

Even more important in practical calculations than the total decay rate are the partial decay rates. In order to obtain a partial decay rate we have to specify the basis vectors $|b\rangle$ in (5.14) or (5.15). For the sake of definiteness we shall give our discussions first in terms of the angular-momentum basis. One can choose

$$|b\rangle = |E_b, \hat{b}\rangle = |E_b, j j_3 \eta\rangle \tag{6.1}$$

if the angular momentum commutes with the internal observables:

$$[L_i, \eta^{op}] = 0. \tag{6.1a}$$

The internal observables η^{op} will in general not agree with the internal observables κ^{op} of which the resonance is an eigenstate with eigenvalue κ_R. We then write (5.15):

$$\Gamma = 2\pi \sum_\eta \int \rho_\eta(E)\, dE\, \delta(E - E_R) \sum_{j j_3} \sum_{l_3} \frac{1}{2l_R + 1} |\langle E j j_3 \eta | V | E_R l_R l_3 \kappa_R\rangle|^2. \tag{6.2}$$

This gives the rate for the decay of R into decay products with any of the internal quantum numbers η. If the detector detects only decay products with one particular value for η, then one measures the quantity

$$\Gamma(R \to \eta) = 2\pi \int \rho_\eta(E)\, dE\, \delta(E - E_R)$$

$$\times \sum_{j j_3} \sum_{l_3} \frac{1}{2l_R + 1} |\langle E j j_3 \eta | V | E_R l_R l_3 \kappa_R\rangle|^2. \tag{6.3}$$

$\Gamma(R \to \eta)$ is called the *partial decay rate* for the decay of R into the state with internal quantum numbers η (decay channel η). Obviously one has

$$\sum_\eta \Gamma(R \to \eta) = \Gamma. \tag{6.4}$$

This statement is identical to the time derivative of (2.10).

Most often V is rotationally symmetric, so that only the term with $j = l_R$ in (6.3) is different from zero.

The weight function $\rho_n(E)$ is related to the normalization of the generalized eigenvectors $|E_b, j\, j_3\, \eta\rangle$:

$$\langle E\, j\, j_3\, \eta\, |E'\, j'\, j'_3\, \eta'\rangle = \delta_{\eta\eta'}(\rho_n(E))^{-1}\delta(E - E')\delta_{jj'}\,\delta_{j_3 j'_3}. \qquad (6.5)$$

Instead of the basis vector (6.1) one may use different basis vectors, e.g., the generalized momentum eigenvectors

$$\langle \mathbf{p}\,\eta\,|\mathbf{p}'\,\eta'\rangle = \delta(\mathbf{p} - \mathbf{p}')\delta_{\eta\eta'} \qquad (6.6)$$

if

$$[P_i, \eta^{\mathrm{op}}] = 0. \qquad (6.6a)$$

Then one has instead of (6.3)

$$\Gamma(R \to \eta) = 2\pi \int d^3p\, \delta(E(p) - E_R)\sum_{l_3}\frac{1}{2l_R + 1}|\langle \mathbf{p}\,\eta\,|V\,|E_R\, l_R\, l_3\, \kappa_R\rangle|^2. \qquad (6.7)$$

Instead of the momenta or angular momenta one can choose any other labels

$$|E_b, \hat{b}\rangle = |E_b, \beta, \eta\rangle. \qquad (6.8)$$

The only requirement is that

$$[\beta^{\mathrm{op}}, \eta^{\mathrm{op}}] = 0.$$

For the basis vectors that one uses in the decay-rate formula one should always choose the most convenient ones, and which the most convenient ones are depends upon the particular problem.[24] Therefore we now discuss the general case.

Let

$$K, \alpha^{\mathrm{op}}, \kappa^{\mathrm{op}} \qquad (6.9)$$

be a complete system of commuting operators for the decaying system, where we have split $\hat{a}^{\mathrm{op}} = (\kappa^{\mathrm{op}}, \alpha^{\mathrm{op}})$ so that the decaying system has a definite value $\kappa = \kappa_R$ for the first set of operators and α^{op} have not been measured. The reduced matrix element is then

$$\langle \hat{a}\| W^R \|\hat{a}'\rangle = \langle \kappa\,\alpha\| W^R \|\kappa'\,\alpha'\rangle = \delta_{\kappa\kappa_R}\,\delta_{\kappa'\kappa_R}\,\delta_{\alpha\alpha'}\frac{1}{\dim \mathcal{R}(\kappa_R)} \qquad (6.10)$$

where $\mathcal{R}(\kappa_R)$ is the eigenspace of κ^{op} with eigenvalue κ_R.

Let

$$K, \beta^{\mathrm{op}}, \eta^{\mathrm{op}} \qquad (6.11)$$

be a complete system of commuting observables chosen so that the detector detects decay products with definite quantum number η and detects states with any values of β. Then one can ask for the partial decay rate for the decay

[24] For instance, if it should happen that (6.6a) is not fulfilled but that

$$\hat{P}_i = P_i M^{-1}$$

commutes with η^{op}, where M is the mass operator (which is a function of the internal quantum numbers), then one would choose the basis vectors $|\hat{p}, \eta\rangle$, where $\hat{p}_i = p_i/m(\eta)$.

of R into the decay channel η. Inserting (6.10) into (5.14) and using eigenvectors of (6.9) and (6.11), one obtains

$$\Gamma = \sum_\eta \Gamma(R \to \eta),\tag{6.12}$$

where the partial decay rates are:

$$\Gamma(R \to \eta) = 2\pi \int \rho_\eta(E)\, dE\, \delta(E - E_R) \overline{\sum_{\beta,\,\alpha}} |\langle E\,\eta\,\beta|V|E_R\,\kappa_R\,\alpha\rangle|^2. \tag{6.13}$$

Here

$$\overline{\sum_{\alpha\beta}} = \frac{1}{\dim \mathscr{R}(\kappa_R)} \sum_\beta \sum_\alpha \tag{6.14}$$

means summing (or integrating) over all values of the labels for the basis vectors of the space of final states and averaging over all initial quantum numbers. One often states this as "summing over the final states and averaging over the initial states."

The formula (6.13) for the decay rates has numerous applications in all branches of quantum physics. Whenever it makes sense to speak of a decaying quantum-mechanical system, (6.13) with suitably chosen basis vectors can be applied. The interaction Hamiltonian or transition operator V depends, of course, upon the particular physical system that one considers and is determined by the algebra of observables. To find the right expression for V is one of the tasks of understanding the physical system.

Problems

1. The decaying state vector $|f^D\rangle$ is defined by

$$|f^D\rangle = \frac{1}{i} \int_0^\infty dE\, |E^+\rangle \left(\frac{\Gamma}{2\pi}\right)^{1/2} \frac{1}{E_R - i\Gamma/2 - E},$$

where $|E^+\rangle$ are generalized eigenvectors of H:

$$H|E^+\rangle = E|E^+\rangle.$$

Calculate the norm of the vector $H|f^D\rangle$ for finite values of $2E_R/\Gamma$ and for $E_R/\Gamma \to \infty$. Interpret the result.

2. Show that the state W_R given by (3.10) with (3.13) corresponds to a state of exponential growth ("capture" state).

3. In this problem we suggest the calculations of the results in Section XXI.5 using a different method.

(a) Define the quantity

$$\mathscr{F}_{E_b}(E, E') = \sum_b \sum_{\hat{a}\hat{a}'} \rho(E)\rho(E')\rho(E_b)\langle E_b\,\hat{b}|V|E\,\hat{a}^+\rangle$$

$$\times \langle E'\,\hat{a}'^+|V|E_b\,\hat{b}\rangle\langle\hat{a}\|W^R\|\hat{a}'\rangle(\rho(E_R))^{-1}. \tag{P.0}$$

Extend this function to complex values ω_b, ω, ω' of the energy and show that, as a consequence of the Hermiticity of V and time-reversal invariance,

$$\mathscr{F}_{\omega_b}(\omega, \omega') = \overline{\mathscr{F}}_{\bar\omega_b}(\varpi, \varpi'), \tag{P.1a}$$

$$\mathscr{F}_{\omega_b}(\omega', \omega) = \mathscr{F}_{\omega_b}(\omega, \omega'). \tag{P.1b}$$

Justify

$$\left\langle E_b, b \left| V \right| E_R + i\frac{\Gamma}{2}, \hat{a}^+ \right\rangle = -\left\langle E_b, b \left| V \right| E_R - i\frac{\Gamma}{2}, \hat{a}^+ \right\rangle^* \qquad (P.2)$$

using the symmetry relation of the S-matrix (Equation XVIII.5.10). [(P.2) is needed for the calculations of the following problems.]

(b)   Use (3.10) with (3.14) to show that the decay rate (XIV.2.18) can be written as

$$\dot{\mathscr{P}}(t) = \dot{\mathscr{P}}_1(t) + \dot{\mathscr{P}}_2(t) \qquad (P.3)$$

with

$$\dot{\mathscr{P}}_1(t) = -i\frac{\Gamma}{2\pi} \iiint dE_b\, dE\, dE'\, \mathscr{F}_{E_b}(E, E')$$

$$\times \frac{e^{-iEt}}{E - E_R + i(\Gamma/2)} \frac{e^{iE't}}{E' - E_R - i(\Gamma/2)} \frac{1}{E' - E_b - i\epsilon}, \qquad (P.4)$$

$$\dot{\mathscr{P}}_2(t) = i\frac{\Gamma}{2\pi} \iiint dE_b\, dE\, dE'\, \mathscr{F}_{E_b}(E, E')$$

$$\times \frac{e^{-iEt}}{E - E_R + i(\Gamma/2)} \frac{e^{iE't}}{E' - E_R - i(\Gamma/2)} \frac{1}{E - E_b + i\epsilon}, \qquad (P.5)$$

and show that

$$\dot{\mathscr{P}}_1(t) = (\dot{\mathscr{P}}_2(t))^*. \qquad (P.6)$$

(c)   Use the approximation (3.7) and show, applying the Titchmarsh theorem, that

$$\dot{\mathscr{P}}(t) = 4\pi^2 \Gamma e^{-\Gamma t/2} \operatorname{Re} \mathscr{F}_{E_R + i(\Gamma/2) - i\epsilon}\left(E_R - i\frac{\Gamma}{2}, E_R + i\frac{\Gamma}{2}\right). \qquad (P.7)$$

(d)   Show that from the condition $\mathscr{P}(t \to \infty) = 1$ it follows that

$$4\pi^2 2 \operatorname{Re} \mathscr{F}_{E_R + i(\Gamma/2) - i\epsilon}\left(E_R - i\frac{\Gamma}{2}, E_R + i\frac{\Gamma}{2}\right) = 1, \qquad (P.8)$$

and obtain (3.40) and (3.41).

4. Assume that the detector has a very good energy resolution $\Delta E_b \ll \Gamma$. Then the decay rate per unit energy as a function of E_b can be measured.

(a)   Using the same method as in Problem 3, show that this decay rate per unit energy is given by

$$\frac{d\dot{\mathscr{P}}}{dE_b}(t = 0) = 2\pi \Gamma \frac{\Gamma}{(E_b - E_R)^2 + (\Gamma/2)^2} \mathscr{F}_{E_b}(E_R - i0, E_b + i0).$$

(b)   Show that for the decaying angular-momentum state (3.16) the decay probability per unit energy is given by

$$\frac{d\dot{\mathscr{P}}(t = 0)}{dE_b} = b(E_b) \sum_b 2\pi \rho^b(E_b) \sum_{l_3} \frac{1}{2l_R + 1} |\langle E_b, b | V | E_R\, l_R\, l_3\, \kappa_R \rangle|^2,$$

where

$$b(E_b) = \frac{\Gamma}{2\pi} \frac{1}{(E_b - E_R)^2 + (\Gamma/2)^2}.$$

$b(E_b)$ is called the natural line width.

Hint: Use (P.2) of Problem 3.

CHAPTER XXII

Quantal Phase Factors and Their Consequences

XXII.1 Introduction

At the beginning of the book we started with simple models, the oscillator model (Section II.3), the rotator model (Sections III.3 and III.4), and the Kepler system (Section VI.4) and we discussed how these systems can be combined (Section III.5) into more complicated systems to give, e.g., the vibrating rotator, which has many realizations in molecular physics. Combining this vibrating rotator further with the one-particle motion in a central field (of which the Kepler system is a particular idealization), we can obtain an understanding of the general features of molecular spectra, a typical example of which is depicted in Figure III.5.3. Thus the dynamics of a complicated system like the molecule could be understood by "putting together" more elementary motions.

We now want to approach the problem of a complicated system from the opposite direction. Instead of starting with elementary systems and synthesizing them into more complicated objects we want to start with a complicated extended object with internal structure and dissect it into its parts. "Parts" does not have to mean constituents like the electrons and nuclei of a molecule (or the protons and neutrons of a nuclei or the quarks of a nucleon) but it can also mean other simpler subsystems, such as certain collective motions (e.g., rotation of the extended object). Often a physical system is divided into two parts because the variables naturally fall into two different classes: the fast variables that change rapidly in time and the slow variables that can be treated as fixed or slowly varying parameters. The electronic variables of a molecule are an example of fast variables, and the nuclear coordinates are an example of slow variables. In this way the molecule is naturally divided into two parts, the fast motion of the electrons and the slow collective vibrations and rotations of the molecule as a whole. In the past this problem was treated in the following way: One first investigated the motion of the fast variables considering the slow variables as *fixed* parameters. After the dynamics of the fast variables has been solved for all values of these fixed parameters, one turns to the dynamics of the slow variables. In this treatment the division into the slow- and fast-moving parts is trivial. These are the ideas that underlie the molecular Born-Oppenheimer approximation[1] that lead to the typical molecular spectra of Chapter III.

The nuclear coordinates are, however, *not fixed* but change in time. And the two motions are not independent but the fast motion effects the slow motion and modifies its dynamics. This follows directly from the Born-Oppenheimer method for molecules if one does not make the conventional approximation of considering the slow variables as fixed. Then, a gauge potential (also called connection) emerges naturally, which can explain a nontrivial phase factor.[2] This was done by Mead et al.[3,4] in 1978. We will discuss this in Chapter XXIII.

Conceptually simpler is the discussion of Hamiltonians that depend on slowly changing parameters. This was done in 1983 in a beautiful paper by M. V. Berry[5] for any quantum system in a slowly changing ("adiabatic") classical environment. Independently he arrived at the same ("Berry") connection and ("Berry") phase that were derived by the Born-Oppenheimer method.[6] Berry's paper showed that a phase factor, which was proven to be unimportant for noncyclic evolutions in 1928[7] and set equal to zero ever since, it not unimportant for adiabatic cyclic evolution.[8] We discuss this in Section 2 and give an application in Section 3.

A generalization of the Berry phase from adiabatic cyclic evolution, which is an approximation, to real cyclic processes uses the Aharonov-Anandan phase.[9] We discuss this in Section 4.

[1] M. Born and J. Oppenheimer, *Ann. Phys.* **84**, 457 (1927).

[2] G. Herzberg and H. C. Longuet-Higgins, *Discuss, Faraday Soc.* **35**, 77 (1963); H. C. Longuet-Higgins, *Proc. Roy. Soc. London Ser.* **A344**, 147 (1975).

[3] C. Mead and D. Truhlar, *J. Chem. Phys.* **70**, 2284 (1979); C. A. Mead, *Chem. Phys.* **49**, 23 and 33 (1980).

[4] See also J. Moody, A. Shapere, and F. Wilczek, *Phys. Rev. Lett.* **56**, 893 (1986); R. Jackiw, *Int. J. Mod. Phys.* **A3**, 285 (1988); A. Bohm, B. Kendrick, and Mark E. Loewe, *J. Quant. Chem.* **41**, 53 (1992).

[5] M. V. Berry, *Proc. Roy. Soc. London Ser.* **A392**, 45 (1984).

[6] For earlier "anticipations" of the Berry phase, see M. Berry, *Physics Today*, Dec., 34 (1990).

[7] V. Fock, *Z. Phys.* **49**, 323 (1928).

[8] This phase was shown to be the holonomy of a fiber bundle with Berry connection; B. Simon, *Phys. Rev. Lett.* **51**, 2167 (1983).

[9] Y. Aharonov and J. Anandan, *Phys. Rev. Lett.* **58**, 1593 (1987); J. Anandan and Y. Aharonov, *Phys. Rev.* **D38**, 1863 (1988).

The geometric phase and the Mead-Berry connection arise in the dissection of complicated quantum systems into simple subsystems. This reduction to the simpler is the basis of understanding in science. In the old drastic approximation the dissection results in the trivial direct product of the states for the two subsystems (Section III.5). Due to the Mead-Berry connection, this is not necessarily so, and the parts of a complicated physical system may turn out to be quite different from what we naively expected.

As the material discussed in this chapter is new we will give more references than in earlier chapters. However our selection is quite arbitrary and far from complete in any sense. The literature on this subject is enormous; fortunately there exist several reviews[10] where a more systematic list of references can be found. As we only mention the underlying mathematics of fiber bundles and use the geometric terms, such as connection, only in the very vague manner of a physicist, we also add a few references to the more mathematical literature.[11]

XXII.2 A Quantum Physical System in a Slowly Changing Environment

XXII.2.a Time Evolution for Parameter-Dependent Hamiltonians

In this subsection we shall discuss dynamical evolution of states for quantum systems whose observables and, in particular, whose Hamiltonians $h(R)$ depend on parameters $R = (X, Y, \dots)$ that describe the environment in which the physical system is immersed (e.g., laboratory setup). We write the parameters here like a three-vector R, but they really can be arbitrary coordinates of differentiable manifold. Evolution of states of such a system is determined by the environmental process, i.e., by the way in which the environmental parameters R change. For a given physical situation, the way in which the parameters change, i.e., the path in the environment's parameter space, must be specified. In specifying the environmental process it is usually convenient to give the environmental parameters a time parameterization $R(t) = (X(t), Y(t), \dots)$. One then obtains from $h(R)$ a time-dependent Hamiltonian $h(t) = h(R(t))$ and speaks of time evolution.

An example of a quantum physical system whose Hamiltonian depends on environmental parameters is a (quantum) magnetic moment **m** in a rotating

[10] A. Shapere and F. Wilczek, *Geometric Phases in Physics* (Review and Reprint volume), World Scientific (1989); C. Alden Mead, "The Geometric Phase in Molecular Systems," *Rev. Mod. Phys.* **64**, 51 (1992); J. Zwanziger, M. Koenig, and A. Pines, "Berry's Phase," *Ann. Rev. Phys. Chem.* **41**, 601 (1990).

[11] A. Uhlmann, "Parallel Transport of Phases," *Lecture Notes of Physics* **379**, 55 (1991). (J. D. Hennig, W. Lücke, and J. Tolar, Eds.), Springer-Verlag (1991); J. E. Avron, L. Sadun, J. Segert, and B. Simon, "Chern Numbers, Quaternions and Berry's Phases," *Commun. Math. Phys.* **124**, 595 (1989).

(classical) magnetic field $\mathbf{B}$ of constant magnitude $B \equiv |\mathbf{B}|$. The parameter-dependent Hamiltonian is given according to (IX.3.1) by

$$H = H_0 - \mathbf{m} \cdot \boldsymbol{B} = H_0 - Bg\frac{e}{2mc}\hat{\boldsymbol{R}} \cdot \boldsymbol{J} = H_0 + b\hat{\boldsymbol{R}} \cdot \boldsymbol{J}. \qquad (2.1)$$

Here H_0 is the Hamiltonian without the magnetic field, which we shall ignore. The quantum system's magnetic moment is given in terms of its angular momentum operator $\boldsymbol{J}$ according to (IX.3.18) by

$$\mathbf{m} = +\frac{e}{2mc}g\boldsymbol{J}, \qquad (2.2)$$

$\hat{\boldsymbol{R}}(t) = \boldsymbol{B}/B$ is the unit vector pointing in the direction of the magnetic field, and $b = -Bg(e/2mc)$ is a constant. The parameter-dependent term of the Hamiltonian is $h(\hat{\boldsymbol{R}}) = b\hat{\boldsymbol{R}} \cdot \boldsymbol{J}$ and the parameter space of the environment is the unit sphere $\{\hat{\boldsymbol{R}}; |\hat{\boldsymbol{R}}| = 1\}$. We can use as the parameters the polar angles $\hat{\boldsymbol{R}} = (\theta, \phi)$ of the unit vectors $\hat{\boldsymbol{R}}$. Environmental processes are described by how the direction $\hat{\boldsymbol{R}}$ of the magnetic field changes. The direction of the magnetic field may change periodically; e.g., it may perform rotations in the $1 - 2$ plane with angular velocity ω, $\hat{\boldsymbol{R}}(t) = \mathbf{e}_1 \cos \omega t - \mathbf{e}_2 \sin \omega t$, or it may run through other closed paths $\boldsymbol{C}$ in the parameter space. In this case the Hamiltonian $h(t) = h(\hat{\boldsymbol{R}}(t))$ returns to its original form as time progresses from $t = 0$ to the period $t = T = 2\pi/\omega$. This system will be the paradigm for our future deliberations.

As the observables of the quantum physical system may depend on the environmental parameters $\boldsymbol{R}$ so may the quantum physical system's space of physical states $\mathscr{H}(\boldsymbol{R})$. This means that it is in general possible to have different spaces of physical states for different values of $\boldsymbol{R}$. We shall assume here, however, that the space of physical states $\mathscr{H}$ is the same for all values of $\boldsymbol{R}$. For any given value of $\boldsymbol{R}$, one may choose an orthonormal basis of eigenvectors $|n; \boldsymbol{R}\rangle$ of the parameter-dependent Hamiltonian $h(\boldsymbol{R})$,[12]

$$h(\boldsymbol{R})|n; \boldsymbol{R}\rangle = E_n(\boldsymbol{R})|n; \boldsymbol{R}\rangle, \qquad (2.3)$$

$$\langle m; \boldsymbol{R}|n; \boldsymbol{R}\rangle = \delta_{m,n}, \qquad (2.4)$$

and write the Hamiltonian according to its spectral resolution (I.4.10d):

$$h(\boldsymbol{R}) = \sum_n E_n(\boldsymbol{R})|n; \boldsymbol{R}\rangle\langle n; \boldsymbol{R}|. \qquad (2.5)$$

Given any environmental process along with a time parameterization $\boldsymbol{R}(t)$, one obtains from the parameter-dependent Hamiltonian $h(\boldsymbol{R})$ in (2.5) a time-dependent Hamiltonian $h(t) = h(\boldsymbol{R}(t))$ along with its spectral resolution (of I.4.10d):

---

[12] We assume here, for simplicity, that the spectrum of $h(\boldsymbol{R})$ is discrete. If in addition to $h(\boldsymbol{R})$ other observables $A_i(\boldsymbol{R})$, $i = 1, \ldots, N - 1$ are needed to form a c.s.c.o. for any given value of $\boldsymbol{R}$, then their eigenvalues $a_i(\boldsymbol{R})$ will also be needed to label the basis vectors $|n; \boldsymbol{R}\rangle = |n, a_1, \ldots, a_{N-1}; \boldsymbol{R}\rangle$.

$$h(R(t)) = \sum_n E_n(R(t))|n; R(t)\rangle \langle n; R(t)|. \tag{2.6}$$

The projection operators

$$\Lambda_n(R(t)) \equiv |n; R(t)\rangle \langle n; R(t)|$$

change in general with time.

We shall assume that the observables are single-valued as functions of R over the whole parameter space of the environment. Single-valuedness of the observables means that if the same value of R occurs more than once (i.e., at different times) during a process, then the observables are the same at each occurrence. In particular, if the environmental process is closed, i.e., if the environmental parameters $R(t)$ traverse a closed path C and return, after some period T, to their original values,

$$C: R(0) \to R(t) \to R(T) = R(0), \tag{2.7}$$

then the Hamiltonian and its eigenvalues and projection operators, which are uniquely defined by (2.5), are the same at $R(T)$ as they are at $R(0)$:

$$h(R(T)) = h(R(0)), \tag{2.8}$$

$$E_n(R(T)) = E_n(R(0)), \tag{2.9}$$

$$|n; R(T)\rangle \langle n; R(T)| = |n; R(0)\rangle \langle n; R(0)|. \tag{2.10}$$

We shall also assume that the observables are continuous and differentiable as functions of R over the whole parameter space of the environment. (See Appendix XII.A.) In particular, the Hamiltonian $h(R)$, its eigenvalues $E_n(R)$, and its projection operators $|n; R\rangle \langle n; R|$ may change smoothly as R changes and the eigenspaces of $h(R)$, i.e, the subspaces of $\mathcal{H}$ onto which the projection operators project, may perform smooth motions within $\mathcal{H}$ as R changes.

The basis vectors $|n; R\rangle$ are determined by (2.3) and (2.4) only up to phase factors and need not be assumed to be single-valued, continuous, or differentiable vector-valued functions of R. We will assume, however, that given any environmental process along with a time parameterization $R(t)$, the basis vectors $|n; R(t)\rangle$ can be chosen to be single-valued functions of time. This means in particular that:

$$|n; R(T)\rangle = |n; R(0)\rangle \qquad \text{for} \quad R(T) = R(0). \tag{2.11}$$

Even when we do not explicitly specify the time parameterization we shall frequently retain the time label in $R(t)$. This notation will help to distinguish that we are considering a given process rather than the whole parameter space and to distinguish between different occurrences of the same value of R along the process. As the normalized eigenvectors $|n; R\rangle$ are only determined up to phase factor, we can define a new system of eigenvectors $|n; R\rangle$ by making phase transformations (also called gauge transformations as shall be explained later),

$$|n; R\rangle \to |n; R\rangle' = e^{i\zeta_n(R)}|n; R\rangle, \tag{2.12}$$

where the $\zeta_n(R)$ are arbitrary real phase angles. Any such basis constitutes just as valid a basis of eigenvectors of the Hamiltonian as the $|n; R\rangle$. We assume that the phase factors $e^{i\zeta_n(R(t))}$ are single-valued functions.

As was mentioned at the end of Chapter XII, for a quantum physical system whose Hamiltonian $h(t) = h(R(t))$ depends on time, it is not possible to give the time evolution by a unitary operator as in (XII.1.15) and (XII.1.16), or as in (XII.1.22) and (XII.1.23). In particular, for a Hamiltonian that depends on time, the expression for the time development operator $U(t)$ given by (XII.1.16) = (XII.1.23) is no longer valid.

Equation (XII.1.15), which gives (in the Schrödinger picture, which we shall use here) the dynamical time dependence of the statistical operator $W(t)$, can, however, be justified for fairly general time evolutions [including time evolutions for which $U(t)$ is not only nonunitary but also noninvertible]. In order to obtain a connection between the time evolution operator $U(t)$ and the Hamiltonian $h(t)$ for a quantum system in a changing environment, we assume that the Schrödinger equation (XII.1.37) remains valid when H is generalized to a time-dependent Hamiltonian $h(t)$. This is a basic assumption of quantum mechanics (a generalization of Va of Chapter XII):

Vd. The time evolution of a (nonconservative) physical system in a changing external environment described by a time-dependent Hamiltonian $h(t)$ is given by the Schrödinger equation:

$$i\hbar \frac{d|\psi(t)\rangle}{dt} = h(t)|\psi(t)\rangle. \tag{2.13}$$

$h(t) = h(R(t))$ is a Hermitian operator that depends on the parameters R of the environment.

The time-evolved state vector $|\psi(t)\rangle$ is defined by the analogue of (XII.1.35) with a general initial state vector:

$$\psi(t) = U^\dagger(t)\psi(0). \tag{2.14}$$

Therefore the assumption (2.13) is equivalent to the assumption that $U^\dagger(t)$ fulfills the Schrödinger equation[13]

$$i\hbar \frac{dU^\dagger(t)}{dt} = h(t)U^\dagger(t). \tag{2.13'}$$

Since $h(t)$ depends on time, (2.13') cannot be integrated to give (the adjoint of) (XII.1.16). One can, however, write (2.13') in the form of an integral equation:

$$U^\dagger(t) = I - \frac{i}{\hbar} \int_0^t h(t')U^\dagger(t') \, dt'. \tag{2.15}$$

---

[13] If $U^\dagger(t)$ is invertible, then $h(t)$ may be given in terms of $U^\dagger(t)$ by

$$h(t) = i\hbar \frac{dU^\dagger(t)}{dt} U^{\dagger-1}(t).$$

If $U(t)$ is unitary, then $h(t)$ is Hermitian (Problems 1 and 2).

Differentiating (2.15) with respect to time gives (2.13′). An expression for $U^\dagger(t)$ in terms of $h(t)$ is then given by successively substituting the right-hand side of (2.15) for the $U^\dagger(t)$ that appears in the integrand:

$$U^\dagger(t) = I + \frac{1}{i\hbar} \int_0^t dt' h(t') + \left(\frac{1}{i\hbar}\right)^2 \int_0^t dt' \int_0^{t'} dt'' h(t')h(t'')U^\dagger(t'')$$

$$= I + \frac{1}{i\hbar} \int_0^t dt' h(t') + \left(\frac{1}{i\hbar}\right)^2 \int_0^t dt' \int_0^{t'} dt'' h(t')h(t'') + \cdots$$

$$+ \left(\frac{1}{i\hbar}\right)^n \int_0^t dt_1 \int_0^{t_1} dt_2 \cdots \int_0^{t_{n-1}} dt_n h(t_1)h(t_2) \ldots h(t_n) + \cdots . \qquad (2.16)$$

Note that the operators appearing in the integrands have decreasing time arguments reading from left to right, $t \geq t' \geq t'' \geq 0, t \geq t_1 \geq t_2 \geq \cdots \geq t_n \geq 0$, and that their order is important since they do not in general commute:

$$[h(t), h(t')] \neq 0 \qquad \text{for } t' \neq t. \qquad (2.17)$$

For the case that $h(t)$ commutes at different times,

$$[h(t), h(t')] = 0 \qquad \text{for all } t, t', \qquad (2.18)$$

the order of the operators in (2.16) is not important and it can be shown (Problem 3) that

$$U^\dagger(t) = \sum_{n=0}^\infty \frac{1}{n!} \left(\frac{1}{i\hbar}\right)^n \int_0^t dt_1 \ldots \int_0^t dt_n h(t_1) \ldots h(t_n) = e^{-i/\hbar \int_0^t dt' h(t')}. \qquad (2.19)$$

According to (II.4.45a), (2.18) means that the projection operators $\Lambda_n(R(t))$ $= |n; R(t)\rangle \langle n; R(t)|$ commute:

$$[\Lambda_n(R(t)), \Lambda_{n'}(R(t'))] = 0 \qquad \text{for all } n, n' \text{ and } t, t'. \qquad (2.20)$$

Together with the assumed continuity of the $\Lambda_n(R(t))$, this means that the projection operators and eigenspaces of $h(t)$ are in fact time-independent with only the eigenvalues $E_n(t)$ depending on time. The projection operators may then be written as $\Lambda_n(R(t)) = \Lambda_n = |n\rangle\langle n|$ with time-independent vectors $|n\rangle$. The resulting spectral resolution (2.6) of $h(t)$ may then be used in (2.19) to give the following spectral resolution of $U^\dagger(t)$:

$$U^\dagger(t) = \sum_n e^{-i/\hbar \int_0^t dt' E_n(t')} |n\rangle\langle n|. \qquad (2.21)$$

If the eigenvalues are also independent of time, i.e., if the Hamiltonian is time-independent, $h(t) = h$, then (2.19) becomes

$$U^\dagger(t) = e^{-ith/\hbar}, \qquad (2.22)$$

which is just the adjoint of (XII.1.16) for the time development operator of a conservative system. For time-independent h, the energy eigenvalues are also time independent and (2.21) becomes:

$$U^\dagger(t) = \sum_n e^{-itE_n/\hbar} |n\rangle\langle n|. \qquad (2.23)$$

But time-independent eigenvalues E_n can also appear if $h(t)$ and $\Lambda_n(R(t))$ depend on time.

For a Hamiltonian that does not commute at different times the expressions (2.19)–(2.23) are not valid. An expression for $U^\dagger(t)$ in terms of $h(t)$ that is valid in general and looks similar to (2.19) is

$$U^\dagger(t) = \sum_{n=0}^{\infty} \frac{1}{n!} \left(\frac{1}{ih}\right)^n \int_0^t dt_1 \dots \int_0^t dt_n \mathcal{T}(h(t_1) \dots h(t_n)) \equiv \mathcal{T}(e^{-i/h \int_0^t dt' h(t')}).$$

(2.24)

In here $\mathcal{T}$ is just a new notation that means to rearrange a product of time-labeled operators such that those operators whose time labels take smaller values appear to the right, e.g.,

$$\mathcal{T}(h(t_1)h(t_2)h(t_3)) = h(t_3)h(t_1)h(t_2) \qquad \text{if } t_3 \geq t_1 \geq t_2. \qquad (2.25)$$

$\mathcal{T}$ is called the *time-ordered product* or $\mathcal{T}$*-product*. We give (2.24) here to emphasize the difference of the general case (2.17) from the solution (2.19) or (2.22). It is the standard expression[14] but only a formal "solution" of (2.15). An expression for the spectral resolution of $U^\dagger(t)$ that is valid also in this more general case is not known. We shall not make use of (2.24); we present it here only to provide an indication that there will be more to (2.14) than just (2.19).

XXII.2.b "Dynamical" Phase and "Geometrical" Phase

For a system whose Hamiltonian $h(t)$ commutes at different times, $U^\dagger(t)$ of (2.21) may be applied to an arbitrary initial state vector $|\psi(0)\rangle$ and the resulting time-evolved state vector

$$|\psi(t)\rangle = U^\dagger(t)|\psi(0)\rangle = \sum_n e^{-i/h \int_0^t dt' E_n(t')}|n\rangle\langle n|\psi(0)\rangle \qquad (2.26)$$

will satisfy the Schrödinger equation (2.13). If the system is initially in the n_0th eigenstate of $h(t)$, then the projection operators $|n\rangle\langle n|$ act on $|\psi(0)\rangle$ according to

$$|n\rangle\langle n|\psi(0)\rangle = |\psi(0)\rangle\delta_{n_0 n} \qquad (2.27)$$

and the sum in (2.26) may be performed to give

$$|\psi(t)\rangle = e^{-i/h \int_0^t dt' E_{n_0}(t')}|\psi(0)\rangle = e^{-i/h \int_0^t dt' E_{n_0}(t')}|n_0; R(0)\rangle. \qquad (2.28)$$

As $|\psi(t)\rangle$ differs from $|\psi(0)\rangle$ only by a phase factor it always spans the n_0th eigenspace of $h(t)$ and the state is stationary. For a stationary state, the phase factor acquired by the state vector,

$$e^{-i/h \int_0^t dt' E_{n_0}(t')}, \qquad (2.29)$$

is called the *dynamical phase factor*.

---

[14] See, e.g., C. Itzykson and J. B. Zuber, *Quantum Field Theory*, section 4-1-4, McGraw-Hill publishers (1980).

We now consider a Hamiltonian that does *not* fulfill (2.18). This is the much more common case for a quantum system in a changing environment (example (2.1) is of this kind). Let us start with a stationary state. Stationary states of conservative systems are always eigenstates (or mixtures of eigenstates) of the Hamiltonian, cf. (XII.1.41) (XII.1.44). For nonconservative systems a pure stationary state is defined (in analogy to (XII.1.41)) by

$$|\psi(t)\rangle \langle \psi(t)| = |\psi(0)\rangle \langle \psi(0)| \qquad \text{for all } t. \tag{2.30a}$$

This means that the state vector of a stationary state of a general Hamiltonian $h(t)$ differs from the initial state vector by at most a time-dependent phase factor:

$$|\psi(t)\rangle = e^{i\zeta_\psi(t)}|\psi(0)\rangle. \tag{2.30b}$$

Substituting this into the Schrödinger equation (2.13), taking the scalar product with $|\psi(t)\rangle$, and assuming $|\psi(t)\rangle$ to be normalized gives

$$-\hbar \frac{d\zeta_\psi(t)}{dt} = \langle \psi(t)|h(t)|\psi(t)\rangle. \tag{2.31}$$

Integrating this and substituting the result into (2.30b) gives

$$|\psi(t)\rangle = e^{-i/\hbar \int_0^t \langle \psi(t')|h(t')|\psi(t')\rangle\, dt'}|\psi(0)\rangle. \tag{2.32}$$

We see that the phase factor for the cyclic time development of a stationary state vector is given by

$$e^{-i/\hbar \int_0^t \langle \psi(t')|h(t')|\psi(t')\rangle\, dt'}. \tag{2.33}$$

For a Hamiltonian that commutes at different times a stationary state is an eigenstate and the expectation value $\langle \psi(t)|h(t)|\psi(t)\rangle$ of the Hamiltonian reduces to an eigenvalue $E_{n_0}(t)$ as in (2.29). For a Hamiltonian that does not commute at different times, i.e., that fulfills (2.17) instead of (2.18), a stationary state is not in general an eigenstate, and vice versa. We will call (2.33) the dynamical phase factor also in case that (2.17) is fulfilled.

In general, the time evolution for a time-dependent Hamiltonian is not stationary, i.e., for a given Hamiltonian $h(t)$ an initial state $\psi(0)$ does not evolve in time according to (2.30a) and (2.30b). As we will show the time development will then not be given by (2.32) but will contain an additional phase factor. But as long as $|\psi(t)\rangle$ is normalized we will call (2.33) the dynamical phase factor.

The set of one-dimensional projection operators $\Lambda(t) = |\psi(t)\rangle \langle \psi(t)|$ of the Hilbert space $\mathcal{H}$ is denoted by $\mathcal{P}(\mathcal{H})$; it is the projective space of (pure) physical states. If the state $\Lambda(t)$ changes in time, then the projection operator traverses a path $\mathscr{C}$ in $\mathcal{P}(\mathcal{H})$:

$$\mathscr{C}: t \to \Lambda(t)$$

We will give particular consideration to an evolution that is not stationary, $\Lambda(t) \neq \Lambda(0)$ but is cyclic. A cyclic evolution of a quantum physical system is an evolution in which the state of the system returns after a period $t = T$ to its original state. This means it is given by a closed path in $\mathscr{P}(\mathscr{H})$:

$$\mathscr{C}: t \to \Lambda(t), \qquad 0 \le t \le T, \qquad \Lambda(T) = \Lambda(0). \tag{2.34}$$

That $\Lambda(t) = |\psi(t)\rangle \langle \psi(t)|$ traverses a closed path in the state space $\mathscr{P}(\mathscr{H})$ does not necessarily mean that the normalized state vector $\psi(t)$, which fulfills the Schrödinger equation (2.13), also traverses a closed path in $\mathscr{H}$. In general the path

$$C: t \to \psi(t) \qquad 0 \le t \le T; \qquad (\psi(t), \psi(t)) = 1 \tag{2.35}$$

is not closed in $\mathscr{H}$ but fulfills

$$\psi(T) = e^{-i\alpha_\psi}\psi(0), \tag{2.36}$$

where α_ψ is a phase angle (real number modulo 2π) depending upon ψ.

If the initial state at $t = 0$ is an eigenstate of $h(R(0))$, i.e., if $\Lambda(0) = |n; R(0)\rangle \langle n; R(0)|$, then $\psi(0)$ can be given (perhaps after a choice of a phase) by

$$\psi(0) = |n; R(0)\rangle. \tag{2.37}$$

For a given Hamiltonian $h(R(t))$ whose parameter changes in time the solutions of the Schrödinger equation (2.13) can give wildly varying paths $\Lambda(t)$ so that an eigenstate at $t = 0$ need not remain an eigenstate at later time. In particular the interaction with the environment can cause the physical system to jump from the nth eigenstate at $t = 0$ into all other eigenstates $|m; R(t)\rangle \langle m; R(t)|, m \ne n$ at a later time t. A very particular situation arises if this is not the case and if the state remains an eigenstate of $h(R(t))$ at all time t with the same energy quantum number n. This means that $|\psi(t)\rangle \langle \psi(t)|$ changes in such a way that at all times t

$$|\psi(t)\rangle \langle \psi(t)| \overset{\substack{\text{adiabatic}\\\downarrow}}{=} |n; R(t)\rangle \langle n; R(t)|. \tag{2.38}$$

This time development is called adiabatic time development. It represents an approximation, not a limiting case, called the adiabatic approximation.[15] It is a generalization of the stationary state-time development (2.30a). The equality (2.38) constitutes an additional condition that we impose upon the solutions of (2.13). For a given Hamiltonian no solution of this kind may exist, and usually there is no $\psi(t)$ with $\psi(0) = |n; R(0)\rangle$, which fulfills (2.38), except when $\psi(t)$ is stationary. But under certain conditions one can find an initial state $\psi(0) = \varphi(0) \ne |n; R(0)\rangle$ that is "close enough" to $|n; R(t)\rangle \langle n; R(t)|$. Then we say that (2.38) is approximately fulfilled and call $\psi(t)$ the adiabatic approximation. We will see later (in Section 4.2) that (2.38) is a good approximation (i.e., the l.h.s. of (2.38) is "close enough" to the r.h.s. of (2.38)) if $h(t)$ or $R(t)$ changes slowly over the period T (it becomes an equality only for $T \to \infty$ or when $h(t)$ becomes independent of time and the time development becomes stationary).

We will now discuss the solution $\psi(t)$ of the Schrödinger equation. At any time t, we can expand $\psi(t)$ with respect to the basis system $|n; R(t)\rangle$:

[15] M. Born and V. Fock, *Zeit. f. Phys.* **51**, 165 (1928).

$$\psi(t) = \sum_m c_m(t)|m; \boldsymbol{R}(t)\rangle = \sum_m a_m(t)e^{-(i/\hbar)\int_0^t E_m(t')dt'}|m; \boldsymbol{R}(t)\rangle. \tag{2.39}$$

In the second equality we have separated the dynamical phase of the expansion coefficients $c_m(t)$. For the special case (2.38) of the adiabatic approximation the expansion (2.39) becomes

$$\psi(t) \overset{\text{adiabatic}}{\underset{\downarrow}{=}} c_n(t)|n, \boldsymbol{R}(t)\rangle = e^{-(i/\hbar)\int_0^t E_n(t')dt'}a_n(t)|n, \boldsymbol{R}(t)\rangle \tag{2.40}$$

where because of (2.37)

$$a_n(0) = 1. \tag{2.37'}$$

This is not an equality but an approximate equality, which we indicate by the qualification "adiabatic" at the equality sign. Inserting (2.40) into the Schrödinger equation (2.13) and using (2.3) we obtain after a little calculation

$$\frac{d}{dt}a_n(t) \equiv \dot{a}_n(t) = -a_n\langle n; \boldsymbol{R}(t)|\left(\frac{d}{dt}|n; \boldsymbol{R}(t)\rangle\right). \tag{2.41}$$

This can be integrated:

$$\int_{a_n(0)}^{a_n(t)} \frac{da_n}{a_n} = -\int_0^t \langle n; \boldsymbol{R}(t)|\frac{d}{dt}|n; \boldsymbol{R}(t)\rangle\, dt \tag{2.42}$$

and leads with (2.37') to:

$$a_n(t) = e^{i\int_{R(0)}^{R(t)} i\langle n; R(t)|(\partial/\partial R^i)|n; R(t)\rangle\, dR^i} \equiv e^{i\gamma_n(t)}. \tag{2.43}$$

The coefficients $c_n(t)$ and $a_n(t)$ in (2.40) are phase factors (their absolute value is one), because $\psi(t)$ is obtained from $\psi(0)$ of (2.37) by a unitary time development (2.14) and is therefore a normalized vector. This justifies the definition of $\gamma_n(t)$ as a real phase angle by (2.43).[16] Note that $\gamma_n(t)$ is only defined up to a multiple of 2π.

The phase angle $\gamma_n(t)$ is defined in terms of an integral over a vector-valued function

$$\boldsymbol{A}^n(\boldsymbol{R}) \equiv i\langle n; \boldsymbol{R}|\boldsymbol{\nabla}|n; \boldsymbol{R}\rangle, \tag{2.44}$$

which is called the Berry vector potential or Berry connection.[17] It is defined as the scalar product of the eigenvectors of $h(t)$ and its derivative with respect to the parameter $\boldsymbol{R}$, $\boldsymbol{\nabla}|n; \boldsymbol{R}\rangle$. For the definition (2.44) of $A(\boldsymbol{R})$ one uses single-valued basis vectors only, whereas (2.43) does not use this requirement. This requirement also does not follow from the physical condition of single valuedness of the observables (2.8), (2.9), and (2.10). Single-valued $|n; \boldsymbol{R}\rangle$ and therewith single-valued $A(\boldsymbol{R})$ can in general not be found on the whole parameter space but only on open subsets (patches). We will discuss this problem in

---

[16] See also Problem 4.

[17] The importance of this vector potential has been discovered two times independently, by C. A. Mead et al. (1979) and by M. V. Berry (1983). We will, therefore, call it also the Mead-Berry potential.

detail in Section 3.1. Here we will assume that the curve C lies in one of these patches and that the $|n; R\rangle$ are single-valued. The integral

$$
\gamma_n(t) = \int_0^t i\langle n; R(t)|\frac{d}{dt}|n; R(t)\rangle \, dt = \int_{R(0)}^{R(t)} i\langle n; R(t)|\frac{\partial}{\partial R^i}|n; R(t)\rangle \, dR^i
$$

$$
= \int_{R(0)}^{R(t)} i\langle n; R(t)|d|n; R(t)\rangle = \int_{R(0)}^{R(t)} A_i^n(R) \, dR^i \equiv \int A^n \qquad (2.45)
$$

is independent of the parameterization as expressed by the third term of (2.45) as line integral of a differential along the curve $t \to R(t)$. For the sake of simplicity we have so far pretended that the environmental parameter is a 3-vector R, then a differential one-form is defined by

$$
A^n \equiv i\langle n; R|\frac{\partial}{\partial R^i}|n; R\rangle \, dR^i = A_i^n \, dR^i = i\langle n; R|d|n; R\rangle, \qquad (2.46)
$$

where A_i^n are the components of the 3-vector (2.44). In general the parameter space is (assumed to be) a differentiable manifold and R^i; $i = 1, 2, \ldots f$; are its arbitrary (local) coordinates. For any covector A^n one defines with the differentials dR^i a differential form $A_i^n \, dR^i$ that is invariant under coordinate changes. Thus $A^n \equiv i\langle n; R|d|n; R\rangle$ is a differential one-form independent of the parameterization.[18] The expressions for the phase angle $\gamma_n(t)$ thus also apply to an f-dimensional parameter space, in which case the Berry vector potential A^n is an f-dimensional covector. If we do not want to emphasize that R is a 3-vector we will also write R in place of R for the f parameters.

Returning to our calculation of $\psi(t)$ we obtain by inserting (2.43) into (2.40) for the state vector in the adiabatic approximation with initial condition (2.37)

$$
\psi(t) = e^{-(i/\hbar)\int_0^t E_n(t')dt'} e^{i\gamma_n(t)} |n; R(t)\rangle. \qquad (2.47)
$$

In addition to the dynamical phase factor (2.29) one obtains the phase factor $e^{i\gamma_n(t)}$ given in terms of the eigenvectors of $h(R(t))$ by (2.45).

According to (2.12) the $|n; R(t)\rangle$ are only determined up to a phase factor. Thus, the additional phase factor in (2.47) can be transformed away by a phase transformation. For this reason the phase factor $e^{i\gamma_n(t)}$ had been ignored for half a century until such phase changes were shown to occur for certain molecular systems[19] and general theoretical considerations demonstrated their universal significance. We first show how γ_n can be eliminated and then discuss when it cannot be eliminated.

From (2.12) it follows that under a phase transformation

$$
A^n(R) \to A'^n(R) = i\langle n; R|'(\nabla|n; R\rangle')
$$

$$
= i\langle n; R|e^{-i\zeta_n(R)}(\nabla e^{i\zeta_n(R)}|n; R\rangle)
$$

$$
= i\langle n; R|\nabla|n; R\rangle + ie^{-i\zeta_n(R)}(\nabla e^{i\zeta_n(R)})
$$

[18] C. J. Isham, *Modern Differential Geometry for Physicists*, World Scientific Publishers (1989).

[19] G. Herzberg and H. C. Longuet-Higgins, Discuss, *Faraday Soc.* **35**, 77 (1963); C. A. Mead and D. G. Truhlar, *J. Chem. Phys.* **70**, 2284 (1979); C. A. Mead, *Chem. Phys.* **49**, 23, 33 (1980).

or

$$A^n(R) \to A'^n(R) = A^n(R) - \nabla \zeta_n(R). \tag{2.48}$$

Thus

$$\gamma_n(t) = \int_{R(0)}^{R(t)} A^n(R) \cdot dR \to \int_{R(0)}^{R(t)} A'^n(R) \cdot dR = \gamma_n(t) - \zeta_n(R(t)) + \zeta_n(R(0)) = \gamma'_n(t). \tag{2.49}$$

If we repeat the calculations that led to (2.47) using the $|n; R\rangle'$ in place of the $|n; R\rangle$, we obtain (2.47) with the primed quantities on the right-hand side:

$$e^{i\gamma'_n(t)}|n; R(t)\rangle' = e^{i\gamma'_n(t)} e^{i\zeta_n(R(t))}|n; R(t)\rangle.$$

As long as $\zeta_n(R(t))$ is an arbitrary single-valued function modulo 2π we can choose it such that the phase factor becomes unity and we obtain in place of (2.47):

$$\psi(t) = e^{-i/\hbar \int_0^t E_n(t') dt'}|n; R(t)\rangle. \tag{2.50}$$

This also fulfills the initial condition (2.37). Since $|n; R\rangle'$ is as valid a basis system (fulfilling (2.3) and (2.4)) as is $|n; R\rangle$, we can use it and thus describe the time development of the state vector by (2.50) with the dynamical phase factor only.

The preceding arguments made use of the fact that $\zeta_n(R(t))$ was arbitrary. If after some period T the environmental parameters return to their original value, $R(T) = R(0)$ so that the state $\Lambda(t)$ evolves along a closed path (2.34) one cannot choose $\zeta_n(R)$ freely to remove γ_n. In this case $\gamma_n(T)$ is given by the line integral over the corresponding closed path C of (2.7):

$$\gamma_n(T) = \oint_C i\langle n; R| \frac{\partial}{\partial R^i} |n; R\rangle \, dR^i = \oint_C A^n(R) \cdot dR. \tag{2.51}$$

Since $e^{i\zeta_n(R)}$ is a single-valued function of R, we must have for $R(T) = R(0)$

$$e^{i\zeta_n(R(T))} = e^{i\zeta_n(R(0))} \quad \text{or} \quad \zeta_n(R(T)) = \zeta_n(R(0)) + 2\pi \cdot \text{integer}. \tag{2.52}$$

Therefore according to (2.49)

$$\gamma_n(T) \to \gamma'_n(T) \equiv \oint_C A'^n(R) \cdot dR = \gamma_n(T) - 2\pi \cdot \text{integer}$$

$$= \oint_C A^n(R) \cdot dR - 2\pi \cdot \text{integer}. \tag{2.53}$$

Thus $\gamma_n(T)$, which is only defined modulo 2π, is invariant with respect to the phase transformation (2.12) of the basis vectors $|n; R\rangle$ and can therefore not be removed. We therefore have

$$\psi(T) = e^{-i/\hbar \int_0^T dt\, E_n(t) + i\gamma_n(T)}|n; R(T)\rangle \tag{2.54}$$

with the $\gamma_n(T)$ given by (2.51), modulo 2π. The phase angle $\gamma_n(T)$ is called the *Berry phase angle*; $e^{i\gamma_n(T)}$ is called the *Berry phase factor*. Though it cannot be removed it could, of course, still be zero if the integral on the r.h.s. of (2.51) turns out to be zero. But as we shall see in Subsection 3.6, there are examples for which this quantity is different from zero.

From (2.48) we see that the Berry vector potential transforms under the phase transformation (2.12) like the vector potential of electrodynamics transforms under a gauge transformation.[20] The set of phase factors $e^{i\zeta_n(R)}$ form the group $U(1)$, they are continuous, differentiable, single-valued functions of the parameters R. We therefore have here a gauge theory with gauge group $U(1)$ and gauge potential $A^n(R)$. This is the reason for which we call the phase transformation (2.12) a $U(1)$-gauge transformation. Whereas $A^n(R)$ is not an invariant with respect to a gauge transformation but transforms like (2.48), the Berry phase is gauge invariant. Precisely, the phase factor $e^{i\gamma_n(T)}$ is $U(1)$ gauge invariant and the phase angle $\gamma_n(T)$ modulo 2π is $U(1)$ gauge invariant (it is only defined modulo 2π). If the parameter space is three-dimensional and the parameter $R = \boldsymbol{R}$ is the three-dimensional coordinate vector, then we have complete analogy with the electromagnetic theory. But the physical meaning of these quantities is different, as the gauge potential (2.44) is defined in terms of the eigenvectors of the Hamiltonian and has nothing to do with electromagnetism. For an f-dimensional parameter space we again have a $U(1)$-gauge theory, but the gauge transformations and gauge potentials now depend upon an f-dimensional parameter R and the $A^n(R)$ have f components. It is called a $U(1)$-gauge theory over and f-dimensional base ($=$ parameter) space.

In analogy to electrodynamics we can define a gauge field

$$F^n_{ij} = \frac{\partial}{\partial R^i} A^n_j - \frac{\partial}{\partial R^j} A^n_i \qquad (2.55a)$$

or (if $\boldsymbol{R}$ is a 3-vector)

$$\boldsymbol{F}^n = \nabla \wedge \boldsymbol{A}^n. \qquad (2.55b)$$

The corresponding 2-form is:

$$F^n \equiv dA^n = \frac{\partial A^n_j}{\partial R^i} dR^i \wedge dR^j = \frac{1}{2} F^n_{ij} dR^i \wedge dR^j. \qquad (2.55c)$$

This gauge field F^n with A^n given by (2.46) is also called Berry curvature. From (2.48) it follows immediately that the gauge field—as in every abelian gauge theory—is gauge invariant:

$$F^n \to F'^n = \nabla \wedge A'^n = F^n. \qquad (2.55')$$

In formula (2.51) Stokes theorem may be used to convert the line integral over the closed path C in the environment's parameter space into a surface integral over *any* surface S enclosed by the path C. For three-dimensional $\boldsymbol{R}$ that is:

$$\gamma_n(T) = \oint_C \boldsymbol{A}^n(\boldsymbol{R}) \cdot d\boldsymbol{R} = \iint_S d\boldsymbol{S} \cdot (\nabla \wedge \boldsymbol{A}^n) \quad \text{mod } 2\pi, \qquad (2.56)$$

[20] J. D. Jackson, *Classical Electrodynamics*, section 6.5.

and for an f-dimensional parameter space

$$\gamma_n(T) = \oint_C i\langle n; R| \frac{\partial}{\partial R^i} |n; R\rangle \, dR^i = \oint_C A^n = \int\int_S dA^n = \int\int_S F^n \quad \text{mod } 2\pi.$$

(2.56')

S is the surface that subtends C and dS denotes the surface element in the parameter space whose direction is normal to the surface S and with the line element dR of C traversed in the right-hand sense with respect to the normal.

Using (2.46) in (2.55') we can also express $\gamma_n(T)$ in the form

$$\gamma_n(T) = \int\int_S F^n = \int d(i\langle n; R|d|n; R\rangle) \quad \text{mod } 2\pi,$$

(2.57)

or, for the three-dimensional case,

$$\gamma_n(T) = \int\int_S dS \cdot F^n = \int\int dS \cdot i\nabla \wedge (\langle n, R|\nabla|n; R\rangle).$$

(2.57')

We have added mod 2π in these equations because $\gamma_n(T)$ is defined only up to an integer times 2π. As $\langle n; R|d|n; R\rangle$ is purely imaginary (see Problem 4) we also write for the connection one-form

$$A^n = -\text{Im}[\langle n; R|d|n; R\rangle]$$

and, for the Berry curvature

$$F^n = dA^n = id(\langle n; R|d|n; R\rangle) = i(d\langle n; R|)d|n; R\rangle,$$

we write

$$F^n = -\text{Im}[d\langle n; R|d|n; R\rangle] = -\text{Im}\left[\left(\frac{\partial}{\partial R^i}\langle n; R|\right)\frac{\partial}{\partial R^j}|n; R\rangle \, dR^i \wedge dR^j\right].$$

For the three-dimensional case this reads

$$A^n = -\text{Im}[\langle n; R|\nabla|n; R\rangle]$$

and

$$F^n = -\text{Im}[(\nabla\langle n; R|) \wedge \nabla|n; R\rangle].$$

(2.58')

Equation (2.57') shows that the Berry phase $\gamma_n(T)$ is analogous to the magnetic flux of the electromagnetic theory.

Equations (2.57) and (2.57') express $\gamma_n(T)$ in terms of gauge-invariant quantities only. It also shows immediately, as does (2.51), that $\gamma_n(T)$ is independent of the time parameterization of the process $t \to R(t)$. That means it is independent of how the closed loop C is traversed (provided it is such that all our conditions are fulfilled, which means, for the case under consideration, that the loop is traversed slowly enough for the adiabatic approximation (2.38) to hold). The Berry phase angle $\gamma_n(T)$ thus does not depend on the dynamics of the quantum system or the details of the environmental changes. It only depends on the path C (with the provision of adiabaticity) and is thus a quantity determined by the geometry of the parameter space. It is therefore also called the geometric phase. We will reserve the name geometric phase for a generalization of (2.51) to nonadiabatic processes and call $\gamma_n(T) = \gamma_n(C)$, expressed by (2.51) in terms of the adiabatically changing energy eigenstates, the Berry phase.

Phase factors such as $e^{i\gamma_n}$ have been dismissed in the past as unimportant. Since the projection operators $|n; R\rangle\langle n; R|$ (or the rays) are the physical quantities, with the (normalized) eigenvectors $|n; R\rangle$ being defined only up to a phase, it appeared plausible that these phases could be chosen such that $e^{i\gamma_n}$ would be unity. As we have seen, it is not always possible to change $e^{i\gamma_n}$ by making a different choice for the phases of the basis vectors. The importance of this fact and of the phase factor $e^{i\gamma_n}$ has only recently been realized.[3, 5]

We have assumed that the observables are continuous, differentiable, single-valued functions of the parameters R. This means that in particular (2.10) holds. The basis vectors $|n; R\rangle$ themselves will in general not be single-valued functions of R over the whole parameter space, and it is usually necessary to use different parameterizations over different patches of the parameter space. Thus, in general (2.10) will not imply

$$|n; R(T)\rangle = |n; R(0)\rangle \qquad \text{for } R(T) = R(0) \qquad (2.59)$$

but only

$$|n; R(T)\rangle = e^{i\zeta_n}|n; R(0)\rangle \qquad \text{for } R(T) = R(0) \qquad (2.60)$$

where $e^{i\zeta_n}$ is a phase factor.

With the phase factor for initial state vector $\psi(0)$ chosen as in (2.37), we thus obtain from (2.54) for the state vector $\psi(T)$ that has developed from the state vector $\psi(0)$ while $R(t)$ evolved through the closed loop C of (2.7):

$$\psi(T) = e^{-(i/\hbar)\int_0^T dt' E_n(t')}e^{i\tilde{\gamma}_n(C)}e^{i\zeta_n}\psi(0). \qquad (2.61)$$

If the parameterization of the basis vectors can be chosen such that the basis vectors $|n; R\rangle$ are single-valued in a domain of the parameter space that contains the closed loop C, then (2.59) can be assumed and $\zeta_n = 0 \pmod{2\pi}$. If (2.59) has been assumed for one basis system, then the single valuedness of the gauge transformation—in particular (2.52)—guarantees that it also holds for the other basis systems.

If $\gamma_n(C)$ is calculated with single-valued basis vectors $|n; R\rangle$ by (2.51), then (2.61) becomes

$$\psi(T) = e^{-(i/\hbar)\int_0^T dt' E_n(t')}e^{i\gamma_n(C)}\psi(0). \qquad (2.62)$$

If $\gamma_n(C)$ is calculated by the first equality of (2.51) with basis vectors that are not single-valued but fulfill (2.60) ($A^n(R)$ is only defined for single-valued $|n; R\rangle$), then we obtain (2.61) instead of (2.62). The result is the same; there is an extra phase angle in addition to the dynamical phase. Only this new phase angle is now called $(\tilde{\gamma}_n + \zeta_n)$, where $\tilde{\gamma}_n$ is calculated with the basis vectors that fulfill (2.60). As it is conventional to call Berry phase that angle calculated with the single-valued basis vectors, (2.59), the Berry phase using (2.60) is given by $(\tilde{\gamma}_n + \zeta_n)$.

To see how nontrivial phases arise when two energy levels become degenerate for some values of the parameter, we express the Berry curvature in terms of the eigenvalues $E_n(R)$. We continue the calculation of F in (2.58). Inserting the complete basis system

$$I = \sum_m |m; R\rangle \langle m; R| \qquad (2.63)$$

we obtain

$$F^n = -\text{Im}\left[\sum_m (\nabla\langle n; R|)|m; R\rangle \wedge \langle m; R|\nabla|n; R\rangle \right]$$

$$= -\text{Im}[\overline{\langle n; R|\nabla|n; R\rangle} \wedge \langle n; R|\nabla|n; R\rangle]$$

$$\quad - \text{Im}\left[\sum_{m \neq n} \overline{\langle m; R|\nabla|n; R\rangle} \wedge \langle m; R|\nabla|n; R\rangle \right].$$

The first term is zero because $\langle n; R|\nabla|n; R\rangle$ is purely imaginary. In the second term we use

$$\langle m; R|\nabla|n; R\rangle = \langle m; R|(\nabla h(R))|n; R\rangle \frac{1}{E_n(R) - E_m(R)} \quad \text{for} \quad n \neq m, \quad (2.64)$$

which will be proven in Problem 5. Then we obtain

$$F^n = -\text{Im}\left[\sum_{m \neq n} \frac{\langle n; R|[\nabla h(R)]|m; R\rangle \wedge \langle m; R|[\nabla h(R)]|n; R\rangle}{[E_n(R) - E_m(R)]^2} \right], \quad (2.65)$$

and the Berry phase angle is given in terms of this quantity:

$$\gamma_n(C) = \iint_S dS \cdot F^n \qquad \text{(modulo } 2\pi\text{)}, \qquad (2.66)$$

where S is any surface in the parameter space that subtends the closed curve C (we write this also as $C = \partial S$ and call ∂S the boundary of S).

The results (2.65) and (2.66) are remarkable because they do not contain any phase factor of the basis vectors $|m; R\rangle$. Whereas (2.45) is only defined for single-valued $|n; R\rangle$ and can therefore be defined only on a patch where single-valued $|n; R\rangle$ exist, in (2.66) and (2.65) single-valuedness is not necessary, and these formulas can also be used for an area where single-valued $|n; R\rangle$ do not exist.

XXII.3 A Spinning Quantum System in a Slowly Changing External Magnetic Field—The Adiabatic Approximation

XXII.3.a Parameterization of the Eigenvectors

As an application of the theory of a quantum physical system in a changing environment we will now discuss in detail a quantum particle with magnetic moment

$$\mathbf{m} = +\frac{e}{2mc} g\mathbf{J} \qquad (3.1)$$

in a magnetic field

$$\mathbf{B} = B\hat{\mathbf{R}}(t) \qquad (3.2)$$

where $\hat{\mathbf{R}}(t)$, $|\hat{\mathbf{R}}(t)| = 1$, is the unit vector in the direction of the magnetic field that rotates slowly ("adiabatically") around the origin. This is the system that we mentioned in the introduction of Section 2. Its Hamiltonian is (cf. Section IX.3) given by

$$h(\hat{R}(t)) = -Bg\frac{e}{2mc}\hat{R}(t)\cdot J = b\hat{R}(t)\cdot J \qquad (3.3)$$

where J is the angular momentum operator of the quantum physical system, $\hat{R}(t)$ is the parameter that describes the changing environment, and $b = -Bg\frac{e}{2mc}$ is a constant. ($\frac{eh}{2mc} = \mu_B$ is the Bohr magneton and g is the Landé factor of the quantum particle. b is a frequency and $b/(-g) \equiv \omega_L$ is called the *Larmor frequency*.)

The parameter space of the environment (the set of all $\hat{R}$) is the unit sphere in two dimensions, S^2. We parameterize S^2 by the polar angles (θ, φ) according to

$$\hat{R} = \hat{R}(\theta, \varphi) = \begin{bmatrix} \sin\theta\cos\varphi \\ \sin\theta\sin\varphi \\ \cos\theta \end{bmatrix}, \qquad \begin{matrix} 0 \le \theta \le \pi \\ 0 \le \varphi < 2\pi \end{matrix}. \qquad (3.4)$$

This parameterization associates unique values of the pair (θ, φ) to each unit vector $\hat{R}$ except for the unit vector

$$\mathbf{e}_3 \equiv \begin{bmatrix} 0 \\ 0 \\ 1 \end{bmatrix} \qquad (3.5)$$

of the north pole and the unit vector $-\mathbf{e}_3$ of the south pole. $\mathbf{e}_3$ and $-\mathbf{e}_3$ are given, respectively, by $\theta = 0$ and $\theta = \pi$ for all values of φ in the range $0 \le \varphi < 2\pi$; the value of φ is not determined when $\hat{R} = \pm\mathbf{e}_3$.

In the space of physical states of the quantum system there exists, for each unit vector $\hat{R}$, a unit ray of normalized eigenvectors

$$e^{i\mu}|k; \hat{R}\rangle, \qquad 0 \le \mu < 2\pi \qquad (3.6)$$

of the Hamiltonian $h(\hat{R})$ with eigenvalue $b\hbar k$, i.e., for each direction $\hat{R}$ of the magnetic field there exists eigenvectors of the operator $\hat{R}\cdot J$ with eigenvalue $\hbar k$ [$= n$ of (2.3)]:

$$\hat{R}\cdot Je^{i\mu}|k; \hat{R}\rangle = \hbar k e^{i\mu}|k; \hat{R}\rangle. \qquad (3.7)$$

Here k is the quantum number for the component of angular momentum along the direction of the external magnetic field. If the vectors $|k; \hat{R}\rangle$ are also eigenvectors of J^2 then the possible values are given by

$$k = -j, -j+1, \ldots, +j; \qquad j = \text{integer or half-integer} \qquad (3.8)$$

where $\hbar^2 j(j+1)$ is the eigenvalue of the angular momentum J^2. We assume here that j and any other additional quantum numbers, which may be connected with the eigenvalues of the rotationally invariant part H_0 of the total Hamiltonian $H^{\text{total}} = H_0 + h(\hat{R})$, have fixed values and we suppress them

throughout the current discussion. In this special case $|k; \hat{R}\rangle$, k, and $b\hbar k$ play
the roles of $|n; R\rangle$, n, and $E_n(R)$ in the general theory of Section 2.

But it may also be that the $|k; \hat{R}\rangle$ are not eigenvectors of J^2 and thus do
not have a definite value of j associated to them. This kind of vectors we will
need in Chapter XXIII, where R becomes a quantum mechanical observable.
If the $|k; \hat{R}\rangle$ are not eigenvectors of J^2, then it is not obvious that k are integer
or half-integer. But we will show that $2k$ will be integer.

The vectors of (3.6) are parameterized by the unit vector $\hat{R}$, which in turn is
described by the polar angles (θ, φ). The vectors (3.6) can therefore also be
parameterized by (θ, φ). This may be done by applying (θ, φ)-dependent
rotations to the eigenvectors $e^{i\mu}|k; \mathbf{e}_3\rangle$ of the component of angular momen-
tum in the direction of the north pole, $J_3 = \mathbf{e}_3 \cdot J$.

There are many (θ, φ)-dependent rotations $\mathscr{R}(\theta, \varphi) \in \mathrm{SO}(3)$ that when applied to $\mathbf{e}_3$ give the arbitrary unit vector $\hat{R}(\theta, \varphi)$. We consider the product
rotation

$$\mathscr{R}(\theta, \varphi) = \mathscr{R}_3(\varphi)\mathscr{R}_2(\theta)\mathscr{R}_3(-\varphi) \tag{3.9}$$

obtained by first applying the rotation

$$\mathscr{R}_3(-\varphi) = \begin{bmatrix} \cos\varphi & \sin\varphi & 0 \\ -\sin\varphi & \cos\varphi & 0 \\ 0 & 0 & 1 \end{bmatrix} = \mathscr{R}_3^{-1}(\varphi) \tag{3.10}$$

by the angle $-\varphi$ around the $\mathbf{e}_3$-axis, and then the rotation

$$\mathscr{R}_2(\theta) = \begin{bmatrix} \cos\theta & 0 & \sin\theta \\ 0 & 1 & 0 \\ -\sin\theta & 0 & \cos\theta \end{bmatrix} \tag{3.11}$$

by the angle θ around the $\mathbf{e}_2$-axis, and then the rotation

$$\mathscr{R}_3(\varphi) = \begin{bmatrix} \cos\varphi & -\sin\varphi & 0 \\ \sin\varphi & \cos\varphi & 0 \\ 0 & 0 & 1 \end{bmatrix} \tag{3.12}$$

by the angle φ around the $\mathbf{e}_3$-axis. The arbitrary unit vector $\hat{R}(\theta, \varphi)$ is then
given by (Problem 6)

$$\mathscr{R}(\theta, \varphi)\mathbf{e}_3 = \mathscr{R}_3(\varphi)\mathscr{R}_2(\theta)\mathscr{R}_3(-\varphi)\mathbf{e}_3 = \mathscr{R}_3(\varphi)\mathscr{R}_2(\theta)\mathbf{e}_3 = \mathscr{R}_3(\varphi)\hat{R}(\theta, 0)$$

$$= \hat{R}(\theta, \varphi) \tag{3.13}$$

where $\mathscr{R}_3(-\varphi)$ does nothing to the unit vector $\mathbf{e}_3$ and where $\mathscr{R}_2(\theta)$ produces
the unit vector $\hat{R}(\theta, 0)$, which lies in the 1-3 plane at an angle θ with respect
to the $\mathbf{e}_3$-axis as shown in Fig. 3.1. The rotation $\mathscr{R}_3(-\varphi)$ has been included
in the definition of $\mathscr{R}(\theta, \varphi)$ in order for the rotation $\mathscr{R}(0, \varphi)$ to be independent
of φ.

Rotations, like any other continuous transformations, are represented in
the space of quantum physical states by unitary operators. The unitary operators that represent the rotations $\mathscr{R}_3(\varphi)$ and $\mathscr{R}_2(\theta)$ are given by

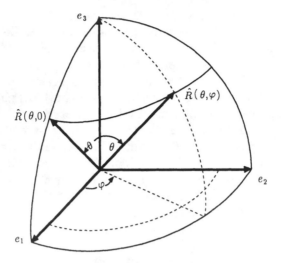

Figure 3.1 Polar coordinates.

$$U_3(\varphi) = e^{-i(\varphi/\hbar)J_3} \equiv I + \frac{\varphi}{i\hbar}J_3 + \frac{1}{2!}\left(\frac{\varphi}{i\hbar}J_3\right)^2 + \cdots, \qquad (3.14)$$

$$U_2(\theta) = e^{-i(\theta/\hbar)J_2} \qquad (3.15)$$

and an analogous expression holds for rotations about the e_1-axis. The product of two or more rotations is represented by the product of the corresponding operators. The rotation $\mathcal{R}(\theta, \varphi)$ of (3.9) is thus represented by the operator

$$U(\theta, \varphi) = U_3(\varphi)U_2(\theta)U_3(-\varphi) = e^{-i(\varphi/\hbar)J_3}e^{-i(\theta/\hbar)J_2}e^{i(\varphi/\hbar)J_3}. \qquad (3.16)$$

We now choose a fixed normalized eigenvector $|k; e_3\rangle$ of $e_3 \cdot J = R(0, \varphi) \cdot J$.

Applying the unitary operator $U(\theta, \varphi)$ to the normalized vector $e^{i\mu}|k; e_3\rangle$ gives the vector

$$e^{i\mu}|k; \theta\varphi\rangle \equiv U(\theta, \varphi)e^{i\mu}|k; e_3\rangle, \qquad 0 \le \mu < 2\pi. \qquad (3.17)$$

This vector is an eigenvector of the operator $\hat{R}(\theta, \varphi) \cdot J$ with eigenvalue $\hbar k$:

$$\hat{R}(\theta, \varphi) \cdot J e^{i\mu}|k; \theta\varphi\rangle = \hbar k e^{i\mu}|k; \theta\varphi\rangle. \qquad (3.18)$$

To prove (3.18) we have to use some properties of the angular momentum operators J_i that follow from their commutation relations (III.2.16). These properties are

$$e^{-i(\theta/\hbar)J_2}J_3 e^{i(\theta/\hbar)J_2} = J_3 \cos\theta + J_1 \sin\theta = R(\theta, 0) \cdot J, \qquad (3.19)$$

$$e^{-i(\varphi/\hbar)J_3}J_1 e^{i(\varphi/\hbar)J_3} = J_1 \cos\varphi + J_2 \sin\varphi, \qquad (3.20)$$

$$e^{-i(\varphi/\hbar)J_3}J_2 e^{i(\varphi/\hbar)J_3} = J_1(-\sin\varphi) + J_2 \cos\varphi. \qquad (3.21)$$

To prove (3.19), (3.20), and (3.21) we use the following general (Baker-Campbell-Hausdorff) formula, which holds for any two linear operators A

and B (Problem 9):

$$e^B A e^{-B} = \sum_{n=0}^{\infty} \frac{1}{n!} \underbrace{[B, [B, \ldots [B, A] \ldots]]}_{n}$$

$$= A + [B, A] + \frac{1}{2!}[B, [B, A]] + \frac{1}{3!}[B, [B, [B, A]]] + \cdots . \quad (3.22)$$

Taking $A = J_1$ and $B = -i(\varphi/\hbar)J_3$ and using the commutation relations $[J_3, J_1] = i\hbar J_2$ and $[J_3, J_2] = -i\hbar J_1$, we obtain

$$e^{-i(\varphi/\hbar)J_3} J_1 e^{i(\varphi/\hbar)J_3} = J_1 + \left(-i\frac{\varphi}{\hbar}\right)[J_3, J_1] + \frac{1}{2!}\left(-i\frac{\varphi}{\hbar}\right)^2 [J_3, [J_3, J_1]]$$

$$+ \frac{1}{3!}\left(-i\frac{\varphi}{\hbar}\right)^3 [J_3, [J_3, [J_3, J_1]]] + \cdots$$

$$= J_1 + \varphi J_2 - \frac{\varphi^2}{2!}J_1 - \frac{\varphi^3}{3!}J_2 + \cdots$$

$$= J_1\left(1 - \frac{\varphi^2}{2!} + \frac{\varphi^4}{4!} - \cdots\right) + J_2\left(\varphi - \frac{\varphi^3}{3!} + \frac{\varphi^5}{5!} - \cdots\right)$$

$$= J_1 \cos \varphi + J_2 \sin \varphi. \quad (3.23)$$

This establishes (3.20); (3.19) and (3.21) are proved in the same way. Using (3.19) and (3.21) we calculate

$$\hbar k e^{i\mu}|k; \mathbf{e}_3\rangle = e^{i\mu} J_3 |k; \mathbf{e}_3\rangle$$

$$= e^{i\mu} J_3 e^{-i(\varphi/\hbar)J_3} e^{i(\theta/\hbar)J_2} e^{i(\varphi/\hbar)J_3} |k; \theta\varphi\rangle$$

$$= e^{i\mu} e^{-i(\varphi/\hbar)J_3} J_3 e^{i(\theta/\hbar)J_2} e^{i(\varphi/\hbar)J_3} |k; \theta\varphi\rangle$$

$$= e^{i\mu} e^{-i(\varphi/\hbar)J_3} e^{i(\theta/\hbar)J_2} (J_3 \cos \theta + J_1 \sin \theta) e^{i(\varphi/\hbar)J_3} |k; \theta\varphi\rangle$$

$$= e^{i\mu}(e^{-i(\varphi/\hbar)J_3} e^{i(\theta/\hbar)J_2} e^{i(\varphi/\hbar)J_3} J_3 \cos \theta$$

$$+ e^{-i(\varphi/\hbar)J_3} e^{i(\theta/\hbar)J_2} J_1 e^{i(\varphi/\hbar)J_3} \sin \theta)|k; \theta\varphi\rangle$$

$$= e^{i\mu} e^{-i(\varphi/\hbar)J_3} e^{i(\theta/\hbar)J_2} e^{i(\varphi/\hbar)J_3} (J_3 \cos \theta$$

$$+ J_1 \sin \theta \cos \varphi + J_2 \sin \theta \sin \varphi)|k; \theta\varphi\rangle$$

$$= e^{i\mu} e^{-i(\varphi/\hbar)J_3} e^{i(\theta/\hbar)J_2} e^{i(\varphi/\hbar)J_3} \hat{R}(\theta, \varphi) \cdot \mathbf{J}|k; \theta\varphi\rangle. \quad (3.24)$$

In the second equality we have used the inverse of (3.17); in the fourth equality we have used (3.19); in the sixth equality we have used (3.20), and in the seventh equality we have used (3.4). Acting on both sides of (3.24) with $U(\theta, \varphi)$ then gives (3.18).

If the value of μ in (3.17) is held fixed at, say, $\mu = 0$, for convenience, then one obtains

$$|k; \theta\varphi\rangle = U(\theta, \varphi)|k; \mathbf{e}_3\rangle = e^{-i(\varphi/\hbar)J_3} e^{-i(\theta/\hbar)J_2} e^{i(\varphi/\hbar)J_3}|k; \mathbf{e}_3\rangle, \quad (3.25)$$

which is a (continuous, differentiable) vector-valued function of (θ, φ). If we want the $|k; \theta, \varphi\rangle$ to be eigenvectors of $\boldsymbol{J}^2$ then $|k; \mathbf{e}_3\rangle$ must also be chosen to be an eigenvector of $\boldsymbol{J}^2$. But we can choose $|k; \mathbf{e}_3\rangle$ to be just an eigenvector of $\hat{\boldsymbol{R}}(0, \varphi) \cdot \boldsymbol{J}$ and not $\boldsymbol{J}^2$. The $|k; \theta\varphi\rangle$ need not be angular momentum eigenvectors.

This vector-valued function gives a unique vector for all $\hat{\boldsymbol{R}}$ except for the south pole $\hat{\boldsymbol{R}} = -\mathbf{e}_3$. At the south pole, $\theta = \pi$, (3.19) becomes

$$e^{-i(\pi/\hbar)J_2} J_3 e^{i(\pi/\hbar)J_2} = -J_3, \tag{3.26}$$

which implies

$$e^{-i(\pi/\hbar)J_2} e^{-i(\varphi/\hbar)J_3} e^{i(\pi/\hbar)J_2} = e^{i(\varphi/\hbar)J_3}. \tag{3.27}$$

This may be used to obtain

$$\begin{aligned} |k; \pi\varphi\rangle &= e^{-i(\varphi/\hbar)J_3} e^{-i(\pi/\hbar)J_2} e^{i(\varphi/\hbar)J_3} |k; \mathbf{e}_3\rangle \\ &= e^{-i(\pi/\hbar)J_2} e^{2i(\varphi/\hbar)J_3} |k; \mathbf{e}_3\rangle \\ &= e^{-i(\pi/\hbar)J_2} e^{2ik\varphi} |k; \mathbf{e}_3\rangle. \end{aligned} \tag{3.28}$$

This shows that at the south pole different vectors of the unit ray are obtained as φ varies in the range $0 \leq \varphi < 2\pi$. φ also varies in the range $0 \leq \varphi < 2\pi$ at the north pole $\mathbf{e}_3 = \hat{\boldsymbol{R}}(0, \varphi)$ but $|k; \theta\varphi\rangle$ is single-valued at the north pole because $|k; 0\varphi\rangle$ does not depend upon φ as a result of the inclusion of the rotation $\mathscr{R}_3(-\varphi)$ in the definition of $\mathscr{R}(\theta, \varphi)$. $|k; \theta\varphi\rangle$ is thus not a vector-valued function over all of $\mathscr{S}^2$ because it does not give a well-defined vector at the south pole $\hat{\boldsymbol{R}} = -\mathbf{e}_3$. It is a single-valued vector function everywhere on S^2 except at the south pole.

A vector-valued function that is well defined at the south pole but not at the north pole is obtained by absorbing the factor $e^{2ik\varphi}$ of (3.28) into the definition of the vector-valued function by setting $\mu = -2k\varphi$ in (3.17) and defining:

$$|k; \theta\varphi\rangle' = e^{-i2k\varphi} |k; \theta\varphi\rangle = e^{-i(\varphi/\hbar)J_3} e^{-i(\theta/\hbar)J_2} e^{-i(\varphi/\hbar)J_3} |k; 00\rangle. \tag{3.29}$$

$|k; \theta\varphi\rangle'$ differs from $|k; \theta\varphi\rangle$ by the gauge transformation $e^{i\zeta(\theta, \varphi)} = e^{-i2k\varphi}$. At the south pole $|k; \theta\varphi\rangle'$ evaluates to a single vector

$$|k; \pi\varphi\rangle' = e^{-i(\pi/\hbar)J_2} |k; \mathbf{e}_3\rangle \tag{3.30}$$

but at the north pole it evaluates to many vectors

$$|k; 0\varphi\rangle' = e^{-2ik\varphi} |k; \mathbf{e}_3\rangle \tag{3.31}$$

of the same unit ray as φ varies in the range $0 \leq \varphi < 2\pi$. The vector $|k; \pi\varphi\rangle'$ is an eigenvector of J_3 with eigenvalue $-k\hbar$ and an eigenvector of $\hat{\boldsymbol{R}}(\pi, \varphi) \cdot \boldsymbol{J} = -\mathbf{e}_3 \cdot \boldsymbol{J} = -J_3$ with eigenvalue $k\hbar$ (Problem 7).

We thus see that two different parameterizations of $|k; \hat{\boldsymbol{R}}\rangle$ are needed. Since $|k; \theta\varphi\rangle$ and $|k; \theta\varphi\rangle'$ differ only by the phase factor $e^{-i2k\varphi}$ the projection operators and subspaces are the same:

$$|k; \theta\varphi\rangle\langle k; \theta\varphi| = |k; \theta\varphi\rangle''\langle k; \theta\varphi|. \tag{3.32}$$

XXII.3.b Berry Connection and Berry Phase—Magnetic Monopole Potentials

In this subsection we want to calculate the vector potentials (Berry connection) (2.44) and the Berry phase (2.51) using the basis vectors (3.25) and (3.29). From the general theory in Section 2 we expect that the Berry connection differs by a gauge transformation (2.48) for the different parameterizations (3.25) and (3.29) but that the Berry phase is independent of the parameterization and depends only on the path on S^2. The quantity that we want to calculate is the one-form of the Berry connection (2.46):

$$A^{k'k}(R^i) = A_i^{k'k} \, dR^i = i\langle k'; R^j | \frac{\partial}{\partial R^i} | k; R^j \rangle \, dR^i. \tag{3.33}$$

We first use the vectors (3.25). As the vectors (3.25) and (3.29) are parameterized by the angles θ, φ we will use for R^i the polar coordinates: $R = (r \sin\theta \cos\varphi, r \sin\theta \sin\varphi, r \cos\theta)$.

We are mainly interested in the diagonal matrix elements $A^{kk} \equiv A^k$ but include in our derivation the off-diagonal matrix elements without much additional effort (we will need them in Chapter XXIII).

If dR^i are the Cartesian components, then $A_i^{k'k}$ are the Cartesian components of the vector

$$A^{k'k}(\theta, \varphi) = i\langle k'; \theta\varphi | \nabla_R | k; \theta\varphi \rangle$$

$$= \langle k'; e_3 | iU^\dagger(\theta, \varphi) \left(e_r \frac{\partial}{\partial r} + e_\theta \frac{1}{r} \frac{\partial}{\partial \theta} + e_\varphi \frac{1}{r \sin\theta} \frac{\partial}{\partial \varphi} \right) U(\theta, \varphi) | k; e_3 \rangle$$

$$= e_r \hat{A}_r^{k'k}(\theta, \varphi) + e_\theta \hat{A}_\theta^{k'k}(\theta, \varphi) + e_\varphi \hat{A}_\varphi^{k'k}(\theta, \varphi). \tag{3.34}$$

e_r, e_θ, e_φ are the unit vectors along the r-direction, the θ-direction, and the φ-direction, respectively, and the standard form for the vector $\mathbf{V}$ has been used. $\hat{A}_r$, $\hat{A}_\theta$, $\hat{A}_\varphi$ are the spherical orthonormal components of A. As $|k; \theta, \varphi\rangle = U(\theta, \varphi)|k; e_3\rangle$ is independent of r

$$\hat{A}_r^{k'k} = 0. \tag{3.35}$$

The connection 1-form therefore is

$$A^{k'k}(\theta, \varphi) = \hat{A}_\theta^{k'k} r \, d\theta + \hat{A}_\varphi^{k'k} r \sin\theta \, d\varphi$$

$$= i\langle k'; \theta\varphi | \frac{\partial}{\partial\theta} | k; \theta\varphi \rangle \, d\theta + i\langle k'; \theta\varphi | \frac{\partial}{\partial\varphi} | k; \theta\varphi \rangle \, d\varphi$$

$$= A_\theta^{k'k} \, d\theta + A_\varphi^{k'k} \, d\varphi \tag{3.36}$$

where

$$A_\theta^{k'k} = r\hat{A}_\theta^{k'k} = \langle k'; e_3 | iU^\dagger(\theta, \varphi) \frac{\partial}{\partial\theta} U(\theta, \varphi) | k; e_3 \rangle \tag{3.37}$$

and

$$A_\varphi^{k'k} = r \sin\theta \hat{A}_\varphi^{k'k} = \langle k'; e_3 | iU^\dagger(\theta, \varphi) \frac{\partial}{\partial\varphi} U(\theta, \varphi) | k; e_3 \rangle \tag{3.38}$$

are the spherical covariant components of A. To obtain these we calculate:

$$iU^\dagger(\theta, \varphi)\frac{\partial U(\theta, \varphi)}{\partial \theta} = ie^{-i(\varphi/\hbar)J_3}e^{i(\theta/\hbar)J_2}e^{i(\varphi/\hbar)J_3}\frac{\partial}{\partial \theta}e^{-i(\varphi/\hbar)J_3}e^{-i(\theta/\hbar)J_2}e^{i(\varphi/\hbar)J_3}$$

$$= \frac{1}{\hbar}e^{-i(\varphi/\hbar)J_3}J_2 e^{i(\varphi/\hbar)J_3}$$

$$= \frac{1}{\hbar}(J_2 \cos \varphi - J_1 \sin \varphi) \qquad (3.39)$$

and

$$iU^\dagger(\theta, \varphi)\frac{\partial U(\theta, \varphi)}{\partial \varphi} = \frac{1}{\hbar}[U^\dagger(\theta, \varphi)J_3 U(\theta, \varphi) - U^\dagger(\theta, \varphi)U(\theta, \varphi)J_3]$$

$$= \frac{1}{\hbar}[-(J_1 \cos \varphi + J_2 \sin \varphi) \sin \theta + J_3(\cos \theta - 1)]. \quad (3.40)$$

Inserting this into (3.37) and (3.38) we obtain:

$$A_\theta^{k'k}(\theta, \varphi) = \frac{1}{\hbar}\langle k'; e_3|(J_2 \cos \varphi - J_1 \sin \varphi)|k; e_3\rangle, \qquad (3.41)$$

$$A_\varphi^{k'k}(\theta, \varphi) = \frac{1}{\hbar}\langle k'; e_3|[-(J_1 \cos \varphi + J_2 \sin \varphi) \sin \theta + J_3(\cos \theta - 1)]|k; e_3\rangle. \qquad (3.42)$$

For the diagonal matrix elements, J_1 and J_2 do not contribute so that we obtain

$$A_\theta^k(\theta, \varphi) = 0, \qquad (3.43)$$

$$A_\varphi^k(\theta, \varphi) = \frac{1}{\hbar}\langle k; e_3|J_3(\cos \theta - 1)|k; e_3\rangle = -k(1 - \cos \theta), \qquad \theta \neq \pi. \quad (3.44)$$

Thus the vector potential A has only a component in the φ-direction, which is given by:

$$\hat{A}^k(\theta, \varphi) = e_\varphi \frac{k(\cos \theta - 1)}{r \sin \theta}, \qquad \theta \neq \pi. \quad (3.45)$$

We now calculate the connection using the vector functions $|n; \hat{R}\rangle'$ of (3.29), which are single-valued everywhere except at the north pole. We obtain in a similar way as above

$$A_\theta'^{k'k}(\theta, \varphi) = \frac{1}{\hbar}'\langle k'; e_3|(J_2 \cos \varphi - J_1 \sin \varphi)|k; e_3\rangle', \qquad (3.46)$$

$$A_\varphi'^{k'k}(\theta, \varphi) = \frac{1}{\hbar}'\langle k'; e_3|[-(J_1 \cos \varphi + J_2 \sin \varphi) \sin \theta$$

$$+ J_3(\cos \theta - 1) + 2k\hbar]|k; e_3\rangle'. \qquad (3.47)$$

The diagonal matrix elements are:

$$A_\theta'^k(\theta, \varphi) = 0, \tag{3.48}$$

$$A_\varphi'^k(\theta, \varphi) = \frac{1}{\hbar}\,'\langle k; e_3|[J_3(\cos\theta - 1) + 2k\hbar]|k; e_3\rangle' = k(\cos\theta + 1), \quad \theta \neq 0, \tag{3.49}$$

and the vector potential is given by:

$$A'^k(\theta, \varphi) = e_\varphi \frac{k(\cos\theta + 1)}{r \sin\theta}, \quad \theta \neq 0. \tag{3.50}$$

According to the general theory of Section 2, we expect from (2.48) and $\zeta_k = -2k\varphi$:

$$A'^k - A^k = -d\zeta_k(\theta, \varphi) = d(2k\varphi) = 2k\,d\varphi \tag{3.51}$$

or for the vectors

$$A'^k - A^k = -\nabla\zeta_k = 2k\frac{1}{r\sin\theta}e_\varphi. \tag{3.52}$$

Comparing (3.49) with (3.44) we have indeed $A_\varphi'^k - A_\varphi^k = 2k$.

We will make use of the off-diagonal matrix elements in a later section and consider here only the case of one given energy eigenstate with the fixed quantum number k.

We now calculate the Berry curvature, for which we use the differential forms because it simplifies the calculations. According to (2.55)

$$F^k = dA^k = \frac{\partial A_\theta^k}{\partial\varphi}d\varphi \wedge d\theta + \frac{\partial A_\varphi^k}{\partial\theta}d\theta \wedge d\varphi, \tag{3.53}$$

where we have already made use of (3.35). Using (3.43) and (3.44) we obtain

$$F^k = -k\sin\theta\,d\theta \wedge d\varphi \equiv F_{\theta\varphi}^k\,d\theta \wedge d\varphi. \tag{3.54}$$

Except for

$$F_{\theta\varphi}^k = -F_{\varphi\theta}^k = -k\sin\theta \tag{3.55}$$

all other spherical covariant components of F_{ij} are zero.

We now use (2.56) to calculate the Berry phase angle

$$\gamma_k(C) = \oint_C A^k = \oint_C i\langle k; \theta(t), \varphi(t)|d|k; \theta(t), \varphi(t)\rangle$$

$$= \int_S F^k = -k\int_S \sin\theta\,d\theta \wedge d\varphi = -k\int_S d\Omega \quad \text{mod } 2\pi \tag{3.56}$$

where S is the surface (any surface) that spans the closed curve C (i.e., for which C is the boundary, $C = \partial S$) and $d\Omega$ is the solid angle element (area element of the unit sphere). We will denote by $\Omega(C)$ the solid angle that C subtends:

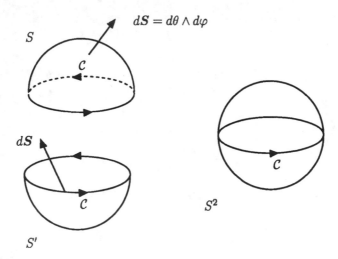

Figure 3.2 The difference of the line integrals of A and A' can be transformed, using Stokes' theorem, into an integral over the closed two-surface $S \cup S'$.

$$\Omega(C) = \int_S \sin \theta \, d\theta \wedge d\varphi. \tag{3.56'}$$

Then (3.56) can be written.

$$\gamma_k(C) = -k\Omega(C) \qquad \text{mod } 2\pi. \tag{3.57}$$

According to (3.56) S is the surface that lies above C when one traverses the curve C (the direction in which C is traversed points along the fingers and the normal of S, i.e., the direction of $d\theta \wedge d\varphi$ points along the thumb of the right hand; cf. Fig. 3.2). Now let us consider the surface S' below C. Then again

$$\gamma_k(C) = \int_{S'} F^k = -k \int_{S'} \sin \theta \, d\theta \wedge d\varphi = -k\left(-\int_{S^2 \setminus S} \sin \theta \, d\theta \wedge d\varphi\right) \tag{3.58}$$

where the direction of the normal of S' and the direction in which C is traversed are again given by the right-hand rule. This means that the normal of S' points into the sphere S^2, $(C = \partial S')$ and the normal of S points out of the sphere S^2, $(C = \partial S)$. In equation (3.58) we denote by $S^2 \setminus S$ the surface with the same area as S' but whose normal points out of the sphere S^2. As the integral of $d\Omega$ over the whole unit sphere is 4π and the integral over S is given by (3.56') we obtain from (3.58)

$$\gamma_k(C) = k\left(\int_{S^2} \sin \theta \, d\theta \wedge d\varphi - \int_S \sin \theta \, d\theta \wedge d\varphi\right)$$

or

$$\gamma_k(C) = +k(4\pi - \Omega(C)) \qquad \text{modulo } 2\pi. \tag{3.59}$$

From this it follows that k is integer or half-integer, because from comparing (3.59) and (3.57) it follows

$$-k\Omega(C) = 4\pi k - k\Omega(C) \qquad \text{mod } 2\pi, \qquad (3.60)$$

which can only be fulfilled if $k = 0, \pm 1/2, \pm 1, \ldots$.[21]

If $F_{\theta\varphi} = -F_{\varphi\theta}$ is the only nonzero spherical covariant component of F then the only nonzero spherical orthonormal component is $\hat{F}_{\theta\varphi} = -\hat{F}_{\varphi\theta}$. The two-form F is given by

$$F = F_{\theta\varphi}\, d\theta \wedge d\varphi = -k \sin \theta\, d\theta \wedge d\varphi = \hat{F}_{\theta\varphi}(r\, d\theta)(r \sin \theta\, d\varphi) = F \cdot dS,$$

and we immediately read off the value of $\hat{F}_{\theta\varphi}$:

$$\hat{F}_{\theta\varphi} = -\frac{k}{r^2} \qquad \text{or} \qquad F = -\frac{k}{r^2}\mathbf{e}_r = -\frac{k}{r^2}\hat{R}(\theta, \varphi) \qquad (3.61)$$

where $\mathbf{e}_r$ is the unit vector in the radial direction, which we denoted by $\hat{R}(\theta, \varphi)$.

The gauge field (3.61) has a very familiar form; it is the form of the magnetic field of a monopole. An imaginary electromagnetic system, with the vector potential given by (3.45) or (3.50) and the magnetic field given by (3.61), was envisioned by Dirac more than 50 years ago[22] and has intrigued several generations of theoreticians and experimentalists.[23] It consists of a charge e and a magnetic monopole with magnetic monopole strength g with the vector R pointing from g to e. The magnetic field $B^{\text{mon.}}$ of this charge-monopole system is given by $B^{\text{mon.}} = g\frac{\hat{R}}{r^2}$ and the electromagnetic vector potential $A^{\text{mon.}}$ is given by (3.45) and (3.50) with $k \to -\frac{eg}{hc}$. For this magnetic monopole, the field and the vector potential are supposed to be electromagnetic.

If one such magnetic monopole exists with magnetic charge g, then all electric charges must be multiples of $2\pi/g$ because k is integer or half-integer (from comparing (3.59) and (3.57)). This would be a remarkable explanation of one of the basic facts in nature. Unfortunately, in spite of all efforts, such electromagnetic type of monopoles have never been found.

The field F and the vector potential A, which we considered in this section, are not electromagnetic but kinematic in nature. They have their origin in a quantum system spinning with fixed angular momentum component k around the axis $\hat{R}(t)$. The analogy between the mechanical systems of this type and

---

[21] The number $C_1 = -\frac{1}{2\pi}\int_{S^2} F$ is called the first Chern number of the monopole bundle ($U(1) = S^1$ bundle over the base space $M = S^2$). It characterizes the bundle and takes the integer $2k$ as shown. If k is the component of angular momentum in an irreducible representation of the J_i then k must be one of the integers or half-integers $-j, -j+1, \ldots +j$. But there is also another proof that shows that k must be integer or half-integer, even if $|k; e_3\rangle$ is not an eigenvector of J^2. R_i are the eigenvalues of the operators X_i, which together with J_i are the generators of the group E_3. Its unitary irreducible representations (see Appendix V.3) are characterized by the numbers (k, ϵ) with $\epsilon^2 = R_i R_i (= 1)$ and $k = $ eigenvalue $\frac{1}{\epsilon}X_i J_i$, where ϵ is any real positive number and k is a positive or negative integer or half-integer including 0. There are no unitary irreducible representations of E_3 for $k \neq 2$ integer.

[22] P. A. M. Dirac, *Proc. Roy. Soc.* **133**, 60 (1931).

[23] P. Goddard and D. I. Olive, *Rep. Prog. Phys.* **41**, 1357 (1978); L. C. Biedenham and J. D. Louck (1974) Chapter 5.2; S. Coleman, *The Unity of the Fundamental Interactions*, H. Zichichi (Ed.) Plenum Press (1983) Erice lecture (1981); A. B. Balachandran, G. Marmo, B. S. Skagerstam, and A. Stern, *Classical Topology and Quantum States Part I*, World Scientific Publishers (1991).

the charge-monopole system have been known for some time.[24] The question is whether these vector potentials, induced by the rapid rotation about the slowly changing axis $\hat{R}(t)$, lead to any physical consequences for the environment. We will see in later sections that when the classical environment (represented by the external magnetic field $B(t)$) is replaced by a quantum system, then the dynamics of this quantum system will be modified by these vector potentials in the same way as the electromagnetic monopole potential would modify the motion of a charged particle.

XXII.4 A Spinning Quantum System in a Precessing External Magnetic Field—The General Cyclic Evolution

XXII.4.a The Exact State Vector

In Section 3 we calculated the Berry phase and the vector potential for a magnetic moment in a classical rotating magnetic field. As we employed an eigenvector of the Hamiltonian using (2.44) and (2.51), phase angle (3.57) and vector potential (3.45) are only the adiabatic approximations for this physical system. We will now consider the special case of this physical system in which the magnetic field precesses uniformly about the three-axis,

$$B(t) = B(\sin \theta \cos \omega t, \sin \theta \sin \omega t, \cos \theta) = B\hat{R}(\theta, \omega t) \qquad (4.1)$$

with B, θ, and $\omega = \varphi/t$ being constants.

This problem has been solved exactly in a series of famous papers (for $k = 1/2$).[25] Knowing the exact solution we do not have to resort to the adiabatic approximation to obtain the geometric phase. We can obtain the total phase angle from the solution of the Schrödinger equation and then subtract the dynamical phase angle, which we defined in (2.33) as $-\frac{1}{\hbar} \int \langle \psi(t)|h(t)|\psi(t)\rangle \, dt$.[26] The geometrical phase, which we obtain in this way for a rapidly precessing magnetic field, will go in the limit of the adiabatic approximation into the Berry phase of Section 3. This geometrical phase is a

---

[24] J. M. Leinaas, *Physica Scripta* **17**, 483 (1978).

[25] I. I. Rabi, N. F. Ramsey, and J. Schwinger, *Rev. Mod. Phys.* **26**, 167 (1954).

[26] The phase given by the dynamics (whose differences are the measurable physical quantities) is, of course, the total phase. Only for historical reasons was (2.33) also called the dynamical phase for time-dependent Hamiltonians. An extra phase factor that appears for time-dependent Hamiltonians was chosen to unity in V. Fock, *Z. Phys.* **49**, 323 (1928). This phase choice was justified by the requirement that the (eigen)vectors $\psi(t)$ have the least possible oscillatory behavior, $\|d\psi(t)/dt\| = \min$. For the adiabatic case, where this requirement leads to $\langle n(t)|\frac{d}{dt}|n(t)\rangle = 0$, the same choice was again made by M. Born and V. Fock, *Z. Phys.* **51**, 165 (1928). Though cyclic evolution was not considered in the proof of Fock, this phase choice was believed always possible until Berry (1984) showed that (according to (2.53)) an extra phase factor can arise in an adiabatic cyclic process. He called this phase geometrical, and we call it the Berry phase in distinction to its generalization introduced by Aharonov and Anandan (1987).

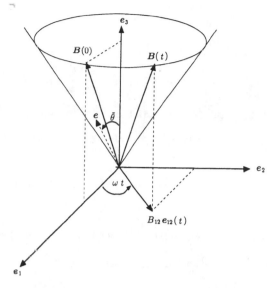

Figure 4.1 The external magnetic field precesses uniformly around a cone of semiangle θ with e_3.

special case[27] of the phase introduced by Aharonov and Anandan.[28] The discussions in this section will help us to understand the meaning of the adiabatic approximation (see Problem 10). We will see that an adiabatic evolution, which is nonstationary, really does not exist. But there exist, under certain conditions, cyclic evolutions that are very close to the adiabatic time development (2.38). The Aharonov-Anandan phase is the geometrical phase for an evolution that really occurs. The adiabatic (Berry) phase is a good approximation of this for a slowly changing environment (slowly changing external magnetic field), but an evolution for which the equality sign in (2.38) holds does not exist.

The precession of the magnetic field[25] is shown in Fig. 4.1. For $t = 0$ the magnetic field lies in the 1-3 plane

$$B(0) = B(\sin \theta, 0, \cos \theta) = (B_{12}, 0, B_3)$$

As t increases the component $B_{12} = B \sin \theta$ rotates in the 1-2 plane

$$B_{12} = B \sin \theta e_{12}(t) = B \sin \theta (e_1 \cos \omega t + e_2 \sin \omega t) \qquad (4.2)$$

so that

$$B(t) = B_{12} e_{12}(t) + B_3 e_3 = B(\sin \theta (e_1 \cos \omega t + e_2 \sin \omega t) + \cos \theta e_3)$$

$$= \frac{b}{(g \frac{e}{2mc})} \hat{R}(\theta, \omega t). \qquad (4.3)$$

[27] S. J. Wang, *Phys. Rev.* **A42**, 5107 (1990).

[28] Y. Aharonov and J. Anandan, *Phys. Rev. Lett.* **58**, 1593 (1987).

The Hamiltonian (3.3) now has the form

$$h(t) = b\hat{R}(t) \cdot J = b(\sin\theta \mathbf{e}_{12}(t) + \cos\theta \mathbf{e}_3) \cdot J$$

$$= b\cos\theta\left(J_3 + \frac{\sin\theta}{\cos\theta}\mathbf{e}_{12}(t) \cdot J\right). \tag{4.4}$$

We rewrite the time-dependent term of the Hamiltonian using (3.20)

$$\mathbf{e}_{12}(t) \cdot J = J_1\cos\omega t + J_2\sin\omega t = e^{-i(t/\hbar)\omega J_3}J_1 e^{i(t/\hbar)\omega J_3}. \tag{4.5}$$

Thus we obtain for (4.4)

$$h(t) = b\cos\theta J_3 + b\sin\theta e^{-i(t/\hbar)\omega J_3}J_1 e^{i(t/\hbar)\omega J_3} = e^{-i(t/\hbar)\omega J_3}h_0 e^{i(t/\hbar)\omega J_3} \tag{4.6}$$

with

$$h_0 = b\hat{R}(\theta, 0) \cdot J = b(\cos\theta J_3 + \sin\theta J_1) = be^{-i(\theta/\hbar)J_2}J_3 e^{i(\theta/\hbar)J_2}. \tag{4.7}$$

The time-dependent Hamiltonian $h(t)$ is periodic with

$$h(0) = h(T) = h_0; \qquad T = 2\pi/\omega. \tag{4.8}$$

Viewed from the laboratory frame the vector B_{12} rotates in the 1-2 plane, and consequently B precesses about the $\mathbf{e}_3$-axis with an angular velocity ω. Viewed from a frame that rotates with $\mathbf{e}_{12}(t)$ about $\mathbf{e}_3$ the magnetic field, B_{12} and B do not change.

We are interested in solutions of the Schrödinger equation

$$i\hbar\frac{\partial\psi}{\partial t} = h(t)\psi(t); \qquad h(t) = b\hat{R}(\theta, \omega t) \cdot J \tag{4.9}$$

that describe cyclic evolutions (2.34):

$$|\psi(T)\rangle\langle\psi(T)| = |\psi(0)\rangle\langle\psi(0)|. \tag{4.10}$$

It is not clear that solutions of (4.9) exist that fulfill the condition (4.10), but for a periodic Hamiltonian, (4.8), we would expect it.

In Section 3 we considered the adiabatic approximation for the Hamiltonian (3.3) of which the Hamiltonian (4.9) is a special case. The adiabatic approximation was defined by (2.38) or (2.40). For this approximation we obtained the Berry phase (3.57), which for the special case (4.1) with the curve $C: t \to \hat{R}(\theta, \omega t)$, is given by:

$$\gamma_k(C) = -k\int_0^{2\pi}\int_0^{\theta}\sin\theta\, d\theta\, d\varphi = -k2\pi(1 - \cos\theta). \tag{4.11}$$

If we choose—as required in the adiabatic approximation—for the initial state $\psi(0)$ the eigenvector (3.25) of the Hamiltonian $h(t) = bR(\theta, \omega t) \cdot J$ with eigenvalue $E_k(t) = bkt$ then we obtain for $\psi(T)$, according to (2.62),

$$\psi(T) = e^{-i2\pi(b/\omega)k}e^{i\gamma_k(C)}\psi(0); \qquad \psi(0) = |k; \theta, 0\rangle. \tag{4.12}$$

The expression for the state vector (2.47) in this case would be adiabatic

$$\psi(t) \overset{\downarrow}{=} e^{-ibtk}e^{i\gamma_k(t)}|k; \theta, \omega t\rangle, \qquad \omega = \frac{2\pi}{T}, \qquad \gamma_k(t) = \gamma_k(C)\frac{t}{T}. \quad (4.13)$$

This formula is another expression of the adiabatic approximation (2.40); it is an approximate equality indicated by the suffix "adiabatic" and, in general, not an equality (except for $\omega = 0$ in which case it is the stationary state of a time-independent Hamiltonian).

We will see that there do *not* exist exact cyclic solutions of (4.9) for which $\psi(0)$ is an eigenvector of the Hamiltonian. But we will show the following: There are other initial states $\psi(0)$ that undergo a cyclic evolution (4.10). If the parameters of the Hamiltonian in (4.9) fulfill the condition $\omega/b \ll 1$ (we assume that $\theta \neq 0$), then these cyclic solutions of (4.9) will be close to the right-hand side of (4.13). The condition $\omega/b \ll 1$ means that the angular frequency of the precession of the external magnetic field ω is very slow compared to the internal (Larmor) frequency b with which the state rotates due to the strength B of the magnetic field (we will explain this later). The adiabatic time development (4.13) is thus a good approximation of an evolution given by (4.9) for a "slow" change of the environment. But (4.13) is a fictitious process that never happens in reality. This explains the need to generalize the Berry phase of Section 2 to the Aharonov-Anandan phase of nonadiabatic cyclic evolution.

To find the exact solutions of (4.9) we proceed in the following way: The magnetic field (4.3) rotates counterclockwise by an angle ωt about the e_3-axis. Instead of rotating the observables (Hamiltonian, etc.) counterclockwise and keeping the state the same (laboratory frame), we can rotate the state $\psi(t)$ clockwise by an angle $(-\omega t)$ about e_3 and keep the observables the same (view the magnetic field from a frame that rotates with $e_{12}(t)$). For the observable quantities, the expectation values, these two points of view give the same results:

$$\langle \psi(t)|h(t)|\psi(t)\rangle = \langle \psi(t)|e^{-i(t/\hbar)\omega J_3}h_0 e^{i(t/\hbar)\omega J_3}|\psi(t)\rangle$$

$$= \langle e^{i(t/\hbar)\omega J_3}\psi(t)|h_0|e^{i(t/\hbar)\omega J_3}\psi(t)\rangle. \quad (4.14)$$

This suggests to make the unitary transformation and define the rotating state $\psi'(t)$ (the state $\psi(t)$ viewed from the rotating frame):

$$\psi(t) \to \psi'(t) = \tilde{U}(t)\psi(t) = e^{i(t/\hbar)\omega J_3}\psi(t); \qquad \tilde{U}(t) = U_3^\dagger(\omega t). \quad (4.15)$$

Inserting $\psi(t) = \tilde{U}^\dagger(t)\psi'(t)$ into the Schrödinger equation (4.9) we obtain the Schrödinger equation for $\psi'(t)$ after a short calculation

$$i\hbar\frac{\partial\psi'}{\partial t} = (-\omega J_3 + e^{i(t/\hbar)\omega J_3}h(t)e^{-i(t/\hbar)\omega J_3})\psi'(t) \equiv h'(t)\psi'(t) \quad (4.16)$$

$\psi'(t)$ thus evolves in time according to the transformed Hamiltonian

$$h'(t) = \tilde{U}h(t)\tilde{U}^\dagger - i\hbar\tilde{U}\frac{\partial\tilde{U}^\dagger}{\partial t} = e^{i(t/\hbar)\omega J_3}h(t)e^{-i(t/\hbar)\omega J_3} - \omega J_3. \quad (4.17)$$

From (4.6), it follows that $h'(t)$ is independent of time (as we expected because in the rotating frame B does not change):

$$h'(t) = h_0 - \omega J_3 = b\left(\cos\theta J_3 + \sin\theta J_1 - \frac{\omega}{b}J_3\right) \equiv H. \tag{4.18}$$

As $h'(t) = H$ is independent of time, the equation (4.16) can be immediately integrated to give, according to (XII.1.35):

$$\psi'(t) = e^{-i(t/\hbar)H}\psi'(0). \tag{4.19}$$

In the same way as $h(t)$ of (4.14) can be related to the angular momentum along $\hat{R}(\theta, \omega t)$ using the angular frequency b, the transformed Hamiltonian H can also be related to the angular momentum along another direction, which we call $\mathbf{e}$ using another angular frequency Ω. We make the ansatz

$$H = \Omega \mathbf{e} \cdot \mathbf{J} \tag{4.20}$$

and obtain Ω and $\mathbf{e}$ from (4.18):

$$\Omega \mathbf{e} \cdot \mathbf{J} = b\left(\left(\cos\theta - \frac{\omega}{b}\right)J_3 + \sin\theta J_1\right). \tag{4.21}$$

Thus the unit vector $\mathbf{e}$ is given by

$$\mathbf{e} = \frac{b}{\Omega}\left(\cos\theta - \frac{\omega}{b}\right)\mathbf{e}_3 + \frac{b}{\Omega}\sin\theta\,\mathbf{e}_1 \equiv \cos\tilde{\theta}\,\mathbf{e}_3 + \sin\tilde{\theta}\,\mathbf{e}_1, \tag{4.22}$$

by which we also define the new angle $\tilde{\theta}$ shown in Fig. 4.1. The new angular frequency Ω is obtained from the normalization condition $\mathbf{e}^2 = 1$:

$$\Omega = b\left(1 + \frac{\omega}{b}\left(-2\cos\theta + \frac{\omega}{b}\right)\right)^{1/2}. \tag{4.23}$$

The angle $\tilde{\theta}$ between the unit vectors $\mathbf{e}$ and $\mathbf{e}_3$ defined by (4.22) is, according to (4.22), given by:

$$\cos\tilde{\theta} = \frac{b}{\Omega}\left(\cos\theta - \frac{\omega}{b}\right), \qquad \sin\tilde{\theta} = \frac{b}{\Omega}\sin\theta. \tag{4.24}$$

For the Hamiltonian (4.9) the angle $\tilde{\theta}$ is, like θ, not a parameter but a constant. $\mathbf{e} = \hat{R}(\tilde{\theta}, 0)$ is a unit vector in the 1-3 plane as shown in Fig. 4.1. It differs from $\hat{R}(\theta, 0)$.

The time development (4.19) with (4.20) represents a "rotation" of the state $\psi'(0)$ by an angle Ωt about the axis $\mathbf{e}$:

$$\psi'(t) = e^{-i(t/\hbar)\Omega\mathbf{e}\cdot\mathbf{J}}\psi'(0). \tag{4.19'}$$

[In the same way

$$\psi(t) = e^{-itb\hat{R}(\theta,0)\cdot\mathbf{J}/\hbar}\psi(0) \tag{4.25}$$

means a rotation of the state vector $\psi(0)$ by an angle bt about the axis $\hat{R}(\theta, 0)$ and

$$\psi(t) = e^{-i(t/\hbar)bJ_3}\psi(0) \qquad (4.26)$$

means a rotation of the state vector $\psi(0)$ by an angle bt about e_3.[29]] Thus we have seen that the time development in the rotating frame by the constant Hamiltonian H means (according to (4.19′)) a rotation by Ωt about e. We now transform back to the laboratory-fixed frame using (4.15) and obtain with (4.19).

$$\psi(t) = e^{-i(t/\hbar)\omega J_3}e^{-i(t/\hbar)H}\psi(0). \qquad (4.27)$$

Thus the time development given by (4.10) consists of a rotation about e by Ωt followed by a rotation about e_3 by ωt:

$$\psi(t) = U^{\dagger}(t)\psi(0) = e^{-i(t/\hbar)\omega J_3}e^{-i(t/\hbar)\Omega e \cdot J}\psi(0), \qquad (4.28)$$

which according to (4.18) can also be written as[30]

$$\psi(t) = e^{-it\omega J_3/\hbar}e^{-i(t/\hbar)(h_0-\omega J_3)}\psi(0). \qquad (4.29)$$

From (4.22) and (4.23) we see that when $\omega \to 0$ (i.e., the precession of the external magnetic field slows down to zero angular velocity) then

$$\Omega \to b \qquad \text{and} \qquad e \to \hat{R}(\theta, 0). \qquad (4.30)$$

Then the time-dependent Hamiltonian $h(t)$ becomes

$$h(t) \to h_0 \to H = b\hat{R}(\theta, 0)\cdot J. \qquad (4.31)$$

In this limit (4.28) and (4.29) goes over into (4.25), which for the special case that the angle θ is chosen $\theta = 0$ becomes (4.26). In this limit we have a constant

---

[29] To understand this statement let us assume that $\psi(t)$ represents a pure state for which at $t = 0$ all magnetic moments are aligned along $e_1 = \hat{R}(\theta = \frac{\pi}{2}, \varphi = 0)$. For this state the probability to measure the observable $|k; \theta = \frac{\pi}{2}, \varphi = 0\rangle\langle k; \theta = \frac{\pi}{2}, \varphi = 0|$ is one:

$$\mathscr{P}_k^{e_1}(t=0) = \langle\psi(0)|k, \frac{\pi}{2}0\rangle\langle k; \frac{\pi}{2}, 0|\psi(0)\rangle = 1.$$

Let us compare this with the probability that for this state $\psi(t)$ at time t all magnetic moments are aligned along $\hat{R}(\theta = \frac{\pi}{2}, \varphi = bt)$. This probability is given by

$$\mathscr{P}_k^{\hat{R}(\pi/2, bt)}(t) = |\langle\psi(t)|k; \theta = \frac{\pi}{2}, \varphi = bt\rangle|^2,$$

which according to (4.26) and (3.25), is given by

$$\mathscr{P}_k^{\hat{R}(\pi/2, bt)}(t) = |\langle\psi(t)|k; \theta = \frac{\pi}{2}, \varphi = bt\rangle|^2$$

$$= |\langle\psi(0)|e^{i(t/\hbar)bJ_3}e^{-i(t/\hbar)bJ_3}|k; \theta = \frac{\pi}{2}, \varphi = 0\rangle|^2$$

$$= |\langle\psi(0)|k; \theta = \frac{\pi}{2}, \varphi = 0\rangle|^2 = \mathscr{P}_k^{e_1}(0) = 1.$$

This means that the probability that at time t all magnetic moments are aligned along $\hat{R}(\frac{\pi}{2}, bt)$ is again one. The state $\psi(0) = |k; \frac{\pi}{2}, 0\rangle$ with all magnetic moments along e_1 has developed at time t (as a consequence of the constant magnetic field B (for $\omega = 0$)) into a state with all magnetic moments rotated by the angle bt, $\psi(t) = |k; \frac{\pi}{2}, \varphi = bt\rangle$.

[30] Note that $e^{-i(t/\hbar)(h_0-\omega J_3)} \neq e^{-i(t/\hbar)h_0}e^{i(t/\hbar)\omega J_3}$ because h_0 and J_3 do not commute.

Hamiltonian (4.31) and if we choose for $\psi(0)$ an eigenstate of h we obtain the stationary state time development of an energy eigenstate

$$\psi(t) = e^{-itbk}|k; \theta, 0\rangle; \quad \psi(0) = |k; \theta, 0\rangle \quad \text{for} \quad \frac{\omega}{b} = 0. \quad (4.32)$$

This is the "extreme adiabatic" limit, namely, that of no change at all of the external environment.

The adiabatic approximation is fulfilled when the external environment and therewith $h(t)$ varies very slowly in time compared to the change of the state $\psi(t)$ in time due to the action of the external magnetic field $\boldsymbol{B}$. This means that the external angular velocity ω (for the precession of $\boldsymbol{B}$) is small compared to the Larmor frequency b (for the rotation of $\psi(t)$):

$$\frac{\omega}{b} \ll 1. \quad (4.33)$$

For the exact cyclic evolution with $\omega \neq 0$, the initial state cannot be an eigenstate of $h(0)$ as we shall now explain: From the exact solution (4.27), (4.28), and (4.29) we see that not any arbitrary initial state $\psi(0)$ can perform a cyclic evolution (4.10) under $h(t)$. In order that (4.28) describe a cyclic evolution, we have to choose an initial state vector $\psi(0)$ that is an eigenvector of the operator

$$U^\dagger(T) = e^{-(i/\hbar)2\pi J_3} e^{-i(T/\hbar)H}; \quad (4.34)$$

we shall now find the eigenvectors and eigenvalues of this operator. We restrict ourselves here to the case $T = \frac{2\pi}{\omega}$, i.e., to the case that T is equal to the period for the environmental changes (4.8). There are other cyclically evolving states with the period $T = \frac{2\pi}{\Omega}$ which, however, are not related to the adiabatic limit (4.12).[28]

As we shall show immediately, H given by (4.18) and $e^{-i2\pi J_3/\hbar}$ commute. Therefore eigenvectors of $U^\dagger(T)$ must be eigenvectors of both $e^{-iTH/\hbar}$ and $e^{-i2\pi J_3/\hbar}$. From this it follows immediately that $|k; \theta, 0\rangle$ cannot be the initial state of a cyclic time evolution because it is an eigenvector of $h(0) = b\hat{\boldsymbol{R}}(\theta, 0) \cdot \boldsymbol{J}$ and cannot be an eigenvector of

$$H = \Omega \boldsymbol{e} \cdot \boldsymbol{J} = \Omega \hat{\boldsymbol{R}}(\tilde{\theta}, 0) \cdot \boldsymbol{J} \quad \text{for} \quad \tilde{\theta} \neq \theta \text{ i.e. for } \omega \neq 0. \quad (4.35)$$

It remains to show that H commutes with $e^{-i2\pi J_3/\hbar}$ or that $[U^\dagger(T), H] = 0$. This means we have to show that

$$U^\dagger(T)HU(T) = e^{-i2\pi J_3/\hbar}He^{+i2\pi J_3/\hbar} = H. \quad (4.36)$$

To do this, we calculate using (4.18) and (3.20) for $\varphi = 2\pi$:

$$e^{-(i/\hbar)2\pi J_3}He^{(i/\hbar)2\pi J_3} = b\left(\cos\theta J_3 + \sin\theta e^{-(i/\hbar)2\pi J_3}J_1 e^{(i/\hbar)2\pi J_3} - \frac{\omega}{b}J_3\right)$$

$$= b\left(\cos\theta J_3 + \sin\theta J_1 - \frac{\omega}{b}J_3\right) = H.$$

Let us denote the eigenvectors of $H = \Omega\hat{\boldsymbol{R}}(\tilde{\theta}, 0) \cdot \boldsymbol{J}$ by ϕ_k where $\hbar k$ again denotes the component of angular momentum along $\boldsymbol{e} = \hat{\boldsymbol{R}}(\tilde{\theta}, 0)$:

$$H\phi_k = \hbar\Omega k\phi_k; \qquad \mathbf{e}\cdot\mathbf{J}\phi_k = \hbar k\phi_k. \tag{4.37}$$

From (3.18) and (3.25) with θ replaced by $\tilde{\theta}$ and φ by 0 follows that ϕ_k is (up to a phase factor if ϕ_k is assumed to be normalized) given by

$$\phi_k = |k; \tilde{\theta}, 0\rangle = U(\tilde{\theta}, 0)|k; \mathbf{e}_3\rangle = e^{-i\tilde{\theta}J_2/\hbar}|k; \mathbf{e}_3\rangle. \tag{4.38}$$

To obtain the eigenvalue of $e^{-i2\pi J_3/\hbar}$ we calculate, using (3.21) for $\varphi = 2\pi$:

$$\begin{aligned} e^{-i2\pi J_3/\hbar}\phi_k &= e^{-i2\pi J_3/\hbar}|k; \tilde{\theta}, 0\rangle = e^{-i2\pi J_3/\hbar}e^{-i\tilde{\theta}J_2/\hbar}|k; 0, 0\rangle \\ &= e^{-i\tilde{\theta}J_2/\hbar}e^{-i2\pi J_3/\hbar}|k; 0, 0\rangle \\ &= e^{-i2\pi k}|k; \tilde{\theta}, 0\rangle. \end{aligned} \tag{4.39}$$

From (4.39) and (4.37) we see that the eigenvalues of $U^\dagger(T)$ are given by:

$$U^\dagger(T)\phi_k = e^{-i2\pi k}e^{-i2\pi(\Omega/\omega)k}\phi_k \equiv e^{-i\alpha_k}\phi_k. \tag{4.40}$$

The eigenvalues of $U^\dagger(T)$ are labeled by k and we denote them by $e^{-i\alpha_k}$. This is, of course, the phase factor in the cyclic evolution (4.10):

$$\psi(T) = U^\dagger(T)\psi(0) = e^{-i\alpha_k}\psi(0) = e^{-i2\pi k}e^{-i2\pi(\Omega/\omega)k}\psi(0), \tag{4.41}$$

where $\psi(0)$ must be, except for a phase factor, one of the eigenvectors ϕ_k.[31]

From comparison with the discussions in Section 3 we see that the ϕ_ks are something very similar to the eigenvectors $|k; \theta, \varphi = 0\rangle$ of $h(0)$ in (3.18) and (3.25). But they are not eigenvectors of $h(0)$, because they are not eigenvectors of $\hat{R}(\theta, \varphi = 0)\cdot\mathbf{J}$ but eigenvectors of $\hat{R}(\tilde{\theta}, \varphi = 0)\cdot\mathbf{J}$. If we choose the relative phase factor between the initial state vector $\psi(0)$ and ϕ_k equal to unity[32] we have the analog of the initial condition (2.37) for the adiabatic approximation

$$\psi(0) = \phi_k = \phi_k(0) = |k; \tilde{\theta}, 0\rangle. \tag{4.42}$$

We have also denoted this eigenvector of (4.37) by $\phi_k(0)$, indicating that it is one of the many vectors in the one-dimensional subspace of the projection operator $|\psi(0)\rangle\langle\psi(0)|$. We say that this vector $\phi_k(0)$ of $\mathcal{H}$ lies above the point $|\psi(0)\rangle\langle\psi(0)|$ of the projective Hilbert space $\mathcal{P}(\mathcal{H})$.

We now have to determine the analog of the eigenvectors $|k; R(t)\rangle = |k; \theta(t), \varphi(t)\rangle$ of (3.25). These vectors $\phi_k(t)$ lie on a curve above the closed curve $\mathscr{C}: t \to |\psi(t)\rangle\langle\psi(t)|$ in $\mathcal{P}(\mathcal{H})$ of (2.34) and thus have to fulfill

$$\phi_k(t) = \text{phase factor}\cdot\psi(t). \tag{4.43a}$$

We also want them to be single-value, i.e., to fulfill the analog of (2.11):

$$\phi_k(T) = \phi_k(0). \tag{4.43b}$$

There are many such vectors, all differing by a phase transformation

$$\phi_k(t) \to \phi_k'(t) = e^{i\zeta_k(t)}\phi_k(t), \qquad \zeta_k(0) = \zeta_k(T) \bmod 2\pi. \tag{4.43c}$$

---

[31] One can show that this result generalizes to any time-dependent Hamiltonian of the form $h(t) = e^{-iAt}h(0)e^{iAt}$. The cyclic initial states are given by the eigenvectors of the time-independent operator $H = h(0) - A$.

[32] Here, as well as in (2.37) we could have made another choice of the relative phase factor between $\psi(0)$ and ϕ_k or $|n; R(0)\rangle$.

One, and by far the most obvious choice for $\phi_k(t)$ is

$$\phi_k(t) \equiv U(\tilde{\theta}, \omega t)|k; \mathbf{e}_3\rangle = |k; \tilde{\theta}, \omega t\rangle = e^{-i(\omega t/\hbar)J_3}e^{-i(\tilde{\theta}/\hbar)J_2}e^{i\omega tk}|k; \mathbf{e}_3\rangle. \quad (4.44)$$

This fulfills the condition (4.43b), as seen immediately from (4.39). To see that it also fulfills the condition (4.43a) we calculate starting with (4.44):

$$\phi_k(t) = e^{i\omega tk}e^{-i(\omega t/\hbar)J_3}\phi_k(0) = e^{i\omega tk}e^{i\Omega tk}e^{-i(\omega t/\hbar)J_3}e^{-i(t/\hbar)\Omega \mathbf{e} \cdot \mathbf{J}}\phi_k(0),$$

where we have used (4.37). Using (4.28) with (4.42) we obtain:

$$\phi_k(t) = e^{i\omega tk}e^{i\Omega tk}\psi(t)$$

or

$$\psi(t) = e^{-i\omega tk}e^{-i\Omega tk}|k; \tilde{\theta}, \omega t\rangle. \quad (4.45)$$

Therewith we have expressed the state vector for cyclic time evolutions in terms of the known quantities ω, Ω (of (4.23)), and $|k; \tilde{\theta}, \omega t\rangle$ (of (4.44)). The cyclic evolutions are, as in the adiabatic case, characterized by the integer or half-integer k, where k is again a component of angular momentum (4.37) but no longer the component along the direction of the changing magnetic field. The equation (4.45) is the generalization of the adiabatic "equality" (4.13) and (4.41) is the generalization of (4.12). However the phase factor in (4.45) and (4.41) is not in the form of a product of the dynamical and the geometrical part as in (4.12) and (4.13). In order to obtain that we have to calculate the dynamical or the geometrical phase factor separately.

XXII.4.b The Dynamical and the Geometrical Phase Factor

The splitting of the phase factor into a "dynamical" and a "geometrical" part can be approached from two different points of view. Either one gives an argument why a certain part is geometrical and obtains the dynamical phase as the difference between the total and geometrical phase, or one defines the dynamical phase and obtains the geometrical phase as the difference between the total and dynamical phase. We will start here with the dynamical phase because we have already defined it in (2.33); the geometrical phase for general cyclic evolution has not yet been defined.[26]

We calculate for the dynamical phase:

$$\langle\psi(t)|h(t)|\psi(t)\rangle = \langle\phi_k|e^{it\Omega \mathbf{e} \cdot \mathbf{J}/\hbar}e^{it\omega J_3/\hbar}h(t)e^{-it\omega J_3/\hbar}e^{-it\Omega \mathbf{e} \cdot \mathbf{J}/\hbar}|\phi_k\rangle$$

$$= \langle\phi_k|e^{it\omega J_3/\hbar}h(t)e^{-it\omega J_3/\hbar}|\phi_k\rangle,$$

where we have used (4.28) and (4.42) and (4.37). Using (4.6) and (4.18) this can be written

$$\langle\psi(t)|h(t)|\psi(t)\rangle = \langle\phi_k|(H + \omega J_3)|\phi_k\rangle. \quad (4.46)$$

The matrix element of H is obtained from (4.37), and it remains to calculate:

$$\langle\phi_k|J_3|\phi_k\rangle = \langle k; 0, 0|e^{i\tilde{\theta}J_2/\hbar}J_3 e^{-i\tilde{\theta}J_2/\hbar}|k; 0, 0\rangle$$

$$= \langle k; 0, 0|(J_3 \cos\tilde{\theta} + J_1 \sin\tilde{\theta})|k; 0, 0\rangle = \hbar k \cos\tilde{\theta}. \quad (4.47)$$

In this calculation (4.38) and (3.19) has been used. Inserting (4.47) and (4.37) into (4.46) we obtain

$$\langle \psi(t)|h(t)|\psi(t)\rangle = \Omega\hbar\left(1 + \frac{\omega}{\Omega}\cos\tilde{\theta}\right)k. \tag{4.48}$$

From this we obtain the dynamical phase angle:

$$-\alpha_k^{\text{dyn}} \equiv -\frac{1}{\hbar}\int_0^T \langle \psi(t)|h(t)|\psi(t)\rangle\, dt$$

$$= -k2\pi\left(\frac{\Omega}{\omega} + \cos\tilde{\theta}\right) = -k\Omega\left(1 + \frac{\omega}{\Omega}\cos\tilde{\theta}\right)T. \tag{4.49}$$

If the geometrical phase for the general cyclic evolution is again, as in the adiabatic approximation (2.62), the difference between the total phase and the dynamical phase

$$-\alpha_k^{\text{geom}} = -\alpha_k + \alpha_k^{\text{dyn}}, \tag{4.50}$$

then we calculate from (4.40) and (4.49) that

$$-\alpha_k^{\text{geom}} = -k2\pi(1 - \cos\tilde{\theta}) = -k2\pi\left(1 + \frac{\omega}{\Omega} - \frac{b}{\Omega}\cos\theta\right). \tag{4.51}$$

This has the same form as (4.11) for the adiabatic case except that θ is replaced by $\tilde{\theta}$ of (4.24).

In the adiabatic approximation (4.33) we can expand the expressions in (4.23) and (4.24) with respect to $\frac{\omega}{b}$:

$$\frac{\Omega}{b} \approx 1 - \frac{\omega}{b}\cos\theta + \left(\frac{\omega}{b}\right)^2\frac{1}{4}(1 - \cos 2\theta),$$

$$\sin\tilde{\theta} \approx \sin\theta + \frac{1}{2}\frac{\omega}{b}\sin 2\theta; \qquad \cos\tilde{\theta} \approx \cos\theta + \frac{1}{2}\frac{\omega}{b}(\cos 2\theta - 1) \tag{4.52}$$

and obtain the adiabatic approximation of the geometrical phase angle (4.51), which we called the Berry phase angle:

$$-\alpha_k^{\text{geom}} \approx -2\pi k(1 - \cos\theta) = \gamma_k^{\text{Berry}}. \tag{4.53}$$

These results indicate that (4.50) with (4.49) is the appropriate choice of the geometrical phase also for the general cyclic evolution. We now want to show that this phase angle can be obtained in the same way as Berry's phase for the adiabatic approximation, and then we shall discuss in what sense it is geometric in nature.

For that purpose we define in analogy to (2.44) and (2.46) the connection one-form

$$A^{\phi_k} = i\langle \phi_k(t)|d|\phi_k(t)\rangle = \langle k; \tilde{\theta}, \omega t|d|k; \tilde{\theta}, \omega t\rangle$$

$$= \langle k; \mathbf{e}_3|U^\dagger(\tilde{\theta}, \omega t)\frac{d}{dt}U(\tilde{\theta}, \omega t)|k; \mathbf{e}_3\rangle\, dt. \tag{4.54}$$

From the results (3.40) (with $\varphi = \omega t$) we obtain immediately

$$A^{\phi_k} = \frac{1}{\hbar} \langle k; \mathbf{e}_3 | [-(J_1 \cos \varphi + J_2 \sin \varphi) \sin \tilde{\theta} + J_3(\cos \tilde{\theta} - 1)] | k; \mathbf{e}_3 \rangle d(\omega t)$$

$$(4.54')$$

or

$$A^{\phi_k} = -k(1 - \cos \tilde{\theta}) \, d(\omega t) = -k\left(1 - \frac{b}{\Omega} \cos \theta + \frac{\omega}{\Omega}\right) d\varphi(t). \quad (4.55)$$

The curvature (field strength) that follows from this connection (gauge potential) is

$$F = dA = -k \frac{(1 - \frac{\omega}{b} \cos \theta)}{(\frac{\Omega}{b})^3} \sin \theta \, d\theta \wedge d\varphi$$

or

$$F_{\theta\varphi} = -k \frac{1 - \frac{\omega}{b} \cos \theta}{[1 - \frac{\omega}{b} \cos \theta + (\frac{\omega}{b})^2]^{3/2}} \sin \theta; \qquad F = -\frac{k}{r^2} \frac{(1 - \frac{\omega}{b} \cos \theta)}{(\frac{\Omega}{b})^3} \hat{R}(\theta, \varphi).$$

$$(4.56)$$

This is again the "monopole" type field as in (3.61) but with modified "monopole strength."

The connection one-form (4.55) has the same form as (3.44) for the adiabatic connection, except that here we have the angle $\tilde{\theta}$, which is not the angle θ for the direction of the magnetic field (parameter of the Hamiltonian).

We can now use the analog of the formula (2.56') and (3.56) for the adiabatic approximation and define a "geometric phase" angle also for general cyclic evolution by

$$\gamma_k(\mathscr{C}) \equiv \gamma_k \equiv \oint_{\mathscr{C}} A^{\phi_k} = \oint i\langle \phi_k(t)|d|\phi_k(t)\rangle. \quad (4.57)$$

Then we obtain with (4.55)

$$\gamma_k(\mathscr{C}) = -k \int_0^T (1 - \cos \tilde{\theta}) \, d(\omega t) = -k2\pi(1 - \cos \tilde{\theta})$$

$$= -2\pi k\left(1 - \frac{b}{\Omega} \cos \theta + \frac{\omega}{\Omega}\right) \quad (4.58)$$

$$\gamma_k(t) = -k2\pi \frac{t}{T}(1 - \cos \tilde{\theta}). \quad (4.58')$$

This agrees with the result (4.51) obtained from (4.50).

The geometrical phase (4.58) is not purely geometrical in the way that (4.53) is, which only depends upon θ and is given solely by the path in the parameter space. The geometrical phase for general cyclic evolution also contains the parameters ω, b of the Hamiltonian. To call (4.49) the dynamical phase and the remainder of the total phase given by (4.50) the geometrical phase has of course only historical reasons. It is the total phase α_k that is given by the Schrödinger equation (4.9). And it is the difference in the total phase that is of physical meaning and can be measured (in interference experiments). The splitting $\alpha_k = \alpha^{\text{dyn}} + \alpha^{\text{geom}}$ given by (4.50) has a meaning only if we specify either α_k^{dyn} or α_k^{geom}. In (4.49) we have specified $\alpha^{\text{dyn}\cdot}$ in accordance with (2.33) and found that $-\alpha_k + \alpha_k^{\text{dyn}}$ agreed with γ_k defined by (4.57). Conversely, if we define α^{geom} by $-\alpha^{\text{geom}} = \gamma_k$ of (4.57) then the so-called dynamical phase is given by (4.49).

For our specific example of a spinning quantum system in a precessing magnetic field we have now solved the problem posed in (2.34) of a general cyclic evolution (nonadiabatic): For a given Hamiltonian $h(t)$ there exist a countable number of cyclic paths in the projective Hilbert space $\mathscr{P}(\mathscr{H})$

$$\mathscr{C}_k: \Lambda_k(0) \to \Lambda_k(t) = |\psi(t)\rangle\langle\psi(t)| \to \Lambda_k(T) = \Lambda_k(0) \qquad (4.59)$$

where k is the number specified by the initial condition (4.42),[33]

$$\Lambda_k(t) = |k; \tilde{\theta}, \omega t\rangle\langle k; \tilde{\theta}, \omega t| = |\phi_k(t)\rangle\langle\phi_k(t)|.$$

These closed paths correspond to the closed curve

$$C: \hat{R}(\theta, \varphi = 0) \to \hat{R}(\theta, \varphi = \omega t) \to \hat{R}(\theta, \varphi = \omega T) = \hat{R}(\theta, \varphi = 0) \quad (4.60)$$

in parameter space S^2 (two-dimensional unit sphere). Above the curve $\mathscr{C}_k$ lies the curve (cf. (2.35)):

$$C_k: \psi(0) \to \psi(t) \to \psi(T) = e^{-i\alpha_k}\psi(0) \qquad (4.61)$$

of state vectors in Hilbert space, where $\psi(t)$ is a solution of the Schrödinger equation with Hamiltonian $h(t)$. We call this the dynamical lift of the curve $\mathscr{C}_k$. This is shown in Fig. 4.2. We now proceed to explain the other features of Fig. 4.2.

Above the curve $\mathscr{C}_k$ also lies the following curve in Hilbert space:

$$C_k^{\text{closed}}: |k; \tilde{\theta}, 0\rangle \to \phi_k(t) = |k; \tilde{\theta}, \omega t\rangle \to |k; \tilde{\theta}, \omega T\rangle = |k; \tilde{\theta}, 0\rangle. \quad (4.62)$$

This is a closed lift of the curve $\mathscr{C}_k$ and in place of (4.62) we could take any other closed lift $\phi_k'(t)$ that differs from (4.62) by a phase (gauge) transformation (4.43c):

$$C_k^{\text{closed}'}: \phi_k'(0) \to \phi_k'(t) \to \phi_k'(T) = \phi_k'(0). \qquad (4.62')$$

In addition to the dynamical lift (4.61) and the closed lifts (4.62), (4.62') we can define another lift given by

$$\tilde{\psi}(t) = e^{i\Omega k(1 + (\omega/\Omega)\cos\tilde{\theta})t}\psi(t) = e^{(i/\hbar)\int_0^t \langle\psi(t)|h(t)|\psi(t)\rangle \, dt}\psi(t). \qquad (4.63)$$

As a consequence of (4.41), (4.50), and (4.63) we have $\tilde{\psi}(T) = e^{-i\alpha_k^{\text{geom}}}\tilde{\psi}(0)$. We call this lift,

$$\tilde{C}_k: \tilde{\psi}(0) \to \tilde{\psi}(t) \to \tilde{\psi}(T) = e^{i\gamma_k}\tilde{\psi}(0), \qquad (4.64)$$

the horizontal lift[34] of $\mathscr{C}_k$. The geometrical phase factor $e^{i\gamma_k}$ is sometimes also

---

[33] To be precise, $\psi(t)$ should also be labeled by k.

[34] These terms have their origin in the underlying differential geometry, which we do not want to explain here any further. For more see reference 18 of Section 2.

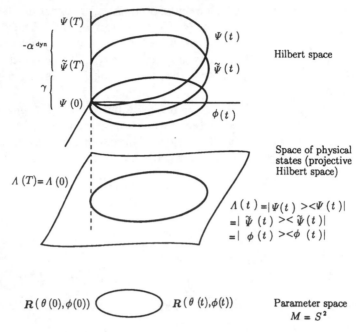

Figure 4.2 Closed path in the parameter space and the space of physical states and its lifts.

called holonomy. The relation between the dynamical lift and the closed lift (4.62) is given by (using $\gamma_k(t) = -2\pi k \frac{t}{T}(1 - \cos\tilde{\theta})$):

$$\psi(t) = e^{-i\omega t k(1 - \cos\tilde{\theta})} e^{-i\omega t k((\Omega/\omega) + \cos\tilde{\theta})} |k; \tilde{\theta}, \varphi\rangle$$

$$= e^{i\gamma_k(t)} e^{-i/\hbar \int_0^t \langle\psi(t')|h(t')|\psi(t')\rangle\, dt'} |k; \tilde{\theta}, \varphi(t)\rangle. \tag{4.65}$$

The concepts introduced in this section for the cyclic evolution (4.59) are the analogs of the concepts in Sections 2 and 3 for the adiabatic approximation. The closed lifts ϕ_k are the analogs of the Hamiltonian eigenvectors $|k; \theta, \varphi\rangle$. The connections A^{ϕ_k} of (4.54) are generalizations of the adiabatic connections (2.46). The geometric phase (4.58) is a generalization of the Berry phase (2.53), which it tends to in the adiabatic approximation (4.33).

The distinction between (4.65) and (4.13) is that in (4.65) we have an equality sign. This means that the state vector $\psi(t)$ "rotates" (in the sense described in footnote 29) along the curve (4.61) described by the curve $t \to (\tilde{\theta}, \omega t)$ in parameter space but *not* by the curve $t \to (\theta, \omega t)$, which is the path for the magnetic field. For slow enough ω (however, before ω becomes zero, which is the stationary state development with $\gamma_k = 0$) these two curves are "close enough" together. The adiabatic time development (2.38) is a fiction. There is really no solution $\psi(t)$ of (4.9) that fulfills this, only nonadiabatic cyclic evolutions (4.65) exist. The geometric (Aharonov-Anandan) phase (4.57) and the general connection (4.54) describe these cyclic evolutions that really exist, the adiabatic

(Berry) connection (2.44) and the adiabatic Berry-phase (2.51) are approximations of them.

XXII.4.c General Cyclic Quantum Evolution

The concepts introduced for the spinning quantum system in a precessing magnetic field can also be introduced for the cyclic evolution of any quantum system with time-dependent Hamiltonian.

This means that we have a Hamiltonian $h(t)$ that has at least one (and then probably a countable number as in the example of Subsection 4.6) cyclic path (2.34) in the projective Hilbert space $\mathscr{P}(\mathscr{H})$:

$$\mathscr{C}: \Lambda(0) \to \Lambda(t) = |\psi(t)\rangle\langle\psi(t)| \to \Lambda(T) = \Lambda(0). \qquad (4.66)$$

By $\psi(t)$ we denote the dynamical lift, i.e., the solution of the Schrödinger equation with Hamiltonian $h(t)$. As a generalization of (4.63) we define another lift of (4.66) by:

$$\tilde{\psi}(t) = e^{1/\hbar \int_0^t \langle\psi(t')|h(t')|\psi(t')\rangle \, dt'} \psi(t). \qquad (4.67)$$

This we call again the horizontal lift. It is a straightforward calculation (Problem 11) to show that $\tilde{\psi}(t)$ fulfills the following equation and initial condition if $\psi(t)$ fulfills the Schrödinger equation (4.9):

$$i\hbar \frac{d\tilde{\psi}(t)}{dt} = (h(t) - \langle\psi(t)|h(t)|\psi(t)\rangle I)\tilde{\psi}(t) \qquad (4.68)$$

$$\tilde{\psi}(0) = \psi(0). \qquad (4.69)$$

Taking the scalar product of (4.68) with $\tilde{\psi}(t)$ or $\psi(t)$ we obtain

$$\langle\tilde{\psi}(t)|\frac{d}{dt}|\tilde{\psi}(t)\rangle = \langle\psi(t)|\frac{d}{dt}|\tilde{\psi}(t)\rangle = 0. \qquad (4.70)$$

The tangent vector $\frac{d}{dt}\tilde{\psi}(t)$ of $\tilde{\psi}(t)$ is orthogonal to $\psi(t)$ and to $\tilde{\psi}(t)$. (This is why $\tilde{\psi}(t)$ is called horizontal in contrast to vertical, which is given by the direction of the "fiber" along $\psi(t)$). The dynamical lift C is uniquely determined by the Hamiltonian $h(t)$, but not uniquely determined by the physical problem. A simple substitution

$$h(t) \to h'(t) = h(t) - \kappa(t)I \qquad \text{with } \kappa(t) \in \mathbb{R} \qquad (4.71)$$

leads to a new Hamiltonian $h'(t)$, which describes the same physics. It has the same closed curves of physical states $t \to \Lambda(t) = |\psi(t)\rangle\langle\psi(t)|$ in the projective Hilbert space $\mathscr{P}(\mathscr{H})$ as $h(t)$ (Problem 13). But $h(t)$ and $h'(t)$ have different dynamical lifts $t \to \psi(t)$ and $t \to \psi'(t)$ (Problem 14). Also, two Hamiltonians $h(t)$ and $h'(t)$ that have the same curves of physical states $t \to |\psi(t)\rangle\langle\psi(t)|$ and $t \to |\psi'(t)\rangle\langle\psi'(t)|$ differ by a multiple of the unit operator $\kappa(t)I$. We are looking for a lift that is unique for a given physical problem, i.e., a lift that is uniquely associated to a closed curve $t \to |\psi(t)\rangle\langle\psi(t)|$ in $\mathscr{P}(\mathscr{H})$. Such a lift happens to be the horizontal lift $\tilde{C}: t \to \tilde{\psi}(t)$, because $\tilde{\psi}(t)$ defined by (4.67) with $h(t)$ and $\tilde{\psi}'(t)$ defined by (4.67) with $h'(t)$ are the same, $\tilde{\psi}'(t) = \tilde{\psi}(t)$ (Problem 15).

We will now discuss the generalizations of (4.62), the closed lifts of the curve $\mathscr{C}$. As seen from (4.57) these are the quantities in terms of which the geometric phase can be calculated if one does not have a solution $\psi(t)$ of the Schrödinger equation. There are various ways to obtain these single-valued vectors $\phi(t) \in \mathscr{H}$.[35] In general, they are curves of local sections. These sections are continuously differentiable maps of an open area in $\mathscr{P}(\mathscr{H})$ into $\mathscr{H}$. We will assume that our curve $\mathscr{C}$ lies in such an open area. (If the curve C in (4.60) $t \to R(\theta(t), \varphi(t))$ includes the south pole then the function (4.44) is not uniquely defined and one has to take another section differing from $|k; \bar{\theta}, \varphi\rangle$ by a gauge transformation $\phi_k(\theta, \varphi) \to \phi_k'(\theta, \varphi) = e^{i\zeta(\theta, \varphi)}\phi_k(\theta, \varphi)$). Then for any closed curve $\mathscr{C}: t \to \Lambda(t)$ one has the single-valued lift

$$t \to \phi(t) = \phi(R(t)), \qquad \Lambda(t) = |\phi(t)\rangle\langle\phi(t)|; \qquad \phi(T) = \phi(0), \quad (4.72a)$$

which is determined only up to the gauge transformation

$$\phi(t) \to \phi'(t) = e^{i\zeta(t)}\phi(t); \qquad e^{i\zeta(T)} = e^{i\zeta(0)}. \tag{4.72b}$$

Since $\psi(t)$, $\tilde{\psi}(t)$ and any $\phi(t)$ are lifts of the same closed curve $\Lambda(t)$, there must exist a phase factor $\omega(t)$ such that

$$\tilde{\psi}(t) = \omega(t)\phi(t). \tag{4.73}$$

We now calculate $\omega(t)$. Taking the derivative of (4.73), we get,

$$\frac{d\tilde{\psi}(t)}{dt} = \frac{d\omega(t)}{dt}\phi(t) + \omega(t)\frac{\phi(t)}{dt}.$$

Taking the scalar product of this with $\tilde{\psi}(t)$, we obtain

$$\omega^*(t)\frac{d\omega(t)}{dt} + |\omega(t)|^2\left(\phi(t), \frac{d\phi(t)}{dt}\right) = \left(\tilde{\psi}(t), \frac{d\tilde{\psi}(t)}{dt}\right) = 0,$$

where the last equality holds because of (4.70). Thus

$$\frac{1}{\omega(t)}\frac{d\omega(t)}{dt} = -\left(\phi(t), \frac{d\phi(t)}{dt}\right).$$

Integrating this we obtain

$$\frac{\omega(t)}{\omega(0)} = e^{-\int_0^t (\phi(t'), d\phi(t')/dt')dt'}. \tag{4.74}$$

This phase factor we call $e^{i\gamma(t)}$ and write (4.73) as

$$\tilde{\psi}(t) = \omega(0)e^{i\int_0^t i(\phi(t'), d\phi(t')/dt')dt'}\phi(t) \equiv \omega(0)e^{i\gamma(t)}\phi(t). \tag{4.75}$$

If we choose the arbitrary constant phase $\omega(0)$ again such that $\tilde{\psi}(0) = \psi(0) = \phi(0)$, i.e., choose $\omega(0) = 1$, then we have

---

[35] The method used in Sections 4.1 and 4.2 can be generalized in a straightforward way for Hamiltonians that are linear functions of generators of any Lie algebra. For more details and for other calculational methods of the closed lifts, see D. J. Moore, *Physics Report* **240**, No. 1 (1992) sects. 3, 4, and 5 and references thereof.

$$\tilde{\psi}(t) = e^{i\gamma(t)}\phi(t), \tag{4.76}$$

and with (4.67)

$$\psi(t) = e^{-i/\hbar \int_0^t (\psi(t'), h(t')\psi(t'))dt'} e^{i\gamma(t)}\phi(t). \tag{4.77}$$

This is the general relation between the solution of the Schrödinger equation $\psi(t)$ and the closed lift $\phi(t)$. The equation (2.47) is the adiabatic approximation of this, and (4.65) is the special case of this for the spinning quantum system in the precessing magnetic field. For the closed path $\mathscr{C}$ in $\mathscr{P}(\mathscr{H})$ the phase angle is (according (4.75))

$$\gamma(T) = \gamma(\mathscr{C}) = \int_0^T i\langle\phi(t)|\frac{d}{dt}|\phi(t)\rangle \, dt = \oint i\langle\phi|d|\phi\rangle \bmod 2\pi \tag{4.78}$$

where ϕ is any of the closed lifts of $\mathscr{C}$. It is independent of the choice of the parameterization (the speed with which $\phi(t)$ traverses its closed path), gauge-invariant, independent of the choice of the Hamiltonian as long as these Hamiltonians describe the same closed path $\mathscr{C}$ in $\mathscr{P}(\mathscr{H})$. It depends upon the closed curve $\mathscr{C}$. It also depends upon the class of Hamiltonians that are connected to each other by the transformation (4.71), but only to the extent that the possible curves $\mathscr{C}$ are determined by $h(t)$. It can thus be considered a "geometric" property of the curve $\mathscr{C}$ in the space of physical states $\mathscr{P}(\mathscr{H})$. Thus the name geometric phase.

The one-form

$$A^\phi = i\langle\phi|d|\phi\rangle \tag{4.79}$$

transforms under a gauge transformation (4.72b) as (Problem 16):

$$A^\phi \to A^{\phi'} = A^\phi - d\zeta. \tag{4.80}$$

This one-form is called the (nonadiabatic) connection form. The formula (4.79) for the connection was derived from the fact that $\tilde{\psi}(t)$ was the horizontal lift fulfilling (4.70), which in turn was uniquely determined by its definition (4.67). This again was the only possible definition of a lift that depends only on the physics of the problem (and not on the arbitrary choice of the Hamiltonian within the class connected by the transformation (4.71)) and which is uniquely determined by the closed curve $\mathscr{C}$ in the space of physical states. The horizontal lift $\tilde{\psi}$ and the connection A of (4.79) are therefore uniquely determined by the requirement that they be a property of the physical quantities $\Lambda(t)$ only.

From (4.76) and (4.72a) it follows immediately that

$$\tilde{\psi}(T) = e^{i\gamma(\mathscr{C})}\psi(0). \tag{4.81}$$

Using (4.67) one obtains then for the cyclic evolution of a state vector:

$$\psi(T) = e^{-i/\hbar \int_0^T \langle\psi(t')|h(t')|\psi(t')\rangle dt'} e^{i\gamma(\mathscr{C})}\psi(0). \tag{4.82}$$

This is the general relation of which (2.62) is the adiabatic approximation.

As the physics is not in ψ (which depends on some arbitrary choice of the Hamiltonian) but in $\tilde{\psi}$ (which is uniquely determined by the physical state

$|\psi(t)\rangle\langle\psi(t)|)$, the so-called geometrical phase $\gamma(\mathscr{C})$ is really the phase that is of physical importance.

Problems

1. Show that the Schrödinger equation $i\hbar\dfrac{d\psi(t)}{dt} = h(t)\psi(t)$ with time-dependent Hamiltonian $h(t)$ is equivalent to the equation $i\hbar\dfrac{dU^{\dagger}(t)}{dt} = h(t)U^{\dagger}(t)$ where $U^{\dagger}(t)$ is the time development operator and $\psi(t)$ is defined in terms of $\psi(0)$ by $\psi(t) = U^{\dagger}(t)\psi(0)$ (analog of (XII 1.35)).

2. Show that if the time development operator $U^{\dagger}(t)$ is invertible, then a time-dependent Hamiltonian may be given in terms of $U^{\dagger}$ by $h(t) = i\hbar\dfrac{dU^{\dagger}}{dt}\cdot U^{\dagger-1}$. Show that $h(t)$ is Hermitian if $U^{\dagger}(t)$ is unitary.

3. Show that if $[h(t), h(t')] = 0$ for all t, t', then $U^{\dagger}(t)$ that fulfills the Schrödinger equation $i\hbar\dfrac{dU^{\dagger}(t)}{dt} = h(t)U^{\dagger}(t)$ can be given in terms of $h(t)$ by

$$U^{\dagger}(t) = e^{-i/\hbar\int_0^t dt' h(t')}.$$

4. Show that as a consequence of the normalization condition for the eigenvectors $|n; R\rangle$ of $h(R)$, the quantity $\langle n; R|\nabla_R|n; R\rangle$ is purely imaginary and the Berry phase angle γ_n is real.

5. Show that for the eigenvectors $|n; R\rangle$ defined by

$$h(R)|n; R\rangle = E_n(R)|n; R\rangle; \qquad \langle n; R|m; R\rangle = \delta_{nm}$$

the following relation holds for $m \neq n$:

$$\langle m; R|\nabla|n; R\rangle = \langle m; R|(\nabla h(R))|n; R\rangle\frac{1}{E_n(R) - E_m(R)}.$$

6. (a) Show that the rotation $\mathscr{R}(\theta, \varphi) = \mathscr{R}_3(\varphi)\mathscr{R}_2(\theta)\mathscr{R}_3(-\varphi)$ transforms the unit vector e_3 in the direction of the north pole into the unit vector $\hat{R} = \hat{R}(\theta, \varphi) = (\sin\theta\cos\varphi, \sin\theta\sin\varphi, \cos\theta)$. $\mathscr{R}_3(\varphi)$ is a rotation around e_3 by an angle φ and $\mathscr{R}_2(\theta)$ is a rotation around e_2 by an angle θ.

   (b) What is the difference between $\mathscr{R}(\theta, \varphi)$ acting upon e_3 and $\mathscr{R}_3(\varphi)\mathscr{R}_2(\theta)$ acting upon e_3.

   (c) Find a rotation $\mathscr{R}'(\theta, \varphi)$ that has the following properties
   i.   It transforms e_3 into the vector $\hat{R}(\theta, \varphi)$.
   ii. The rotation $\mathscr{R}'(\theta = \pi, \varphi)$ is independent of φ.

7. Show that the vector

$$|k; \theta = \pi, \varphi\rangle' = e^{i(\varphi/\hbar)J_3}e^{i(\pi/\hbar)J_2}e^{i(\varphi/\hbar)J_3}|k; 00\rangle$$

is an eigenvector of $\hat{R}(\theta = \pi, \varphi)\cdot J$ and of J_3. Calculate the eigenvalues of these two operators.

8. Using the standard formulas of vector analysis calculate $F = curl\ A$ when A is

given by

$$A = \mathbf{e}_\varphi \frac{k(\cos\theta + 1)}{r\sin\theta}$$

where $\mathbf{e}_\varphi$ is the normalized unit vector along the φ-direction.

9. Verify the relation

$$e^{iB}Ae^{-iB} = A + i[B, A] + \frac{i^2}{2!}[B, [B, A]] + \cdots$$

where B and A are linear operators.
Hint: Use the Taylor series for the function $F(x) = e^{ixB}Ae^{-ixB}$.

10. Find a Hamiltonian $h(t)$ that has nonstationary $(|\psi(t)\rangle\langle\psi(t)| \neq |\psi(0)\rangle\langle\psi(0)|)$ solutions of the Schrödinger equation $i\hbar\frac{d}{dt}\psi(t) = h(t)\psi(t)$, which also obey the adiabaticity condition $|\psi(t)\rangle\langle\psi(t)| = |n; R(t)\rangle\langle n; R(t)|$, where $h(t)|n; R(t)\rangle = \epsilon_n(t)|n; R(t)\rangle$, and $\psi(0) = |n; R(0)\rangle$.

11. Show that the vector

$$\tilde\psi(t) = \exp\left(\frac{i}{\hbar}\int_0^t \langle\psi(t')|h(t')|\psi(t')\rangle\, dt'\right)\psi(t)$$

obeys the equation

$$i\hbar\frac{d}{dt}\tilde\psi(t) = (h(t) - \langle\psi(t)|h(t)|\psi(t)\rangle I)\tilde\psi(t)$$

if $\psi(t)$ obeys the Schrödinger equation.

12. Use the Schrödinger equation for time-dependent Hamiltonian postulated in (2.13) as one of the basic assumptions of quantum mechanics to derive the time evolution equation in differential form for the pure state $W(t) = |\psi(t)\rangle\langle\psi(t)|$ and compare the result with Equation (XII 1.33) for time-independent Hamiltonians.

13. Show that the closed path $\mathscr{C}: t \to |\psi(t)\rangle\langle\psi(t)|$ in the space of physical states $\mathscr{P}(\mathscr{H})$ generated by a Hamiltonian $h(t)$ is not affected by a change of the Hamiltonian $h(t) \to h'(t) = h(t) - \kappa(t)I$ where $\kappa(t) \in \mathbb{R}$.

14. Show that the dynamical lift of the closed path $\mathscr{C}$ in $\mathscr{P}(\mathscr{H})$ changes under the substitution $h(t) \to h'(t) - \kappa(t)I$ and calculate the phase difference between the dynamical lift $\psi(t)$ belonging to $h(t)$ and the dynamical lift $\psi'(t)$ belonging to $h'(t)$.

15. Show that the horizontal lift $\tilde C$ of a closed path $\mathscr{C}$ in $\mathscr{P}(\mathscr{H})$, defined in terms of the dynamical lift $\psi(t)$ by

$$\tilde\psi(t) = \exp\left(\frac{i}{\hbar}\int_0^t \langle\psi(t')|h(t')|\psi(t')\rangle\, dt'\right)\psi(t),$$

is not affected by the substitution

$$h(t) \to h'(t) = h(t) - \kappa(t)I.$$

16. Find the transformation of the connection one-form $A^{\phi_k} \equiv i\langle\phi_k|d|\phi_k\rangle$ when the section ϕ_k undergoes the phase transformation $\phi_k(t) \to \phi'_k(t) = e^{i\zeta_k(t)}\phi_k(t)$.

A Quantum Physical System in a Quantum Environment—The Gauge Theory of Molecular Physics

XXIII.1 Introduction

In the preceeding chapter we considered a complicated physical system that consisted of a quantum physical system (spinning particle with magnetic moment) and a classical physical system (external time-dependent magnetic field); we called the latter environment. Now we want to consider a complicated physical system that is entirely quantum mechanical. To understand it, we will divide it into simpler parts, which are again quantum systems. A natural division for a large class of quantum systems is the separation into a fast-moving part and a slow-moving part. The slowly moving part usually consists of the collective motion of the quantum system as a whole. The paradigm of this procedure is the Born-Oppenheimer method in molecular physics. The swiftly moving electron or electrons correspond to the quantum system of the previous chapter and the slowly moving nuclei, which determine the collective motion of the molecule as a whole, take the place of the quantum mechanical environment.

The conventional Born-Oppenheimer[1] approximation used an assumption that did not allow for any of the effects connected with the Berry phase. Though most of the structure of molecular quantum mechanics based on the Born-Oppenheimer approximation remains valid today, there are molecular systems that exhibit geometric phase effects. Indeed, it was in molecular physics where the effect of the geometric phase was first noticed as sign

---

[1] M. Born and J. Oppenheimer, *Ann. Phys.* **84**, 457 (1927).

ambiguities of the electronic wave function.[2] If one relaxes the conventional assumption, the Born-Oppenheimer method leads directly to the Berry vector potential.[3] In the following, we give the general quantum mechanical derivation, which also introduces the gauge theoretical framework that is the basis of the new Born-Oppenheimer procedure.

XXIII.2 The Hamiltonian of the Diatomic Molecule

In this section we give a brief introduction to the structure of molecules. This leads to the standard form of the Hamiltonian that is the starting point of the Born-Oppenheimer method. In order to avoid unnecessary complications, we want to restrict ourselves here to diatomic molecules. In the classical constituent picture the diatomic molecule consists of two nuclei and numerous electrons, all interacting with each other; the potential energy is given by the sum of the Coulomb potentials between each pair of particles. We use r and p or r_i, p_i $(i = 1, 2, \ldots N_e)$ to denote the electronic coordinates and momenta and R, P or R_j, P_j $(j = 1, 2)$ to denote the coordinates and momenta of the nuclei. We let T and U denote kinetic energy and potential energy, respectively; the energy E for a diatomic molecule has the form

$$E = T_{\text{nuclei}} + T_{\text{electrons}} + U_{\text{nuclei}} + U_{\text{nuclei-electrons}} + U_{\text{electrons}}. \qquad (2.1)$$

In (2.1) the first two terms are, respectively, the kinetic energies of the nuclei and electrons,

$$T_{\text{nuclei}} = \frac{1}{2m_1} P_1^2 + \frac{1}{2m_2} P_2^2,$$

$$T_{\text{electrons}} = \frac{1}{2m_e} \sum_{i=1}^{N_e} p_i^2. \qquad (2.2)$$

The three potential energy terms in (2.1) represent, respectively, the Coulomb potential energy due to the interaction between the two nuclei, the Coulomb potential energy resulting from the interaction among the nuclei and electrons, and the Coulomb potential energy arising from the interaction among the electrons. The three potential energy terms are as follows:

$$U_{\text{nuclei}} = \frac{k(Z_1 e)(Z_2 e)}{|R_1 - R_2|}, \qquad (2.3a)$$

$$U_{\text{nuclei-electron}} = \sum_{i=1}^{N_e} \sum_{j=1}^{2} \frac{k(Z_j e)(-e)}{|r_{ei} - R_j|}, \qquad (2.3b)$$

[2] G. Herzberg and H. C. Longuet-Higgins, Discuss. *Faraday Soc.* **35**, 77 (1963); H. C. Longuet-Higgins, *Proc. Roy. Soc. London* Ser. **A344**, 147 (1975).

[3] C. Mead and D. Truhlar, *J. Chem. Phys.* **70**, 2284 (1978); C. A. Mead, *Chem. Phys.* **49**, 23, 33 (1980).

$$U_{\text{electron}} = \sum_{j=i+1}^{N_e} \sum_{i=1}^{N_e} \frac{k(-e)(-e)}{|r_{ei} - r_{ej}|}. \tag{2.3c}$$

The charges on the two nuclei are $Z_1 e$ and $Z_2 e$, respectively.

When the correspondence principle is employed to obtain the Hamiltonian H from the classical energy E of (2.1) by replacing the classical observables with operators, the resulting eigenvalue equation

$$H|\psi^E\rangle = E|\psi^E\rangle \tag{2.4}$$

or its position re$_1^{\cdot}$ $_2$sentation, the $N = 2 + N_e$ body Schrödinger equation

$$\langle R_1, R_2, r_1 \ldots r_{N_e}|H|\psi^E\rangle = E\langle R_1, \ldots r_{N_e}|\psi^E\rangle, \tag{2.4'}$$

is very complicated and cannot be solved exactly.

To arrive at a simpler picture (which can also provide some guidance for suitable approximation of the quantum N-body problem (2.4)) we now consider not the constituents but the types of motion that are occurring in the molecule. The first step in this process is to consider the motion of the electrons and nuclei separately because they differ greatly from each other. The light, swiftly moving electrons whirl around the much heavier nuclei which, from the point of view of the electrons, are almost stationary. Thus, to a good approximation, the electrons move in the presence of two almost stationary nuclei, and their system is described approximately by:

$$E_{\text{electrons}} \cong T_{\text{electrons}} + U_{\text{nuclei-electrons}} + U_{\text{electrons}} \tag{2.5}$$

The explicit expressions for the potential energy terms are given in (2.3) and contain the slowly changing positions R_j of the nuclei as parameters.

Since the light electrons instantaneously follow the motion of the heavy nuclei, the slowly changing nuclear variables are essentially identical with the variables for the molecule as a whole. The variables that describe the slow motion of the molecule as a whole are called the collective variables.

Often, what is of interest is the electronic spectra caused by the excitation of *one* electron, while the remaining $N_e - 1$ electrons remain in the ground state. Then it is possible to go to a further simplified picture in which the least tightly bound electron moves in a potential $U(r)$. Here r is the coordinate of the single excited electron, and $U(r)$ is the potential resulting from the smeared-out negative charge distribution of the other $N_e - 1$ electrons and the positive charge distribution of the two almost stationary nuclei. It also depends parametrically on the relative coordinates of the nuclei R: $U(r) = U(r; R)$ and does not have any of the forms in (2.3). It can be guessed only approximately. The energy of the electron for this model, which (except for an additive constant that represents the energy of the remaining $N_e - 1$ electrons) approximates (2.5), is

$$E_e = \frac{1}{2m_e}p_e^2 + U(r; R). \tag{2.6}$$

The quantum analog of (2.6), in which the p_i and r_i are operators fulfilling the

Heisenberg commutation relations

$$[\hat{p}_i, \hat{r}_j] = \frac{\hbar}{i} \delta_{ij} 1,$$

and in which the position of the nuclei R is considered as a parameter, describes the electronic energy spectra of molecules with a single excited electron. These energy spectra are similar to electronic spectra of atoms. The energy value and the state of this quantum system (one electron) are labeled by the quantum number n and some other quantum numbers. If the spin of the electron is also taken into consideration (spin orbit coupling), then (2.6) will contain an additional term, which for the time being we can imagine as being contained in $U(r; R)$.

Now let us consider the molecule from the point of view of the nuclei: The (classical) electrons, moving about in complicated orbits, create a smeared-out charge distribution. This smeared-out charge distribution gives rise to a potential $E_n^e(R)$ of the orbiting electrons, which is an average of $U_{\text{nuclei-electrons}}$ of (2.3b), with the average being taken over all possible electron orbits and depending on the quantum number of the electronic state. The average potential $U_n(R)$ experienced by the nuclei is thus

$$U_n(R) = E_n^e(R) + U_{\text{nuclei}}(R), \tag{2.7}$$

where R is the relative position of the nuclei $R = R_2 - R_1$.

The second term $U_{\text{nuclei}}(R)$ is the Coulomb repulsion term (2.3a). Thus if it were not for the potential $E_n^e(R)$ created by the whirling electrons, then the two positively charged nuclei would repel each other. Depending on the net effect of the electron's motion, the sum $U_n(R)$ can have various shapes for different molecules and different electronic states n. If $U_n(R)$ decreases uniformly with increasing R, one has repulsion between the two atoms and no stable molecules will result. In many cases, especially in the electronic ground states, the potential energy curve has a minimum at an equilibrium nuclear separation, as shown in Fig. 2.1. As a consequence, the nuclei remain bound to each other, as long as their energy does not exceed the dissociation energy D_e. If one expands $U_n(R)$ about R_e (see Section 7), one obtains as the leading term the parabolic potential curve depicted by the dashed line in Fig. 2.1. Near $R = x_e$—called the harmonic region—this curve is a good qualitative approximation. As a consequence of this shape for the potential energy, the internuclear axis will perform harmonic oscillations.

The energy for the relative motion of the nuclei is therefore

$$E_{\text{nuclei}} = T_{\text{nuclei}} + U_n(R) = T_{\text{nuclei}} + E_n^e(R) + U_{\text{nuclei}}(R) \tag{2.8}$$

where

$$T_{\text{nuclei}} = \frac{P^2}{2\mu}; \quad \mu = \frac{m_1 m_2}{m_1 + m_2} = \text{reduced mass}; \quad P = \frac{m_2 P_1 - m_1 P_2}{m_1 + m_2} \tag{2.9}$$

and where $U_n(R)$ depends on the orbiting electrons through the potential $E_n^e(R)$. (Each energy eigenstate of electron Hamiltonian given by (2.5) has a

Figure 2.1 Typical potential energy curve for bound states of a diatomic molecule. It shows the classical potential energy of the nuclei as a function of the internuclear distance. This function is the same as the induced scalar potential $\varepsilon_n(x)(x = R)$, in the Schrödinger equation for the collective (nuclear) motion, Equation (3.25). As the consequence of its approximate parabolic shape, the collective energy levels are the vibrational excitations $v = 0, 1, 2, 3. \ldots$ As discussed in Section 5.7, each of these vibrational levels splits into doubly degenerate rotational bands (not shown here).

charge density that creates a different potential. Thus $U_n(R)$ depends on the quantum numbers n of the electronic states.)

This picture, which separates the electronic motion from the motion of the nuclei, is the essence of the Born-Oppenheimer approximation of molecular physics. The expression (2.6) (or in the general case (2.5)) leads to the electronic spectra and (2.8) leads to the energy spectra associated with the collective motions of the molecule (which is essentially identical to the motion of the nuclei, because the much lighter electrons follow the nuclei almost instantaneously).

There are three types of collective motions:

1. Translation of the center of mass of the molecule (which is essentially the center of mass of the nuclei).
2. Rotation of the molecule about an axis through its center of mass.
3. Vibration along the internuclear axis.

The translation of the center of mass will lead to a continuum of energy values similar to the translational motion of a free mass point. We will ignore the translational energy (as in (2.8)) and consider the diatomic molecule from a reference system in which the center of mass is at rest.

The pure rotational motion of a diatomic molecule is obtained if the two nuclei remain at a fixed distance R_0 from each other and if the electrons stay in a state with fixed electronic quantum numbers. (Classically, we would say the electrons are in a stable configuration.) Then the classical picture is that of a rigid dumbbell, as shown in Fig. III.2.1, which rotates about a direction perpendicular to R_0. The energy spectra of molecules caused by rotational motion are called rotational spectra or rotational bands. They are discussed in Section III.4.

In order for the distance between the nuclei to remain fixed at R_0, the potential energy $U_n(R)$ must have a minimum at $R = R_0$, implying that R_0 is the equilibrium separation between the two nuclei.

The total energy of the molecule is the sum of the energy of the collective motion (2.8) and the electronic motion (2.6) (or (2.5) in the more general situation).

$$E = E_{\text{nuclei}} + E_e = T_{\text{nuclei}} + U_n(R) + E_e \qquad (2.10)$$

This is—except for an additive constant including the center of mass energy— the same or approximately the same as (2.1).

From (2.10) we make the transition to the quantum theory by replacing the classical observables by quantum mechanical operators. For the quantum mechanical Hamilton operator, we then obtain:

$$H = \frac{P^2}{2\mu} + h(\hat{p}, \hat{r}, R) \qquad (2.11)$$

where

$$h(\hat{p}, \hat{r}, R) = \frac{\hat{p}^2}{2m_e} + V(R, \hat{r}) = H_e + U(R). \qquad (2.12)$$

$U(R)$ is the operator obtained from $U_n(R)$ and H_e is the operator obtained from E_e of (2.6) (or E_{electron} of (2.5)). $V(R, \hat{r})$, obtained from the combination of $U(\hat{r}, R)$ and $U_n(R)$, is a complicated function of the operators R and $\hat{r}$ (and possibly some other operators suppressed here, such as spin). The operators $\hat{p}$ and $\hat{r}$ describe the fast (electronic) degrees of freedom and are called the fast variables. The operators P and R describe the slow degrees of freedom of the nuclei or the molecule as a whole. They are called the slow (collective) variables. In the quantum theory, the fast and slow variables are operators, and $h(\hat{p}, \hat{r}, R)$ describes a quantum system in a quantum environment.

In an intermediate semiclassical theory one can consider the slow variables as slowly changing classical parameters. Then (2.12) is a parameter-dependent (with parameter R) Hamiltonian describing a quantum system (electron) in a slowly changing classical environment (molecule as a whole). If the motion of the (classical) nuclei is slow enough then this may not change the quantum number of the electron state, and we have an adiabatic time evolution. To this intermediate theory the results of the previous chapter apply. The eigenvectors and eigenvalues of (2.12) depend on R as in (XXII.2.3). The evolution is cyclic (if the internuclear distance $R(t)$ returns after a period T to its original position) and Berry connection (XXII.2.44) and Berry phase (XXII.2.51) are obtained in terms of the eigenvectors $|n; R\rangle$ of the fast electronic Hamiltonian $h(R)$ of (2.12).

In the quantum theory all observables are operators in the space of physical states. The physical system is a combination of two systems: the system of the fast or electronic motions and the system of the slow or nuclear motions, which—according to the preceding—gives the collective motions of the molecule as a whole. Denoting the space for the fast motion by $\mathscr{H}^{\text{fast}}$ and the space for the slow motion by $\mathscr{H}^{\text{slow}}$, the space of physical states of the quantum mechanical molecule is, according to the basic assumption IVa of Section III.5, the direct product space

$$\mathscr{H} = \mathscr{H}^{\text{slow}} \otimes \mathscr{H}^{\text{fast}}. \tag{2.13}$$

The spaces $\mathscr{H}^{\text{slow}}$ and $\mathscr{H}^{\text{fast}}$ are in general infinite dimensional but often one obtains a good approximation if one restricts oneself to a finite-dimensional subspace of $\mathscr{H}^{\text{fast}}$. We will call $\mathscr{N} = \dim \mathscr{H}^{\text{fast}}$, which can be one, finite, or infinite. The Hamiltonians (2.11) and (2.12) and the observables $\hat{p}$, $\hat{r}$, P, R, and the other observables, such as electron spin, which we have not explicitly mentioned, are operators in $\mathscr{H}$.

XXIII.3 The Born-Oppenheimer Method

Now that the problem has been set up, we can proceed to the solution, which is to solve the eigenvalue problem

$$H\psi^E = E\psi^E \tag{3.1}$$

where H is the Hamiltonian (2.11). This can, of course, only be done approximately. In the Born-Oppenheimer approximation[1] one first solves the eigenvalue problem for the fast Hamiltonian (2.12) and obtains the eigenvalues $\varepsilon_n(x)$ as functions of the eigenvalues x_i of the operator R_i (slow position, i.e., relative position of the nuclei).[4] In order to obtain these $\varepsilon_n(x)$, which we will not calculate explicitly, we introduce the following complete systems of basis vectors of the space $\mathscr{H}$ in (2.13).

The $|x_i, r_i \ldots\rangle$ are generalized eigenvectors of the operators R_i and $\hat{r}_i$,

$$R_i|x_i, r_i \ldots\rangle = x_i|x_i, r_i \ldots\rangle \tag{3.2a}$$

$$\hat{r}_i|x_i, r_i \ldots\rangle = r_i|x_i, r_i \ldots\rangle. \tag{3.2b}$$

These eigenvectors are the standard (δ-function normalized) position eigenvectors.

The basis vectors $|x_i, \ldots n\rangle$ are generalized eigenvectors of the operators R_i and $h(R_i)$. They fulfill the equations,

$$R_i|x_i, \ldots n\rangle = x_i|x_i, \ldots n\rangle \tag{3.3a}$$

$$h(R)|x_i, \ldots n\rangle = \varepsilon_n(x)|x_i, \ldots n\rangle, \tag{3.3b}$$

and are normalized in the standard way,

$$\langle x_i', \ldots n'|x_j, \ldots n\rangle = \delta^3(x_i' - x_i)\delta_{n'n}. \tag{3.4}$$

---

[4] In the conventional Born-Oppenheimer approach R itself is the classical parameter; here we treat immediately the fully quantum mechanical version.

Notice that $h(R)$ is an operator that acts in both $\mathcal{H}^{\text{fast}}$ and $\mathcal{H}^{\text{slow}}$. The existence of these simultaneous eigenvectors follows from the fact that $[R, h(R)] = 0$. The label n stands for the set of electronic quantum numbers, all or part of which label the electronic energy value ε_n. The dots stand for the other quantum numbers that are needed in addition to x_i and n to label the basis vectors completely. The eigenvalue $\varepsilon_n(x)$ is obtained by solving (3.3b) for each value of x.

Each vector of the basis system in (3.3) can be expanded with respect to the basis system in (3.2) (we omit the ... from now on),

$$|x_i, n\rangle = \int d^3x' \int d^3r' |x_i', r_j'\rangle \langle x_i', r_j'|x_i, n\rangle. \tag{3.5}$$

The transition matrices between these two basis systems are given by

$$\langle x_i', r_j'|x_i, n\rangle = \delta^3(x_i' - x_i)\langle r_j'|n(x_i)\rangle. \tag{3.6}$$

This equation is a statement and a definition. It states that the x-dependence is given by the δ-function, and it defines $\langle r_j|n(x_i)\rangle \equiv \langle r_j|n\rangle_{x_i}$ (the reduced transition matrix), which for every r_j and n is a function of x_i. This reduced transition matrix defines a basis system $\{|n(x_i)\rangle; n = \text{all electronic quantum numbers}\}$ in $\mathcal{H}^{\text{fast}}$ for every value of x_i. With these basis vectors, the basis vectors (3.3b) of $\mathcal{H}$ can be represented as

$$|x, n\rangle = |x\rangle \otimes |n(x_i)\rangle, \tag{3.7a}$$

where according to (3.4),

$$\langle n(x)|m(x)\rangle = \delta_{nm} \tag{3.7b}$$

and where $|x\rangle = |x \ldots\rangle$ is the basis system of (generalized, δ-function normalized) eigenvectors of R for the space $\mathcal{H}^{\text{slow}}$. Comparing with the definition of the direct product (III.5.3), the basis system of the second factor of (3.7a) depends on the eigenvalue of the (generalized) basis vector of the first factor, because the quantum numbers x_i and n are, according to (3.3b), not independent of each other.

We can now define the operator $h(x)$ as the operator in $\mathcal{H}^{\text{fast}}$ obtained from the operator $h(R)$ in $\mathcal{H}^{\text{slow}} \otimes \mathcal{H}^{\text{fast}}$ by replacing the operator R by the number x. From (3.4) and (3.7a) it follows that

$$h(R)|x\rangle \otimes |n(x)\rangle = |x\rangle \otimes h(x)|n(x)\rangle$$
$$= \varepsilon_n(x)|x\rangle \otimes |n(x)\rangle. \tag{3.8}$$

Consequently, the operator $h(x)$, which depends upon the parameter x, has the eigenvector $|n(x)\rangle$ with eigenvalue $\varepsilon_n(x)$. Both of these also depend on the parameter[5] x,

$$h(x)|n(x)\rangle = \varepsilon_n(x)|n(x)\rangle. \tag{3.9}$$

---

[5] The value of the parameter x in both operators and vector must be the same. $h(x)$ applied to $|n(x')\rangle$, with $x' \neq x$, is a complicated expression obtained by expanding $|n(x')\rangle$ with respect to the basis system $|n(x)\rangle$ in $\mathcal{H}^{\text{fast}}$.

The symbol $n(x)$ in $|n(x)\rangle$ does not mean that the value of the quantum number changes as a function of x, but that the physical meaning of the quantum number changes, i.e., the fast observable with eigenvalue n is different for different values of x (see Eq. (XXII.3.7), where the eigenvalue k remains the same but the observable $\hat{R} \cdot J$ changes with the parameter $\hat{R}$).

Unlike the x and r of (3.2), the x and n of (3.3) are not independent quantum numbers. As a consequence of this, the matrix elements $\langle x, n|\mathcal{O}|\psi\rangle$ of operators $\mathcal{O}$ do not have the familiar form. In particular, whereas for the basis vectors (3.2) we have,

$$\langle x_i, r_j|P|\psi\rangle = \frac{1}{i}\nabla_x\langle x_i, r_i|\psi\rangle \tag{3.10}$$

for any state vector $|\psi\rangle$, for the basis vectors (3.3) a similar expression does not hold. This means that the following is in general *not* true:

$$\langle x_i, n|P|\psi\rangle = \frac{1}{i}\nabla_x\langle x_i, n|\psi\rangle. \tag{3.11}$$

Instead one derives for the matrix element of $P = P \otimes I^{\text{fast}}$:

$$\langle x, n|P|\psi^E\rangle = \langle x, n|P \sum_{n'} \int d^3x'|x', n'\rangle\langle x'n'|\psi^E\rangle$$

$$= (\langle x| \otimes \langle n(x)|)(P \otimes I^{\text{fast}}) \sum_{n'} \int d^3x'(|x'\rangle \otimes |n'(x')\rangle)\langle x', n'|\psi^E\rangle.$$

Here we have used (3.7a) and the fact that $|x, n\rangle$ is a complete system of basis vectors in $\mathcal{H} = \mathcal{H}^{\text{slow}} \otimes \mathcal{H}^{\text{fast}}$:

$$\sum_{n=1}^{\mathcal{N}} \int d^3x|x, n\rangle\langle x, n| = I.$$

The operator P acts only in $\mathcal{H}^{\text{slow}}$ and we obtain

$$\langle x, n|P|\psi^E\rangle = \langle n(x)| \sum_{n'} \int d^3x'|n'(x')\rangle\langle x|P|x'\rangle\langle x', n'|\psi^E\rangle.$$

The position representation of P in $\mathcal{H}^{\text{slow}}$, is (according to the three-dimensional version of (II.7.53) given by $\frac{1}{i}\nabla$. Thus we obtain[6]

---

[6] Precisely what we calculate is:

$$\int dx'\langle x|P|x'\rangle\varphi(x') = \int dx'\left(\frac{1}{i}\frac{d}{dx}\delta(x - x')\right)\varphi(x')$$

$$= \int dx'\frac{-1}{i}\delta'(x' - x)\varphi(x') = \frac{-1}{i}(-1)\varphi'(x)$$

where $\varphi(x)$ is a well-behaved function. In this derivation we have used results from section I.2 of *Generalized Functions*, Vol. 1, by I. M. Gelfand and G. E. Shilov.

$$\langle x, n|P|\psi^E\rangle = \langle n(x)| \int d^3x' \sum_{n'} |n'(x')\rangle \left[\frac{-1}{i}\nabla_{x'}\delta^3(x' - x)\right]\langle x', n'|\psi^E\rangle$$

$$= \frac{1}{i} \sum_{n'} \langle n(x)|\nabla_x|n'(x)\rangle\langle x, n'|\psi^E\rangle. \tag{3.12}$$

The differentiation ∇_x applies to both factors to the right of it: $|n'(x)\rangle$ and $\langle x, n'|\psi^E\rangle$.

If we execute the differentiation using the product rule we obtain

$$\langle x, n|P|\psi^E\rangle$$

$$= \sum_{n'} \left(\langle n(x)|n'(x)\rangle\frac{1}{i}\nabla\langle x, n'|\psi^E\rangle + \frac{1}{i}(\langle n(x)|\nabla|n'(x)\rangle)\langle x, n'|\psi^E\rangle\right)$$

$$= \sum_{n'} \left(\langle n(x)|n'(x)\rangle\frac{1}{i}\nabla\langle x, n'|\psi^E\rangle - i(\langle n(x)|\nabla|n'(x)\rangle)\langle x, n'|\psi^E\rangle\right).$$

Using the orthogonality (3.7b) we write this as

$$\langle x, n|P|\psi^E\rangle = \sum_{n'} \left(\frac{1}{i}\delta^{nn'}\nabla - A^{nn'}(x)\right)\langle x, n'|\psi^E\rangle$$

$$= \sum_{n'} \frac{1}{i}D^{nn'}\langle x, n'|\psi^E\rangle. \tag{3.13}$$

Thus if n is the eigenvalue of an x-dependent observable, the matrix element of the operator P for an arbitrary $\psi \in \mathscr{H}$ is *not* given by the derivative (3.11) but by the more complicated expression (3.13). The following quantities appeared naturally in the derivation of (3.13):

$$A^{nn'}(x) \equiv i\langle n(x)|\nabla|n'(x)\rangle \tag{3.14}$$

$$\frac{1}{i}D^{nn'} \equiv \frac{1}{i}\delta^{nn'}\nabla - A^{nn'}(x). \tag{3.15}$$

These are $\mathscr{N}$-dimensional quadratic matrices where $\mathscr{N} = \dim \mathscr{H}^{\text{fast}}$ (in the exact theory $\mathscr{N} = \infty$). The D^{nm} of (3.15) is called the gauge covariant derivative or just covariant derivative. The A^{nm} of (3.14) is called the (non-Abelian for $\mathscr{N} \neq 1$) gauge potential. The nomenclature has its origin in the gauge theory that underlies the new Born-Oppenheimer description (which does not use the conventional Born-Oppenheimer approximation). This will be discussed in detail in Sections 4 and 5.

The gauge potential (3.14) (for $\mathscr{N} = 1$) was first introduced by Mead et al.[3] It has the same form as the Berry vector potential that we defined in (XXII.2.44) and that we used in (XXII.3.33) for the quantum system in a classical environment. The fast quantum system in the molecule is the electron(s); the place of the slowly changing classical environmental parameters is taken in the quantum theory of the molecule by the eigenvalues of the observable R for the internuclear axis. In Section XXII.2 the adiabaticity assumption restricted us to one fixed value of the quantum number n, and we considered only the diagonal matrix elements. Here, the gauge potential is an $\mathcal{N} \times \mathcal{N}$ matrix where the value of $\mathcal{N}$ is determined by the required approximation. The Mead-Berry connection (3.14) describes many observable effects, which could not be explained in the conventional approximation.

In the same way one derived (3.13), one can derive for the matrix element of $\boldsymbol{P}^2 = \boldsymbol{P}^2 \otimes I^{\text{fast}}$

$$\langle x, n | \boldsymbol{P}^2 | \psi^E \rangle = -\sum_{n'} \langle n(x) | \nabla_x^2 | n'(x) \rangle \langle x, n' | \psi^E \rangle$$

$$= \sum_{n'} \sum_{l} \left(\frac{1}{i} \delta^{nl} \nabla_x - A^{nl} \right) \cdot \left(\frac{1}{i} \delta^{ln'} \nabla_x - A^{ln'} \right) \langle x, n' | \psi^E \rangle. \quad (3.16)$$

If we compare (3.11) and (3.13) we see the difference between the usual or naively expected representations of the momentum operator and the new representation of the momentum operator. The difference arises from the x-dependence of the eigenvectors $|n(x)\rangle$. If these basis vectors are independent of x, then the $A^{nn'}(x)$ are zero and the momentum operator is represented by the ordinary derivative. If the dependence of these eigenvectors on the slow variable x is not neglected then the momentum operator is represented by the covariant derivative. The standard assumption of the ordinary Born-Oppenheimer approximation is that $|n(x)\rangle$ changes little as a function of x so that the derivative of it can be neglected.

We will now write the two eigenvalue equations (3.1) and (3.9) in the form of differential equations.

The eigenvector $|\psi^E\rangle$ of the molecular Hamiltonian in (3.1) can be expanded with respect to the basis system (3.3):

$$|\psi^E\rangle = \sum_n \int d^3x' |x', n\rangle \langle x', n | \psi^E \rangle. \quad (3.17)$$

The sum extends over all values of the electronic quantum numbers n, which are in general infinite, but often a small number of $\mathcal{N}$ terms in (3.17) provides a good approximation (as mentioned in the paragraph following equation (2.13)).

The wave function of the energy eigenstate $|\psi^E\rangle$ is the scalar product of it with the basis vectors (3.1),

$$\psi^E(x, r) = \langle x, r | \psi^E \rangle. \quad (3.18)$$

We define the following quantities:

$$\psi_n^E(x) \equiv \langle x, n | \psi^E \rangle \quad (3.19)$$

and

$$\phi_n(x, r) \equiv \langle r | n(x) \rangle. \quad (3.20)$$

If we take the scalar product of (3.17) with the basis vectors $|x, r\rangle$ and use (3.6), we obtain

$$\psi^E(x, r) = \sum_n \phi_n(x, r) \psi_n^E(x). \quad (3.21)$$

The $\phi_n(x, r)$ are called the electronic eigenfunctions. The $\psi_n^E(x)$ are called the nuclear wave function, though they are not wave functions in the usual sense in the slow space $\mathcal{H}^{\text{slow}}$. (Usually by wave function in the position representation in the space $\mathcal{H}^{\text{slow}}$, one means the component $\langle x|\psi\rangle$ of the vector $\psi \in \mathcal{H}^{\text{slow}}$ with respect to the basis system $|x\rangle$, whereas the $\psi_n^E(x)$ are defined by (3.19).)

If we take the scalar product of (3.9) with the position eigenkets $|r\rangle$ of $\mathcal{H}^{\text{fast}}$ and use (2.12) we obtain

$$\langle r|h(x)|n(x)\rangle = \langle r|[\frac{\hat{p}^2}{2m_e} + V(x, \hat{r})]|n(x)\rangle$$

$$= \left[-\frac{\nabla_r^2}{2m_e} + V(x, r)\right]\langle r|n(x)\rangle = \varepsilon_n(x)\langle r|n(x)\rangle. \quad (3.22)$$

Using the notation (3.20) we can write this as:

$$\left[-\frac{\nabla_r^2}{2m_e} + V(x, r)\right]\phi_n(x, r) = \varepsilon_n(x)\phi_n(x, r). \quad (3.23)$$

This is the Schrödinger equation for the electronic wave function, which depends parametrically on the eigenvalue x of the slow observable R.

If we take the scalar product of (3.1) with the basis vectors $|x, n\rangle$ of (3.16) using (2.11) and (3.3b), we find

$$\langle x, n|\frac{P^2}{2\mu}|\psi^E\rangle + \varepsilon_n(x)\psi_n^E(x) = E\psi_n^E(x). \quad (3.24)$$

The first matrix element in this equation is given by (3.16), and we obtain:

$$\sum_m \left[\frac{1}{2\mu}\sum_l(-i\delta^{nl}\nabla_x - A^{nl})\cdot(-i\delta^{lm}\nabla_x - A^{lm}) + \varepsilon_n(x)\delta^{nm}\right]\psi_m^E(x) = E\psi_n^E(x).$$

$$(3.25)$$

This is a system of an infinite number of coupled differential equations.

If in place of (3.13) we use the relation (3.11), which is in general not correct and is equivalent to the assumption that the basis vectors $|n(x)\rangle$ are independent of x, then we derive in place of (3.25) the following equation for every value n of the electronic quantum number:

$$\left[\frac{-1}{2\mu}\nabla_x^2 + \varepsilon_n(x)\right]\psi_n^E(x) = E\psi_n^E(x). \quad (3.26)$$

Equation (3.26) is the conventional Born-Oppenheimer approximation for the Schrödinger equation of the slow motion.

We shall first discuss the solution of the conventional Born-Oppenheimer problem, which uses (3.23) and (3.26).

One first solves Equation (3.23) for all values of the parameter x. This is done using various approximation methods and empirical assumptions.[7] In

[7] For example, A. C. Hurley, *Introduction to the Electron Theory of Small Molecules*, Academic Press, NY (1976).

this way the eigenvalues $\varepsilon_n(x)$ are obtained as functions of $x = |x|$. The function $\varepsilon_n(x)$ is then treated as a potential function in the Schrödinger equation (3.26). For this reason one calls $\varepsilon_n(x)$ the induced scalar potential. It is the potential that is induced by the electrons in the nth electronic state in the dynamical equation for the slow or collective motion. Comparing (2.12) with (2.10) we see that $\varepsilon_n(x)$ corresponds to the classical potential function $U_n(R) + E_e$. We expect to find as the solution of Equation (3.23) that $\varepsilon_n(x)$ is of the form given in Fig. 5.1. This function has a minimum at the position $x = x_e$, called the equilibrium position, and behaves like an anharmonic oscillator in the neighborhood of x_e.

We shall not discuss here the calculation of $\varepsilon_n(x)$ by solving (3.23) using various approximation methods but shall assume that this has been done and that $\varepsilon_n(x)$ is given to us for the values of the electronic quantum number n with which we are concerned. We shall then use this potential function $\varepsilon_n(x)$ for the solution of the Schrödinger equations (3.26) or (3.25) of the collective (which is essentially the same as the nuclear) motion of the molecule. We shall discuss more details in the example in the next section. Here we shall describe the general picture.

If we take (3.26) as the Schrödinger equation for the energy spectrum E of the collective motion (conventional Born-Oppenheimer approximation) then all these energy values belong to one electronic state with a definite value of the electronic quantum number such as $n = A$ or $n = B$, and the energy values E of the molecule are therefore labeled by this electronic quantum number $E_{n=A}$ or $E_{n=B}$. In addition there is the quantum number N, which distinguishes the different solutions of (3.26) for a fixed value of n. Thus the electronic energy level $E_{n=A}$ splits into sublevels which are the different solutions of Equation (3.26) for $n = A$. Such a splitting is shown in Fig. III.5.3 where N is the vibrational quantum number called n in Section III.5 (which is not to be mistaken for the electronic quantum number called n in this section) and the rotational quantum number called l in Section III.5. In this way one obtains the standard picture of the molecular spectra as shown in Fig. III.5.3: Each electronic state $n = A$, $n = B$... splits into vibrational excitations and then into rotational bands. As all eigenvalues of (3.26) belong to one electronic quantum number $n = A$ determined by the choice of $\varepsilon_A(x)$ (from (3.23)), there is only one electronic state contributing to the eigenvector ψ^E in (3.17); thus

$$\psi^E = \psi^{E_A} = \int d^3x' |x', n = A\rangle \langle x', A|\psi^{E_A}\rangle \qquad (3.27)$$

and all other expansion coefficients in (3.17) are zero

$$\langle x, n|\psi^E\rangle = 0 \qquad \text{for } n \neq A. \qquad (3.28)$$

In contrast to the conventional Born-Oppenheimer approximation (3.26) the exact[8] Schrödinger equation (3.25) is an infinite-dimensional matrix differential equation. In addition to the induced scalar potentials $\varepsilon_n(x)$, it contains

---

[8] It is exact if the $\varepsilon_n(x)$ are exactly known.

induced vector potentials $A^{nm}(x)$ that couple all electronic quantum numbers to one given energy value E (the sum in ((3.17) is infinite).

The system of coupled differential equations (3.25) is of no practical value unless the infinite sums can be replaced by finite sums with a small number of terms $\mathcal{N}$. This means that (3.17) and (3.21) can be approximated by a sum of $\mathcal{N}$ terms.[9] The simplest case is, of course, the case (3.27) with $\mathcal{N} = 1$. More common is the case that two or three electronic states contribute. Of special interest is the case where two electronic quantum numbers n_1 and n_2 with nonzero contribution to (3.17) and (3.21) correspond to the same potential curve $\varepsilon_{n_1}(x) = \varepsilon_{n_2}(x)$. In these approximations, which are often very good, (3.25) is the $\mathcal{N}$-dimensional matrix equation

$$\sum_m^{\mathcal{N}} \left[\sum_l^{\mathcal{N}} \frac{-1}{2\mu} D^{nl} \cdot D^{lm} + \tilde{\varepsilon}^{nm}(x) \right] \psi_m^E(x) = E \psi_n^E(x), \tag{3.29}$$

where

$$\tilde{\varepsilon}^{nm}(x) \equiv \varepsilon_n(x)\delta^{nm} + \frac{1}{2\mu} \sum_{l'=\mathcal{N}+1}^{\infty} A^{nl'} \cdot A^{l'm}. \tag{3.30}$$

The covariant derivatives D^{nl} in (3.29) and the Berry connections $A^{nl}(x)$ in (3.29) are $\mathcal{N}$-dimensional matrices defined again by (3.15) and (3.14) with n, $n' = 1, 2, \ldots \mathcal{N}$. The Schrödinger equation (3.29) defines the Hamiltonian that governs the motion of the slow system which, in the position representation of the slow system, has the form:

$$H_{\text{eff}}^{nm}(x) \equiv \langle n(x)|H|m(x)\rangle = \frac{-1}{2\mu} \sum_l D^{nl} \cdot D^{lm} + \tilde{\varepsilon}^{nm}(x). \tag{3.31}$$

It is called the effective Hamiltonian and is an $\mathcal{N}$-dimensional quadratic matrix. Its rows and columns are labeled by the fast quantum numbers and its elements are differential operators acting on the functions $\psi_n^E(x) = \langle x, n|\psi^E\rangle$. (Note that the $\psi_n^E(x)$ are not wave functions in the slow space $\mathcal{H}^{\text{slow}}$ by which one means the component $\langle x|\psi\rangle$ with respect to the basis system $|x\rangle$).

Equations (3.25) and (3.29) have been written in a form reminiscent of a gauge theory with A defined by (3.14) being the gauge potential and D defined by (3.15) being the gauge-covariant derivative. In the (infinite) matrix notation, Equation (3.29) becomes

$$\left[\frac{-1}{2\mu} D \cdot D + \tilde{\varepsilon}(x) \right] \psi^E(x) = E \psi^E(x) \tag{3.32}$$

where D are the matrices in (3.15), $\psi^E(x)$ is a column matrix whose rows are labeled by n, and $\tilde{\varepsilon}(x)$ is the matrix whose matrix elements are given by equation (3.30)). We will now show that the new Born-Oppenheimer description, which does not use the conventional Born-Oppenheimer approximation, leads in a natural way to a gauge theory.

---

[9] This is called the Born-Huang approximation; cf. footnote 7.

XXIII.4 Gauge Theories

In Equations (3.14) and (3.15), we introduced quantities that appear in every gauge theory, and the Schrödinger equations for the collective motion (3.29) and (3.32) have been written in a gauge-covariant way. The reason for this is that the Born-Oppenheimer description (if one does *not* use the conventional drastic Born-Oppenheimer approximation) leads in a natural way to a gauge theory. That the appearance of the Berry phase is a feature of an underlying gauge theory was already indicated in Section XXII.2 preceeding the definition (XXII.2.55). There we mentioned the analogy between the electromagnetic quantities of Maxwell's theory and the Berry vector potential (connection) and Berry gauge field (curvature). With the Born-Oppenheimer wave function for the collective motion we have now also introduced the matter field of a gauge theory. For $\mathcal{N} \neq 1$ we have a non-Abelian gauge theory. Before we show that the *new* Born-Oppenheimer description of molecules leads in a natural way to a gauge theory, we will present a brief review[10] of the general features of a gauge theory.

We start from the gauge theory of electromagnetism. The Hamiltonian for a particle of charge e in an electromagnetic field

$$E^{el} = -\nabla A_0^{el} - \frac{1}{c}\frac{\partial A^{el}}{\partial t}, \qquad B^{el} = \nabla \wedge A^{el} \tag{4.1}$$

is given by

$$H^{el} = \frac{1}{2m}\left(P - \frac{eA^{el}}{c}\right)^2 + eA_0^{el} \tag{4.2}$$

where A_0^{el} and A^{el} are the electromagnetic scalar and vector potentials, respectively. In place of the usual canonical momentum P, the Hamiltonian contains the covariant (also called gauge-covariant, kinematical, or mechanical) momentum

$$\Pi = P - \frac{e}{c}A^{el}, \tag{4.3}$$

which fulfills the commutation relation

$$[\Pi_i, \Pi_j] = i\hbar \frac{e}{c}\epsilon_{ijk}B_k^{el}. \tag{4.4}$$

The Schrödinger equation for the Hamiltonian (4.2) is

$$\left(i\hbar\frac{\partial}{\partial t} - eA_0^{el}\right)\psi = \frac{1}{2m}\left(\frac{\hbar}{i}\nabla - \frac{e}{c}A^{el}\right)^2\psi. \tag{4.5}$$

---

[10] Readers familiar with this subject can skip this section. For further studies consult, e.g., C. Quigg, *Gauge Theories of the Strong, Weak and Electromagnetic Interactions*, Benjamin/Cummings Publ. Co. (1983); L. O'Raifeartaigh, *Group Structure of Gauge Theories*, Cambridge (1986).

As is well known, the electric and magnetic fields E^{el} and B^{el} are unaltered if the potentials are transformed in the following way

$$A^{el} \to A^{el} + \nabla\alpha \equiv A'^{el} \equiv A^{(\alpha)el} \tag{4.6a}$$

$$A_0^{el} \to A_0^{el} - \frac{1}{c}\frac{\partial\alpha}{\partial t} \equiv A_0'^{el} \equiv A_0^{(\alpha)el} \tag{4.6b}$$

where $\alpha(x, t)$ is a well-behaved (continuous infinitely differentiable single-valued) function of x and t. The notation $A^{(\alpha)el}$, $A_0^{(\alpha)el}$ indicates that $A' = A^{(\alpha)}$ is the vector potential obtained from a given A by the transformation with $\alpha(x, t)$.

In order that the Schrödinger equation (4.5) also remains unaltered, the wave function $\psi(x, t)$ has to undergo the (phase) transformation:

$$\psi(x, t) \to \psi'(x, t) = e^{+i(e/hc)\alpha(x, t)}\psi(x, t). \tag{4.7}$$

Because the combination of the two transformations (4.6) and (4.7) leads to:

$$\left(\frac{\hbar}{i}\nabla - \frac{e}{c}A'^{el}\right)\psi' = e^{+i(e/hc)\alpha}\left(\frac{\hbar}{i}\nabla - \frac{e}{c}A^{el}\right)\psi \tag{4.8a}$$

and

$$\left(i\hbar\frac{\partial}{\partial t} - eA_0'^{el}\right)\psi' = e^{+i(e/hc)\alpha}\left(i\hbar\frac{\partial}{\partial t} - eA_0^{el}\right)\psi, \tag{4.8b}$$

and therefore

$$\frac{1}{2m}\left(\frac{\hbar}{i}\nabla - \frac{e}{c}A'^{el}\right)^2\psi' = e^{+i(e/hc)\alpha}\frac{1}{2m}\left(\frac{\hbar}{i}\nabla - \frac{e}{c}A^{el}\right)^2\psi, \tag{4.8c}$$

so that the Schrödinger equation (4.5) is the same in terms of $A_0'^{el}$, A'^{el}, ψ' and in terms of A_0^{el}, A^{el}, ψ.

The combination of the two transformations (4.6) and (4.7) is called a gauge transformation and A_0^{el}, A^{el} is called the gauge potential. In particular, with the interpretation of the preceding quantities as the electromagnetic scalar and vector potential and of ψ as the wave function of a charged particle (4.6) and (4.7) is called the electromagnetic gauge transformation.

If one defines, in analogy to the mechanical (or covariant) momentum (4.3), the gauge-covariant derivative D and D_t by

$$\frac{1}{i}D = \frac{1}{i}D(A) \equiv \frac{1}{i}\nabla - \frac{e}{\hbar c}A \tag{4.9a}$$

$$iD_t = iD_t(A_0) \equiv i\frac{\partial}{\partial t} - \frac{e}{\hbar}A_0 \tag{4.9b}$$

then one can write the Schrödinger equation (4.5) as

$$i\hbar D_t\psi = \frac{-\hbar^2}{2m}D^2\psi. \tag{4.10}$$

The transformations (4.8) can be written in terms of the covariant derivatives as:

$$\frac{1}{i}D(A')e^{+i(e/\hbar c)\alpha}\psi(x, t) = e^{+i(e/\hbar c)\alpha}\frac{1}{i}D(A)\psi(x, t) \tag{4.11a}$$

$$iD_t(A_0')e^{+i(e/\hbar c)\alpha}\psi(x, t) = e^{+i(e/\hbar c)\alpha}iD_t(A_0)\psi(x, t) \tag{4.11b}$$

where $D(A')(D_t(A_0'))$ is the covariant derivative defined as in (4.9) but in terms of the transformed vector (scalar) potential $A'(A_0')$. Since $\psi(x, t)$ is an arbitrary wave function the transformation (4.11) of the covariant derivative can also be written as

$$D(A') \equiv D(A^{(\alpha)}) = e^{+i(e/\hbar c)\alpha}D(A)e^{-i(e/\hbar c)\alpha}. \tag{4.12}$$

In order to generalize this gauge theory, we introduce the following notation

$$U(\alpha(x, t)) = e^{+i(e/\hbar c)\alpha(x,t)}. \tag{4.13}$$

Then the transformation (4.7) is written

$$\psi(x, t) \rightarrow \psi'(x, t) = U(\alpha(x, t))\psi(x, t) \tag{4.7'}$$

and the transformation (4.11) is written

$$D(A')\psi' = U(\alpha)(D(A)\psi) \tag{4.11'}$$

or

$$D(A') = U(\alpha)D(A)U^{-1}(\alpha). \tag{4.12'}$$

The transformation of the gauge potential (4.6) can then be written as:

$$A' = UAU^{-1} + i\frac{\hbar c}{e}U\nabla U^{-1}$$

$$= A + \nabla\alpha. \tag{4.6'}$$

The transformations (4.7) form a group,[11] which is called the one-dimensional unitary group and is denoted by U(1). As the product of two U(1) transformations fulfills

$$U(\alpha_1)\cdot U(\alpha_2) = U(\alpha_1 + \alpha_2) = U(\alpha_2)\cdot U(\alpha_1), \tag{4.14}$$

U(1) is called an Abelian group.[12] The inverse of $U(\alpha)$ is given by

$$U^{-1}(\alpha) = U(-\alpha) = U^\dagger(\alpha). \tag{4.15}$$

The electromagnetic gauge theory is therefore called an Abelian gauge theory. The group U(1), whose parameter α depends on x, t, is called the gauge group.

[11] A set of elements is called a group if it has the following properties:

1. For any two $g; h \in G$ there is defined the product $g\cdot h \in G$.
2. $(g\cdot h)\cdot f = g\cdot(h\cdot f) = g\cdot h\cdot f$ for any three $g, h, f \in G$.
3. There exists an $e \in G$ such the $e\cdot g = g\cdot e = g$ for every $g \in G$. e is called the unit element.
4. For any $g \in G$ there exists and $h \in G$ such that $h\cdot g = g\cdot h = e$. h is called the inverse of g.

[12] A group is called Abelian if $g\cdot h = h\cdot g$ for all $g, h \in G$.

The concept of an Abelian gauge theory has been generalized to non-Abelian or Yang-Mills gauge theories. In place of the group U(1) of phase transformations (4.11) and (4.7), one takes an arbitrary Lie group G (usually a compact and often a simple group). We will restrict ourselves here to the unitary groups in $\mathcal{N}$ dimensions, U($\mathcal{N}$) and we will explain Yang-Mills gauge theories using the example of SU(2).[13]

The ingredients of a gauge theory are

1. a parameter space M (differentiable manifold). M is usually a subset of the three-dimensional Euclidian space $\mathbb{R}^3$ or the Minkowski space with elements labeled by x and t. But it can also be another manifold whose elements $x \in M$ we call "parameter."
2. the gauge group G whose elements, $g = g(x)$, are smooth functions of the parameter $x \in M$.
3. the matter fields $\psi(x)$, which are smooth functions of $x \in M$ and that transform according to a unitary representation $G \ni g \to U(g)$ of G:[14]

$$\psi(x) \to \psi'(x) = U(g(x))\psi(x), \qquad (4.16)$$

or in matrix notation

$$\psi'_n(x) = U_{nm}(g(x))\psi_m(x) \qquad (4.16')$$

where $\psi_n(x); n = 1, 2, \ldots \mathcal{N}$; are the orthonormal $((\psi_n(x), \psi_m(x)) = \delta_{nm})$ basis vectors of the representation space.
4. the gauge potentials $A_i(x) = (A_i^{nm}(x))$, which are quadratic matrices of the same dimension $\mathcal{N}$ as the dimension of the representation space and which are postulated to transform under gauge transformations (4.16) according to

$$A'_i(x) = U(g(x))A_i(x)U^{-1}(g(x)) + \frac{i}{g}(\partial_i U(g(x)))U^{-1}(g(x)) \qquad (4.17)$$

---

[13] U(2) is the set of 2×2 matrices with complex elements that fulfill the condition (unitarity) $U \cdot U^{\dagger} = U^{\dagger} \cdot U = 1$, where $U^{\dagger}$ is the transposed complex conjugate of the matrix U and $\cdot$ denotes matrix multiplication. With the matrix product $\cdot$ as the group product, these matrices form a group. The subset of unitary matrices with det $U = 1$ is also a group called SU(2). U($\mathcal{N}$) is the group of $\mathcal{N} \times \mathcal{N}$ unitary matrices.

[14] The definition of a representation of a group G is as follows: A correspondence (map)

$$g \to U(g)$$

that associates to every $g \in G$ a linear operator $U(g)$ in a linear space Φ is called a linear representation, if the following conditions are fulfilled
a) $U(g_1 \cdot g_2) = U(g_1) \cdot U(g_2)$, for all $g_1, g_2 \in G$
b) $U(e) = 1$, where 1 is the unit operator in Φ and e is the unit element of the group. Φ is called the representation space and the dimension of Φ is called the dimension of the representation. The dimension can be finite or infinite.
   A representation is called unitary if the representation space Φ is a Hilbert space (scalar product space) and
c) $(U(g)\phi, U(g)\psi) = (\phi, \psi)$ for all $\phi, \psi \in \Phi$ and every $g \in G$.
   The $g \to U(g)$ in (4.16) can be an arbitrary (usually finite-dimensional) representation of a group G. In our applications in this section we will use it only when $g \to U(g)$ is the self-representation that means the group G is the group of unitary operators U($\mathcal{N}$) in the $\mathcal{N}$-dimensional Hilbert space Φ (where $\mathcal{N}$ can also be ∞).

(or in vector notation

$$A'(x) = UAU^{-1} - \frac{i}{g}(\nabla U)U^{-1}). \tag{4.17'}$$

In this postulated transformation property a constant g occurs that has a dimension, dim $g = (\dim x \cdot \dim A)^{-1}$. It is needed to adjust the dimension of $\frac{\partial}{\partial x^i}$ and the dimension of A_i, which in general will be different from that of $\frac{\partial}{\partial x^i}$ if this A_i has already been given a physical interpretation as, e.g., in the electromagnetic gauge theory. This constant g is called the coupling constant of the gauge theory; it is characteristic of the gauge theory under consideration.

With the gauge potential one defines the gauge-covariant derivatives:

$$D^i(A) \equiv \partial^i + igA^i(x) \tag{4.18}$$

(or in vector notation

$$-D = -\nabla + igA(x) \tag{4.18'}$$

where D, ∇, and A are the vectors with the Cartesian components $-D^i$, $-\partial^i = -\frac{\partial}{\partial x_i}$ and A^i, respectively.[15] The transformation property of the gauge-covariant derivative follows from the postulated transformation property of the $A^i(x)$. It can be shown immediately using (4.17') (see Problem 3) that

$$D(A') \equiv \nabla - igA' = U(g(x))D(A)U^{-1}(g(x)) \tag{4.19}$$

and

$$D(A)\psi \to D(A')\psi' = U(g(x))D(A)\psi. \tag{4.20}$$

The means that the covariant derivative $D\psi$ of ψ transforms in the same way as ψ itself. The gauge potentials with their postulated transformation property (4.17) have, indeed, been introduced with the intention to obtain a derivative that has the transformation property (4.20) (the ordinary derivative $\nabla\psi$ does not possess this property). For the special case of an Abelian gauge theory for which A and U are 1×1 matrices we obtain by inserting (4.13) into (4.17'):

$$A'(x) = A(x) + \frac{1}{g}\frac{e}{\hbar c}\nabla\alpha(x). \tag{4.6''}$$

This is identical to (4.6) for $g = \frac{e}{\hbar c}$. The electromagnetic gauge theory is therefore the special case of a general gauge theory with gauge group U(1) and coupling constant $\frac{e}{\hbar c}$.

---

[15] This confusing notation and the change of sign in (4.17') and (4.18') have their origins in the use of the metric $\eta_{\mu\nu} = (1, -1, -1, -1)$ for the Minkowski space as a consequence of which

$$\partial^\mu \equiv \frac{\partial}{\partial x_\mu} = \left\{\frac{\partial}{\partial ct}; -\nabla\right\}$$

and

$$\partial_\mu \equiv \frac{\partial}{\partial x^\mu} = \left\{\frac{\partial}{\partial ct}; \nabla\right\}.$$

It is of no further consequence for the remainder of this chapter.

In analogy to the electromagnetic field (or the magnetic field in the three-dimensional, t-independent version) (4.1),

$$F_{ij} = F^{ij} = \epsilon_{ijk} B^k = \partial^j A^i - \partial^i A^j = \frac{\partial}{\partial x^j} A_i - \frac{\partial}{\partial x^i} A_j, \tag{4.21}$$

one can define the non-Abelian gauge field

$$F_{ij}^{nm} = \frac{1}{ig} [D_j, D_i]^{nm} = \frac{\partial}{\partial x^j} A_i^{nm} - \frac{\partial}{\partial x^i} A_j^{nm} - ig[A_i^{nl}, A_j^{lm}]. \tag{4.22}$$

The brackets in (4.22) mean the commutator of the $\mathcal{N} \times \mathcal{N}$ matrices and this involves a summation over all values $l = 1, 2, \ldots \mathcal{N}$:

$$[A_i^{nl}, A_j^{lm}] \equiv \sum_{l=1}^{\mathcal{N}} A_i^{nl} A_j^{lm} - \sum_{l=1}^{\mathcal{N}} A_j^{nl} A_i^{lm}. \tag{4.23}$$

From the transformation property (4.17) of the gauge potentials, one can calculate that the transformed gauge field tensor (i.e., the gauge field tensor defined with the transformed gauge potentials (4.17)):

$$F_{ij}'(x) \equiv \frac{1}{ig} [D_j(A'), D_i(A')] \tag{4.24}$$

is obtained from the original F_{ij} of (4.22) by

$$F_{ij}'(x) = U(g(x)) F_{ij}(x) U^{-1}(g(x)) \quad \left(F_{ij}'^{nm} = \sum_{l,k=1}^{\mathcal{N}} U^{nl}(g) F_{ij}^{lk} U^{-1km}(g) \right). \tag{4.25}$$

This means that the gauge field is not gauge-invariant as for the Abelian gauge theory, but is gauge-covariant; it transforms like the matrix of an operator under the basis transformation (4.17′). The non-Abelian gauge potential has the more complicated transformation property (4.17), which is a combination of the matrix transformation property (4.25) and the vector potential transformation property (4.6).

From the definition (4.22) of F_{ij} we see that it can happen that $F_{ij} = 0$ even though $A_i \neq 0$. A gauge theory for which F_{ij} is identically zero is called trivial. One can show (cf. Problem 5) that for a trivial gauge theory the gauge potential A_i is given by

$$A_i^{nm}(x) = -i \sum_l (\partial_i V^{\dagger nl}(x)) V^{lm}(x) = i \sum_l V^{\dagger nl}(x)(\partial_i V^{lm}(x)) \tag{4.26}$$

where $V(x)$ is a unitary matrix. A gauge potential of the form (4.26) is called pure gauge.

XXIII.5 The Gauge Theory of Molecular Physics

In order to devise the gauge theory of molecular physics we have to choose for the ingredients of the abstract gauge theory defined in Section 4 the concrete physical quantities of the molecule we discussed in Section 3 and show that they fulfill the defining relations (4.16), (4.17), etc. of the abstract quantities.

We will distinguish two cases. The first case is the exact description for which the sum in the expansions (3.17) and (3.21) extends over a complete system of fast basis vectors. As $\mathscr{H}^{\text{fast}}$ is in general infinite-dimensional, the sums in (3.17), (3.21), and (3.25) are also infinite. In this case the gauge theory will turn out to be trivial, i.e., the gauge covariant field F is zero. The second case, which is the only case of practical importance, is when the sum in (3.17) and (3.21) extends over an incomplete system $n = 1, 2 \ldots \mathscr{N}$, where $\mathscr{N}$ is a small number. This is an approximation, the Born-Huang approximation and in this case F is in general nonzero. In the second case we will single out the special subcase that

$$\varepsilon_1(x) = \varepsilon_2(x) = \cdots \varepsilon_n(x) = \cdots \varepsilon_{\mathscr{N}}(x) \equiv \varepsilon(x) \tag{5.1}$$

or where this degeneracy is at least approximately correct.

In either case the gauge theory ingredients are given by the same physical quantities, and their equations are similar. The main difference is that the dimension of the matrices and summation in the matrix equations are in the first case $\mathscr{N} = \dim \mathscr{H}^{\text{fast}}(=\infty)$ and in the second case $\mathscr{N} = $ small number.

The matter field in the gauge theory of molecular physics is the so-called nuclear wave function $\psi_n^E(x)$ and the gauge transformation is defined by the unitary matrix $U(x) \in U(\mathscr{N})$, which depends on the parameter of the slow-motion x:

$$\psi_n^E(x) \to \psi_n'^E(x) = \sum_l U^{nl}(x)\psi_l^E(x). \tag{5.2}$$

The gauge potentials of molecular physics are the $\mathscr{N} \times \mathscr{N}$ matrices defined in (3.14). They are defined in terms of the basis vectors for the fast motion but occur in the equations for the slow motion, e.g., in (3.29). The covariant derivatives of the molecular gauge theory are then the ones defined in the usual way (3.15) in terms of these Mead-Berry potentials (3.14).

We first have to show that the A defined by (3.14) are really gauge potentials, i.e., have the transformation property (4.17) with respect to the transformation (5.2). We show now that this follows from the postulated principle that the wave function of the total system $\psi^E(x, r) = \langle x, r | \psi^E \rangle$ not be affected by the gauge transformation:

$$\psi^E(x, r) = \sum_n \phi_n(x, r)\psi_n^E(x) = \sum_n \phi_n'(x, r)\psi_n'^E(x). \tag{5.3}$$

For this to be fulfilled it follows from (5.2) that the electronic wave functions $\phi_n(x, r)$ have to transform according to

$$\phi_n(x, r) \to \phi_n'(x, r) = \sum_l \phi_l(x, r)U^{\dagger ln}(x). \tag{5.4}$$

This means that the basis vectors $|n(x)\rangle$ of $\mathscr{H}^{\text{fast}}$ transform according to

$$|n(x)\rangle \to |n(x)\rangle' = \sum_l |l(x)\rangle U^{\dagger ln}(x). \tag{5.5}$$

If we insert (5.5) into (3.14), the definition of the $A^{nm}(x)$, we obtain (omitting the summation symbol and summing over equal indices $n, m = 1 \dots \mathcal{N}$):

$$A^{nm}(x) \to A'^{nm}(x) = U^{nn'}(x) i \langle n'(x) | (\nabla_x | m'(x) \rangle U^{\dagger m'm}(x))$$

$$= U^{nn'}(x) i \langle n'(x) | (\nabla_x | m'(x) \rangle) U^{\dagger m'm}(x)$$

$$+ U^{nn'}(x) i \langle n'(x) | m'(x) \rangle (\nabla_x U^{\dagger m'm}(x))$$

and, using (3.14) and (3.7b),

$$A'^{nm}(x) = U^{nn'}(x) A^{n'm'}(x) U^{\dagger m'm}(x) + i U^{nn'}(x) (\nabla_x U^{\dagger n'm}(x)). \tag{5.6}$$

This is the desired transformation property (4.17). From this, as in the general gauge theory, it follows that D of (3.15) transforms in the desired way:

$$D\psi^E \to D'\psi'^E = U D\psi^E \tag{5.7}$$

$$\frac{1}{i} D \to \frac{1}{i} D' = U \frac{1}{i} D U^\dagger. \tag{5.8}$$

The gauge covariant field F_{ij}^{nm} is defined in the standard way (4.22) in terms of the Mead-Berry potentials (3.14):

$$F_{ij}^{nm} = \partial_i A_j^{nm} - \partial_j A_i^{nm} - i [A_i^{nl}, A_j^{lm}] = i [D_i^{nl}, D_j^{lm}] \tag{5.9}$$

where the commutator of the two matrices in the second term is defined as

$$[A_i^{nl}, A_j^{lm}] = \sum_{l=1}^{\mathcal{N}} (A_i^{nl} A_j^{lm} - A_j^{nl} A_i^{lm}) \tag{5.10}$$

and the sum extends over $\mathcal{N} = \dim \mathcal{H}^{\text{fast}}$ (in the first exact case) or over a small number of values (in the second case).

We now want to consider the effect of the $U(\mathcal{N})$ gauge transformations on the Schrödinger equation for the collective motion (3.29). Using (5.2) and (5.8) in (3.29) we obtain (after multiplying with U from the left):

$$\left(\frac{-1}{2\mu} D' \cdot D' + U \tilde{\varepsilon} U^\dagger \right) \psi'^E = E \psi'^E. \tag{5.11}$$

In this equation D' is the gauge-transformed D, i.e., the covariant derivative calculated form (3.14) (3.15) using the transformed basis vectors $|n(x)\rangle'$; these have been proved to be connected with the original D (calculated using the original $|n(x)\rangle$) by (5.8). The ψ'^E are the gauge-transformed ψ^E of (5.2). In order that (5.11) have the same form in terms of the transformed quantities as (3.29) in terms of the original quantities, the term $U\tilde{\varepsilon}U^\dagger$ must be $\tilde{\varepsilon}'$, the gauge transformed $\tilde{\varepsilon}$, i.e., the matrix (3.30) calculated in terms of the transformed basis vectors $|n(x)\rangle'$.

The matrix $\tilde{\varepsilon}$ of (3.30) consists of two terms. The second term is not present in the exact case (for which, usually, $U(\mathcal{N}) = U(\infty)$). We will, therefore, discuss the exact case first. The matrix $\tilde{\varepsilon}(x)$ is then diagonal with the matrix elements

$$\tilde{\varepsilon}^{nm} = \varepsilon_n(x) \delta^{nm}. \tag{5.12}$$

In this case, the summation in these sums extends over a complete system of quantum numbers n for which the following completeness relation holds:

$$\sum_{n=1}^{\infty} |n(x)\rangle\langle n(x)| = I^{\text{fast}}. \tag{5.13}$$

Using this completeness relation and inserting (3.14) into (5.9) one can show by a straightforward calculation (Problem 6):

$$F_{ij}^{nm} = 0. \tag{5.14}$$

This means that in the exact case the gauge theory is trivial. From this it follows that A^{nm} is pure gauge and can be expressed in terms of a unitary (infinite) matrix V by

$$A_i^{nm} = -i \sum_l (\partial_i V^{nl}(x)) V^{\dagger lm}(x) = i \sum_l V^{nl}(x)(\partial_i V^{\dagger lm}(x)). \tag{5.15}$$

It can be shown by a straightforward calculation (Problem 7) starting with the definition (3.14) of A^{nm} that the matrix V is given by

$$V^{nm}(x) = \langle n(x)|m(e_3)\rangle \tag{5.16}$$

where e_3 is the direction of the space-fixed z-axis or any other *fixed* value x_0 of the parameter x.

Since A^{nm} is given by (5.15), one can obtain $A'^{nm} = 0$ in (5.11) by choosing for U in (5.5) $U = V$. However, in this gauge, the transformed matrix

$$\tilde{\varepsilon}' = U\tilde{\varepsilon}U^\dagger \tag{5.17}$$

has the matrix elements

$$\tilde{\varepsilon}'^{n'm'} = V^{n'n}\tilde{\varepsilon}^{nm}V^{\dagger mm'} = \sum_l V^{n'l}\tilde{\varepsilon}_l V^{\dagger lm'} \tag{5.18}$$

and if $\tilde{\varepsilon}$ in the original gauge was diagonal, then in the new gauge, in which $A'^{nm} = 0$, it will no more be diagonal. This is a reflection of the fact that in the old gauge the basis vectors $|n(x)\rangle$ are eigenvectors of the electronic Hamiltonian $h(x)$ and in the new gauge the transformed basis vectors $|n(x)\rangle'$ of (5.5) are in general no more eigenvectors of $h(x)$.

In this new gauge Equation (5.11) will have the form

$$\sum_m \left[-\frac{\nabla_x^2}{2\mu}\delta^{nm} + \sum_l V^{nl}(x)\varepsilon_l(x)V^{\dagger lm}(x) \right] \psi_m'^E(x) = E\psi_n'^E(x), \tag{5.19}$$

which is a somewhat simpler looking system of equations but is still a coupled system of (infinite) differential equations. It is of no practical use.

We shall now discuss the second case, the Born-Huang approximation by $\mathcal{N}$ terms in the electronic energy eigenfunctions.

In the spirit of the Born-Oppenheimer description we want to use as basis vectors $|n(x)\rangle$ of $\mathcal{H}^{\text{fast}}$ only eigenvectors of the fast Hamiltonian $h(x)$. We shall therefore allow in (5.5) only gauge transformations that transform one eigenvector of $h(x)$ into another eigenvector of $h(x)$. The U($\mathcal{N}$) transform in (5.5) must therefore transform within the same energy eigenspace $\mathcal{H}_\varepsilon \subset \mathcal{H}^{\text{fast}}$ of $h(x)$. This means we have the subcase (5.1). (The other subcase in which U($\mathcal{N}$)

is not a symmetry transformation of $h(x)$ leads to nondiagonal $\tilde{\varepsilon}$ and nontrivial A.)

If all the $\varepsilon_n(x)$ are nondegenerate, then the largest gauge group is U(1). Usually for diatomic molecules the $\varepsilon_n(x)$ are doubly degenerate (Λ-doublets, corresponding to electrons orbiting or spinning clockwise or counterclockwise around the internuclear axis). In this case the gauge group is U(2). The sum in (5.2), (5.3), (5.4), (5.5), etc. extends now only over the $\mathcal{N}$ different electronic quantum numbers n with the same eigenvalue (5.1). In particular the system of basis vectors that occur in (5.13) is incomplete, fulfilling

$$\sum_{l=1}^{\mathcal{N}} |l(x)\rangle\langle l(x)| = \Lambda_{\mathcal{H}_\varepsilon}(x) \tag{5.20}$$

where $\Lambda_{\mathcal{H}_\varepsilon}$ is the projection operator onto the $\mathcal{N}$-dimensional subspace $\mathcal{H}_\varepsilon$. From (5.20) it can no longer be proven that (5.14) holds. Therefore, the $\mathcal{N}$-dimensional gauge potential A^{nm} defined by (3.14) with eigenvectors $|n(x)\rangle$ of the fast Hamiltonian $h(x)$ cannot in general be made to vanish by a U($\mathcal{N}$) gauge transformation.

The matrix $\tilde{\varepsilon}$ in (3.29) is now more complicated and contains the second term of (3.30) where the summation over l' extends over all those electronic quantum numbers, $l' = \mathcal{N} + 1, \mathcal{N} + 2 \cdots \infty$, which do not belong to the electronic energy $\varepsilon(x)$. We will write (3.30) now as

$$\tilde{\varepsilon}^{nm}(x) = \varepsilon(x)\langle n(x)|m(x)\rangle + \frac{i^2}{2\mu} \sum_{l=\mathcal{N}+1,\ldots}^{\infty} \langle n(x)|(\nabla|l(x)\rangle)\cdot\langle l(x)|\nabla|m(x)\rangle \tag{5.21}$$

$$= \varepsilon(x)\delta^{nm} + \frac{i^2}{2\mu}\tilde{\tilde{\varepsilon}}^{nm} \qquad n, m = 1, 2, \ldots \mathcal{N}. \tag{5.22}$$

As the vectors $|l(x)\rangle$ in (5.21) of the subspace orthogonal to $\mathcal{H}_\varepsilon$ are not affected by our U($\mathcal{N}$) gauge transformation we can take any basis system for them and use the projection operator

$$\tilde{\Lambda} = 1 - \Lambda_{\mathcal{H}_\varepsilon} = \sum_{l=\mathcal{N}+1,\cdots}^{\infty} |l(x)\rangle\langle l(x)|. \tag{5.23}$$

Then we can write $\tilde{\tilde{\varepsilon}}$ (using $\langle n(x)|\nabla|l(x)\rangle = -(\nabla\langle n(x)|)|l(x)\rangle$):

$$\tilde{\tilde{\varepsilon}}^{nm} = -\sum_{l=\mathcal{N}+1\cdots}^{\infty} (\nabla\langle n(x)|)\cdot|l(x)\rangle\langle l(x)|\nabla|m(x)\rangle = -(\nabla\langle n(x)|)\cdot\tilde{\Lambda}(\nabla|m(x)\rangle). \tag{5.24}$$

Thus (5.22) can be written

$$\tilde{\varepsilon}^{nm}(x) = \varepsilon(x)\langle n(x)|m(x)\rangle + \frac{1}{2\mu}(\nabla\langle n(x)|)\tilde{\Lambda}(\nabla|m(x)\rangle). \tag{5.25}$$

These matrix elements depend on the basis vectors $|n(x)\rangle$; $n = 1, \cdots \mathcal{N}$ only. We consider the gauge-transformed matrix $\tilde{\varepsilon}'$, i.e., the matrix (5.25) calculated in terms of the transformed $|n(x)\rangle'$:

$$\tilde{\varepsilon}'^{nm}(x) = \varepsilon(x)' \langle n(x)|m(x)\rangle' + \frac{1}{2\mu}(\mathbf{\nabla}'\langle n(x)|)\tilde{\Lambda}(\mathbf{\nabla}|m(x)\rangle'). \tag{5.26}$$

We now want to express these matrix elements in terms of the original matrix $\tilde{\varepsilon}^{nm}$ as we have done for the A' in (5.6). We first consider

$$\tilde{\tilde{\varepsilon}}'^{nm} \equiv -(\mathbf{\nabla}'\langle n(x)|)\tilde{\Lambda}(\mathbf{\nabla}|m(x)\rangle'). \tag{5.27}$$

Using (5.5) we calculate

$$\tilde{\tilde{\varepsilon}}'^{nm} = -(\mathbf{\nabla} U^{n'n}(x)\langle n(x)|)\tilde{\Lambda}(\mathbf{\nabla}|m(x)\rangle U^{\dagger mm'}(x))$$

$$= -[U^{n'n}(\mathbf{\nabla}\langle n(x)|)\tilde{\Lambda}(\mathbf{\nabla}|m(x)\rangle)U^{\dagger mm'} + (\mathbf{\nabla} U^{n'n})\langle n|\tilde{\Lambda}(\mathbf{\nabla}|m\rangle)U^{\dagger mm'}$$

$$+ U^{n'n}(\mathbf{\nabla}\langle n|)\tilde{\Lambda}|m\rangle(\mathbf{\nabla} U^{\dagger mm'}) + (\mathbf{\nabla} U^{n'n})\langle n|\tilde{\Lambda}|m\rangle(\mathbf{\nabla} U^{\dagger mm'})].$$

The last three terms on the r.h.s. are zero because $\tilde{\Lambda}$ is orthogonal to the $|n\rangle$ and $|m\rangle$ that occur in the sums (and in the sum of (5.5)). Thus we obtain

$$\tilde{\tilde{\varepsilon}}'^{n'm'} = - \sum_{n,m=1}^{\mathcal{N}} U^{n'n}(\mathbf{\nabla}\langle n(x)|)\tilde{\Lambda}(\mathbf{\nabla}|m(x)\rangle)U^{\dagger mm'}$$

$$= - \sum_{n,m=1}^{\mathcal{N}} U^{n'n} \sum_{l=\mathcal{N}+1}^{\infty} (\mathbf{\nabla}\langle n|)|l\rangle\langle l|(\mathbf{\nabla}|m\rangle)U^{\dagger mm'}, \tag{5.28}$$

which means

$$\tilde{\tilde{\varepsilon}}'^{n'm'} = U^{n'n}\tilde{\tilde{\varepsilon}}^{nm}U^{\dagger mm'}. \tag{5.29}$$

The first term in (5.21) fulfills a relation like this trivially; therefore we have

$$\tilde{\varepsilon}'^{n'm'}(x) = U^{n'n}\tilde{\varepsilon}^{nm}(x)U^{\dagger mm'}. \tag{5.30}$$

This is the analog of (5.6) and (5.8) for the matrix $\tilde{\varepsilon}$. But this is also exactly what the gauge-transformed matrix has to be in order that the second term, $U\tilde{\varepsilon}U^{\dagger}$, in (5.11) becomes $\tilde{\varepsilon}'$. Thus (5.11) becomes

$$\left(-\frac{1}{2\mu}\mathbf{D}'\cdot\mathbf{D}' + \tilde{\varepsilon}'(x)\right)\psi'^{E} = E\psi'^{E}. \tag{5.31}$$

Therewith we have shown that the Schrödinger equation (3.29), (3.32) of the slow motion is independent of the choice of the basis vectors of $\mathcal{H}^{fast}$. It has the same form in terms of the eigenvectors $|n(x)\rangle$ and in terms of the gauge-transformed eigenvectors $|n(x)\rangle'$. Note that this is only the case if (5.1) holds.

Therewith we have seen that in the case of an $\mathcal{N}$-fold degenerate electronic energy value we have a U($\mathcal{N}$) gauge theory for the slow collective (nuclear) motion, which leaves the Schrödinger equation unaltered. In this gauge theory the matter field is given by the so-called nuclear wave function (3.19) and the gauge potential is given by the Mead-Berry connection (3.14). This gauge theory is in general nontrivial.

In the case where $\mathcal{N} = 1$ the operator matrices D_i^{nm}, A_i^{nm}, and H_{eff}^{nm} become just operators. We call n_A the value of the electronic quantum number, which is associated with the eigenvalue E. The wave functions $\langle x, n|\psi^E\rangle = \psi_n^E(x)$ in (3.19) and (3.21) fulfill (3.28) as in the old Born-Oppenheimer approximation.

For $n = n_A$, the Schrödinger equation (3.29) becomes

$$\left[\frac{-1}{2\mu} D \cdot D + \tilde{\varepsilon}(x)\right]\psi^E(x) = E\psi^E(x)$$

$$\left[\frac{-1}{2\mu}\left(\frac{\nabla_x}{i} - A\right)^2 + \tilde{\varepsilon}(x)\right]\psi^E(x) = E\psi^E(x) \tag{5.32}$$

where we have defined

$$A \equiv A^{n_A n_A} \tag{5.33}$$

$$D \equiv D^{n_A n_A} \tag{5.34}$$

$$\psi^E \equiv \psi^E_{n_A} \tag{5.35}$$

$$\tilde{\varepsilon}(x) \equiv \varepsilon_{n_A}(x) + \frac{1}{2\mu}\sum_{l \neq n_A} A^{n_A l}(x) \cdot A^{l n_A}(x) \tag{5.36}$$

$$\tilde{\varepsilon}(x) = \varepsilon_{n_A}(x) + \frac{1}{2\mu}(\nabla_x\langle n_A(x)|) \cdot (\nabla_x|n_A(x)\rangle) - \frac{1}{2\mu}A^{n_A n_A}(x) \cdot A^{n_A n_A}(x). \tag{5.37}$$

Equation (5.32) is the new Schrödinger equation for the slow collective degrees of freedom that correspond to the one electronic state with quantum numbers $n = n_A$. This new Schrödinger equation uses the same assumption for the approximation of the wave function (3.27), (3.28) as the conventional Born-Oppenheimer equation but it does not ignore the x-dependence of the eigenvectors $|n(x)\rangle$. As a consequence, this new Schrödinger equation differs from the conventional Born-Oppenheimer approximation (3.26) by two additional contributions. The first is the Mead-Berry potential A, which is induced by the fast electronic motion. Since the curvature, which for $\mathcal{N} = 1$ is given by

$$F_{ij} = F^{n_A n_A}_{ij} = \partial_i A_j - \partial_j A_i, \tag{5.38}$$

is in general not zero, we are not assured that there exists a transformation by which $A = A^{n_A n_A}$ can be gauged away. This induced vector potential can make an important contribution and change the results of the conventional Born-Oppenheimer approximation. It modifies the momenta and the new "effective" momentum operator $\frac{1}{i}D$ fulfills the new commutation relation

$$\left[\frac{1}{i}D_i, \frac{1}{i}D_j\right] = iF_{ij}, \tag{5.39}$$

which is a special case of (5.9). The second new contribution is the modification of the scalar potential by the second term in (5.36). In our example discussed in the following section we will show that this additional term in the scalar potential just contributes an irrelevant constant. The solutions of Equation (5.32) have to fulfill the following auxiliary conditions, which follow from (3.29) for $n \neq n_A$

$$\sum_l D^{nl} \cdot D^{l n_A}\psi^E(x) = 0. \tag{5.40}$$

In the case that will be of interest to us in the following section, $\varepsilon_n(x)$ is doubly degenerate ($\mathcal{N} = 2$, $\varepsilon_n = \varepsilon_{n_+} = \varepsilon_{n_-}$) with the two states corresponding to electrons orbiting (or spinning) clockwise and counterclockwise around the internuclear axis (doubly degenerate Λ levels). We now have a 2×2 matrix equation given by (for $n = n_+, n_-$)

$$\sum_{m=n_+,n_-} \left[\sum_{l=n_+,n_-} \frac{-1}{2\mu} \boldsymbol{D}^{nl} \cdot \boldsymbol{D}^{lm} + \tilde{\varepsilon}^{nm}(x) \right] \psi_m^E(x) = E\psi_n^E(x) \qquad (5.41)$$

where

$$\tilde{\varepsilon}^{nm}(x) = \varepsilon_n(x)\delta^{nm} + \frac{1}{2\mu} \sum_{l \neq n_+,n_-} \boldsymbol{A}^{nl} \cdot \boldsymbol{A}^{lm}. \qquad (5.42)$$

The covariant momenta are now 2×2 matrices whose commutator gives a nonzero field strength (curvature). We will evaluate these quantities in Section 7 for the example of doubly degenerate Λ-levels of diatomic molecules.

XXIII.6 The Electronic States of Diatomic Molecules

In order to apply the new Born-Oppenheimer method to the diatomic molecule we have to discuss the properties of the electronic quantum numbers n in more detail.[16]

The classification of the electronic states of the diatomic molecules is in one respect similar to the classification of atomic energy levels, but in other respects quite different. The Rydberg series (VI.5.12) or (VII.2.24), which dominate the atomic spectra, are unimportant in molecular spectroscopy. So the principal quantum number of atomic physics is not relevant. The orbital angular momentum of the electrons is also not a good quantum number, because in a diatomic molecule one does not have spherical symmetry but only axial symmetry about the internuclear axis $\hat{\boldsymbol{R}}$ (the operator of electronic angular momentum square, $L_{\mathrm{elec.}}^2$, does not commute with the Hamiltonian). But $L_{\mathrm{elec.}} \cdot \hat{\boldsymbol{R}}$, and $(L_{\mathrm{elec.}} \cdot \hat{\boldsymbol{R}})^2$ commutes with h and is thus a good quantum number.

One, therefore, classifies the electronic states of a diatomic molecule according to the value of

$$\Lambda = +\sqrt{\text{eigenvalue } (\boldsymbol{L} \cdot \hat{\boldsymbol{R}})^2}. \qquad (6.1)$$

The quantum number Λ can take the values

$$\Lambda = 0, 1, 2, 3 \cdots \qquad (6.2)$$

and the corresponding molecular states are called

[16] Readers familiar with this subject can skip this section. For more details, consult, e.g., G. Herzberg (1966) Vol. 1.

$$\Sigma, \Pi, \Delta, \Phi \cdots \qquad (6.2')$$

states (in analogy to the $S, P, D, F \ldots$ designation for atoms).

Since $L_{\text{elec.}} \cdot \hat{R}$ can have two eigenvalues, $+\Lambda$ and $-\Lambda$ for a given energy level (which depends only on Λ), the $\Pi, \Delta, \Phi \cdots$ states are doubly degenerate and the Σ states are nondegenerate. (The physical eigenstates are not eigenvectors of $L_{\text{elec.}} \cdot \hat{R}$ but eigenvectors of parity, as discussed in Section V.4. The physical states are thus linear combinations of two eigenvectors of $L_{\text{elec.}} \cdot \hat{R}$ with eigenvalue $+\Lambda$ and $-\Lambda$).

The electron spin cannot be neglected in the classification of the electronic spectra of molecules. The electronic spin $S_{\text{elec.}}$ is the sum of the spins of each electron, the corresponding quantum number S can therefore be any integer or half-integer number, depending on whether the number of electrons is even or odd.

The eigenvalue of $S_{\text{elec.}} \cdot \hat{R}$ is denoted by Σ and can take the following values

$$\Sigma = -S, -S + 1, -S + 2, \ldots +S. \qquad (6.3)$$

Σ and Λ are not always well-defined separately. It can be that the spin-orbit coupling is so large that $L_{\text{elec.}}$ and $S_{\text{elec.}}$ are coupled to form the total electronic angular momentum $J = L_{\text{elec.}} + S_{\text{elec.}}$ (this is called Hund's coupling case c). Then only

$$\Omega = \sqrt{\text{eigenvalue of } (J \cdot \hat{R})^2} \qquad (6.4)$$

is a good quantum number. We will consider the case where Λ and Σ separately are good quantum numbers (called Hund's coupling case a). Then

$$\Omega = |\Lambda + \Sigma|, \qquad (6.5)$$

and for a fixed-value Λ (with $\Lambda \geq S$), Ω takes the $2S + 1$ values

$$\Omega = \Lambda + S, \Lambda + S - 1, \ldots \Lambda - S. \qquad (6.6)$$

The corresponding molecular states are denoted $^{2S+1}(\Lambda)_{\Omega}$, where (Λ) stands for the symbols (6.2′) of the values (6.2) for Λ. Thus, e.g., the electronic states with $\Lambda = 1$ and $S = 1$ are denoted by $^{3}\Pi_{0}, {}^{3}\Pi_{1}, {}^{3}\Pi_{2}$ for the value $\Omega = 0(\Sigma = -1), \Omega = 1(\Sigma = 0)$ and $\Omega = 2(\Sigma = +1)$, respectively. They form the $^{3}\Pi$ multiplet. The quantum numbers $n_a \equiv {}^{2S+1}(\Lambda)_{\Omega}$ usually specify the electronic energy, i.e., the potential energy curves $\varepsilon_n(x)$. Sometimes, when the molecule has a center of inversion (homonuclear molecules like O_2), the additional quantum number g (gerade) and u (ungerade) appear with values $+1$ and -1 of this molecular-fixed-inversion operator (replacement of all electron coordinates by their negatives). Then the quantum numbers n_a that specify the potential energy curves $\varepsilon_n(x)$ are $n_a = {}^{2S+1}(\Lambda)_{\Omega}$, g or u: $\varepsilon_n(x) = \varepsilon_{n_a}(x)$.

In addition there are the degeneracy quantum numbers n_b; $n = (n_a; n_b)$, which the electronic energy does not depend on. For the diatomic molecule this is the intrinsic parity with the eigenvalue $+1$ or -1 (Section V.4, in particular (V.4.21)). The intrinsic parity is the eigenvalue π of the parity operator U_P (space-fixed-inversion operator) divided by $(-1)^j$, where j is the total angular momentum of the molecule, i.e., the angular momentum of the electronic and collective motions combined. The total angular momentum j is always (as a consequence of isotropy of space) a good quantum number, and its component j_3 is also a good quantum number.

The total electronic angular momentum j_{el}, i.e., the eigenvalue $j_{el}(j_{el} + 1)$ of J^2, is not a good quantum number. The reason for this is that the electrons do not have spherical symmetry with respect to a center (like in atoms), but only axial symmetry with respect to the internuclear axis R.

The basis vectors in the electronic space $\mathcal{H}^{fast}$ are thus labeled by the following quantum numbers

$$|n(x)\rangle = |n_a; n_b(x)\rangle = |^{2S+1}(\Lambda), \Omega; \pm(x)\rangle. \tag{6.7}$$

The parameter x is the eigenvalue of R, and its unit vector will be denoted by $\hat{x} = (\sin\theta \cos\varphi, \sin\theta \sin\varphi, \cos\theta)$.

The energy values $\varepsilon_{n_a}(x)$ depend on $n_a = {}^{2S+1}(\Lambda)_\Omega$ and the degeneracy quantum number, $n_b = +1$ or -1, is the parity or intrinsic parity. Except for $\Lambda = 0$ (Σ states), when the intrinsic parity is either $+1$ or -1, the energy values have double degeneracy. This means both parity values $+$ and $-$ occur. Instead of using the parity as degeneracy quantum number, one can use the eigenvalue $k = +\Omega$ or $k = -\Omega$ of the operator $\hat{x} \cdot J$. This means instead of choosing a basis consisting of the parity eigenvectors, a basis of eigenvectors of the operator $\hat{x} \cdot J$ can be chosen. We denote these eigenvectors as $|n_a, k(\theta, \varphi)\rangle$ (θ and φ are the polar coordinates of the internuclear axis). For brevity we drop the labels n_a

$$\hat{x} \cdot J|k(\theta, \varphi)\rangle = k|k(\theta, \varphi)\rangle. \tag{6.8}$$

The parity eigenvectors are linear combinations of these vectors

$$|^{2S+1}(\Lambda)\Omega^{\pm}(\theta, \varphi)\rangle \equiv |\pi = \pm; (\theta, \varphi)\rangle$$

$$= \frac{1}{\sqrt{2}}(|k = \Omega(\theta, \varphi)\rangle \pm |k = -\Omega(\theta, \varphi)\rangle). \tag{6.9}$$

We omit all the labels $^{2S+1}(\Lambda)_u^g$ that are fixed in a given multiplet and we choose the notation $k = \pm\Omega = \pm|\Sigma + \Lambda|$. In general, Ω can take any of its possible values: $(0, \frac{1}{2}, 1, \frac{3}{2}, 2 \ldots)$; for a given electronic energy level it has a fixed value. In order to have a concrete example we choose $\Sigma = 0$ and $\Lambda = 1$ (i.e., $(\Lambda) = \Pi$), then $k = \pm 1$. The value $k = \Omega$ corresponds to an electron rotating counterclockwise around the internuclear axis and $k = -\Omega$ corresponds to a clockwise-rotating electron. The diatomic molecule is thus a realization of the model of a dumbbell with flywheel (of Section V.4).

XXIII.7 The Monopole of the Diatomic Molecule

The vectors (6.8) are familiar from Section XXII.3, where the vectors (XXII.3.7) were eigenvectors of a Hamiltonian describing a magnetic moment in a rotating magnetic field. The vectors (6.8) are electronic energy eigenvectors but the eigenvalue k is now not an eigenvalue connected with the electronic energy $\varepsilon_{n_a}(x)$ as was the case in Section XXII.3. Otherwise, the mathematics is the same as in Section XXII.3, and the physics is related. In place of the precessing external magnetic field we have the magnetic field in the direction of the internuclear axis created by the rapidly rotating electronic charges. In place of the magnetic moment of the spinning quantum system in Section XXII.3 we have here the magnetic moment connected with the angular momentum of the electrons. Thus, in a certain sense, the diatomic dumbbell molecule with flywheel is the quantum analog of the quantum system in a slowly changing classical environment of Sections XXII.3 and XXII.4. It is a quantum system in a slowly changing quantum environment. The results here will, therefore, be reminiscent of the results we had obtained in Sections XXII.3 and XXII.4.

The calculation of the Berry connection proceeds in the same way as in Section XXII.3; the only difference is that k now takes the two values $\pm \Omega$ and we obtain 2×2 matrices for the non-Abelian connections. We start with the vector $|k(\mathbf{e}_3)\rangle$, which is an eigenvector of $\hat{\mathbf{x}} \cdot \boldsymbol{J}$ when the internuclear axis points along the space-fixed three-axis $\mathbf{e}_3$:

$$\mathbf{e}_3 \cdot \boldsymbol{J}|k(\mathbf{e}_3)\rangle = J_3|k(\mathbf{e}_3)\rangle = k|k(\mathbf{e}_3)\rangle. \tag{7.1}$$

Note that $|k(\mathbf{e}_3)\rangle$ does not have to be—and in general is not—an eigenvector of $\boldsymbol{J}^2$ (because the total electronic angular momentum is not a good quantum number). It may only be an eigenvector of the electronic angular momentum component J_3. Starting with the vector $|k(\mathbf{e}_3)\rangle$, we define—as in Section XXII.3—a vector-valued function of (θ, φ) by

$$|k(\theta, \varphi)\rangle \equiv U(\theta, \varphi)|k(\mathbf{e}_3)\rangle = e^{-i\varphi J_3}e^{-i\theta J_2}e^{i\varphi J_3}|k(\mathbf{e}_3)\rangle. \tag{7.2}$$

This defines a unique vector for all (θ, φ) *except* at $\theta = \pi$ (south pole). The other parameterization is given by

$$|k(\theta, \varphi)\rangle' \equiv e^{-i\varphi J_3}e^{-i\theta J_2}e^{-i\varphi J_3}|k(\mathbf{e}_3)\rangle = e^{-i2k\varphi}|k(\theta, \varphi)\rangle, \tag{7.3}$$

which defines a unique vector for all (θ, φ) *except* at $\theta = 0$ (north pole).

With these vectors we can evaluate the Berry connection

$$\tilde{A}^{k'k} = A_i^{k'k}\,dx^i = A_\theta^{k'k}\,d\theta + A_\varphi^{k'k}\,d\varphi = i\langle k'(\theta, \varphi)|d|k(\theta, \varphi)\rangle \tag{7.4}$$

where $A_i^{k'k}$ are the Cartesian components of the vector (3.14). We use here the spherical covariant components of this vector potential:

$$A_\theta^{k'k} = i\langle k'(\theta, \varphi)|\frac{\partial}{\partial\theta}|k(\theta, \varphi)\rangle \tag{7.5}$$

$$A_\varphi^{k'k} = i\langle k'(\theta, \varphi)|\frac{\partial}{\partial\varphi}|k(\theta, \varphi)\rangle. \tag{7.6}$$

The $A_i^{k'k}(i = \theta, \varphi)$ are two-by-two matrices ($\tilde{A}$ is a matrix-valued differential form). The calculation was done in Section XXII.3 with the result (XXII.3.41), (XXII.3.42), which we write here as (with $\hbar = 1$):

$$A_\theta^{k'k} = e^{-i\varphi k'}e^{i\varphi k}\langle k'(\mathbf{e}_3)|J_2|k(\mathbf{e}_3)\rangle \tag{7.7}$$

$$A_\varphi^{k'k} = \langle k'(\mathbf{e}_3)|[-\sin\theta(\cos\varphi J_1 + \sin\varphi J_2) + (\cos\theta - 1)J_3]|k(\mathbf{e}_3)\rangle. \tag{7.8}$$

If the $|k(\mathbf{e}_3)\rangle$ were eigenvectors of J^2, i.e., if they were basis vectors of an irreducible representation of $SO(3)_{J_i}$, the matrix elements in (7.7) and (7.8) would be well known (III.3.22) and (III.3.23). But even if the $|k(\mathbf{e}_3)\rangle$ are not $SO(3)$-basis vectors, the matrix elements are easily calculated (from the $SO(2)_{J_3}$-vector operator property of J_1 and J_2). If $|k - k'| \neq 1$ (i.e., if $|k| \equiv \Omega \neq \frac{1}{2}$) one obtains (for $\theta \neq \pi$):

$$A_\theta^{k'k} = 0 \tag{7.9}$$

$$A_\varphi^{k'k} = \begin{matrix} & k = +\Omega & k = -\Omega \\ k' = +\Omega \\ k' = -\Omega \end{matrix}\begin{pmatrix} -\Omega(1 - \cos\theta) & 0 \\ 0 & +\Omega(1 - \cos\theta) \end{pmatrix} = -\Omega(1 - \cos\theta)\sigma_3^{k'k}. \tag{7.10}$$

Using the parameterization (7.3) one calculates in the same way (for $\theta \neq 0$):

$$A_\theta'^{k'k} = 0 \tag{7.11}$$

$$A_\varphi'^{k'k} = k\delta_{k'k}(1 + \cos\theta) = \Omega(1 + \cos\theta)\sigma_3^{k'k}. \tag{7.12}$$

For $|k| \equiv \Omega = \frac{1}{2}$ the calculation is now a little more complicated because for $|k' - k| = 1$ the operators J_1 and J_2 will have nonzero off-diagonal matrix elements. One can show (Problem 8) that $\tilde{A}$ can be expressed in terms of a real parameter κ, which is defined by the matrix element $\langle k(\mathbf{e}_3)|J_2|k(\mathbf{e}_3)\rangle$ as

$$\langle k'(\mathbf{e}_3)|J_2|k(\mathbf{e}_3)\rangle = \frac{\kappa}{2}\sigma_2^{k'k}. \tag{7.13}$$

The result in terms of this arbitrary parameter is:

$$A_\theta^{k'k} = \begin{matrix} & k = +\frac{1}{2} & k = -\frac{1}{2} \\ k' = +\frac{1}{2} \\ \\ k' = -\frac{1}{2} \end{matrix}\begin{bmatrix} 0 & \dfrac{-ie^{-i\varphi}}{2}\kappa \\ \dfrac{+ie^{i\varphi}}{2}\kappa & 0 \end{bmatrix} \tag{7.14}$$

$$A_\varphi^{k'k} = \begin{matrix} & k = +\frac{1}{2} & k = -\frac{1}{2} \\ k' = +\frac{1}{2} \\ \\ k' = -\frac{1}{2} \end{matrix}\begin{bmatrix} \dfrac{-(1 - \cos\theta)}{2} & \dfrac{-\sin\theta e^{-i\varphi}}{2}\kappa \\ \dfrac{-\sin\theta e^{i\varphi}}{2}\kappa & \dfrac{(1 - \cos\theta)}{2} \end{bmatrix}. \tag{7.15}$$

The spherical orthonormal components of the vector A (denoted by $A =$

$a_x \mathbf{e}_x + a_\theta \mathbf{e}_\theta + a_\varphi \mathbf{e}_\varphi)$ are defined by $\tilde{A} = a_\theta x \, d\theta + a_\varphi x \sin \theta \, d\varphi$. By comparison with Equations (7.9) and (7.10), we see that

$$a_x^{k'k} = 0 \tag{7.16}$$

$$a_\theta^{k'k} = 0 \tag{7.17}$$

$$a_\phi^{k'k} = \frac{-\Omega(1 - \cos \theta)}{x \sin \theta} \sigma_3^{k'k}. \tag{7.18}$$

The first equality holds because the vectors (7.2) and (7.3) are independent of the radius x. From the connection—either in the gauge given by (7.9) and (7.10) or in the gauge given by (7.11) and (7.12)—we can calculate the curvature. The curvature is the matrix valued two-form[17]

$$\tilde{F}^{k'k} = F_{ij}^{k'k} \, dx^i \wedge dx^j = F_{x\theta}^{k'k} \, dx \wedge d\theta + F_{x\varphi}^{k'k} \, dx \wedge d\varphi + F_{\theta\varphi} \, d\theta \wedge d\varphi \tag{7.19}$$

where F_{ij} are its Cartesian components given by (5.9) in terms of the matrices A_i and $F_{x\theta}, F_{x\varphi}$, and $F_{\theta\varphi}$ are the covariant spherical components given in terms of the matrices A_θ, A_φ by

$$F_{\theta\varphi} = \partial_\theta A_\varphi - \partial_\varphi A_\theta - i[A_\theta, A_\varphi]. \tag{7.20}$$

It is easily seen that all other components $F_{x\varphi}, F_{x\theta}$ of $\tilde{F}$ except for $F_{\theta\varphi} = -F_{\varphi\theta}$ are zero. The $F_{\theta\varphi}$ component of the curvature calculated from (7.9) and (7.10) (or (7.11) and (7.12)) is the matrix (for $\Omega \neq \frac{1}{2}$)

$$F_{\theta\phi}^{k'k} = -\Omega \sin \theta \sigma_3^{k'k}. \tag{7.21a}$$

Thus the curvature two-form for our case is:

$$\tilde{F}^{k'k} = F_{\theta\varphi}^{k'k} \, d\theta \wedge d\varphi = -\Omega \sin \theta \sigma_3^{k'k} \, d\theta \wedge d\varphi. \tag{7.21b}$$

For the case $\Omega = \frac{1}{2}$, a separate calculation shows that (Problem 8)

$$F_{\theta\varphi}^{k'k} = -\frac{1}{2}(1 - \kappa^2) \sin \theta \sigma_3^{k'k}. \tag{7.22}$$

In the derivation of (7.21) the commutator on the r.h.s. of (7.20) is zero (because A_θ is zero). However, in the derivation of (7.22) this commutator gives a nonzero contribution. For $\Omega \neq \frac{1}{2}$, $\tilde{F}$ is always different from zero (except in the trivial case $\Omega = 0$). For $\Omega = \frac{1}{2}$, the curvature $\tilde{F}$ can vanish when $\kappa = 1$. According to the definition of κ given by (7.13), the case $\kappa = 1$ means that the $\{|k(\mathbf{e}_3)\rangle; k = \frac{1}{2}, -\frac{1}{2}\}$ span the irreducible representation space of the rotation group generated by the electron angular momentum operators J_1, J_2, and J_3 with electron angular momentum $\frac{1}{2}$.

The Berry connection and the Berry curvature that we have obtained for the diatomic molecule with fixed angular momentum component k along the

---

[17] The curvature is the covariant exterior derivative of the connection one-form $\tilde{F}^{k'k} = D\tilde{A}^{k'k} \equiv d\tilde{A}^{k'k} + (\tilde{A} \wedge \tilde{A})^{k'k}$. See, e.g., C. J. Isham, *Modern Differential Geometry for Physicists*, World Scientific (1989) or C. Nash and S. Sen, *Topology and Geometry for Physicists*, Academic Press (1983).

internuclear is given by the same expressions that we obtained in Section XXII.3. The difference between the molecule and the quantum system of Section XXII.3 is that the fast electron state that we consider here is not a pure state described by a one-dimensional subspace of the space of electron states, but a mixture (c.f. Section II.4) described by a two-dimensional subspace spanned by the vectors $|k(\theta, \varphi)\rangle$ with $k = \Omega$ and $k = -\Omega$. As a consequence of this, the connection $\tilde{A}$ (or A) and curvature $\tilde{F}$ (or F) are 2×2 matrices.

Instead of using the basis vectors $|k(\theta, \varphi)\rangle$ for this two-dimensional space, the parity basis vectors $|\pi = \pm; (\theta, \varphi)\rangle$ of (6.9) can be used. Then $\tilde{F}$ is no longer diagonal but is given by

$$F^{\pi'\pi} = \sum_{k,k'=\pm\Omega} \langle \pi'|k'\rangle F^{k'k}\langle k|\pi\rangle, \qquad (7.23)$$

where $\langle k|\pi \rangle$ is the transition matrix of the transformation (6.9) given by

$$\langle k|\pi \rangle = \begin{array}{c} \\ k = +\Omega \\ k = -\Omega \end{array} \overset{\begin{array}{cc} \pi = + & \pi = - \end{array}}{\begin{bmatrix} \dfrac{1}{\sqrt{2}} & \dfrac{1}{\sqrt{2}} \\ \dfrac{1}{\sqrt{2}} & -\dfrac{1}{\sqrt{2}} \end{bmatrix}}. \qquad (7.24)$$

Performing the transformation (7.23) gives

$$F^{\pi'\pi}_{\theta\varphi} = -\Omega \sin \theta \sigma_1^{\pi'\pi} = -\Omega \sin \theta \begin{pmatrix} 0 & 1 \\ 1 & 0 \end{pmatrix}. \qquad (7.25)$$

The electronic energy eigenstate is a mixture of electron states rotating clockwise and counterclockwise around the internuclear axis. Equivalently, the state can be considered a mixture of positive and negative parity eigenstates.

The vector field F that corresponds to the two-form component (7.21) (and similarly for (7.22) and (7.25)) has only a component in the x-direction (direction of internuclear axis). In order to obtain this vector we use in (7.21b) the orthonormal spherical components $\hat{F}_{\theta\varphi}$, $\hat{F}_{x\theta}$, $\hat{F}_{x\varphi}$:

$$\tilde{F} = \hat{F}_{x\theta} \, dx \wedge (x \, d\theta) + \hat{F}_{x\varphi} \, dx \wedge (x \sin \theta \, d\varphi) + \hat{F}_{\theta\varphi}(x \, d\theta) \wedge (x \sin \theta \, d\varphi). \qquad (7.26)$$

Comparing this with (7.21b) we obtain:

$$\hat{F}_{\theta\varphi} = -\Omega \frac{1}{x^2}\sigma_3, \qquad \hat{F}_{x\theta} = 0, \qquad \hat{F}_{x\varphi} = 0$$

or

$$F = -\Omega\sigma_3 \frac{x}{x^3}. \qquad (7.27)$$

The matrix elements of A and the matrix elements of F are identical with the vector potential and the field strength of a charge-monopole system if the quantum number k (component of electronic angular momentum along the internuclear axis) is replaced by $-\frac{eg}{4\pi}$ where e is the electric charge and g the magnetic monopole strength. This is most apparent from (7.27), the vector form of $\tilde{F}$. Equation (7.27) is the standard expression for the magnetic field of a monopole if $\Omega \to |\frac{eg}{4\pi}|$.

Thus we have derived the monopole vector potential and field from the rapid motion of the electrons around the internuclear axis of a diatomic molecule. This "magnetic monopole" is of course not of the electromagnetic kind as originally imagined by Dirac[18] but has its sole origin in the fast motion of the electrons. This monopole is characterized by the motion constant k, in place of the electromagnetic constants (e, g). These "motional monopoles" of the diatomic dumbbell with flywheel always come in pairs with opposite motion constants $|k|$ and $-|k|$ or, equivalently, in pairs with opposite parity. The parity eigenstates are linear combinations (6.9) of these monopoles.

We shall now discuss the effective Hamiltonian of our system. The gauge-covariant momenta, defined by

$$\pi^{k'k} \equiv \frac{1}{i} D^{k'k} = \delta^{k'k} P - A^{k'k}, \tag{7.28}$$

are two-dimensional matrices like the A and F whose rows and columns are labeled by k' and k (the eigenvalues of the operator $\hat{x} \cdot J$). The effective Hamiltonian of (3.31) is also a 2×2 matrix

$$H_{eff} = \frac{1}{2\mu} \pi \cdot \pi + \tilde{\varepsilon}(x) \tag{7.29}$$

with

$$\tilde{\varepsilon}^{k'k}(x) = \varepsilon_\Omega(x)\delta^{k'k} + \frac{1}{2\mu} \sum_{l \neq \pm\Omega} A^{k'l} \cdot A^{lk}. \tag{7.30}$$

The multiplication of the π is matrix multiplication (there is also a sum over the three vector components in (7.29)).

The matrix elements of the two-dimensional matrices π_i ($i = 1, 2, 3$) are operators in the space of the slow motion. For the case $|k| \neq \frac{1}{2}$ the matrices π_i are diagonal but their matrix elements are noncommuting operators. For $|k| = \frac{1}{2}$ their matrix elements are noncommuting operators, and from (7.14) and (7.15) we see that they are nondiagonal matrices as well. In both cases, the operator matrices π_i fulfill the commutation relations (4.22), which in our case (7.21a) and (7.22) are given by

$$[\pi_i, \pi_j] = -i\epsilon_{ijl} \frac{x_l}{x^3} \Omega \begin{pmatrix} 1 & 0 \\ 0 & -1 \end{pmatrix} \qquad |k| \neq \frac{1}{2} \tag{7.31}$$

and

[18] P. Dirac. *Proc. Roy. Soc. London Ser.* **A133**, 60 (1931); P. Goddard and D. I. Olive, *Rep. Prog. Phys.* **41**, 1357 (1978); S. Coleman "The Unity of the Fundamental Interactions," H. Zichichi (editor) (Erice lectures 1981), Plenum Press (1983).

$$[\pi_i, \pi_j] = -i\epsilon_{ijl}\frac{x_l}{x^3}\frac{1}{2}(1 - \kappa^2)\begin{pmatrix} 1 & 0 \\ 0 & -1 \end{pmatrix} \qquad |k| = \frac{1}{2}. \qquad (7.32)$$

If the matrix elements of the 2×2 matrices π_i are labeled not by the quantum numbers k but by the quantum numbers π (eigenvalues of intrinsic parity) then their commutation relations are similar to (7.31) and (7.32) with σ_3 on the r.h.s. replaced by σ_1 as in (7.25). If we restrict ourselves to a subspace of the space of electronic states with a fixed value for the operator $\hat{x} \cdot J$ (e.g. to $k' = k = |k|$), then the π_i are just operators and not operator matrices. Their commutation rel. ..ons are (for $|k| \neq \frac{1}{2}$):

$$[\pi_i, \pi_j] = -i\epsilon_{ijl}\frac{x_l}{x^3}k, \qquad (7.33)$$

which are the standard commutation relations for a charge-monopole system.

We now consider the scalar potential given by (7.30). The first term $\varepsilon_\Omega(x)$, which is the scalar potential for the diatomic molecule in a definite electronic energy level (the Born-Oppenheimer eigenvalue), is specified by the value of Ω and the other fixed electronic quantum numbers $^{2S+1}(\Lambda)$. In many cases $\varepsilon_\Omega(x)$ describes a potential curve containing a deep attractive well (bonding) as shown in Fig. 2.1. In this case it is well approximated near the equilibrium value $x = x_e$ of the internuclear separation by a harmonic oscillator potential $\varepsilon_\Omega(x) \approx \frac{f}{2}(x - x_e)^2$, leading to vibrational bound states.

We shall now sketch the calculation of the second term in (7.30). The result is given in (7.40). Using the same techniques that were used in the evaluation of (5.24) and (5.37), one obtains

$$\sum_{l\neq\pm\Omega} A^{k'l} \cdot A^{lk} = (\nabla\langle k'(\theta, \varphi)|)\nabla|k(\theta, \varphi)\rangle - \sum_{l=\pm\Omega} A^{k'l} \cdot A^{lk}. \qquad (7.34)$$

The first term is calculated using the following result, which can be obtained by a straightforward calculation from (XXII.3.39) and (XXII.3.40):

$$\nabla U^\dagger(\theta, \varphi) \cdot \nabla U(\theta, \varphi) = \frac{1}{x^2}\frac{\partial U^\dagger}{\partial\theta}\frac{\partial U}{\partial\theta} + \frac{1}{x^2\sin^2\theta}\frac{\partial U^\dagger}{\partial\varphi}\frac{\partial U}{\partial\varphi}$$

$$= \frac{1}{x^2}\left[J_1^2 + J_2^2 - \frac{1 - \cos\theta}{\sin^2\theta}J_3^2\right]$$

$$+ \frac{1 - \cos\theta}{x^2\sin\theta}\{J_3, J_1\cos\varphi + J_2\sin\varphi\}. \qquad (7.35)$$

The second term of (7.34), which involves a summation over the two matrix indices $l = \pm|k|$ only, is calculated from (XXII.3.41) and (XXII.3.42):

$$\sum_{l=\pm\Omega} A^{k'l}A^{lk} = \frac{1}{x^2}\sum_{l=\pm\Omega}\left(j_1^{k'l}j_1^{lk} + j_2^{k'l}j_2^{lk} + j_3^{k'l}j_3^{lk}\frac{(1 - \cos\theta)^2}{\sin^2\theta}\right) \qquad (7.36)$$

where j_i^{lk} are the 2×2 matrices:

$$j_i^{lk} = \langle l(\mathbf{e}_3)|J_i|k(\mathbf{e}_3)\rangle; \qquad l, k = +\Omega \text{ or } -\Omega. \qquad (7.37)$$

Note that the calculation of the matrix elements of the operators $J_i J_j$ in (7.35) involves a summation over all infinite electronic quantum numbers, not just $l = \pm\Omega$:

$$\langle k'(\mathbf{e}_3)|J_i J_j|k(\mathbf{e}_3)\rangle = \sum_l \langle k'(\mathbf{e}_3)|J_i|l(\mathbf{e}_3)\rangle \langle l(\mathbf{e}_3)|J_j|k(\mathbf{e}_3)\rangle.$$

Also note that the vectors $|k(\mathbf{e}_3)\rangle$ are not eigenvectors of $\mathbf{J}^2$ (not basis vectors of an $SO(3)_{J_i}$ representation).

We first consider the case $\Omega \neq \frac{1}{2}$. The matrix elements of the anti-commutator in (7.35) are zero, the matrix elements of the first term are not. Taking the matrix elements of (7.35) and inserting this and (7.36) into (7.34) one obtains

$$\sum_{l \neq \pm\Omega} A^{k'l} \cdot A^{lk} = \frac{1}{2\mu x^2} \langle k'(\mathbf{e}_3)|(J_1^2 + J_2^2)|k(\mathbf{e}_3)\rangle. \tag{7.38}$$

The matrix elements on the r.h.s. cannot be explicitly evaluated using the matrix elements of the generators of $SO(3)$ ($\langle k'(\mathbf{e}_3)|\mathbf{J}^2|k(\mathbf{e}_3)\rangle$ is not $j_{el}(j_{el} + 1)$). However, they can be expressed in terms of an $SO(2)_{J_3}$ reduced matrix element Δ of the $SO(2)_{J_3}$-scalar operator $J_1^2 + J_2^2$ (using the Wigner-Eckart theorem for $SO(2)$) and the $SO(2)$ Clebsch-Gordan coefficient $\delta_{k'k}$):

$$\langle k'(\mathbf{e}_3)|(J_1^2 + J_2^2)|k(\mathbf{e}_3)\rangle = \Delta \delta^{k'k}. \tag{7.39}$$

The factor Δ can of course be different for different values of the (fixed) quantum number $\Omega = |k|$; $\Delta = \Delta(\Omega)$. Thus, we can write the second term of equation (7.30) as

$$\frac{1}{2\mu} \sum_{l \neq \pm\Omega} A^{k'l} \cdot A^{lk} = \frac{1}{2\mu x^2} \Delta \delta^{k'k}. \tag{7.40}$$

For the case $\Omega = \frac{1}{2}$, the matrix elements of (7.37) are given by the Pauli matrices, $j_i = \frac{\kappa}{2}\sigma_i$, $i = 1, 2$. A separate calculation shows that in this case Δ in (7.40) has to be replaced by $\Delta + \frac{\kappa^2}{2}$ where κ is the constant introduced in (7.13).

Equation (7.40) shows that this term is an inverse-square potential originating from an inverse-cube repulsive force centered at $x = 0$.[19] For future reference we rewrite (7.30) (for both cases $|k| \neq \frac{1}{2}$ and $|k| = \frac{1}{2}$) as

$$\tilde{\varepsilon}^{k'k}(x) = \tilde{\varepsilon}_\Omega(x)\delta^{k'k}, \tag{7.41}$$

where

$$\tilde{\varepsilon}_\Omega(x) = \varepsilon_\Omega(x) + \frac{1}{2\mu x^2}\Delta. \tag{7.42}$$

Therewith the effective Hamiltonian of the diatomic dumbbell with flywheel can be given by:

$$H_{\text{eff}} = \frac{1}{2\mu}\boldsymbol{\pi} \cdot \boldsymbol{\pi} + \frac{1}{2\mu x^2}\Delta + \varepsilon_\Omega(x). \tag{7.43}$$

If the last two terms in (7.43) were zero, i.e., $\tilde{\varepsilon}_\Omega(x) = 0$ then the Hamiltonian

---

[19] M. V. Berry and R. Lim, *J. Phys. A: Math. Gen.* **23**, L655 (1990).

would be the Hamiltonian of a charge-monopole system:

$$H = H^{\text{mon.}} = \frac{1}{2\mu} \pi \cdot \pi. \tag{7.44}$$

The radial motion of the charge-monopole system is free. If $\tilde{\varepsilon}_\Omega(x)$ would be a Coulomb potential, then (7.43) would be the Hamiltonian of a dyon system (two mass points with charge and magnetic monopole strength). In our case of the diatomic molecule (dumbbell with flywheel) the effective Hamiltonian (7.43) has the same induced vector potential as the monopole or dyon system but the induced scalar potential is given by (7.42) where $\tilde{\varepsilon}_\Omega(x)$ is close to an oscillator potential. Thus this system has the same angular motion as that of a magnetic monopole and a radial motion that is oscillatory.

We now investigate the property of the effective Hamiltonian for the "slow" system with the electronic potential energy curve $\tilde{\varepsilon}_{n_a}(x) = \tilde{\varepsilon}_\Omega(x)$. To $\tilde{\varepsilon}_\Omega(x)$ belong the two electronic quantum numbers $k = +\Omega$, $k = -\Omega$ and the electronic energy eigenspace $\mathcal{H}_{\varepsilon_\Omega} \subset \mathcal{H}^{\text{fast}}$ is two-dimensional. The subspace $\mathcal{H}^{\text{slow}} \otimes \mathcal{H}_{\varepsilon_\Omega} \subset \mathcal{H} = \mathcal{H}^{\text{slow}} \otimes \mathcal{H}^{\text{fast}}$ is thus the direct sum

$$\mathcal{H}^{\text{slow}} \otimes \mathcal{H}_{\varepsilon_\Omega} \equiv \mathcal{H}^\Omega = \mathcal{H}^{k=\Omega} \oplus \mathcal{H}^{k=-\Omega}. \tag{7.45}$$

In this space the effective Hamiltonian $H^\Omega_{\text{eff}} = \langle k'(\hat{x})|H|k(\hat{x})\rangle = \frac{1}{2\mu}\pi^2 - \tilde{\varepsilon}_\Omega(R)$ is a 2×2 matrix, and the slow observables R (whose eigenvalues are the parameters x) and $\pi = P - A(R)$ are also 2×2 matrices (not necessarily diagonal) whose elements are operators in $\mathcal{H}^\Omega$.

The commutation relations of R_i with R_j, and R_i with π_j are as usual

$$[R_i, R_j] = 0 \qquad [\pi_i, R_j] = i\delta_{ij}. \tag{7.46}$$

The commutation relations of the π_i are the anomalous commutation relations (7.32).

We now want to construct the angular momentum operator of the dumbbell molecule with flywheel. This is different from the angular momentum operator $J = L_{\text{elec.}} + S_{\text{elec.}}$, which we discussed earlier (and in Section XXII.3) and which was the angular momentum of the electronic motion. The angular momentum that we want to construct now is the angular momentum of the whole system, fast and slow. It is, therefore, the generator of rotations. As a consequence of the isotropy of space, this operator $\mathcal{J}_i$ must therefore commute with the Hamiltonian of the molecule:

$$[H, \mathcal{J}_i] = 0. \tag{7.47}$$

Further it must fulfill the standard commutation relations:

$$[\mathcal{J}_i, \mathcal{J}_j] = i\epsilon_{ijl}\mathcal{J}_l \tag{7.48}$$

and the position R_i and momentum π_i operators must be vector operators (of Section V.3) with respect to the angular momentum:

$$[\mathcal{J}_i, R_j] = i\epsilon_{ijl}R_l \tag{7.49}$$

$$[\mathcal{J}_i, \pi_j] = i\epsilon_{ijl}\pi_l. \tag{7.50}$$

We want to construct $\mathcal{J}_i$ in terms of the position operator R, the momentum operator π, and the other physical quantities that we constructed for the collective motion. We expect that this angular momentum will not be the operator $R \wedge \pi$ (or the operator $R \wedge P$), because these operators have a zero component of angular momentum along the internuclear axis, $\hat{R}, (R \wedge \pi) = 0$. We know that the molecule in the state characterized by Ω has a nonzero component of angular momentum along the internuclear axis due to the fast intrinsic motion of the electrons (the flywheel motion):

$$\hat{R} \cdot \mathcal{J} = k = \pm \Omega = \hat{R} \cdot J. \tag{7.51}$$

From the theory of Dirac's magnetic monopole, we would expect that

$$\mathcal{J}_i = \epsilon_{ijl} R_j \pi_l - R_i \frac{1}{2} \epsilon_{mnl} R_m F_{nl} \tag{7.52}$$

is the right expression for the angular momentum operator. That means the operator

$$\mathcal{T}_i^{(k)} = \epsilon_{ijl} R_j \pi_l^{(k)} + k \frac{R_i}{R} \tag{7.53}$$

is the angular momentum operator in each subspace $\mathcal{H}^k$. It can be shown by direct calculation that this operator indeed fulfills the conditions (7.47)–(7.50). From these results, we conclude that the dynamics of a diatomic molecule in a doubly degenerate Λ level is the same as that of the hypothetical charge-monopole system, except that the radial interaction is not free as in (7.44), but given by the scalar potential $\tilde{\varepsilon}_\Omega(x)$ in (7.43).

To derive the properties of this physical system (diatomic molecule in a fixed electronic state) we separate the radial and angular motion. From (7.52) we obtain by a straightforward calculation using the commutation relations (7.33) and (7.46)

$$\mathcal{J}^2 = R^2 \pi^2 + iR \cdot \pi - (R \cdot \pi)^2 + k^2. \tag{7.54}$$

Since the radial component of A $(a_x = \hat{R} \cdot A)$ is equal to zero (7.16), we have

$$R \cdot \pi = R \cdot P. \tag{7.55}$$

Thus we can replace the π is the second and third term on the r.h.s. of (7.54) with P. Then we obtain from (7.54)

$$\pi^2 = \frac{1}{R^2} (\mathcal{J}^2 - k^2) + \frac{1}{R^2} ((R \cdot P)^2 - iR \cdot P). \tag{7.56}$$

The last term on the r.h.s is the square of the radial momentum operator (VII.2.2):

$$P_{\text{rad}} = \frac{1}{2} \left\{ \frac{R_i}{R}, P_i \right\}. \tag{7.57}$$

P_{rad} is the component of P along the direction R. The anticommutator $\{\frac{R_i}{R}, P_i\}$ $= \frac{R_i}{R} P_i + P_i \frac{R_i}{R}$ takes into account that R_i and P_i are noncommuting operators.

(a)

(b)

(c)

Figure 7.1 Comparison of physical systems of two mass-points m_1, and m_2 with magnetic monopole vector potentials. (a) the charge-monopole system with electric charge e and magnetic monopole strength g with no force in the radial direction. (b) the dyon system consisting of two mass-points m_i with electric charge e_i and magnetic monopole strength g_i and with a force in the radial direction given by the $\frac{1}{R}$ Coulomb potential. (c) the vibrating dumbbell with flywheel. In contrast to systems (a) and (b), the monopole potential is not of electromagnetic origin but caused by the fast flywheel motion with fixed angular momentum component k around the dumbbell axis. The radial interaction is not free but is given by the Born-Oppenheimer potential curve $\varepsilon_\Omega(R)$, which is approximated by an oscillator potential.

P_{rad} is the operator conjugate to the radius operator R ($[R, P_{\text{rad}}] = i l$), and $\frac{P_{\text{rad}}^2}{2\mu}$ is the operator for the kinetic energy of the radial motion (μ is the reduced mass of the two atoms). Taking the square of (7.57) gives

$$P_{\text{rad}}^2 = \frac{1}{R^2}\left((R \cdot P)^2 - i(R \cdot P)\right). \tag{7.58}$$

From this we see that (7.56) can be written (after dividing by 2μ) as

$$\frac{1}{2\mu}\pi^2 = \frac{1}{2\mu R^2}(\mathscr{J}^2 - k^2) + \frac{1}{2\mu}P_{\text{rad}}^2. \tag{7.59}$$

If we insert this into (7.43) we obtain for the effective Hamiltonian

$$H_{\text{eff}} = \frac{1}{2\mu R^2}(\mathscr{J}^2 - k^2 + \Delta) + \frac{1}{2\mu}P_{\text{rad}}^2 + \varepsilon_\Omega(R). \tag{7.60}$$

This Hamiltonian is best understood if we compare it with the two standard Hamiltonians connected with magnetic monopoles. This is illustrated in Fig. 7.1.

The charge-monopole system consists of a mass m_1 with magnetic monopole strength g and another mass m_2 with electric charge e. Its Hamiltonian is given by (7.44) with $k = -\frac{eg}{4\pi}$. Using (7.56) this can be expressed as:

$$H^{\text{mon.}} = \frac{1}{2\mu x^2}(\mathscr{J}^2 - k^2) + \frac{1}{2\mu}P_{\text{rad}}^2. \tag{7.61}$$

The dyon system[20] consists of two particles of masses m_1 and m_2 each of which has electric charge e_1 and e_2, respectively, and magnetic monopole strength g_1 and g_2, respectively. The Hamiltonian is given by

$$H^{\text{dyon}} = \frac{1}{2\mu R^2}(\mathscr{J}^2 - k^2) + \frac{1}{2\mu}P_{\text{rad}}^2 + \frac{1}{4\pi}(e_1e_2 + g_1g_2)\frac{1}{R}$$

with

$$k = -\frac{e_1g_2 - e_2g_1}{4\pi}. \tag{7.62}$$

This dyon Hamiltonian differs from the monopole Hamiltonian (7.61) by the Coulomb potential term for the radial motion; for the charge-monopole system the radial motion is free.

The diatomic dumbbell with flywheel is also a physical system that combines rotational motion, given (approximately) by the first term of (7.60), and radial motion given by the term $\frac{1}{2\mu}P_{\text{rad}}^2 + \varepsilon_\Omega(x)$. The rotational motion is again the same as for the monopole and dyon system, except that the k is not of electromagnetic origin but comes from the fast flywheel motion. The radial motion is now different and given by the induced scalar potential $\varepsilon_\Omega(x)$, which is often (approximately) oscillatory. We will now discuss the properties for the molecule in detail.

The scalar potential ε_Ω, Fig. 2.1, has a minimum at the position x_e and is approximated by an oscillator potential given by

$$\varepsilon_\Omega \approx \frac{f}{2}X^2 + cX^3 \ldots, \qquad X = R - x_e. \tag{7.63}$$

The instantaneous moment of inertia is defined by $I = \mu R^2$ and is therefore written as

$$I = \mu R^2 = \mu(x_e + X)^2 = \mu x_e^2 + 2\mu x_e X + \mu X^2 \tag{7.64}$$

where X (and R) is the operator conjugate to P_{rad}. Introducing the constant $\omega = \sqrt{\frac{f}{\mu}}$, we define the new operators

$$q = \sqrt{\mu\omega}X, \qquad p = \frac{1}{\sqrt{\mu\omega}}P_{\text{rad}} \tag{7.65}$$

and so

$$I = I_e\left(1 + \frac{1}{\sqrt{\mu\omega}x_e}q\right)^2 \qquad I_e = \mu x_e^2 \tag{7.66}$$

and

$$I^{-1} = \frac{1}{I_e}\left(1 - \gamma q + \frac{3}{4}\gamma^2 q^2 - \cdots\right) \qquad \gamma = \frac{2}{\sqrt{\mu\omega}x_e}. \tag{7.67}$$

[20] J. Schwinger, *Science* **165**, 757 (1969).

Inserting this result into (7.60) we obtain

$$H_{\text{eff}} = \frac{1}{2I_e}(\mathcal{J}^2 - k^2 + \Delta) + \frac{\omega}{2}(p^2 + q^2) + H' \qquad (7.68)$$

where H' is treated as a perturbation and given by

$$H' = -\frac{\gamma q}{2I_e}(\mathcal{J}^2 - k^2 + \Delta) + \frac{3\gamma^2 q^2}{8I_e}(\mathcal{J}^2 - k^2 + \Delta) - \dots \qquad (7.69)$$

We will not discuss this correction term H' any further.

The effective Hamiltonian for the slow motion of the diatomic molecule in a doubly degenerate Λ-level is thus given by the sum of the Hamiltonian for a radial vibrator

$$H_{\text{rad vib}} = \frac{\omega}{2}(p^2 + q^2) \qquad (7.70)$$

and a rotator

$$H_{\text{rot}} = \frac{1}{2I_e}(\mathcal{J}^2 - (\hat{\boldsymbol{R}} \cdot \boldsymbol{\mathcal{J}})^2 + \Delta) = \frac{1}{2I_e}(\mathcal{J}^2 - k^2 + \Delta). \qquad (7.71)$$

Thus we derived that the collective motion of the diatomic molecule is in first approximation that of a vibrating rotator as described in Section II.5. The eigenvalues of the radial vibrator Hamiltonian (7.70) are obtained by (II.3.30). The eigenvalues of the rotator (with flywheel and therefore symmetrical top) are obtained by (V.4.55). The operators $\mathcal{J}_i$ and R_i generate the enveloping algebra of E_3 (corresponding to the J_i and Q_i in Section V.4). Its irreducible representation spaces are derived in the Appendix to Section V.3 and are applied to describe the spectrum of the rotator (and the symmetric top) in Section V.4. These irreducible representation spaces $\mathcal{H}(k, x)$ are characterized by the two numbers k and x. These two numbers are related to the eigenvalues of the Casimir operators $R_i \mathcal{J}_i = xk$ and $R_i R_i = x^2$, of which x^2 can take any real value and k can only be integer or half-integer (which we derived already by completely different means in (3.60)). $\mathcal{H}(k, x)$ is the space of physical states of the rotational motion (E_3 is the spectrum-generating group of the rotator). The spectrum of (7.71) is given by the property of the space $\mathcal{H}(k, x)$.

Since the energy levels are defined only up to an arbitrary constant, the k^2 and Δ terms in (7.71) can be ignored as fixed constants. For $k = 0$ one obtains from (7.71) the ordinary rotator spectrum

$$E_{\text{rot}} = \frac{1}{2I_e}j(j + 1) \qquad j = 0, 1, 2, 3 \dots . \qquad (7.72)$$

For $k^2 > 0$ one obtains the spectrum

$$E_{\text{rot}} = \frac{1}{2I_e}j(j + 1) \qquad j = k, k + 1, k + 2 \dots . \qquad (7.73)$$

Equation (7.68) with (7.70) and (7.71) is the standard result, including the flywheel term $(\hat{\boldsymbol{R}} \cdot \boldsymbol{\mathcal{J}})^2$ in H_{rot}, which is usually obtained by phenomenological arguments from the assumption that the electronic angular momentum J is pointed along the internuclear axis R, $(J = \pm \Omega \hat{R})$. We have shown here that it can be derived (without any further phenomenological assumption) directly from the Schrödinger equation if one does not make the drastic assumption of the Born-Oppenheimer approximation (which is that x is a fixed value). If the collective coordinates and momenta are treated as quantum mechanical observables, the effective Hamiltonian for the slow motion retains terms that can be expressed in terms of a non-Abelian gauge potential. The induced gauge potential has its origin in the fast (electronic) motion and has the same form as Berry's connection. If the fast motion is the flywheel motion (of the electrons) about the body axis, the Berry connection is identical with the vector potential of Dirac's hypothetical monopole—once the electromagnetic constants $\frac{eg}{hc}$ are replaced by the motion constant $|k|$. As a consequence, the effective Hamiltonian of the slow motion is governed, in addition to the radial potential, by a monopole Hamiltonian.[21] This monopole Hamiltonian leads to the experimentally well-known rotator spectrum (7.68) with two-fold degeneracy (Λ-type doubling). These well-known results are thus a manifestation of the Berry phase.

Although we used the diatomic molecule as an example, our arguments made very little use of the fact that the complicated system was a molecule. If we have a quantum system that we can visualize as a bead that can slide along a rod that rotates slowly in space, we will come to the same conclusions: The rapid rotation of the bead about the rod induces into the dynamics of the slow motion of the bead along the rotating rod a vector potential of the same form as that of a magnetic monopole. Quantum systems of this kind exist in many areas of physics. Thus, even though magnetic monopoles of the electromagnetic kind probably do not exist, physical systems with the same dynamics as that of magnetic monopoles are abundant. These physical systems are the slow "parts" of a complicated physical system in which the fast "part" is a rapid rotation.

Problems

1. Show that the covariant momentum of electromagnetism

$$\Pi = P - \frac{e}{c} A^{el}(X)$$

fulfills the commutation relation

$$[\Pi_i, \Pi_j] = i\hbar \frac{e}{c} \epsilon_{ijk} B_k^{el}$$

where the canonical momentum P_i and position X_j fulfill the usual Heisenberg commutation relations and where $B^{el} = \nabla_X \wedge A^{el}(X)$.

---

[21] It has been known for some time that a bead that is rapidly rotating about an axis along which it can freely slide has the same dynamics as a magnetic monopole in the field of an electric charge; J. M. Leinaas, *Physica Scripta* 17, 483 (1978).

2. (a) Show that the Schrödinger equation

$$\left(i\hbar\frac{\partial}{\partial t} - eA_0^{el}\right)\psi = \frac{1}{2m}\left(\frac{\hbar}{i}\nabla - \frac{e}{c}A^{el}\right)^2\psi$$

remains unaltered when A_0^{el}, A^{el}, and ψ undergo the following gauge transformations

$$A^{el} \rightarrow A^{(\alpha)el} \equiv A^{el} + \nabla\alpha(x, t)$$

$$A_0^{el} \rightarrow A_0^{(\alpha)el} \equiv A_0^{el} - \frac{1}{c}\frac{\partial\alpha}{\partial t}$$

$$\psi \rightarrow \psi^\alpha \equiv e^{if}\psi$$

where $f(x)$ is a function that depends on α. Calculate f.

(b) Obtain the transformation property of the gauge-covariant derivative

$$\frac{1}{i}D \equiv \frac{1}{i}\nabla - \frac{e}{\hbar c}A$$

under the preceding gauge transformations, i.e., find the operator $U(\alpha)$ in the transformation

$$D(A^\alpha) = U(\alpha)D(A)U^{-1}(\alpha).$$

3. Obtain the transformation property of the covariant derivative of the matter field $D\psi$ from the transformation property of the gauge potential A with respect to a non-Abelian gauge transformation.

4. Show that the gauge transformations

$$\psi_n^E(x) \rightarrow \psi_n^{\prime E}(x) = \sum_{l=1}^{\mathcal{N}} U^{nl}(x)\psi_l(x)$$

of the collective wave function leave the $\mathcal{N} \times \mathcal{N}$ matrix Schrödinger equation for $\psi_n^E(x)$ unchanged. Hint: Use that the total wave function $\psi^E(x, r) = \sum_{n=1}^{\mathcal{N}}\phi_n(x, r)\psi_n^E(x)$ should not be affected by this gauge transformation.

5. Show that for a gauge field with property $F_{ij}^{nm} = 0$ the gauge potential A_i^{nm} is given by $A_i^{nm}(x) = iV^{\dagger nl}(x)\partial_i V^{lm}(x)$, where $V^{lm}(x)$ is a unitary matrix.

6. If one allows for the gauge transformations of molecular physics the set of (in general infinite-dimensional) unitary transformations of the basis vectors of $\mathcal{H}^{fast}$:

$$|n(x)\rangle \rightarrow |n(x)\rangle' = \sum_l |l(x)\rangle U^{\dagger ln}(x),$$

where the sum extends over all values of the electronic quantum numbers $\sum_l |l(x)\rangle\langle l(x)| = I^{fast}$, then one obtains a trivial gauge theory: $F_{ij}^{nm} = 0$. Show this.

7. In a trivial gauge theory $A^{nm} = i\langle n(x)|\nabla|m(x)\rangle$ is pure gauge:

$$A^{nm}(x) = i\sum_l V^{\dagger nl}(x)(\nabla V^{lm}(x)).$$

Obtain a unitary matrix $V^{nm}(x)$ that gives this A^{nm}.

8. (a) Calculate the Berry connection

$$A_\theta^{k'k} = i\langle k'(\theta, \varphi)| \frac{\partial}{\partial\theta} |k(\theta, \varphi)\rangle$$

$$A_\varphi^{k'k} = i\langle k'(\theta, \varphi)| \frac{\partial}{\partial\varphi} |k(\theta, \varphi)\rangle$$

for the diatomic molecule with $\Omega = |k| = \frac{1}{2}$ (component of angular momentum along the internuclear axis) and express it in terms of the real parameter κ defined by $\langle k'(e_3)|J_2|k(e_3)\rangle = \frac{\kappa}{2}\sigma_2^{k'k}(\sigma_i = $ Pauli matrix).

(b) Show that the Berry curvature (covariant spherical components) for this case is given by

$$F_{\theta\varphi}^{k'k} = -\frac{1}{2}(1 - \kappa^2) \sin \theta \sigma_3^{k'k}.$$

Epilogue

The purpose of a physical theory is fulfilled if it provides a mathematical image of some domain of reality that allows us to relate experimental data and foresee new situations by making mathematical deductions. In this book we have restricted ourselves to this program and have not mentioned any of the further-reaching implications of quantum mechanics. Yet quantum mechanics has affected all scientific and even general human thinking. And when presented in its full generality, as done in this book, it effortlessly reveals the two facts whose lessons reach far beyond the boundaries of physics.

Classical science is based on two assumptions: (1) the deterministic nature of predictions, and (2) the atomistic nature of understanding. Quantum mechanics teaches us to revise both.

Classical theories are deterministic. The laws of classical physics are constructed in such a way that if the initial values of a system's dynamical variables are given, then their precise values can be calculated for any later time. These laws were not simply derived solely from experience, and then accepted as the basis of scientific philosophy and general thinking. Rather, and perhaps to a greater degree, these laws have been extricated from nature because they were in accord with the prevailing philosophical idea that nothing can be without a cause.

Probability statements in classical physics are always associated with insufficient knowledge, i.e., they are statements about the observer's knowledge and not about the physical system, which according to the principles of classical physics can be known to unlimited accuracy. In quantum theory (as described in Section II.5 and elsewhere in the text) one can only say with what probability certain values can be expected, even if one knows the state as well as possible, i.e., even if the system is in a pure state. Thus in quantum theory statements are inherently probabilistic; the occurrence of probability functions is not just a consequence of the observer's insufficient knowledge, but a property attributed to the physical systems themselves. Quantum predictions of experimental results are statements of how a microphysical process shows up in the macrophysical domain. These traces of microphysical processes in the macrophysical domain, the only source of human knowledge about such processes, do not obey deterministic laws. Earlier traces of a microphysical process do not determine later traces uniquely, but only probabilistically. Quantum theory teaches us that there are inherent limitations to human knowledge.

The second point, that of the profoundly holistic nature (of the understanding) of quantum physical systems, is not often emphasized, even though it is an obvious consequence of the quantum-mechanical description of physical systems. Although holism has already become rather widely accepted in other disciplines (e.g., psychology), it has been resisted by the physicists, who seem to be influenced by the success of atomism in classical physics. The quantum physical system is a structured whole described by the mathematical structure of an algebra of operators. From the laws of the combination of quantum physical systems (Section III.5), it follows that there are observables—of the combination of the two subsystems (described by $\mathcal{H}_1 \otimes \mathcal{H}_2$) that are incompatible with all observables of either subsystem (described by $\mathcal{H}_1$ or $\mathcal{H}_2$).

Thus, in quantum physics there exist holistic properties that cannot be obtained as combinations of the properties of the subsystems. In this sense the whole is not the sum of the parts.

In the atomistic approach understanding comes from the reduction of the complex system to simpler subsystems by ever finer separations until one comes to the ultimate constituents. In quantum physics the presence of holistic properties prevents this reduction process, and the notion of ultimate constituents loses its meaning. Atomism belongs to classical physics. A quantum physical system such as a molecule cannot be fully understood by dissecting it into nuclei and electrons, although, in the tradition of our scientific heritage, it is tempting to do this. What one arrives at in this way, however, is only the classical analogue of the quantum physical system, as in the Kepler system of proton and electron for the classical analogue of the hydrogen atom. An electron in an atom "*is*" something different from an electron in a linear accelerator, and the whole picture of the electron can only be displayed by giving its different aspects as they are mathematically described by the various basis systems in the space of physical states.

The visual picture that one usually requires for the process of understanding is in quantum physics not the geometrical picture of the object, but the picture of its image in the space of physical states. The reduction from the more complex to the simpler is performed not on the physical object, leading to simpler constituent objects, but on the space of physical states, leading to the irreducible subspaces for ever simpler structures. At every stage of this reduction one still has a whole picture describing all aspects by the various basis systems of the subspaces. But within a subspace the structure is simpler and describes a narrower domain of physics.

Dissecting a quantum physical system may destroy it. Therefore a quantum physical system (such as the CO molecule) cannot be understood only atomistically (as a di-atom) but is often more adequately understood holistically and functionally (as a vibrator rotator).

Atomism has been a great achievement of the past, and most technology is based on it. But quantum theory has revealed its limitations and shown that even for simple systems a holistic method is needed also.

Bibliography

The cited literature has been chosen rather arbitrarily. I have not made a systematic search for the most suitable list of books for further or supplementary reading, and mention just those I happen to have come across. Many of these books I have used myself.

The books and a few review articles are separated into several categories and then listed in alphabetical order; a few comments are added here and there.

1 Foundations of Quantum Mechanics

P. A. M. Dirac, *The Principles of Quantum Mechanics*, Clarendon Press, Oxford, 1958 (fourth edition).
Written by one of the creators of quantum mechanics many years ago, this is still modern and one of the greatest books written on this subject.

J. M. Jauch, *Foundations of Quantum Mechanics*, Addison-Wesley, Reading, Mass., 1968.
This text is concerned with the conceptual foundations of quantum mechanics and contains practically no applications. Furthermore, it differs from the present presentation by starting from the lattice structure of quantum mechanics. Still, it is recommended even to those who do not want to learn lattice theory.

G. Ludwig, *Grundlagen der Quantenmechanik*, Springer, Berlin, 1954.

J. von Neumann, *Mathematical Foundations of Quantum Mechanics*, Springer, Berlin, 1932; Princeton University Press, Princeton, 1955.
This is the first book written on the Hilbert-space formulation of quantum mechanics developed by its author.

2 Scattering Theory

A. I. Baz, Ya. B. Zeldovich, A. M. Perelomov, *Scattering Reactions and Decay in Non-relativistic Quantum Mechanics*, Israel Program for Scientific Translations, Jerusalem, 1969.

L. Fonda, G. C. Ghirardi, A. Rimini, Decay Theory of Unstable Quantum Systems, *Reports on Progress in Physics*, **41**, 587 (1978).

M. L. Goldberger, K. M. Watson, *Collision Theory*, Wiley, New York, 1964.

R. G. Newton, *Scattering Theory of Waves and Particles*. McGraw-Hill, New York, 1966.
In particular, for the formal theory of scattering this book is highly recommended.

H. M. Nussenzveig, *Causality and Dispersion Relations*, Academic Press, New York, 1972.

A. G. Sitenko, *Lectures in Scattering Theory*, Pergamon Press, New York, 1971.

K. Smith, *The Calculation of Atomic Collision Processes*, Wiley-Interscience, New York, 1971.

John R. Taylor, *Scattering Theory*, Wiley, New York, 1972.
This is a very clearly written book, which contains more material than the second part of the present book, however, it is written in a somewhat different spirit. It is highly recommended.

3 Theory of Angular Momentum and Group Theory

L. C. Biedenharn, J. D. Louck, *Angular Momentum in Quantum Physics*, Addison-Wesley, Reading, Mass., 1979.

A. R. Edmonds, *Angular Momentum in Quantum Mechanics*, Princeton University Press, Princeton, 1957.

I. M. Gelfand, R. A. Minlos, Z. Ja. Shapiro, *Representations of the Rotation Group and of the Lorentz Group*, Pergamon Press, New York, 1963.

M. A. Naimark, *Linear Representations of the Lorentz Group*, Pergamon Press, New York, 1964.
The first part of this book gives an introduction to the theory of group representations and discusses the rotation group in full detail. This is a book written by a mathematician for physicists, very readable, and mathematically rigorous. Some of the material of Chapter I is contained in this book.

M. Hamermesh, *Group Theory*, Addison-Wesley, Reading, Mass., 1962.

L. Michel, Applications of Group Theory to Quantum Physics: Algebraic Aspects; and

L. O'Raifeartaigh, Unitary Representations of Lie Groups in Quantum Mechanics.
Both published in *Group Representations in Mathematics and Physics* (Ed. V. Bargmann), Springer Lecture Notes in Physics, Vol. 6, 1970.
The two reviews by Michel and O'Raifeartaigh discuss many applications of group theory to quantum physics.

M. E. Rose, *Elementary Theory of Angular Momentum*, Wiley, New York, 1957.

4 Experimental Subjects

N. L. Alpert, W. E. Keiser, H. A. Szymanski, *Theory and Practice of Infrared Spectroscopy*, Wiley, New York, 1970.

A. Barbaro–Galtieri, Baryon Resonances, in *Advances in Particle Physics*, Vol. 2 (Eds. R. L. Cook, R. E. Marshak) Interscience, New York, 1968.

R. P. Bauman, *Absorption Spectroscopy*, Wiley, New York, 1962.

G. Herzberg, *Molecular Spectra and Molecular Structure*, D. van Nostrand, New York, 1966 (in particular Vol. 1).
This book teaches more than just molecular physics. It is one of the most beautiful books on physics and is highly recommended to every student.

H. S. W. Massey, E. H. S. Burhop, H. B. Gilbody, *Electronic and Ionic Impact Phenomena*, Clarendon Press, Oxford, 1969 (in particular Vol. 1: *Collisions of Electrons with Atoms*).

5 Mathematical Material

Milton Abramowitz and I. A. Stegun, *Handbook of Mathematical Functions*, Dover, New York, 1965.

A. Bohm, *The Rigged Hilbert Space and Quantum Mechanics*, Springer Lecture Notes in Physics, Vol. 78, 1978.

A. Bohm, M. Gadella, *Dirac Kets, Gamow Vectors and Galfand Triplets*, Springer Lecture Notes in Physics, Vol. 348 (1989).

I. M. Gel'fand and G. P. Shilov, *Generalized Functions*, Vols. 1, 2, 4, Academic Press, New York, 1964.

A. Lichnerowicz, *Linear Algebra and Analysis*, Holden–Day, San Francisco, 1967.

K. Maurin, *Methods of Hilbert Spaces*, Polish Scientific Publishers, Warsaw, 1967.

K. Maurin, *General Eigenfunction Expansions and Unitary Representations of Topological Groups*, Polish Scientific Publishers, Warsaw, 1968.

V. I. Smirnov, *A Course of Higher Mathematics*, Pergamon Press, New York, 1964 (in particular Vol. III, Part 1, on Linear Algebra, and Vol. III, Part 2, on Functions of a Complex Variable).

G. N. Watson, *Theory of Bessel Functions*, Cambridge University Press, Cambridge, 1958.

6 History of Quantum Mechanics

F. Hund, *Geschichte der Quantentheorie*, BI Wissenschaftsverlag, Berlin, 1975.

Max Jammer, *The Conceptual Development of Quantum Mechanics*, McGraw-Hill, New York, 1966.

7 Quantum Mechanics, Textbooks and Atomic Physics

G. Baym, *Lectures on Quantum Mechanics*, Benjamin, New York, 1969.

H. A. Bethe, E. E. Salpeter, *Quantum Mechanics of One- and Two-Electron Atoms*, Springer-Verlag, New York, 1957.

D. I. Blokhintsev, *Quantum Mechanics*, Reidel, Dordrecht, 1964.

R. H. Dicke, J. W. Wittke, *Introduction to Quantum Mechanics*, Addison-Wesley, Reading, Mass., 1960.

R. P. Feynman, R. B. Leighton, M. Sands, *The Feynman Lectures on Physics*, Vol. 3, *Quantum Mechanics*, Addison-Wesley, Reading, Mass., 1965.

S. Gasiorowicz, *Quantum Physics*, Wiley, New York, 1974.

D. T. Gillespie, *A Quantum Mechanics Primer*, Intex Publisher, 1970.

K. Gottfried, *Quantum Mechanics*, Benjamin, New York, 1966.

L. D. Landau, E. M. Lifshitz, *Quantum Mechanics*, Pergamon Press, New York, 1958.

H. J. Lipkin, *Quantum Mechanics, New Approaches to Selected Topics*, North-Holland, Amsterdam, 1973.

E. Merzbacher, *Quantum Mechanics*, Wiley, New York, 1961.

A. Messiah, *Quantum Mechanics*, Vols. 1, 2, Interscience, New York, 1961.

M. Mizushima, *Quantum Mechanics of Atomic Spectra and Atomic Structure*, Benjamin, New York, 1970.

M. Sargent, M. O. Scully, W. E. Lamb, *Laser Physics*, Addison-Wesley, Reading, Mass., 1974.

L. T. Schiff, *Quantum Mechanics*, McGraw-Hill, New York, 1968 (third edition).

8 Prerequisites from Classical Physics

V. D. Barger, M. Olsson, *Classical Mechanics: A Modern Perspective*, McGraw-Hill, New York, 1973.

A. O. Barut, *Electrodynamics and Classical Theory of Fields and Particles*, Macmillan, New York, 1964.

H. C. Corben, *Classical and Quantum Theories of Spinning Particles*, Holden-Day, San Francisco, 1968.

H. Goldstein, *Classical Mechanics*, Addison-Wesley, Reading, Mass., 1980 (second edition).

J. D. Jackson, *Classical Electrodynamics*, Wiley, New York, 1975 (second edition).

F. Rohrlich, *Classical Charged Particles*, Addison-Wesley, Reading, Mass., 1965.

Index